CAMBRIDGE LIBRARY COLLECTION

Books of enduring scholarly value

Mathematical Sciences

From its pre-historic roots in simple counting to the algorithms powering modern desktop computers, from the genius of Archimedes to the genius of Einstein, advances in mathematical understanding and numerical techniques have been directly responsible for creating the modern world as we know it. This series will provide a library of the most influential publications and writers on mathematics in its broadest sense. As such, it will show not only the deep roots from which modern science and technology have grown, but also the astonishing breadth of application of mathematical techniques in the humanities and social sciences, and in everyday life.

The Scientific Papers of Sir George Darwin

Sir George Darwin (1845-1912) was the second son and fifth child of Charles Darwin. After studying mathematics at Cambridge he read for the Bar, but soon returned to science and to Cambridge, where in 1883 he was appointed Plumian Professor of Astronomy and Experimental Philosophy. His family home is now the location of Darwin College. His work was concerned primarily with the effect of the sun and moon on tidal forces on Earth, and with the theoretical cosmogony which evolved from practical observation: he formulated the fission theory of the formation of the moon (that the moon was formed from still-molten matter pulled away from the Earth by solar tides). He also developed a theory of evolution for the Sun–Earth–Moon system based on mathematical analysis in geophysical theory. This volume of his collected papers covers figures of equilibrium of rotating liquid and geophysical investigations.

Cambridge University Press has long been a pioneer in the reissuing of out-of-print titles from its own backlist, producing digital reprints of books that are still sought after by scholars and students but could not be reprinted economically using traditional technology. The Cambridge Library Collection extends this activity to a wider range of books which are still of importance to researchers and professionals, either for the source material they contain, or as landmarks in the history of their academic discipline.

Drawing from the world-renowned collections in the Cambridge University Library, and guided by the advice of experts in each subject area, Cambridge University Press is using state-of-the-art scanning machines in its own Printing House to capture the content of each book selected for inclusion. The files are processed to give a consistently clear, crisp image, and the books finished to the high quality standard for which the Press is recognised around the world. The latest print-on-demand technology ensures that the books will remain available indefinitely, and that orders for single or multiple copies can quickly be supplied.

The Cambridge Library Collection will bring back to life books of enduring scholarly value (including out-of-copyright works originally issued by other publishers) across a wide range of disciplines in the humanities and social sciences and in science and technology.

The Scientific Papers of Sir George Darwin

Volume 3: Figures of Equilibrium of Rotating Liquid and Geophysical Investigations

George Howard Darwin

Cambridge University Press

CAMBRIDGE UNIVERSITY PRESS

Cambridge New York Melbourne Madrid Cape Town Singapore São Paolo Delhi

Published in the United States of America by Cambridge University Press, New York

www.cambridge.org
Information on this title: www.cambridge.org/9781108004459

This edition first published 1910
This digitally printed version 2009

ISBN 978-1-108-00445-9

SCIENTIFIC PAPERS

CAMBRIDGE UNIVERSITY PRESS

London: FETTER LANE, E.C.

C. F. CLAY, Manager

Edinburgh: 100, PRINCES STREET

Berlin: A. ASHER AND CO.

Leipzig: F. A. BROCKHAUS

New York: G. P. PUTNAM'S SONS

Bombay and Calcutta: MACMILLAN AND CO., Ltd.

SCIENTIFIC PAPERS

BY

SIR GEORGE HOWARD DARWIN,
K.C.B., F.R.S.
FELLOW OF TRINITY COLLEGE
PLUMIAN PROFESSOR IN THE UNIVERSITY OF CAMBRIDGE

VOLUME III

FIGURES OF EQUILIBRIUM
OF
ROTATING LIQUID
AND
GEOPHYSICAL INVESTIGATIONS

Cambridge:
at the University Press
1910

Cambridge:
PRINTED BY JOHN CLAY, M.A.
AT THE UNIVERSITY PRESS.

PREFACE

THE papers here collected together treat of the figure and of the movement of an actual or an ideal planet or satellite. I have failed to devise a short title for this volume which should describe exactly the scope of the subjects considered, and the title on the back of the book can only be held to apply strictly to three-quarters of the whole.

The first three papers fall somewhat further outside the proper meaning of the abridged title than do any of the others, for they are devoted to the mathematical solution of a geological problem. The second paper is indeed only a short note on a controversy long since dead; and the third is of little value.

The discussion of the amount of the possible changes in the position of the earth's axis of rotation, resulting from subsidences and upheavals, has some interest, but the conclusions arrived at in my paper are absolutely inconsistent with the sensational speculations as to the causes and effects of the glacial period which some geologists have permitted themselves to make.

At the end of this first paper there will be found an appendix containing an independent investigation by Lord Kelvin of the subject under discussion. He was one of the referees appointed by the Royal Society to report upon my paper, and he seemed to find that on these occasions the quickest way of coming to a decision was to talk over the subject with the author himself—at least this was frequently so as regards myself. Our discussion in this instance is very memorable to me, since it was the means of bringing me into intimate relationship with Lord Kelvin. The first two volumes of these collected papers may be regarded as the scientific outcome of our conversation of the year 1877; but for me at least science in this case takes the second place in importance.

The next four papers fall in more exactly with the proper meaning of my title, since they deal with the theory of the figure of the earth and of the planets. The ninth paper contains an attempt to determine the figures of a liquid planet and of its satellite when they are very close together. During the time that I was at work at this, M. Poincaré's great memoir, "Sur l'équilibre d'une masse fluide animée d'un mouvement de rotation," was not available to guide me. When it appeared I found therein the enunciation of the fundamental principles of stability, and as my work was not quite finished I kept it back for some time with the object of applying those principles to my problem. However I made a mistake in the attempted application, and the erroneous portion of the work is now suppressed. The failure to determine the stability of these figures of equilibrium deprives the work of much of the interest which it might have possessed. The present value of the paper is moreover yet further diminished by the fact that substantially the same problem, inclusive of the determination of stability, is solved in the last paper in the volume by means of far more appropriate analytical methods.

The next four papers are devoted to an extension of M. Poincaré's results as to figures of equilibrium, and he must be regarded as the presiding genius —or shall I say my patron saint—in this volume, just as Lord Kelvin was for the two preceding ones.

It was, I think, M. Poincaré who pointed out the resemblance to a pear of his conjectural drawing of the new figure of equilibrium which he had discovered. A name is generally convenient in such a case, and I have called it the pear-shaped figure, although when actual numerical values became available for drawing the figure, the resemblance to a pear was seen to have become much less striking.

The pear-shaped figure being derived from Jacobi's ellipsoid, its form must be defined by means of ellipsoidal harmonic deformations of a certain ellipsoid. Accordingly the solution of the problem demanded a thorough working knowledge of these harmonic functions of Lamé.

Ellipsoidal harmonic analysis has been used effectively in various analytical investigations—as for example in this very discovery by M. Poincaré of the new form of rotating liquid; but to the best of my belief no one had previously made systematic use of the method for numerical work. My first task, therefore, was the codification of the functions in a form convenient for practical use. It is not for me to pronounce on the merits of my scheme, yet it may be claimed that at least it attained its object by providing straightforward rules, whereby the computation could be carried out to any required degree of accuracy.

Numerical values were found in this way for the speed of rotation and for the form of the pear-shaped figure. Further than this, its stability was, as I believe, established. Yet it should be stated, in contradiction to this last result, that M. Liapounoff claims to have proved that the figure is unstable; and neither of us is as yet able to reconcile his result with that of the other. In this volume I naturally state my own point of view.

I cannot refrain from drawing attention to the *tour de force* whereby M. Poincaré has performed the apparently impossible task of applying harmonic analysis so as to take into account the thickness of the layer which constitutes the departure from the standard form of reference. As a verification of the complicated analysis applicable to the ellipsoid, I have, as appears below, applied this method to the cases of Jacobi's and Maclaurin's ellipsoids, using respectively spheroidal and spherical harmonic analysis.

In the last paper in this volume ellipsoidal harmonic analysis is applied to the extension and verification of the work of another great Frenchman, Edouard Roche. It has been already pointed out that this paper affords a far more complete solution of the problems proposed, than that attempted in the ninth paper.

Immediately after this preface, as in volumes I. and II., there is given a chronological list of my papers, corrected and extended up to the present time.

In conclusion I wish once more to acknowledge the admirable care bestowed by the printers and readers of the Cambridge University Press on the production of this book. Their task was rendered somewhat more difficult from the fact that several of the papers, especially the twelfth, have undergone extensive revision.

G. H. DARWIN.

December, 1909.

CONTENTS

CHRONOLOGICAL LIST OF PAPERS WITH REFERENCES TO THE VOLUMES IN WHICH THEY ARE OR WILL BE CONTAINED.

YEAR	TITLE AND REFERENCE	Volume in collected papers
1875	On two applications of Peaucellier's cells. London Math. Soc. Proc., 6, 1875, pp. 113—114.	IV
1875	On some proposed forms of slide-rule. London Math. Soc. Proc., 6, 1875, p. 113.	IV
1875	The mechanical description of equipotential lines. London Math. Soc. Proc., 6, 1875, pp. 115—117.	IV
1875	On a mechanical representation of the second elliptic integral. Messenger of Math., 4, 1875, pp. 113—115.	IV
1875	On maps of the World. Phil. Mag., 50, 1875, pp. 431—444.	IV
1876	On graphical interpolation and integration. Brit. Assoc. Rep., 1876, p. 13.	IV
1876	On the influence of geological changes on the Earth's axis of rotation. Roy. Soc. Proc., 25, 1877, pp. 328—332; Phil. Trans., 167, 1877, pp. 271—312.	III
1876	On an oversight in the *Mécanique Céleste*, and on the internal densities of the planets. Astron. Soc. Month. Not., 37, 1877, pp. 77—89.	III
1877	A geometrical puzzle. Messenger of Math., 6, 1877, p. 87.	IV
1877	A geometrical illustration of the potential of a distant centre of force. Messenger of Math., 6, 1877, pp. 97—98.	IV
1877	Note on the ellipticity of the Earth's strata. Messenger of Math., 6, 1877, pp. 109—110.	III
1877	On graphical interpolation and integration. Messenger of Math., 6, 1877, pp. 134—136.	IV
1877	On a theorem in spherical harmonic analysis. Messenger of Math., 6, 1877, pp. 165—168.	IV
1877	On a suggested explanation of the obliquity of planets to their orbits. Phil. Mag., 3, 1877, pp. 188—192.	III
1877	On fallible measures of variable quantities, and on the treatment of meteorological observations. Phil. Mag., 4, 1877, pp. 1—14.	IV

YEAR	TITLE AND REFERENCE	Volume in collected papers
1878	On Professor Haughton's estimate of geological time. Roy. Soc. Proc., 27, 1878, pp. 179—183.	III
1878	On the bodily tides of viscous and semi-elastic spheroids, and on the Ocean tides on a yielding nucleus. Roy. Soc. Proc., 27, 1878, pp. 419—424; Phil. Trans., 170, 1879, pp. 1—35.	II
1878	On the precession of a viscous spheroid. Brit. Assoc. Rep., 1878, pp. 482—485.	omitted
1879	On the precession of a viscous spheroid, and on the remote history of the Earth. Roy. Soc. Proc., 28, 1879, pp. 184—194; Phil. Trans., 170, 1879, pp. 447—538.	II
1879	Problems connected with the tides of a viscous spheroid. Roy. Soc. Proc., 28, 1879, pp. 194—199; Phil. Trans., 170, 1879, pp. 539—593.	II
1879	Note on Thomson's theory of the tides of an elastic sphere. Messenger of Math., 8, 1879, pp. 23—26.	II
1879	The determination of the secular effects of tidal friction by a graphical method. Roy. Soc. Proc., 29, 1879, pp. 168—181.	II
1880	On the secular changes in the elements of the orbit of a satellite revolving about a tidally distorted planet. Roy. Soc. Proc., 30, 1880, pp. 1—10; Phil. Trans., 171, 1880, pp. 713—891.	II
1880	On the analytical expressions which give the history of a fluid planet of small viscosity, attended by a single satellite. Roy. Soc. Proc., 30, 1880, pp. 255—278.	II
1880	On the secular effects of tidal friction. Astr. Nachr., 96, 1880, col. 217—222.	omitted
1881	On the tidal friction of a planet attended by several satellites, and on the evolution of the solar system. Roy. Soc. Proc., 31, 1881, pp. 322—325; Phil. Trans., 172, 1881, pp. 491—535.	II
1881	On the stresses caused in the interior of the Earth by the weight of continents and mountains. Phil. Trans., 173, 1882, pp. 187—230; Amer. Journ. Sci., 24, 1882, pp. 256—269.	II
1881	(Together with Horace Darwin.) On an instrument for detecting and measuring small changes in the direction of the force of gravity. Brit. Assoc. Rep., 1881, pp. 93—126; Annal. Phys. Chem., Beibl. 6, 1882, pp. 59—62.	I
1882	On variations in the vertical due to elasticity of the Earth's surface. Brit. Assoc. Rep., 1882, pp. 106—119; Phil. Mag., 14, 1882, pp. 409—427.	I
1882	On the method of harmonic analysis used in deducing the numerical values of the tides of long period, and on a misprint in the Tidal Report for 1872. Brit. Assoc. Rep., 1882, pp. 319—327.	omitted
1882	A numerical estimate of the rigidity of the Earth. Brit. Assoc. Rep., 1882, pp. 472—474; § 848, Thomson and Tait's Nat. Phil. second edition.	I

YEAR	TITLE AND REFERENCE	Volume in collected papers
1883	Report on the Harmonic analysis of tidal observations. Brit. Assoc. Rep., 1883, pp. 49—117.	I
1883	On the figure of equilibrium of a planet of heterogeneous density. Roy. Soc. Proc., 36, pp. 158—166.	III
1883	On the horizontal thrust of a mass of sand. Instit. Civ. Engin. Proc., 71, 1883, pp. 350—378.	IV
1884	On the formation of ripple-mark in sand. Roy. Soc. Proc., 36, 1884, pp. 18—43.	IV
1884	Second Report of the Committee, consisting of Professors G. H. Darwin and J. C. Adams, for the harmonic analysis of tidal observations. Drawn up by Professor G. H. Darwin. Brit. Assoc. Rep., 1884, pp. 33—35.	omitted
1885	Note on a previous paper. Roy. Soc. Proc., 38, pp. 322—328.	II
1885	Results of the harmonic analysis of tidal observations. (Jointly with A. W. Baird.) Roy. Soc. Proc., 39, pp. 135—207.	omitted
1885	Third Report of the Committee, consisting of Professors G. H. Darwin and J. C. Adams, for the harmonic analysis of tidal observations. Drawn up by Professor G. H. Darwin. Brit. Assoc. Rep., 1885, pp. 35—60.	I
1886	Report of the Committee, consisting of Professor G. H. Darwin, Sir W. Thomson, and Major Baird, for preparing instructions for the practical work of tidal observation; and Fourth Report of the Committee, consisting of Professors G. H. Darwin and J. C. Adams, for the harmonic analysis of tidal observations. Drawn up by Professor G. H. Darwin. Brit. Assoc. Rep., 1886, pp. 40—58.	I
1886	Presidential Address. Section A, Mathematical and Physical Science. Brit. Assoc. Rep., 1886, pp. 511—518.	IV
1886	On the correction to the equilibrium theory of tides for the continents. I. By G. H. Darwin. II. By H. H. Turner. Roy. Soc. Proc., 40, pp. 303—315.	I
1886	On Jacobi's figure of equilibrium for a rotating mass of fluid. Roy. Soc. Proc., 41, pp. 319—336.	III
1886	On the dynamical theory of the tides of long period. Roy. Soc. Proc., 41, pp. 337—342.	I
1886	Article 'Tides.' (Admiralty) Manual of Scientific Inquiry, pp. 53—91.	I
1887	On figures of equilibrium of rotating masses of fluid. Roy. Soc. Proc., 42, pp. 359—362; Phil. Trans., 178A, pp. 379—428.	III
1887	Note on Mr Davison's Paper on the straining of the Earth's crust in cooling. Phil. Trans., 178A, pp. 242—249.	IV
1888	Article 'Tides.' Encyclopaedia Britannica.	Certain sections in I
1888	On the mechanical conditions of a swarm of meteorites, and on theories of cosmogony. Roy. Soc. Proc., 45, pp. 3—16; Phil. Trans., 180A, pp. 1—69.	IV

YEAR	TITLE AND REFERENCE	Volume in collected papers
1889	Second series of results of the harmonic analysis of tidal observations. Roy. Soc. Proc., 45, pp. 556—611.	omitted
1889	Meteorites and the history of Stellar systems. Roy. Inst. Rep., Friday, Jan. 25, 1889.	omitted
1890	On the harmonic analysis of tidal observations of high and low water. Roy. Soc. Proc., 48, pp. 278—340.	I
1891	On tidal prediction. Bakerian Lecture. Roy. Soc. Proc., 49, pp. 130—133; Phil. Trans., 182A, pp. 159—229.	I
1892	On an apparatus for facilitating the reduction of tidal observations. Roy. Soc. Proc., 52, pp. 345—389.	I
1896	On periodic orbits. Brit. Assoc. Rep., 1896, pp. 708—709.	omitted
1897	Periodic orbits. Acta Mathematica, 21, pp. 101—242, also (with omission of certain tables of results) Mathem. Annalen, 51, pp. 523—583.	IV
	[by S. S. Hough. On certain discontinuities connected with periodic orbits. Acta Math., 24 (1901), pp. 257—288.]	IV
1899	The theory of the figure of the Earth carried to the second order of small quantities. Roy. Astron. Soc. Month. Not., 60, pp. 82—124.	III
1900	Address delivered by the President, Professor G. H. Darwin, on presenting the Gold Medal of the Society to M. H. Poincaré. Roy. Astron. Soc. Month. Not., 60, pp. 406—415.	IV
1901	Ellipsoidal harmonic analysis. Roy. Soc. Proc., 68, pp. 248—252; Phil. Trans., 197A, pp. 461—557.	III
1901	On the pear-shaped figure of equilibrium of a rotating mass of liquid. Roy. Soc. Proc., 69, pp. 147, 148; Phil. Trans., 198A, pp. 301—331.	III
1902	Article 'Tides.' Encyclopaedia Britannica, supplementary volumes.	Certain sections in I
1902	The stability of the pear-shaped figure of equilibrium of a rotating mass of liquid. Roy. Soc. Proc., 71, pp. 178—183; Phil. Trans., 200A, pp. 251—314.	III
1903	On the integrals of the squares of ellipsoidal surface harmonic functions. Roy. Soc. Proc., 72, p. 492; Phil. Trans., 203A, pp. 111—137.	III
1903	The approximate determination of the form of Maclaurin's spheroid. Trans. Amer. Math. Soc., 4, pp. 113—133.	III
1903	The Eulerian nutation of the Earth's axis. Bull. Acad. Roy. de Belgique (Sciences), pp. 147—161.	IV
1905	The analogy between Lesage's theory of gravitation and the repulsion of light. Roy. Soc. Proc., 76A, pp. 387—410.	IV
1905	Address by Professor G. H. Darwin, President. Brit. Assoc. Rep., 1905, pp. 3—32.	IV
1906	On the figure and stability of a liquid satellite. Roy. Soc. Proc., 77A, pp. 422—425; Phil. Trans., 206A, pp. 161—248.	III

YEAR	TITLE AND REFERENCE	Volume in collected papers
1908	Further consideration of the figure and stability of the pear-shaped figure of a rotating mass of liquid. Roy. Soc. Proc., 80A, pp. 166—7; Phil. Trans., 208A, pp. 1—19.	III
1908	Antarctic Tidal Observations. I. Observations of the 'Discovery.' National Antarctic Expedition 1901—4, Physical Observations, pp. 3—12.	I
	II. Observations of the 'Scotia' (a short note). Ibid. p. 16.	omitted
1908	Further note on Maclaurin's Spheroid. Trans. Amer. Math. Soc., vol. 9, No. 1, pp. 34—38.	III
1908	'Bewegung der Hydrosphäre' (jointly with Mr S. S. Hough). Encyklopädie der Mathematischen Wissenschaften, VI. 1, B. 81 pp.	IV
1908	The Rigidity of the Earth. Atti del IV congresso internat. dei matematici (Roma, 6—11 Aprile 1908), Vol. III. Sez. III. A, pp. 5—11; and 'Scientia,' Rivista di Scienza, Vol. V. (1909), N. X—2. 13 pp.	IV
1909	The Genesis of Double Stars. Essay XXVIII. of 'Darwin and Modern Science,' pp. 543—564.	IV
1909	A theory of the evolution of the solar system. Internationale Wochenschrift, 3 Jahrgang Nr 30, 24 Juli 1909, pp. 921—932.	IV
1909	On certain families of Periodic Orbits. Roy. Astron. Soc. Month. Not., 70, 1909, pp. 108—145.	IV
1910	Reduction of the tidal observations made in the Antarctic expedition of the 'Nimrod.'	IV
	Unpublished Article 'Tides.' Encyclopaedia Britannica, new edition to be published hereafter (by permission of the proprietors).	Certain sections in I

ERRATA.

Vol. I., p. 377, line 14 from foot of page,
for 28 *read* 27.

Vol. III., p. 81, line 20, in the last term of U_e' *for* $\frac{18}{35}$ *read* $\frac{8}{35}$.

" p. 84, in the second of equations (12)
for $-\frac{24}{35}hP_4$ *read* $-\frac{36}{35}hP_4$.

" p. 95, in the first term of the first equation in § 8 the factor $(1-h)$ is omitted; *read* therefore

$$\tfrac{4}{3}\pi w\frac{a^3}{r}(1-h)(1+\mu).$$

" p. 394, line 4 from foot of page,
for $\mathfrak{A}_1{}^1$ *read* $\mathsf{A}_1{}^1$.

" p. 402, line 12 from foot of page,
for $EF'+F'F-FF'$ *read* $EF'+E'F-FF'$.

1.

ON THE INFLUENCE OF GEOLOGICAL CHANGES ON THE EARTH'S AXIS OF ROTATION.

[*Philosophical Transactions of the Royal Society*, Part I. Vol. 167 (1877), pp. 271—312.]

THE subject of the fixity or mobility of the earth's axis of rotation in that body, and the possibility of variations in the obliquity of the ecliptic, have from time to time attracted the notice of mathematicians and geologists. The latter look anxiously for some grand cause capable of producing such an enormous effect as the glacial period. Impressed by the magnitude of the phenomenon, several geologists have postulated a change of many degrees in the obliquity of the ecliptic and a wide variability in the position of the poles on the earth; and this, again, they have sought to refer back to the upheaval and subsidence of continents.

Mr John Evans, F.R.S.*, the late President of the Geological Society, in an address delivered to that Society, has recurred to this subject at considerable length. After describing a system of geological upheaval and subsidence, evidently designed to produce a maximum effect in shifting the polar axis, he asks:—"Would not such a modification of form bring the axis of figure about 15° or 20° south of the present, and on the meridian of Greenwich—that is to say, midway between Greenland and Spitzbergen? and would not, eventually, the axis of rotation correspond in position with the axis of figure?

"If the answer to these questions is in the affirmative, then I think it must be conceded that even minor elevations within the tropics would produce effects corresponding to their magnitude, and also that it is unsafe to assume that the geographical position of the poles has been persistent throughout all geological time†."

* [Subsequently Sir John Evans, K.C.B.]

† *Quart. Journ. Geol. Soc.*, 1876, XXXII. *Proc.*, p. 108.

On the few occasions on which this subject has been referred to by mathematicians, the adequacy of geological changes to produce effects of such amount has been denied. Amongst others, the Astronomer Royal and Sir William Thomson have written briefly on the subject*, but, as far as I know, the subject has not hitherto been treated at much length.

The following paper is an attempt to answer the questions raised by Mr Evans; but as I have devoted a section to the determination of the form of continent and sea which would produce a maximum effect in shifting the polar axis, I have not taken into consideration the configuration proposed by him.

The general plan of this paper is to discuss the following problems:—

First. The precession and nutation of a body slowly changing its shape from internal causes, with especial reference to secular alterations in the obliquity of the ecliptic.

Second. The changes in the position of the earth's axis of symmetry, caused by any deformations of small amount.

Third. The modifications introduced by various suppositions as to the nature of the internal changes accompanying the deformations.

In making numerical application of the results of the previous discussions to the case of the earth, it has of course been necessary to betake one's self to geological evidence; but the vagueness of that evidence has precluded any great precision in the results.

In conclusion I must mention that, since this paper has been in manuscript, Sir William Thomson, in his Address to the Mathematical Section of the British Association at Glasgow, has expressed his opinion on this same subject. He there shortly states results in the main identical with mine, but without indicating how they were arrived at.

The great interest which this subject has recently been exciting both in England and America, coupled with the fact that several of my results are not referred to by Sir William Thomson, induces me to persist in offering my work to the Royal Society.

I. Precession of a Spheroid slowly changing its Shape.

I begin the investigation by discussing the precession and nutation of an ellipsoid of revolution slowly and uniformly changing its shape. The changes are only supposed to continue for such a time, that the total changes in the principal moments of inertia are small compared to the difference between the greatest and least moments of inertia of the ellipsoid in its initial state.

For brevity, I speak of the ellipsoid as the earth; and shall omit some parts of the investigation, which are irrelevant to the problem under discussion.

* In papers referred to below.

The changes are supposed to proceed from internal causes, and to be any whatever; and in the application made they will be supposed to go on with a uniform velocity.

§ 1. *The Equations of Motion.*

M. Liouville has given the equations of motion about a point of a body which is slowly changing its shape from internal causes*; these equations, he says, are only applicable to the case of the point being fixed or moving uniformly in a straight line. They may, however, be extended to the motion of the earth about its centre of inertia, because the centrifugal force due to the orbital motion and the unequal orbital motion will not add anything to the moments of the impressed forces. These equations are, in fact, an extension of Euler's equations for the motion of a rigid body, which are ordinarily applied to the precessional problem. To make them intelligible I reproduce the following from Mr Routh's *Rigid Dynamics*†, where the proof is given more succinctly than in the original:—

"Let x, y, z be the coordinates of any particle of mass m at the time t, referred to axes fixed in space. Then we have the equation of motion

$$\Sigma m\left(x\frac{d^2y}{dt^2} - y\frac{d^2x}{dt^2}\right) = \text{N} \quad\text{.........................(1)}$$

and two similar equations.

"Let
$$h_3 = \Sigma m\left(x\frac{dy}{dt} - y\frac{dx}{dt}\right) \quad\text{.........................(2)}$$

with similar expressions for h_1, h_2.

"Then the equation (1) becomes

$$\frac{dh_3}{dt} = \text{N} \quad\text{..................................(3)}$$

"Let the motion be referred to three rectangular axes Ox', Oy', Oz' moving in any manner about the origin O. Let α, β, γ be the angles these three axes make with the fixed axis of z. Now h_3 is the sum of the products of the mass of each particle into twice the projection on the plane of xy of the area of the surface traced out by the radius vector of that particle drawn from the origin. Let h_1', h_2', h_3' be the corresponding 'areas' described on the planes $y'z'$, $z'x'$, $x'y'$ respectively. Then by a known theorem proved in Geometry of Three Dimensions, the sum of the projections of h_1', h_2', h_3' on xy is equal to h_3; therefore

$$h_3 = h_1'\cos\alpha + h_2'\cos\beta + h_3'\cos\gamma \quad\text{...................(4)}$$

* Liouville's *Journ. Math.*, 2me série, t. III., 1858, p. 1.

† Page 150, edit. of 1860, but omitted in later editions. L, M, N are the couples of the impressed forces about the axes.

"Since the fixed axes are quite arbitrary, let them be taken so that the moving axes are passing through them at the time t. Then

$$h_1' = h_1, \quad h_2' = h_2, \quad h_3' = h_3$$

and by the same reasoning, as in Arts. 114 and 115, we can deduce from equation (4) that

$$\frac{dh_3}{dt} = \frac{dh_3'}{dt} - h_1'\theta_2 + h_2'\theta_1 \quad \text{.........................(5)}$$

where θ_1, θ_2, θ_3 are the angular velocities of the axes with reference to themselves. Hence the equations of motion of the system become

$$\left.\begin{aligned} \frac{dh_1}{dt} - h_2\theta_3 + h_3\theta_2 &= \mathrm{L} \\ \frac{dh_2}{dt} - h_3\theta_1 + h_1\theta_3 &= \mathrm{M} \\ \frac{dh_3}{dt} - h_1\theta_2 + h_2\theta_1 &= \mathrm{N} \end{aligned}\right\} \quad \text{.........................(6)}$$

"These equations may be put under another form which is more convenient. Let x', y', z' be the coordinates of the particle m referred to the moving axes, and let

$$\mathrm{H}_3 = \Sigma m \left(x' \frac{dy'}{dt} - y' \frac{dx'}{dt}\right).$$

"Since the fixed axes coincide with these at the time t, we have $x = x'$, $y = y'$, and by Art. 114,

$$\left.\begin{aligned} \frac{dx}{dt} &= \frac{dx'}{dt} + \theta_2 z - \theta_3 y \\ \frac{dy}{dt} &= \frac{dy'}{dt} + \theta_3 x - \theta_1 z \end{aligned}\right\}$$

Therefore $\qquad h_3' = \mathrm{H}_3 + \mathrm{C}\theta_3 - \mathrm{E}\theta_1 - \mathrm{D}\theta_2$*

and by similar reasoning

$$h_1' = \mathrm{H}_1 + \mathrm{A}\theta_1 - \mathrm{F}\theta_2 - \mathrm{E}\theta_3$$
$$h_2' = \mathrm{H}_2 + \mathrm{B}\theta_2 - \mathrm{D}\theta_3 - \mathrm{F}\theta_1$$

"Hence the general equation of motion becomes

$$\frac{d}{dt}(\mathrm{C}\theta_3 - \mathrm{E}\theta_1 - \mathrm{D}\theta_2 + \mathrm{H}_3) + \mathrm{F}(\theta_2^2 - \theta_1^2) + (\mathrm{B} - \mathrm{A})\,\theta_1\theta_2 + \mathrm{E}\theta_2\theta_3 - \mathrm{D}\theta_1\theta_3 + \theta_1\mathrm{H}_2 - \theta_2\mathrm{H}_1 = \mathrm{N} \quad \text{.....(7)}$$

and two similar equations.

"Let the moving axes be so chosen as to coincide with the principal axes at the time t. Then $\mathrm{D} = 0$, $\mathrm{E} = 0$, $\mathrm{F} = 0$, and the equations become,"

$$\frac{d}{dt}(\lambda_1\theta_1 + \mathrm{H}_1) - (\lambda_2 - \lambda_3)\,\theta_2\theta_3 + \theta_2\mathrm{H}_3 - \theta_3\mathrm{H}_2 = \mathrm{L}$$

and two similar equations; where λ_1, λ_2, λ_3 (replacing the A, B, C of

* A, B, C, D, E, F are, as usual, the moments and products of inertia.

Mr Routh) are the three principal moments of inertia, and are functions of the time.

In order to apply these equations to the present problem, we must consider the meaning of the quantities θ_1, θ_2, θ_3. A system of particles may be made to pass from any one configuration to any other by means of the rotation of the system as a whole about any axis through any angle, and a subsequent displacement of every particle in a straight line to its ultimate position. Of all the axes and all the angles about and through which the preliminary rotation may be made, there is one such that the sum of the squares of the subsequent paths is a minimum. By analogy with the method of least squares this rotation may be said to be that which most nearly represents the passage of the system from one configuration to the other. If the two configurations differ by little from one another, and if the best representative rotation be such that the curvilinear path of any particle is large compared to its subsequent straight path, the system may be said to be rotating as a rigid body, and at the same time slowly changing its shape. Now this is the case we have to consider in a slow distortion of the earth.

Divide the time into a number of equal small intervals τ, and in the first interval let the earth be rigid, and let each pair of its principal axes rotate about the third (with angular velocities double those with which they actually rotate). At the end of that interval suppose that each pair has rotated about the third through angles $2\omega_1\tau$, $2\omega_2\tau$, $2\omega_3\tau$. Then reduce the earth to rest, and during the next interval let the matter constituting the earth flow (with velocities double those with which it actually flows) so that the pairs of principal axes have, at the end of the interval, rotated with respect to the third ones through the angles $-2\alpha\tau$, $-2\beta\tau$, $-2\gamma\tau$. Lastly let $2\theta_1\tau$, $2\theta_2\tau$, $2\theta_3\tau$ be the rotations of each pair of axes about the third by which they could have been brought directly from their initial to their final positions in the time 2τ.

Therefore, by the principle of superposition of small motions,

$$\theta_1 = \omega_1 - \alpha, \quad \theta_2 = \omega_2 - \beta, \quad \theta_3 = \omega_3 - \gamma$$

Now supposing these two processes to go on simultaneously with their actual velocities, instead of in alternate intervals of time with double velocities, it is clear that θ_1, θ_2, θ_3 are "the angular velocities of the axes with reference to themselves"; ω_1, ω_2, ω_3 are the component angular velocities of the earth considered as a rigid body; and $-\alpha$, $-\beta$, $-\gamma$ are the component angular velocities of the principal axes relatively to the earth, arising from the supposed continuous distortion of that body.

With respect to the other quantities involved in the equations of motion:—

Let C, A be the principal moments of inertia of the earth initially when t is zero; and at any time t, let

$$\lambda_1 = A + at, \quad \lambda_2 = A + bt, \quad \lambda_3 = C + ct$$

We here suppose that the changes in the earth are so slow that terms depending on the higher powers of t may be neglected.

Lastly the quantities H_1, H_2, H_3 are respectively twice the areas conserved on the planes of $\theta_2\theta_3$, $\theta_3\theta_1$, $\theta_1\theta_2$ by the motion of the earth relative to these axes. If the earth were rigid, they would all be zero, because there would be no motion relative to the principal axes: thus ω_1, ω_2, ω_3 do not enter into these quantities. Now the motion which *does* take place may be analysed into two parts. Divide the time into a number of equal small elements τ, and in the first of them let the matter constituting the earth flow (with a velocity double that with which it actually flows); this motion will conserve double-areas on the planes of $\theta_2\theta_3$, $\theta_3\theta_1$, $\theta_1\theta_2$, which we may call $2\mathfrak{H}_1\tau$, $2\mathfrak{H}_2\tau$, $2\mathfrak{H}_3\tau$. In the next interval of time let each pair of axes rotate round the third (with angular velocities double those with which they actually rotate), so that at the end of the interval they have turned through the angles $-2\alpha\tau$, $-2\beta\tau$, $-2\gamma\tau$. Now since during this second interval the axes have rotated in a negative direction through the solid, therefore the solid has rotated in a positive direction with reference to the axes. Remembering then that λ_1, λ_2, λ_3 are the principal moments of inertia, the double-areas conserved on the three planes in this second interval are $2\lambda_1\alpha\tau$, $2\lambda_2\beta\tau$, $2\lambda_3\gamma\tau$. Hence if $2H_1\tau$, $2H_2\tau$, $2H_3\tau$ be the double areas conserved in this double interval of time, we have $2H_1\tau = 2\mathfrak{H}_1\tau + 2\lambda_1\alpha\tau$, $2H_2\tau = 2\mathfrak{H}_2\tau + 2\lambda_2\beta\tau$, $2H_3\tau = 2\mathfrak{H}_3\tau + 2\lambda_3\gamma\tau$.

Therefore if we now suppose the two processes to go on simultaneously with their actual velocities, instead of in alternate elements of time with double velocities, and if we substitute for λ_1, λ_2, λ_3 their values in terms of A, a, t, &c., we get

$$H_1 = (A + at)\,\alpha + \mathfrak{H}_1, \quad H_2 = (A + bt)\,\beta + \mathfrak{H}_2, \quad H_3 = (C + ct)\,\gamma + \mathfrak{H}_3$$

where $\mathfrak{H}_1$, $\mathfrak{H}_2$, $\mathfrak{H}_3$, denote those parts of the double areas conserved, which depend only on the internal motions accompanying the change of shape.

Then, if the changes proceed with uniform velocity, a, α, $\mathfrak{H}$, &c. are all constant.

Corresponding also to the equations of motion are the geometrical equations

$$\frac{d\theta}{dt} = \theta_2 \cos\phi + \theta_1 \sin\phi$$

$$\frac{d\psi}{dt} \sin\theta = -\theta_1 \cos\phi + \theta_2 \sin\phi$$

$$\frac{d\phi}{dt} + \frac{d\psi}{dt} \cos\theta = \theta_3$$

In figure 1, A, B, C are the axes, about which the moments of inertia are

$\lambda_1, \lambda_2, \lambda_3$; XY is the ecliptic; and the meaning of the other symbols is sufficiently indicated*.

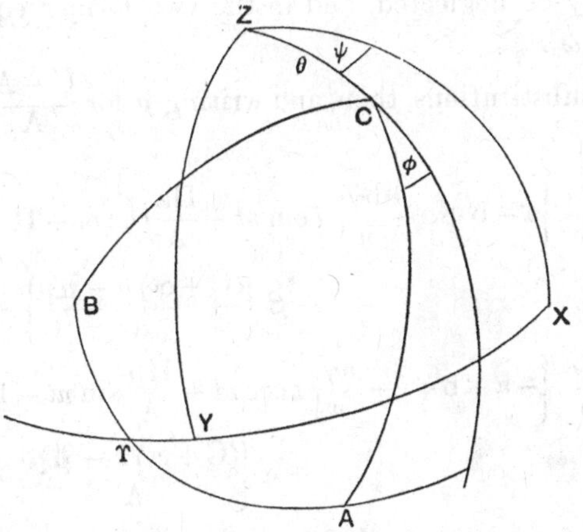

FIG. 1.

Substituting, then, for the various symbols in the original equations of motion, it will be found that

$$A\frac{d\omega_1}{dt} - (A' - C)\,\omega_2\omega_3 = L - t\left\{a\frac{d\omega_1}{dt} - (b - c)\,\omega_2\omega_3 + b\gamma\omega_2 - c\beta\omega_3\right\}$$
$$- a\omega_1 - \{A'\gamma + \mathfrak{H}_3\}\,\omega_2 + \{C\beta + \mathfrak{H}_2\}\,\omega_3 + \beta\mathfrak{H}_3 - \gamma\mathfrak{H}_2$$

and two similar equations†.

Now the terms on the right-hand side are always very small compared to $A\frac{d\omega_1}{dt}$, because the time will not run on until they have become large; hence approximate values may be substituted therein.

Let the angular velocity of rotation of the earth be $-n$, and let Π cosec θ be the precession of the equinoxes; then in the small terms the following substitutions may be made:—

$$\phi = -(nt + e), \quad \omega_1 = -\Pi\cos(nt + e), \quad \omega_2 = -\Pi\sin(nt + e), \quad \omega_3 = -n$$

and the e may be omitted for brevity.

Further, N (depending on the attractions of the sun and moon) is very small; and a consideration of the third equation of motion shows that, when integrated, it leads to $\omega_3 = -n +$ terms, which are very small during the limited period under consideration. And if these terms were substituted on the left-hand side of the two former equations, they would be still further

* [I have followed Mr Routh in using "perverted" coordinate axes, but it does not seem worth while to revert to the more desirable usage by redrawing the figure and by changing the signs of many of the terms.]

† The A is written A′ in two places, where it may be taken to stand for B; and then the other equations may be found by cyclic changes of letters and suffixes.

diluted by multiplication by the small quantities $\dfrac{C-A}{A}$ and Π. Hence the third equation may be neglected, and in the two former equations $-n$ may be substituted for ω_3.

Making these substitutions, then, and writing μ for $\dfrac{C-A}{A}n$, the equations become

$$\frac{d\omega_1}{dt} - \mu\omega_2 = \frac{L}{A} - \frac{\Pi n}{A}\left\{a-b+c-\frac{b\gamma}{n}\right\} t \sin nt + \frac{\Pi a}{A}\cos nt + \Pi\left\{\gamma+\frac{\mathfrak{H}_3}{A}\right\}\sin nt$$

$$- \beta\left\{\frac{(C+ct)n - \mathfrak{H}_3}{A}\right\} - \frac{\mathfrak{H}_2}{A}(n+\gamma)$$

$$\frac{d\omega_2}{dt} + \mu\omega_1 = \frac{M}{A} - \frac{\Pi n}{A}\left\{-a+b+c-\frac{a\gamma}{n}\right\} t \cos nt + \frac{\Pi b}{A}\sin nt - \Pi\left(\gamma+\frac{\mathfrak{H}_3}{A}\right)\cos nt$$

$$+ \alpha\left\{\frac{(C+ct)n - \mathfrak{H}_3}{A}\right\} + \frac{\mathfrak{H}_1}{A}(n+\gamma)$$

Then neglecting γ/n compared to unity, putting $C+ct=A$ in the small terms, and only retaining the more important terms,

$$\frac{d\omega_1}{dt} - \mu\omega_2 = \frac{L}{A} - \frac{\Pi n}{A}\{a-b+c\}\,t\sin nt + \frac{\Pi a}{A}\cos nt + \Pi\left\{\gamma+\frac{\mathfrak{H}_3}{A}\right\}\sin nt - n\beta$$

$$\frac{d\omega_2}{dt} + \mu\omega_1 = \frac{M}{A} + \frac{\Pi n}{A}\{-a+b+c\}\,t\cos nt + \frac{\Pi b}{A}\sin nt - \Pi\left\{\gamma+\frac{\mathfrak{H}_3}{A}\right\}\cos nt + n\alpha^*$$

These are the required equations of motion, and in integrating them they may be treated as linear.

§ 2. *Inequalities independent of the Impressed Forces.*

First, then, suppose that $L = M = 0$.

Integrate the equations, and neglect μ compared with n, and we have

$$\omega_1 = \frac{\Pi}{A}(a-b+c)\,t\cos nt + \frac{\Pi}{An}(b-c)\sin nt - \frac{\Pi}{n}\left(\gamma+\frac{\mathfrak{H}_3}{A}\right)\cos nt$$

$$+ \frac{n}{\mu}\left(\alpha+\frac{\mathfrak{H}_1}{A}\right) + F\cos\mu t - G\sin\mu t$$

$$\omega_2 = \frac{\Pi}{A}(-a+b+c)\,t\sin nt + \frac{\Pi}{An}(c-a)\cos nt - \frac{\Pi}{n}\left(\gamma+\frac{\mathfrak{H}_3}{A}\right)\sin nt$$

$$+ \frac{n}{\mu}\left(\beta+\frac{\mathfrak{H}_2}{A}\right) - F\sin\mu t - G\cos\mu t$$

---

* If we wish to treat a, α, $\mathfrak{H}_1$, &c. as variable, we have only to add to the right-hand sides of these equations $\dfrac{\Pi}{A}t\cos nt\,\dfrac{da}{dt} - \dfrac{1}{A}\dfrac{d\mathfrak{H}_1}{dt}$, and $\dfrac{\Pi}{A}t\sin nt\,\dfrac{db}{dt} - \dfrac{1}{A}\dfrac{d\mathfrak{H}_2}{dt}$ respectively. If we put $L=M=0$ and neglect $\mathfrak{H}_1$, $\mathfrak{H}_2$, $\mathfrak{H}_3$, these equations will be found to be identical with the equations (2) given by Sir W. Thomson in App. C. I had not noticed until it was pointed out by him, how nearly applicable my equations were to the case of varying velocities of distortion.—April 26, 1877.

The last two terms in ω_1 and ω_2 represent the complementary function; and the values of F and G must be determined from the initial conditions.

Now it will be shown later that α, β, γ are comparable with $\frac{\text{a}}{\text{C}-\text{A}}$, $\frac{\text{b}}{\text{C}-\text{A}}$, $\frac{\text{c}}{\text{C}-\text{A}}$; hence the terms in the second lines are much more important than those in the first. Thus in determining the values of F and G we may neglect the first lines.

Initially, the instantaneous axis coincides with the axis of greatest moment of inertia; so that when $t=0$, $\omega_1=\omega_2=0$, and therefore

$$\text{F} = -\frac{n}{\mu}\left(\alpha + \frac{\mathfrak{H}_1}{\text{A}}\right),\quad \text{G} = \frac{n}{\mu}\left(\beta + \frac{\mathfrak{H}_2}{\text{A}}\right)$$

The terms in F and G represent an inequality of 306 days period.

§ 3. *The Inequality of 306-Days Period*.*

I have worked out the values of $\mathfrak{H}_1$ and $\mathfrak{H}_2$ in two supposed cases of elevation, under certain suppositions as to the nature of the internal movements of the earth. In one of them I found $\frac{\mathfrak{H}_2}{\beta\text{A}} = \frac{1}{141}$, and $\mathfrak{H}_1 = 0$; and in the other $\mathfrak{H}_1 = \mathfrak{H}_2 = 0$. In order not to interrupt the thread of the argument, the calculation is given in Appendix A; it will also be more intelligible after the latter part of this paper has been read. In the general case the same kind of proportion will subsist between $\mathfrak{H}_1$ and $\text{A}\alpha$, $\mathfrak{H}_2$ and $\text{A}\beta$, and we may therefore, without serious error, neglect the former compared with the latter.

Thus, as far as concerns the present inequality,

$$\omega_1 = \frac{n\alpha}{\mu}(1-\cos\mu t) - \frac{n\beta}{\mu}\sin\mu t$$

$$\omega_2 = \frac{n\alpha}{\mu}\sin\mu t + \frac{n\beta}{\mu}(1-\cos\mu t)$$

* I have thought it necessary to discuss this inequality fully, both on account of its intrinsic interest, and because it has been referred to by the Astronomer Royal [Sir George Airy] and Sir William Thomson.

The former says (*Athenæum*, Sept. 22, 1860):—"Now, let us suppose the earth not absolutely rigid, but that there is susceptibility to change of form, either from that degree of yielding or fracture to which most solid substances are liable, or from the hydrostatic pressure of internal fluid. This, as I conceive, puts an end to all supposition of change of axis. The first day's whirl would again make the axis of rotation to be a principal axis, and the position of the axis would then be permanent."

But Sir George Airy is here speaking of the effect of the elevation of a mountain mass in about latitude 45°, by something like a gaseous explosion. This supposition is not at all in accordance with the belief of geologists, whereas a gradual elevation is so.

Sir W. Thomson, on the other hand, says (*Trans. Geol. Soc. Glasgow*, 1874, Vol. XIV., p. 312):—"In the present condition of the earth, any change in the axis of rotation could not be permanent, because the instantaneous axis would travel round the principal axis of the solid in a period of 296 days. . . . . In very early geologic ages, if we suppose the earth to have been plastic, the yielding of the surface might have made the new axis a principal axis. But certain it is that the earth at present is so rigid that no such change is possible." And he adds that practical rigidity has prevailed throughout geologic history.

On account of this inequality the greatest angular distance (in radians) of the instantaneous axis from the pole is $2(\alpha^2+\beta^2)^{\frac{1}{2}}/\mu$. It will appear from the latter part of this paper that, if the elevation of a large continent proceeds at the rate of two feet in a century, $(\alpha^2+\beta^2)^{\frac{1}{2}}$ may be about $\frac{1}{100}''$ per annum, and μ is 360° in 306 days; whence it follows that the greatest angle made by the instantaneous axis with the axis of figure is comparable with $\frac{1}{376}''$, a quantity beyond the power of observation. On the score of these terms the instantaneous axis will therefore remain sensibly coincident with the axis of figure.

They will, moreover, produce no secular alteration in the obliquity of the ecliptic, nor in the precession, because they will appear as periodic in $d\theta/dt$ and $\sin\theta\, d\psi/dt$, with arguments n and $n\pm\mu$.

Now although this inequality is so small, it nevertheless is of interest.

If we map, on a tangent plane to the earth at its initial pole, the relative motion of the instantaneous axis and the pole of figure, we get, as the equation to the curve,

$$x=\frac{\alpha}{\mu}(1-\cos\mu t)-\frac{\beta}{\mu}\sin\mu t$$

$$y=\frac{\alpha}{\mu}\sin\mu t+\frac{\beta}{\mu}(1-\cos\mu t)$$

If t be eliminated from these equations, we get

$$\left(x-\frac{\alpha}{\mu}\right)^2+\left(y+\frac{\beta}{\mu}\right)^2=\frac{\alpha^2+\beta^2}{\mu^2}$$

Thus the relative motion is a circle, passing through the origin, and touching a line inclined to the axis of y at an angle $\tan^{-1}\alpha/\beta$. Therefore the instantaneous axis describes a circle passing through the pole of figure every 306th day; and this circle touches the meridian, along which the axis of figure is travelling with uniform velocity, in consequence of the geological deformation of the earth.

The motion of the instantaneous axis in the earth is a prolate cycloid.

§ 4. *Adjustments to a Form of Equilibrium.*

If the earth were a viscous fluid there is no doubt but that the pole of figure would tend to displace itself towards the instantaneous axis, whose mean position would be the centre of the circle above referred to.

But Sir William Thomson has shown* that the earth is sensibly rigid; and in any case the earth is not a viscous fluid, properly so called, although it may be slightly plastic.

* In his Address to the British Association, 1876, he states that the argument derived from precession (Thomson and Tait's *Natural Philosophy*, p. 691) is fallacious; he adduces, however, a number of cogent arguments on this point.

M. Tresca has shown that all solids are plastic under sufficiently great stresses, but that, until a certain magnitude of stress is reached, the solid refuses to flow*. Now in the case of a very small inequality like this, the stresses introduced by the want of coincidence of the instantaneous axis with the axis of figure are very small, even when at their maximum; and every 306th day they are zero. It seems, therefore, extremely improbable that the stresses can be great enough to bring the earth into what M. Tresca calls the state of fluidity; and therefore it is unlikely that there can be any adaptation of the earth's form to a new form of equilibrium in consequence thereof.

In all the other inequalities introduced, whether arising from the first three terms above given in ω_1 and ω_2, or arising from the impressed forces, to be treated hereafter, the centre of the positions of the instantaneous axis is coincident with the pole of figure, and therefore there can hardly be any adaptation of figure eccentric to the axis of greatest moment to balance the stresses introduced by centrifugal force.

It would appear probable that, whilst a geological change is taking place, the earth is practically rigid for long periods. But as the earth comes to depart more and more from a form of equilibrium, the stresses due to the mutual gravitation of the parts, and to the rotation, increase gradually, until they are sufficiently great to cause the solid matter to flow. A rough kind of adjustment to a form of equilibrium would then take place. The existence of continents, however, shows that this adjustment does not take place by the subsidence of the upheaved part; and as this adaptation of form would be produced by an entirely different cause from that to which the upheaval was due, that upheaval would probably persist independently of the approximate adoption of a new form of equilibrium by the earth.

M. Tresca's experiments on the punching of metals would lead one to believe that the change would take place somewhat suddenly, and would in fact be by an earthquake, or a succession of earthquakes. On each of these occasions the tendency would be to adjust the form to one of equilibrium about the instantaneous axis. Now the principal axis λ_3 has (in consequence of the postulated deformation) been travelling along the meridian in longitude $\pi + \tan^{-1} \alpha/\beta$, measured from the plane containing λ_1 and λ_3.

The earthquake will take place when, to the stresses due to mutual gravitation, are superadded the maximum stresses due to centrifugal force; that is to say, when the instantaneous axis is at its greatest distance from λ_3, the axis of greatest moment of inertia. At the instant of the earthquake the principal axis will be moved towards the position of the instantaneous axis. And as the circle described by the instantaneous axis touches the meridian of displacement of the principal axis, therefore the principal axis will be carried

* "Sur l'écoulement des Corps Solides," *Mém. des Sav.*, tom. XVIII.

by the adjustment towards the centre of the circle described by the instantaneous axis, and therefore perpendicular to the meridian of displacement.

Thus, if the adjusting earthquakes take place at long intervals, the motion of the principal axis will not deviate sensibly from continuing along the meridian, along which it would travel in consequence merely of the geological deformation. If, however, the adjustments are frequent, the path of λ_3 will diverge sensibly from the meridian along which it started. If the readjustments become infinitely frequent and infinitely small, there is a continuous flow of the matter of the earth, which is always seeking to bring back the earth's figure to one of equilibrium, from which figure it is also supposed to be continuously departing under the action of internal forces. In this state the earth may be considered as formed of a stiff viscous fluid.

According to these ideas, at each adjustment λ_1, λ_2, λ_3 will be suddenly reduced to nearly their primitive values, A, A, C; but α, β, γ depend on the rate of accession and diminution of matter at various parts of the earth, and remain constant. The only effect, then, is that each adjusting earthquake must be taken as a new epoch.

As far as I can see, it seems quite possible that the earth may be sensibly rigid to the tidally deforming influences of the sun and moon, and yet may bring itself back from any considerable departure from a form of equilibrium to approximately that form. It therefore seems worth while to consider the case of the adjustments being continuous, whilst the deformation is also continuous.

§ 5. *Adjustments to the Form of Equilibrium continuous**.

I therefore propose to consider geometrically, but not dynamically, the paths of the instantaneous axis, and of the principal axis, when the earth is viscous and continuously deformed by internal forces. It is supposed that the velocities of flow of the matter of the earth are so small that inertia may be neglected, and that the displacements are so small that the principle of the superposition of small motions is applicable.

As before the paths of the instantaneous and principal axes may be mapped on a tangent plane to the spheroid, at the extremity of the primitive pole, the mean radius of the spheroid being taken as unity.

In consequence of the continuous deformation, the principal axis travels with a linear velocity (on the map) $-\sqrt{(\alpha^2+\beta^2)}$ along the meridian of longitude $\tan^{-1}\alpha/\beta$. Take this meridian as axis of x, and measure y, so that the angular velocity μ is from x towards y, and call $\sqrt{(\alpha^2+\beta^2)}$, u.

Then the principal axis λ_3 moves along the axis of x with a uniform linear velocity $-u$, and, from dynamical principles, the instantaneous axis I moves round the instantaneous position of λ_3 with a uniform angular velocity μ.

* [Compare § 4, Paper 1, "On the tides of viscous...spheroids...," Vol. II., p. 11.]

But because of the earth's viscosity, λ_3 always tends to approach I. The stresses introduced in the earth by the want of coincidence of λ_3 with I vary as λ_3I. Also the amount of flow of a viscous fluid, in a small interval of time, varies jointly as that interval and the stress. Hence the linear velocity (on the map), with which λ_3 approaches I, varies as λ_3I (equal to r suppose). Let this velocity be νr, where ν depends on the viscosity of the earth, diminishing as the viscosity increases.

Thus the principal axis describes a sort of curve of pursuit on the map; it is animated with a constant velocity $-u$ parallel to x, and with a velocity νr towards I, which rotates round it with a uniform angular velocity μ.

The motion of I, relative to λ_3, is that of a point moving with a constant velocity u parallel to x, rotating round a fixed point with a constant angular velocity μ, and moving towards that point with a velocity νr.

Let ξ, η be the relative coordinates of I with respect to λ_3, and x, y the coordinates of λ_3. Then the differential equations which give the above motions are:—

$$\frac{d\xi}{dt} = u - \nu\xi - \mu\eta \quad \text{.............................(1)}$$

$$\frac{d\eta}{dt} = -\nu\eta + \mu\xi \quad \text{.............................(2)}$$

$$\frac{dx}{dt} = -u + \nu\xi \quad \text{.............................(3)}$$

$$\frac{dy}{dt} = \nu\eta \quad \text{.............................(4)}$$

If (1) and (2) be integrated, and the constants determined so that, when $t=0$, $\xi=\eta=0$ (which expresses that initially λ_3 and I are coincident), it will be found that

$$\xi = \frac{u}{\nu^2+\mu^2}\left\{\nu\left(1-e^{-\nu t}\cos\mu t\right) + \mu e^{-\nu t}\sin\mu t\right\}$$

$$\eta = \frac{u}{\nu^2+\mu^2}\left\{\mu\left(1-e^{-\nu t}\cos\mu t\right) - \nu e^{-\nu t}\sin\mu t\right\}$$

These give the path of I relative to λ_3. It may be seen to be a spiral curve diminishing with more or less rapidity, according as the earth is less or more viscous. If $\nu=0$, it becomes the circle found above from the dynamical equations.

Substitute in (3) and (4) for ξ and η; integrate, and determine the constants, so that when $t=0$, $x=y=0$. It will then be found that

$$x = -\frac{u}{\nu^2+\mu^2}\left[\nu\left(\nu^2-\mu^2\right) + \mu^2 t + \frac{\nu}{\nu^2+\mu^2}\left\{-\left(\nu^2-\mu^2\right)e^{-\nu t}\cos\mu t + 2\mu\nu e^{-\nu t}\sin\mu t\right\}\right]$$

$$y = \frac{u}{\nu^2+\mu^2}\left[-2\mu\nu^2 + \mu\nu t + \frac{\nu}{\nu^2+\mu^2}\left\{2\mu\nu e^{-\nu t}\cos\mu t + \left(\nu^2-\mu^2\right)e^{-\nu t}\sin\mu t\right\}\right]$$

These give the path of λ_3 on the map. It may be seen to be a cycloidal curve, in which the radius of the rolling circle diminishes with more or less rapidity, according as the earth is less or more viscous.

After some time $e^{-\nu t}$ becomes very small, and the motion is steady; and then $\xi = u\nu/(\mu^2+\nu^2)$, $\eta = u\mu/(\nu^2+\mu^2)$, or I is fixed, relatively to λ_3, at a distance $u/(\nu^2+\mu^2)^{\frac{1}{2}}$ from it, and on the meridian, measured from the axis of x, in longitude $\tan^{-1}\mu/\nu$. This point is the centre of the above-mentioned spiral curve.

If ν be very small (or the earth nearly rigid) this meridian differs by little from the axis of y. But it may be that ν is so small that $e^{-\nu t}$ has not time to become insensible before the geological changes cease. This case corresponds very nearly to the hypothesis, in the last section, of adjusting earthquakes.

If the earth be very mobile, or ν large, $\xi = u/\nu$, $\eta = 0$.

Again, with respect to the path of λ_3, when the motion has become steady,

$$x = -\frac{u\nu(\nu^2-\mu^2)}{\nu^2+\mu^2} - \frac{\mu^2 u}{\nu^2+\mu^2}t$$

$$y = -\frac{2\mu u\nu^2}{\nu^2+\mu^2} + \frac{\mu u\nu}{\nu^2+\mu^2}t$$

and eliminating t, $\nu x + \mu y = -u\nu^2$.

That is to say, when the motion is steady, λ_3 moves parallel to the meridian of longitude $\pi - \tan^{-1}\nu/\mu$, and distant from it $u\nu^2/(\nu^2+\mu^2)^{\frac{1}{2}}$ on the negative side. This straight line is the degraded form of the above-mentioned cycloidal curve.

If the earth is nearly rigid this path does not differ sensibly from the axis of x; if very mobile, it is nearly perpendicular to the axis of x, and a long way from the origin. In this last case the solution becomes nugatory, except as showing that the very small inequality of 306 days would be capable of disturbing and quite altering the path of the principal axis, as arising merely from geological changes on the surface of the earth.

In the case contemplated by the Astronomer Royal, where the elevation is explosive, u must be put equal to zero, and the constants of integration so determined, that when $t=0$, $\xi = \mathrm{R}$ suppose, and $\eta = x = y = 0$. It will then be found that when the agitation has subsided, $x = \mathrm{R}\nu^2/\mu^2$, $y = \mathrm{R}\nu/\mu$, or the pole of figure will have taken up a position on one side of the meridian, along which it was initially propelled by the explosion.

It thus seems probable that during the consolidation of the earth there was a great instability in the position of the principal axis, and therefore also of the axis of rotation which followed it.

§ 6. *Secular alteration in the obliquity of the Ecliptic, resulting from terms independent of the Impressed Forces.*

To return to the main line of the inquiry:—If the values of ω_1 and ω_2, found in § 2, be substituted in the geometrical equations for $d\theta/dt$ and $\sin\theta d\psi/dt$ (see § 1) a number of periodic terms will arise, and these terms have diurnal and semidiurnal periods, but their amplitudes are so small that they have no practical interest.

The only thing which concerns us is to inquire whether there can be any secular change in the obliquity of the ecliptic.

Select, then, only terms in $\sin nt$ in ω_1, and in $\cos nt$ in ω_2, and substitute in the geometrical equation $d\theta/dt = -\omega_1 \sin nt + \omega_2 \cos nt$, and reject periodic terms. It will then be found that

$$\frac{d\theta}{dt} = -\frac{\Pi}{2An}(a + b - 2c)$$

§ 7. *Terms dependent on the Impressed Forces.*

It now remains to consider the effect of the impressed forces on the precession and obliquity of the ecliptic.

The equations of motion are reduced to

$$\frac{d\omega_1}{dt} + \frac{C-A}{A}\omega_3\omega_2 = \frac{L}{A}$$

$$\frac{d\omega_2}{dt} - \frac{C-A}{A}\omega_3\omega_1 = \frac{M}{A}$$

$$\frac{d\omega_3}{dt} = \frac{N}{A}$$

If we write $L + \delta L$, $M + \delta M$, δN for L, M, N, and indicate by L and M the couples caused by the attractions of the sun and moon on the protuberant parts of the earth before it has begun to change its shape, then L and M only cause the ordinary precession and nutations. For the present problem it is therefore only necessary to consider the effects of δL, δM, δN, which arise from the change of shape of the earth.

It follows, from the same arguments that were used in § 1, that the change in the earth's angular velocity of rotation due to δN will only have a very small effect on ω_1 and ω_2; so that, as far as is now important, ω_3 may be put equal to $-n$ in the first two equations, which may then be written

$$\frac{d\omega_1}{dt} - \mu\omega_2 = \frac{\delta L}{A}$$

$$\frac{d\omega_2}{dt} + \mu\omega_1 = \frac{\delta M}{A}$$

Now δL and δM are the changes in L and M, when $A + at$, $A + bt$, $C + ct$ are written for A, A, and C respectively. If δL, δM be thus formed, and the equations integrated, it will be found that the principal terms, arising from the sun's attraction, are nine both in $d\theta/dt$ and $\sin\theta\, d\psi/dt$; the same number of terms arise in the precession and nutation with respect to the plane of the lunar orbit, and these would have to be referred to the ecliptic. Sixteen out of the eighteen terms represent, however, only very small nutations, and the only terms of any interest are those which give rise to a secular change in the obliquity of the ecliptic. These terms may be picked out without reproducing the long calculation above referred to, for they arise entirely out of the constant couple acting about the equinoctial line, which gives rise to the uniform precession.

Now this constant couple is $C\Pi n$; whence

$$L = C\Pi n \sin nt, \qquad M = -C\Pi n \cos nt$$

And since Π involves $(C - A)/C$, therefore

$$\delta L = -C\Pi n \frac{b - c}{C - A} t \sin nt, \quad \delta M = -C\Pi n \frac{c - a}{C - A} t \cos nt$$

If these be substituted in the equations of motion and the equations integrated, and only terms in $\sin nt$ in ω_1 and those in $\cos nt$ in ω_2 be retained, we get

$$\omega_1 = -\frac{\Pi}{n}\frac{b - c}{C - A}\sin nt, \quad \omega_2 = -\frac{\Pi}{n}\frac{c - a}{C - A}\cos nt$$

Substituting in the geometrical equation $d\theta/dt = -\omega_1 \sin nt + \omega_2 \cos nt$ and rejecting periodic terms,

$$\frac{d\theta}{dt} = \frac{\Pi}{2n}\frac{a + b - 2c}{C - A}$$

§ 8. *General result with respect to the Obliquity of the Ecliptic.*

It was found in § 6 that the secular rate of change of θ, as due to the internal changes in the earth, was $-\frac{\Pi}{2n} \cdot \frac{a + b - 2c}{A}$. Since $C - A$ is small compared to A, this term is small compared with the term found at the end of § 7. Hence, finally, taking all the terms together, we get the approximate result,

$$\frac{d\theta}{dt} = \frac{\Pi}{2n} \cdot \frac{a + b - 2c}{C - A}$$

and for small changes in the obliquity, insufficient to affect Π materially,

$$\theta = i + \frac{\Pi}{2n} \cdot \frac{a + b - 2c}{C - A} t$$

This equation has been obtained on the supposition that the change in the earth's form never becomes so great that at, bt, ct exceed small fractions of $C - A$; a condition which is satisfied in the case of such geological changes as those of which we have any cognizance at present.

It will appear from a comparison with results given hereafter, that $\frac{a+b-2c}{C-A}t$ cannot ever exceed two or three degrees; and since $\Pi/2n$ is a very small fraction, it follows that *the obliquity of the ecliptic must have remained sensibly constant throughout geological history**. *Also the instantaneous axis of rotation must always have remained sensibly coincident with the principal axis of figure, however the latter may have wandered in the earth's body.*

It has hitherto been assumed that the change of form and the angular velocities of the principal axes in the earth's body are uniform. But the preceding investigation shows clearly that no material change would be brought about by supposing the changes to proceed with varying velocities. This being so, dynamical considerations may be dismissed henceforth; and accordingly the next part of this paper will be devoted to the kinematical question, as to the change in position of the earth's axis of figure as due to geological changes.

* During the Glacial Period there must have been heavy ice-caps on one or both poles of the earth. The above equation will give the disturbance of the obliquity of the ecliptic produced thereby.

I will take what I believe is the most extreme view held by any geologist. Mr Belt is of opinion that an enormous ice-sheet, which was thickest in about lat. 70° N. and S., descended from both poles down to lat. 45°; the amount of ice was so great that the sea stood some 2000 feet lower than now throughout the unfrozen regions between lat. 45° N. and S.

Suppose that the whole of this equatorial region was sea, and that the water contained in 2000 feet of depth of this sea was gradually piled on the polar regions in the form of ice. Then the effect in diminishing C and increasing A cannot be so great as if the whole of this mass were subtracted actually from the equator and piled actually on the poles. The latter supposition will then give a superior limit to the amount of alteration in the obliquity of the ecliptic. I have calculated this alteration by means of the above formula, taking the numerical data used later in this paper, and taking the specific gravity of water to that of surface-rock as 4 to 11. I find, then, that the superior limit to the increase of the obliquity of the ecliptic would be 0''·00045; that is to say, the position of the arctic circle cannot have been shifted so much as half an inch. And this is an accumulated effect, and the matter is distributed in the most favourable manner possible.

In this case the amount of matter displaced is enormous, and is placed in the most favourable position for affecting the obliquity; hence, *à fortiori*, geological changes in the earth cannot have sensibly affected the obliquity.

But although this equation leads to no startling results in the geological history of the earth, I hope to show in a future paper that it may have some bearing on the very remote history of the earth and of the other planets [see Paper 3, p. 51]. In consequence of a mistake in the work it was erroneously stated in the abstract of this paper in the *Proceedings* that the change in the position of the arctic circles might amount to 3 inches, instead of to half an inch.

The various assumptions made above will incidentally be justified in the course of the work.

For some remarks of Sir William Thomson on this part of the paper see Appendix C.

II. The Principal Axes of the Earth.

§ 9. *Preliminary Assumptions.*

It is assumed at first that, in consequence of some internal causes, the earth is undergoing a deformation, but that there is no disturbance of the strata of equal density, and that there is no local dilatation or contraction in any part of the body. The cases at present excluded will be considered later.

The result of this assumption is, that the volume of the body remains constant, and that the parts elevated or depressed above or below the mean surface of the ellipsoid have the same density as the rest of the surface. Such changes of form must, of course, be produced by a very small flow of the solid matter of the earth. Since the whole volume remains the same, this hypothesis may be conveniently called that of incompressibility; although, if the matter of the earth flowed quite incompressibly, there would be some slight dislocation of the strata of equal density.

It is immaterial for the present purpose what may be the forces which produce, and the nature of, this internal flow; but it was assumed in the dynamical investigation that the forces were internal, and that the flow proceeded with uniform velocity.

After deformation the body may be considered as composed of the original ellipsoid, together with a superposed layer of matter, which is positive in some parts and negative in others. The condition of constancy of volume necessitates that the total mass of this layer should be zero. If we take axes with the origin at the centre of the ellipsoid and symmetrical thereto, and let $hF(\theta, \phi)$ represent the depth of the layer at the point θ, ϕ, the condition of incompressibility is expressed by the integral of $F(\theta, \phi)$ over the surface of the ellipsoid being zero. Then by varying h, elevations and depressions of various magnitudes may be represented.

§ 10. *Moments and Products of Inertia after Deformation.*

Before the deformation:—

Let A, C be the principal moments of inertia of the earth; a, b its semi-axes; M its mass; $\mathfrak{D}$ its mean density; ρ its surface density; and c its mean radius, so that $3c = 2a + b$; and let the earth's centre of inertia be at the origin.

After the deformation:—

Let a, b, c, D, E, F be the moments and products of inertia of the above ideal shell of matter about the axes; x_1, y_1, z_1 the coordinates of the earth's centre of inertia.

Then, since the ellipticity of the earth is small, the integrals may be taken over the surface of a sphere of radius c, instead of over the ellipsoid. Therefore,

$$\mathrm{a} = h\rho c^4 \iint \mathrm{F}(\theta, \phi) \sin\theta \, (\sin^2\theta \sin^2\phi + \cos^2\theta) \, d\theta \, d\phi$$

$$\mathrm{M}x_1 = h\rho c^3 \iint \mathrm{F}(\theta, \phi) \sin^2\theta \cos\phi \, d\theta \, d\phi$$

$$\mathrm{M} = \tfrac{4}{3}\pi \mathfrak{D} c^3$$

and other integrals of a like nature for b, c, D, E, F, y_1, z_1.

Since $\iint \mathrm{F}(\theta, \phi) \sin\theta \, d\theta \, d\phi = 0$, therefore $\mathrm{a} + \mathrm{b} + \mathrm{c} = 0$

If A be the moment of inertia of the body, after deformation, about an axis parallel to x, through x_1, y_1, z_1,

$$A = \mathrm{A} + \mathrm{a} - \mathrm{M}(y_1^2 + z_1^2)$$

Now a varies as h/c, whilst $\mathrm{M}(y_1^2 + z_1^2)$ varies as $(h/c)^2$. But the greatest elevation or depression to be treated of is about two miles, whilst the mean radius c is about 4000 miles; hence h/c cannot exceed about $\frac{1}{2000}$, and accordingly the term $\mathrm{M}(y_1^2 + z_1^2)$ is negligible compared to a. Whence $A = \mathrm{A} + \mathrm{a}$.

In like manner, the terms introduced in the other moments and products of inertia by the shifting of the earth's centre of inertia are negligible compared to the direct changes. Thus it may be supposed that the centre of inertia remains fixed at the origin, and that the moments and products of inertia of the earth after deformation are A + a, A + b, C + c, D, E, F.

§ 11. *General Theorem with respect to Principal Axes.*

A general theorem will now be required to determine the position of the principal axes after the deformation.

Take as axes the principal axes of a body about which its moments of inertia are A, B, C. Let the body undergo a small deformation, which turns the principal axes through small angles α, β, γ about the axes of reference, and makes the new principal moments A′, B′, C′. And let the moments and products about the axes of reference become in consequence $A+a$, $B+b$, $C+c$, D, E, F. Then it is required to find α, β, γ in terms of these last quantities.

Let l, m, n be the direction cosines of any line through the origin, and let them remain unaltered by the deformation. Let I be the moment of inertia about this line after deformation. Let $l+\delta l$, $m+\delta m$, $n+\delta n$ be the direction cosines of the line with respect to the new principal axes. Then, by a well-known theorem,

$$\delta l = \gamma m - \beta n, \quad \delta m = \alpha n - \gamma l, \quad \delta n = \beta l - \alpha m$$

Now

$$I = (A+a)\, l^2 + (B+b)\, m^2 + (C+c)\, n^2 - 2Dmn - 2Enl - 2Flm$$

But it is also equal to

$$A'(l+\delta l)^2 + B'(m+\delta m)^2 + C'(n+\delta n)^2$$

and by substituting for δl, δm, δn, this is equal to

$$A'l^2 + B'm^2 + C'n^2 - 2mn\,(C'-B')\,\alpha - 2ln\,(A'-C')\,\beta - 2lm\,(B'-A')\,\gamma$$

to the first order of small quantities.

This expression must be identical with the former for all values of l, m, n; hence putting $l=1$, $m=n=0$, $A'=A+a$, and similarly $B'=B+b$, $C'=C+c$. Wherefore also

$$\alpha = \frac{D}{C'-B'} = \frac{D}{C-B} \text{ nearly}$$

and

$$\beta = \frac{E}{A-C}, \quad \gamma = \frac{F}{B-A}$$

and these are the required expressions for α, β, γ.

If, however, $B=A$, γ becomes infinite, and the solution is nugatory: but since, under this condition, all axes in the plane of xy were originally principal axes, the axes of reference may always be so chosen that F is zero absolutely; and then

$$\alpha = \frac{D}{C-A}, \quad \beta = -\frac{E}{C-A}, \quad \gamma = 0$$

Therefore the new principal axis C′ is inclined to the old C at a small angle $(D^2+E^2)^{\frac{1}{2}}/(C-A)$, and is displaced along the meridian, whose longitude, measured from the plane of xz, is $\pi + \tan^{-1} D/E$. This is the case to be dealt with in the present problem. The positions of the other principal axes will be of no interest.

§ 12. *Application of preceding Theorem.*

To solve the problem numerically in any particular case, it will be necessary to find the integrals

$$D = h\rho c^4 \iint F(\theta, \phi) \sin^2\theta \cos\theta \sin\phi \, d\theta d\phi$$

$$E = h\rho c^4 \iint F(\theta, \phi) \sin^2\theta \cos\theta \cos\phi \, d\theta d\phi$$

If $D/h\rho c^4$ and $E/h\rho c^4$ be called d and e, then d and e stand for the above integrals, which depend on the distribution of surface-matter in continents and seas.

It will be convenient to use a foot as the unit for measuring h, and seconds of arc for the measurement of the inclination i of the new principal axis to the old. For this purpose the value of the coefficient $\rho c^4/(C - A)$ may be calculated once for all. Let its value when multiplied by the appropriate factors for the use of the above units be called K*. Now

$$C - A = \tfrac{2}{3}(\epsilon - \tfrac{1}{2}m) Ma^2 = \tfrac{2}{3}(1 + \tfrac{2}{3}\epsilon)(\epsilon - \tfrac{1}{2}m) Mc^2$$

Then if we take $\epsilon = \cdot 0033439$, being the mean of the values given by Colonel A. R. Clarke, $m = 1/289\cdot 66$, and $c = 20{,}899{,}917$ feet†, $M = \tfrac{4}{3}\pi \mathfrak{D} c^3$, and $\mathfrak{D}/\rho = 2$, we get

$$C - A = \tfrac{8}{3}\pi\rho c^4 \times \cdot 0010809 \times 20{,}899{,}917$$

and

$$K = 1\cdot 08986$$

If, in accordance with Thomson and Tait, $\mathfrak{D}/\rho = 2\cdot 1$, $K = 1\cdot 0380$, but I shall take K as 1·090. Then we have $i'' = Kh\sqrt{(d^2 + e^2)}$, where $K = 1\cdot 090$, h being measured in feet, and i'' being the angular change in the position of the principal axis of greatest moment of inertia of the earth, due to a deformation given by $hF(\theta, \phi)$ all over the surface of the spheroid.

The angle $\dfrac{a + b - 2c}{C - A} t$ is clearly of the same order of magnitude as i, as it was assumed to be in Part I.

* I have to thank Prof. J. C. Adams for his help with respect to the numerical data, and for having discussed several other points with me.

† See Thomson and Tait, *Natural Philosophy*, pp. 648, 651.

III. FORMS OF CONTINENTS AND SEAS WHICH PRODUCE THE MAXIMUM DEFLECTION OF THE POLAR AXIS.

§ 13. *Conditions under which the Problem is treated.*

On the hypothesis of incompressibility, the effect of a deformation in deflecting the pole is exactly equivalent to the removal of a given quantity of matter from one part of the earth's surface to another. But as no continent exceeds a few thousand feet in average height, the removal is restricted by the condition that the hollows excavated, and the continents formed, shall nowhere exceed a certain depth and height. The areas of present continents and seas, and their heights and depths, give some idea of the amount of matter at disposal, as will be shown hereafter. It is interesting, therefore, to determine what is the greatest possible deflection of the pole which can be caused by the removal of given quantities of matter from one part of the earth to another, subject to the above condition as to height and depth.

§ 14. *Problem in Maxima and Minima.*

This involves the following problem:—To remove a given quantity of matter from one part of a sphere to another, the layers excavated or piled up not being greater than k in thickness, so as to make $\sqrt{(D^2 + E^2)}$ a maximum, the axes being so chosen as to make $F = 0$.

If D', E' be the products of inertia referred to other axes having the same origin and axis of z as before, it may easily be shown, from the fact that $D^2 + E^2 = D'^2 + E'^2$, that $D^2 + E^2$ is greatest and equal to E'^2 for that distribution of matter which makes $D' = 0$ and E' a maximum.

The problem is thus reduced to the following:—Rectangular axes are drawn at the centre of a sphere of radius c; it is required to effect the above-described removal of matter, so that the product of inertia about a pair of planes through z, and inclined to xz at 45° on either side, shall be a maximum, subject to the above condition as to depth, k being small compared to c. For convenience, I refer to the plane xy as the equator, to xz as prime meridian, from which longitudes ψ are measured from x towards y, and to θ the colatitude. These must not be confused with the terrestrial equator, longitude, and latitude.

A little consideration shows that the seas and continents must be of uniform depth k, that there must be two of each, that they must all be of the same shape, must be symmetrical with respect to the equator, and that the continents must be symmetrical with respect to the prime meridian, and the seas with respect to meridians 90° and 270°.

Also the total product of inertia P, produced by this distribution, is 16 times that produced by the part of one continent lying in the positive octant of space; and the mass of matter removed is 8 times the mass of this same portion of one continent.

The problem is, therefore, to find the outline of the continent, so that P may be a maximum, subject to the condition that the mass is given.

Take the surface-density of the sphere as unity, and let the mass removed be given as an elevation of a height k over a fraction q of the whole sphere's surface; so that the mass removed from hollow to continent is $4\pi c^2 kq$. Then it may easily be shown that

$$P = 4kc^4 \int_0^{\frac{1}{2}\pi} \sin^3\theta \sin 2\psi \, d\theta$$

and

$$q = \frac{1}{\pi}\int_0^{\frac{1}{2}\pi} 2\psi \sin\theta \, d\theta$$

where ψ is a function of θ to be determined. Then writing ω for 2ψ, and μ for $\cos\theta$, we have to make

$$\int_0^1 \{(1-\mu^2)\sin\omega - \omega\cos^2\alpha\}\, d\mu \text{ a maximum}$$

for it will be seen later that $-c^2\cos^2\alpha$ is a proper form for the constant, to be introduced according to the principles of the Calculus of Variations. This leads at once to

$$(1-\mu^2)\cos\omega = \cos^2\alpha$$

or

$$\sin^2\theta\cos 2\psi = \cos^2\alpha$$

That is to say, the outline of the continent is the sphero-conic formed by the intersection with the sphere of the cone, whose Cartesian equation is

$$y^2(1+\cos^2\alpha) + z^2\cos^2\alpha = x^2\sin^2\alpha$$

Reverting to the expressions for P and q, altering the variable of integration, and the limits, so as to exclude the imaginary parts of the integrals, we have as the equation to find α

$$q = \frac{1}{\pi}\int_0^{\alpha} \cos\chi \cos^{-1}\left(\frac{\cos\alpha}{\cos\chi}\right)^2 d\chi$$

and

$$P = 4kc^4 \int_0^{\alpha} \cos\chi \sqrt{(\cos^4\chi - \cos^4\alpha)}\, d\chi$$

These integrals are reducible to elliptic functions; but in order not to interrupt the argument, I give the reduction in Appendix B. If $\cos 2\gamma = \cos^2\alpha$, the result is that

$$\pi q = \sqrt{2}\,\frac{\cos 2\gamma}{\cos\gamma}\,[\Pi^1(-2\sin^2\gamma) - F^1]$$

or

$$q = 1 - \frac{2}{\pi}\{E^1 F - F^1(F-E)\}$$

and
$$\frac{P}{kc^4} = \tfrac{8}{3}\sqrt{2}\cos\gamma\,[E^1 - \cos 2\gamma F^1]$$

where the modulus of the complete functions E^1, F^1, Π^1 is $\tan\gamma$, and where E, F have a modulus $\cos\alpha/\cos\gamma$ and an amplitude $\frac{1}{2}\pi - \gamma$.

It will be observed that α is the semi-length of the continent in latitude, and γ the semi-breadth in longitude.

From these expressions I have constructed the following Table:—

Semi-breadth of continent (γ)	Semi-length of continent (α)	Fraction of surface elevated or depressed (q)	Product of inertia $\left(\frac{P}{kc^4}\right)$
°	° ′		
0	0 0	·0000	·0000
5	7 5	·0054	·0672
10	14 13	·0216	·2628
15	21 28	·0486	·5697
20	28 55	·0867	·9603
25	36 42	·1362	1·3981
30	45 0	·1979	1·8399
35	54 12	·2732	2·2371
40	65 22	......	......
45	90 0	·5000	2·6667

§ 15. *Application of preceding problem to the case of the Earth.*

In the application to the case of the earth, what has been called, for brevity, the equator (EE in fig. 2) must be taken as a great circle, passing through a point in *terrestrial* latitude 45°.

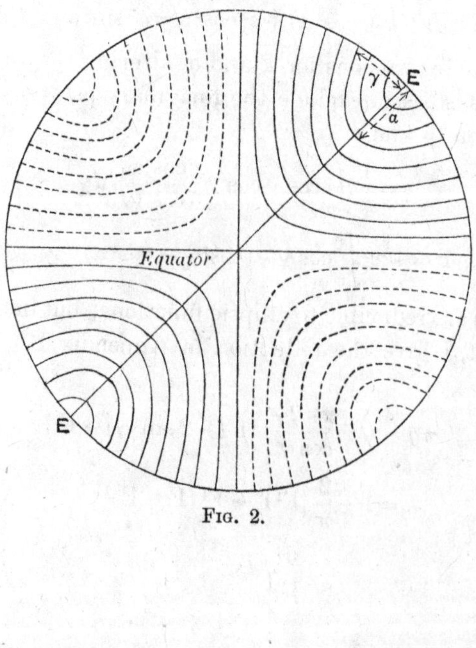

FIG. 2.

Figure 2 gives the stereographic projection of the forms of continents and seas, the firm lines showing continents, and the broken ones seas, when covering various fractions of the whole surface; α and γ are indicated on one of the continents. The other hemisphere is the same as this figure, when seen in a looking-glass. It will be observed that the limiting case is when the two continents fill up two quarters of the earth, and the two seas the other two.

It is clear that the greatest deflection of the polar axis which can be produced by the elevation of continents of height k and having a total area $4\pi c^2 q$, and the depression of similar seas, will be PK.

A numerical Table of results will be given below, formed by interpolation in the above Table.

IV. On Geological Changes on the Earth's Surface.

§ 16. *The points to be considered.*

It is now necessary to consider what kind and amount of superficial changes are brought about in the earth's shape by such geological changes as are believed to have taken place. The points to be determined are:—

i. Over what extent of the earth's surface is there evidence of consentaneous subsidence, or upheaval, during any one period.

ii. What is the extreme vertical amount of that subsidence or upheaval.

iii. How the sea affects the local excesses and deficiencies of matter on the earth's surface.

iv. How marine and aërial erosion affect the distribution of the excess or deficiency of matter.

v. The possibility of wide-spread deformations of the earth, which approximately carry the level surfaces with them.

The object of this discussion is to find what areas and amounts of elevation and subsidence on a sealess and rainless globe are equivalent, as far as moving the principal axis, to those which obtain on the earth. These areas and effects will be referred to as "effective areas and amounts of elevation or subsidence."

It is probable that during the elevation or subsidence of any large area, the change proceeds at unequal rates in different parts; probably one part falls or rises more quickly than another, and then the latter gains on the former. But it has been shown, in the dynamical part of this paper, that the axis of rotation sensibly follows the axis of figure. Hence it is immaterial by what course the earth changes its configuration, provided the changes do not proceed by large impulses, a supposition which may be certainly excluded. The essential point is, to compare the final and initial distributions of matter, after and before a period of large geographical change.

§ 17. *Areas of subsidence and elevation.*

When a new continent is being raised above the sea, there is no certainty as to the extent to which areas in the adjoining seas partake in the elevation; even in the case of S. America, where the area of elevation is supposed to be abruptly limited towards the west, the line of 15,000 feet depth lies a long way from the coast.

As soon, moreover, as the land is raised above the sea, the rivers begin washing away its surface, and the sea eats into its coasts. The materials of the land are carried away, and deposited in the surrounding seas. Thus to form a continent of 1000 feet in height, perhaps entails an elevation of the surface of from 3000 to 4000 feet, and all the matter of the additional 2000 to 3000 feet is deposited in the sea. This tends to make the adjoining seas shallower, and to cause some increase to the area of the land. Therefore in a sealess globe the effect must be represented by a greater area of elevation and a less height.

The bed of a deep sea is hardly at all subject to erosion, and therefore the tendency seems to be to make the negative features of an ocean-bottom more pronounced than the positive features of mountain-ranges, at least in the parts very remote from land.

The areas, then, of existing continents may not be a due measure of the areas of effective elevation; we can only say that the latter may considerably exceed the former. The direct evidence as to the extent of the earth's surface over which there has been a general movement during any one period, is also very meagre. It appears certain that very large portions of S. America have undergone a general upward movement within a recent geological period; but there is no certainty whatever as to the limits of this area, nor as to whether the beds of the adjoining seas have partaken to any extent of this general movement. Thus the case of S. America is of scarcely any avail in determining the point in question. The presence of deep ocean up to the Chilian coast seems, however, to make it probable that areas of elevation are more or less abruptly divided from those of rest or subsidence.

There is only one area of large extent in which we possess fairly well-marked evidence of a general subsidence; and this is the area embracing the Coral islands of the Pacific Ocean. The evidence is derived from the structure of the Coral islands, and is confirmed in certain points by the geographical distribution of plants and animals. Some naturalists are of opinion that there is evidence of the existence of a previous continent; others (and amongst them my father, Mr Charles Darwin) that there existed there an archipelago of islands. In this dearth of precise information, only a rough estimate of area is possible.

My father, who has especially attended to the subject of the subsidence of the Pacific islands, has marked for me, on the map given in his work on Coral Reefs, a large area which he believes to have undergone a general subsiding motion. This area runs in a great band from the Low Archipelago to the Caroline Islands, and embraces the greater number of the islands coloured dark-blue in his map. The boundary may be defined as passing through:—

	N.							S.							
Lat. ...	3	5	15	22	18	10	5	5	15	25	30	18	15	10	8
Long...	150	140	150	165	180	165	150	135	120	120	135	150	165	180	165
	E.					W.									E.

He also marked a smaller area, embracing New Caledonia, the S.E. corner of New Guinea, and the N.E. coast of Australia.

It is noteworthy that the former large area consists of sea more than 15,000 feet deep, except in patches round some of the islands, where it appears to be from 10,000 to 15,000 feet deep*.

I marked these areas on a globe, and cut out a number of pieces of paper to fit them, and then weighed them. By this method I determined that the former area was ·055 of the whole surface of the globe, and the latter was ·01; the two together were therefore ·065.

It thus appears that we have some evidence of an area of between 5 and 7 per cent. of the globe having undergone a general motion of subsidence within a late geological period. But between this area and the coast of S. America there is a vast and deep ocean, and nothing whatever is known with respect to the movements of its bed. Hence it is quite possible that the area which has really sunk, in this quarter of the globe, is considerably larger than the one above spoken of.

On the whole, then, perhaps from ·05 to ·1 of the whole surface may at various times have partaken of a consentaneous movement, so as to convert deep sea into land, and *vice versâ*.

Besides this kind of general movement, there have certainly been many more or less local rises and falls, but this small oscillation is not fitted to produce any sensible effect on the position of the earth's axis.

* See frontispiece-map to Wallace's *Geographical Distribution of Animals*.

§ 18. *Amount of Elevation, and the effects of Water.*

Humboldt has shown that the mean height of the present continents is a little less than 1100 feet from the sea-level*. But this, of course, does not give the limit to the amount of change of level. On the other hand, there are perhaps 50,000 to 80,000 feet of superposed strata at most places on the earth; but neither does this give the indication required, because the surface must have risen and fallen many times during the deposition of these strata.

But, as before pointed out, the actual upward or downward movement of land is by no means the same as its effective elevation or subsidence; for erosion causes the effective to be far slower than the actual. And the actual upward or downward movement of an ocean-bed is different from the effective; for the sea-water will flow off or in from the adjoining seas. The specific gravity of water is about one-third of that of surface rock, and the local loss or gain of matter is the actual loss or gain of surface rock, less the mass of the sea-water admitted or displaced. Thus the effective downward or upward movement of a sea-bed is about $\frac{2}{3}$ of the actual; of this a more accurate estimate will be given presently.

It is fortunately not important to trace the series of changes through their course; and in order to avoid the complication of doing so, the way seems to be to estimate the amount of transference of matter entailed in the conversion of a deep ocean into a continent of the present mean height.

Suppose, then, that an ocean area of 15,000 feet in depth were gradually elevated, and that the final result, notwithstanding erosion, were a continent of 1100 feet in height. Conceive a prism, the area of whose section is unity, running vertically upwards from what was initially the ocean-bed. Initially this prism contained 15,000 feet of sea-water, and finally it contains 16,100 feet of rock; so that the local gain of matter, on this unit of area of the earth's surface, is the difference between the masses of this prism, initially and finally.

Now 1·02 is the specific gravity of sea-water, and 2·75 that of surface rock; therefore the same local gain of matter, in a sealess globe, would be given by an elevation of

$$16{,}100 - \tfrac{102}{275} \text{ of } 15{,}000 = 10{,}436 \text{ feet}$$

That is to say, 10,436 feet has been the effective elevation.

I therefore adopt 10,000 feet as the effective elevation equivalent to the conversion of deep ocean into a continent; and in the examples given hereafter, where I find the deflection of the pole for various forms and sizes of continent, I shall give the results of such an assumed conversion.

* Sir J. Herschel seems to have doubled the height through a misconception of Humboldt's meaning. The mean height of the land is in English feet: Europe, 671; N. America, 748; Asia, 1132; S. America, 1151. See a letter to *Nature*, by Mr J. Carrick Moore, April 18th, 1872.

§ 19. *Wide-spread Deformations of the Earth.*

It has hitherto been assumed that the elevation of land would not affect the sea-level; but there can be no doubt but that elevations, such as those already spoken of, would do so to the extent of, say, a hundred feet. In so far, then, as this is the case, the elevation would be masked from the eyes of geologists. But if the change of form were a gradual rising over a very wide area, the level surfaces would approximately follow the form of the rocky surface. For instance, the elliptical form of the equator carries the ocean level with it; the amount of this ellipticity is such that the difference between the longest and shortest equatorial radii is 6378 feet*. So long, however, as these bulges remain equatorial they cannot affect the position of the principal axis, even should they vary in amount from time to time. But this kind of deformation, if not symmetrical with respect to the equator, would alter the position of the principal axis, without leaving any trace whatever of elevation or depression for geologists to discover.

The discrepancy which is found between the ellipticity of the earth, as deduced from various arcs of meridian, is, I presume, attributable to real inequalities in the earth's form, and not entirely to errors of observation and to the elliptical form of the equatorial section. It seems, moreover, quite possible that these wide-spread inequalities may have varied from time to time.

Hence, even if the deposit of strata in the sea did not produce a continual shifting of the weights on the earth's surface, and even if geologists should ultimately come to the conclusion that there has never been any consentaneous elevation and depression of very large continents relative to the sea-level, but that the oscillations of level have always been local, it would by no means follow that the earth's axis has remained geographically fixed.

V. Numerical Application to the Case of the Earth.

§ 20. *Continents and Seas of Maximum Effect.*

As far as I can learn, geologists are not of opinion that there is any more reason why upheavals and subsidences should take place at one part of the earth's surface than at another. It is accordingly of interest to suppose the elevations and depressions to take place in the most favourable places for shifting the axis of figure. The area over which a consentaneous change may take place is also a matter of opinion.

The theorem in maxima and minima in Part III. makes it easy to construct a table from which that area may be selected which seems most

* Thomson and Tait, *Natural Philosophy*, p. 648.

probable to geologists. The following Table is formed by interpolation in the Table in § 14; the first column gives the fraction of the earth's surface over which an elevation is supposed to take place, a depression over an equal area taking place simultaneously. The second column gives the angular shift in the earth's axis of figure, due to 10,000 feet of effective elevation; as was shown in Part IV., this would convert a deep ocean into a continent. If 10,000 feet be thought too high an estimate, the last column may be reduced in any desired proportion. Lastly, fig. 2 shows the forms of these continents and seas of maximum effect.

Area of elevation or subsidence, as fraction of Earth's surface	Deflection of pole for 10,000 feet effective elevation
·001	$2\frac{1}{4}'$
·005	$11\frac{1}{3}'$
·01	$22\frac{1}{2}'$
·05	$1°\ 46\frac{1}{2}'$
·1	$3°\ 17'$
·15	$4°\ 33\frac{2}{3}'$
·2	$5°\ 36\frac{2}{3}'$
·5	$8°\ 4\frac{1}{2}'$

N.B. The area of Africa is about ·059, and of S. America about ·033 of the Earth's surface.

§ 21. *Examples of other forms of Continent.*

I will now apply the preceding work to a few cases where the continents and seas do not satisfy the condition of giving the maximum effect.

Figures 3, 4, 5, and 6 represent the shapes of the continents as projected stereographically. The shaded parts represent areas of elevation, the dotted parts those of depression; and in the shelving continents and seas the contour lines are roughly indicated. P′ shows the new position of the pole. In every case here given $d=0$ and $F=0$.

Fig. 3. $F(\theta, \phi) = \sin 2\theta \cos 2\phi$, from $\theta = 0$ to π, and $\phi = -\frac{1}{4}\pi$ to $+\frac{1}{4}\pi$, and zero over the rest of the globe.

$$e = 2\int_0^{\pi}\int_{-\frac{1}{4}\pi}^{\frac{1}{4}\pi} \sin^3\theta \cos^2\theta \cos 2\phi \cos\phi \, d\theta d\phi = \tfrac{16}{45}\sqrt{2}$$

$$i'' = Khe = {\cdot}5480h$$

If the effective elevation or depth in the middle of continent or sea be 10,000 feet, $PP' = 1°\ 31\frac{1}{3}'$.

This is the form of continent for which $\mathfrak{H}_2$ is worked out in Appendix A.

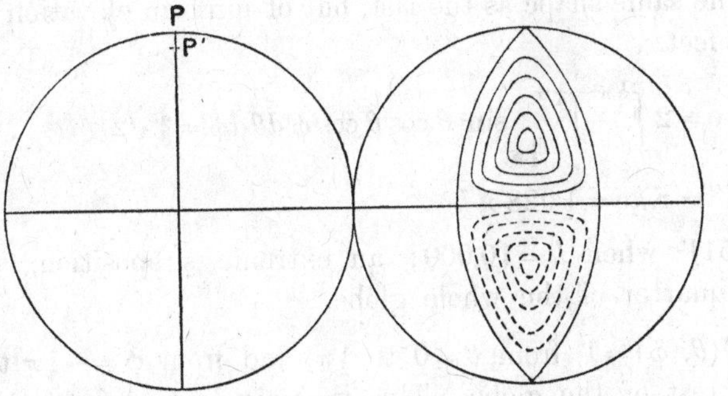

FIG. 3.

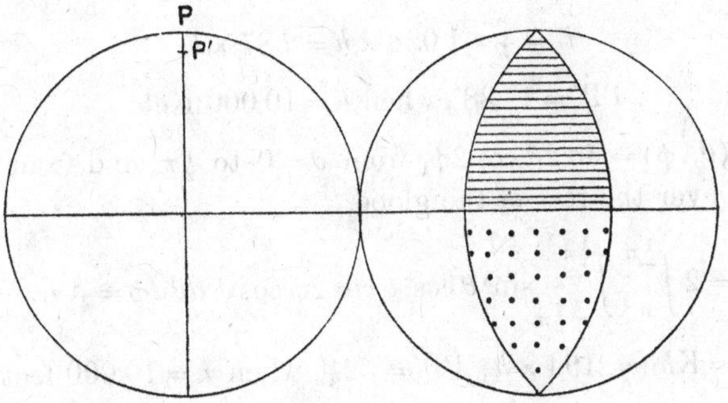

FIG. 4.

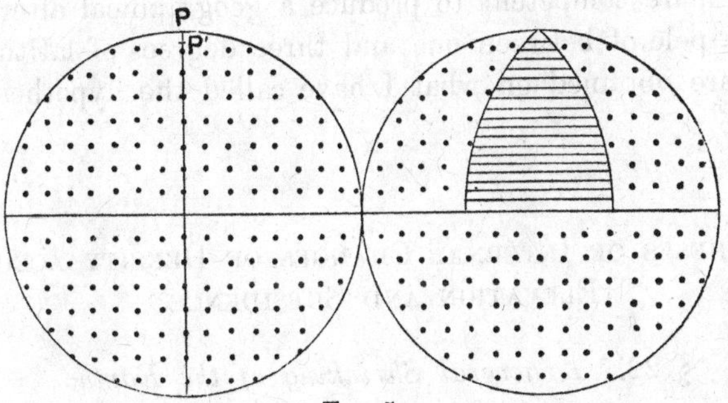

FIG. 5.

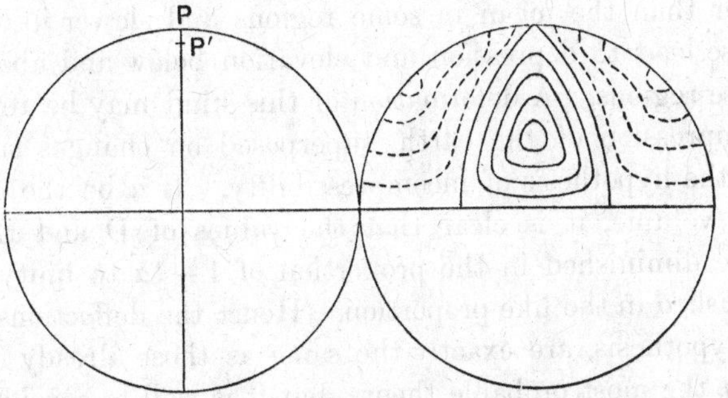

FIG. 6.

Fig. 4. The same shape as the last, but of uniform elevation and depression of 10,000 feet.

$$e = 2\int_0^{\frac{1}{2}\pi}\int_{-\frac{1}{4}\pi}^{\frac{1}{4}\pi} \sin^2\theta \cos\theta \cos\phi \, d\theta d\phi = \tfrac{2}{3}\sqrt{2}$$

$$i'' = Khe = 1{\cdot}028 \times h$$

$PP' = 2° \; 51\frac{1}{4}'$ when $h = 10{,}000$; an extreme supposition, as the area affected is a quarter of the whole globe.

Fig. 5. $F(\theta, \phi) = 1$, from $\theta = 0$ to $\frac{1}{2}\pi$, and from $\phi = -\frac{1}{4}\pi$ to $+\frac{1}{4}\pi$, and $-\frac{1}{7}$ over the rest of the globe. This is equivalent to $F(\theta, \phi) = \frac{8}{7}$ within the above limits.

Then
$$i'' = \tfrac{4}{7} \times 1{\cdot}028 \times h = {\cdot}587 \times h$$
and
$$PP' = 1° \; 38', \text{ when } h = 10{,}000 \text{ feet}$$

Fig. 6. $F(\theta, \phi) = \sin 2\theta \cos 2\phi$, from $\theta = 0$ to $\frac{1}{2}\pi$, and from $\phi = -\frac{1}{2}\pi$ to $+\frac{1}{2}\pi$, and zero over the rest of the globe.

$$e = 2\int_0^{\frac{1}{2}\pi}\int_{-\frac{1}{2}\pi}^{\frac{1}{2}\pi} \sin^3\theta \cos^2\theta \cos 2\phi \cos\phi \, d\theta d\phi = \tfrac{8}{45}$$

$$i'' = Khe = {\cdot}194 \times h; \quad PP' = 32\tfrac{1}{3}', \text{ when } h = 10{,}000 \text{ feet}$$

On the whole, then, it appears that continents, such as those with which we have to deal, are competent to produce a geographical alteration in the position of the pole of between one and three degrees of latitude. But all these results are obtained on what I have called the hypothesis of incompressibility.

VI. Hypotheses of Internal Changes of Density accompanying Elevation and Subsidence.

§ 22. *A general Shrinking of the Earth.*

It may be supposed that the earth is gradually shrinking, but that it shrinks quicker than the mean in some regions and slower in others. This would of course lead to depression and elevation below and above the mean surface in those regions. A deformation of this kind may be represented as a uniform compression of the earth, superposed on changes such as those considered on the hypothesis of incompressibility. If α be the coefficient of contraction of volume, it is clear that the values of D and E, as already found, must be diminished in the proportion of $1 - \frac{2}{3}\alpha$ to unity, and $C - A$ must be diminished in the like proportion. Hence the deflections of the polar axis, on this hypothesis, are exactly the same as those already found. This seems, perhaps, the most probable theory, but it is well to consider others.

The redistribution of matter caused by the erosion of continents will clearly produce the same effect as deformations on the theory of incompressibility.

§ 23. *Changes of Internal Density producing Elevation.*

In discussing the above hypothesis, I shall confine myself to the case of the upheaval or subsidence being of uniform height over given areas, and shall make certain other special assumptions. This will considerably facilitate the analysis, and will give sufficient insight into the extent to which previous results will be modified.

I assume, then, that the elevation of the surface is produced by a swelling of the strata contained between distances r_1 and r_2 from the centre of the globe and immediately under the area of elevation, and that the coefficient of cubical expansion α is constant throughout the intumescent portion.

This will cause a fracture of the strata of equal density, and will produce a discontinuity such as that shown in figure 7, where the dotted circle of radius r_2 indicates the upper boundary of the swelling strata before their intumescence.

But the shift of the earth's axis, caused by this kind of deformation, will differ insensibly from what would obtain if there were a more or less abrupt flexure of the strata of equal density at the boundaries of the intumescent volume and of the area of elevation.

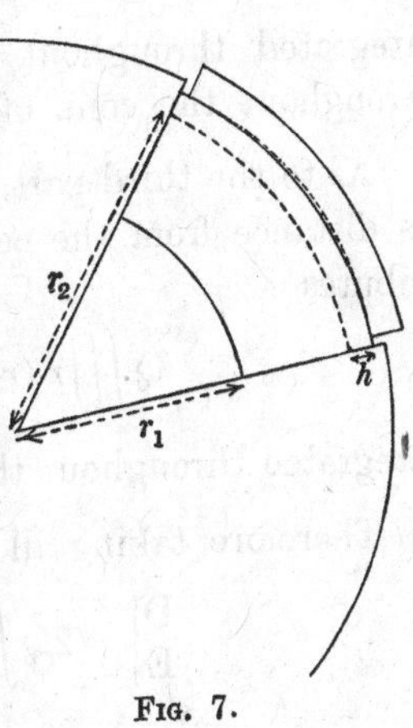

FIG. 7.

Suppose, as before, that h is the height to which the continent is raised above the surface; then we require to know α in terms of h.

Before intumescence, let r, θ, ϕ be the coordinates of any point within the intumescent volume; and suppose that r becomes $r+u$, whilst θ and ϕ, of course, remain constant.

The equation of continuity is easily found to be

$$\frac{du}{dr} + \frac{2u}{r} = \alpha$$

of which the integral is $ur^2 = \frac{1}{3}\alpha r^3 + \beta$.

If β be determined, so that when $r = r_1$, $u = 0$,

$$u = \tfrac{1}{3}\alpha \frac{r^3 - r_1^3}{r^2}$$

But when $r=r_2$, $u=h$, the elevation of the surface; therefore

$$\alpha=\frac{3h}{r_2}\frac{1}{1-r_1^3/r_2^3}$$

the required expression for α in terms of h.

Also, before intumescence, Laplace's law of internal density held good, viz. $Q\sin qr/r$, therefore afterwards the density of the stratum distant $r+u$ from the centre is $Q(1-\alpha)\sin qr/r$.

Now the propositions given in Part II., as to the change in the position of the earth's axis, remain true here also; and the only difference is that the products of inertia D and E must now be expressed by different integrals.

After intumescence the earth may be conceived to consist of:—*first*, itself as it was before; *secondly*, of *negative* matter, of which the law of density is $Q\sin qr/r$, throughout the space bounded by $r=r_1$, $r=c$, and the cone of elevation; and, *thirdly*, of the matter which formerly lay within this space, in the configuration attained by it after intumescence.

The first part clearly contributes nothing to D and E; and the second contributes

$$-Q\iiint r^3\sin qr\sin^2\theta\cos\theta\begin{Bmatrix}\sin\phi\\ \cos\phi\end{Bmatrix}dr\,d\theta\,d\phi$$

integrated throughout the above space, that is from $r=r_1$ to $r=c$, and throughout the cone of elevation.

As to the third part, the mass of any element remains unchanged, whilst its distance from the centre has become $r+u$. Hence the third part contributes

$$Q\iiint r(r+u)^2\sin qr\sin^2\theta\cos\theta\begin{Bmatrix}\sin\phi\\ \cos\phi\end{Bmatrix}dr\,d\theta\,d\phi$$

integrated throughout the above space.

Therefore, taking all together, and treating u as small,

$$\left.\begin{matrix}D\\ E\end{matrix}\right\}=2Q\iiint ur^2\sin qr\sin^2\theta\cos\theta\begin{Bmatrix}\sin\phi\\ \cos\phi\end{Bmatrix}dr\,d\theta\,d\phi$$

Therefore
$$\frac{D}{d}=\frac{E}{e}=2Q\int ur^2\sin qr\,dr$$

where d and e have the same meanings as before, in Part II. § 12.

Now this last integral divides itself into two parts: first, from $r=c$ to $r=r_2$, $u=h$; and, secondly, from $r=r_2$ to $r=r_1$, $u=\frac{1}{3}\alpha(r^3-r_1^3)/r^2$.

Therefore

$$\frac{D}{d}=\frac{E}{e}=2Qh\int_{r_2}^{c}r^2\sin qr\,dr+\tfrac{2}{3}Q\alpha\int_{r_1}^{r_2}(r^3-r_1^3)\sin qr\,dr$$

If the value of α be substituted, and the integrations effected, it will be found that

$$\frac{D}{2d\rho hc^4} = \frac{E}{2e\rho hc^4} = -\frac{\cot qc}{qc} + \frac{2}{(qc)^2}\frac{S}{\sin qc} + \frac{3c}{r_2}\cdot\frac{1}{(qc)^2}\frac{1}{1 - r_1^3/r_2^3}\frac{T}{\sin qc}$$
$$= U \text{ suppose}$$

where S stands for the expression $\frac{r}{c}\sin qr + \cos qr$, taken between the limits c and r_2, and T for the expression $\frac{r^2}{c^2}\sin qr + 2\frac{r}{c}\cdot\frac{1}{qc}\cos qr - \frac{2}{(qc)^2}\sin qr$, taken between the limits r_2 and r_1.

Substituting in the expression $i = \sqrt{(D^2 + E^2)}/(C - A)$, and using the coefficient K, we get

$$i'' = 2KUh\sqrt{(d^2 + e^2)}$$

It must be noticed that this investigation is applicable as much to subsidence caused by internal compression as it is to elevation; and the word intumescence is used to cover both phenomena. In the case of subsidence h is negative.

Now on the hypothesis of incompressibility it was shown that

$$i'' = Kh\sqrt{(d^2 + e^2)}$$

Hence, on the present hypotheses, the estimated deflection of the pole must be diminished in the proportion of 2U : 1.

Taking $qc = 141°$ (which makes $\mathfrak{D}/\rho = 2$, very nearly), I have calculated the values of 2U, when $\frac{r_1}{c} = \frac{399}{400}$, and $\frac{r_2}{c} = \frac{79}{80}, \frac{9}{10}, \frac{3}{4}, \frac{1}{2}, 0$. If the earth's radius be taken as 4000 miles, this gives, that the superficial strata for 10 miles in thickness do not swell, but are merely heaved up, and that the lower surface of the intumescent volume is at the various distances from the earth's surface given in the first column of the following Table. The second column gives 2U, or the factor by which previous results would have to be diminished on the present hypothesis. The third column gives the so diminished value of 1° of deflection of the pole.

Depth below surface of bottom of intumescent volume, in miles, $(c - r_2)$	Factor of diminution of former results, (2U)	A deflection of 1° would be reduced to (2U × 1°)
50	·0126	46″
400	·1011	6′ 4″
1000	·2731	16′ 23″
2000	·5171	31′ 2″
4000	·6721	40′ 20″

The last row, of course, indicates that the intumescence extends quite down to the centre of the earth.

This Table shows that if elevation is due to the swelling of strata at all near the surface, the alteration in the position of the polar axis would be reduced to quite an insignificant amount. The alleged deficiency of density under the Himalayas affords some slight evidence that it is so, at least occasionally. I believe, also, that Mr Mallet is of opinion that the centre of disturbance of earthquake-shocks is not at a greater distance than 30 miles below the surface*. It does not, of course, follow from this evidence that there may not be elevations of both kinds going on, some being approximately superficial phenomena, and others probably due to unequal shrinking of the earth as a whole. The latter kind would be likely to produce more extensive deviations from the external form of equilibrium than the former.

On the whole, then, it appears that the deflection of the polar axis cannot exceed that which was found in the case of incompressibility, and it may possibly be considerably less. The complete want of knowledge of the internal movements only allows us to state a superior limit to the change which might be produced by any one upheaval or subsidence.

VII. Summary and Conclusion.

§ 24. *Summary.*

For the sake of those who do not read mathematics, I will shortly recapitulate the chief results arrived at.

The change in the obliquity of the ecliptic caused by any gradual deformation of the earth's shape of small amount is very small. Even so great a redistribution of weights on the earth's surface as is entailed by immense polar ice-caps during the Glacial Period, cannot have altered the obliquity by so much as $\frac{1}{2200}$ of a second of arc; and this is the most favourable redistribution of weights for producing this effect. Thus throughout geological history the obliquity of the ecliptic must have remained sensibly constant. And, further, when the earth undergoes any such deformation, the axis of rotation follows, and remains sensibly coincident with the principal axis of figure.

It thus only remains to consider the change in the geographical position of the poles caused by the deformation.

The principal axes at the centre of inertia of a body are three lines mutually perpendicular, and their position is entirely determined by the shape of the body. Hence if a nearly spherical body be slightly deformed,

* Referred to at second hand by Mr Carruthers, *Trans. New-Zeal. Inst.*, Vol. VIII., p. 363.

the extremities of these principal axes will move from their original positions and describe paths on the surface of the body, which may be shortly described as the paths of the principal axes. In the case of the earth, as geologically deformed, it is only of interest to consider the path of one of these axes, which is, in common parlance, the earth's axis.

If the earth be sensibly rigid, or should only readjust itself to an approximate form of equilibrium at long intervals (as maintained in Part I.), the geographical path of the axis is very nearly the same as is due merely to the geological deformation of the earth's shape; but if the earth be more or less plastic, or should readjust itself frequently to an approximate form of equilibrium, the dynamical reactions introduced are such as more or less to modify the geographical path of the axis. In the case of great plasticity these reactions would suffice to entirely alter the character of the path. It seems probable that during the consolidation of the earth there was great instability in the geographical position of the poles. Throughout the rest of the investigation suppositions of plasticity are set aside, and the hypothesis of sensible rigidity is adhered to.

Formulæ for the change in the geographical position of the pole due to any small deformation are found in Part II.

On the assumption that the internal density of the earth remains unchanged by the deformation, the forms of continent and depression which produce the greatest deflection of the poles, for the transport of a given quantity of matter from one part of the earth's surface to another, are then investigated. These forms are shown, projected stereographically, in fig. 2 (p. 24).

Part IV. gives what evidence I have been able to collect of the areas and amounts of deformation to which the earth may have been subjected in geological history; but as the discussion is not mathematical, it seems unnecessary to give an abstract thereof.

Part V. gives numerical applications of the preceding theorems to the case of the earth, on the assumption that the internal density is unaltered by the deformation. From this it appears that the poles may have been deflected from 1° to 3° in *any one geological period*; but the reader is referred back to that part for details.

If upheaval and subsidence of the surface are due to a shrinking of the earth as a whole, but to a more rapid shrinking in some regions than others, the deflection of the poles is the same as that found where there is no disturbance of the strata of equal density.

But if the upheaval and subsidence are due to local intumescence and contraction of the strata underneath the rising or falling areas, the previous numerical estimates must be largely reduced; for the extent of this reduction the reader is referred to the Table in § 23 (p. 35).

It thus appears that the deflection of the poles first given is a superior limit to that which is possible.

§ 25. *Conclusion.*

There remain, in conclusion, one or two miscellaneous points to be referred to.

In a letter to Sir Charles Lyell read before the Geological Society*, Sir John Herschel has pointed out that the isothermal strata near the surface of the earth must approximately follow the solid surface. Therefore, when a thick stratum is deposited at the bottom of the ocean, the primitive bottom is gradually warmed and expands. There is thus a tendency for the upheaval of sea-beds, on which a large amount of matter has been deposited; but this kind of upheaval certainly falls within the case of superficial intumescence, and could therefore affect the geographical position of the poles but little more than would be due merely to the weight of the deposited stratum. It must be noticed, moreover, that the weight of the deposited stratum would tend to compress the primitive sea-bed, and might counteract the expansion due to rise of temperature.

If the earth were absolutely rigid the pole could never have wandered more than from 1° to 3° from its primitive position, whatever geological changes were successively to take place; because the new pole could never be brought to a greater distance from its original position, by any fresh distribution of the matter forming the continents, than the maximum for this amount of matter arranged in continents of a like height.

But it was maintained in Part I. that from time to time the earth makes a kind of rough adjustment to a figure of equilibrium. If this adjustment is, as seems probable, by an earthquake, it will take place with reference to the axis of rotation at the instant of the earthquake. Now there exists in erosion and marine deposits a cause of terrestrial deformation which is certainly independent of such adjustments; and it seems probable that the causes of geological upheaval and subsidence are so also. We have therefore clearly a state of things in which the pole may wander indefinitely from its primitive position. On this hypothesis, as in successive periods the continents have risen and fallen, the pole may have worked its way, in a devious course, some 10° or 15° away from its geographical position at consolidation, or may have made an excursion of smaller amount and have returned to near its old position. May not the Glacial Period, then, have been only apparently a period of great cold? If at that period the N. pole stood somewhere where Greenland now stands, would not the whole of Europe and a large part of N. America have been glaciated? And if the N. pole retreated to its present

* *Proc. Geol. Soc.*, Vol. II., p. 549.

position, would it not leave behind it the appearance of a very cold climate having prevailed in those regions?

But although such a cumulative effect is possible with respect to the geographical position of the pole, none such is possible with respect to the obliquity of the ecliptic.

Now this kind of wandering of the poles would of course require extensive and numerous deformations, and it is hard to see how there can have been a shifting of the surface weights sufficient to produce it, without frequent changes in the geographical distribution of land and water. If, then, geologists are right in supposing that where the continents now stand they have always stood, would it not be almost necessary to give up any hypothesis which involved a very wide excursion of the poles?

APPENDIX A. (See p. 9.)

To calculate $\mathfrak{H}_1$ and $\mathfrak{H}_2$ in a supposed case of elevation and subsidence.

Take the case of § 21 (fig. 3), where the elevation is given by $ht \sin 2\theta \cos 2\phi$, from $\theta = 0$ to π, and from $\phi = -\frac{1}{4}\pi$ to $\frac{1}{4}\pi$, and zero over the rest of the sphere. Suppose that the internal motion is entirely confined to the quarter of the sphere defined by the above limits of θ and ϕ, that radial particles are always radial, and that the motion is entirely meridional.

Let $\theta + \vartheta$ be the disturbed colatitude of the point θ, ϕ. Then the equation of continuity, which expresses that the volume of the elementary pyramid $\frac{1}{3}c^3 \sin \theta d\theta d\phi$ remains constant, when θ becomes $\theta + \vartheta$, is

$$\frac{d}{d\theta}(\vartheta \sin \theta) + \frac{3ht}{c} \sin \theta \sin 2\theta \cos 2\phi = 0$$

the integral of which is

$$\vartheta \sin \theta + \frac{2h}{c} t \cos 2\phi \sin^3 \theta = \text{a constant}$$

and since ϑ is zero, when $\phi = \pm \frac{1}{4}\pi$, for all values of t, $\vartheta = -\frac{2h}{c} t \cos 2\phi \sin^2 \theta$,

and $$\frac{d\vartheta}{dt} = -\frac{2h}{c} \cos 2\phi \sin^2 \theta$$

Hence H_2, twice the area conserved on the plane of xz, is

$$\iiint \rho r^2 \sin \theta \, dr \, d\theta \, d\phi \, . \, \frac{r^2 d\vartheta}{dt} \cos \phi$$

taken from $r = 0$ to c, $\theta = 0$ to π, $\phi = -\frac{1}{4}\pi$ to $+\frac{1}{4}\pi$.

If the sphere be taken as homogeneous,

$$H_2 = -\frac{2h}{c}\rho\iiint r^4 \sin^3\theta \cos 2\phi \cos\phi \, dr\, d\theta\, d\phi$$

$$= -\tfrac{16}{45}\sqrt{2}\, h\rho c^4 = -\frac{4\sqrt{2}}{15\pi}\mathrm{M}hc$$

H_1 and H_3 are both clearly zero.

The above value of H_2 is larger than what it would be in the case of the earth, if Laplace's law of internal density were true, because the external layers have been taken too heavy, and the internal too light. But taking that law of density, $\mathrm{A} = \frac{1}{3}\mathrm{M}c^2$ very nearly.

Hence $\dfrac{H_2}{\mathrm{A}} = -\dfrac{4\sqrt{2}}{5\pi}\dfrac{h}{c}$.

If we let the time run on until the highest point of the continent has risen one foot, so that $\dfrac{ht}{c} = \dfrac{1}{20{,}900{,}000}$, then $\dfrac{H_2 t}{\mathrm{A}} = -\dfrac{4\sqrt{2}}{5\pi}\dfrac{1}{20{,}900{,}000}$.

But reference to § 21 (fig. 3) shows that $i'' = \cdot 5480h$, or in the present notation,

$$\beta t = \cdot 5 \times \frac{\pi}{648{,}000} \text{ nearly}$$

Therefore $$\frac{H_2}{\mathrm{A}\beta} = -\frac{8 \times 648\sqrt{2}}{104{,}500\pi^2} = -\tfrac{1}{141} \text{ nearly}$$

But generally, since the angular velocities α, β, γ of the moving axes, to which $\mathfrak{H}_1, \mathfrak{H}_2, \mathfrak{H}_3$ refer, are very small, therefore

$$\mathfrak{H}_1 = H_1, \quad \mathfrak{H}_2 = H_2, \quad \mathfrak{H}_3 = H_3$$

to the first order of small quantities, within the limited period to which the investigation applies. So that in this particular case,

$$\frac{\mathfrak{H}_2}{\mathrm{A}\beta} = -\tfrac{1}{141} \text{ nearly, and } \mathfrak{H}_1 = \mathfrak{H}_3 = 0$$

And, besides, this value of $\mathfrak{H}_2/\mathrm{A}\beta$ is larger than it ought to be, because $\mathfrak{H}_2$ was calculated on an assumed homogeneity of the earth. This, then, justifies the conclusion in the text on p. 9.

In the elevation and subsidence given by $ht \sin 2\theta \sin 2\phi$ from $\theta = 0$ to $\frac{1}{2}\pi$, and from $\phi = -\frac{1}{2}\pi$ to $\frac{1}{2}\pi$, H_1 and H_2 are clearly zero, under a like supposition as to the nature of the internal motions accompanying upheavals.

APPENDIX B. (See p. 23.)

To reduce the integrals $\int_0^a \cos\chi \cos^{-1}\frac{\cos^2 \alpha}{\cos^2\chi}\,d\chi$ *and* $\int_0^a \cos\chi\,\sqrt{(\cos^4\chi - \cos^4\alpha)}\,d\chi$ *to elliptic functions.*

Call the former A and the latter B.

Integrating A by parts,

$$A = -\int_0^a \sin\chi\, d\left(\cos^{-1}\frac{\cos^2\alpha}{\cos^2\chi}\right)$$

Put $x = \sin\chi$, and $\cos 2\gamma = \cos^2\alpha$, then we get

$$A = 2\cos 2\gamma \int_0^{\sqrt{2}\sin\gamma} \left(\frac{1}{1-x^2} - 1\right)\frac{dx}{\sqrt{(x^4 - 2x^2 + \sin^2 2\gamma)}}$$

and if $x = \sqrt{2}\sin\gamma\sin\phi$, this becomes

$$\sqrt{2}\,\frac{\cos 2\gamma}{\cos\gamma}\{\Pi^1(-2\sin^2\gamma) - F^1\}, \text{ where the modulus is } \tan\gamma$$

Again, integrating B by parts,

$$B = \int_0^a \sin\chi \cdot \frac{4\cos^3\chi\sin\chi\, d\chi}{2\sqrt{(\cos^4\chi - \cos^4\alpha)}}$$

$$= 2\int_0^{\sqrt{2}\sin\gamma} \frac{(1-x^2)\,x^2\,dx}{\sqrt{(x^4 - 2x^2 + \sin^2 2\gamma)}}$$

But B is also $= \int_0^{\sqrt{2}\sin\gamma} \frac{\sin^2 2\gamma - x^2 - (1-x^2)\,x^2}{\sqrt{(x^4 - 2x^2 + \sin^2 2\gamma)}}\,dx$ from the expression before partial integration. Multiplying the latter expression by 2 and adding to the former,

$$3B = 2\int_0^{\sqrt{2}\sin\gamma} \frac{\sin^2 2\gamma - x^2}{\sqrt{(\sin^2 2\gamma - 2x^2 + x^4)}}\,dx$$

and substituting the above value for x,

$$\tfrac{3}{2}B = \frac{\sin^2 2\gamma}{\sqrt{2}\cos\gamma}F^1 + \sqrt{2}\cos\gamma\,(E^1 - F^1)$$

Whence $B = \frac{2}{3}\sqrt{2}\cos\gamma\,[E^1 - \cos 2\gamma\, F^1]$, the modulus being $\tan\gamma$.

B may be calculated from this form by means of the tables in Legendre's *Fonctions Elliptiques*, tom. II. But A is not yet in a form adapted for numerical calculation.

The parameter $-2\sin^2\gamma$ of Π^1 is negative and numerically greater than the square of the modulus; therefore Π^1 falls within Legendre's second class (*op. cit.* tom. I. p. 72). Now it is shown by Legendre (tom. I. p. 138) that

$$\frac{b^2\sin\theta\cos\theta}{\Delta(b,\theta)}[\Pi^1(n,c) - F^1(c)]$$
$$= \tfrac{1}{2}\pi + F^1(c)\,F(b,\theta) - E^1(c)\,F(b,\theta) - F^1(c)\,E(b,\theta)$$

In this case θ will be found to be $\frac{1}{2}\pi - \gamma$,

$$\frac{b^2 \sin\theta \cos\theta}{\Delta(b, \theta)} = \tfrac{1}{2}\sqrt{2} \cdot \frac{\cos 2\gamma}{\cos\gamma}, \quad \text{and } b = \frac{\cos\alpha}{\cos\gamma}$$

whence

$$A = \pi - 2\{E^1F - F^1(F - E)\}$$

where the moduli of F and E are $\cos\alpha/\cos\gamma$, and their amplitude $\frac{1}{2}\pi - \gamma$.

From this form A may be calculated numerically.

APPENDIX C. (Added April 1877.)

Sir William Thomson, who was one of the referees requested by the Royal Society to report on this paper, has remarked that the subject of Part I. may also be treated in another manner.

The following note contains his solution, but some slight alterations have been made in a few places.

The axis of resultant moment of momentum remains invariable in space whatever change takes place in the distribution of the earth's mass; or, in other words, the normal to the invariable plane is not altered by internal changes in the earth.

Now suppose a change to take place so slowly that the moment of momentum round any axis of the motion of any part of the earth relatively to any other part may be neglected compared to the resultant moment of momentum of the whole*; or else suppose the change to take place by sudden starts, such as earthquakes. Then, on either supposition (except during the critical times of the sudden changes, if any), the component angular velocities of the mass relatively to fixed axes, coinciding with the positions of its principal axes at any instant, may be written down at once from the ordinary formulæ, in terms of the direction-cosines of the normal to the invariable plane with reference to these axes, and in terms of the moments of inertia round them, which are supposed to be known.

Hence we find immediately the angular velocity and direction of the motion of that line of particles of the solid which at any instant coincides with the normal to the invariable plane at the origin. This is equal and opposite to the angular velocity with which we see the normal to the invariable plane travelling through the solid, if we, moving with the solid, look upon the solid as fixed. Let, at any instant, x, y, z be the direction-cosines of the normal to the invariable plane relatively to the principal axes; and let A, B, C be the principal moments of inertia at that instant. Let h

* This is equivalent to neglecting $\mathfrak{H}_1$, $\mathfrak{H}_2$, $\mathfrak{H}_3$ of Part I.; by which Sir W. Thomson is of opinion that nothing is practically lost.

be the constant moment of momentum (or twice the area conserved on the invariable plane).

Consider axes fixed relatively to the solid in the positions of the principal axes at any instant, but not moving with them, if they are being shifted in virtue of changes in the distribution of portions of the solid.

The component angular velocities of the rest of the universe are, relatively to these axes, hx/A, hy/B, hz/C; and therefore, if N be the point in which the normal to the invariable plane at the origin cuts a sphere of unit radius, the components parallel to these axes of the velocity of N relatively to them are

$$yz\left(\frac{h}{\mathrm{C}}-\frac{h}{\mathrm{B}}\right),\quad zx\left(\frac{h}{\mathrm{A}}-\frac{h}{\mathrm{C}}\right),\quad xy\left(\frac{h}{\mathrm{B}}-\frac{h}{\mathrm{A}}\right)^*$$

Now, suppose that by slow continuous erosion and deposition the positions of the principal axes change slowly and continuously relatively to the solid.

Let ϖ, ρ, σ be the components round the axes (which, of course, are always mutually at right angles) of the angular velocity of the actual solid relatively to an ideal solid moving with the principal axes†. Then the component velocities relatively to this ideal solid of the point of the body coinciding at any instant with N are

$$z\rho - y\sigma,\quad x\sigma - z\varpi,\quad y\varpi - x\rho$$

and the components parallel to the principal axes of the velocity of N relatively to these axes are dx/dt, dy/dt, dz/dt. Hence we have

$$\left.\begin{aligned}\frac{dx}{dt} &= yz\left(\frac{h}{\mathrm{C}}-\frac{h}{\mathrm{B}}\right)-(z\rho-y\sigma)\\ \frac{dy}{dt} &= zx\left(\frac{h}{\mathrm{A}}-\frac{h}{\mathrm{C}}\right)-(x\sigma-z\varpi)\\ \frac{dz}{dt} &= xy\left(\frac{h}{\mathrm{B}}-\frac{h}{\mathrm{A}}\right)-(y\varpi-x\rho)\end{aligned}\right\}\quad\ldots\ldots\ldots\ldots(1)‡$$

These three equations give $x\frac{dx}{dt}+y\frac{dy}{dt}+z\frac{dz}{dt}=0$, and therefore they are equivalent to two independent equations to determine two of the three unknown quantities x, y, z as functions of t, the three fulfilling the condition $x^2+y^2+z^2=1$, and it being understood that ϖ, ρ, σ are given functions of the time.

* The angular velocity of the rest of the universe relatively to the earth being opposite to the angular velocity of the earth relatively to the rest of the universe, the components of the former round the axes x, y, z are taken as in the negative direction, *i.e.* from z to y, x to z, y to x.

† ϖ, ρ, σ are the same as $-\alpha$, $-\beta$, $-\gamma$ of Part I.

‡ These equations are the same as those given by me in Part I., p. 8.

To apply these equations to the questions proposed as to the earth's axis, let the normal to the invariable plane be very nearly coincident with the axis of greatest moment of inertia C. Let O be the point where the axis C cuts the earth's surface, and let OX, OY be parallel to the axes A and B. Then $z = 1$; and if the earth's radius be taken as unity, x and y will be the coordinates relatively to OX, OY of the point P in which the normal to the invariable plane cuts the surface.

Putting therefore $z = 1$ in the preceding equations, we find for the determination of x, y that

$$\left.\begin{aligned} \frac{dx}{dt} &= -\,yh\left(\frac{1}{\mathrm{B}} - \frac{1}{\mathrm{C}}\right) - \rho + y\sigma \\ \frac{dy}{dt} &= \quad xh\left(\frac{1}{\mathrm{A}} - \frac{1}{\mathrm{C}}\right) + \varpi - x\sigma \end{aligned}\right\} \quad \ldots\ldots\ldots\ldots\ldots\ldots(2)$$

or

$$\left.\begin{aligned} &\frac{dx}{dt} + ay = u \\ &\frac{dy}{dt} - bx = v \\ \text{where}\quad & a = h\left(\frac{1}{\mathrm{B}} - \frac{1}{\mathrm{C}}\right) - \sigma, \quad b = h\left(\frac{1}{\mathrm{A}} - \frac{1}{\mathrm{C}}\right) - \sigma \\ \text{and}\quad & u = -\rho, \quad v = \varpi \end{aligned}\right\} \quad \ldots\ldots\ldots\ldots\ldots(3)$$

In these equations we are to regard a, b, u, v as given functions of the time.

Eliminating y, we have

$$\frac{d}{dt}\left(\frac{1}{a}\frac{dx}{dt}\right) + bx = \frac{d}{dt}\left(\frac{u}{a}\right) - v \quad \ldots\ldots\ldots\ldots\ldots\ldots(4)$$

which is a linear equation, from which x may be found by integration; and then, by the first of equations (3),

$$y = \frac{1}{a}\left(u - \frac{dx}{dt}\right) \quad \ldots\ldots\ldots\ldots\ldots\ldots\ldots\ldots(5)$$

If $\mathrm{B} = \mathrm{A}$, the presence of σ in the equations would merely mean that the axes of x and y revolve with an angular velocity σ; and so we lose nothing of interest with reference to the terrestrial problem by supposing $\sigma = 0$. If, then, we take A and B constant, equation (4) becomes,

$$\left.\begin{aligned} &\frac{d^2x}{dt^2} + \omega^2 x = \frac{du}{dt} - av \\ \text{where}\quad &\omega^2 = ab \end{aligned}\right\} \quad \ldots\ldots\ldots\ldots\ldots\ldots\ldots(6)$$

To integrate this according to the method of variation of parameters, put

$$x = \mathrm{P}\cos\omega t + \mathrm{Q}\sin\omega t \quad \ldots\ldots\ldots\ldots\ldots\ldots\ldots\ldots(7)$$

and
$$\frac{dx}{dt} = -P\omega \sin \omega t + Q\omega \cos \omega t \quad \text{.....................(8)}$$

so that
$$\frac{dP}{dt} \cos \omega t + \frac{dQ}{dt} \sin \omega t = 0$$

We find then

$$\left.\begin{aligned} P &= -\frac{1}{\omega}\int\left(\frac{du}{dt} - av\right)\sin \omega t\, dt \\ Q &= \frac{1}{\omega}\int\left(\frac{du}{dt} - av\right)\cos \omega t\, dt \end{aligned}\right\} \quad \text{.....................(9)}$$

For the case considered in Part I., where u and v are constant,

$$P = -\frac{av}{\omega^2}\cos \omega t + C, \quad Q = -\frac{av}{\omega^2}\sin \omega t + C'$$

and therefore by (7)

$$x = -\frac{av}{\omega^2} + C \cos \omega t + C' \sin \omega t \quad \text{.....................(10)}$$

The solution expressed in equations (5), (7), (8), (9) is convenient for discontinuous as well as for continuously varying and constant values of u and v.

Consider, then, the case of $u = 0$ and $v = 0$, except at certain instants when u and v have infinite values, so that $\int_{T'}^{T} u dt$ and $\int_{T'}^{T} v dt$ express the components of a single abrupt change in the position of the instantaneous axis; where T and T′ denote any instants before and after the instant of the change, but so that the interval does not include more than one abrupt change.

Therefore, if t_0 be the instant of the change

$$\left.\begin{aligned} \int_{T'}^{T} v \sin \omega t\, dt &= \sin \omega t_0 \int_{T'}^{T} v dt \\ \int_{T'}^{T} v \cos \omega t\, dt &= \cos \omega t_0 \int_{T'}^{T} v dt \end{aligned}\right\} \quad \text{.....................(11)}$$

Hence the part of x depending on v vanishes at the instant immediately after the abrupt change when $t = t_0$. Also we have by integration by parts,

$$\left.\begin{aligned} \int \frac{du}{dt} \sin \omega t\, dt &= u \sin \omega t - \omega \int u \cos \omega t\, dt \\ \int \frac{du}{dt} \cos \omega t\, dt &= u \cos \omega t + \omega \int u \sin \omega t\, dt \end{aligned}\right\} \quad \text{..............(12)}$$

And, therefore, taking the integrals between the prescribed limits, since $u=0$ both when $t=\mathrm{T}$ and when $t=\mathrm{T}'$, we have

$$\left.\begin{aligned}\int \frac{du}{dt}\sin\omega t\,dt &= -\omega\cos\omega t_0\int_{\mathrm{T}'}^{\mathrm{T}} udt\\ \int \frac{du}{dt}\cos\omega t\,dt &= \omega\sin\omega t_0\int_{\mathrm{T}'}^{\mathrm{T}} udt\end{aligned}\right\}\quad\ldots\ldots\ldots\ldots(13)$$

Using these in (9) and (7) we find, at the instant after the abrupt change,

$$x=\int_{\mathrm{T}'}^{\mathrm{T}} udt \quad\ldots\ldots\ldots\ldots(14)$$

and similarly

$$y=\int_{\mathrm{T}'}^{\mathrm{T}} vdt \quad\ldots\ldots\ldots\ldots(15)$$

which of course might be deduced from (8) and (5).

2.

ON PROFESSOR HAUGHTON'S ESTIMATE OF GEOLOGICAL TIME.

[*Proceedings of the Royal Society*, XXVII. (1878), pp. 179—183.]

In a paper recently read before the Royal Society*, Professor Haughton has endeavoured by an ingenious line of argument to give an estimate of the time which may have elapsed in the geological history of the earth. The results attained by him are, if generally accepted, of the very greatest interest to geologists, and on that account his method merits a rigorous examination. The object, therefore, of the present note is to criticise the applicability of his results to the case of the earth; and I conceive that my principal criticism is either incorrect, and will meet its just fate of refutation, or else is destructive of the estimate of geological time.

Professor Haughton's argument may be summarised as follows:—The impulsive elevation of a continent would produce a sudden displacement of the earth's principal axis of greatest moment of inertia. Immediately after the earthquake, the axis of rotation being no longer coincident with the principal axis, will, according to dynamical principles, begin describing a cone round the principal axis, and the complete circle of the cone will be described in about 306 days. Now, the ocean not being rigidly connected with the nucleus, a 306-day tide will be established, which by its friction with the ocean bed will tend to diminish the angle of the cone described by the instantaneous axis round the principal axis: in other words, the "wabble" set up by the earthquake will gradually die away.

Then by means of Adams and Delaunay's estimate of the alteration of the length of day, which is attributed to tidal friction, Professor Haughton obtains a numerical value for the frictional effect of the residual tidal current. He then applies this to the 306-day tide, and deduces the time required to reduce a "wabble" of given magnitude to any given extent.

* "Notes on Physical Geology. No. III. On a New Method of finding Limits to the Duration of certain Geological Periods." *Proc. Roy. Soc.*, Vol. XXVI., pp. 534—546 (December 20, 1877).

He is of opinion that if, at the present time, the instantaneous axis of rotation of the earth were describing a circle of more than 10 feet in diameter at the earth's surface, then the phenomenon could not escape detection by modern astronomical instruments. From the absence of any such inequality he concludes, after numerical calculation, "if Asia and Europe were manufactured *per saltum*, causing a sudden displacement of the axis of figure through 69 miles, that this event cannot have happened at an epoch less than 641,000 years before the present time, and that this event may have occurred at an epoch much more remote."

He then passes on to consider the case where the elevation takes place by a number of smaller impulses instead of by one large one. He treats first the case of "69 geological convulsions, each of which displaced the axis of figure through one mile," and where "the radius of the wabble" is "reduced from one mile to 5 feet in the interval between each two successive convulsions"; and, secondly, the case where "the increase of this radius is exactly destroyed by friction during each wabble, so that the radius of 5 feet continues constant."

In the first case he finds that the total time occupied by the manufacture of Europe and Asia is 27½ millions of years, and also that "no geological change, altering the position of the axis of figure through one mile, can have taken place within the past 400,000 years." And in the second case, he finds that the same elevation would occupy 4,170 millions of years. A little lower he adds: "It is extremely improbable that the continent of Asia and Europe was formed *per saltum*, and therefore our minor limit of time is probably far short of the reality."

It appears from these passages that Professor Haughton is of opinion that a succession of smaller impulses at short intervals will necessarily increase the radius of the "wabble"; but it is not very clear to me whether he means that the radius of the "wabble" would be the same by whatever series of impulses the principal axis was moved from one position to another. Now, I conceive that it is by no means necessary that a second impulse succeeding a first should augment the radius of the "wabble"; it might, indeed, annihilate it. I admit that by properly timed impulses the radius of the "wabble" might be made as great as if the whole change took place by a single convulsion. But where the impulses take place at hazard there will be a certain average effect on the radius of the "wabble," which, as far as I can see, Professor Haughton makes no attempt to determine. It seems, therefore, an unjustifiable assumption that sufficient time must elapse between the successive impulses to reduce the radius of the "wabble" to 5 feet, for if the impulses took place more frequently they might tend to some extent to counteract one another. If this assumption is unjustifiable, then Professor Haughton's estimate of time falls with it.

In my paper on the "Influence of Geological Changes on the Earth's Axis of Rotation*," I have considered the effects of a slow continuous distortion of the earth. The results there attained would, of course, have been identical, had I considered the effects of a series of infinitely small and infinitely frequent earthquakes. I presume Professor Haughton will agree with me in thinking this supposition more consonant with geological science than the larger earthquakes which he postulates.

I will now show, from the results of my paper, that *without calling in any effects whatever of tidal friction*, Asia and Europe might have been gradually upheaved in 19,200 years, without leaving any "wabble" sufficiently large to be detected astronomically, and, moreover, that at no time during the elevation could the "wabble" have been detected had astronomers been in existence to make observations; and further, that under certain not improbable suppositions, this estimate of time may be largely reduced. Let α be the angular velocity of the principal axis relatively to the solid earth, arising from the continuous elevation of the continent; n the earth's angular velocity of rotation; C, A the greatest and least principal moments of inertia of the earth; and $\mu = (C - A)\, n/A$.

Then, in section 2 of my paper, I show that the extremity of the instantaneous axis describes a circle at the earth's surface in 306 days, and that this circle passes through the extremity of the principal axis, and touches the meridian along which the principal axis is travelling with velocity α in consequence of the postulated geological change. Strictly speaking, the curve described by the instantaneous axis, is a trochoid, because the circle travels in the earth along with the principal axis; but the motion of the circle is so slow compared with that of the instantaneous axis along its arc, that it is more convenient to say that the instantaneous axis describes a circle which slowly changes its position. It must be noticed that this circle is unlike the "wabble" considered by Dr Haughton, inasmuch as the extremity of the principal axis lies on its arc instead of being at its centre. It is also shown in the same section that the diameter of the circle is equal to $2\alpha/\mu$.

I will now suppose that the geological changes begin suddenly from rest, and proceed at such a rate that the variations in the position of the principal axis are imperceptible to astronomical observation. I will suppose, therefore, that the extremity of the instantaneous axis is never more than 5 feet distant from the extremity of the principal axis. Now, 5 feet at the earth's surface, subtends very nearly $0''\cdot05$ at the earth's centre, and, therefore, to find α on this supposition, $2\alpha/\mu$ must be put equal to $0''\cdot05$.

μ is an angular velocity of 360° in 306 days, and if we wish to express α

* *Phil. Trans.*, Vol. 167, Part I., p. 271. [Paper 1.]

in seconds of arc per annum, μ must be expressed in those units, and 0″·05 must be expressed in circular measure. Thus

$$\alpha = \tfrac{1}{2} \times \cdot 05 \times \frac{\pi}{648000} \times 360 \times 60 \times 60 \times \frac{365 \cdot 25}{306}$$

$$= \frac{18 \cdot 263}{306} \pi = \tfrac{3}{16} \text{ very nearly.}$$

Therefore, α is an angular velocity of 1° (or 69 miles) in 19,200 years.

But, according to Professor Haughton, 69 miles is the displacement of the earth's principal axis, due to the elevation of Europe and Asia; hence, at this rate of elevation, Europe and Asia would have been heaved up in 19,200 years.

Now, if the elevation be supposed to stop suddenly, the instantaneous axis cannot, at the time of the stoppage, be more than 5 feet distant from the axis of figure, and it may even be coincident with it. Therefore the stoppage cannot set up a "wabble" of more than 10 feet in diameter, and it may set up none at all. But even this maximum "wabble" of 10 feet, would, according to Professor Haughton, be imperceptible, and *à fortiori* the circle of 5 feet in diameter, described in the course of the elevation, would be imperceptible.

On any of the following suppositions, the elevation might be much more rapid, without increasing the residual "wabble":—

(1) The stoppage of the elevation to take place at a time when the instantaneous axis is separated from the principal axis by a small angle.

(2) The elevation partly counterbalanced by simultaneous elevations in other parts of the world, so that the upheaval of Europe and Asia would not displace the pole of figure by so much as 69 miles.

(3) The elevation partly or altogether produced by the intumescence of the strata immediately underlying those continents. (See Part VI. of my paper above referred to.)

(4) The elevation not uniform but more rapid in the earlier portion of the time, so that the magnitude of the "wabble" would be reduced by the friction of the 306-day tide; for we are by no means compelled to believe that that inequality of motion must always have remained as small as it is at present.

It appears to me, from these considerations, that the continents of Europe and Asia might have been elevated in very much less than 20,000 years, and yet leave no record of the fact in the present motion of the earth. Therefore, if my solution of the problem is correct, it is certain that Professor Haughton's method can give us no clue to the times which have elapsed in the geological history of the earth.

3.

ON A SUGGESTED EXPLANATION OF THE OBLIQUITY OF PLANETS TO THEIR ORBITS*.

[*Philosophical Magazine*, III. (1877), pp. 188—192.]

IN a former paper† I have shown that if θ be the obliquity to the ecliptic of a planet which is slowly changing its shape, so that its principal moments of inertia at the time t are $A + at$, $A + bt$, $C + ct$, then, so long as at, bt, ct remain small compared with $C - A$,

$$\frac{d\theta}{dt} = \frac{\Pi}{2n} \frac{a + b - 2c}{C - A}$$

$\Pi \operatorname{cosec} \theta$ being the precession of the equinoxes, and $-n$ the rotation of the planet. This equation will hold true for long periods, if all the quantities on the right hand are treated as functions of the time; and if $a = b$ it may be written

$$\frac{d\theta}{dt} = -\frac{\Pi}{n} \frac{\frac{d}{dt}(C - A)}{C - A}$$

In the case of the earth,

$$\frac{6\pi^2}{n} \left\{ \frac{1}{T^2} + \frac{1}{T'^2} \frac{1 - \frac{3}{2}\sin^2 i}{1 + \nu} \right\} \frac{C - A}{C} = \frac{\Pi}{\sin\theta\cos\theta} = \frac{p}{n} \frac{C - A}{C}, \text{ suppose}$$

where T, T′ are the year and month, ν is the ratio of the earth's mass to the moon's, and i is the inclination of the lunar orbit to the ecliptic. In the corresponding function for any other planet there will be a term for each satellite, and $1 - \frac{3}{2}\sin^2 i$ will be replaced by a certain function called λ by Laplace.

* [I attach very little importance to this paper, but a reference is given to it in Vol. II., p. 456, and it is now included for the sake of completeness.]

† "On the Influence of Geological Changes on the Earth's Axis of Rotation." [Paper 1.]

The equation may now be written

$$\frac{Cn}{p}\frac{d\theta}{dt}\log\tan\theta = -\frac{1}{n}\frac{d}{dt}(C-A)$$

The object of the present note is to apply this equation to the supposition that the planets were originally nebulous masses, and contracted symmetrically under the influence of the mutual gravitation of their parts. This application involves a large assumption, viz. that the precession of a nebulous mass is nearly the same as though it were rigid. In defence thereof I can only quote Sir W. Thomson, who says, "Now although the full problem of precession and nutation, and what is now necessarily included in it—tides, in a continuous revolving liquid spheroid, whether homogeneous or heterogeneous, has not yet been coherently worked out, I think I see far enough towards a complete solution to say that precession and nutations will be practically the same in it as in a solid globe, and that the tides will be practically the same as those of the equilibrium theory*."

I therefore once for all make this assumption.

The coefficient p depends solely on the orbit of the planet and of its satellites, and during the contraction of the mass will have been constant, or very nearly so. To determine the other quantities involved, we have the three following principles:—

(1) The conservation of angular momentum.

(2) The constancy of mass of the planet.

(3) That the form of the planet is one of equilibrium.

(1) is expressed by the equation $Cn = H$, a constant; and, if a, ρ be the mean radius and density of the planet at any time, (2) by $\frac{4}{3}\pi\rho a^3 = M$, the mass. Then, if the law of internal density during contraction be that of Laplace, viz. $Q \sin qr/r$, if k be the ratio of the surface-density to the mean density, e the ellipticity of the surface, and m the ratio of the centrifugal force at the distance a to mean pure gravity, the third principle gives†

$$\frac{5m}{2e} = \frac{(qa)^2}{3k(qa-1)} - 3k$$

Also

$$C = \frac{2}{3}\left\{1 + \frac{6}{(qa)^2}(k-1)\right\} Ma^2$$

$$C - A = \frac{2}{3}\left(e - \frac{1}{2}m\right) Ma^2$$

$$m = \frac{3n^2}{4\pi\mu\rho}$$

* Address to Section A. of the British Association at Glasgow, *Nature*, September 14, 1876, p. 429.

† Compare Thomson and Tait's *Natural Philosophy*, § 824 (14), § 827 (20).

Hence (1), (2), and (3) lead to the following equations:—

$$\tfrac{2}{3}\left\{1+\frac{6}{(qa)^2}(k-1)\right\}Ma^2n = H \quad\ldots\ldots\ldots\ldots\ldots\ldots(4)$$

$$\rho a^3 = \frac{3M}{4\pi} \quad\ldots\ldots\ldots\ldots\ldots\ldots(5)$$

$$\frac{n^2}{4\pi\mu\rho}\left\{\frac{5}{\dfrac{(qa)^2}{3k(qa-1)}-3k}-1\right\}Ma^2 = C - A \quad\ldots\ldots\ldots\ldots(6)$$

If during contraction qa remains constant, and if the coefficient of Ma^2n in (4) be called γ, and that of $Ma^2n^2/4\pi\mu\rho$ in (6) be called β, then it will be found that

$$\frac{1}{n}\frac{d}{dt}(C-A) = -\frac{H\beta}{\gamma}\frac{1}{12\pi\mu\rho^2}\frac{d\rho}{dt}$$

Hence, remembering that $Cn = H$,

$$\frac{d}{d\rho}\log\tan\theta = \frac{p\beta}{12\pi\mu\gamma\rho^2}$$

Integrate, and let D, I be the present values of ρ and θ; then

$$\log\frac{\tan\theta}{\tan I} = \frac{p\beta}{12\pi\mu D\gamma}\left(1-\frac{D}{\rho}\right)$$

If we assume that qa has always the same value as it now has in the case of the earth*,

$$\gamma = \cdot 3344, \quad \beta = \cdot 9507, \quad \text{and} \quad \frac{\beta}{\gamma} = 2\cdot 8433$$

If during contraction the planet were always homogeneous, the factor β/γ would be replaced by $\frac{15}{4}$, or 3·75.

Let K stand for 2·8433, or 3·75, as the case may be; let

$$Q = \frac{1}{T^2} + \Sigma\frac{\lambda}{T'^2(1+\nu)}$$

let P be the periodic time of a pendulum of length equal to the present mean radius of the planet, swinging under mean pure gravity. Then

$$\frac{p}{2\pi\mu D} = \tfrac{1}{6}P^2Q$$

and the equation becomes

$$\log\frac{\tan\theta}{\tan I} = \tfrac{1}{6}KP^2Q\left(1-\frac{D}{\rho}\right)$$

* In determining the precessional constants of Jupiter and Saturn, Laplace assumed that their law of internal density was the same as that of the earth. The assumption is, I believe, unjustifiable; but it will give sufficiently good results for the present purpose. The limiting value of β/γ, when the surface-density is infinitely small, and if the Laplacian law still holds good, is 1·99. [See Paper 5].

This equation shows that as ρ diminishes θ diminishes, and when ρ is infinitely small θ is zero. That is to say, if a nebulous mass is rotating about an axis nearly perpendicular to the plane of its orbit, its equator tends to become oblique to its orbit as it contracts.

In the case of the earth, $P^2Q = 8{\cdot}5577 \times 10^{-8}$; and taking the present obliquity of the ecliptic as 23° 28′, the equation may be written

$$\text{Log}_{10} \tan \theta = 9{\cdot}63761 - \frac{1{\cdot}7612}{10^8} \cdot \frac{D}{\rho}$$

On the hypothesis of homogeneity, 1·7612 must be replaced by 2·3229.

The extreme smallness of the coefficient of D/ρ shows that the earth must have had nearly the same obliquity even when its matter was rare enough to extend to the moon. But if it can be supposed that the moon parted from the earth without any abrupt change in the obliquity of the planet to the ecliptic, then from that epoch backwards the function Q would have had only one term, viz. $1/T^2$, and P^2Q would be $2{\cdot}5750 \times 10^{-8}$. The coefficient of D/ρ in the above equation would be reduced to $5{\cdot}30 \times 10^{-9}$, or $7{\cdot}00 \times 10^{-9}$, according to whichever value of K is taken. This being granted, it follows that when the diameter of the earth was 1000 times as large as at present, the obliquity to the ecliptic was only a few minutes.

This somewhat wild speculation can hardly be said to receive much support from the cases of other planets; but it is not thereby decisively condemned. In all the planets up to and inclusive of Jupiter, the expression Q will have to be reduced to its first term $1/T^2$, because the satellites are rather near their primaries. Hence one would expect that the obliquities of the planets to their orbits would diminish as we go away from the sun. It is believed (but the observations seem doubtful) that Mercury and Venus are very oblique to their orbits; and Mars has an obliquity nearly the same as that of the earth. The region of the asteroids is a blank; and then we come to Jupiter, with a very small obliquity.

The next in order is Saturn: and his case is unfavourable; for he is slightly more oblique to his orbit than is the earth. Nevertheless it must be observed that he has a large number of satellites, and some are very remote from him, and his mean density is very small; hence, if the satellites can have affected the obliquity in any case, one would expect them to have done so in that of Saturn.

No light whatever is thrown on the case of Uranus, whose axis is said to lie nearly in the plane of his orbit.

4.

NOTE ON THE ELLIPTICITY OF THE EARTH'S STRATA.

[*Messenger of Mathematics*, VI. (1877), pp. 109, 110.]

IN Pratt's *Figure of the Earth*, the following expression is given for the ellipticity e of the stratum of mean radius a, viz.

$$\frac{e}{\epsilon} = \frac{\tan q\mathrm{a} - q\mathrm{a}}{\tan qa - qa} \frac{\left(1 - \dfrac{3}{q^2 a^2}\right) \tan qa + \dfrac{3}{qa}}{\left(1 - \dfrac{3}{q^2 \mathrm{a}^2}\right) \tan q\mathrm{a} + \dfrac{3}{q\mathrm{a}}}$$

where ϵ, a are the ellipticity and mean radius of the Earth's surface. A similar expression, but with a different notation, will be found in Thomson and Tait's *Nat. Phil.*, § 824 (9). It is proposed to reduce this to a simpler form. In what follows, D' denotes the mean density, and ρ' the stratum or surface density of the spheroid of mean radius a, excluding the parts of the earth which lie outside this spheroid; D, ρ, a have parallel meanings for the whole earth.

Laplace's law for the density of the stratum a may be written $\rho \dfrac{\mathrm{a} \sin q\mathrm{a}}{a \sin qa}$; then it may easily be shown that

$$\frac{D}{\rho} = \frac{3}{(q\mathrm{a})^2} \left(1 - \frac{q\mathrm{a}}{\tan q\mathrm{a}}\right)$$

And if (following Pratt) $z = 1 - \dfrac{q\mathrm{a}}{\tan q\mathrm{a}}$

$$\frac{3z}{(q\mathrm{a})^2} = \frac{D}{\rho}$$

If $z' = 1 - \dfrac{qa}{\tan qa}$, it is clear also that

$$\frac{3z'}{(qa)^2} = \frac{D'}{\rho'}$$

Hence
$$\tan q\text{a} - q\text{a} = \frac{z}{1-z} q\text{a}$$

$$\left(1 - \frac{3}{q^2\text{a}^2}\right) \tan q\text{a} + \frac{3}{q\text{a}} = \frac{1 - D/\rho}{1-z} q\text{a}$$

and two parallel expressions with italics and accents.

Substituting in the expression for $\frac{e}{\epsilon}$, we get

$$\frac{e}{\epsilon} = \frac{z}{z'} \frac{1 - D'/\rho'}{1 - D/\rho} = \frac{\text{a}^2}{a^2} \frac{1 - \rho'/D'}{1 - \rho/D}$$

If in the case of the earth $\frac{\rho}{D} = \frac{1}{2}$,

$$e = 2\text{a}^2\epsilon \frac{1 - \rho'/D'}{a^2}$$

But whether this proportion of D to ρ holds good or not, it follows that the ellipticity of any stratum is proportional to the attraction of a particle, whose mass is proportional to $1 - \rho'/D'$, placed at the earth's centre at the distance of the stratum in question.

ρ'/D' diminishes as we go away from the centre, and therefore $1 - \rho'/D'$ increases from zero to $1 - \rho/D$; on the other hand a^2 increases rapidly, and therefore one would be led to expect (as is the fact) that the ellipticity increases on going away from the centre.

The preceding expression also leads to another curious expression for the ellipticity.

If we add a small stratum of thickness δa to the spheroid of mean radius a, the mean density of course falls, and it is very easily shown that

$$\delta D' = -3 \frac{D'}{a} \left(1 - \frac{\rho'}{D'}\right) \delta a$$

whence it follows at once, that

$$e \propto - \frac{d \log D'}{d(a^2)}$$

5.

ON AN OVERSIGHT IN THE *MÉCANIQUE CÉLESTE*, AND ON THE INTERNAL DENSITIES OF THE PLANETS.

[*Royal Astronomical Society, Monthly Notices*, XXXVII. (1877), pp. 77—89.]

IN the following paper an endeavour is made to point out an inconsistency, which appears to have escaped the notice of Laplace, in his determination of the precessional constants of the planets *Jupiter* and *Saturn*. From this I have been led on to speculate on the law of internal density of those planets, and of *Mars*, and to make some reference to the ellipticities of *Mercury* and *Venus*.

§ 1. *Laplace's Law of the Internal Density of the Planets.*

In the investigation of the figure of the Earth, Laplace assumed that, in molten rock, the hydrostatic pressure *plus* a constant varies as the square of the density. The result of this assumption is that, after the consolidation of a planet, the density of any stratum of mean radius r is given by the law $\frac{F}{r} \sin \frac{r\theta}{a}$, where a is the mean radius of the surface, and θ and F are constants.

Throughout the rest of this paper, besides the foregoing, the following notation is used:—

α, β, the equatoreal and polar radii;

ϵ the ellipticity of the surface;

m the ratio of the centrifugal force of the planet's rotation at the distance a to the mean pure gravity;

f the ratio of the mean to the surface density;

C, A, the greatest and least principal moments of inertia of the planet;

P or $(C-A)/C$ the precessional constant; and

M the mass.

Other symbols will be defined as they arise.

Then the results of the Laplacian law may be embodied in the following equations*:—

$$\tfrac{1}{3}f = \frac{1}{\theta^2} - \frac{\cot\theta}{\theta} \quad \text{.............................(1)}$$

$$\frac{5m}{2\epsilon} = \frac{f\theta^2}{3(f-1)} - \frac{3}{f} \quad \text{.........................(2)}$$

$$C - A = \tfrac{2}{3}(\epsilon - \tfrac{1}{2}m)\,Ma^2 \quad \text{......................(3)}$$

$$C = \tfrac{2}{3}\left\{1 - \frac{6(f-1)}{f\theta^2}\right\} Ma^2 \quad \text{.. ....(4)}$$

$$\frac{\epsilon - \tfrac{1}{2}m}{P} = 1 - \frac{6(f-1)}{f\theta^2} \quad \text{.........................(5)}$$

Now, if (following Thomson and Tait) we take for the Earth $f = 2{\cdot}1$, which corresponds with $\theta = 144°$, we have

$$P = 1{\cdot}994\,(\epsilon - \tfrac{1}{2}m) \quad \text{...........................(6)}$$

$$\frac{5m}{2\epsilon} = 2{\cdot}562 \quad \text{............ ....(7)}$$

Hence, in any planet where the law of density is the same as in the Earth (i.e., $f = 2{\cdot}1$, $\theta = 144°$), we should find (7) satisfied, and (6) would give the precessional constant.

In order to illustrate the Laplacian law, the following table has been constructed, giving the values of f and $5m/2\epsilon$ for every 10° of θ, with the omission of a few at the early part, which are rather troublesome to calculate, and are of little value.

θ	f	$5m/2\epsilon$	θ	f	$5m/2\epsilon$
0°	1·0000	2·000	90°	1·2159	2·165
10°			100°	1·2879	2·213
20°	1·0082		110°	1·3827	2·270
30°	1·0188		120°	1·5109	2·338
40°	1·0341	2·029	130°	1·6922	2·422
50°	1·0548	2·046	140°	1·9657	2·525
60°	1·0817	2·067	150°	2·4225	2·652
70°	1·1161	2·094	160°	3·3363	2·813
80°	1·1600	2·126	170°	6·0750	3·019
90°	1·2159	2·165	180°	infin.	3·290

This table shows how $5m/2\epsilon$ increases, as the planet passes from homogeneity to infinitely small surface density; $5m/2\epsilon$ can never be less than 2, nor greater than 3·290. Although it is not strictly involved in the subject

* See Thomson and Tait's *Natural Philosophy*, §§ 824, 827. Equation (3) is independent of Laplace's assumption.

of this paper, I may point out that the first column of the table may be employed to calculate the ellipticity of any internal stratum. I have shown elsewhere* that the ellipticity e of any stratum of mean radius $\theta' a/\theta$ is given by

$$\frac{e}{\epsilon} = \left(\frac{\theta}{\theta'}\right)^2 \frac{f(f'-1)}{f'(f-1)}$$

where f, θ refer to the surface, and f' is the value corresponding to θ' in the table. For example, to find the ellipticity of the Earth's stratum of mean radius $\frac{5}{12}a$, we must take $\theta' = 60°$, because $\theta = 144°$; then from the table $f' = 1{\cdot}0817$, and

$$e = \epsilon \left(\tfrac{12}{5}\right)^2 \cdot \frac{2{\cdot}1 \times {\cdot}0817}{1{\cdot}0817 \times 1{\cdot}1} = \frac{\epsilon}{1{\cdot}204}$$

$$= \tfrac{1}{357}, \text{ when } \epsilon = \tfrac{1}{297}$$

§ 2. *Jupiter's Precessional Constant.*

To determine the precessional constant Laplace uses the following argument:—"If we suppose the densities of the strata of Jupiter and the Earth, at distances proportional to the diameters of these planets, to be in a constant ratio to each other....In this hypothesis, if both planets be fluid, their ellipticities will be, as in [2068 *k*, etc., III. v. § 43] proportional to the respective values of ϕ†, corresponding to each of them; or to their ellipticities, if they be homogeneous. If we suppose the ratio to obtain in their actual state, and we have seen in [2069, III. v. § 43] that this is nearly conformable to observation, *then the values of* $(2C - A - B)/C$ *will be, for each of these planets, respectively proportional to the ellipticities corresponding to the case of homogeneity.* These ellipticities by the same article [2068″], are as ·10967000 to 0·00433441‡." He uses ·00291193 as the Earth's precessional constant, and thence deduces $(2C - A - B)/C = {\cdot}14735$, or in my notation $P = {\cdot}07368$.

By the theory of the perturbations of the satellites, he finds

$$\epsilon - \tfrac{1}{2}m = {\cdot}0219013\S$$

this value does not involve any assumption as to the law of internal density, except, I conceive, that the strata of equal density must be nearly spherical‖.

Then by the periodic time of the Fourth Satellite, and its observed distance from *Jupiter*, he finds $m = {\cdot}0987990$; whence $\epsilon = {\cdot}0713008$§.

* [Paper 4, p. 55.] The notation is, however, different.

† The m of this paper.

‡ Bowditch, Trans. of *Méc. Cél.* VIII. vii., § 23.

§ *Ib.* VIII. ix., § 27.

‖ See Pratt's *Figure of Earth*, Arts. 90—94.

Using these values of m and ϵ, we get $5m/2\epsilon = 3{\cdot}4642$. Now Laplace does not seem to have noticed that these values are not only incompatible with the identity of the law of internal density in the *Earth* and *Jupiter*, but are also incompatible with that form of law at all; for, as appears from the table, $5m/2\epsilon$ can never be greater than 3·290.

My attention was first drawn to the point by observing that, if there were this identity of law of density, the precessional constant of *Jupiter* ought to be nearly twice $\epsilon - \frac{1}{2}m$, or ·0438; whereas Laplace, by a different method, founded on the same assumption, finds it to be ·0737. This of course indicated that the assumption was untenable.

In view of this discrepancy, it will be well to go over Laplace's work again by the light of later, and presumably better, observations.

The distance of the Fourth Satellite is open to some doubt, but the following has seemed to me the best value attainable. Bessel gives the apparent distance at the planet's mean distance as the angle 498″·8663*.

M. Kaiser gives, as the result of a long series of his own observations when united with those of Bessel, that the polar and equatoreal diameters of *Jupiter* subtend at the same distance 35″·170 and 37″·563 respectively†.

Hence the mean distance D of the satellite is 26·5616 equatoreal radii; or $D = 26{\cdot}5616\,a$; T the satellite's periodic time is 16·68902 m. s. days, t *Jupiter's* sidereal day is ·4135 m. s. days. Therefore

$$m = \left(\frac{T}{t}\right)^2 \left(\frac{a}{D}\right)^3 (1-\epsilon)$$

$$= {\cdot}0869258\,(1-\epsilon) = m'(1-\epsilon) \text{ suppose}$$

Professor J. C. Adams informs me that M. Damoiseau has recalculated the values of $\epsilon - \frac{1}{2}m$, and of the masses of the satellites, basing his work on better observations than those which were available to Laplace‡. He has taken, however, certain coefficients from Laplace, which ought to have been recalculated. This work has been completed by Professor Adams himself, and he has obtained the following value for $\epsilon - \frac{1}{2}m$, of which he has, with very great kindness, allowed me to make use, viz.

$$\epsilon - \tfrac{1}{2}m = {\cdot}0216623\,\S$$

* *Astron. Untersuch. Bestim. der Masse des Jupiter*, Königsberg, 1842.

† *Astron. Nachr.* 48, p. 111.

‡ *Tables des Satellites de Jupiter.* Bur. des Long., Bachelier, Paris, 1836.

§ The corresponding masses of the Satellites multiplied by 1000 are found by him to be:

·283113
·232355
·812453
·214880

They seem to bear hardly any relation to the ordinarily received values.

Substituting then for m, we have, for the ellipticity of *Jupiter*,

$$\epsilon = \frac{\frac{1}{2}m' + \cdot 0216623}{1 + \frac{1}{2}m'} = \cdot 06241 = \frac{1}{16\cdot 022}$$

and $$m = m'(1 - \epsilon) = \cdot 081501\,^*$$

Laplace remarks† that the ellipticity of *Jupiter* may be found by this method with greater accuracy than by the best observations. It is interesting, therefore, to observe how closely this value agrees with that of all the later observations. The following are the values given for the reciprocal of the ellipticity by—Secchi, 16·06 [*Ast. Nach.* 43, p. 142]; Kaiser, (i) 15·36, (ii) 15·98 [*Ast. Nach.* 45, p. 211]; Bessel and Kaiser, 15·70; Bessel, 15·73; Secchi, 15·99 [*Ast. Nach.* 48, p. 111]; Schmidt, 15·6 [*Ast. Nach.* 65, p. 102]; Main, 16·84 [*Month. Not. Ast. Soc.* 16, p. 142]. The reductions were in several cases made by myself.

It is, nevertheless, a curious coincidence that the value of ϵ found by Laplace agreed well with the older observations of Struve, who found $\epsilon = 1/13\cdot 74$ [*Ast. Nach.* 45, p. 211].

Using the above value of m, I find (following Plana‡) that on the hypothesis of homogeneity $\alpha/\beta = 1\cdot 1164$ (compare with 1·116515 of Plana), and therefore the homogeneous ellipticity is ·10405. But the homogeneous ellipticity of the Earth is 1/230·433 (Plana); and the Earth's precessional constant is ·003272; hence, following Laplace's method, we should have for *Jupiter*,

$$P = \cdot 10405 \times 230\cdot 433 \times \cdot 003272$$
$$= \cdot 078451$$

But if Laplace's assumption as to the internal density of *Jupiter* were justifiable, we ought to have by equation (6),

$$P = 1\cdot 994\,(\epsilon - \tfrac{1}{2}m) = 1\cdot 994 \times \cdot 02166$$
$$= \cdot 043$$

Thus these new values of the quantities involved leave as wide a discrepancy as before between the two values of P. Therefore, Laplace's method cannot be justified.

§ 3. *The Internal Density of Jupiter.*

But this unjustifiability may be seen in another way, for the new values of m and ϵ give $5m/2\epsilon = 3\cdot 2646$; and reference to equation (7) shows that this differs widely from the corresponding value for the Earth. It is not, however, inconsistent with the Laplacian form of law, as was the same function when Laplace's values of m and ϵ were used. In fact, I find, after some rather

* This value may be compared with ·08163 found by Plana, *Astr. Nachr.* 36, p. 155.

† *Méc. Cél.*, VIII. ix., § 27.

‡ All future references to M. Plana are to his paper in *Astr. Nachr.* 36, p. 154.

troublesome arithmetic, that $5m/2\epsilon = 3{\cdot}2646$ corresponds to $\theta = 179^\circ\, 11'\, 20''$ nearly, and to $f = 68$.

Using these values of θ and f, I find by equation (5) that *Jupiter's* precessional constant $= 2{\cdot}528\,(\epsilon - \frac{1}{2}m) = {\cdot}0548$; and this I take to be far nearer the truth than the value assigned by Laplace.

On account of the uncertainty in the determination of the quantity m, the above values of θ and f can, of course, have no claim to precision. In fact, if following Herschel, Loomis, and others, we take $D/\alpha = 26{\cdot}99835$, we should find $\epsilon = 1/16{\cdot}52$, $m = {\cdot}0778$, and $5m/2\epsilon = 3{\cdot}211$. This last corresponds to θ about $177^\circ\, 42'$, and f about 24.

The value of the precessional constant will, however, be but little changed, being $2{\cdot}489\,(\epsilon - \frac{1}{2}m) = {\cdot}0539$.

Now it seems reasonable *à priori* to assume that the law of internal density within *Jupiter* is of the same *nature* as in the Earth, and from that assumption it follows that the surface density of *Jupiter* is vastly less than the mean: for it must, I think, be admitted that numerical values can be assigned to m and ϵ with some precision.

But the true meaning of this result would seem to be that the Laplacian law of density is not exact for *Jupiter*, but that that planet must be very much denser in the centre than at the surface. Is it not possible that *Jupiter* may still be in a semi-nebulous condition, and may consist of a dense central part with no well-defined bounding surface? Does not this view accord with the remarkable cloudy appearance of the disk, and the remarkable belts?

§ 4. *Saturn.*

The preceding method cannot be applied very satisfactorily to the case of *Saturn*, on account of the uncertainty in all the quantities which enter into the determination of m and ϵ; but the balance of evidence appears to me decidedly in favour of his surface density being far less than the mean.

m may be best determined from the motion of *Japetus*.

There seems almost complete unanimity as to his periodic time, and I take $T = 79{\cdot}3296$ m. s. days.

Two values are assigned for the sidereal day of *Saturn*, viz., ·4370 (Hansen), and ·4278 (Herschel) m. s. days, the latter being more generally accepted; I take then $t = {\cdot}430$.

There is a wide difference of opinion as to the mean distance D of *Japetus*, the estimates varying from $57{\cdot}4\alpha$ (Jacob, *Mem. Astr. Soc.* Vol. XXVIII.) to $64{\cdot}359\alpha$ (used by Herschel, Loomis, and others, and most generally accepted). I take $D = 62\alpha$.

The ellipticity seems fairly well determined at $\frac{1}{11}$*.

Using these quantities in the formula

$$m = \left(\frac{T}{t}\right)^2 \left(\frac{\alpha}{D}\right)^3 (1-\epsilon)$$

I find $$m = \cdot131270 - \cdot61056\delta t - \cdot00635\delta\left(\frac{D}{\alpha}\right) - \cdot14281\delta\epsilon$$

(i) If all the small variations are zero, and $\epsilon = \frac{1}{11}$, we get $\frac{5m}{2\epsilon} = 3\cdot610$.

(ii) Loomis, *Pract. Astron.*, Tables 33, 34,

$$t = \cdot437, \quad \frac{D}{\alpha} = 64\cdot359, \quad \epsilon = \tfrac{1}{10}$$

$$\delta t = \cdot007, \quad \delta\frac{D}{\alpha} = 2\cdot359, \quad \delta\epsilon = \cdot01909$$

then $$m = \cdot1093, \quad \frac{5m}{2\epsilon} = 2\cdot732$$

(iii) Hind's *Solar System*, pp. 103–4,

$$t = \cdot4278, \quad \frac{D}{\alpha} = 60\cdot3436, \quad \epsilon = \frac{1}{10\cdot28} \text{ according to Bessel}$$

$$\delta t = -\cdot0022, \quad \delta\frac{D}{\alpha} = -1\cdot6564, \quad \delta\epsilon = \cdot00637$$

then $$m = \cdot1422, \quad \frac{5m}{2\epsilon} = 3\cdot66$$

If we take $\epsilon = \frac{1}{9\cdot23}$ from the *Greenwich Observations*, $\frac{5m}{2\epsilon}$ will be slightly smaller.

(iv) Captain Jacob's *Madras Observations* give $\frac{D}{\alpha} = 57\cdot4$, and $\delta\frac{D}{\alpha} = -4\cdot6$, this with $\epsilon = \frac{1}{11}$ and all the other small variations zero, gives $m = \cdot160$ and $\frac{5m}{2\epsilon}$ very nearly $4\frac{1}{4}$.

No doubt other values might be assigned to $\frac{5m}{2\epsilon}$ with equal plausibility, but I think (ii) and (iv) contain extreme values in the two directions, and the truth probably lies at some intermediate point. It is noticeable that

* I find the following in a paper by Mr Grant (*Monthly Notices of the Royal Astronomical Society*, Vol. XIII., p. 195), where he reduces observations of apparent ellipticity to their real values, viz., Lassell 1/11·05, Main 1/9·227, De La Rue 1/10·611. The mean of all Bessel's observations would seem to be 1/10·20, but Mr Hind (see below) assigns a slightly different value as Bessel's result.

Sir W. Herschel found 1/10·368 (see p. 79 of above vol.), but 1/10·384 is also attributed to him on the same page.

even the value (ii) indicates a larger ratio of mean to surface density than in the Earth, namely 2·809 (corresponding to $\theta = 155° \ 14'$), whilst all the other values are far outside the limits of admissibility under the Laplacian law.

I have tried also to find m from the motion of *Titan*. Taking

$$\frac{D}{a} = 21, \quad T = 15{\cdot}9454, \quad t = {\cdot}43, \quad \epsilon = \tfrac{1}{11}$$

I find

$$m = {\cdot}13648 - {\cdot}1174\delta t - {\cdot}0195\delta \frac{D}{a} - {\cdot}1485\delta\epsilon$$

If the small increments are zero, $5m/2\epsilon = 3{\cdot}75$; and if the various possible values be assigned to them, the results will be found to be very similar to those given above.

The value of m assigned by Laplace is ·165970*, and this, together with the ellipticities $\frac{1}{9}$ and $\frac{1}{12}$, gives $5m/2\epsilon$ equal to 3·734 and 4·979 respectively, either value quite incompatible with the Laplacian law of density. The ellipticity of the planet had apparently not been observed in his time, and the numbers used by him involve an ellipticity of 1/68·6, which we now know is far too small.

Laplace makes *Saturn's* precessional constant $\frac{1}{2}$ of ·27934 or ·13967 (VIII. xvii. § 37), but this value is inconsistent with any of those given above for m and ϵ.

Taking $\epsilon = \frac{1}{11}$, $m = {\cdot}1313$, we have $\epsilon - \frac{1}{2}m = {\cdot}0252$. It can hardly be supposed that *Saturn* is more nearly homogeneous than is the Earth, and if the law of density were the same we should, by (6), have $P = {\cdot}050$.

On the other hand, if the Laplacian law of density were to hold good, and if the surface density were zero—of course an ideal case†—we should have by (5),

$$\frac{\epsilon - \frac{1}{2}m}{P} = 1 - \frac{6}{\pi^2} = \frac{1}{2{\cdot}5505}$$

or

$$P = 2{\cdot}5505 \times {\cdot}0252 = {\cdot}064$$

I conceive then, that *Saturn's* precessional constant must be very nearly equal to ·06, since the former of these values is certainly too small, whilst the latter is a little too large.

* *Méc. Cél.*, VIII. xvii., § 36.

† It may be observed, that to be thoroughly consistent we should have in this case, $P = {\cdot}663m = {\cdot}873\epsilon$, because $5m/2\epsilon = 3{\cdot}290$; but the method is here only used to find the limiting value of the coefficient of $\epsilon - \frac{1}{2}m$.

§ 5. *Mars.*

Mars rotates in $24^h\ 37^m\ 22^s{\cdot}6$, or 1·025956 m. s. days, according to the elaborate observations of M. Kaiser*. His density is ·948 of that of the Earth†. The sidereal day is ·997270 m. s. days, and for the Earth m is $\frac{1}{289{\cdot}66}$; hence for *Mars*,

$$m = \frac{1}{\cdot 948}\left(\frac{\cdot 997270}{1{\cdot}025956}\right)^2 \frac{1}{289{\cdot}66}$$

$$= \cdot 0034409 = \frac{1}{290{\cdot}6}\ \ddagger$$

With respect to the ellipticity of the planet the most various estimates are given. Thus, for the reciprocal of the ellipticity I find, according to Loomis (*Pract. Astr.*, Table 33) 50; Main (*Month. Not. Ast. Soc.* XVI., p. 142) 62; Kaiser (Guillemin, *Le Ciel*, p. 257) 118; Arago, doubtfully, 32; Bessel failed to detect ellipticity in observations at Königsberg; and lastly, from observations of Dr Winnecke, to which I shall recur below, 225 (*Ast. Nach.* 48, p. 97). According to Dr Winnecke, M. Kaiser doubts whether Mr Main's result represents any corresponding reality.

With an ellipticity of $\frac{1}{225}$, we have $5m/2\epsilon = 1{\cdot}93$, and with the other values of ellipticity, a very much smaller value.

Now a reference to the table in § 1 will show that not only is even the largest of these quantities incompatible with the Laplacian law of density, but they are also incompatible with the homogeneity of the planet. We should require the planet to decrease in density towards the centre, or actually to be hollow; and on any theory of original fluidity such a state of things is almost inconceivable.

The wide discrepancy between various observations shows that the results are to be very little depended on; and for the following reasons, I venture to think that all the above values of the ellipticity are almost worthless, but that Dr Winnecke's is by far the nearest to the truth. He gives§ as the equatoreal and polar apparent diameters, $9''{\cdot}227$ and $9''{\cdot}186$, subject to mean errors of $0''{\cdot}045$ and $0''{\cdot}032$ respectively.

* See a paper by M. Julius Schmidt, *Astr. Nachr.*, 82, p. 333.

† Guillemin, *Le Ciel*, p. 257. The other details with respect to *Mars* are given on the authority of M. Kaiser, and this I presume is so also.

‡ Plana gives 1/288·84, using slightly different data; with Herschel's value of the density ·72, $m = 1/220{\cdot}7$.

§ *Astr. Nachr.*, 48, p. 97. The observations are said to be reduced for "Phase und Refraction," but I do not see it expressly stated that they are corrected for the Earth's Martian declination.

From these measurements I find that the *probable error* of the ratio of the axes is ·0050441. The ratio itself is in decimals ·9955564, and therefore the ratio of the axes is ·9955564 ± ·0050441. That is to say, it is an even chance that the ratio of the axes lies between the 1·0006 and ·990512. In other words, it results from Dr Winnecke's observations that it is an even chance that the figure of *Mars* lies between a prolate spheroid, with an ellipticity of $\frac{1}{1667}$, and an oblate one with an ellipticity of $\frac{1}{105}$; the observed values give an ellipticity of $\frac{1}{225}$.

Dr Winnecke himself says, "Es sprechen diese Messungen entschieden gegen eine Abplattung des *Mars*, die für das Berliner Instrument messbar wäre."

Now the telescope shows that *Mars* has an external physical constitution exceedingly like that of the Earth, and therefore there is a strong probability that its internal structure is somewhat the same. Seeing then that the observations of the best observers permit such very wide limits of error, and that the value of m can be assigned with some precision, I submit that there is a far better chance of being near the truth in trusting to indirect evidence than to direct observation, for the determination of the ellipticity. Assuming then that the law of internal density is the same as in the Earth, I find an ellipticity of $\frac{1}{298}$, and this, I venture to think, is nearer the truth than any of the above-quoted values derived from observation*.

§ 6. *Mercury, Venus.*

M. Plana assigns as the values of m for these planets, 1/325·82 and 1/251·54 respectively; and following the arguments used in the case of *Mars* (although they have not here equal force), we should assign to *Mercury* an ellipticity of about $\frac{1}{330}$, and to *Venus* of about $\frac{1}{260}$.

* [This paper was written before the discovery of the Martian Satellites. I find from Asaph Hall's *Observations and Orbits of the Satellites of Mars* (Washington Government Printing Office, 1878), that the mean distance of Deimos is 32″·3541, and of Phobos 12″·9531; their periodic times are respectively $1^d{\cdot}262429$ and $0^d{\cdot}3189244$.

If we adopt Winnecke's value of the equatoreal radius of Mars, viz., 4″·6135, we find that Deimos furnishes for m the value $\frac{1}{228}(1-\epsilon)$, and Phobos gives $\frac{1}{229}(1-\epsilon)$. Hence m must be very nearly equal to 1/228, agreeing closely with Herschel's value given above.

Now from the table in § 1 we see that $5m/2\epsilon$ must lie between 2 and 3·290. Hence the reciprocal of ϵ must lie between 184 and 300.

From the general similarity of Mars to the Earth, it seems reasonable to suppose that $5m/2\epsilon$ is about 2·525, this would make the ellipticity of Mars 1/230, agreeing nearly with Winnecke's value. November, 1908.]

§ 7. *Conclusion.*

The results arrived at may be summed up as follows:

The values assigned by Laplace to the precessional constants of *Jupiter* and *Saturn* cannot be maintained; and the objections to them remain equally strong when later observations are consulted.

Professor Adams's calculation of $\epsilon - \frac{1}{2}m$ gives, as the ellipticity of *Jupiter*, 1/16·02.

The surface density of *Jupiter* is far less than the mean, and he may perhaps still be in a semi-nebulous condition.

There is a considerable probability that the like is true of *Saturn*.

The value of *Jupiter's* precessional constant is about ·0548; and that of *Saturn* probably about ·06. No dependence is to be placed on the observations of the ellipticity of *Mars*, because of their wide limits of error. Indirect evidence of the ellipticity seems safer, and we are thereby led to assign to him an ellipticity of $\frac{1}{298}$. [But observations of the satellites indicate an ellipticity of about $\frac{1}{230}$.] In the cases of *Mercury* and *Venus*, we have only indirect evidence to rely on, and these ellipticities are probably about $\frac{1}{330}$ and $\frac{1}{260}$.

In conclusion, I must add that if in any case I have underrated or neglected the work of important observers, or have overrated the worth of other observations, my excuse must be my previous slight acquaintance with observational astronomy.

POSTSCRIPT.—The method of this paper may perhaps give some idea of the amount of difference which might be expected to be found between the apparent equatoreal and polar diameters of the Sun.

To find m, the Earth may be treated as a solar satellite, and the same formula as before applied. d/D is here the Sun's mean apparent radius (which I take as 961″·82) in circular measure.

$T = 365{\cdot}2569$ m. s. days.

t, according to Carrington, is 25·38 m. s. days.

But Mr Christie (to whom I owe my thanks for his help in this matter) informs me that the true period may perhaps be 26 or 27 days. The following results are therefore given in duplicate; in those marked with an accent t is taken as 27.

Then $m = {\cdot}00002100$, or $m' = {\cdot}00001829$.

If the law of internal density were the same as in the Earth, we should have,

$$\epsilon = \cdot 00002048, \quad \text{or} \quad \epsilon' = \cdot 00001784$$

These correspond to differences of apparent diameter of $0''\cdot 0394$ and $0''\cdot 0343$ respectively. If the Laplacian law of density were to hold good, and if the surface density were zero, we should have, $\epsilon = \cdot 00001596$, or $\epsilon' = \cdot 00001390$, which correspond to differences of $0''\cdot 0307$ and $0''\cdot 0267$ respectively.

Mr Christie informs me that the probable error in the determination of the Sun's apparent diameter is not less than $0''\cdot 1$, and it appears that the observed difference of diameters is also $0''\cdot 1$.

May we not conclude that the difference of the apparent diameters is probably less than $0''\cdot 04$, but perhaps greater than $0''\cdot 03$?

6.

ON THE FIGURE OF EQUILIBRIUM OF A PLANET OF HETEROGENEOUS DENSITY.

[*Proceedings of the Royal Society*, XXXVI. (1884), pp. 158—166.]

THE problem of the figure of the earth has, so far as I know, only received one solution, namely, that of Laplace*. His solution involves an hypothesis as to the law of compressibility of the matter forming the planet, and a solution involving another law of compressibility seems of some interest, even although the results are not perhaps so conformable to the observed facts with regard to the earth as those of Laplace†.

The solution offered below was arrived at by an inverse method, namely, by the assumption of a form for the law of the internal density of the planet, and the subsequent determination of the law of compressibility. One case of the solution gives us constant compressibility, and another gives the case where the modulus of compressibility varies as the density, as with gas.

It would be easy to fabricate any number of distributions of density, any one of which would lead to a law of compressibility equally probable with that of Laplace; but the solution of Clairaut's equation for the ellipticity of the internal strata of equal density seems in most cases very difficult. Indeed, it is probable that Laplace formulated his law because it made the equation in question integrable, and because it was not improbable from a physical point of view.

* Since this paper was presented I have seen a reference to a paper by the late M. Edouard Roche, in Vol. I. of the *Memoirs of the Academy of Montpellier* (1848), in which the problem is solved, when the rate of increase of the density varies as the square of the radius. See Tisserand, *Comptes Rendus*, 23rd April, 1883.

† Laplace's hypothetical law of compressibility arises from a law of internal density for which the problem had previously been worked out, as an example, by Legendre. See Todhunter's *History of the Figure of the Earth*, Vol. II., pp. 117 and 337.

The following notation will be adopted:—

For an internal stratum of equal density let

r be the radius vector of any point,

a the mean radius of the stratum,

e the ellipticity,

w the density,

ϕ colatitude from the axis of rotation,

p the hydrostatic pressure at the point r, ϕ.

For the surface let $\mathfrak{r}$, $\mathfrak{a}$, $\mathfrak{e}$, $\mathfrak{w}$ denote the similar things.

Let M be the mass of the planet, ρ its mean density, ω the angular velocity of rotation, m the ratio $\omega^2/\frac{4}{3}\pi\rho$.

Let k be the ratio of the density of the stratum a to the mean density of all the matter situated inside that stratum, and $\mathfrak{k}$ the surface value of k*.

Let C, A be the greatest and least principal moments of inertia of the planet about axes through its centre of inertia.

Let $æ$ be the ellipticity which the surface would have if the planet were homogeneous with density ρ, so that $æ = \frac{5}{4}m$.

The condition that the surface of the planet is a level surface is satisfied by

$$\mathrm{C} - \mathrm{A} = \tfrac{2}{3}\mathrm{M}\mathfrak{a}^2(\mathfrak{e} - \tfrac{1}{2}m) \quad \text{.........................(1)}$$

The condition that the internal surfaces are also surfaces of equilibrium demands that e should satisfy Clairaut's equation

$$\left(\frac{d^2e}{da^2} - 6\frac{e}{a^2}\right)\int_0^a wa^2da + 2wa^2\left(\frac{de}{da} + \frac{e}{a}\right) = 0 \quad \text{..............(2)}$$

It may be proved from (2), and the consideration that w must diminish as a increases, that e cannot have a maximum or minimum value.

Also it may be shown that the constants introduced in the integral of this equation must be such that

$$\frac{æ}{\mathfrak{e}} = \frac{5m}{4\mathfrak{e}} = \frac{1}{2\mathfrak{e}\mathfrak{a}}\frac{d}{da}(ea^2) \quad \text{.........................(3)}$$

when a is put equal to $\mathfrak{a}$ after differentiation.

The mean density is given by

$$\mathfrak{a}^3\rho = 3\int_0^{\mathfrak{a}} wa^2da \quad \text{.........................(4)}$$

And
$$\mathfrak{k} = \frac{\mathfrak{w}}{\rho}$$

* k, $\mathfrak{k}$, are the reciprocals of f, $\mathfrak{f}$, according to the notation adopted in Thomson and Tait's *Natural Philosophy* (edit. of 1883), § 824.

Neglecting the ellipticity of the strata, we have the moment of inertia about any diameter of the planet given by

$$C = \tfrac{8}{3}\pi \int_0^{\mathfrak{a}} wa^4 da \text{ ....(5)}$$

The ratio of (1) to (5) gives the precessional constant. The pressure and density are connected by the equation

$$\int_a^{\mathfrak{a}} \left(\frac{1}{w}\frac{dp}{da} + 4\pi wa\right) da + \frac{4\pi}{a}\int_0^a wa^2 da = 0 \text{.................(6)}$$

Now if ϖ be a function such that $wdp = d\varpi$, the differentiation of (6) leads to

$$\frac{d\varpi}{da} + \frac{4\pi}{a^2}\int_0^a wa^2 da = 0 \text{ ....(7)}$$

and a second differentiation to

$$\frac{d^2}{da^2}(\varpi a) + 4\pi wa = 0 \text{ ....(8)}$$

It is well known that Laplace assumed that the modulus of compressibility of rock varies as the square of the density. Since this modulus is wdp/dw, Laplace's hypothesis makes ϖ proportional to w, and the equation (8) is at once soluble.

After the determination of w as a function of a, the solution of all the other equations follows.

In this paper I propose to find a new solution, and to compare the results with those of Laplace.

In order to simplify the analysis let the unit of length be equal to the mean radius $\mathfrak{a}$ of the planet, and the unit of time be such that the surface density $\mathfrak{w}$ of the planet is also unity.

Now let us assume that the law of internal density is

$$w = a^{-n} \text{ ....(9)}$$

Then the mean density of all the matter lying inside the stratum a is $a^{-n}/(1 - \tfrac{1}{3}n)$. Hence, by definition we have

$$k = 1 - \tfrac{1}{3}n \text{ ....(10)}$$

Thus we see that k is a constant for all strata, and therefore also for the surface. In Laplace's theory k is variable. With our assumed law of density and the special units, ρ the mean density is equal to the reciprocal of k.

It is clear that n must be positive, otherwise heavier strata lie above lighter, and it must be less than 3 in order to avoid infinite mass at the centre of the planet.

Now let us find the law connecting pressure and density, and the modulus of compressibility.

Equation (7) becomes

$$\frac{d\varpi}{da} = -\tfrac{4}{3}\pi \frac{a^{1-n}}{1-\frac{1}{3}n}$$

and by definition of ϖ and the assumption (9),

$$\frac{dp}{da} = -\tfrac{4}{3}\pi \frac{a^{1-2n}}{1-\frac{1}{3}n}$$

Integrating this, with the condition that the pressure vanishes at the surface, we have

$$p = \frac{2\pi}{3(1-n)(1-\frac{1}{3}n)}[1 - a^{2(1-n)}]$$

$$= \frac{2\pi}{3(1-n)(1-\frac{1}{3}n)}[1 - w^{-2(1-n)/n}]$$

whence the modulus of compressibility is

$$w\frac{dp}{dw} = \tfrac{4}{3}\pi \frac{1}{n(1-\frac{1}{3}n)} w^{-2(1-n)/n}$$

The case of $n=1$ is interesting; it gives a constant modulus of compressibility equal to 2π, and the law of pressure $p = 2\pi \log w$.

If n be less than unity the compressibility, or reciprocal of the modulus, increases with the density, which is of course physically improbable. If n be greater than unity and less than 3, the compressibility becomes less the greater the density. The assumed law probably does not give such good results as those of Laplace, because the decrease of compressibility with increasing density is not sufficiently rapid.

The range of $n=3$ to $n=1$ gives the results which possess most physical interest.

In comparing results with those of Laplace there will be occasion to express the modulus as a length; that is to say, we are to find the length of a column of unit section whose weight (referred to the surface gravity of the planet) is equal to the force specified in the modulus.

Now if g be gravity

$$g = \tfrac{4}{3}\pi\rho = \tfrac{4}{3}\pi \frac{1}{1-\frac{1}{3}n}$$

Hence the modulus is $\frac{g}{n} w^{2(n-1)/n} = g\mathfrak{a}\mathfrak{w} \times \frac{1}{n}\left(\frac{w}{\mathfrak{w}}\right)^{2(n-1)/n}$, the units $\mathfrak{a}$, $\mathfrak{w}$ being reintroduced to give the expression the proper dimensions. Now $g\mathfrak{a}\mathfrak{w}$ is a pressure, and therefore the length of the modulus is $\frac{\mathfrak{a}}{n}\left(\frac{w}{\mathfrak{w}}\right)^{2(n-1)/n}$. Thus the surface matter has a length modulus equal to $\mathfrak{a}/n$.

Now let us find the ellipticity of the internal strata.

Substituting for w from (9) in (2), we have

$$a\frac{d^2e}{da^2} + 2(3-n)\frac{de}{da} - 2n\frac{e}{a} = 0$$

If the solution be assumed of the form $e = ca^{\beta}$, β must satisfy

$$\beta(\beta-1) + 2(3-n)\beta - 2n = 0$$

whence

$$\beta = -(\tfrac{5}{2} - n) \pm \sqrt{\{(\tfrac{5}{2})^2 - n(3-n)\}}$$

Now $n(3-n)$ is a maximum when it is equal to $(\frac{3}{2})^2$, and therefore the square root can never become imaginary.

From the sign of the last term in the equation for β, it is clear that one of the values of β is negative. Hence to avoid infinite ellipticity at the centre, the c corresponding to the negative root must be zero. Hence the solution of Clairaut's equation (2) is

$$e = \mathfrak{e}a^{n - \frac{5}{2} + \sqrt{[(\frac{5}{2})^2 - n(3-n)]}}$$

The surface value of $\frac{1}{\mathfrak{e}}\frac{d}{da}(ea^2)$ is clearly $n - \frac{1}{2} + \sqrt{\{(\frac{5}{2})^2 - n(3-n)\}}$. Thus from (3) we have

$$\frac{\oe}{\mathfrak{e}} = \tfrac{1}{2}(n - \tfrac{1}{2}) + \tfrac{1}{2}\sqrt{\{(\tfrac{5}{2})^2 - n(3-n)\}}$$

Then substituting for w from (9) in (5),

$$\mathrm{C} = \frac{8\pi}{3(5-n)}$$

And since $\mathrm{M} = \frac{4}{3}\pi\rho = \frac{4}{3}\pi\frac{1}{1-\frac{1}{3}n}$, we have for the precessional constant

$$\frac{\mathrm{C}-\mathrm{A}}{\mathrm{C}} = \frac{5-n}{3-n}(\mathfrak{e} - \tfrac{1}{2}m)$$

Now let us collect these results, and express them in terms of k instead of n. The solution is

$$w = a^{-3(1-k)}$$

And the mass inside any radius a is $\frac{4\pi}{3k}a^{3k}$

$$p = \tfrac{2}{3}\pi\frac{w^{(4-6k)/(3-3k)} - 1}{k(2-3k)}$$

$$w\frac{dp}{dw} = \tfrac{4}{3}\pi\frac{w^{(4-6k)/(3-3k)}}{3k(1-k)}$$

And when $k = \frac{2}{3}$, $p = 2\pi\log w$, $w\frac{dp}{dw} = 2\pi$, $w = a^{-1}$

The length of the modulus at the surface is $1/3\,(1-k)$ of the planet's radius.

$$e = \mathfrak{e}\, a^{\frac{1}{2} - 3k + 3\sqrt{[(\frac{5}{6})^2 - k(1-k)]}}$$

$$\frac{\text{æ}}{\mathfrak{e}} = \tfrac{5}{4} - \tfrac{3}{2}k + \tfrac{3}{2}\sqrt{\{(\tfrac{5}{6})^2 - k(1-k)\}}$$

$$\mathrm{C} = \frac{3k}{2+3k} \cdot \tfrac{2}{3}\mathrm{M}\mathfrak{a}^2$$

$$\frac{\mathrm{C}-\mathrm{A}}{\mathrm{C}} = \frac{2+3k}{3k}\,(\mathfrak{e} - \tfrac{1}{2}m)$$

Any value from unity to an infinitely small value may be assigned to k, that is to say, we may have any arrangement of density from homogeneity to infinitely small surface density, but if k be greater than $\frac{2}{3}$ the compressibility increases with the density, which is physically improbable.

The infinite density and infinite pressure, which occur in this solution actually *at* the centre, may be avoided by imagining the centre occupied by a rigid spherical homogeneous nucleus, of very small radius δa, and of density $1/k\delta a^{3(1-k)}$.

We have to compare this solution with Laplace's.

For this case k is not constant, and its surface value is $\mathfrak{k}$.

Let $\vartheta = a/\kappa$, where κ is a constant, being the arbitrary constant introduced in this solution; and let θ be the surface value of ϑ*.

The solution is
$$w = \frac{\theta}{\vartheta}\,\frac{\sin\vartheta}{\sin\theta}$$

And the mass inside any radius a is $\dfrac{4\pi}{3k}a^{3k}$

$$p = 2\pi\kappa^2(w^2 - 1)$$

$$w\frac{dp}{dw} = 4\pi\kappa^2 w^2$$

$$k = \frac{\vartheta^2}{3(1 - \vartheta\cot\vartheta)}$$

The length of the modulus at the surface is $1/(1-\theta\cot\theta)$ or $3\mathfrak{k}\theta^{-2}$ of the planet's radius.

$$e = \mathfrak{e}\,\frac{\theta^2}{\vartheta^2}\,\frac{1-k}{1-\mathfrak{k}}$$

$$\frac{\text{æ}}{\mathfrak{e}} = \frac{\theta^2}{6(1-\mathfrak{k})} - \tfrac{3}{2}\mathfrak{k}$$

$$\mathrm{C} = [1 - 6(1-\mathfrak{k})\,\theta^{-2}]\,\tfrac{2}{3}\mathrm{M}\mathfrak{a}^2$$

$$\frac{\mathrm{C}-\mathrm{A}}{\mathrm{C}} = \frac{1}{1 - 6(1-\mathfrak{k})\,\theta^{-2}}\,(\mathfrak{e} - \tfrac{1}{2}m)$$

* See Thomson and Tait's *Natural Philosophy*, 1883, § 824.

The following table gives the numerical values of the solution, together with columns for comparison with the results of Laplace's theory, for various values of the ratio of surface to mean density. The case of $\mathfrak{k} = 0$ gives the planet infinite mass at the centre, and the values are only inserted in order to complete the series.

$\mathfrak{k}$	Length mod. at surface in terms of a as unity	$e = \mathfrak{e}a^x$ where $x=$	$\frac{æ}{\mathfrak{e}}$	Laplace $\frac{æ}{\mathfrak{e}}$	$\frac{C-A}{C(\mathfrak{e}-\frac{1}{2}m)}$	Laplace $\frac{C-A}{C(\mathfrak{e}-\frac{1}{2}m)}$	$p \propto (w^y \sim 1)$ where $y=$
1·0	∞	0·000	1·000	*1·00*	1·667	*1·67*	$-\infty$
·9	3·333	0·132	1·066	*1·04*	1·741	*1·71*	−4·667
·8	1·667	0·293	1·147	*1·09*	1·833	*1·77*	−1·333
·7	1·111	0·488	1·244	*1·15*	1·952	*1·83*	−0·067
·6	1÷1·2	0·722	1·361	*1·21*	2·111	*1·90*	0·333
·5	1÷1·5	1·000	1·500	*1·27*	2·333	*1·98*	0·667
·4	1÷1·8	1·322	1·661	*1·34*	2·667	*2·07*	0·889
·3	1÷2·1	1·688	1·844	*1·41*	3·222	*2·17*	1·048
·2	1÷2·4	2·093	2·047	*1·48*	4·333	*2·28*	1·167
·1	1÷2·7	2·532	2·266	*1·56*	7·667	*2·40*	1·259
·0	1÷3·0	3·000	2·500	*1·65*	∞	*2·55*	1·333
·667	1·000	0·562	1·281	*1·17*	2·000	*1·85*	$p \propto \log w$
·333	1÷2·0	1·562	1·781	*1·39*	3·000	*2·13*	1·000

Note.—The values in the two columns applicable to Laplace's theory were found by graphical interpolation from a series of values given in [Paper 5, p. 58], or Thomson and Tait, *Natural Philosophy* (1883), § 824′.

In Laplace's theory $p \propto (w^2 - 1)$, and the modulus of compressibility $\propto w^2$. In the present theory the modulus $\propto w^y$.

The value $\mathfrak{k} = {\cdot}667$ corresponds to constant compressibility, and $\mathfrak{k} = {\cdot}333$ to gaseous compressibility.

One of the grounds on which Laplace's solution is held to be satisfactory is that if we take the value of æ, as determined by the known angular velocity and mean density of the earth, and the value of $\mathfrak{e}$ as determined by geodesy, and find the value of $\mathfrak{k}$, the ratio of surface to mean density, which corresponds with the ratio æ$/\mathfrak{e}$, this same value of $\mathfrak{k}$ is found to give a proper value to the coefficient of $\mathfrak{e} - \frac{1}{2}m$, so as to obtain the observed precessional constant. To be more precise, m is found to be 1/289·66, which gives æ $= 1/231{\cdot}7$, and $\mathfrak{e}$ has been found to be approximately 1/295. These give æ$/\mathfrak{e} = 1{\cdot}273$, and this corresponds with $\mathfrak{k} = 1/2{\cdot}06 = {\cdot}49$. This value of $\mathfrak{k}$, with the same values of $\mathfrak{e}$ and m, gives the precessional constant as ·0033, and Leverrier and Serret give its value as ·00327.

Now it appears remarkable that almost as good a correspondence is obtainable from my solution. The value æ$/\mathfrak{e} = 1{\cdot}273$ corresponds with $\mathfrak{k} = {\cdot}675$, and when $\mathfrak{k} = {\cdot}675$ the coefficient of $\mathfrak{e} - \frac{1}{2}m$ in the precessional constant is 1·99, which gives the same precessional constant ·0033.

This value of $\mathfrak{k}$ corresponds very nearly with constant modulus of compressibility, and with pressure determined by $p = 2\pi \log w$.

It is claimed in favour of the Laplacian hypothesis that it corresponds to a surface density which is nearly a half of the mean density of the earth, and that we know that average rock has a density of about 2·8. Also it is pointed out in Thomson and Tait's *Natural Philosophy* that the length modulus of compressibility of the surface rock is about 1/4·4 of the earth's radius, which is very nearly the observed length modulus of iron.

These conditions are not well satisfied by the present solution, for the surface density is found to be ·675, or 1/1·48 of the mean density of the planet, whence the specific gravity at the surface is 3·7; and the length modulus at the surface is equal to the planet's radius. It is to be admitted that this density is large, and that the substance is also highly incompressible.

Thus in these respects the Laplacian hypothesis has the advantage. It seems to me, however, that too much stress should not be laid on these arguments. We know nothing of the materials of the earth, excepting for a mile or two in thickness from the surface, hence it is not safe to argue confidently as to the degree of compressibility of the interior. There seems reason to believe that there is a deficiency in density under great mountain ranges, and this would agree with the hypothesis that our continents are a mere intumescence of the surface layers.

According to this view we might expect to find a rather sudden change in density within a few miles of the surface. Now in any theory of the earth's density such a sudden change in the thin shell on the surface could not be taken into account, and the numerical value for the surface density should be taken from below the intumescent layer if it exists. Hence it is not unreasonable to say that a solution of the problem, which gives a higher surface density than that of rock, lies near the truth. I do not maintain that my solution is as likely as that of Laplace, but it is not to be condemned at once because it does not satisfy these conditions as to the density and compressibility of rock.

The two cases which are given at the foot of the above table each possess an interest, the first of constant compressibility, because it corresponds with the case of the earth, and the second of modulus of compressibility varying as the density, because this is the gaseous law.

With constant compressibility the internal ellipticity varies as the ·562 power of a, or nearly as the square root of the radius; with gaseous compressibility it varies as the 1·562 power of a, or nearly as the square root of the cube of the radius.

A numerical comparison of the case of constant compressibility with Laplace's solution for $\mathfrak{k} = \frac{1}{2}$ gives the following results:—

$a =$	0	$\frac{1}{4}$a	$\frac{1}{2}$a	$\frac{3}{4}$a	a
(Laplace) $\frac{e}{\epsilon} =$	0	·812	·844	·902	1·000
(Constant compress.) $\frac{e}{\epsilon} =$	0	·459	·693	·851	1·000

Thus the Laplacian solution attributes much higher ellipticity to the internal strata. The solution with constant compressibility in fact gives so large a proportion of the mass in the central region, that attraction has a greater influence compared with rotation, than in the solution of Laplace.

POSTSCRIPT.—If, as is not improbable, the increase of density in the interior of the earth is due rather to the heavier materials falling down to the centre than to great pressure compressing the material until it has a high density, then the determination of a modulus of compressibility would be fallacious, and it would be more logical to leave the expressions for the pressure and the density both as functions of the radius, without proceeding to eliminate the radius and to form an expression for the modulus of compressibility. I owe this suggestion to a conversation with Sir William Thomson.

7.

THE THEORY OF THE FIGURE OF THE EARTH CARRIED TO THE SECOND ORDER OF SMALL QUANTITIES.

[*Monthly Notices of the Royal Astronomical Society*, LX. (1900), pp. 82—124.]

INTRODUCTION.

As far as I know, Airy was the first to include quantities of the second order in investigating the theory of the Earth's figure; his paper is dated 1826, and is published in Part III. of the *Philosophical Transactions of the Royal Society* for that year.

He gave the formula for gravity which I have obtained below (§ 6 (40)). Our results would be *literatim* identical but that my e is expressed by $e \div (1 - e)$ in his notation, and that I denote by $-f$ the quantity which he wrote as A. He also established equations, equivalent to my (13) and (14), which express the identity of the surfaces of equal density with the level surfaces. He remarked that these may be reduced to the form of differential equations, but he did not give the results, since he found himself unable to solve them, even for an assumed law of internal density. I have succeeded in solving these equations in this paper.

Airy further concluded that the Earth's surface must be depressed below the level of the true ellipsoid in middle latitudes. He gave no numerical estimate of this depression, but expressed the opinion that it must be very small.

In the second volume of his *Höhere Geodäsie*, Dr Helmert has also investigated the formula for gravity to the second order of small quantities. The expression for gravity which has been compared with the results of pendulum experiments by Dr Helmert was taken as having no term dependent on the fourth power of the sine of the latitude. The results of the experiments are somewhat irregular, and there was no apparent advantage in the inclusion of such a term; accordingly Dr Helmert assumed that such a term is actually evanescent, and pointed out that this implies that the Earth's surface is elevated above the true ellipsoid, instead of being depressed below it, in middle latitudes. There can, I think, be no doubt that there should be depression, and it therefore seems as if it would be safer to adopt such a formula as that given below in § 6 (41) in future reductions of pendulum experiments.

In volume XIX. (1889, pp. E, 1–84) of the *Annals of the Observatory of Paris* M. Callandreau has carried out an elaborate investigation of the problems considered in this paper. The publication of my work might, indeed, have been unnecessary were it not that my procedure is, as I think, simpler than his, and that my formulæ are presented in a more tractable shape. I have, however, in some respects, as for instance in the numerical solution of the differential equations, carried the work somewhat further than he has done; but, on the other hand, he has considered several interesting points on which I do not touch. Our two methods differ in detail from first to last, and it would be rather troublesome to compare them from point to point. I have then been satisfied with the knowledge that we are travelling along parallel roads. M. Callandreau has also written a short but valuable note on the same subject in the *Bulletin Astronomique* for 1897. I refer to this paper in § 12.

Lastly Professor Wiechert has published an important memoir on the distribution of masses in the interior of the Earth in the *Transactions* of the Royal Society of Sciences of Göttingen (1897, pp. 221–243). He has there adduced weighty arguments in favour of the hypothesis that the Earth consists of an iron nucleus with a superstratum of rock. He also has taken into account quantities of the second order, and has calculated interesting numerical results corresponding to his theory. I refer to this paper in §§ 10, 12, and in the summary.

The first part of my paper contains the mathematical investigation, and this is followed by a summary and discussion of results.

MATHEMATICAL INVESTIGATION.

§ 1. *The Moments of Inertia and the Potential of a homogeneous Spheroid.*

The equation to the surface of a homogeneous oblate ellipsoid of revolution of density w whose semi-axes are a and $a(1-e)$ is

$$r^2\left(\frac{\cos^2\theta}{(1-e)^2}+\sin^2\theta\right)=a^2$$

where r is the radius vector and θ the colatitude measured from the axis of revolution.

If the cubes and higher powers of e are neglected, the equation may be written

$$r = a\{1 - e\cos^2\theta - \tfrac{3}{2}e^2\sin^2\theta\cos^2\theta\}$$

Now let us consider a spheroid of which the equation is

$$r = a\{1 - e\cos^2\theta + (f - \tfrac{3}{2}e^2)\sin^2\theta\cos^2\theta\}$$

where f is of the same order of magnitude as e^2.

This surface has ellipticity e, and the excess of its radius vector over that of the true ellipsoid is $af\sin^2\theta\cos^2\theta$. The maximum excess occurs in colatitude 45°, and it amounts to $\tfrac{1}{4}af$.

I now introduce the zonal harmonics

$$P_2 = \tfrac{3}{2}\cos^2\theta - \tfrac{1}{2}, \quad P_4 = \tfrac{35}{8}\cos^4\theta - \tfrac{15}{4}\cos^2\theta + \tfrac{3}{8}$$

so that $\cos^2\theta = \tfrac{2}{3}P_2 + \tfrac{1}{3}, \quad \sin^2\theta\cos^2\theta = -\tfrac{8}{35}P_4 + \tfrac{2}{21}P_2 + \tfrac{2}{15}$

Accordingly, the equation to the spheroid may be written

$$r = a\{1 - \tfrac{1}{3}e - \tfrac{1}{5}e^2 + \tfrac{2}{15}f - \tfrac{2}{3}(e + \tfrac{3}{14}e^2 - \tfrac{1}{7}f)P_2 + \tfrac{8}{35}(\tfrac{3}{2}e^2 - f)P_4\} \quad \text{...(1)}$$

This form of the equation will be needed hereafter, but for the present it is convenient to regard the body as consisting of a homogeneous ellipsoid of density w, and with semi-axes a and $a(1-e)$, together with the excess above the ellipsoid of a body of which the equation is

$$r = a\{1 + \tfrac{2}{15}f + \tfrac{2}{21}fP_2 - \tfrac{8}{35}fP_4\} \quad \text{..................(2)}$$

The developments will only be carried to the order e^2, and therefore we may regard this excess as a layer of surface density

$$wa(\tfrac{2}{15}f + \tfrac{2}{21}fP_2 - \tfrac{8}{35}fP_4)$$

distributed over the surface of a sphere of radius a.

The mass of the excess is clearly $4\pi wa^3(\tfrac{2}{15}f)$, and the mass of the ellipsoid is $\tfrac{4}{3}\pi wa^3(1-e)$; hence the mass of the spheroid is given by

$$M = \tfrac{4}{3}\pi wa^3(1 - e + \tfrac{2}{5}f) \quad \text{.........................(3)}$$

If ρ, θ, ϕ be the polar coordinates of a point whose Cartesian coordinates are x, y, z we have

$$x^2 + y^2 = \tfrac{2}{3}\rho^2(1 - P_2), \quad \begin{Bmatrix} x^2 + z^2 \\ y^2 + z^2 \end{Bmatrix} = \tfrac{2}{3}\rho^2(1 + \tfrac{1}{2}P_2) \pm \tfrac{1}{2}\rho^2\sin^2\theta\cos 2\phi$$

Now let C', A', denote the moments of inertia of the shell (2) about the axes of z, and of x or of y, and we have

$$\begin{Bmatrix} C' \\ A' \end{Bmatrix} = \tfrac{2}{3}\iiint w\rho^4 \begin{Bmatrix} 1 - P_2 \\ 1 + \tfrac{1}{2}P_2 \end{Bmatrix} \sin\theta\, d\theta\, d\phi$$

integrated throughout the layer comprised between the surface (2) and the sphere a.

Then, since $\int \rho^4 d\rho = \frac{1}{5}a^5 \{\frac{2}{3}f + \frac{10}{21}fP_2 - \frac{8}{7}fP_4\}$, and since the integrals over the surface of a sphere of a spherical harmonic, and also of the product of two harmonics of different orders vanish, we have

$$C' = 4\pi wa^5 \int_0^\pi (\tfrac{2}{15}f - \tfrac{2}{21}fP_2^2) \sin\theta \, d\theta$$

$$A' = 4\pi wa^5 \int_0^\pi (\tfrac{2}{15}f + \tfrac{1}{21}fP_2^2) \sin\theta \, d\theta$$

Also $\int_0^\pi P_2^2 \sin\theta \, d\theta = \frac{2}{5}$, and therefore

$$C' = \tfrac{8}{15}\pi wa^5 (\tfrac{4}{7}f), \quad A' = \tfrac{8}{15}\pi wa^5 (\tfrac{5}{7}f)$$

Now denoting the moments of inertia of the homogeneous ellipsoid by C'', A'', we have

$$C'' = \tfrac{8}{15}\pi wa^5 (1 - e), \quad A'' = \tfrac{8}{15}\pi wa^5 (1 - 2e + \tfrac{3}{2}e^2)$$

The sums $C' + C''$, $A' + A''$, give the moments of inertia of the spheroid; so that

$$\left.\begin{aligned} C = \tfrac{8}{15}\pi wa^5 (1 - e + \tfrac{4}{7}f), \quad A &= \tfrac{8}{15}\pi wa^5 (1 - 2e + \tfrac{3}{2}e^2 + \tfrac{5}{7}f) \\ C - A &= \tfrac{8}{15}\pi wa^5 (e - \tfrac{3}{2}e^2 - \tfrac{1}{7}f) \end{aligned}\right\} \quad \ldots(4)$$

It is, in the next place, necessary to evaluate the potential of the spheroid both internally and externally, and I will begin by considering the part contributed by the shell defined by (2). It consists of a spherical shell of mass $\frac{8}{15}\pi wa^3 f$ and radius a, together with surface density $\frac{2}{21}wafP_2 - \frac{8}{35}wafP_4$. Accordingly, if U_e', U_i' denote the external and internal potentials

$$U_e' = \tfrac{8}{15}\pi wa^3 \frac{f}{r} + \tfrac{4}{5}\pi \,.\, \tfrac{2}{21}waf \,.\, \frac{a^4}{r^3} P_2 - \tfrac{4}{9}\pi \,.\, \tfrac{18}{35}waf \,.\, \frac{a^6}{r^5} P_4$$

$$U_i' = \tfrac{8}{15}\pi wa^2 f + \tfrac{4}{5}\pi \,.\, \tfrac{2}{21}waf \,.\, \frac{r^2}{a} P_2 - \tfrac{4}{9}\pi \,.\, \tfrac{8}{35}waf \,.\, \frac{r^4}{a^3} P_4$$

We have now to find the external and internal potentials of the ellipsoid, which may be denoted by U_e'', U_i''.

It is well known that the external potential of an ellipsoid of revolution of semi-major axis a, *eccentricity* η, and mass M' is

$$U_e'' = \frac{M'}{r}\left\{1 - \tfrac{1}{5}\frac{a^2\eta^2}{r^2} P_2 + \tfrac{3}{35}\frac{a^4\eta^4}{r^4} P_4 \ldots\right\}$$

Also if $\eta = \sin\gamma$, the internal potential is

$$U_i'' = \frac{3M'\gamma}{\sin\gamma} - \tfrac{3}{4}\frac{M'}{a^3 \sin^3\gamma}(\gamma - \sin\gamma\cos\gamma)(x^2 + y^2) - \tfrac{3}{2}\frac{M'}{a^3\sin^3\gamma}(\tan\gamma - \gamma) z^2$$

Now

$$\tan\gamma - \gamma = \tfrac{1}{3}\eta^3 (1 + \tfrac{9}{10}\eta^2 + \tfrac{45}{56}\eta^4 + \ldots)$$

$$\gamma - \sin\gamma\cos\gamma = \tfrac{2}{3}\eta^3 (1 + \tfrac{3}{10}\eta^2 + \tfrac{9}{56}\eta^4 + \ldots)$$

Also $$(x^2+y^2)=\tfrac{2}{3}r^2(1-P_2),\quad z^2=\tfrac{2}{3}r^2(\tfrac{1}{2}+P_2)$$

Therefore

$$U_i''=\tfrac{3}{2}\frac{M'}{a}(1+\tfrac{1}{6}\eta^2+\tfrac{3}{40}\eta^4\ldots)-\tfrac{1}{2}\frac{M'r^2}{a^3}(1+\tfrac{1}{2}\eta^2+\tfrac{3}{8}\eta^4+\ldots)$$
$$-\tfrac{1}{5}\frac{M'r^2}{a^3}(\eta^2+\tfrac{15}{14}\eta^4\ldots)P_2$$

But $\eta^2=2e(1-\frac{1}{2}e)$ and $M'=\frac{4}{3}\pi wa^3(1-e)$, and on substitution in the above formulæ it will be found that

$$U_e''=\tfrac{4}{3}\pi w\frac{a^3}{r}(1-e)-\tfrac{8}{15}\pi w\frac{a^5}{r^3}e(1-\tfrac{3}{2}e)P_2+\tfrac{16}{35}\pi w\frac{a^7}{r^5}e^2P_4$$

$$U_i''=2\pi wa^2(1-\tfrac{2}{3}e-\tfrac{1}{5}e^2)-\tfrac{2}{3}\pi wr^2-\tfrac{8}{15}\pi wr^2(e+\tfrac{9}{14}e^2)P_2$$

The external potential U_e of the spheroid is equal to $U_e'+U_e''$, and the internal potential U_i is equal to $U_i'+U_i''$. For application to the problem of the figure of the Earth the potential of rotation must be added, and if ω be the angular velocity of rotation, this potential is

$$\tfrac{1}{2}\omega^2r^2\sin^2\theta\quad\text{or}\quad\tfrac{1}{3}\omega^2r^2(1-P_2)$$

Now let V_e, V_i be the external and internal potentials, inclusive of rotation, and we have

$$\left.\begin{aligned}V_e&=\tfrac{4}{3}\pi w\frac{a^3}{r}(1-e+\tfrac{2}{5}f)-\tfrac{8}{15}\pi w\frac{a^5}{r^3}(e-\tfrac{3}{2}e^2-\tfrac{1}{7}f)P_2\\&\qquad+\tfrac{16}{35}\pi w\frac{a^7}{r^5}(e^2-\tfrac{2}{9}f)P_4+\tfrac{1}{3}\omega^2r^2(1-P_2)\\V_i&=2\pi wa^2(1-\tfrac{2}{3}e-\tfrac{1}{5}e^2+\tfrac{4}{15}f)-\tfrac{2}{3}\pi wr^2\\&\qquad-\tfrac{8}{15}\pi wr^2(e+\tfrac{9}{14}e^2-\tfrac{1}{7}f)P_2-\tfrac{32}{315}\pi w\frac{r^4}{a^2}fP_4+\tfrac{1}{3}\omega^2r^2(1-P_2)\end{aligned}\right\}\quad(5)$$

The first term of V_i is independent of r, θ and is only inserted in order that V may be a continuous function at the surface of the spheroid.

It will be supposed hereafter that the heterogeneous Earth is built up of layers of density w, bounded externally and internally by spheroids defined by $a+\delta a$, $e+\delta e$, $f+\delta f$ and a, e, f. It is obvious that the potential of such a body may be written down from (5) by replacing each term by an integral.

If a, e, f denote the superficial values of a, e, f we shall have such integrals as

$$\int_0^{\mathrm{a}} w\frac{d}{da}[a^5(e-\tfrac{3}{2}e^2-\tfrac{1}{7}f)]\,da\quad\text{and}\quad\int_a^{\mathrm{a}} w\frac{d}{da}[e+\tfrac{9}{14}e^2-\tfrac{1}{7}f]\,da$$

For the sake of brevity I shall write these

$$\int_0^{\mathrm{a}} wd[a^5(e-\tfrac{3}{2}e^2-\tfrac{1}{7}f)]\quad\text{and}\quad\int_a^{\mathrm{a}} wd[e+\tfrac{9}{14}e^2-\tfrac{1}{7}f]$$

It may be well to note here that for the heterogeneous Earth

$$\left.\begin{aligned} M &= \tfrac{4}{3}\pi\int_0^a wd\,[a^3(1-e+\tfrac{2}{5}f)] \\ C &= \tfrac{8}{15}\pi\int_0^a wd\,[a^5(1-e+\tfrac{4}{7}f)] \\ A &= \tfrac{8}{15}\pi\int_0^a wd\,[a^5(1-2e+\tfrac{3}{2}e^2+\tfrac{5}{7}f)] \\ C-A &= \tfrac{8}{15}\pi\int_0^a wd\,[a^5(e-\tfrac{3}{2}e^2-\tfrac{1}{7}f)] \end{aligned}\right\}\quad\ldots\ldots\ldots\ldots(6)$$

§ 2. *Heterogeneous Planet. The surfaces of equal density are level surfaces.*

It will now be supposed that the substance of which the heterogeneous planet is formed is plastic enough to allow the surfaces of equal density to be level surfaces.

It is necessary to write down the potential of the planet at any internal point, and for this purpose I find it better to introduce a new parameter in place of the ellipticity e. This parameter is h, defined by

$$h = e - \tfrac{25}{14}e^2 - \tfrac{1}{7}f \quad\ldots\ldots\ldots\ldots(7)$$

I do not quite understand why this substitution should lead to simplification, but I may remark that $h = e - 2e^2 - \frac{1}{7}(f - \frac{3}{2}e^2)$, and that $f - \frac{3}{2}e^2$ is the coefficient of $\sin^2\theta\cos^2\theta$ in the equation to the spheroid. Thus the existence of the fraction $\frac{25}{14}$ is in some sense explained.

It will be found that the equation (1) to the stratum a, in terms of this parameter, becomes

$$r = a\{1 - \tfrac{1}{3}h - \tfrac{167}{210}h^2 + \tfrac{3}{35}f - \tfrac{2}{3}h(1+2h)P_2 + \tfrac{8}{35}(\tfrac{3}{2}h^2 - f)P_4\}\quad\ldots(8)$$

Also

$$\left.\begin{aligned} C &= \tfrac{8}{15}\pi\int_0^a wd\,[a^5(1-h-\tfrac{25}{14}h^2+\tfrac{3}{7}f)] \\ C-A &= \tfrac{8}{15}\pi\int_0^a wd\,[a^5(h+\tfrac{2}{7}h^2)] \end{aligned}\right\}\quad\ldots\ldots\ldots\ldots(9)$$

Let us now for brevity write

$$\left.\begin{aligned} S_0 &= \int_0^a wd\,[a^3(1-e+\tfrac{2}{5}f)] &&= \int_0^a wd\,[a^3(1-h-\tfrac{25}{14}h^2+\tfrac{9}{35}f)] \\ S_2 &= \int_0^a wd\,[a^5(e-\tfrac{3}{2}e^2-\tfrac{1}{7}f)] &&= \int_0^a wd\,[a^5(h+\tfrac{2}{7}h^2)] \\ S_4 &= \int_0^a wd\,[a^7(e^2-\tfrac{2}{3}f)] &&= \int_0^a wd\,[a^7(h^2-\tfrac{2}{3}f)] \\ T_2 &= \int_a^a wd\,[e+\tfrac{9}{14}e^2-\tfrac{1}{7}f] &&= \int_a^a wd\,[h+\tfrac{17}{7}h^2] \\ T_4 & &&= \int_a^a wd\left[\frac{f}{a^2}\right] \end{aligned}\right\}(10)$$

The terms of the second order in S_0 are not required, so that, in fact, I take $S_0 = \int_0^a wd\,[a^3(1-h)]$.

The formulæ (5) and (10) enable us to write down the potential at any internal point as follows:—

$$\frac{3V}{4\pi} = \frac{S_0}{r} - \tfrac{2}{5}\frac{S_2 P_2}{r^3} + \tfrac{12}{35}\frac{S_4 P_4}{r^5} - \tfrac{2}{5}r^2 T_2 P_2 - \tfrac{8}{105}r^4 T_4 P_4 + \frac{\omega^2}{4\pi} r^2 (1 - P_2) \quad \text{.........(11)}$$

If it be assumed that the equipotential surfaces are also surfaces of equal density, the equation $V =$ constant must be reducible to the form (8). Hence, if in (11) we attribute to r the value (8), the coefficients of P_2 and of P_4 must vanish. In effecting this substitution we attribute to r its full value in the term of the lowest order, namely S_0/r; in terms of the first order, namely those involving S_2, T_2, and ω^2, we may put $a(1 - \frac{1}{3}h - \frac{2}{3}hP_2)$ for r; and in the terms of the second order, namely, those involving S_4 and T_4, we simply put a for r. A consideration of (11) shows that there are several functions of r, P_2, P_4, which will have to be evaluated, and it is obvious that the expressions in question, when developed to the required order, will involve P_2^2. But

$$P_2^2 = \tfrac{1}{5} + \tfrac{2}{7}P_2 + \tfrac{18}{35}P_4$$

and it will be found by aid of this formula that, to the required order of approximation,

$$\left.\begin{aligned}
\frac{a}{r} &= 1 + \tfrac{1}{3}h + \tfrac{209}{210}h^2 - \tfrac{3}{35}f + \tfrac{2}{3}h\left(1 + \tfrac{20}{7}h\right)P_2 + \tfrac{8}{35}\left(f - \tfrac{1}{2}h^2\right)P_4 \\
\frac{a^3}{r^3}P_2 &= \tfrac{2}{5}h + \left(1 + \tfrac{11}{7}h\right)P_2 - \tfrac{24}{35}hP_4 \\
\frac{r^2}{a^2}P_2 &= -\tfrac{4}{15}h + \left(1 - \tfrac{22}{21}h\right)P_2 - \tfrac{24}{35}hP_4 \\
\frac{r^2}{a^2}(1 - P_2) &= 1 - \tfrac{2}{5}h - \left(1 + \tfrac{2}{7}h\right)P_2 + \tfrac{24}{35}hP_4
\end{aligned}\right\} \text{...(12)}$$

By aid of (12) it may be shown that the conditions that the level surfaces shall be surfaces of equal density are

$$\frac{S_0}{a}\left(h + \tfrac{20}{7}h^2\right) - \tfrac{3}{5}\frac{S_2}{a^3}\left(1 + \tfrac{11}{7}h\right) - \tfrac{3}{5}a^2 T_2\left(1 - \tfrac{22}{21}h\right) - \frac{3\omega^2}{8\pi}a^2\left(1 + \tfrac{2}{7}h\right) = 0 \quad (13)$$

$$\frac{S_0}{a}\left(f - \tfrac{1}{2}h^2\right) - \tfrac{9}{5}\frac{S_2}{a^3}h + \tfrac{3}{2}\frac{S_4}{a^5} + \tfrac{6}{5}a^2 T_2 h - \tfrac{1}{3}a^4 T_4 + \frac{3\omega^2}{4\pi}a^2 h = 0 \text{.........(14)}$$

Since $\omega^2 a^3/S_0$ is a quantity of the same order as h, (13) involves terms of two orders, but (14) consists entirely of terms of the second order.

If the terms of the second order be omitted, (13) becomes

$$\frac{S_0}{a}h - \tfrac{3}{5}\frac{S_2}{a^3} - \tfrac{3}{5}a^2 T_2 - \frac{3\omega^2 a^2}{8\pi} = 0 \text{.....................(15)}$$

This may be used for the elimination of T_2 and of ω^2 from (14); for multiplying it by $2h$ and adding it to (14) we have

$$\frac{S_0}{a}(f+\tfrac{3}{2}h^2)-3\frac{S_2}{a^3}h+\tfrac{3}{2}\frac{S_4}{a^5}-\tfrac{1}{3}a^4T_4=0 \quad \text{..............(16)}$$

The conditions for the identity of the two surfaces are then (13) and (16).

In order to obtain the differential equations to be satisfied by h and f, it is necessary to eliminate the S and T integrals by differentiation, and the results may be much simplified by the use of the approximate form (15) of (13), and of its derivatives.

Our first task is, then, to pursue the approximate equation (15).

The equation (15) may be written in the form

$$\frac{S_0}{a^3}h-\tfrac{3}{5}\frac{S_2}{a^5}-\tfrac{3}{5}T_2-\frac{3\omega^2}{8\pi}=0$$

To the same degree of approximation we have $S_0=3\int_0^a wa^2da$, so that $\frac{dS_0}{da}=S_0'=3wa^2$. It follows, therefore, that

$$w=\tfrac{1}{3}\frac{S_0'}{a^2} \quad \text{..............................(17)}$$

On differentiating our equation and using (17) we find

$$S_2=S_0a^3\left(\frac{h}{a}-\tfrac{1}{3}\frac{dh}{da}\right) \quad \text{..........................(18)}$$

Differentiating again, and effecting some reductions, we have

$$S_0\frac{d^2h}{da^2}=\frac{6S_0}{a^2}h-2S_0'\left(\frac{dh}{da}+\frac{h}{a}\right) \quad \text{......................(19)}$$

It would be easy, by means of (18), to eliminate S_2 from the terms of the second order in (13), but I prefer to postpone that elimination for the present; we can, however, at once eliminate it from (16). Effecting this elimination, and repeating (13), our conditions are

$$\frac{S_0}{a}(h+\tfrac{20}{7}h^2)-\tfrac{3}{5}\frac{S_2}{a^3}(1+\tfrac{11}{7}h)-\tfrac{3}{5}a^2T_2(1-\tfrac{22}{21}h)-\frac{3\omega^2}{8\pi}a^2(1+\tfrac{2}{7}h)=0$$

$$\frac{S_0}{a^5}\left(f-\tfrac{3}{2}h^2+ah\frac{dh}{da}\right)+\tfrac{3}{2}\frac{S_4}{a^9}-\tfrac{1}{3}T_4=0 \quad \text{......(20)}$$

The equation (19) has not been used yet, but it will prove useful hereafter.

§ 3. *The Differential Equation for f.*

By differentiating the second of (20) and using (17) and (19), I find

$$S_0 a^5 \left\{\frac{df}{da} - 5\frac{f}{a} + \tfrac{27}{2}\frac{h^2}{a} - 7h\frac{dh}{da} + a\left(\frac{dh}{da}\right)^2\right\} - \tfrac{27}{2} S_4 = 0 \quad \ldots\ldots(21)$$

Differentiating again, and again using (17) and (19), I find

$$S_0 a^5 \left\{\frac{d^2 f}{da^2} - 20\frac{f}{a^2} + 12\frac{h^2}{a^2} + 4\frac{hdh}{ada} - \left(\frac{dh}{da}\right)^2\right\}$$

$$+ S_0' a^6 2 \left\{\frac{df}{ada} + \frac{2f}{a^2} - 4\frac{h^2}{a^2} - 6\frac{hdh}{ada} - 3\left(\frac{dh}{da}\right)^2\right\} = 0 \quad \ldots(22)$$

I now introduce a new symbol w_0, which is to denote the mean density of all the matter lying inside the spheroid defined by a; then

$$S_0 = \int_0^a w da^3 = w_0 a^3$$

$$S_0' = 3wa^2$$

Thus $\dfrac{S_0'}{S_0} = \dfrac{3}{a}\dfrac{w}{w_0}$.

In obtaining this result the ellipticity of the spheroid has been neglected, but for the present this approximation suffices, and (22) is then easily reducible to the form

$$\frac{d^2 f}{da^2} + 6\frac{w}{w_0}\frac{df}{ada} - \left(20 - 6\frac{w}{w_0}\right)\frac{f}{a^2} + 12\left(1 - \frac{w}{w_0}\right)\frac{h^2}{a^2} + \left(4 - 18\frac{w}{w_0}\right)\frac{hdh}{ada}$$

$$- \left(1 + 9\frac{w}{w_0}\right)\left(\frac{dh}{da}\right)^2 = 0 \quad \ldots(23)$$

This is the differential equation for *f*, and I shall in § 9 solve it on the assumption of a certain law of internal density of the Earth.

§ 4. *The Differential Equation for h.*

I now return to the first of (20), and divide it by $a^2(1 - \frac{22}{21}h)$, so that it becomes

$$\frac{S_0}{a^3}(h + \tfrac{82}{21}h^2) - \tfrac{3}{5}\frac{S_2}{a^5}(1 + \tfrac{55}{21}h) - \tfrac{3}{5}T_2 - \frac{3\omega^2}{8\pi}(1 + \tfrac{4}{3}h) = 0 \quad \ldots\ldots(24)$$

As a preliminary to the differentiation which will eliminate the T_2 integral I indicate certain transformations.

We have

$$S_0 = \int_0^a w d\,[a^3(1 - h)]$$

$$S_0' = 3wa^2\left(1 - h - \tfrac{1}{3}a\frac{dh}{da}\right)$$

Hence $$w = \frac{S_0'}{3a^2}\left(1 + h + \tfrac{1}{3}a\frac{dh}{da}\right) \quad\text{.........................(25)}$$

If S_2, T_2 as defined in (10) be differentiated, and if the value of w as given in (25) be used, it will be found that

$$3\frac{dS_2}{da} = \frac{S_0'}{a^2}\left\{5h + a\frac{dh}{da} + \tfrac{45}{7}h^2 + \tfrac{68}{21}ah\frac{dh}{da} + \tfrac{1}{3}\left(a\frac{dh}{da}\right)^2\right\}$$

and

$$3(1 + \tfrac{55}{21}h)\frac{dS_2}{da} = \frac{S_0'}{a^2}\left\{5h + a\frac{dh}{da} + \tfrac{410}{21}h^2 + \tfrac{41}{7}ah\frac{dh}{da} + \tfrac{1}{3}\left(a\frac{dh}{da}\right)^2\right\} \quad (26)$$

$$3\frac{dT_2}{da} = -\frac{S_0'}{a^2}\left\{\frac{dh}{da} + \tfrac{41}{7}h\frac{dh}{da} + \tfrac{1}{3}a\left(\frac{dh}{da}\right)^2\right\}$$

On differentiating (24) and using (26) and (18) it will be found that

$$S_2 + S_0 a^3\left\{-\frac{h}{a} + \tfrac{1}{3}\frac{dh}{da} - \tfrac{9}{7}\frac{h^2}{a} + \tfrac{76}{63}h\frac{dh}{da} + \tfrac{11}{63}a\left(\frac{dh}{da}\right)^2\right\} - \frac{\omega^2 a^6}{6\pi}\frac{dh}{da} = 0 \quad\text{...(27)}$$

On differentiating (27) and again using (26) and (19) in the small terms it will be found that

$$S_0\left\{\frac{d^2h}{da^2} - 6\frac{h}{a^2} + 14\frac{h^2}{a^2} + \tfrac{66}{7}\frac{hdh}{ada} + \tfrac{40}{7}\left(\frac{dh}{da}\right)^2\right\}$$
$$+ S_0'\left\{2\frac{dh}{da} + 2\frac{h}{a} - \tfrac{14}{3}\frac{h^2}{a} - \tfrac{52}{21}ah\frac{dh}{da} - \tfrac{26}{21}a\left(\frac{dh}{da}\right)^2\right\}$$
$$-\frac{3\omega^2a^2}{\pi S_0}(S_0 - \tfrac{1}{3}S_0'a)\left(\frac{h}{a} + \frac{dh}{da}\right) = 0 \quad\text{...(28)}$$

Since $S_0 = w_0 a^3(1 - h)$, it follows from (25) that

$$\frac{S_0'}{S_0} = \frac{3}{a}\frac{w}{w_0}\left(1 - \tfrac{1}{3}a\frac{dh}{da}\right) \quad\text{.........................(29)}$$

I shall later denote by m the ratio of equatorial centrifugal force to equatorial gravity, and by an extension of this notation I will now write

$$m = \frac{\omega^2 a^3}{\frac{4}{3}\pi S_0} \quad\text{..................................(30)}$$

By means of (29) and (30) equation (28) becomes

$$\frac{d^2h}{da^2} + 6\frac{w}{w_0}\frac{dh}{ada} - \left(1 - \frac{w}{w_0}\right)\left[6\frac{h}{a^2} - 14\frac{h^2}{a^2} - \tfrac{66}{7}\frac{hdh}{ada} - \tfrac{40}{7}\left(\frac{dh}{da}\right)^2\right.$$
$$\left. + 4m\left(\frac{h}{a^2} + \frac{dh}{ada}\right)\right] = 0 \quad\text{...(31)}$$

This is the differential equation to be satisfied by h. If the terms of the second order be omitted it is the equation for the ellipticity of internal strata as given in any treatise on the theory of the figure of the Earth.

The transformation by which h was substituted for e has enabled us to obtain an equation for h in which f is not involved. Moreover h has been chosen as such a function of e that all the latter terms are multiplied by the simple factor $1 - w/w_0$, instead of by various more complex functions of the density, as was the case in the differential equation for f.

§ 5. *Radau's form of the Differential Equation of the Ellipticity.*

I shall now include the terms of the second order in M. Radau's very remarkable transformation of the equation of the last section.

Since
$$w_0 a^3 (1-h) = \int_0^a w d\,[a^3(1-h)]$$

it follows that
$$\frac{w}{w_0} = 1 + \tfrac{1}{3}\left(1 + \tfrac{1}{3} a \frac{dh}{da}\right) \frac{a dw_0}{w_0 da} \quad \text{......................(32)}$$

The equation (31) may therefore be written
$$\frac{d^2h}{da^2} + 6\frac{dh}{ada} + 2\frac{adw_0}{w_0 da}\left(\frac{dh}{ada} + \frac{h}{a^2}\right)$$
$$- \tfrac{1}{3}\frac{adw_0}{w_0 da}\left\{14\frac{h^2}{a^2} + \tfrac{52}{7}\frac{hdh}{ada} + \tfrac{26}{7}\left(\frac{dh}{da}\right)^2 - 4m\left(\frac{dh}{ada} + \frac{h}{a^2}\right)\right\} = 0$$

The last term here includes all the terms of the second order.

If we write
$$\eta = \frac{adh}{hda}$$
$$\frac{d^2h}{da^2} = \frac{h}{a^2}\left(a\frac{d\eta}{da} - \eta + \eta^2\right), \quad \frac{dh}{ada} = \frac{h}{a^2}\eta$$

Thus the equation becomes
$$a\frac{d\eta}{da} + 5\eta + \eta^2 + 2\frac{adw_0}{w_0 da}(1+\eta)$$
$$- \tfrac{4}{21}\frac{adw_0}{w_0 da} h \left\{18 + \tfrac{13}{2}(1+\eta)^2 - 7\frac{m}{h}(1+\eta)\right\} = 0$$

Now it is easy to prove that
$$a\frac{d\eta}{da} = \frac{2\sqrt{(1+\eta)}}{w_0 a^4}\frac{d}{da}[w_0 a^5 \sqrt{(1+\eta)}] - 2\left(5 + \frac{adw_0}{w_0 da}\right)(1+\eta)$$

Therefore the equation may be written
$$\frac{2\sqrt{(1+\eta)}}{w_0 a^4}\frac{d}{da}[w_0 a^5 \sqrt{(1+\eta)}]$$
$$= 10\left(1 + \tfrac{1}{2}\eta - \tfrac{1}{10}\eta^2\right) + \tfrac{4}{21}\frac{adw_0}{w_0 da} h \left\{18 + \tfrac{13}{2}(1+\eta)^2 - 7\frac{m}{h}(1+\eta)\right\}$$

In the terms of the second order we may put $\frac{a dw_0}{w_0 da}$ equal to $-3\left(1-\frac{w}{w_0}\right)$; therefore

$$\frac{d}{da}[w_0 a^5 \sqrt{(1+\eta)}] = \frac{5w_0 a^4}{\sqrt{(1+\eta)}}\left\{1+\tfrac{1}{2}\eta - \tfrac{1}{10}\eta^2 - \tfrac{2}{35}h\left(1-\frac{w}{w_0}\right)\left[18+\tfrac{13}{2}(1+\eta)^2 - 7\frac{m}{h}(1+\eta)\right]\right\} \quad \ldots(33)$$

This is M. Radau's equation, with the inclusion of the terms of the second order. These terms have been determined by M. Callandreau, but the form in which he gives them appears to me more complicated than the above.

Let us consider the function within { } on the right-hand side of the equation (33), and in the first place omitting the terms of the second order consider the function $\frac{(1+\frac{1}{2}\eta-\frac{1}{10}\eta^2)}{\sqrt{(1+\eta)}}$.

It is equal to unity when η is zero, rises to a maximum of 1·00074 when $\eta=\frac{1}{3}$, and only falls to ·8 when $\eta=3$. Now it has been proved that η is necessarily less than 3, and is positive*. In all the cases which are likely to prove of practical interest η is very much less than 3. In the case of the Earth, for example, η is equal to about ·56 at the surface, and vanishes at the centre. Now when $\eta=·56$ this function is equal to ·99971. Therefore between the centre and the surface it rises from unity to 1·00074 and then falls to ·99971. It is obvious that any kind of average value of the function, estimated over the range from centre to surface, can at most differ from unity in the fourth place of decimals. When the terms of the second order are included I think the average value is yet nearer unity than when they are omitted, for in such a case as that of the Earth $(1+\frac{1}{2}\eta-\frac{1}{10}\eta^2)\div\sqrt{(1+\eta)}$ is greater than unity throughout the greater part of its range, and, although I cannot prove it absolutely, I believe that

$$18+\tfrac{13}{2}(1+\eta)^2-7\frac{m}{h}(1+\eta) \quad \text{or} \quad 18-\tfrac{49}{26}\frac{m^2}{h^2}+\tfrac{13}{2}\left(1+\eta-\tfrac{7}{13}\frac{m}{h}\right)^2$$

is always positive. If this is the case the whole function is on the average nearer unity than when the terms of the second order are omitted.

Supposing, then, that $1+\lambda$ denotes a proper mean value of the function

$$\left\{1+\tfrac{1}{2}\eta-\tfrac{1}{10}\eta^2-\tfrac{2}{35}h\left(1-\frac{w}{w_0}\right)\left[18+\tfrac{13}{2}(1+\eta)^2-7\frac{m}{h}(1+\eta)\right]\right\}(1+\eta)^{-\frac{1}{2}}$$

estimated over the whole range from centre to surface, we must have

$$\mathrm{w}_0 \mathrm{a}^5\sqrt{(1+\eta_1)} = 5(1+\lambda)\int_0^{\mathrm{a}} w_0 a^4 da \quad \ldots\ldots\ldots\ldots\ldots(34)$$

* Professor Helmert tells me that he doubts the universal validity of the proof of this. I have to thank him for valuable information given me while writing this paper.

where w_0, a, η_1 denote the superficial values of those quantities; and we may feel sure that $1+\lambda$ will not differ from unity until we come at least to the fourth place of decimals.

I shall in § 12 attempt to use this remarkable result for evaluating the ellipticity of the Earth's surface from the Precessional Constant.

§ 6. *Gravity at the Earth's Surface.*

In order to render this presentation of the theory of the Earth's figure more complete I shall now go on to find the theoretical expression for gravity, although the same investigation is to be found in various other places.

At the surface of the planet the integrals T_2, T_4 vanish, and the potential (11) becomes

$$\frac{3V}{4\pi} = \frac{S_0}{r} - \frac{2}{5}\frac{S_2P_2}{r^3} + \frac{12}{35}\frac{S_4P_4}{r^5} + \frac{\omega^2 r^2}{4\pi}(1-P_2)$$

In this section, as elsewhere, the superficial values of the various quantities are denoted by Roman in place of Italic type.

Through the vanishing of T_2, T_4 the equations (20), which denote that the surface is a level surface, become

$$\frac{\mathrm{S}_0}{\mathrm{a}}(\mathrm{h}+\tfrac{20}{7}\mathrm{h}^2) - \tfrac{3}{5}\frac{\mathrm{S}_2}{\mathrm{a}^3}(1+\tfrac{11}{7}\mathrm{h}) - \frac{3\omega^2}{8\pi}\mathrm{a}^2(1+\tfrac{2}{7}\mathrm{h}) = 0$$

$$\frac{\mathrm{S}_0}{\mathrm{a}}(\mathrm{f}+\tfrac{3}{2}\mathrm{h}^2) - 3\frac{\mathrm{S}_2}{\mathrm{a}^3}\mathrm{h} + \tfrac{3}{2}\frac{\mathrm{S}_4}{\mathrm{a}^5} = 0$$

It must be observed that the mass M of the spheroid is equal to $\frac{4}{3}\pi \mathrm{S}_0$; and we deduce

$$\left.\begin{aligned} \mathrm{S}_2 &= \frac{5\mathrm{Ma}^2}{4\pi}\left\{\mathrm{h}+\tfrac{9}{7}\mathrm{h}^2 - \tfrac{1}{2}\frac{\omega^2\mathrm{a}^3}{\mathrm{M}}(1-\tfrac{9}{7}\mathrm{h})\right\} \\ \mathrm{S}_4 &= \frac{\mathrm{Ma}^4}{2\pi}\left\{\tfrac{7}{2}\mathrm{h}^2 - \mathrm{f} - \tfrac{5}{2}\frac{\omega^2\mathrm{a}^3}{\mathrm{M}}\mathrm{h}\right\} \end{aligned}\right\} \qquad \text{.............(35)}$$

The potential may therefore be written

$$V = \mathrm{M}\left[\frac{1}{r} - \tfrac{2}{3}\frac{\mathrm{a}^2}{r^3}\left\{\mathrm{h}+\tfrac{9}{7}\mathrm{h}^2 - \tfrac{1}{2}\frac{\omega^2\mathrm{a}^3}{\mathrm{M}}(1-\tfrac{9}{7}\mathrm{h})\right\}P_2 \right.$$
$$\left. + \tfrac{8}{35}\frac{\mathrm{a}^4}{r^5}\left\{\tfrac{7}{2}\mathrm{h}^2 - \mathrm{f} - \tfrac{5}{2}\frac{\omega^2\mathrm{a}^3}{\mathrm{M}}\mathrm{h}\right\}P_4 + \tfrac{1}{3}\frac{\omega^2\mathrm{a}^3}{\mathrm{M}}\frac{r^2}{\mathrm{a}^3}(1-P_2)\right]$$

At the equator $r=\mathrm{a}$, $P_2 = -\tfrac{1}{2}$, $P_4 = \tfrac{3}{8}$, and equatorial gravity, say g_e, is equal to $-dV/dr$.

It is easy then by differentiation to show that

$$g_e = \frac{\mathrm{M}}{\mathrm{a}^2}\left[1+\mathrm{h}+\tfrac{39}{14}\mathrm{h}^2 - \tfrac{3}{7}\mathrm{f} - \tfrac{3}{2}\frac{\omega^2\mathrm{a}^3}{\mathrm{M}}(1+\tfrac{2}{7}\mathrm{h})\right] \qquad \text{...........(36)}$$

Now let m denote the ratio of equatorial centrifugal force to equatorial gravity, so that

$$m = \frac{\omega^2 a}{g_e}$$

It follows that

$$\frac{\omega^2 a^3}{M} = m\left[1 + h + \tfrac{39}{14}h^2 - \tfrac{3}{7}f - \tfrac{3}{2}\frac{\omega^2 a^3}{M}(1 + \tfrac{9}{7}h)\right]$$

$$= m\,[1 + h - \tfrac{3}{2}m + \tfrac{39}{14}h^2 - \tfrac{27}{14}mh + \tfrac{9}{4}m^2 - \tfrac{3}{7}f] \quad \text{.............(36)}$$

The function which naturally arises in the consideration of figures of equilibrium of rotating fluid is the ratio of ω^2 to the density. But

$$M = \tfrac{4}{3}\pi w_0 a^3 (1 - h - \tfrac{25}{14}h^2 + \tfrac{9}{35}f)$$

hence
$$\frac{3\omega^2}{4\pi w_0} = m\,[1 - \tfrac{3}{2}m + \tfrac{9}{4}m^2 - \tfrac{3}{7}mh - \tfrac{6}{35}f]$$

If m were an ideally perfect parameter in which to express our results, h and f should have entirely disappeared from this equation, since m should only depend on ω^2 and w_0, and should be perfectly independent of the figure of the planet.

However, for our present purpose it suffices to take

$$\frac{\omega^2 a^3}{M} = m\,(1 + h - \tfrac{3}{2}m), \quad \text{or} \quad \frac{3\omega^2}{4\pi w_0} = m\,(1 - \tfrac{3}{2}m) \quad \text{........(36)}$$

To this order there is no objection to the use of m, and I bow to custom in continuing to use it.

The potential of the planet may now be written

$$V = M\left[\frac{1}{r} - \tfrac{2}{3}\frac{a^2}{r^3}(h - \tfrac{1}{2}m + \tfrac{9}{7}h^2 + \tfrac{1}{7}mh + \tfrac{3}{4}m^2)\,P_2 + \tfrac{8}{35}\frac{a^4}{r^5}(\tfrac{7}{4}h^2 - \tfrac{5}{2}mh - f)\,P_4 + \tfrac{1}{3}m\,(1 + h - \tfrac{3}{2}m)\frac{r^2}{a^3}(1 - P_2)\right] \quad \text{...(37)}$$

It will perhaps be more convenient here to reintroduce the ellipticity instead of h. I observe, then, that $h + \tfrac{9}{7}h^2$ is equal to $e - \tfrac{1}{2}e^2 - \tfrac{1}{7}f$, and that throughout the rest of the expression e may be written for h.

The quantity 𝔡 used by Dr Helmert in his *Geodesy* (Vol. II. pp. 77—85) is the same as my $\tfrac{7}{2}e^2 - \tfrac{5}{2}me - f$, and my results will be found to agree with his.

We have, then,

$$\left.\begin{aligned} V &= M\left[\frac{1}{r} - \tfrac{2}{3}\frac{a^2}{r^3}(e - \tfrac{1}{2}m - \tfrac{1}{2}e^2 + \tfrac{1}{7}me + \tfrac{3}{4}m^2 - \tfrac{1}{7}f)\,P_2 + \tfrac{8}{35}\frac{a^4}{r^5}(\tfrac{7}{2}e^2 - \tfrac{5}{2}me - f)\,P_4 + \tfrac{1}{3}m(1 + e - \tfrac{3}{2}m)\frac{r^2}{a^3}(1 - P_2)\right] \\ g_e &= \frac{M}{a^2}[1 + e - \tfrac{3}{2}m + e^2 - \tfrac{27}{14}me + \tfrac{9}{4}m^2 - \tfrac{4}{7}f] \end{aligned}\right\} \quad \text{...(38)}$$

Clairaut's ratio is that of the excess of polar above equatorial gravity to equatorial gravity. I follow Dr Helmert in denoting this ratio by $\mathfrak{b}$; so that

$$\mathfrak{b} = \frac{g_p - g_e}{g_e}$$

At the pole $r = a(1 - e)$, $P_2 = P_4 = 1$. If $-dV/dr$ be found and these values introduced, we get

$$g_p = \frac{M}{a^2}(1 + m + \tfrac{6}{7}me - \tfrac{3}{2}m^2 - \tfrac{6}{7}f)$$

Then
$$g_p - g_e = \frac{M}{a^2}(\tfrac{5}{2}m - e - e^2 + \tfrac{39}{14}me - \tfrac{15}{4}m^2 - \tfrac{2}{7}f)$$

and since to the first order $g_e = \frac{M}{a^2}(1 + e - \frac{3}{2}m)$, we have

$$\mathfrak{b} = \tfrac{5}{2}m - e - \tfrac{17}{14}me - \tfrac{2}{7}f \quad \text{.......................(39)}$$

The form of the potential shows that the general expression for gravity must be

$$g = g_e[1 + \mathfrak{b}\cos^2\theta + \alpha\sin^2\theta\cos^2\theta]$$

where it remains to determine α.

This may be written

$$g = g_e[1 + \tfrac{1}{3}\mathfrak{b} + \tfrac{2}{15}\alpha + (\tfrac{2}{3}\mathfrak{b} + \tfrac{2}{21}\alpha)P_2 - \tfrac{8}{35}\alpha P_4]$$

The value of α is then determinable by finding the coefficient of P_4 in the expression for g.

Now $g^2 = \left(\frac{dV}{dr}\right)^2 + \left(\frac{dV}{rd\theta}\right)^2$, where after differentiation the value (1) is attributed to r.

Suppose that $-\frac{dV}{dr} = \frac{M}{a^2}[G_0 + G_1 + G_2]$, $\frac{dV}{rd\theta} = \frac{M}{a^2}H_1$, where the suffixes denote the orders of the several terms; then it is easy to prove that $g = \frac{M}{a^2}\left[G_0 + G_1 + G_2 + \frac{1}{2}\frac{H_1^2}{G_0}\right]$.

Since G_0 is equal to unity as far as the order zero,

$$g = -\frac{dV}{dr} + \frac{M}{a^2}(\tfrac{1}{2}H_1^2)$$

In order to find H_1 the terms of the second order in (38) are to be dropped, and we find

$$\frac{dV}{rd\theta} = \frac{M}{a^2} \cdot 2e\sin\theta\cos\theta$$

Therefore
$$H_1 = 2e\sin\theta\cos\theta$$

and
$$\tfrac{1}{2}H_1^2 = 2e^2\sin^2\theta\cos^2\theta$$
$$= 2e^2(\tfrac{2}{15} + \tfrac{2}{21}P_2 - \tfrac{8}{35}P_4)$$

We are only concerned with the term in P_4, and this portion of the transverse component of gravity is $-\frac{16}{35}\frac{M}{a^2}e^2P_4$.

It is next required to find the term in P_4 in $-\frac{dV}{dr}$. After the differentiation of V we shall require to determine the term in P_4 in the following functions, namely, $\frac{a^2}{r^2}$, $\frac{a^4}{r^4}P_2$, $\frac{a^6}{r^6}P_4$, $\frac{r}{a}(1-P_2)$.

Now
$$\frac{a^2}{r^2} = -\tfrac{16}{35}(\tfrac{3}{2}e^2 - f)P_4 + 3(\tfrac{2}{3}e)^2P_2^2 + \ldots = \tfrac{16}{35}fP_4 + \ldots$$
$$\frac{a^4}{r^4}P_2 = \tfrac{8}{3}eP_2^2 + \ldots = \tfrac{8}{3}e \cdot \tfrac{18}{35}P_4 + \ldots \qquad = \tfrac{48}{35}eP_4 + \ldots$$
$$\frac{a^6}{r^6}P_4 \qquad = P_4$$
$$\frac{r}{a}(1-P_2) = \tfrac{2}{3}eP_2^2 + \ldots = \tfrac{2}{3}e \cdot \tfrac{18}{35}P_4 + \ldots = \tfrac{12}{35}eP_4 + \ldots$$

It follows that the term in P_4 in $-dV/dr$ has a coefficient
$$\frac{M}{a^2}[\tfrac{16}{35}f - 2(e - \tfrac{1}{2}m)\tfrac{48}{35}e + \tfrac{8}{7}(\tfrac{7}{2}e^2 - \tfrac{5}{2}me - f) - \tfrac{2}{3}m \cdot \tfrac{12}{35}e]$$

The term is therefore
$$-\tfrac{8}{35}\frac{M}{a^2}(3f - \tfrac{11}{2}e^2 + \tfrac{15}{2}me)P_4$$

Then adding the transverse component, it appears that the whole term is
$$-\tfrac{8}{35}\frac{M}{a^2}(3f - \tfrac{7}{2}e^2 + \tfrac{15}{2}me)P_4$$

But the coefficient of P_4 was shown to be $-\frac{8}{35}g_e\alpha$; and since $g_e = M/a^2$ to the order zero, the coefficient in the expression for g is given by
$$\alpha = 3f - \tfrac{7}{2}e^2 + \tfrac{15}{2}me$$

In the expression for gravity the geocentric colatitude θ is used, and it remains to introduce the true colatitude λ, which is connected with θ by the formulæ
$$\theta = \lambda + 2e\sin\lambda\cos\lambda$$
and
$$\cos^2\theta = \cos^2\lambda - 4e\sin^2\lambda\cos^2\lambda$$

Thus finally
$$\left.\begin{aligned} g &= g_e[1 + \mathfrak{b}\cos^2\lambda - (\tfrac{5}{2}me - \tfrac{1}{2}e^2 - 3f)\sin^2\lambda\cos^2\lambda] \\ \text{where}\quad g_e &= \frac{M}{a^2}[1 + e - \tfrac{3}{2}m + e^2 - \tfrac{27}{14}me + \tfrac{9}{4}m^2 - \tfrac{4}{7}f] \\ \mathfrak{b} &= \tfrac{5}{2}m - e - \tfrac{17}{14}me - \tfrac{2}{7}f \end{aligned}\right\} \quad \ldots\ldots(40)$$

I shall show in § 9 that f is probably about $-\cdot 00000205$; m is $\frac{1}{288\cdot 41}$, or $\cdot 0034672$, and if e be taken as $\frac{1}{298}$

$$\tfrac{5}{2}\mathrm{me} - \tfrac{1}{2}\mathrm{e}^2 - 3\mathrm{f} = \cdot 0000295$$

Thus $$g = g_e\{1 + \mathfrak{b}\cos^2\lambda - \cdot 0000295\sin^2\lambda\cos^2\lambda\} \quad\text{..........(41)}$$

The results of the pendulum experiments which have been made up to the present time are hardly sufficiently numerous or concordant amongst themselves to make it worth while to take into account this small term. If it represented a detectable inequality in gravity the residuals given on page 240 of volume II. of Helmert's *Geodesy* would vary more or less as the square of the sine of twice the latitude; but they are quite irregular, being as follows:—

Latitude	Residuals	Latitude	Residuals
5°	$-\cdot 0000139$	45°	$-\cdot 0000015$
15°	$-$ 054	55°	$-$ 73
25°	$+$ 246	65°	$-$ 62
35°	$+$ 008	75°	$+$ 40

A great mass of material now awaits reduction, and we shall no doubt soon have from Dr Helmert far more accurate results for gravity and for the ellipticity of the Earth's figure than any that have been obtained up to the present time.

§ 7. *The superficial values of the rates of increase of h and f.*

The rates of increase of *h* and *f* are given in (27) and (21); whence

$$\frac{a\,dh}{da} - 3\mathrm{h} - \tfrac{27}{7}\mathrm{h}^2 + \tfrac{76}{21}\mathrm{h}\left(\frac{a\,dh}{da}\right) + \tfrac{11}{21}\left(\frac{a\,dh}{da}\right)^2 - \frac{\omega^2 a^3}{2\pi S_0}\left(\frac{a\,dh}{da}\right) + \frac{S_2}{S_0 a^2} = 0$$

$$\frac{a\,df}{da} - 5\mathrm{f} + \tfrac{27}{2}\mathrm{h}^2 - 7\mathrm{h}\left(\frac{a\,dh}{da}\right) + \left(\frac{a\,dh}{da}\right)^2 - \tfrac{27}{2}\frac{S_4}{S_0 a^4} = 0$$

But $$S_0 = \frac{3}{4\pi}M$$

$$S_2 = \frac{5Ma^2}{4\pi}\left[\mathrm{h} + \tfrac{9}{7}\mathrm{h}^2 - \tfrac{1}{2}\frac{\omega^2 a^3}{M}(1 - \tfrac{2}{7}\mathrm{h})\right]$$

$$S_4 = \frac{Ma^4}{2\pi}\left[\tfrac{7}{2}\mathrm{h}^2 - \mathrm{f} - \tfrac{5}{2}\frac{\omega^2 a^3}{M}\mathrm{h}\right]$$

Whence

$$\frac{a\,dh}{da} + 2\mathrm{h} + \tfrac{18}{7}\mathrm{h}^2 + \tfrac{76}{21}\mathrm{h}\left(\frac{a\,dh}{da}\right) + \tfrac{11}{21}\left(\frac{a\,dh}{da}\right)^2 - \tfrac{5}{2}\frac{\omega^2 a^3}{M}\left(1 - \tfrac{2}{7}\mathrm{h} + \tfrac{4}{15}\frac{a\,dh}{da}\right) = 0$$

$$\frac{a\,df}{da} + 4\mathrm{f} - 18\mathrm{h}^2 - 7\mathrm{h}\left(\frac{a\,dh}{da}\right) + \left(\frac{a\,dh}{da}\right)^2 + \tfrac{45}{2}\frac{\omega^2 a^3}{M}\mathrm{h} = 0$$

But $\frac{\omega^2 a^3}{M} = m(1 - h - \frac{3}{2}m)$, so that the last term of the first of these equations may be written $-\frac{5}{2}m\left(1 - \frac{2}{7}h - \frac{3}{2}m + \frac{4}{15}\frac{adh}{da}\right)$, and the last term of the second is $+\frac{45}{2}mh$. The first equation then shows that to the first order $\frac{adh}{da} = \frac{5}{2}m - 2h$. If this be used to eliminate dh/da from the terms of the second order in the first equation, and completely from the second, we get

$$\left.\begin{aligned} \frac{adh}{da} &= \frac{5}{2}m - 2h + \frac{18}{7}h^2 - \frac{41}{7}mh - \frac{75}{14}m^2 \\ \frac{adf}{da} &= -4f + 5mh - \frac{25}{4}m^2 \end{aligned}\right\} \quad \ldots\ldots\ldots\ldots(42)$$

The second of these may be combined with the first in the simple form

$$\frac{ad}{da}(f + \frac{5}{2}mh) + 4f = 0$$

The parameter h denotes $e - \frac{25}{14}e^2 - \frac{1}{7}f$; whence it may be shown that

$$\frac{ade}{da} = \frac{5}{2}m - 2e - e^2 + \frac{53}{14}me - \frac{25}{4}m^2 - \frac{2}{7}f$$

§ 8. *The Figure of a Homogeneous Mass with a small Nucleus.*

In order to obtain a preliminary estimate of the magnitude of f in the case of the Earth I shall first consider the case of a homogeneous mass with a nucleus at the centre of finite mass, but of infinitely small linear dimensions.

If we suppose the mass of the nucleus to be μ times that of the fluid, and that the form of surface is defined by the parameters a, h, f, it is clear from (10) and (11) that the external potential of the whole is

$$V = \frac{4}{3}\pi w\frac{a^3}{r}(1+\mu) - \frac{8}{15}\pi w\frac{a^5}{r^3}(h + \frac{2}{7}h^2)P_2 + \frac{16}{35}\pi w\frac{a^7}{r^5}(h^2 - \frac{9}{9}f)P_4 + \frac{1}{3}\omega^2 r^2(1 - P_2)$$

The equation to the surface will be of the form (8), and the transformations (12) hold good.

The conditions that the surface shall be level are, as before, that in V the coefficients of P_2 and P_4 shall vanish when r has the value (8).

These conditions are

$$\frac{2}{3}(1+\mu)a^2h(1-h)(1+\frac{20}{7}h) - \frac{2}{5}a^2h(1+\frac{2}{7}h)(1+\frac{11}{7}h) - \frac{\omega^2 a^2}{4\pi w}(1+\frac{2}{7}h) = 0$$

$$8(1+\mu)a^2(f - \frac{1}{2}h^2) - \frac{72}{5}a^2h^2 + \frac{8}{3}a^2(\frac{9}{2}h^2 - f) + \frac{6\omega^2 a^2}{\pi w}h = 0$$

whence
$$h\left(1+\tfrac{11}{7}h\right)=\tfrac{15}{16}\frac{\omega^2}{\pi w}\frac{1}{1+\frac{5}{2}\mu} \quad\text{.....................(43)}$$

$$f\left(1+\tfrac{3}{2}\mu\right)=\left(\tfrac{6}{5}+\tfrac{3}{4}\mu\right)h^2-\frac{9\omega^2 h}{8\pi w}$$

To the first order of small quantities $\dfrac{\omega^2}{\pi w}=\tfrac{16}{15}\left(1+\tfrac{5}{2}\mu\right)h$, so that the second equation becomes

$$f=\frac{-\frac{9}{4}\mu h^2}{1+\frac{3}{2}\mu} \quad\text{..............................(43)}$$

If μ vanishes f also vanishes, and the surface is a true ellipsoid, as obviously should be the case.

Let us apply these formulæ (43) to the case of the Earth. It is known that $\omega^2 a^3/M$ is about $\frac{1}{288}$, and it may be denoted by m, although the meaning of that symbol is slightly changed from that which it bears elsewhere.

Then
$$m=\frac{\omega^2}{\frac{4}{3}\pi w(1+\mu)}$$

and
$$h\left(1+\tfrac{11}{7}h\right)=\tfrac{5}{4}m\frac{1+\mu}{1+\frac{5}{2}\mu}$$

$$f=-\left(\tfrac{5}{4}m\right)^2\frac{\frac{9}{4}\mu(1+\mu)^2}{\left(1+\frac{3}{2}\mu\right)\left(1+\frac{5}{2}\mu\right)^2}$$

The equation which determines the value of μ which makes f a maximum is

$$\frac{1}{\mu}-\frac{3}{2+\mu}+\frac{2}{1+\mu}-\frac{10}{2+5\mu}=0$$

This reduces to the quadratic $\mu^2-\frac{1}{4}\mu-\frac{1}{2}=0$, of which the positive root is

$$\mu=\tfrac{1}{8}\left[1+\sqrt{33}\right]=\cdot 84307$$

The departure from the true ellipsoidal figure is greatest when the nucleus forms $\frac{843}{1843}$ or ·457 of the whole mass, and in this case

$$f=-\cdot 2946\left(\tfrac{5}{4}m\right)^2$$

Taking $\frac{5}{4}m=\frac{1}{231}$

$$f=-\cdot 0000055$$

The Earth's equatorial radius is about 6,378,000 metres, and the depression of the surface below the true ellipsoid in latitude 45° is $\frac{1}{4}af$, which gives 8·8 metres for this ideal case.

It is clear that the depression in actuality must be considerably less than this, and we shall hereafter see reason to believe that it is about one-third of this maximum value.

§ 9. *Evaluation of the actual departure of the Earth's Figure from true Ellipticity.*

In order to determine the superficial value of f it is necessary to make some definite hypothesis as to the law of internal density.

The theory of the Earth's figure has been worked out according to several hypotheses as to the internal density. Two of these may be described as more prominent than others; they are the hypotheses of Laplace and of Roche. The results derived from both these theories are conformable to our knowledge, and as Roche's hypothesis seems more tractable than the other I shall adopt it.

Roche then supposes that the mean density w_0 of all the matter lying inside equatorial radius a is expressed by the formula

$$w_0 = \rho\left[1 - \mathrm{k}\left(\frac{a}{\mathrm{a}}\right)^2\right]$$

The mean density of the whole Earth is clearly given by $\mathrm{w}_0 = \rho(1-\mathrm{k})$.

Since $w = w_0 - \frac{1}{3}a\frac{dw_0}{da}$, it follows that

$$w = \rho\left[1 - \tfrac{5}{3}\mathrm{k}\left(\frac{a}{\mathrm{a}}\right)^2\right] \quad\text{.........................(44)}$$

This is only true as a first approximation, but it suffices for the present. I now put

$$x = \mathrm{k}\left(\frac{a}{\mathrm{a}}\right)^2$$

and change the independent variable from a to x in the differential equations for f and h, the latter only being taken to the first order of small quantities.

The equations (31) and (23) then become

$$\left.\begin{aligned} &x(1-x)\frac{d^2h}{dx^2} + \tfrac{1}{2}(7-11x)\frac{dh}{dx} - h = 0 \\ &x(1-x)\frac{d^2f}{dx^2} + \tfrac{7}{2}(1-x)\left(\frac{df}{dx} - \frac{f}{x}\right) - 2x\left(\frac{df}{dx} + \tfrac{1}{2}\frac{f}{x}\right) + Q = 0 \\ \text{where}\quad &Q = 2h^2 + 13h\left(x\frac{dh}{dx}\right) + 16\left(x\frac{dh}{dx}\right)^2 - \frac{1}{x}\left[7\left(x\frac{dh}{dx}\right) + 10\left(x\frac{dh}{dx}\right)^2\right] \end{aligned}\right\} \text{...(45)}$$

The first of these is Roche's equation, and I proceed to find the solution*.

* Tisserand, *Méc. Cél.*, Vol. II. A full account of the researches of Legendre, Laplace, Roche, Radau, Tisserand, Callandreau, and of others is given in this work.

If it be assumed that

$$h = E\sum_0^\infty H_n x^n$$

where $H_0 = 1$, it is easy to prove that

$$H_{n+1} = \frac{n^2 + \frac{9}{2}n + 1}{n^2 + \frac{9}{2}n + \frac{7}{2}} H_n$$

Then if (n) denotes $\dfrac{n^2 + \frac{9}{2}n + 1}{n^2 + \frac{9}{2}n + \frac{7}{2}}$

$$h = E[1 + (0)x + (0)(1)x^2 + (0)(1)(2)x^3 + \ldots]$$

Translating this into numbers I find

$$\frac{h}{E} = 1 + \frac{2}{7}x + \frac{13}{3^2.7}x^2 + \frac{52}{3^3.11}x^3 + \frac{47}{3^3.11}x^4 + \frac{658}{3^4.5.11}x^5 + \frac{31913}{3^5.5.11.17}x^6 + \ldots \quad \ldots\ldots(46)$$

From this I make the following series of deductions:—

$$\frac{h^2}{E^2} = 1 + \frac{4}{7}x + \frac{218}{3^2.7^2}x^2 + \frac{6812}{3^3.7^2.11}x^3 + \frac{20045}{3^4.7^2.11}x^4 + \frac{3896}{3^5.5.7}x^5 + \frac{24210052}{3^6.5.7.11^2.17}x^6 + \ldots \quad \ldots(47)$$

$$\frac{h}{E^2}\left(x\frac{dh}{dx}\right) = \frac{2}{7}x + \frac{218}{3^2.7^2}x^2 + \frac{3406}{3^2.7^2.11}x^3 + \frac{40090}{3^4.7^2.11}x^4 + \frac{1948}{3^5.7}x^5 + \frac{24210052}{3^5.5.7.11^2.17}x^6 + \ldots \quad \ldots(48)$$

$$\frac{1}{E}\left(x\frac{dh}{dx}\right) = \frac{2}{7}x + \frac{26}{3^2.7}x^2 + \frac{52}{3^2.11}x^3 + \frac{188}{3^3.11}x^4 + \frac{658}{3^4.11}x^5 + \frac{63826}{3^4.5.11.17}x^6 + \ldots \quad \ldots\ldots(49)$$

$$\frac{1}{E^2}\left(x\frac{dh}{dx}\right)^2 = \frac{4}{7^2}x^2 + \frac{104}{3^2.7^2}x^3 + \frac{20540}{3^4.7^2.11}x^4 + \frac{4960}{3^4.7.11}x^5 + \frac{251176}{3^5.7.11^2}x^6 + \ldots(50)$$

By means of these developments I find that the function Q in (45) is given by

$$\frac{Q}{E^2} = \frac{256}{3^2.7^2}x + \frac{7024}{3^2.7^2.11}x^2 + \frac{13120}{3^2.7^2.11}x^3 + \frac{578128}{3^5.7^2.11}x^4 + \frac{116120320}{3^5.5.7.11^2.17}x^5 + \ldots$$

Then, assuming $f = E^2\sum_1^\infty K_n x^n$, and substituting in the second of (45), we get

$$\Sigma\left[n\left(n + \tfrac{9}{2}\right)K_{n+1} - \left(n^2 + \tfrac{9}{2}n - \tfrac{5}{2}\right)K_n\right]x^n + Q = 0$$

or

$$\left(\tfrac{11}{2}K_2 - 3K_1\right)x + \left(13K_3 - \tfrac{21}{2}K_2\right)x^2 + \left(\tfrac{45}{2}K_4 - 20K_3\right)x^3 + \left(34K_5 - \tfrac{63}{2}K_4\right)x^4 + \left(\tfrac{95}{2}K_6 - 45K_5\right)x^5 + \ldots + Q = 0$$

The coefficients of the several powers of x are to be equated to zero, and the successive equations solved. Carrying out this process I find

$$\left.\begin{aligned}
K_2 &= \frac{6}{11}K_1 - \frac{512}{3^2.7^2.11} &&= \cdot 54545K_1 - \cdot 15545 \\
K_3 &= \frac{63}{11.13}K_1 - \frac{12400}{3^2.7^2.11.13} &&= \cdot 44056K_1 - \cdot 19663 \\
K_4 &= \frac{56}{11.13}K_1 - \frac{167424}{3^4.7^2.11.13} &&= \cdot 39161K_1 - \cdot 29499 \\
K_5 &= \frac{882}{11.13.17}K_1 - \frac{11668616}{3^5.7^2.11.13.17} &&= \cdot 36281K_1 - \cdot 40312 \\
K_6 &= \frac{15876}{11.13.17.19}K_1 - \frac{15778709488}{3^5.5.7^2.11^2.13.17.19} &&= \cdot 34372K_1 - \cdot 52164 \\
&\text{Extrapolated } K_7 &&= \cdot 330K_1 \;\; - \cdot 635
\end{aligned}\right\} \quad \text{.........(51)}$$

The coefficient K_1 remains indeterminate as yet, and I must now show how it is to be found.

The second of equations (20), (21), and (23) are as follows:—

$$f - \tfrac{3}{2}h^2 + ah\frac{dh}{da} + \frac{3}{2w_0a^7}S_4 - \frac{a^2}{3w_0}T_4 = 0 \text{ ....} (a)$$

$$\frac{df}{da} - \frac{5f}{a} + \frac{27}{2}\frac{h^2}{a} - 7h\frac{dh}{da} + a\left(\frac{dh}{da}\right)^2 - \frac{27}{2w_0a^8}S_4 = 0 \text{ ....} (b)$$

$$\frac{d^2f}{da^2} + 6\frac{w}{w_0}\frac{df}{ada} - \left(20 - 6\frac{w}{w_0}\right)\frac{f}{a^2} + 12\left(1 - \frac{w}{w_0}\right)\frac{h^2}{a^2} + \left(4 - 18\frac{w}{w_0}\right)\frac{hdh}{ada}$$
$$- \left(1 + 9\frac{w}{w_0}\right)\left(\frac{dh}{da}\right)^2 = 0 \text{ ....} (c)$$

In obtaining (b) from (a) we multiplied (a) by w_0/a^2, differentiated, and divided by w_0/a^2. When x is independent variable we multiply by $(1-x)/x$, perform on the result the operation $x^{\frac{1}{2}}\dfrac{d}{dx}$, and divide by $(1-x)/x$.

Supposing all the functions to be expressed in series of powers of x, the equation (a) may be taken to be

$$P_1x + P_2x^2 + \ldots + P_nx^n + \ldots = 0 \text{ ....} (a')$$

It must be understood that the P's are numbers and not spherical harmonics. Then the equation (b) would be

$$(P_2 - P_1)x^{\frac{3}{2}} + (2P_3 - P_2 - P_1)x^{\frac{5}{2}} + (3P_4 - P_3 - P_2 - P_1)x^{\frac{7}{2}}$$
$$+ (4P_5 - P_4 - P_3 - P_2 - P_1)x^{\frac{9}{2}} + \ldots = 0 \text{ ....} (b')$$

Again, (c) was derived from (b) by multiplication by w_0a^8, differentiation, and division by w_0a^8. When x is independent variable we multiply by $x^4(1-x)$, perform the operation $x^{\frac{1}{2}}\dfrac{d}{dx}$, and divide by $x^4(1-x)$. Hence (c) must be equivalent to

$$11(P_2-P_1)x+(2.13P_3-15P_2-11P_1)x^2+(3.15P_4-19P_3-15P_2-11P_1)x^3$$
$$+(4.17P_5-21P_4-19P_3-15P_2-11P_1)x^4+\ldots=0\ldots(c')$$

Now the coefficients in f were evaluated by equating to zero the coefficients of the successive powers of x in (c) or (c'). Therefore

$$P_2-P_1=0,\quad 2.13P_3-15P_2-11P_1=0$$

and so forth.

These equations are satisfied by

$$P_1=P_2=P_3=\ldots=P_n$$

but one of the coefficients, say P_1, remains indeterminate.

The equation (b') is satisfied by the same condition; but in order that (a') may be satisfied it is necessary not only that all the P's should be equal, but that they should also vanish.

Accordingly (a) affords us one more condition from which K_1 will be determinable.

When x is the independent variable, the equation (a) may be written

$$f-\tfrac{3}{2}h^2+2h\left(\frac{xdh}{dx}\right)+\tfrac{3}{2}\frac{1}{x^{\frac{7}{2}}(1-x)}\int_\theta^x(1-\tfrac{5}{3}x)\,d\,[x^{\frac{7}{2}}(h^2-\tfrac{2}{9}f)]$$
$$-\frac{x}{3(1-x)}\int_x^{\mathrm{k}}(1-\tfrac{5}{3}x)\,d\left(\frac{f}{x}\right)=0$$

Since f/E^2 is expressible by a series beginning with K_1x, we shall be able to find K_1 by developing this equation in a series, but only carrying the development out as far as the first term.

I drop the factor E^2 for the sake of brevity.

Then, $f=K_1x$; $h^2=1+\tfrac{4}{7}x$, from (47); $h^2-\tfrac{2}{9}f=1+\tfrac{4}{7}x-\tfrac{2}{9}K_1x$;

$$\frac{d}{dx}[x^{\frac{7}{2}}(h^2-\tfrac{2}{9}f)]=x^{\frac{5}{2}}(\tfrac{7}{2}+\tfrac{18}{7}x-K_1x)$$

$$(1-\tfrac{5}{3}x)\frac{d}{dx}[x^{\frac{7}{2}}(h^2-\tfrac{2}{9}f)]=x^{\frac{5}{2}}(\tfrac{7}{2}-\tfrac{137}{42}x-K_1x)$$

$$\int_0^x(1-\tfrac{5}{3}x)\,d\,[x^{\frac{7}{2}}(h^2-\tfrac{2}{9}f)]=x^{\frac{7}{2}}(1-\tfrac{137}{189}x-\tfrac{2}{9}K_1x)$$

whence $$\frac{3}{2x^{\frac{7}{2}}(1-x)}\int_0^x(1-\tfrac{5}{3}x)\,d\,[x^{\frac{7}{2}}(h^2-\tfrac{2}{9}f)]=\tfrac{3}{2}+\tfrac{26}{63}x-\tfrac{1}{3}K_1x$$

Also $$-\tfrac{3}{2}h^2+2h\left(\frac{xdh}{dx}\right)=-\tfrac{3}{2}-\tfrac{2}{7}x$$

Therefore as far as the first power of x the equation is

$$K_1 x + \tfrac{8}{63}x - \tfrac{1}{3}K_1 x - \tfrac{1}{3}x\int_0^{\mathrm{k}}(1-\tfrac{5}{3}x)\,d\left(\frac{f}{x}\right) = 0$$

Thus, reintroducing the factor E^2,

$$K_1 = -\tfrac{4}{21} + \frac{1}{2E^2}\int_0^{\mathrm{k}}(1-\tfrac{5}{3}x)\,d\left(\frac{f}{x}\right)$$

But
$$\frac{f}{E^2 x} = K_1 + K_2 x + \ldots + K_n x^{n-1} + \ldots$$

Therefore

$$(1-\tfrac{5}{3}x)\frac{d}{dx}\left(\frac{f}{E^2 x}\right) = K_2 + (2K_3 - \tfrac{5}{3}K_2)x + \ldots + [(n+1)K_{n+2} - \tfrac{5}{3}nK_{n+1}]x^n + \ldots$$

and
$$\frac{1}{E^2}\int_0^{\mathrm{k}}(1-\tfrac{5}{3}x)\,d\left(\frac{f}{x}\right) = K_2\mathrm{k} + (K_3 - \tfrac{5}{6}K_2)\,\mathrm{k}^2 + \ldots$$
$$+\left(K_{n+2} - \frac{5n}{3(n+1)}K_{n+1}\right)\mathrm{k}^{n+1} + \ldots$$

On substituting for the K's their values, I find

$$K_1 + \tfrac{4}{21} = K_1(\cdot 27273\mathrm{k} - \cdot 00699\mathrm{k}^2 - \cdot 04895\mathrm{k}^3 - \cdot 06335\mathrm{k}^4 - \cdot 07002\mathrm{k}^5 - \cdot 0737\mathrm{k}^6 \ldots)$$
$$-(\cdot 07773\mathrm{k} + \cdot 03355\mathrm{k}^2 + \cdot 03826\mathrm{k}^3 + \cdot 01719\mathrm{k}^4 - \cdot 00793\mathrm{k}^5 - \cdot 04475\mathrm{k}^6 \ldots)$$

To find k we have the equations

$$2\frac{\Sigma n A_n \mathrm{k}^n}{\Sigma A_n \mathrm{k}^n} = 2\frac{\mathrm{k}d\mathrm{h}}{\mathrm{h}d\mathrm{k}} = \frac{\mathrm{a}d\mathrm{h}}{\mathrm{h}d\mathrm{a}}$$
$$= \tfrac{5}{2}\frac{\mathrm{m}}{\mathrm{h}} - 2 + \tfrac{18}{7}\mathrm{h} - \tfrac{4}{7}\mathrm{m} - \tfrac{75}{14}\frac{\mathrm{m}^2}{\mathrm{h}}$$
$$\mathrm{h} = \mathrm{e} - \tfrac{25}{14}\mathrm{e}^2 - \tfrac{1}{7}\mathrm{f}$$

The solution is virtually contained in the table of § 12 below*. From this it appears that when e is $\frac{1}{297}$, k is ·464.

Now with k = ·464 I find

$$K_1(\cdot 11426) - \cdot 0464 = K_1 + \tfrac{4}{21}$$

whence on substitution in (51)

$$K_1 = -\cdot 2674$$
$$K_2 = -\cdot 3013$$
$$K_3 = -\cdot 3144$$
$$K_4 = -\cdot 3997$$
$$K_5 = -\cdot 5002$$
$$K_6 = -\cdot 6136$$

and $K_7 = -\cdot 72$, by extrapolation.

* In making the computations I treated f as zero.

With these values for the K's, and with $\mathrm{k} = \cdot 464$,

$$\mathrm{f} = -\cdot 2632 E^2$$

But with $\mathrm{k} = \cdot 46$, $\frac{\mathrm{h}}{E} = 1\cdot 20465$, and with $\mathrm{k} = \cdot 47$, $\frac{\mathrm{h}}{E} = 1\cdot 21193$; whence with $\mathrm{k} = -\cdot 464$, $E = \frac{1}{358\cdot 5}$.

Hence finally

$$\left.\begin{aligned} \mathrm{f} &= -\cdot 00000205 \\ \tfrac{1}{4}\mathrm{af} &= -3\cdot 26 \text{ metres} \end{aligned}\right\} \qquad (52)$$

This result shows that the Earth's surface is $3\frac{1}{4}$ metres below the ellipsoid in latitude 45°.

I have already used this value of f in the evaluation of gravity in equation (41) § 7.

M. Callandreau has not solved his differential equation which corresponds with mine, but he concludes that the depression in latitude 45° must be less than 5 metres *.

§ 10. *The departure from true Ellipticity according to Professor Wiechert's hypothesis.*

Professor Wiechert † has adduced forcible arguments in favour of the hypothesis that the Earth consists of an iron nucleus, of approximately uniform density, with a superposed layer of rock. He concludes that the nucleus must occupy about four-fifths of the radius. He considers the conditions that both nucleus and surface may be level, and gives valuable numerical tables.

The method of this paper permits us to give the conditions from which his tables were computed somewhat more succinctly than he does. I will therefore give my results in outline.

Let ρ', ρ be the densities of the nucleus and of the superficial layer; let a', e' or h', f' define the figure of the nucleus, and a, e or h, f that of the surface; also let ρ_0 be the mean density of the whole.

It is clear that

$$\left.\begin{aligned} \rho_0 &= \rho + (\rho' - \rho)\left(\frac{a'}{a}\right)^3 (1 + h - h') \\ \text{whence} \quad \frac{\rho'}{\rho_0} &= \frac{\rho}{\rho_0} - \left(1 - \frac{\rho}{\rho_0}\right)\left(\frac{a}{a'}\right)^3 (1 - h + h') \end{aligned}\right\} \qquad (53)$$

* "Ann. de l'Obs. de Paris," *Mémoires*, t. XIX., 1889, p. E. 51.

† "Ueber die Massenvertheilung im Innern der Erde," *Nachr. K. Gesell. zu Göttingen*, 1896–7, p. 221.

The mean density ρ_0 may be taken as known, but if the values of ρ and of a'/a be assumed, the value of ρ' can only be rigorously found after the determination of h, h'. As a first approximation, very near the truth however, we may take $h = h'$, so that

$$\frac{\rho'}{\rho_0} = \frac{\rho}{\rho_0} - \left(1 - \frac{\rho}{\rho_0}\right)\left(\frac{a}{a'}\right)^3$$

The integrals S_0, S_2, S_4, T_2, T_4, defined in (10) are only required when $a = a'$ at the boundary of the nucleus, and when $a = a$ at the surface.

When $a = a'$

$$S_0 = \rho' a'^3 (1 - h'); \quad S_2 = \rho' a'^5 h' (1 + \tfrac{2}{7}h'); \quad S_4 = \rho' a'^7 (h'^2 - \tfrac{2}{9}f')$$

$$T_2 = \rho (h - h')(1 + \tfrac{17}{7}h + \tfrac{17}{7}h'); \quad T_4 = \rho \left(\frac{f}{a^2} - \frac{f'}{a'^2}\right)$$

Similarly, the integrals corresponding to $a = a$ may be written down. If ρ' be then eliminated by means of (53), it will be found that they are

$$S_0 = \rho_0 a^3 (1 - h)$$

$$S_2 = \rho_0 a^3 \left[\left(1 - \frac{\rho}{\rho_0}\right) h' (1 - h + \tfrac{9}{7}h') a'^2 + \frac{\rho}{\rho_0} h (1 + \tfrac{2}{7}h) a^2\right]$$

$$S_4 = \rho_0 a^3 \left[\left(1 - \frac{\rho}{\rho_0}\right)(h'^2 - \tfrac{2}{9}f') a'^4 + \frac{\rho}{\rho_0}(h^2 - \tfrac{2}{9}f) a^4\right]$$

$$T_2 = 0, \quad T_4 = 0$$

The conditions that the surfaces a' and a may be level surfaces are given in (13) and (16), with the above values for the S and T integrals. I find that a considerable simplification in the expression results from considering $h(1 + \tfrac{11}{7}h)$, $h'(1 + \tfrac{11}{7}h')$ as the unknown quantities instead of simply h and h'. Accordingly I write

$$k = h(1 + \tfrac{11}{7}h), \quad k' = h'(1 + \tfrac{11}{7}h')$$

It may be remarked in passing that $k = e - \tfrac{3}{14}e^2 - \tfrac{1}{7}f$.

Without giving details I may state that the condition (13) leads to the two equations

$$\left.\begin{aligned} k'\left(\frac{\rho'}{\rho_0} + \frac{3}{2}\frac{\rho}{\rho_0}\right) - k\left(\frac{3}{2}\frac{\rho}{\rho_0}\right) &= \frac{15}{16}\frac{\omega^2}{\pi\rho_0} + \frac{1}{7}\frac{\rho}{\rho_0}(k - k')(9k - 5k') \\ -\frac{3}{2}k'\left(\frac{a'}{a}\right)^2\left(1 - \frac{\rho}{\rho_0}\right) + k\left(\frac{5}{2} - \frac{3}{2}\frac{\rho}{\rho_0}\right) &= \frac{15}{16}\frac{\omega^2}{\pi\rho_0} + \frac{3}{7}\left(1 - \frac{\rho}{\rho_0}\right)k'(k - k')\left(\frac{a'}{a}\right)^2 \end{aligned}\right\} \quad (54)$$

We might first neglect the small terms on the right, and use the approximate value of ρ' on the left, solve the equations and then proceed to a second approximation. The approximate solution

$$\frac{k}{(2\rho' + 3\rho)a^2 + 3(\rho_0 - \rho)a'^2} = \frac{k'}{5\rho_0 a^2} = \frac{\dfrac{15\omega^2}{8\pi\rho_0}}{(5\rho_0 - 3\rho)(2\rho' + 3\rho)a^2 - 9\rho(\rho_0 - \rho)a'^2}$$

is, however, very near the truth, because the small terms on the right not only depend on the squares of the ellipticity, but also involve $k - k'$, which is itself much smaller than either k or k'.

In these equations we have, of course, $\frac{15\omega^2}{8\pi\rho_0} = \frac{5}{2}m\left(1 - \frac{3}{2}m\right)$.

Now turning to the condition (16), in which it is clearly permissible to write k, k', for h, h', I find

$$f\left(\frac{2\rho'}{\rho} + 1\right) - f'\left(\frac{a'^2}{a}\right) = 0$$

$$-f\left(1 - \frac{1}{3}\frac{\rho}{\rho_0}\right) + \frac{1}{3}f'\left(1 - \frac{\rho}{\rho_0}\right)\left(\frac{a'}{a}\right)^4 = \frac{3}{2}\left(1 - \frac{\rho}{\rho_0}\right)\left\{k - k'\left(\frac{a'}{a}\right)^2\right\}$$

From this

$$f = \frac{-\frac{9}{2}a^2(ka^2 - k'a'^2)^2}{\left(3 + 2\frac{\rho}{\rho_0 - \rho}\right)a^6 - \frac{a'^6}{1 + 2\rho'/\rho}} \quad \text{.................(55)}$$

If ρ' were infinite and a' zero, but $\rho' a'^3 = \mu\rho a^3(1 - h)$, we ought to come back on the result for the nucleus of infinitely small dimensions in § 8; and this is so.

In one of the cases considered by Professor Wiechert he took $\rho = 3{\cdot}2$, $\rho' = 8{\cdot}206$, $a'/a = {\cdot}78039$, and found the ellipticity of the surface to be $\frac{1}{297}$. The case corresponds closely with that of the Earth. He also computed a certain function from which my f might be derived. I find, however, by computation directly from (55), that $f = -{\cdot}00000175$, and $\frac{1}{4}af = -2{\cdot}79$ metres.

With Roche's hypothesis I found $f = -{\cdot}00000205$, and $\frac{1}{4}af = -3{\cdot}26$ metres. Thus the two hypotheses lead to nearly the same result.

§ 11. *Solution of the Differential Equation for h.*

I propose to solve the equation (31) with Roche's hypothesis, and thus to estimate the contribution of the terms of the second order to the value of h.

As in § 9, x is to be the independent variable.

The expression for w in (44) is no longer sufficient, but a term of the first order must be included in it.

It was proved in (32) that

$$w = w_0\left[1 + \frac{1}{3}\left(1 + \frac{1}{3}a\frac{dh}{da}\right)\frac{a\,dw_0}{w_0\,da}\right]$$

Now $w_0 = \rho(1 - x)$ and $a\frac{d}{da} = 2x\frac{d}{dx}$, so that

$$w = \rho\left[1 - \frac{5}{3}x - \frac{4}{9}x\left(x\frac{dh}{dx}\right)\right]$$

and

$$w_0 - w = \frac{2}{3}\rho x\left[1 + \frac{2}{3}\left(x\frac{dh}{dx}\right)\right] \quad \text{.....................(55)}$$

When x is introduced as independent variable in (31), and when the equation is divided by 4, the terms of the first order assume the form given in the first of (45). Then by aid of (55) the terms of the second order are easily determined, and the equation will be found to assume the following form:—

$$\left.\begin{aligned} x(1-x)\frac{d^2h}{dx^2}+\tfrac{1}{2}(7-11x)\frac{dh}{dx}-h+P-R=0 \\ \text{where}\qquad P=\tfrac{7}{3}h^2+\tfrac{52}{21}\left[h\left(x\frac{dh}{dx}\right)+\left(x\frac{dh}{dx}\right)^2\right] \\ R=\tfrac{2}{3}m\left(h+2x\frac{dh}{dx}\right) \end{aligned}\right\}\quad\text{.........(56)}$$

In the last of these m denotes $\dfrac{3\omega^2}{4\pi w_0}$; this is equal to $\dfrac{3\omega^2}{4\pi \mathrm{w}_0}\cdot\dfrac{\mathrm{w}_0}{w_0}$ or $\mathrm{m}\dfrac{1-\mathrm{k}}{1-x}$. Accordingly

$$R=\tfrac{2}{3}\mathrm{m}(1-\mathrm{k})\frac{h+2x\dfrac{dh}{dx}}{1-x}\quad\text{......................(56)}$$

By means of the series (47), (48), and (49) I find

$$P=E^2\left\{\frac{7}{3}+\frac{100}{7^2}x+\frac{23890}{3^3.7^3}x^2+\frac{149084}{3^4.7^2.11}x^3+\frac{4134965}{3^5.7^3.11}x^4+\frac{11540024}{3^6.5.7^2.11}x^5+\frac{1658730884}{3^7.7^2.11^2.17}x^6+\dots\right\}$$

or

$$P=E^2\{\tfrac{7}{3}+2{\cdot}0408x+2{\cdot}5796x^2+3{\cdot}4149x^3+4{\cdot}5100x^4+5{\cdot}8738x^5+7{\cdot}5248x^6+9{\cdot}45x^7+\dots\}$$

the last term being extrapolated.

Now we saw in the last section that for the Earth the reciprocal of E was 358·5. As a fact I have here used the value 361·8, and it does not seem worth while to recompute on account of this small change in E.

Then

$$P=E\{{\cdot}006450+{\cdot}005641x+{\cdot}007131x^2+{\cdot}009440x^3+{\cdot}012467x^4+{\cdot}016237x^5+{\cdot}020801x^6+{\cdot}0261x^7+\dots\}$$

When the fractions in (46) and (49) are expressed in decimals, I find

$$h+2x\frac{dh}{dx}=E\{1+{\cdot}85714x+1{\cdot}03175x^2+1{\cdot}22559x^3+1{\cdot}42424x^4+1{\cdot}62469x^5+1{\cdot}82597x^6+2{\cdot}028x^7+\dots\}$$

And

$$\frac{h+2x\dfrac{dh}{dx}}{1-x}=E\{1+1{\cdot}85714x+2{\cdot}88889x^2+4{\cdot}11448x^3+5{\cdot}53872x^4+7{\cdot}16341x^5+8{\cdot}98938x^6+11{\cdot}016x^7+\dots\}$$

This last must be multiplied by $\frac{2}{5}\mathrm{m}(1-\mathrm{k})$ in order to find R.

Now $\mathrm{m} = \cdot 00346$, $\mathrm{k} = \cdot 464$; so that the factor is $\cdot 001236$.

Then

$$R = E\{\cdot 001236 + \cdot 002296x + \cdot 003572x^2 + \cdot 005087x^3 + \cdot 006848x^4 + \cdot 008857x^5 + \cdot 011114x^6 + \cdot 0136x^7 + \ldots\}$$

Finally

$$\frac{P-R}{E} = \cdot 005214 + \cdot 003345x + \cdot 003559x^2 + \cdot 004353x^3 + \cdot 005619x^4 + \cdot 007380x^5 + \cdot 009687x^6 + \cdot 0125x^7 + \ldots$$

I now assume as before, $h = E\sum_0^\infty H_n x^n$

so that the differential equation (56) gives

$$\Sigma\left[(n^2 + \tfrac{9}{2}n + \tfrac{7}{2})H_{n+1} - (n^2 + \tfrac{9}{2}n + 1)H_n\right]x^n + \frac{P-R}{E} = 0$$

When the coefficients of the successive powers of x are equated to zero the first equation is

$$\tfrac{7}{2}H_1 - H_0 = -\cdot 005214$$

If we assume $H_0 = 1$, which is clearly permissible,

$$H_1 = \tfrac{2}{7} - \cdot 001490$$

Then from the successive equations I find

$$H_2 = \frac{13}{3^2 . 7} - \cdot 001448$$

$$H_3 = \frac{52}{3^3 . 11} - \cdot 001444$$

$$H_4 = \frac{47}{3^3 . 11} - \cdot 001473$$

$$H_5 = \frac{658}{3^4 . 5 . 11} - \cdot 001524$$

$$H_6 = \frac{31913}{3^5 . 5 . 11 . 17} - \cdot 001594$$

$$H_7 = \frac{583552}{3^5 . 5 . 11 . 17 . 19} - \cdot 001698$$

The first part of each of these coefficients is the corresponding coefficient in the first approximation, given in (46). If, therefore, we denote by δh the correction to be applied to the first approximation as arising from the terms of the second order, we have

$$\delta h = -E\{\cdot 00149x + \cdot 00145x^2 + \cdot 00144x^3 + \cdot 00147x^4 + \cdot 00152x^5 + \cdot 00159x^6 + \cdot 00170x^7 + \ldots\}$$

By an empirical summation of this series, we may assert that a fair approximation to the result is given by

$$\delta h = -\cdot 0015 \frac{Ex}{1-x}$$

But $\frac{2}{3}\frac{x}{1-x} = 1 - \frac{w}{w_0}$, so that the correction becomes

$$\delta h = -\cdot 00225 E\left(1 - \frac{w}{w_0}\right)$$

or since E is about $\frac{1}{360}$,

$$\delta h = -\cdot 0000062\left(1 - \frac{w}{w_0}\right)$$

The superficial value of h corresponding to $e = \frac{1}{297}$ is ·003349, and

$$\frac{\delta h}{h} = -\cdot 0019\left(1 - \frac{w}{w_0}\right)$$

It is, however, better to look at this correction from another point of view. If h_0 denotes h as derived from the first approximation, we have, by means of the empirical solution,

$$h = h_0 - \cdot 0015 \frac{Ex}{1-x}$$

so that

$$x\frac{dh}{dx} = x\frac{dh_0}{dx} - \cdot 0015 \frac{Ex}{(1-x)^2}$$

We have seen in § 9 that k in Roche's hypothesis is derived from the equation

$$\frac{2\Sigma nA_n\mathrm{k}^n}{\Sigma A_n\mathrm{k}^n} = \tfrac{5}{2}\frac{\mathrm{m}}{\mathrm{h}} - 2 + \tfrac{18}{7}\mathrm{h} - \tfrac{41}{7}\mathrm{m} - \tfrac{75}{14}\frac{\mathrm{m}^2}{\mathrm{h}}$$

It now appears that, when terms of the second order are included in the differential equation for h, the left-hand side of this equation becomes

$$\frac{2\left(\Sigma nA_n\mathrm{k}^n - \cdot 0015\frac{\mathrm{k}}{(1-\mathrm{k})^2}\right)}{\Sigma A_n\mathrm{k}^n - \cdot 0015\frac{\mathrm{k}}{1-\mathrm{k}}}$$

Now I found that when $\mathrm{k} = \cdot 464$, $\Sigma A_n\mathrm{k}^n = 1\cdot 20756$, $\Sigma nA_n\mathrm{k}^n = \cdot 33666$. Therefore

$$\Sigma nA_n\mathrm{k}^n - \cdot 0015\frac{\mathrm{k}}{(1-\mathrm{k})^2} = \Sigma nA_n\mathrm{k}^n\left(1 - \frac{150}{33666}\cdot\frac{\cdot 464}{(\cdot 536)^2}\right)$$

$$= \Sigma nA_n\mathrm{k}^n(1 - \cdot 00720)$$

$$\Sigma A_n\mathrm{k}^n - \cdot 0015\frac{\mathrm{k}}{1-\mathrm{k}} = \Sigma A_n\mathrm{k}^n\left(1 - \frac{150}{120756}\cdot\frac{464}{536}\right)$$

$$= \Sigma A_n\mathrm{k}^n(1 - \cdot 00108)$$

Since $\frac{1-\cdot 00720}{1-\cdot 00108}=1-\cdot 00612=\cdot 99388$, it follows that the left-hand side of the equation for k is

$$\cdot 994\left(\frac{2\Sigma n A_n k^n}{\Sigma A_n k^n}\right)$$

The principal effect of the terms of the second order will therefore be slightly to alter the value of k which satisfies the observed conditions. In every hypothesis as to the internal density there is some parameter derivable from observation, and a similar investigation would show that its value is but slightly affected by the terms of the second order. It is clear, then, that there is not much to be gained by pursuing this investigation further.

§ 12. *The Moments of Inertia, the Precessional Constant, and the Ellipticity of the Earth.*

It was shown in (6) of § 1 that

$$C=\tfrac{8}{15}\pi\int_0^a wd\left[a^5\left(1-e+\tfrac{4}{7}f\right)\right]$$

If the terms of the second order be omitted we may drop f and replace e by h. Also to the first order $S_2=\int_0^a wd\,(a^5 h)$; and by (35)

$$S_2=\frac{5M}{4\pi}a^2\left[h-\tfrac{1}{2}m\right]$$

Hence
$$C=\tfrac{8}{3}\pi\int_0^a wa^4 da-\tfrac{2}{3}Ma^2\left(h-\tfrac{1}{2}m\right)$$

It was shown in (32) that

$$w=w_0+\tfrac{1}{3}a\frac{dw_0}{da}+\tfrac{1}{9}a^2\frac{dh}{da}\frac{dw_0}{da}$$

so that
$$\int_0^a wa^4 da=\int_0^a w_0 a^4 da+\tfrac{1}{3}\int_0^a a^5 dw_0+\tfrac{1}{9}\int_0^a a^6\frac{dh}{da}dw_0$$

If the middle term in this expression be integrated by parts we have a term $-\tfrac{5}{3}\int_0^a w_0 a^4 da$, which will fuse with the first term. Therefore

$$\int_0^a wa^4 da=\tfrac{1}{3}w_0 a^5-\tfrac{2}{3}\int_0^a w_0 a^4 da+\tfrac{1}{9}\int_0^a a^6\frac{dh}{da}dw_0$$

and

$$C=\tfrac{8}{9}\pi w_0 a^5\left[1-\frac{2}{w_0 a^5}\int_0^a w_0 a^4 da+\frac{1}{3w_0 a^5}\int_0^a a^6\frac{dh}{da}dw_0\right]-\tfrac{2}{3}Ma^2\left(h-\tfrac{1}{2}m\right)$$

But to the order of approximation here adopted

$$\tfrac{8}{9}\pi w_0 a^5 = \tfrac{2}{3} M a^2 (1 + h)$$

Therefore

$$C = \tfrac{2}{3} M a^2 \left[1 - \frac{2(1+h)}{w_0 a^5}\int_0^a w_0 a^4 da + \frac{1}{3 w_0 a^5}\int_0^a a^6 \frac{dh}{da} dw_0 + \tfrac{1}{2} m\right]$$

We now make use of M. Radau's transformation. It was proved in (34), § 5, that

$$w_0 a^5 \sqrt{(1 + \eta_1)} = 5(1 + \lambda)\int_0^a w_0 a^4 da$$

where $1 + \lambda$ is a mean value of a certain function.

Therefore $$-\frac{2(1+h)}{w_0 a^5}\int_0^a w_0 a^4 da = -\tfrac{2}{5}\frac{(1+h)\sqrt{(1+\eta_1)}}{1+\lambda}$$

Now $$\int_0^a a^6 \frac{dh}{da} dw_0 = w_0 a^6 \frac{dh}{da} - \int_0^a w_0 \left(6a^5 \frac{dh}{da} + a^6 \frac{d^2 h}{da^2}\right) da$$

But h satisfies the differential equation

$$\frac{d^2 h}{da^2} + 6\frac{w}{w_0}\frac{dh}{a\,da} - 6\left(1 - \frac{w}{w_0}\right)\frac{h}{a^2} = 0$$

and $$\frac{dw_0}{da} = -\frac{3}{a}(w_0 - w)$$

Therefore

$$\int_0^a w_0 \left(6a^5 \frac{dh}{da} + a^6 \frac{d^2 h}{da^2}\right) da = 6\int_0^a (w_0 - w)\left(h a^4 + a^5 \frac{dh}{da}\right) da$$

$$= 6\int_0^a (w_0 - w) h a^4 da - 2\int_0^a a^6 \frac{dh}{da} dw_0$$

Therefore

$$\int_0^a a^6 \frac{dh}{da} dw_0 = -w_0 a^5 h \eta_1 + 6\int_0^a (w_0 - w) h a^4 da$$

Now let

$$\left.\begin{aligned} \Delta_1 &= \tfrac{2}{5}(1+h)\sqrt{(1+\eta_1)}\frac{\lambda}{1+\lambda} \\ \Delta_2 &= -\tfrac{1}{3} h \eta_1 + \frac{2}{w_0 a^5}\int_0^a (w_0 - w) h a^4 da \\ \Delta &= \Delta_1 + \Delta_2 \end{aligned}\right\} \quad \text{..............(57)}$$

and we have $$C = \tfrac{2}{3} M a^2 \left[1 - \tfrac{2}{5}(1+h)\sqrt{(1+\eta_1)} + \tfrac{1}{2} m + \Delta\right] \quad \text{...........(57)}$$

The only part of the expression for C which involves the law of internal density is Δ, and we shall see that Δ is very small compared with unity.

M. Callandreau has established a formula for C with which, as far as I can make out, mine agrees*.

It is clear that the two parts of Δ originate in quite different ways, Δ_1 depending on the equation for the internal ellipticity, and Δ_2 depending on the terms of the second order.

It is first necessary to evaluate λ from the equation

$$1+\lambda = \frac{\frac{1}{5}\mathrm{w}_0 \mathrm{a}^5 \sqrt{(1+\eta_1)}}{\int_0^{\mathrm{a}} w_0 a^4 da} \quad \text{......................(58)}$$

where $$1+\eta_1 = \tfrac{5}{2}\frac{\mathrm{m}}{\mathrm{h}} - 1 + \tfrac{18}{7}\mathrm{h} - \tfrac{41}{7}\mathrm{m} - \tfrac{75}{14}\frac{\mathrm{m}^2}{\mathrm{h}}$$

It is not hard to obtain a close approximation to λ when Laplace's hypothesis as to the density is adopted, but that hypothesis is not so tractable as Roche's, and I therefore adopt the latter.

According to Roche

$$w_0 = \rho(1-x), \quad \text{where } x = \mathrm{k}\left(\frac{a}{\mathrm{a}}\right)^2$$

Therefore $$\int_0^{\mathrm{a}} w_0 a^4 da = \tfrac{1}{2}\frac{\rho \mathrm{a}^5}{\mathrm{k}^{\frac{5}{2}}}\int_0^{\mathrm{k}} (1-x)\, x^{\frac{3}{2}} dx$$

$$= \tfrac{1}{5}\rho \mathrm{a}^5 (1 - \tfrac{5}{7}\mathrm{k})$$

Accordingly $$1+\lambda = \frac{1-\mathrm{k}}{1-\frac{5}{7}\mathrm{k}}\sqrt{(1+\eta_1)}$$

Since $\mathrm{w}_0 = \rho(1-\mathrm{k})$, and with close approximation, $\mathrm{w} = \rho(1-\frac{5}{3}\mathrm{k})$, we have $\rho = (\frac{5}{2}\mathrm{w}_0 - \frac{3}{2}\mathrm{w})$, $\mathrm{k} = \dfrac{3(\mathrm{w}_0 - \mathrm{w})}{5\mathrm{w}_0 - 3\mathrm{w}}$ very nearly. Hence we may also write

$$1+\lambda = \frac{\sqrt{(1+\eta_1)}}{1+\frac{3}{7}(1-\mathrm{w}/\mathrm{w}_0)} \dagger \quad \text{....................(59)}$$

Therefore $$\Delta_1 = \tfrac{2}{5}(1+\mathrm{h})\left[\sqrt{(1+\eta_1)} - 1 - \tfrac{3}{7}(1-\mathrm{w}/\mathrm{w}_0)\right] \quad \text{...........(60)}$$

* *Bulletin Astronomique*, Vol. xiv., 1897, p. 217.

† The approximation here consists in neglecting the variation of ellipticity in evaluating the density w from the mean density w_0. In Laplace's hypothesis $w = \rho\frac{\kappa}{a}\sin\frac{a}{\kappa}$. If we neglect the variation of ellipticity in determining the mean density w_0 from the density w, it will be found that

$$1+\lambda = \tfrac{1}{15}\frac{\mathrm{a}^2}{\kappa^2}\frac{\sqrt{(1+\eta_1)}}{1-\mathrm{w}/\mathrm{w}_0}$$

Since $\dfrac{\mathrm{w}_0}{\mathrm{w}} = \dfrac{\kappa^2}{\mathrm{a}^2}\left(1 - \dfrac{\mathrm{a}}{\kappa}\cot\dfrac{\mathrm{a}}{\kappa}\right)$, κ may be regarded as a function of w_0/w, and λ is seen to be a function of w_0/w and of η_1, as in the text.

It is now required to evaluate Δ_2.

With Roche's hypothesis

$$\frac{2}{\text{w}_0 \text{a}^5}\int_0^{\text{a}} (w_0 - w)\, h a^4 da = \tfrac{2}{3}\frac{\rho}{\text{w}_0} \cdot \frac{1}{\text{k}^{\frac{5}{2}}}\int_0^{\text{k}} h x^{\frac{5}{2}} dx$$

I write the series (46) $\text{h} = E\sum_0^\infty A_n x^n$, and then

$$\frac{2}{\text{w}_0 \text{a}^5}\int_0^{\text{a}} (w_0 - w)\, h a^4 da = \tfrac{2}{3}\frac{E\rho}{\text{w}_0 \text{k}^{\frac{5}{2}}} \sum \int_0^{\text{k}} A_n x^{n+\frac{5}{2}} dx$$

$$= \tfrac{4}{3}\frac{E\text{k}}{1-\text{k}} \sum \frac{1}{2n+7} A_n \text{k}^n$$

But $\text{h} = E\Sigma A_n \text{k}^n$, and therefore

$$\Delta_2 = \tfrac{1}{3}\text{h}\left[-\eta_1 + \frac{4\text{k}\Sigma \frac{1}{2n+7} A_n \text{k}^n}{(1-\text{k})\,\Sigma A_n \text{k}^n}\right]$$

or

$$\Delta_2 = \tfrac{1}{3}\text{h}\left[-\eta_1 + 6\left(1 - \frac{\text{w}}{\text{w}_0}\right)\frac{\Sigma \frac{1}{2n+7} A_n \text{k}^n}{\Sigma A_n \text{k}^n}\right] \quad \text{.........(61)}$$

In order to estimate the magnitude of the correction Δ I have computed it, and its two constituents, in the following table:—

Table of Results according to Roche's Hypothesis.

$$\left(m = \frac{1}{288{\cdot}41}\right).$$

		·1	·2	·3	·4	·46	·47	·5	·55
	k								
Ratio of mean to surface density	w_0/w =	1·08	1·2	1·4	1·8	2·314	2·446	3·0	5·4
Reciprocal of ellipticity ...	$\frac{1}{e}$ =	239·3	248·2	262·2	280·7	295·7	298·6	308·2	327·5
	$h \times 10^6$ =	4148	4001	3788	3541	3362	3329	3227	3037
Moment of inertia, uncorrected	$C \div \frac{2}{3}Ma^2$ =	·58728	·57137	·55088	·52360	·50245	·49847	·48557	·46036
	$\lambda \times 10^6$ =	47	184	387	548	516	493	422	99
First correction	$\Delta_1 \times 10^6$ =	19	79	174	262	258	248	218	54
Second correction ...	$\Delta_2 \times 10^6$ =	−2	−10	−28	−63	−95	−102	−125	−173
Ratio of total correction to moment of inertia ...	$\frac{2}{3}\frac{Ma^2\Delta}{C} \times 10^6$ =	29	120	265	380	322	293	192	−258

It will be observed that λ and Δ_1 reach maxima when k is ·4, whereas $-\Delta_2$ increases throughout as k increases. The last line gives the factor of augmentation, diminished by unity, of the uncorrected value of C. The columns k = ·46 and ·47 are those which correspond to the case of the Earth most closely, and they show that C must be augmented by a factor 1·0003 when Δ is taken into account.

We are now in a position to give the formula for the Precessional Constant $\frac{C-A}{C}$.

It appears from (9), (10), (35), and (36) that

$$C-A=\tfrac{2}{3}Ma^2\left[h-\tfrac{1}{2}m+\tfrac{9}{7}h^2+\tfrac{1}{7}mh+\tfrac{3}{4}m^2\right] \quad \text{.........(61)}$$

So that, dividing (61) by (57),

$$\left.\begin{aligned} \frac{C-A}{C} &= \frac{h-\frac{1}{2}m+\frac{9}{7}h^2+\frac{1}{7}mh+\frac{3}{4}m^2}{1-\frac{2}{5}(1+h)\sqrt{(1+\eta_1)}+\frac{1}{2}m+\Delta} \\ \text{where} \quad 1+\eta_1 &= \tfrac{5}{2}\frac{m}{h}-1+\tfrac{18}{7}h-\tfrac{41}{7}m-\tfrac{75}{14}\frac{m^2}{h} \end{aligned}\right\} \quad \text{.........(62)}$$

We have reason to believe that the term Δ may be allowed for by first treating it as zero, and afterwards multiplying the result by ·9997.

The Precessional Constant is known with a high degree of accuracy, and I cannot but think that this investigation shows that it may be used for determining the actual ellipticity of the Earth's surface with perhaps as little error as by any other method. The uncertainty is, indeed, of a different kind, being dependent on our ignorance of the interior of the Earth. I reduce the formula (62) to numbers in the following manner. I assume a definite value for the ellipticity, namely $e_0=\frac{1}{299}=$ ·00334448. Then, with $f=-$ ·00000205, h_0 is computed and found to be ·00332480. I take also $m=$ ·0034672.

Now let N_0, D_0 be the numerator and denominator of (62), with these values of h, m, and with Δ put equal to zero. Then, if $N=N_0+\delta N$, $D=D_0+\delta D$, $h=h_0+\delta h$ be the true values of those quantities, it is clear that

$$N=N_0+(1+\tfrac{18}{7}h+\tfrac{1}{7}m)\,\delta h$$

$$D=D_0-\tfrac{2}{5}\sqrt{(1+\eta_1)}\,\delta h+\tfrac{1}{5}\frac{(1+h)}{\sqrt{(1+\eta_1)}}\left(\tfrac{5}{2}\frac{m}{h^2}-\tfrac{18}{7}-\tfrac{75}{14}\frac{m^2}{h^2}\right)\delta h$$

From these we may compute δN and δD.

I find then, $\quad N_0=$ ·00161608, $\quad N=N_0(1+624{\cdot}4\,\delta h)$

$D_0=$ ·4991436, $\quad D=D_0(1+248{\cdot}0\,\delta h)$

Then $N_0/D_0=$ ·00323771, and since the denominator should be augmented by 1·0003, it follows that the corrected value of $N_0/D_0=$ ·00323674, and

$$\frac{C-A}{C}=\frac{N}{D}=\cdot 00323674\,(1+376{\cdot}4\,\delta h)$$

The most generally accepted value of the precessional constant is ·003272, and this exceeds our corrected N_0/D_0 by ·00003526. Therefore

$$\delta h=\frac{\cdot 00003526}{376{\cdot}4\times\cdot 0032366}=\cdot 00002895$$

But h_0 was ·00332480, so that h = ·0033537. If $\frac{25}{14}h^2 + \frac{1}{7}f$ be added to h, we obtain e; the result is

$$e = \cdot 0033734 = \frac{1}{296\cdot 4} \, *$$

I have, in fact, repeated the computation with this value of e, and find $\frac{N_0}{D_0} = \cdot 003273$, which only differs from the Precessional Constant by unity in the sixth place of decimals. Oppolzer† adopts the value ·003261 for the Precessional Constant, and this leads to $\frac{1}{297\cdot 5}$ as the value of the ellipticity.

Dr Wiechert has considered the hypothesis that the Earth consists of an iron nucleus with a superstratum of rock. With the Precessional Constant at ·003272 it may be concluded from his table that the ellipticity is $\frac{1}{297\cdot 3}$.

These results, then, point to an ellipticity between $\frac{1}{296}$ and $\frac{1}{298}$, and they agree well with the results of all the methods of determining the ellipticity except that of the pendulum. Dr Helmert's‡ result from the pendulum is $\frac{1}{299\cdot 26 \pm 1\cdot 26}$, and is certainly slightly smaller than the value found here.

In the paper above referred to M. Callandreau has used the Precessional Constant for evaluating the ellipticity of the Earth, which he finds to be $\frac{1}{297\cdot 4}$. The agreement of my work with his is thus very satisfactory.

* As it is desirable that the ellipticity of the earth should be evaluated with all the accuracy possible, it may be well to advert to the augmentation of ellipticity which is due to the direct actions of the Moon and the Sun.

The tide-generating potential of each of these bodies contains a term which is independent of the time, and to which there must correspond a permanent tide.

If i be the obliquity of the ecliptic; and m, c, e_0 the mass, mean distance and eccentricity of orbit for the Moon; while m', c', e_0' represent the same for the Sun; the tide-generating potential contains the term,

$$\left[\frac{m}{c^3}(1+\tfrac{3}{2}e_0^2) + \frac{m'}{c'^3}(1+\tfrac{3}{2}e_0'^2)\right]\tfrac{3}{4}(1-\tfrac{3}{2}\sin^2 i)\, r^2 (\tfrac{1}{3}-\cos^2\theta)$$

This is equal to

$$(1\cdot 46035)\,\tfrac{3}{4}\frac{m}{c^3}(1-\tfrac{3}{2}\sin^2 i + \tfrac{3}{2}e_0^2)\, r^2(\tfrac{1}{3}-\cos^2\theta)$$

The ellipticity of the Earth which corresponds with this term is

$$(1\cdot 46035)\,\frac{\frac{3m}{4M}\left(\frac{a}{c}\right)^3(1-\tfrac{3}{2}\sin^2 i+\tfrac{3}{2}e_0^2)}{1-\tfrac{3}{5}w/w_0} = \frac{\cdot 00000004708}{1-\tfrac{3}{5}w/w_0}$$

If we take $w_0/w = 2\cdot 1$, the ellipticity is ·00000006588. The meaning of this is that the Earth's surface is 28 cm. lower at the poles, and 14 cm. higher at the equator, than would be the case if the Sun and the Moon were obliterated. This term may therefore be safely omitted.

† Helmert, *Höhere Geodäsie*, Vol. II., p. 437.

‡ *Ib.*, Vol. II., p. 241.

SUMMARY AND DISCUSSION.

The space in the neighbourhood of an oblate ellipsoid of revolution may be divided into three regions by two spheres touching it internally and externally. It is clearly possible to express the potential of such a solid homogeneous ellipsoid by series of spherical harmonics which are convergent both inside the smaller sphere and outside the larger one, but for the space between them the convergency is uncertain. In the treatment of the attractions of spheroids by means of spherical harmonics, it is usual to assume the ellipticity to be small, so that the region of possible divergency becomes negligible.

But this method is no longer certainly justifiable when we seek to carry the development as far as the squares of small quantities, and proof is needed of the applicability of the series within the middle region of space. If, having found our two series expressive of the potential for internal and for external space respectively, we determine the form of the surface inside the middle region at which these two potentials are continuous as far as the second order of small quantities, we find that the surface in question is that of the ellipsoid itself. The two series then form a continuous function at the surface of the ellipsoid, and they obviously satisfy the differential equation of the potential both inside and outside the ellipsoid. It follows that although the series were determined by a process which is open to doubt as respects the middle region, they lead to results which are trustworthy as far as the second order of small quantities.

There is, however, another method of finding the potential of the ellipsoid, by which the difficulty as to convergency is avoided. It is well known that the potential of a solid ellipsoid is expressible in a series of spherical harmonics, and that the series is convergent up to the surface, at least for all ellipsoids with eccentricity less than $\frac{1}{2}$. Also there is a rigorous expression in finite terms for the internal potential, involving only the second harmonic. If only the second power of the ellipticity be retained, these two expressions are found to be identical with those derived from integration, and the question of convergency is settled.

If an oblate ellipsoid of revolution be slightly distorted in any way, the deformation being of the order of the square of the ellipticity, the additional terms in the potential may obviously be expressed by the ordinary formulæ of spherical harmonic analysis.

In the theory of the Earth's figure it is unnecessary to contemplate the existence of any other departure from true ellipticity than one expressible by a zonal harmonic of the fourth order. Three parameters are needed to express the surface of such a spheroid. If a and b denote the equatorial and polar radii, the first parameter is a the equatorial radius; the second is the ellipticity e defined as being equal to $(a-b)/a$; and it seems convenient to take as the third a quantity proportional to the elevation of the surface of the spheroid above that of the true ellipsoid in latitude 45°. If the third parameter be denoted by f, the elevation in question is $\frac{1}{4}af$.

I have found, however, that there is a gain in simplicity by displacing the ellipticity e from its position as the second parameter, and by substituting for it a parameter h, which is defined by

$$h = e - \tfrac{25}{14}e^2 - \tfrac{1}{7}f$$

To the first order of small quantities it is clearly immaterial whether h or e be used, but the advantage of the change is found to arise when quantities of the second order are retained in the developments.

When the external and internal potentials of a homogeneous spheroid defined by a, h, f are known, it is easy to express by means of integrals the potential of a heterogeneous body built up by a succession of layers, each of which has its external and internal surfaces defined as a spheroid of the kind under consideration. When the object in view is the study of the figure of a planet, the potential corresponding to a uniform angular velocity must of course be added. The processes explained here are carried out in §§ 1 and 2.

It is generally assumed that the matter of which the Earth is formed is sufficiently plastic to permit the condition of hydrostatic equilibrium to be satisfied throughout the mass. Even if this condition is not rigorously correct, it must be nearly so.

In § 2 the condition of hydrostatic equilibrium is determined. It is expressed by two equations, of which the first corresponds to that ordinarily given for the ellipticity of internal strata, but it contains also terms of the second order. The second equation consists entirely of terms of the second order, and involves the parameter f. The advantage due to the substitution of h for e is now apparent, for the first equation is entirely independent of f. These two equations involve integrals which are eliminated by integration, and we obtain two differential equations of the second order for h and for f; the equation for h does not involve f (§§ 3, 4).

M. Radau has shown that it is possible to reduce the differential equation for the ellipticity to one of the first order, at least to a close degree of approximation. In § 5 his process is carried out with the retention of quantities of the second order. It appears that the approximation is even better than when only the terms of the first order are retained.

The formula for gravity is obtained in § 6. It was given for the first time by Airy, as already remarked in the Introduction. The results of pendulum experiments do not appear as yet to be sufficiently numerous or consistent, *inter se*, to render a revision of the value of the ellipticity of the Earth's surface practicable by aid of this more accurate formula for gravity.

The rates of change of h and of f at the earth's surface are determined in § 7. As a preliminary to the evaluation of the departure of the Earth's figure from true ellipticity, I then consider the figure of a homogeneous mass of fluid with a small heavy nucleus at its centre (§ 8). The departure is found to be greatest when the nucleus contributes ·457, and the fluid ·543 of the whole mass. For such a planet, with the same mean density, size, and length of day as the Earth, it is found that the surface is 8·8 metres below the true ellipsoid in latitude 45°. In the actual Earth the departure will certainly be less than this maximum.

The evaluation of f for a heterogeneous planet is only possible when the law of internal density is known. I have, then, adopted Roche's hypothesis, according to which the mean density of all the matter lying inside any surface of equal density is less than the central density by an amount which varies as the square of the equatorial radius of that surface. In the notation of the paper the law is expressed by $w_0 = \rho [1 - k (a/a)^2]$. It is found in § 9, by some rather tedious analysis and computation, that the surface is depressed in latitude 45° by $3\frac{1}{4}$ metres.

Dr Wiechert has maintained that the Earth probably consists of an iron nucleus with a rocky superstratum. His theory leads to the conclusion that the depression in latitude 45° amounts to 2·75 metres (§ 10). The close agreement between the results of such diverse hypotheses as those of Roche and of Wiechert appears to justify us in maintaining with confidence that the level surface is depressed below the true ellipsoid by 3 metres in latitude 45°.

A solution of the differential equation for h in § 11 does not lead to results of much general interest, and I refer the reader to that section for details.

It has been stated in the Introduction that M. Callandreau has treated these problems by methods which differ somewhat from mine. He has concluded, but without definitely solving the differential equation, that the depression in latitude 45° must be less than 5 metres.

In § 12 formulæ are found for the moment of inertia of the Earth about its axis of rotation, and thence for the Precessional Constant. Similar formulæ have been found by M. Callandreau in the papers referred to above, but I think that my formulæ are somewhat more succinct and tractable than his.

In the various theories of the figure of the Earth which have been propounded up to recent times, the value of the Precessional Constant has always been appealed to as the test whereby the correctness of the hypothesis as to internal density may be tried. But it appears from M. Radau's remarkable investigation that the Earth's moment of inertia about the axis of rotation is in reality nearly independent of the law of internal density; and the difference between the greatest and least moments of inertia depends rigorously on superficial data. Accordingly, the value of the Precessional Constant may be inferred with a considerable degree of accuracy from the form of the surface, and it affords evidence of little weight as to the law of internal density. But although from this point of view the comparison of this constant with theory becomes almost nugatory, yet the very result which shows its uselessness in one respect points out its utility in another, for we may now appeal to it as affording the means for an independent evaluation of the ellipticity of the Earth's surface.

In § 12 an estimate is made of the amount by which the moment of inertia of the earth is affected by the law of internal density. A formula is found for the moment of inertia, which consists of the sum of two parts, the first being dependent only on superficial data, and the second on the law of internal density. From the numerical table of results given in that section it appears that if the moment of inertia C be computed as though the second part were non-existent, it must be multiplied by the factor 1·0003 in order to take into account the neglected portion. Since the moment of inertia C occurs in the denominator of the Precessional Constant, it is obvious that the uncorrected value should be multiplied by ·9997. Proceeding from these results and from the value of the constant, it appears that the ellipticity of the Earth's surface must be about $\frac{1}{297}$. M. Callandreau has arrived also at the same conclusion, and it is confirmed by Dr Wiechert, although his suggested law of internal density differs very widely from that adopted by other investigators. This estimate of the ellipticity agrees well with that derived from all the other methods, except that of the pendulum, from which it is concluded that the ellipticity is about $\frac{1}{299}$. It may be hoped that the various results may be brought into closer agreement with one another when the great mass of pendulum results now accumulated has been reduced.

It has been contended by Tisserand that the ellipticity of the Earth's surface is greater than any value which it is possible to reconcile with the existence of internal hydrostatic equilibrium; and with the values of the constants used by him, this is certainly so. But it seems to me that when the terms of the second order are included, and when more recent data are employed, there is but little evidence in favour of this conclusion. If Tisserand were correct, it would indicate that the internal layers of the earth are more elliptic than is consistent with the present angular velocity of rotation. On the other hand, Dr Wiechert seeks to show that his iron nucleus is deficient in ellipticity. His argument does not, however, carry conviction to my mind, as the data seem to be too uncertain for any such conclusion. It is, I think, preferable to maintain that nothing can, as yet, be decided on this point.

8.

ON JACOBI'S FIGURE OF EQUILIBRIUM FOR A ROTATING MASS OF FLUID.

[*Proceedings of the Royal Society*, XLI. (1887), pp. 319—336.]

I AM not aware that any numerical values have ever been determined for the axes of the ellipsoids, which are figures of equilibrium of a rotating mass of fluid*.

In the following paper the problem is treated from the point of view necessary for reducing the formulæ to a condition for computation, and a table of numerical results is added.

Let a, b, c be the semi-axes of a homogeneous ellipsoid of unit density; let the origin be at the centre and the axes of x, y, z be in the directions a, b, c.

* The following list of papers bearing on this subject is principally taken from a report to the British Association, 1882, by W. M. Hicks:—

Jacobi, *Acad. des Sciences*, 1834; Liouville, *Journ. École Polytech.*, Vol. XIV., p. 289; Ivory, *Phil. Trans.*, 1838, Part I., p. 57; Pontécoulant, *Syst. du Monde*, Vol. II. The preceding are proofs of the theorem, and in more detail we have:—C. O. Meyer, *Crelle*, Vol. XXIV., p. 44; Liouville, *Liouville's Journ.*, Vol. XVI., p. 241; a remarkable paper by Dirichlet and Dedekind, *Borchardt's Journ.*, Vol. LVIII., pp. 181 and 217; Riemann, *Abh. K. Ges. Wiss. Göttingen*, Vol. IX., 1860, p. 3; Brioschi, *Borchardt's Journ.*, Vol. LIX., p. 63; Padova, *Ann. della Sc. Norm. Pisa*, 1868–9 (being Dirichlet and Riemann's work with additions); Greenhill, *Proc. Camb. Phil. Soc.*, Vol. III., p. 233 and Vol. IV., p. 4; Lipschitz, *Borchardt's Journ.*, Vol. LXXVIII., p. 245; Hagen, *Schlömilch's Zeitsch. Math.*, Vol. XXIV., p. 104; Betti, *Ann. di Matem.*, Vol. X., p. 173 (1881); Thomson and Tait's *Natural Philosophy* (1883), Part II., § 778; a very important paper by Poincaré, *Acta Mathem.*, 7, 3 and 4 (1885).

[I have been criticised in respect to this paper by S. Krüger (*Nieuw Archief voor Wiskunde*, 2nd Series, 3rd Part, van Leeuwen, 89 Hoogewoerd, 1896), on "Ellipsoidale Evenwichtsvormen," in a Thesis for the doctorate of Leiden, because I wrote it in ignorance of certain previous work, especially of a paper by Plana (*Astr. Nachr.*, 36, n. 851, c. 169). But as it appears that Plana gave a number of numerical results which were wholly wrong, a knowledge of that paper would have caused me much further trouble.

See also R. Kaibara, "On the Jacobian Ellipsoid," *Tokyo Sugaku-Buturigakkwai Kizi*, 2nd Series, Vol. IV., No. 5 (1907).]

Then if we put

$$A^2 = a^2 + u, \quad B^2 = b^2 + u, \quad C^2 = c^2 + u$$

and

$$\Psi = \int_0^\infty \frac{du}{ABC} \quad \ldots\ldots\ldots\ldots\ldots\ldots\ldots\ldots\ldots\ldots(1)$$

it is known* that the potential of the ellipsoid at an internal point x, y, z is given by

$$V = \pi abc \left[\Psi + \frac{x^2}{a}\frac{\partial\Psi}{\partial a} + \frac{y^2}{b}\frac{\partial\Psi}{\partial b} + \frac{z^2}{c}\frac{\partial\Psi}{\partial c} \right] \quad \ldots\ldots\ldots\ldots\ldots(2)$$

Now let us introduce a new notation, and let

$$c = a\cos\gamma, \quad \kappa = \sqrt{\frac{a^2 - b^2}{a^2 - c^2}}, \quad \text{and} \quad \kappa'^2 = 1 - \kappa^2$$

Let an auxiliary angle β be defined by

$$\sin\beta = \kappa\sin\gamma \quad \ldots\ldots\ldots(3)$$

Then $\quad b = a\sqrt{(1 - \kappa^2\sin^2\gamma)} = a\cos\beta$

Also let $\quad A^2 = u + a^2 = \dfrac{a^2 - c^2}{\sin^2\theta} = a^2\dfrac{\sin^2\gamma}{\sin^2\theta}$

whence $\quad B^2 = u + b^2 = \dfrac{a^2\sin^2\gamma}{\sin^2\theta}(1 - \kappa^2\sin^2\theta)$

$$C^2 = u + c^2 = \frac{a^2\sin^2\gamma}{\sin^2\theta}\cos^2\theta \quad \ldots\ldots\ldots(4)$$

and $\quad du = -\dfrac{2a^2\sin^2\gamma}{\sin^3\theta}\cos\theta\, d\theta$

Thus $\quad \displaystyle\int_0^\infty du = 2a^2\sin^2\gamma \int_0^\gamma \frac{\cos\theta}{\sin^3\theta}\, d\theta = 2a^2\sin^2\gamma \int_0^\gamma \frac{\cos\gamma}{\sin^3\gamma}\, d\gamma$

Lastly, let $\quad \Delta = \sqrt{(1 - \kappa^2\sin^2\gamma)}$

and in accordance with the usual notation for elliptic integrals let

$$F = \int_0^\gamma \frac{d\gamma}{\Delta}, \quad E = \int_0^\gamma \Delta\, d\gamma \quad \ldots\ldots\ldots\ldots\ldots\ldots\ldots\ldots\ldots(5)$$

Then we have the following transformations:—

$$\begin{aligned}
\Psi &= \int_0^\infty \frac{du}{ABC} = \frac{2}{a\sin\gamma}F \\
-\frac{\partial\Psi}{a\,\partial a} &= \int_0^\infty \frac{du}{A^3BC} = \frac{2}{a^3\sin^3\gamma}\int_0^\gamma \frac{\sin^2\gamma}{\Delta}\, d\gamma = \frac{2}{a^3\kappa^2\sin^3\gamma}(F - E) \\
-\frac{\partial\Psi}{b\,\partial b} &= \int_0^\infty \frac{du}{AB^3C} = \frac{2}{a^3\sin^3\gamma}\int_0^\gamma \frac{\sin^2\gamma}{\Delta^3}\, d\gamma \\
-\frac{\partial\Psi}{c\,\partial c} &= \int_0^\infty \frac{du}{ABC^3} = \frac{2}{a^3\sin^3\gamma}\int_0^\gamma \frac{\tan^2\gamma}{\Delta}\, d\gamma
\end{aligned} \quad \ldots(6)$$

* Thomson and Tait's *Natural Philosophy* (1883), § 494, *l*. The form in which the formula is here given is slightly different from that in (8), (11), (15) of §§ 494, *k*, *l*.

It remains to reduce the last two of (6) to elliptic integrals.

The following are known transformations in the theory of elliptic integrals, viz.:—

$$\int_0^\gamma \frac{d\gamma}{\Delta^3} = \frac{1}{\kappa'^2}\mathrm{E} - \frac{\kappa^2 \sin\gamma\cos\gamma}{\kappa'^2\Delta} \quad\text{.......................(7)}$$

$$\int_0^\gamma \frac{\tan^2\gamma}{\Delta} d\gamma = \frac{\Delta\tan\gamma}{\kappa'^2} - \frac{1}{\kappa'^2}\mathrm{E} \quad\text{.......................(8)}$$

Hence $$\int_0^\gamma \frac{\sin^2\gamma}{\Delta^3} d\gamma = \frac{1}{\kappa^2}\int_0^\gamma \frac{1-\Delta^2}{\Delta^3} d\gamma = \frac{1}{\kappa^2\kappa'^2}\mathrm{E} - \frac{\sin\gamma\cos\gamma}{\kappa'^2\Delta} - \frac{1}{\kappa^2}\mathrm{F} \quad\text{........(9)}$$

In the present case $\Delta = \cos\beta$.

Thus (8) and (9) enable us to complete the required transformation of (6) to elliptic integrals.

Substituting from (6) (8) (9) in the expression

$$\mathrm{V} = \tfrac{3}{4}m\left\{\Psi + \frac{x^2}{a}\frac{\partial\Psi}{\partial a} + \frac{y^2}{b}\frac{\partial\Psi}{\partial b} + \frac{z^2}{c}\frac{\partial\Psi}{\partial c}\right\}$$

where $$m = \tfrac{4}{3}\pi abc = \tfrac{4}{3}\pi a^3\cos\beta\cos\gamma$$

we have

$$\mathrm{V} \div \tfrac{3}{4}m = \frac{2}{a\sin\gamma}\mathrm{F} + \frac{2}{a^3\sin^3\gamma}\left[\frac{x^2}{\kappa^2}(\mathrm{E}-\mathrm{F}) + y^2\left(\frac{\sin\gamma\cos\gamma}{\kappa'^2\cos\beta} + \frac{\mathrm{F}}{\kappa^2} - \frac{\mathrm{E}}{\kappa^2\kappa'^2}\right)\right.$$
$$\left. + \frac{z^2}{\kappa'^2}(\mathrm{E} - \tan\gamma\cos\beta)\right] \quad\text{......(10)}$$

Now suppose the ellipsoid to be rotating about the axis of z with an angular velocity ω, and let us choose the axes a, $a\cos\beta$, $a\cos\gamma$, and the angular velocity ω, so that the surface may be a surface of equilibrium.

For this purpose $\mathrm{V} + \tfrac{1}{2}\omega^2(x^2+y^2) =$ constant, must be identical with

$$\frac{x^2}{a^2} + \frac{y^2}{a^2\cos^2\beta} + \frac{z^2}{a^2\cos^2\gamma} = 1$$

In (10) we have V in the form

$$\mathrm{V} = Lx^2 + My^2 + Nz^2 + P \quad\text{.......................(11)}$$

whence $$a^2(L + \tfrac{1}{2}\omega^2) = a^2(M + \tfrac{1}{2}\omega^2)\cos^2\beta = a^2N\cos^2\gamma$$

Hence
$$\left.\begin{aligned} L - M + N\cos^2\gamma\tan^2\beta &= 0 \\ \tfrac{1}{2}\omega^2 &= N\cos^2\gamma - L \\ \text{or}\quad \tfrac{1}{2}\omega^2\sin^2\beta &= M\cos^2\beta - L \end{aligned}\right\} \quad\text{.....................(12)}$$

There are two kinds of solutions of these equations (12).

First, since

$$\left.\begin{aligned} L &= \pi bc\frac{\partial\Psi}{\partial a} = -2\pi\frac{\cos\beta\cos\gamma}{\sin^3\gamma}\int_0^\gamma \frac{\sin^2\gamma}{\Delta} d\gamma \\ M &= \pi ac\frac{\partial\Psi}{\partial b} = -2\pi\frac{\cos\beta\cos\gamma}{\sin^3\gamma}\int_0^\gamma \frac{\sin^2\gamma}{\Delta^3} d\gamma \end{aligned}\right\} \quad\text{...........(13)}$$

it is obvious that $L - M$ vanishes when $\kappa = 0$.

And since when κ vanishes, β also vanishes, the equation

$$L - M + N \cos^2 \gamma \tan^2 \beta = 0$$

is satisfied by

$$\kappa = 0, \quad \beta = 0$$

That is to say there is a solution of the problem which makes $a = b$.

Thus there is a solution which gives us an ellipsoid of revolution.

When $\kappa = 0$, we have also $\beta = 0$, $\Delta = 1$, and

$$L = -\frac{2\pi \cos \gamma}{\sin^3 \gamma} \int_0^\gamma \sin^2 \gamma d\gamma = \frac{\pi \cos \gamma}{\sin^3 \gamma} (\sin \gamma \cos \gamma - \gamma)$$

$$N = -\frac{2\pi \cos \gamma}{\sin^3 \gamma} \int_0^\gamma \tan^2 \gamma d\gamma = \frac{\pi \cos \gamma}{\sin^3 \gamma} (2\gamma - 2 \tan \gamma)$$

Therefore

$$\tfrac{1}{2} \omega^2 = N \cos^2 \gamma - L$$

$$= \frac{\pi}{\tan^3 \gamma} [2\gamma - 2 \tan \gamma - (1 + \tan^2 \gamma)(\sin \gamma \cos \gamma - \gamma)]$$

$$= \frac{\pi}{\tan^3 \gamma} [\gamma (3 + \tan^2 \gamma) - 3 \tan \gamma] \quad \ldots\ldots\ldots\ldots(14)^*$$

and the eccentricity of the ellipsoid of revolution is $\sin \gamma$.

To find the other solution when κ is not zero, we have by comparison between (10) and (11),

$$\left.\begin{aligned} L \frac{\kappa^2 \sin^3 \gamma}{2\pi \cos \beta \cos \gamma} &= \mathrm{E} - \mathrm{F} \\ M \frac{\kappa^2 \sin^3 \gamma}{2\pi \cos \beta \cos \gamma} &= \frac{\kappa^2 \sin \gamma \cos \gamma}{\kappa'^2 \cos \beta} + \mathrm{F} - \frac{\mathrm{E}}{\kappa'^2} \\ N \frac{\kappa^2 \sin^3 \gamma}{2\pi \cos \beta \cos \gamma} &= \frac{\kappa^2}{\kappa'^2} (\mathrm{E} - \tan \gamma \cos \beta) \end{aligned}\right\} \ldots\ldots\ldots(15)$$

Hence the first of (12) gives

$$-(2\mathrm{F} - \mathrm{E}) + \frac{\mathrm{E}}{\kappa'^2} + \frac{\kappa^2}{\kappa'^2} \tan^2 \beta \cos^2 \gamma (\mathrm{E} - \tan \gamma \cos \beta) - \frac{\kappa^2 \sin \gamma \cos \gamma}{\kappa'^2 \cos \beta} = 0$$

or

$$\frac{\mathrm{E}}{\kappa'^2} [1 + (\kappa \tan \beta \cos \gamma)^2] - (2\mathrm{F} - \mathrm{E}) - \frac{\kappa}{\kappa'^2} \tan \beta \cos \gamma (1 + \sin^2 \beta) = 0 \ldots(16)$$

In order to adapt this for computation, we may introduce the auxiliary angles defined by

$$\tan \zeta = \kappa \tan \beta \cos \gamma, \quad \tan \delta = \sin \beta \ldots\ldots\ldots(17)$$

and the equation becomes

$$\frac{\mathrm{E}}{\kappa'^2} \sec^2 \zeta - (2\mathrm{F} - \mathrm{E}) - \frac{\sec^2 \delta}{\kappa'^2} \tan \zeta = 0 \ldots\ldots\ldots(18)$$

* Compare with Thomson and Tait's *Natural Philosophy*, § 771 (3); or any other work which gives a solution of the problem.

The second of (12) gives

$$\frac{\omega^2}{4\pi}\frac{\kappa^2\sin^3\gamma}{\cos\beta\cos\gamma}=\frac{\kappa^2}{\kappa'^2}\cos^2\gamma\,(\mathrm{E}-\tan\gamma\cos\beta)-(\mathrm{E}-\mathrm{F})$$

whence

$$\frac{\omega^2}{4\pi\cos\beta\cos\gamma}=\frac{\mathrm{F}-\mathrm{E}}{\kappa^2\sin^3\gamma}+\frac{\mathrm{E}\cos^2\gamma}{\kappa'^2\sin^3\gamma}-\frac{\cos\beta\cos\gamma}{\kappa'^2\sin^2\gamma}$$

$$\frac{\omega^2}{4\pi}=\cot\beta\operatorname{cosec}\beta\cot\gamma\,(\mathrm{F}-\mathrm{E})+\cot^3\gamma\cos\beta\frac{\mathrm{E}}{\kappa'^2}-\frac{\cos^2\beta\cot^2\gamma}{\kappa'^2}\quad\text{...(19)}$$

Some of the subsequent computations were, however, actually made from a formula deduced from the third of (12), which leads to

$$\frac{\omega^2}{4\pi}=\cot\beta\cot\gamma\operatorname{cosec}^3\beta\,(1+\cos^2\beta)(\mathrm{F}-\mathrm{E})-\frac{\kappa^2}{\kappa'^2}\cot^3\beta\cot\gamma\operatorname{cosec}\beta\mathrm{E}$$
$$+\frac{\cot^2\beta\cot^2\gamma}{\kappa'^2}\quad\text{...(20)}$$

By subtracting (20) from (19) we can deduce (16); hence it follows that (19) and (20) lead to identical results. Most of the subsequent results were computed from both (19) and (20), thus verifying the solution of (18).

[Since the date at which this paper was written I have, however, obtained a much better form for the expression for ω^2, as follows:—

If we retain the partial differentials of Ψ in the expression for V, it is clear that

$$\mathrm{V}+\tfrac{1}{2}\omega^2(x^2+y^2)=\text{constant}$$

is another form of the equation

$$\frac{x^2}{a^2}+\frac{y^2}{b^2}+\frac{z^2}{c^2}=1$$

Hence $$a^2\left(\frac{\partial\Psi}{a\partial a}+\frac{\omega^2}{2\pi abc}\right)=b^2\left(\frac{\partial\Psi}{b\partial b}+\frac{\omega^2}{2\pi abc}\right)=c^2\frac{\partial\Psi}{c\partial c}$$

These two equations may be written in the forms

$$\left.\begin{aligned}\frac{\omega^2}{2\pi}&=\frac{bc}{a}\left(c\frac{\partial\Psi}{\partial c}-a\frac{\partial\Psi}{\partial a}\right)\\ \frac{c^2(a^2-b^2)}{a^2b^2}\frac{\partial\Psi}{c\partial c}&=\frac{\partial\Psi}{b\partial b}-\frac{\partial\Psi}{a\partial a}\end{aligned}\right\}\quad\text{..................(21)}$$

If the differentials of Ψ be expressed as integrals by means of (6), we have

$$\left.\begin{aligned}\frac{\omega^2}{4\pi}&=\frac{\cos\beta\cos\gamma}{\sin^3\gamma}\left[\int_0^\gamma\frac{\sin^2\gamma}{\Delta}d\gamma-\cos^2\gamma\int_0^\gamma\frac{\tan^2\gamma}{\Delta}d\gamma\right]\\ \sin^2\gamma\cos^2\gamma\int_0^\gamma\frac{\tan^2\gamma}{\Delta}d\gamma&=\cos^2\beta\int_0^\gamma\frac{\sin^4\gamma}{\Delta^3}d\gamma\end{aligned}\right\}\quad\text{.....(22)}$$

But for the present I retain the differentials of Ψ as in (21).

Since V is the potential of an ellipsoid of unit density $\nabla^2 V = -4\pi$.

Hence
$$\frac{\partial \Psi}{a\partial a} + \frac{\partial \Psi}{b\partial b} + \frac{\partial \Psi}{c\partial c} = -\frac{2}{abc} \quad \text{.......................(23)}$$

Also since Ψ is a homogeneous function of degree -1 in a, b, c

$$a\frac{\partial \Psi}{\partial a} + b\frac{\partial \Psi}{\partial b} + c\frac{\partial \Psi}{\partial c} = -\Psi \quad \text{.....................(24)}$$

We may now eliminate the differentials of Ψ from the four equations (21), (23), (24) by means of a determinant or otherwise.

The result of the elimination is

$$\frac{\omega^2}{2\pi} = \frac{abc\,\Psi\left(\frac{1}{a^2} + \frac{1}{b^2} + \frac{1}{c^2}\right) - 6}{(a^2 + b^2)\left(\frac{1}{a^2} + \frac{1}{b^2} + \frac{1}{c^2}\right) - 6} \quad \text{....................(27)}$$

Introducing the notation involving the angles γ and β, and expressing Ψ in terms of the elliptic integral F, we have

$$\frac{\omega^2}{4\pi} = \frac{\frac{\cos\beta\cos\gamma}{\sin\gamma}F - \frac{3}{1+\sec^2\beta+\sec^2\gamma}}{1+\cos^2\beta - \frac{6}{1+\sec^2\beta+\sec^2\gamma}} \quad \text{.. ....(28)}$$

This formula is much better than those given in (19) and (20) from which I computed the results given hereafter.

The formulæ (18) and (19) are suitable for finding the solution, except when κ is small or $\sin^{-1}\kappa$ is nearly 90°, when the elliptic integrals are awkward to use. I shall therefore find approximate forms for these cases*.

The formula (28) will always enable us to compute ω when κ has been determined, and therefore it is only the determination of κ which need be considered. In both the cases, namely where κ is small and where it is nearly equal to unity, I proceed from the second of equations (22).

First when κ is small:—

For the sake of brevity I write

$$p = \cos\gamma, \quad q = \sin\gamma$$

It is obvious that $1/\Delta$ and $1/\Delta^3$ may be expanded in powers of κ, so that

$$\frac{1}{\Delta} = 1 + \frac{1}{2}\kappa^2 q^2 + \frac{1\,.\,3}{2\,.\,4}\kappa^4 q^4 + \ldots$$

$$\frac{1}{\Delta^3} = 1 + \frac{3}{2}\kappa^2 q^2 + \frac{3\,.\,5}{2\,.\,4}\kappa^4 q^4 + \ldots$$

* [This portion of the paper has been rewritten and improved.]

When these expressions are introduced into the second of (22) we have two kinds of definite integrals which may be evaluated as follows:—

$$\int_0^\gamma q^{2n}\, d\gamma = -\frac{1}{2n} q^{2n-1} p + \frac{2n-1}{2n}\int_0^\gamma q^{2n-2}\, d\gamma$$

$$\int_0^\gamma \frac{q^{2n}}{p}\, d\gamma = \frac{q^{2n-1}}{p} - (2n-1)\int_0^\gamma q^{2n-2}\, d\gamma$$

Now noting that $\cos^2\beta = 1 - \kappa^2 q^2$, we have the means of obtaining approximate values of both sides of (22). I find in this way that the equation for determining κ is as follows:—

$$10q^2 + 3 - \frac{\gamma}{pq}(3 + 8p^2q^2) + \kappa^2\left[-2q^4 + \tfrac{11}{2}q^2 + \tfrac{15}{4} - \frac{\gamma}{pq}\left(\tfrac{15}{4} + 3q^2 - 6q^4\right)\right]$$

$$+ \kappa^4\left[-\tfrac{7}{8}q^6 - \tfrac{35}{16}q^4 + \tfrac{295}{64}q^2 + \tfrac{525}{128} - \frac{\gamma}{pq}\left(\tfrac{525}{128} + \tfrac{15}{8}q^2 - \tfrac{45}{8}q^4\right)\right] + \ldots = 0 \quad \ldots(29)$$

If κ be zero we have

$$\gamma = \frac{pq(3 + 10q^2)}{3 + 8p^2q^2} = \frac{\sin\gamma\cos\gamma\,(3 + 10\sin^2\gamma)}{3 + 8\sin^2\gamma\cos^2\gamma}$$

This may also be written in the form

$$\gamma = \frac{\sin 2\gamma - \frac{5}{16}\cos 4\gamma}{1 - \frac{1}{4}\cos 4\gamma}$$

when it becomes more convenient for solution by trial and error.

The solution is $\gamma = 54^\circ\ 21'\ 27''$

If we write $\tan\gamma = f$, as in Thomson and Tait's *Natural Philosophy*, § 778′, this equation becomes

$$\frac{\tan^{-1} f}{f} = \frac{1 + \frac{13}{3}f^2}{1 + \frac{14}{3}f^2 + f^4}$$

which is the equation (9), § 778′ of that work.

The ellipsoid of revolution of which the eccentricity is $\sin 54^\circ\ 21'\ 27''$ belongs to the revolutional series of figures of equilibrium, and is the starting point of the Jacobian series of figures. As shown by Lord Kelvin, it is the flattest revolutional figure which is dynamically stable. The Jacobian figures of equilibrium are initially stable, and as stated by M. Poincaré*, there is for this value of γ a crossing point of the two series, and an exchange of stabilities.

When κ is small the result must be computed from (29), but as κ increases with great rapidity as γ increases, it would be necessary to take in higher powers of κ than the fourth to obtain accurate results when γ exceeds $54^\circ\ 21'\ 27''$ by more than a degree.

* *Acta Mathem.*, 7 (1886).

When γ only exceeds 54° 21′ 27″ (or say δ) by a few minutes of arc, it is easy to prove that

$$\gamma - \delta = \tfrac{1}{4}\kappa^2 \sin\delta\cos\delta$$

Whence $$\kappa^2 = 10^{\cdot 9266821}\sin(\gamma - 54° 21' 27'')$$

These approximate formulæ have been used to confirm the results to be given hereafter for the early values in the table, where the use of the elliptic integrals became troublesome.

We now turn to the second case where the elliptic integrals again become troublesome to use, namely where κ is nearly unity and $\sin^{-1}\kappa$ is nearly 90°. It happens that in these cases $\kappa'^2\tan^2\gamma$ is a moderately small fraction and since

$$\Delta^2 = 1 - \kappa^2\sin^2\gamma = \cos^2\gamma\,(1 + \kappa'^2\tan^2\gamma)$$

it is possible to develope $1/\Delta$ and $1/\Delta^3$ in powers of $\kappa'\tan\gamma$.

In § 11 of the last paper in the present volume it became necessary to develope the integrals with which we are now concerned in this way.

In that paper I write

$$\mathfrak{A}_1{}^1 = \frac{\kappa\cos^2\beta}{\sin^2\gamma}\int_0^\gamma \frac{\sin^2\gamma}{\Delta^3}d\gamma,\quad \mathfrak{A}_1 = \frac{\kappa}{\sin^2\gamma}\int_0^\gamma \frac{\sin^2\gamma}{\Delta}d\gamma,\quad \mathsf{A}_1{}^1 = \frac{\kappa\cos^2\gamma}{\sin^2\gamma}\int_0^\gamma \frac{\tan^2\gamma}{\Delta}d\gamma$$

The equations (22) for determining Jacobi's ellipsoid and the angular velocity may be written in this new notation, as follows:—

$$\mathfrak{A}_1{}^1 - \mathsf{A}_1{}^1 - (\cos^2\gamma + \kappa'^2\sin^2\gamma)(\mathfrak{A}_1 - \mathsf{A}_1{}^1) = 0$$

$$\frac{\omega^2}{4\pi} = \frac{\cos\beta\cot\gamma}{\kappa}(\mathfrak{A}_1 - \mathsf{A}_1{}^1)$$

Now when κ is nearly unity and $\kappa'\tan\gamma$ small we have

$$\mathfrak{A}_1{}^1 - \mathsf{A}_1{}^1 = \frac{\kappa\kappa'^2}{\sin\gamma}(\sigma_0 - \sigma_1\kappa'^2 + \sigma_2\kappa'^4 - \ldots)$$

$$\mathfrak{A}_1 - \mathsf{A}_1{}^1 = \frac{\kappa}{\sin\gamma}(\tau_0 - \tau_1\kappa'^2 + \tau_2\kappa'^4 - \ldots)$$

where the σ's and τ's are certain functions of γ defined in (38) of the paper referred to.

Hence we find

$$\left.\begin{aligned} \kappa'^2(\sigma_0 + \tau_1\cos^2\gamma - \tau_0\sin^2\gamma) &= \tau_0\cos^2\gamma + \kappa'^4(\sigma_1 + \tau_2\cos^2\gamma - \tau_1\sin^2\gamma) \\ &\quad - \kappa'^6(\sigma_2 + \tau_3\cos^2\gamma - \tau_2\sin^2\gamma)\ldots \\ \frac{\omega^2}{4\pi} &= \frac{\cos\beta\cos\gamma}{\sin^2\gamma}[\tau_0 - \tau_1\kappa'^2 + \tau_2\kappa'^4 - \ldots] \end{aligned}\right\}\ldots(30)$$

From the first of these it is easy to find κ'^2 for a given value of γ by successive approximation, and then either from the second we find $\omega^2/4\pi$, or else we may use equation (28) of the present paper.

The entries corresponding to $\gamma = 80°$ and $85°$ in the table given below have been recomputed in this way for the present volume.

We shall return later to equation (30) which will afford the value of ω^2 for very long ellipsoids.]

Besides the angular velocity and the axes of the ellipsoid, the other important functions are the moment of momentum, the kinetic energy of rotation, and the intrinsic energy of the mass. In order to express these numerically we must adopt a unit of length, and it will be convenient to take a, where

$$\mathrm{a}^3 = abc = a^3 \cos\beta \cos\gamma$$

Thus

$$a = \mathrm{a}\,(\sec\beta \sec\gamma)^{\frac{1}{3}}$$

Let σ be the density of the fluid which has hitherto been treated as unity, and let $(\frac{4}{3}\pi\sigma)^{\frac{3}{2}}\mathrm{a}^5\mu$, $(\frac{4}{3}\pi\sigma)^2\mathrm{a}^5\epsilon$ be the moment of momentum and kinetic energy, then

$$(\tfrac{4}{3}\pi\sigma)^{\frac{3}{2}}\mathrm{a}^5\mu = \tfrac{1}{5}m(a^2+b^2)\,\omega = \tfrac{4}{15}\pi\sigma\mathrm{a}^5(\sec\beta\sec\gamma)^{\frac{2}{3}}(1+\cos^2\beta)(4\pi\sigma)^{\frac{1}{2}}\left(\frac{\omega^2}{4\pi\sigma}\right)^{\frac{1}{2}}$$

Thus

$$\mu = \tfrac{1}{5}\sqrt{3}\,(\sec\beta\sec\gamma)^{\frac{2}{3}}(1+\cos^2\beta)\left(\frac{\omega^2}{4\pi\sigma}\right)^{\frac{1}{2}} \quad\text{..........(31)}$$

The function (31) is the quantity which will be tabulated.

Again

$$(\tfrac{4}{3}\pi\sigma)^2\mathrm{a}^5\epsilon = \tfrac{1}{2}(\tfrac{4}{3}\pi\sigma)^{\frac{3}{2}}\mathrm{a}^5\mu\omega = \tfrac{1}{2}\sqrt{3}\,(\tfrac{4}{3}\pi\sigma)^2\mathrm{a}^5\mu\left(\frac{\omega^2}{4\pi\sigma}\right)^{\frac{1}{2}}$$

so that

$$\epsilon = \tfrac{1}{2}\mu\sqrt{3}\left(\frac{\omega^2}{4\pi\sigma}\right)^{\frac{1}{2}} \quad\text{..........................(32)}$$

The function (32) is the quantity which will be tabulated.

Thus in the tables the unit of moment of momentum is taken as $(\frac{4}{3}\pi\sigma)^{\frac{3}{2}}\mathrm{a}^5$, or $m^{\frac{3}{2}}\mathrm{a}^{\frac{1}{2}}$, and the unit of energy as $(\frac{4}{3}\pi\sigma)^2\mathrm{a}^5$ or m^2/a.

It remains to evaluate the intrinsic energy, or the energy required to expand the ellipsoid against its own gravitation, into a condition of infinite dispersion.

If dt be an element of volume, then this energy is

$$-\tfrac{1}{2}\iiint V\sigma\,dt$$

integrated throughout the ellipsoid.

This will be denoted by $(\frac{4}{3}\pi\sigma)^2\mathrm{a}^5(i-1)$, or $m^2\mathrm{a}^{-1}(i-1)$, so that i will be positive.

Now $V = Lx^2 + My^2 + Nz^2 + P$, and if we denote by A, B, C, the principal moments of inertia of the ellipsoid, we have

$$\iiint x^2\sigma\,dt = \tfrac{1}{2}(B+C-A) = \tfrac{1}{5}ma^2$$

and similarly, $\iiint y^2 \sigma \, dt = \frac{1}{5} m b^2, \quad \iiint z^2 \sigma \, dt = \frac{1}{5} m c^2$

Also

$$\iiint \sigma \, dt = m$$

Hence
$$\frac{m^2}{\mathrm{a}}(i-1) = \tfrac{1}{10} m \left[La^2 + Mb^2 + Nc^2 + 5P\right]$$
$$= \tfrac{1}{10} m a^2 (\sec\beta \sec\gamma)^{\frac{2}{3}} \left[L + M\cos^2\beta + N\cos^2\gamma + 5Pa^{-2}\right]$$

But if we take the values of L, M, N given in (15), and note that

$$P = \pi a^3 \cos\beta \cos\gamma \,.\, \frac{2}{a \sin\gamma} \mathrm{F}$$

it easily follows that

$$L + M\cos^2\beta + N\cos^2\gamma + Pa^{-2} = 0$$

Hence
$$\frac{m^2}{\mathrm{a}}(i-1) = -\tfrac{2}{5} m a^2 (\sec\beta\sec\gamma)^{\frac{2}{3}} \,.\, Pa^{-2}$$
$$= -\tfrac{2}{5} m a^2 (\sec\beta\sec\gamma)^{\frac{2}{3}} \,.\, \tfrac{3}{4}\frac{m}{a^2} \,.\, \frac{2}{a\sin\gamma}\mathrm{F}$$
$$= -\tfrac{3}{5}\frac{m^2}{\mathrm{a}} \frac{(\cos\beta\cos\gamma)^{\frac{1}{3}}}{\sin\gamma}\mathrm{F}$$

Therefore
$$i = 1 - \tfrac{3}{5}\frac{(\cos\beta\cos\gamma)^{\frac{1}{3}}}{\sin\gamma}\mathrm{F} \quad\text{.................................(33)}$$

For a sphere γ becomes infinitely small, and F becomes equal to γ, so that $\mathrm{F}/\sin\gamma = 1$. Thus $i - 1 = -\frac{3}{5}$. Therefore the exhaustion of energy of a sphere of radius a is $\frac{3}{5} m^2/\mathrm{a}$; which is the known result. For an ellipsoid of revolution $\kappa = 0$, and $\beta = 0$, and $\mathrm{F} = \gamma$; so that

$$i = 1 - \tfrac{3}{5}\gamma \frac{\cos^{\frac{1}{3}}\gamma}{\sin\gamma}$$

The function (33) is the quantity tabulated below. It seemed preferable to tabulate a positive quantity, and it is on this account that the intrinsic energy corresponding to the infinitely long ellipsoid is entered as unity.

Having now obtained all the necessary formulæ, we may proceed to consider the solution of the problem.

We have to solve

$$\frac{\sec^2\zeta}{\kappa'^2}\mathrm{E} - (2\mathrm{F} - \mathrm{E}) - \frac{\tan\zeta\sec^2\delta}{\kappa'^2} = 0 \quad\text{.................(34)}$$

where $\tan\zeta = \kappa\tan\beta\cos\gamma, \quad \tan\delta = \sin\beta = \kappa\sin\gamma$

and $\mathrm{F} = \int_0^\gamma \frac{d\gamma}{\Delta}, \quad \mathrm{E} = \int_0^\gamma \Delta\, d\gamma$

The axes of the ellipsoid are

$$\frac{a}{\mathrm{a}} = (\sec\beta\sec\gamma)^{\frac{1}{3}}, \quad \frac{b}{\mathrm{a}} = \frac{a}{\mathrm{a}}\cos\beta, \quad \frac{c}{\mathrm{a}} = \frac{a}{\mathrm{a}}\cos\gamma \quad\text{...........(35)}$$

If e_1, e_2, e_3 are the eccentricities of the sections through *ca*, *cb*, *ab* respectively, we have

$$e_1 = \sin\beta, \quad e_2 = \sin\gamma, \quad e_3 = \kappa' \sin\gamma \sec\beta \quad \text{..............(36)}$$

Having obtained the solution, we have to compute

$$\frac{\omega^2}{4\pi\sigma} = \cot\beta \operatorname{cosec}\beta \cot\gamma\,(\mathrm{F}-\mathrm{E}) + \frac{\cot^3\beta\cos\beta}{\kappa'^2}\,\mathrm{E} - \frac{\cos^2\beta\cot^2\gamma}{\kappa'^2} \text{* ...(37)}$$

or the preferable formula (28) may be used.

Then we next compute μ and ϵ and i from the formulæ (31), (32), (33).

The functions F and E are tabulated in Table IX of the second volume of Legendre's *Traité des Fonctions Elliptiques*, in a table of double entry for $\sin^{-1}\kappa$ and γ for each degree.

The solution of (34) by trial and error was laborious, as it was necessary to work with all the accuracy attainable with logarithms of seven figures.

The method adopted was to choose an arbitrary value of γ, and then by trial and error to find two values of $\sin^{-1}\kappa$ one degree apart, one of which made the left-hand side of (34) positive, and the other negative.

The smallest value of γ is 54° 21′ 27″, but after that value integral degrees for γ were always chosen.

The solutions for $\gamma = 55°$ and 57° could not be found very exactly from the elliptic integrals with logarithms of only seven figures, but the solutions were confirmed by the approximate formulæ (29). The solution for $\gamma = 80°$ was confirmed by the approximate formulæ and that for $\gamma = 85°$ was only computed therefrom, since when $\gamma = 80°$ the approximate formula gave nearly identical results with the exact one.

The solution obtained is embodied in the table on the next page. The first three columns give the auxiliary angles γ, $\sin^{-1}\kappa$, β, from which the remaining results are computed.

* As stated above, some of the computations were actually made from the formula (20).

Solutions of Jacobi's Problems*.

Auxiliary angles			Axes			Eccentricities of sections			Ang. vel.	Mom. of momentum	Energy		
γ	$\sin^{-1}\kappa$	β	a/a	b/a	c/a	(ac) e_1	(bc) e_2	(ab) e_3	$\omega^2/4\pi\sigma$	μ	Kinetic ϵ	Intrinsic i	Total E
54° 21′ 27″	0° 0′	0° 0′	1·1972	1·1972	·6977	·81267	·81267	·0000	·09356	·30375	·08046	·41495	·49541
55°	$17\frac{3}{4}$°	$14\frac{1}{4}$°	1·216	1·179	·698	·819	·806	·246	·0935	·304	—	—	—
57°	$34\frac{3}{4}$°	$28\frac{1}{2}$°	1·279	1·123	·696	·839	·784	·478	·093	·306	—	—	—
60°	49° 7′	40° 54′	1·3831	1·0454	·6916	·8660	·7500	·6547	·09060	·3134	·0817	·4188	·5005
65°	64° 19′	54° 46′	1·6007	·9235	·6765	·9063	·6807	·8168	·08295	·3407	·0850	·4394	·5244
70°	74° 12′	64° 43′	1·899	·8111	·6494	·9397	·5991	·9042	·07047	·3920	·0901	·4489	·5390
75°	81° 4′	72° 36′	2·346	·7019	·6072	·9659	·5014	·9542	·0536	·4809	·0964	·4808	·5772
80°	85° 49′	79° 10′	3·1294	·5881	·5434	·9848	·3822	·9822	·03307	·6387	·1006	·533	·634
85°	88° 48′	84° 52′	5·0406	·4516	·4393	·9962	·2313	·9960	·01293	1·0087	·0993	·645	·744
90°	90°	90°	∞	·0	·0	1·000	·000	1·000	·000	∞	·000	1·000	1·000

N.B.—*The moment of momentum of the system is* $(\frac{4}{3}\pi\sigma)^{\frac{3}{2}}a^5 . \mu$, *or* $m^{\frac{3}{2}}a^{\frac{1}{2}}\mu$; *the kinetic energy is* $(\frac{4}{3}\pi\sigma)^2a^5 . \epsilon$, *or* $m^2a^{-1}\epsilon$, *and the intrinsic energy is* $(\frac{4}{3}\pi\sigma)^2a^5 . (i-1)$, *or* $m^2a^{-1}(i-1)$, *but in the above table unity has been added to make the results positive;* $E = \epsilon + i$.

* [In the paper referred to in the note at the beginning of this paper, M. R. Kaibara has computed the Jacobian ellipsoids by Weierstrass's method of treating the elliptic integrals, and has then interpolated results which are intended to be exactly comparable with mine. But there must have been some degree of inaccuracy in the interpolations, because the product of the three axes should always come to unity, and in every case Kaibara's product exceeds unity. In most of the entries this discrepancy is so small that it would not affect the last significant figure in the axes, but in the last three entries (for which my results have been recomputed for this volume) the discrepancy becomes considerable. The following table exhibits the discrepancies expressed in units of the fourth place of decimals. It must be borne in mind that several of my results for the axes (marked with asterisks) do not claim to go beyond three places of decimals.

(Discrepancies in units of the 4th place of decimals.)

γ	a/a D − K	b/b D − K	c/c D − K	abc/a³ D	abc/a³ K	$\omega^2/4\pi\sigma$ D − K
54° 21′ 27″	0	0	+1			0
55°	− 2*	+ 3*	+4*			− 2
57°	+ 3*	+ 2*	+5*			− 7*
60°	+ 9	+12	−4			− 3
65°	− 3	+ 2	0			0
70°	− 19*	+ 3	0	1·00026	1·00089	− 1
75°	− 91*	+ 5	+2	·99985	1·00268	− 4
80°	+ 72	− 1	−1	1·00001	1·00066	−41
85°	−117	+ 2	+1	·99999	1·00178	− 5

The internal evidence seems therefore to be adverse to the correctness of M. Kaibara's last three entries.]

As a graphical result is much more intelligible than a numerical one, I have given two figures, showing the three principal sections in two cases, namely, where $\gamma = 60°$, and $\gamma = 75°$. For these figures a is taken as 2 cm., so that the volume of fluid is $\frac{4}{3}\pi \times 2^3$ cubic cm.

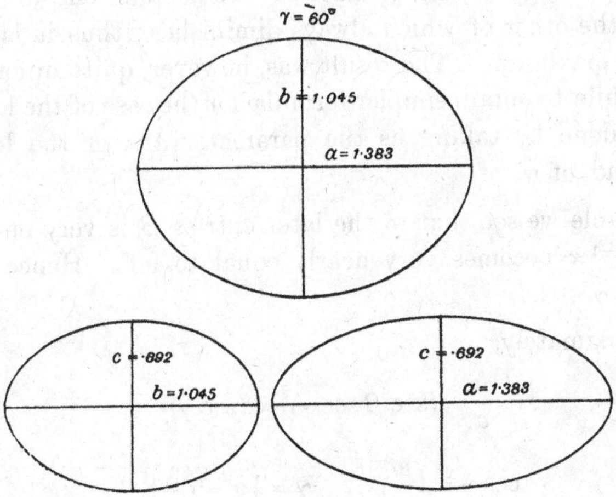

FIG. 1.

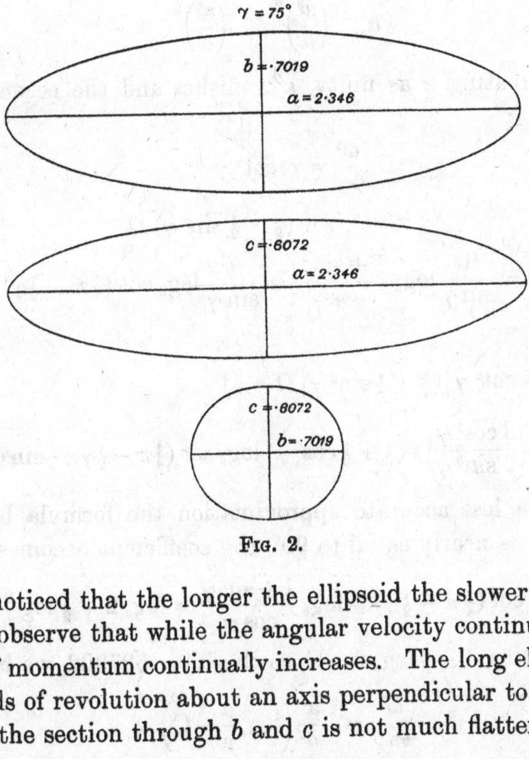

FIG. 2.

It will be noticed that the longer the ellipsoid the slower it rotates. It is interesting to observe that while the angular velocity continually diminishes, the moment of momentum continually increases. The long ellipsoids are very nearly ellipsoids of revolution about an axis perpendicular to that of rotation. Thus in fig. 2 the section through b and c is not much flattened.

The most remarkable point is that there is a maximum of kinetic energy when a/a is about 3, or when the length of the ellipsoid is about five times its diameter. However, notwithstanding this maximum of kinetic energy, the total energy always increases with the length of the ellipsoid.

The kinetic energy is the product of two factors, one of which always increases, and the other of which always diminishes; thus it is obvious that it must have a maximum. The result was, however, quite unforeseen, and it seems worth while to obtain simpler formulæ for the case of the long ellipsoids. This may be done by taking as the parameter a/a, or the length of the ellipsoid, instead of γ.

From the table we see that in the later entries β is very nearly equal to γ, and that $\sin^{-1}\kappa$ becomes very nearly equal to 90°. Hence we may put $\kappa = 1$, and $\beta = \gamma$.

Thus, approximately,

$$\frac{a}{\mathrm{a}} = (\sec\beta\sec\gamma)^{\frac{1}{3}} = (\sec\gamma)^{\frac{2}{3}}$$

and

$$\cos\gamma = \left(\frac{\mathrm{a}}{a}\right)^{\frac{3}{2}}, \quad \gamma = \tfrac{1}{2}\pi - \left(\frac{\mathrm{a}}{a}\right)^{\frac{3}{2}}$$

The axes of the ellipsoid are

$$a, \quad \left(\frac{\mathrm{a}^3}{a}\right)^{\frac{1}{2}}, \quad \left(\frac{\mathrm{a}^3}{a}\right)^{\frac{1}{2}}$$

[As we are treating κ as unity, κ' vanishes and the second equation of (34) becomes

$$\frac{\omega^2}{4\pi} = \tau_0 \cot^2\gamma$$

Now

$$\tau_0 = -\tfrac{3}{2} + (\tfrac{3}{2} - \tfrac{1}{2}\sin^2\gamma)\,\Omega$$

where

$$\Omega = \frac{1}{\sin\gamma}\log_e\frac{1+\sin\gamma}{\cos\gamma} = \frac{1}{\sin\gamma}\log_e\cot(\tfrac{1}{4}\pi - \tfrac{1}{2}\gamma)$$

Therefore

$$\frac{\omega^2}{4\pi\sigma} = \cot^2\gamma\,[(\tfrac{3}{2} - \tfrac{1}{2}\sin^2\gamma)\,\Omega - \tfrac{3}{2}]$$

$$= \frac{3\cos^2\gamma}{2\sin^3\gamma}[\tfrac{2}{3}(1 + \tfrac{1}{2}\cos^2\gamma)\log_e\cot(\tfrac{1}{4}\pi - \tfrac{1}{2}\gamma) - \sin\gamma]$$

If we adopt a less accurate approximation the formula becomes much simpler. Since γ is nearly equal to 90°, the coefficient becomes $\tfrac{3}{2}(\mathrm{a}/a)^3$, and

$$\log_e\cot(\tfrac{1}{4}\pi - \tfrac{1}{2}\gamma) = \log_e\frac{1+\sin\gamma}{\cos\gamma} = \log_e 2 + \tfrac{3}{2}\log\frac{a}{\mathrm{a}}$$

Therefore writing $1 - \tfrac{2}{3}\log_e 2 = \mathrm{C}$, so that $\mathrm{C} = {\cdot}537902$, we have

$$\frac{\omega^2}{4\pi\sigma} = \tfrac{3}{2}\frac{\mathrm{a}^3}{a^3}\left(\log_e\frac{a}{\mathrm{a}} - \mathrm{C}\right)$$

If we put $a/\text{a} = 5{\cdot}04064$ this formula gives $\omega^2/4\pi\sigma = 0{\cdot}012645$. The full value in the preceding tables was 0·01293; thus even with so short an ellipsoid as this the results agree within about 2 per cent.

For the moment of momentum we have

$$\mu = \tfrac{1}{5}\sqrt{3}\,(\sec\beta\sec\gamma)^{\frac{2}{3}}\,(1+\cos^2\beta)\left(\frac{\omega^2}{4\pi\sigma}\right)^{\frac{1}{2}}$$

$$= \frac{3}{5\sqrt{2}}\,\frac{a^{\frac{1}{2}}}{\text{a}^{\frac{1}{2}}}\left(1+\frac{\text{a}^3}{a^3}\right)\left(\log_e\frac{a}{\text{a}} - \text{C}\right)^{\frac{1}{2}}$$

The limit of this is infinite, when a is infinite.]

Again
$$\epsilon = \tfrac{1}{2}\mu\sqrt{3}\left(\frac{\omega^2}{4\pi\sigma}\right)$$

$$= \tfrac{9}{20}\,\frac{\text{a}}{a}\left[1+\frac{\text{a}^3}{a^3}\right]\left[\log_e\frac{a}{\text{a}} - \text{C}\right]$$

Now the function $\dfrac{\text{a}}{a}\left(\log_e\dfrac{a}{\text{a}} - \text{C}\right)$ has a maximum, when

$$\log_e\frac{a}{\text{a}} = 1 + \text{C} = 1{\cdot}5379$$

that is when $a/\text{a} = 4{\cdot}65$. It is probable that this approximation is not a very close one, but it shows that there is actually a maximum of kinetic energy.

Since when $\kappa = 1$, the elliptic integral

$$\text{F} = \log_e\cot\left(\tfrac{1}{4}\pi - \tfrac{1}{2}\gamma\right) = \tfrac{3}{2}\left[\log_e\frac{a}{\text{a}} + \tfrac{2}{3}\log_e 2\right]$$

we have

$$i = 1 - \tfrac{3}{5}\,\frac{(\cos\beta\cos\gamma)^{\frac{1}{3}}}{\sin\gamma}\,\text{F} = 1 - \tfrac{9}{10}\left(\frac{\text{a}}{a}\right)^3\left[\log_e\frac{a}{\text{a}} + \tfrac{2}{3}\log_e 2\right]$$

If we like we may express these several results in terms of the minor and major axes of the ellipsoid, for $b = c = \text{a}^{\frac{3}{2}}/a^{\frac{1}{2}}$, and therefore $\text{a}^3 = c^2 a$.

Thus
$$\left.\begin{aligned}
\frac{\omega^2}{4\pi\sigma} &= \frac{c^2}{a^2}\left[\log_e\frac{a}{c} - \tfrac{3}{2}\text{C}\right]\\
\mu &= \frac{3^{\frac{1}{2}}}{5}\left(\frac{a}{c}\right)^{\frac{1}{3}}\left[1+\frac{c^2}{a^2}\right]\left[\log_e\frac{a}{c} - \tfrac{3}{2}\text{C}\right]^{\frac{1}{2}}\\
\epsilon &= \tfrac{3}{10}\left(\frac{c}{a}\right)^{\frac{2}{3}}\left(1+\frac{c^2}{a^2}\right)\left(\log_e\frac{a}{c} - \tfrac{3}{2}\text{C}\right)\\
i &= 1 - \tfrac{3}{5}\,\frac{a^2}{c^2}\log_e\frac{2a}{c}
\end{aligned}\right\}$$

[POSTSCRIPT.—If throughout the series of Jacobian ellipsoids we keep the moment of momentum constant and equal, say, to h, and throw the change on to the density σ, we have

$$\tfrac{4}{3}\pi a^3 \sigma = m; \quad m^{\frac{3}{2}} a^{\frac{1}{2}} \mu = h; \quad \text{say } \Omega = \frac{\omega^2}{4\pi\sigma}$$

where μ and Ω have the numerical values tabulated above.

It is easy then to show that

$$\sigma = \left(\frac{3m^{10}}{4\pi h^6}\right) \mu^6; \quad \omega = \left(\sqrt{3}\,\frac{m^5}{h^3}\right) \Omega^{\frac{1}{2}} \mu^3$$

the coefficients in brackets being constants.

The following values indicate the march of the angular velocity throughout the series:—

$\gamma =$	$54^\circ\ 21'\ 27''$	60°	70°	80°
$\Omega^{\frac{1}{2}} \mu^3 =$	·0086	·0093	·0160	·0488

Instead of diminishing throughout the series, the angular velocity always increases.]

9.

ON FIGURES OF EQUILIBRIUM OF ROTATING MASSES OF FLUID.

[*Philosophical Transactions of the Royal Society*, Vol. 178, A (1887), pp. 379—428.]

IN a previous paper* I remarked that there might be reason to suppose that the earliest form of a satellite might not be annular. Whether or not the present investigation does actually help us to understand the working of the nebular hypothesis, the idea there alluded to was the existence of a dumb-bell shaped figure of equilibrium, such as is shown in the figures at the end of this paper. These figures were already drawn when a paper by M. Poincaré appeared, in which, amongst other things, a similar conclusion was arrived at. My paper was accordingly kept back in order that an attempt might be made to apply the important principles enounced by him to this mode of treatment of the problem. The results of that attempt are, for reasons explained below, given in the Appendix†.

The subject of figures of equilibrium of rotating masses of fluid is here considered from a point of view so wholly different from that of M. Poincaré that, notwithstanding his priority and the greater completeness of his work, it still appears worth while to present this paper.

The method of treatment here employed is simple of conception; but it is unfortunate that, to carry out the idea, a very formidable array of analysis is necessary.

In the last section a summary will be found of the principal conclusions, in which analysis is avoided.

* *Phil. Trans.*, Part II., 1881, p. 534 [Vol. II., p. 457, footnote].

† [The attempt failed, and the greater part of the Appendix is omitted.]

§ 1. *Formulæ of Spherical Harmonic Analysis.*

Let there be two sets of rectangular axes, as shown in fig. 1; and let z be measured from o to O, whilst Z is measured from O to o; let $r^2 = x^2 + y^2 + z^2$, $R^2 = X^2 + Y^2 + Z^2$; and let $c = oO$.

Then $$x + X = 0, \quad y + Y = 0, \quad z + Z = c \qquad (1)$$

Let w_i, W_i, denote the solid zonal harmonics of degree i of the coordinates x, y, z, and X, Y, Z, respectively.

Now we shall require to express the solid zonal and certain tesseral harmonics of negative degrees with respect to the origin O as solid zonal and tesseral harmonics of positive degrees with respect to the origin o, and *vice versâ*; moreover, the results will have to be applied to a sphere of radius a with centre o, and to a sphere of radius A with centre O. This last clause is introduced in order to explain the introduction of the symbols a, A, in this place.

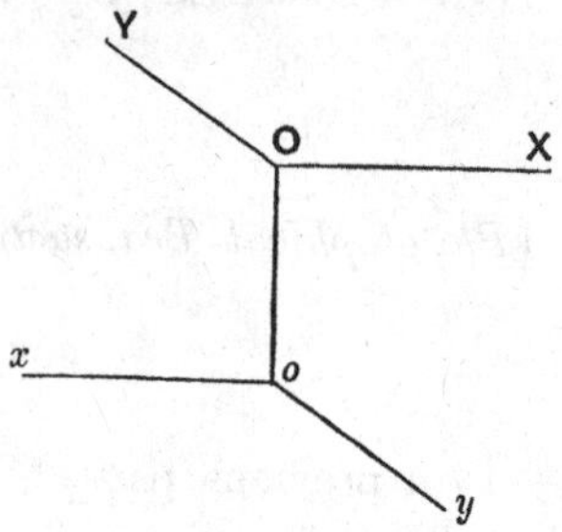

FIG. 1.

The formulæ required will be called "transference formulæ," because they are to be used in shifting the origin from one point to the other.

The obvious symmetry of our axes is such that every transference formula from O to o has its exact counterpart for transference from o to O; thus a second symmetrical formula with capital and small letters interchanged will generally be left unwritten. When necessary, θ, ϕ, will be written for co-latitude and longitude with regard to x, y, z; and Θ, Φ, for the same with respect to X, Y, Z.

Then, since $$R^2 = r^2 + c^2 - 2rc \cos\theta$$

we have the usual expansion in zonal harmonics

$$\frac{c}{R} = \sum_{k=0}^{k=\infty} \frac{w_k}{c^k} \qquad (2)$$

The usual formula for the derivation of the zonal harmonic of negative degree $i+1$ from $1/R$ is

$$\frac{(-)^i}{i!} \frac{d^i}{dZ^i} \frac{1}{R} = \frac{W_i}{R^{2i+1}} \qquad (3)$$

Hence, on differentiating (2) i times with respect to Z, or, which is the same thing, with respect to $-z$, we have, from (3),

$$c \frac{W_i}{R^{2i+1}} = \frac{1}{i!} \sum_{k=0}^{k=\infty} \frac{d^i}{dz^i} \frac{w_k}{c^k}$$

But $$\frac{d^i}{dz^i} w_k = k(k-1) \dots (k-i+1) w_{k-i} = \frac{k!}{k-i!} w_{k-i}$$

Hence $$c^i \frac{W_i}{R^{2i+1}} = \frac{1}{c} \sum_{k=0}^{k=\infty} \frac{k!}{i!\,\overline{k-i}!} \frac{w_{k-i}}{c^{k-i}}$$

In interpreting this formula, it will be noted that, if i is less than k, the term vanishes: hence the summation runs from $k=\infty$ to $k=i$; it is therefore better to write $k+i$ for k, and we thus obtain

$$\frac{c^i W_i}{R^{2i+1}} = \frac{1}{c} \sum_{k=0}^{k=\infty} \frac{\overline{k+i}!}{i!\,k!} \left(\frac{a}{c}\right)^k \frac{w_k}{a^k} \quad \text{.........................(4)}$$

This is the first transference formula by which the solid zonal harmonic of degree $-i-1$ with respect to O is expressed as a series of solid harmonics of positive degree with respect to o. The formula (4) includes (2) as the particular case where $i=0$. The right-hand side of (4) is convergent for r less than a. A similar formula, convergent for r greater than a, is easily obtainable, but with this we shall not concern ourselves.

It remains to find the transference formula for certain tesseral harmonics.

If we put $$\rho = \tfrac{1}{4}(x^2+y^2) \quad \text{.................................(5)}$$

the general expression for the zonal harmonic is

$$w_i = \Sigma (-)^k \frac{i!}{k!^2 . \overline{i-2k}!} z^{i-2k} \rho^k \quad \text{.....................(6)}$$

where the summation extends from $k=0$ to $k=\tfrac{1}{2}i$ or $\tfrac{1}{2}(i-1)$.

From (6) we have

$$\frac{dw_i}{d\rho} = \Sigma (-)^k \frac{i!}{k!^2 . \overline{i-2k}!} k z^{i-2k} \rho^{k-1} \quad \text{..................(7)}$$

Now, since $r^2 = z^2 + 4\rho$, we have

$$r^2 \frac{dw_i}{d\rho} = \Sigma (-)^k i! \left[\frac{-(k+1)}{\overline{k+1}!^2 . \overline{i-2k-2}!} + \frac{4k}{k!^2 . \overline{i-2k}!}\right] z^{i-2k} \rho^k$$

$$= \Sigma (-)^k i! \frac{[4k(k+1) - (i-2k)(i-2k-1)]}{(k+1) . k!^2 . \overline{i-2k}!} z^{i-2k} \rho^k \quad \text{........(8)}$$

Also $$2(2i+1) w_i = \Sigma (-)^k i! \frac{2(2i+1)(k+1)}{(k+1) . k!^2 . \overline{i-2k}!} z^{i-2k} \rho^k \quad \text{...........(9)}$$

Subtracting (9) from (8), and simplifying the difference, we have

$$r^2 \frac{dw_i}{d\rho} - 2(2i+1) w_i = \Sigma (-)^{k+1} \frac{(i+1)(i+2) . i!}{(k+1) . k!^2 . \overline{i-2k}!} z^{i-2k} \rho^k$$

$$= \Sigma (-)^{k+1} \frac{\overline{i+2}!}{\overline{k+1}!^2 . \overline{i+2-2k-2}!} (k+1) z^{i+2-2k-2} \rho^{k+1-1}$$

$$= \frac{d}{d\rho} w_{i+2} \quad \text{..(10)}$$

the last transformation being derived from (7) with $i+2$ in place of i, and $k+1$ in place of k.

Differentiate (10) with respect to ρ, and notice that $dr^2/d\rho = 4$, and we have

$$r^2 \frac{d^2 w_i}{d\rho^2} - 2(2i-1)\frac{dw_i}{d\rho} = \frac{d^2 w_{i+2}}{d\rho^2}$$

Then, with $i+2$ in place of i,

$$r^2 \frac{d^2 w_{i+2}}{d\rho^2} - 2(2i+3)\frac{dw_{i+2}}{d\rho} = \frac{d^2 w_{i+4}}{d\rho^2} \quad \text{.................(11)}$$

Now
$$\frac{d}{d\rho}\frac{w_i}{r^{2i+1}} = \frac{1}{r^{2i+3}}\left\{r^2\frac{dw_i}{d\rho} - 2(2i+1)w_i\right\}$$

$$= \frac{1}{r^{2i+3}}\frac{dw_{i+2}}{d\rho} \text{ by (10)}$$

Differentiating again,

$$\frac{d^2}{d\rho^2}\frac{w_i}{r^{2i+1}} = \frac{1}{r^{2i+5}}\left\{r^2\frac{d^2 w_{i+2}}{d\rho^2} - 2(2i+3)\frac{dw_{i+2}}{d\rho}\right\}$$

$$= \frac{1}{r^{2i+5}}\frac{d^2}{d\rho^2} w_{i+4} \text{ by (11)}$$

or
$$\frac{1}{r^{2i+1}}\frac{d^2 w_{i+2}}{d\rho^2} = \frac{d^2}{d\rho^2}\frac{w_{i-2}}{r^{2i-3}} \quad \text{..........................(12)}$$

But since $\rho = \frac{1}{4}(x^2+y^2)$, it follows that, in operating on a function involving x and y only in the form x^2+y^2,

$$\frac{d}{dx} = \tfrac{1}{2}x\frac{d}{d\rho}, \quad \frac{d}{dy} = \tfrac{1}{2}y\frac{d}{d\rho}, \quad \text{and} \quad x\frac{d}{dx} - y\frac{d}{dy} = \tfrac{1}{2}(x^2-y^2)\frac{d}{d\rho}$$

Also
$$\frac{d^2}{dx^2} = \frac{1}{2}\frac{d}{d\rho} + \tfrac{1}{4}x^2\frac{d^2}{d\rho^2}, \quad \frac{d^2}{dy^2} = \frac{1}{2}\frac{d}{d\rho} + \tfrac{1}{4}y^2\frac{d^2}{d\rho^2}$$

so that
$$\frac{d^2}{dx^2} - \frac{d^2}{dy^2} = \tfrac{1}{4}(x^2-y^2)\frac{d^2}{d\rho^2}$$

Now let us put
$$\delta^2 = \frac{d^2}{dx^2} - \frac{d^2}{dy^2} \quad \text{..........................(13)}$$

Then
$$\delta^2 = \tfrac{1}{4}(x^2-y^2)\frac{d^2}{d\rho^2}$$

and therefore (12) may be written

$$\frac{1}{r^{2i+1}}\delta^2 w_{i+2} = \delta^2\frac{w_{i-2}}{r^{2i-3}} \quad \text{..........................(14)}$$

These expressions in (14) are obviously solid tesseral harmonics.

The transference formula required is for $\dfrac{\delta^2 W_{i+2}}{R^{2i+1}}$.

By the formula (4) we have

$$\frac{c^{i-2}W_{i-2}}{R^{2i-3}} = \frac{1}{c}\sum_{k=0}^{k=\infty}\frac{k+i-2!}{i-2!\,k!}\frac{w_k}{c^k}$$

Operating on both sides by δ^2, and applying (14), we have

$$\frac{c^{i-2}}{R^{2i+1}}\,\delta^2 W_{i+2}=\frac{1}{c}\sum_{k=0}^{k=\infty}\frac{\overline{k+i-2}\,!}{\overline{i-2}\,!\,k\,!}\,\frac{\delta^2 w_k}{c^k}\quad\text{..................(15)}$$

Now the general formula (6) for the zonal harmonic shows us that $d^2w_k/d\rho^2$ is zero when $k=0, 1, 2, 3$, and hence $\delta^2 w_k$ vanishes for the same values of k. Thus the summation in (15) is from $k=\infty$ to $k=4$, or, if we write $k+2$ for k, from ∞ to 2. Hence (15) gives

$$\frac{c^{i}}{R^{2i+1}}\,\delta^2 W_{i+2}=\frac{1}{c}\sum_{k=2}^{k=\infty}\frac{\overline{k+i}\,!}{\overline{i-2}\,!\,\overline{k+2}\,!}\left(\frac{a}{c}\right)^k\frac{\delta^2 w_{k+2}}{a^k}\quad\text{...........(16)}$$

This is the second transference formula required.

We observe that the transference of a negative zonal harmonic gives us positive zonals, and that tesseral harmonics of the type $\delta^2 W_{i+2}/R^{2i+1}$ give us harmonics of the type $\delta^2 w_{k+2}$.

§ 2. *The Mutual Influence of two Spheres of Fluid without Rotation.*

Imagine two approximately spherical masses of fluid of unit density, with their centres at the origins o and O respectively, and with mean radii a and A respectively.

We shall find that each exercises on the other certain forces, one part of which has as potential a solid zonal harmonic of the first degree. This part of the force must remain essentially unbalanced in the supposed system, but we shall see hereafter that it is balanced by the rotation to be afterwards imposed on the system.

Meanwhile it will be supposed that it is annulled in some way, and we shall content ourselves with finding the mutual influence of the spheroids, and the outstanding term of the first degree of harmonics.

Let us assume that the equations, referred to our two origins, of the surfaces of the two spheroids, when they mutually perturb one another, are

$$\left.\begin{aligned}\frac{r}{a}&=1+\left(\frac{A}{a}\right)^3\sum_{i=2}^{i=\infty}\frac{2i+1}{2i-2}\left(\frac{a}{c}\right)^{i+1}h_i\,r^{-i}w_i\\ \frac{R}{A}&=1+\left(\frac{a}{A}\right)^3\sum_{i=2}^{i=\infty}\frac{2i+1}{2i-2}\left(\frac{A}{c}\right)^{i+1}H_iR^{-i}W_{-i}\end{aligned}\right\}\quad\text{.........(17)}$$

The h's and H's are unknown coefficients, to be determined.

We have now to find the potential at any point in space.

The mass of the spheroid o is $\frac{4}{3}\pi a^3$, and its potential is $\frac{4}{3}\pi a^3/r$.

The potential due to the departure from sphericity, represented by the term in h_i in the first of (17), is

$$\frac{4\pi A^3}{3c}\,\frac{3h_i}{2i-2}\left(\frac{a}{c}\right)^i\left(\frac{a}{r}\right)^{i+1}\frac{w_i}{r^i}\quad\text{.........................(18)}$$

This is written in a form convenient for passing to the case of $r=a$. It may also be written in the form

$$\tfrac{4}{3}\pi A^3 \left(\frac{a}{c}\right)^{2i+1} \frac{3h_i}{2i-2} \frac{c^i w_i}{r^{2i+1}} \quad \text{......................(19)}$$

when it is in a suitable form for application of the transference formula (4).

We shall now introduce two new symbols, namely,

$$\gamma = \left(\frac{a}{c}\right)^2, \quad \Gamma = \left(\frac{A}{c}\right)^2 \quad \text{......................(20)}$$

Then (19) may be written

$$\tfrac{4}{3}\pi A^3 \left(\frac{a}{c}\right)^3 \frac{3h_i \gamma^{i-1}}{2i-2} \frac{c^i w_i}{r^{2i+1}}$$

and, of course, the similar potential with the other origin is

$$\tfrac{4}{3}\pi a^3 \left(\frac{A}{c}\right)^3 \frac{3H_i \Gamma^{i-1}}{2i-2} \frac{c^i W_i}{R^{2i+1}} \quad \text{......................(21)}$$

The whole potential at any point of space consists of the potentials of the two spheres and of the inequalities on each. The potential of the inequalities of the sphere o may be written in the form (18), and of sphere O in the form (21).

Thus the whole potential is

$$\tfrac{4}{3}\pi a^2 \cdot \frac{a}{r} + \frac{4\pi A^3}{3c} \sum_{k=2}^{k=\infty} \frac{3h_k}{2k-2} \left(\frac{a}{c}\right)^k \left(\frac{a}{r}\right)^{k+1} \frac{w_k}{r^k} \quad \text{........(22–i.)}$$

$$+ \frac{4\pi A^3}{3c} \cdot \frac{c}{R} + \frac{4\pi a^3}{3} \left(\frac{A}{c}\right)^3 \sum_{i=2}^{i=\infty} \frac{3H_i \Gamma^{i-1}}{2i-2} \frac{c^i W_i}{R^{2i+1}} \quad \text{..........(22–ii.)}$$

The first line of (22) refers to origin o, the second to origin O, and to this latter half the transference formula (4) must be applied.

Now apply (2) to the first term of the second line, and (4) to one term of the series in the second term, and we have

$$\frac{4\pi A^3}{3c} \frac{c}{R} = \frac{4\pi A^3}{3c} \sum_{k=0}^{k=\infty} \left(\frac{a}{c}\right)^k \frac{w_k}{a^k}$$

and

$$\left[\frac{4\pi a^3}{3} \left(\frac{A}{c}\right)^3 \frac{3H_i \Gamma^{i-1}}{2i-2}\right] \frac{c^i W_i}{R^{2i+1}} = \frac{4\pi A^3}{3c} \left(\frac{a}{c}\right)^3 \frac{3H_i \Gamma^{i-1}}{2i-2} \sum_{k=0}^{k=\infty} \frac{k+i\,!}{k\,!\,i\,!} \left(\frac{a}{c}\right)^k \frac{w_k}{a^k}$$

Thus the second line of (22) when transferred is

$$\frac{4\pi A^3}{3c} \left[\sum_{k=0}^{k=\infty} \left(\frac{a}{c}\right)^k \frac{w_k}{a^k} + \tfrac{3}{2} \left(\frac{a}{c}\right)^3 \sum_{i=2}^{i=\infty} \sum_{k=0}^{k=\infty} \frac{k+i\,!}{k\,!\,i\,!} \frac{\Gamma^{i-1}}{i-1} H_i \left(\frac{a}{c}\right)^k \frac{w_k}{a^k}\right] \quad \text{...(22–ii.)}$$

Then (22–i.) and (22–ii.) together constitute the potential now entirely referred to origin o.

We want to choose the h's and H's so that each spheroid may be a level surface, save as to the outstanding term of the first degree.

In order that (17) may be a level surface, when we substitute for r its value (17) in (22), the whole potential must be constant. In effecting this substitution, we may put $r=a$ in the small terms, but in $\frac{4}{3}\pi a^3/r$ we must give it the full value (17).

The constancy of the potential is secured by making the coefficient of each harmonic term vanish separately—excepting the first harmonic, which remains outstanding by supposition.

We may consider each harmonic term by itself.

As far as concerns the term involving w_k, we have, from (22–i.) and (22–ii.), as the value of the potential,

$$\frac{4}{3}\pi a^2 \frac{a}{r} + \frac{4\pi A^3}{3c}\left[\frac{3h_k}{2k-2}\left(\frac{a}{c}\right)^k \left(\frac{a}{r}\right)^{k+1} \frac{w_k}{r^k} + \left(\frac{a}{c}\right)^k \frac{w_k}{a^k} + \tfrac{3}{2}\left(\frac{a}{c}\right)^3 \left(\frac{a}{c}\right)^k \frac{w_k}{c^k} \sum_{i=2}^{i=\infty} \frac{k+i\,!}{k\,!\,i\,!} \frac{\Gamma^{i-1}}{i-1} H_i\right]$$

and the value of r which must make this constant is

$$\frac{r}{a} = 1 + \left(\frac{A}{a}\right)^3 \frac{2k+1}{2k-2} \left(\frac{a}{c}\right)^{k+1} h_k \frac{w_k}{r^k}$$

but in the small terms inside [ ] we may put $r=a$ simply.

Make, therefore, the substitution, and equate the coefficient of w_k to zero. On dividing that coefficient by $\frac{4\pi A^3}{3c}\cdot\left(\frac{a}{c}\right)^k$, we find

$$-\frac{2k+1}{2k-2} h_k + \frac{3h_k}{2k-2} + 1 + \tfrac{3}{2}\left(\frac{a}{c}\right)^3 \sum_{i=2}^{i=\infty} \frac{k+i\,!}{k\,!\,i\,!} \frac{\Gamma^{i-1}}{i-1} H_i = 0$$

Therefore $$h_k = 1 + \tfrac{3}{2}\left(\frac{a}{c}\right)^3 \sum_{i=2}^{i=\infty} \frac{k+i\,!}{k\,!\,i\,!} \frac{\Gamma^{i-1}}{i-1} H_i \quad\text{..................(23)}$$

and, by symmetry,

$$H_r = 1 + \tfrac{3}{2}\left(\frac{A}{c}\right)^3 \sum_{i=2}^{i=\infty} \frac{r+i\,!}{r\,!\,i\,!} \frac{\gamma^{i-1}}{i-1} h_i \quad\text{..................(24)}$$

Multiplying both sides of (24) by the coefficient of H_r in (23), we have

$$\tfrac{3}{2}\left(\frac{a}{c}\right)^3 \frac{k+r\,!}{k\,!\,r\,!} \frac{\Gamma^{r-1}}{r-1} H_r = \tfrac{3}{2}\left(\frac{a}{c}\right)^3 \frac{k+r\,!}{k\,!\,r\,!} \frac{\Gamma^{r-1}}{r-1} + (\tfrac{3}{2})^2 \left(\frac{a}{c}\right)^3 \left(\frac{A}{c}\right)^3 \sum_{i=2}^{i=\infty} \frac{r+i\,!\,r+k\,!}{r\,!\,i\,!\,k\,!\,r\,!} \frac{\gamma^{i-1}}{i-1} \frac{\Gamma^{r-1}}{r-1} h_i$$

Performing $\sum_{r=2}^{r=\infty}$ on both sides, and substituting from (23),

$$h_k - 1 = \tfrac{3}{2}\left(\frac{a}{c}\right)^3 \sum_{r=2}^{r=\infty} \frac{k+r\,!}{k\,!\,r\,!} \frac{\Gamma^{r-1}}{r-1} + (\tfrac{3}{2})^2 \left(\frac{a}{c}\right)^3 \left(\frac{A}{c}\right)^3 \sum_{r=2}^{r=\infty} \sum_{i=2}^{i=\infty} \frac{r+i\,!\,r+k\,!}{r\,!\,i\,!\,k\,!\,r\,!} \frac{\gamma^{i-1}}{i-1} \frac{\Gamma^{r-1}}{r-1} h_i \quad\text{...(25)}$$

Now let

$$\left.\begin{aligned}(k, \Gamma) &= \sum_{r=2}^{r=\infty} \frac{k+r\,!}{k\,!\,r\,!}\,\frac{\Gamma^{r-1}}{r-1}\\ [k, i, \Gamma] &= \sum_{r=2}^{r=\infty} \frac{r+i\,!\,r+k\,!}{r\,!\,i\,!\,k\,!\,r\,!}\,\frac{\Gamma^{r-1}}{r-1}\end{aligned}\right\}\quad\ldots\ldots\ldots\ldots\ldots\ldots(26)$$

And (25) may be written

$$h_k = 1 + \tfrac{3}{2}\left(\frac{a}{c}\right)^3 (k, \Gamma) + (\tfrac{3}{2})^2 \left(\frac{a}{c}\right)^3 \left(\frac{A}{c}\right)^3 \sum_{i=2}^{i=\infty} [k, i, \Gamma]\,\frac{\gamma^{i-1}}{i-1}\,h_i \quad\ldots(27)$$

By imparting to k all integral values from 2 upwards, we get a system of linear equations for the determination of the h's, and it will appear below that as many of them may be found numerically as may be desired.

We now have to consider the series (26).

Let
$$\beta = \frac{\gamma}{1-\gamma}, \qquad B = \frac{\Gamma}{1-\Gamma}$$

and denote the operations

$$\frac{1}{\lambda\,!}\,\frac{d^\lambda}{d\gamma^\lambda}\,\gamma^\lambda \text{ or } \frac{1}{\lambda\,!}\,\frac{d^\lambda}{d\Gamma^\lambda}\,\Gamma^\lambda \text{ by } E^\lambda$$

Consider the function $\gamma^{-1}E^\lambda \,.\, \gamma \log(1+\beta)$.

Now
$$\log(1+\beta) = -\log(1-\gamma) = \sum_{r=2}^{r=\infty} \frac{\gamma^{r-1}}{r-1}$$

Therefore

$$E^k \,.\, \gamma \log(1+\beta) = \frac{1}{k\,!}\,\frac{d^k}{d\gamma^k} \sum_{r=2}^{r=\infty} \frac{\gamma^{k+r}}{r-1} = \sum_{r=2}^{r=\infty} \frac{k+r\,!}{k\,!\,r\,!}\,\frac{\gamma^r}{r-1}$$

Thus
$$(k, \gamma) = \frac{1}{\gamma} E^k \,.\, \gamma \log(1+\beta), \quad (k, \Gamma) = \frac{1}{\Gamma} E^k \,.\, \Gamma \log(1+B) \ldots\ldots(28)$$

Next consider the function $\gamma^{-1}E^kE^i \,.\, \gamma \log(1+\beta)$.

As before,
$$E^i \,.\, \gamma \log(1+\beta) = \sum_{r=2}^{r=\infty} \frac{i+r\,!}{i\,!\,r\,!}\,\frac{\gamma^r}{r-1}$$

and
$$E^kE^i \,.\, \gamma \log(1+\beta) = \frac{1}{k\,!}\,\frac{d^k}{d\gamma^k} \sum_{r=2}^{r=\infty} \frac{i+r\,!}{i\,!\,r\,!}\,\frac{\gamma^{k+r}}{r-1}$$
$$= \sum_{r=2}^{r=\infty} \frac{i+r\,!\,k+r\,!}{i\,!\,r\,!\,r\,!\,k\,!}\,\frac{\gamma^r}{r-1}$$

Hence

$$[k, i, \gamma] = \frac{1}{\gamma} E^kE^i \,.\, \gamma \log(1+\beta), \quad [k, i, \Gamma] = \frac{1}{\Gamma} E^kE^i \,.\, \Gamma \log(1+B) \ldots(29)$$

We must now develop the symbolical sums of the series in (28) and (29). The following theorems are obvious:—

$$\frac{d^n}{d\gamma^n}\,\gamma^p = \frac{p\,!}{p-n\,!}\,\gamma^{p-n}, \quad \frac{d^n}{d\gamma^n} \log(1+\beta) = \frac{n-1\,!}{(1-\gamma)^n}$$

$$\frac{d^n}{d\gamma^n}(1-\gamma)^{-p} = \frac{p+n-1\,!}{p-1\,!}(1-\gamma)^{-p-n}$$

Then, by their aid, we have from Leibnitz's theorem

$$\frac{d^k}{d\gamma^k}\gamma^{k+1}\log(1+\beta)=\sum_{t=0}^{t=k}\frac{k\,!}{t\,!\,k-t\,!}\frac{d^t}{d\gamma^t}\gamma^{k+1}\frac{d^{k-t}}{d\gamma^{k-t}}\log(1+\beta)$$
$$=\sum_{t=0}^{t=k}\frac{k\,!}{t\,!\,k-t\,!}\,\frac{k+1\,!\,k-t-1\,!}{k-t+1\,!}\,\frac{\gamma^{k-t+1}}{(1-\gamma)^{k-t}}$$

in which we interpret $(-1)!/(1-\gamma)^0$ as $\log(1+\beta)$.

Thus $$(k,\gamma)=\sum_{t=0}^{t=k}\frac{1}{(k-t+1)(k-t)}\frac{k+1\,!}{t\,!\,k-t\,!}\beta^{k-t}\quad\ldots\ldots\ldots\ldots(30)$$

with $\beta^0/0=\log(1+\beta)$.

Again

$$\frac{1}{i\,!\,k\,!}\frac{d^i}{d\gamma^i}\gamma^i\frac{d^k}{d\gamma^k}\{\gamma^{k+1}\log(1+\beta)\}$$
$$=\frac{1}{i\,!}\sum_{t=0}^{t=k}\frac{k+1\,!\,k-t-1\,!}{t\,!\,k-t\,!\,k-t+1\,!}\frac{d^i}{d\gamma^i}\frac{\gamma^{i+k-t+1}}{(1-\gamma)^{k-t}}$$
$$=\gamma\sum_{t=0}^{t=k}\sum_{r=0}^{r=i}\frac{k+1\,!\,k-t-1\,!\,i+k-t+1\,!\,i+k-r-t-1\,!}{t\,!\,k-t\,!\,r\,!\,i-r\,!\,k-t+1\,!\,k-t-1\,!\,i+k-r-t+1\,!}\beta^{i+k-r-t}$$

Hence

$$[k,i,\gamma]=\sum_{t=0}^{t=k}\sum_{r=0}^{r=i}\frac{1}{(i+k-r-t+1)(i+k-r-t)}$$
$$\times\frac{k+1\,!\,i+k-t+1\,!}{k-t+1\,!}\cdot\frac{\beta^{i+k-r-t}}{t\,!\,k-t\,!\,r\,!\,i-r\,!}\ldots(31)$$

In (30) and (31) the infinite series are replaced by finite series.

From the form of the series it is obvious that the result must be symmetrical with respect to k and i, so that $[k, i, \gamma]=[i, k, \gamma]$, but this is not obvious on the face of the formula (31).

We shall find, therefore, the symmetrical form of (31) for the first few terms.

If $t=k$, $r=i$, we obviously have

$$\textit{First term}=(k+1)(i+1)\log(1+\beta)$$

The second term arises from $t=k$, $r=i-1$, and $t=k-1$, $r=i$. The two corresponding values of (31) will be found to add together, and we get

$$\textit{Second term}=\tfrac{1}{4}(k+1)(i+1)[2(i+k)+ik]\beta$$

The third term arises from $t=k$, $r=i-2$; $t=k-1$, $r=i-1$; $t=k-2$, $r=i$, and we find

$$\textit{Third term}=\frac{1}{3\,.\,2}(k+1)(i+1)\left\{\frac{i(i-1)+k(k-1)}{2\,!}\right.$$
$$\left.+\frac{ik(i+2)(k+2)}{2\,!\,3\,!}+\frac{ik(i+k+1)}{2\,!^2}\right\}\beta^2$$

A symmetrical form for further terms may be obtained by writing (31) first with i before k and then with k before i, and taking half the sum of the two results. In computing these coefficients it is a useful check to compute from both unsymmetrical forms, when the identity of results verifies the computation.

The following Tables have been computed from (30) and (31). The numbers are the coefficients of the quantities at the heads of the columns for the values of k and i written in the first column. The series (k, γ) is terminable with β^k, and the series $[k, i, \gamma]$ is terminable with β^{k+i}.

In $[k, i, \gamma]$ the coefficients have only been computed as far as β^6, so that the last which is given completely is $[2, 4, \gamma]$; however, with such values of β as we require, the series are carried far enough to give numerical results with sufficient accuracy.

TABLE of (k, γ).

	$\operatorname{Log}(1+\beta)$	$+\beta$	$+\beta^2$	$+\beta^3$	$+\beta^4$	$+\beta^5$
$k=2$	3	3	$\frac{1}{2}$	...	...	...
$k=3$	4	6	2	$\frac{1}{3}$	...	...
$k=4$	5	10	5	$1\frac{2}{3}$	$\frac{1}{4}$	...
$k=5$	6	15	10	5	$1\frac{1}{2}$	$\frac{1}{5}$

TABLE of $[k, i, \gamma]$.

	$\operatorname{Log}(1+\beta)$	$+\beta$	$+\beta^2$	$+\beta^3$	$+\beta^4$	$+\beta^5$	$+\beta^6$	$+\beta^7$
$k=2,\ i=2$	9	27	$18\frac{1}{2}$	8	$1\frac{1}{2}$	...	...	...
$k=2,\ i=3$	12	48	46	31	12	2	...	...
$k=2,\ i=4$	15	75	$92\frac{1}{2}$	85	$50\frac{1}{4}$	17	$2\frac{1}{2}$	...
$k=2,\ i=5$	18	108	163	190	$151\frac{1}{2}$	$77\frac{3}{5}$	23	&c.
$k=3,\ i=3$	16	84	108	103	63	22	$3\frac{1}{3}$	...
$k=3,\ i=4$	20	130	210	260	219	118	$36\frac{2}{3}$	&c.
$k=3,\ i=5$	24	186	362	552	594	$434\frac{4}{5}$	206	&c.
$k=4,\ i=4$	25	200	400	625	$687\frac{3}{4}$	514	$248\frac{1}{3}$	&c.
$k=4,\ i=5$	30	285	680	1285	$1750\frac{1}{2}$	1681	1110	&c.
$k=5,\ i=5$	36	405	1145	2585	4272	$5098\frac{3}{5}$	4345	&c.

We must now go back and determine the value of the outstanding potential of the first degree of harmonics, which will be annulled when rotation is imposed on the system. The potential is given in (22–i.) and (22–ii.); (22–i.) contributes nothing, and (22–ii.) gives us, for $k = 1$,

$$\frac{4\pi A^3}{3c}\left[\frac{a}{c} + \tfrac{3}{2}\left(\frac{a}{c}\right)^4 \sum_{i=2}^{i=\infty} \frac{i+1!}{1!\,i!}\frac{\Gamma^{i-1}}{i-1} H_i\right]\frac{w_1}{a}$$

Thus, if we call u_1, U_1 the outstanding potential of the first degree, when referred to the two origins respectively, we have

$$\left.\begin{aligned} u_1 &= \frac{4\pi A^3}{3c}\left[1 + \tfrac{3}{2}\left(\frac{a}{c}\right)^3 \sum_{i=2}^{i=\infty} \frac{i+1}{i-1}\Gamma^{i-1} H_i\right]\frac{a}{c}\cdot\frac{w_1}{a} \\ U_1 &= \frac{4\pi a^3}{3c}\left[1 + \tfrac{3}{2}\left(\frac{A}{c}\right)^3 \sum_{i=2}^{i=\infty} \frac{i+1}{i-1}\gamma^{i-1} h_i\right]\frac{A}{c}\cdot\frac{W_1}{A} \end{aligned}\right\} \quad\text{......(32)}$$

§ 3. *The Potential due to Rotation.*

Intermediate between the two origins o and O take a third Q, and take the axes of ξ and η parallel to those of x and y, and that of ζ identical with that of z. Let $Qo = d$, $QO = D$.

Then suppose that the system of the two spheroids is in uniform rotation about the axes of ξ with an angular velocity ω.

The potential Ω of the centrifugal force is given by

$$\Omega = \tfrac{1}{2}\omega^2(\eta^2 + \zeta^2) \quad\text{..............................(33)}$$

But

$$\left.\begin{aligned} z &= \zeta + d, & Z &= D - \zeta, \quad d + D = c \\ y &= \eta & Y &= -\eta \\ x &= \xi & X &= -\xi \end{aligned}\right\} \quad\text{..................(34)}$$

Hence

$$\Omega = \tfrac{1}{2}\omega^2(y^2 + z^2 - 2zd + d^2)$$
$$= \tfrac{1}{2}\omega^2\left[-\tfrac{1}{2}(x^2 - y^2) + \tfrac{1}{3}(z^2 - \tfrac{1}{2}x^2 - \tfrac{1}{2}y^2) + \tfrac{2}{3}(x^2 + y^2 + z^2) - 2zd + d^2\right]$$

Then, remembering that

$$w_2 = z^2 - \tfrac{1}{2}x^2 - \tfrac{1}{2}y^2, \quad w_1 = z$$

and if we put

$$q_2 = x^2 - y^2, \quad Q_2 = X^2 - Y^2$$

we have

$$\Omega = -\tfrac{1}{4}\omega^2 q_2 + \tfrac{1}{6}\omega^2 w_2 - \omega^2 d w_1 + \tfrac{1}{3}\omega^2 r^2 + \tfrac{1}{2}\omega^2 d^2 \quad\text{..........(35)}$$

Similarly the rotation potential, when developed with reference to the other origin O, is

$$\Omega = -\tfrac{1}{4}\omega^2 Q_2 + \tfrac{1}{6}\omega^2 W_2 - \omega^2 D W_1 + \tfrac{1}{3}\omega^2 R^2 + \tfrac{1}{2}\omega^2 D^2 \quad\text{........(36)}$$

The last terms of (35) and (36) are constants, and the term in r^2, and that in R^2 are symmetrical about each origin, and so the corresponding forces can produce no departure from sphericity in either mass; thus these terms may be dropped. Next we have in (35) and (36) the outstanding potentials $-\omega^2 dw_1$ and $-\omega^2 DW_1$, which will be annulled by other similar terms, and so need not be considered now. We are left, therefore, with the terms in q_2 and w_2, or in Q_2 and W_2. The q_2 is a sectorial harmonic, the w_2 a zonal, and it will be convenient to treat them separately. We shall begin with the zonal term.

§ 4. *Disturbance due to the Zonal Harmonic Rotational Term.*

The potential whose effects we are to consider is $\frac{1}{6}\omega^2 w_2$ or $\frac{1}{6}\omega^2 W_2$, according to the origin which we are considering.

If an isolated spheroid of fluid of unit density be rotating with angular velocity ω, the ellipticity of the spheroid is $15\omega^2/16\pi$; therefore we put

$$\epsilon = \frac{15\omega^2}{16\pi} \quad \text{.................................(37)}$$

Let us assume, for the equations to the two spheroids,

$$\left.\begin{aligned} \frac{r}{a} &= 1 + \tfrac{1}{3}\epsilon\frac{w_2}{r^2} + \left(\frac{A}{a}\right)^3 \sum_{i=2}^{i=\infty} \frac{2i+1}{2i-2}\left(\frac{a}{c}\right)^{i+1} l_i \frac{w_i}{r^i} \\ \frac{R}{A} &= 1 + \tfrac{1}{3}\epsilon\frac{W_2}{R^2} + \left(\frac{a}{A}\right)^3 \sum_{i=2}^{i=\infty} \frac{2i+1}{2i-2}\left(\frac{A}{c}\right)^{i+1} L_i \frac{W_i}{R^i} \end{aligned}\right\} \quad \text{.........(38)}$$

where l_i, L_i, are unknown coefficients which are to be determined. We now have to determine the potentials at any point of the inequalities (38) on the two spheroids.

The potential of the inequality $\frac{1}{3}\epsilon w_2/r^2$ in the first of (38) is

$$\tfrac{4}{5}\pi a^3 \,.\, a^2 \,.\, \tfrac{1}{3}\epsilon\frac{w_2}{r^5} = \tfrac{1}{4}\omega^2 a^2 \left(\frac{a}{r}\right)^3 \frac{w_2}{r^2} \quad \text{......................(39)}$$

The similar inequality in the second of (38) gives us

$$\tfrac{4}{5}\pi A^3 \,.\, A^2 \,.\, \tfrac{1}{3}\epsilon\frac{W_2}{R^5} = \frac{4\pi A^3}{3c}\,\tfrac{1}{5}\epsilon\left(\frac{A}{c}\right)^2 . \frac{c^3 W_2}{R^5} \quad \text{..............(40)}$$

The term in l_k in the first of (38) gives us, as in § 2,

$$\frac{4\pi A^3}{3c}\,\frac{3l_k}{2k-2}\left(\frac{a}{c}\right)^k \left(\frac{a}{r}\right)^{k+1} \frac{w_k}{r^k} \quad \text{........................(41)}$$

The term in L_i in the second of (38) gives us, as in § 2,

$$\tfrac{4}{3}\pi a^3 \left(\frac{A}{c}\right)^3 \frac{3L_i\Gamma^{i-1}}{2i-2}\,\frac{c^i W_i}{R^{2i+1}} \quad \text{........................(42)}$$

The potential due to rotation is $\frac{1}{6}\omega^2 w_2$ or $\frac{1}{6}\omega^2 W_2$, being the second term of (35) or (36); this term we find it convenient to write

$$\tfrac{1}{6}\omega^2 a^2 \left(\frac{r}{a}\right)^2 \frac{w_2}{r^2} \quad \text{..................................(43)}$$

The sums of the several terms (39), (40), (41), (42), and (43) are to be regarded as the potential of perturbing forces by which the spheroid a or the spheroid A is disturbed, and the arbitrary constants l and L are to be so chosen that each may be a figure of equilibrium.

We may consider the spheroid a by itself, and the solution for it will afford us the solution for the spheroid A by symmetry. In order to find the disturbance, the formulæ (40) and (42) must be transferred.

Now by (4), with $i=2$,

$$\frac{4\pi A^3}{3c}\,\tfrac{1}{5}\epsilon\left(\frac{A}{c}\right)^2\frac{c^3 W_2}{R^5}=\frac{4\pi A^3}{3c}\,\tfrac{1}{5}\epsilon\left(\frac{A}{c}\right)^2\sum_{k=0}^{k=\infty}\frac{k+2\,!}{2\,!\,k\,!}\left(\frac{a}{c}\right)^k\frac{w_k}{a^k}\quad\text{...(40')}$$

And again, by (4),

$$\frac{4\pi a^3}{3c}\left(\frac{A}{c}\right)^3\frac{3L_i\Gamma^{i-1}}{2i-2}\,\frac{c^i W_i}{R^{2i+1}}=\frac{4\pi A^3}{3c}\,\tfrac{3}{2}\left(\frac{a}{c}\right)^3\sum_{k=0}^{k=\infty}\frac{k+i\,!}{i\,!\,k\,!}\,\frac{\Gamma^{i-1}}{i-1}\,L_i\left(\frac{a}{c}\right)^k\frac{w_k}{a^k}\text{...(42')}$$

Then (39), (40′), the sum of (41) from $k=\infty$ to $k=2$, the sum of (42′) from $i=\infty$ to $i=2$, and (43) together constitute the disturbing potential, all now referred to the origin o.

In order to find the disturbance of the spheroid a, we add the perturbing potential to $\frac{4}{3}\pi a^3/r$, give r its value (38) in this term, put $r=a$ in the perturbing potential, and make the whole potential constant by equating to zero the coefficient of each harmonic term.

We will begin by putting $r/a=1+\frac{1}{3}\epsilon w_2/r^2$, and considering only the perturbing potentials (39) and (43). We have then, for the coefficient of w_2/r^2,

$$-\tfrac{4}{3}\pi a^2\,.\,\tfrac{1}{3}\epsilon+\tfrac{1}{4}\omega^2a^2+\tfrac{1}{6}\omega^2a^2$$

Now, with the value of ϵ in (37),

$$-\tfrac{4}{3}\pi a^2\,.\,\tfrac{1}{3}\epsilon=-\tfrac{5}{12}\omega^2a^2\quad\text{and}\quad-\tfrac{5}{12}+\tfrac{1}{4}+\tfrac{1}{6}=0$$

Hence the coefficient of w_2/r^2 vanishes, and the term in ϵ in (38) has been properly chosen to satisfy the perturbing potentials (39) and (43).

Following the similar process with the remaining terms of (38), and equating to zero the coefficient of w_k, we have from (40′), (41), and (42′),

$$-\frac{2k+1}{2k-2}\,l_k+\frac{3l_k}{2k-2}+\tfrac{1}{5}\epsilon\left(\frac{A}{c}\right)^2\frac{k+2\,!}{k\,!\,2\,!}+\tfrac{3}{2}\left(\frac{a}{c}\right)^3\sum_{i=2}^{i=\infty}\frac{k+i\,!}{k\,!\,i\,!}\,\frac{\Gamma^{i-1}}{i-1}\,L_i=0$$

whence
$$l_k=\tfrac{1}{5}\epsilon\left(\frac{A}{c}\right)^2\frac{k+2\,!}{k\,!\,2\,!}+\tfrac{3}{2}\left(\frac{a}{c}\right)^3\sum_{i=2}^{i=\infty}\frac{k+i\,!}{k\,!\,i\,!}\,\frac{\Gamma^{i-1}}{i-1}\,L_i\quad\text{........(44)}$$

By symmetry, the condition that the spheroid A may be a level surface is

$$L_r=\tfrac{1}{5}\epsilon\left(\frac{a}{c}\right)^2\frac{r+2\,!}{r\,!\,2\,!}+\tfrac{3}{2}\left(\frac{A}{c}\right)^3\sum_{i=2}^{i=\infty}\frac{r+i\,!}{r\,!\,i\,!}\,\frac{\gamma^{i-1}}{i-1}\,l_i\quad\text{...........(45)}$$

Multiply both sides of (45) by $\frac{3}{2}\left(\frac{a}{c}\right)^3 \frac{r+k\,!}{r\,!\,k\,!}\frac{\Gamma^{r-1}}{r-1}$, and perform $\sum_{r=2}^{r=\infty}$ on the whole, and substitute from (44); and we have

$$l_k - \tfrac{1}{5}\epsilon\left(\frac{A}{c}\right)^2 \frac{k+2\,!}{k\,!\,2\,!} = \tfrac{1}{5}\epsilon\left(\frac{a}{c}\right)^2 \tfrac{3}{2}\left(\frac{a}{c}\right)^3 \sum_{r=2}^{r=\infty} \frac{r+2\,!\,r+k\,!}{2\,!\,r\,!\,r\,!\,k\,!}\frac{\Gamma^{r-1}}{r-1}$$
$$+ (\tfrac{3}{2})^2\left(\frac{a}{c}\right)^3\left(\frac{A}{c}\right)^3 \sum_{i=2}^{i=\infty}\sum_{r=2}^{r=\infty} \frac{r+i\,!\,r+k\,!}{r\,!\,i\,!\,r\,!\,k\,!}\frac{\gamma^{i-1}}{i-1}\frac{\Gamma^{r-1}}{r-1}\,l_i \text{.........(46)}$$

Introducing the notation (26) for the series involved in (46), we have

$$l_k = \tfrac{1}{5}\epsilon\left(\frac{A}{c}\right)^2\left\{\tfrac{1}{2}(k+1)(k+2) + \tfrac{3}{2}\left(\frac{a}{A}\right)^2\left(\frac{a}{c}\right)^3 [k, 2, \Gamma]\right\}$$
$$+ (\tfrac{3}{2})^2\left(\frac{a}{c}\right)^3\left(\frac{A}{c}\right)^3 \sum_{i=2}^{i=\infty} [k, i, \Gamma]\frac{\gamma^{i-1}}{i-1}\,l_i \text{.........(47)}$$

Each value of k gives a similar equation, and there is a similar series of equations with small and large letters interchanged.

Now put

$$\left.\begin{aligned} l_k &= \tfrac{1}{10}\epsilon\left(\frac{A}{c}\right)^2 (k+1)(k+2)\lambda_k \\ L_i &= \tfrac{1}{10}\epsilon\left(\frac{a}{c}\right)^2 (k+1)(k+2)\Lambda_k \end{aligned}\right\} \text{..................(48)}$$

and (47) becomes

$$\lambda_k = 1 + \frac{3}{(k+1)(k+2)}\left(\frac{a}{A}\right)^2\left(\frac{a}{c}\right)^3 [k, 2, \Gamma]$$
$$+ (\tfrac{3}{2})^2\left(\frac{a}{c}\right)^3\left(\frac{A}{c}\right)^3 \sum_{i=2}^{i=\infty} \frac{(i+1)(i+2)}{(k+1)(k+2)}[k, i, \Gamma]\frac{\gamma^{i-1}}{i-1}\lambda_i \text{......(49)}$$

We attribute to k in (49) all values from ∞ to 2, and thus find a series of equations for the λ's. A similar series of equations holds for the Λ's.

We must now find the outstanding potential of the first degree of harmonics. No such term exists in (39), (41), (43), but it arises entirely out of (40′) and (42′). If we write v_1 for the outstanding potential, we have clearly

$$v_1 = \frac{4\pi A^3}{3c}\left\{\tfrac{1}{5}\epsilon\left(\frac{A}{c}\right)^2 \frac{3\,!}{2\,!\,1\,!}\frac{a}{c}\cdot\frac{w_1}{a} + \tfrac{3}{2}\left(\frac{a}{c}\right)^3 \sum_{i=2}^{i=\infty}\frac{i+1\,!}{i\,!\,1!}\frac{\Gamma^{i-1}}{i-1}L_i\frac{a}{c}\frac{w_1}{a}\right\}$$

whence

$$v_1 = \frac{4\pi A^3}{3c}\tfrac{1}{10}\epsilon\left(\frac{a}{c}\right)^2\left\{6\left(\frac{A}{a}\right)^2 + \sum_{i=2}^{i=\infty}\frac{(i+1)^2(i+2)}{i-1}\Gamma^{i-1}\Lambda_i\right\}\frac{a}{c}\cdot\frac{w_1}{a} \text{...(50)}$$

and, by symmetry,

$$V_1 = \frac{4\pi a^3}{3c}\tfrac{1}{10}\epsilon\left(\frac{A}{c}\right)^2\left\{6\left(\frac{a}{A}\right)^2 + \sum_{i=2}^{i=\infty}\frac{(i+1)^2(i+2)}{i-1}\gamma^{i-1}\lambda_i\right\}\frac{A}{c}\cdot\frac{W_1}{A} \text{...(51)}$$

§ 5. *Disturbance due to the Sectorial Harmonic Rotational Term.*

In (35) and (36) we have found this term to be $-\frac{1}{4}\omega^2 q_2$ or $-\frac{1}{4}\omega^2 Q_2$.

We have already observed that, if the operation $\frac{d^2}{dx^2}-\frac{d^2}{dy^2}$ or δ^2 be performed on w_i, the result vanishes when $i=1, 2, 3$.

Now, by (6), $$w_4 = \sum_{k=0}^{k=2} (-)^k \frac{4\,!}{k\,!^2\, 4-2k\,!} z^{4-2k} \rho^k$$

$$= z^4 - \frac{4\,!}{1\,!^2\, 2\,!} z^2 \rho + \frac{4\,!}{2\,!^2\, 0\,!} \rho^2$$

Hence $\frac{1}{4} d^2 w_4/d\rho^2 = 3$, and, since $\delta^2 w_4 = \frac{1}{4}(x^2-y^2)\, d^2 w_4/d\rho^2$, it follows that

$$q_2 = x^2 - y^2 = \tfrac{1}{3}\delta^2 w_4, \quad \text{and} \quad Q_2 = \tfrac{1}{3}\delta^2 W_4 \quad \text{.............(52)}$$

Hence the sectorial rotational term is $-\frac{1}{12}\omega^2 \delta^2 w_4$ or $-\frac{1}{12}\omega^2 \delta^2 W_4$; this potential is of the second order of sectorial harmonics.

Now, with ϵ as defined in (37), let us assume as the equations to the two surfaces,

$$\left.\begin{aligned} \frac{r}{a} &= 1 - \tfrac{1}{6}\epsilon \frac{\delta^2 w_4}{r^2} - \left(\frac{A}{a}\right)^3 \sum_{i=2}^{i=\infty} \frac{2i+1}{2i-2} \left(\frac{a}{c}\right)^{i+1} m_i \frac{\delta^2 w_{i+2}}{r^i} \\ \frac{R}{A} &= 1 - \tfrac{1}{6}\epsilon \frac{\delta^2 W_4}{R^2} - \left(\frac{a}{A}\right)^3 \sum_{i=2}^{i=\infty} \frac{2i+1}{2i-2} \left(\frac{A}{c}\right)^{i+1} M_i \frac{\delta^2 W_{i+2}}{R^i} \end{aligned}\right\} \quad \text{......(53)}$$

We have now to determine the potentials of the inequalities on the two spheroids expressed by (53).

The potential of the inequality $-\frac{1}{10}\epsilon\,\delta^2 w_4/r^2$ in the first of (53) is

$$-\tfrac{4}{5}\pi a^3 . a^2 . \tfrac{1}{6}\epsilon \frac{\delta^2 w_4}{r^5} = -\tfrac{1}{8}\omega^2 a^2 \left(\frac{a}{r}\right)^3 \frac{\delta^2 w_4}{r^2} \quad \text{.............(54)}$$

The potential of the similar inequality in the second of (53) is

$$-\frac{4\pi A^3}{3c} \cdot \left(\frac{A}{c}\right)^2 \tfrac{1}{10}\epsilon \frac{c^3 \delta^2 W_4}{R^5} \quad \text{......................(55)}$$

The term in m_k in the first of (53) gives us

$$-\frac{4\pi A^3}{3c} \frac{3m_k}{2k-2} \left(\frac{a}{c}\right)^k \left(\frac{a}{r}\right)^{k+1} \frac{\delta^2 w_{k+2}}{r^2} \quad \text{................(56)}$$

The term in M_i in the second of (53) gives us

$$-\frac{4\pi A^3}{3c} \left(\frac{a}{c}\right)^3 \frac{3M_i}{2i-2} \Gamma^{i-1} \frac{c^i \delta^2 W_{i+2}}{R^{2i+1}} \quad \text{................(57)}$$

Lastly, the sectorial term itself is

$$-\tfrac{1}{12}\omega^2 a^2 \left(\frac{r}{a}\right)^2 \frac{\delta^2 w_4}{r^2} \quad \text{........................(58)}$$

The sums of the several terms (54), (55), (56), (57), and (58) are to be regarded as the potential of perturbing forces by which the spheroid a, or the spheroid A, is disturbed, and the arbitrary constants m, M, are to be so chosen that they may each be figures of equilibrium. We may consider the spheroid a by itself, and the solution for it will afford the solution for the spheroid A by symmetry. In order to find the disturbance, the formulæ (55) and (57) must be transferred. For this purpose we require the second transference formulæ.

By (16), with $i = 2$, we have for (55)

$$-\frac{4\pi A^3}{3c}\left(\frac{A}{c}\right)^2 \tfrac{1}{10}\epsilon \frac{c^3\delta^2 W_4}{R^5} = -\frac{4\pi A^3}{3c}\tfrac{1}{10}\epsilon\left(\frac{A}{c}\right)^2 \sum_{k=2}^{k=\infty} \frac{k+2\,!}{0\,!\,k+2\,!}\left(\frac{a}{c}\right)^k \frac{\delta^2 w_{k+2}}{a^k} \quad \ldots(55')$$

And by (16) we have for (57)

$$-\frac{4\pi A^3}{3c}\left(\frac{a}{c}\right)^3 \frac{3M_i}{2i-2}\Gamma^{i-1}\frac{c^i\delta^2 W_{i+2}}{R^{2i+1}}$$

$$= -\frac{4\pi A^3}{3c}\tfrac{3}{2}\left(\frac{a}{c}\right)^3 \sum_{k=2}^{k=\infty} \frac{k+i\,!}{i-2\,!\,k+2\,!}\frac{\Gamma^{i-1}}{i-1}M_i\left(\frac{a}{c}\right)^k \frac{\delta^2 w_{k+2}}{a^k} \quad \ldots(57')$$

Then (54), (55′), the sum of (56) from $k=\infty$ to $k = 2$, the sum of (57′) from $i = \infty$ to $i = 2$, and (58) together constitute the disturbing potential, all now referred to the origin o.

In order to find the disturbance of the spheroid a, we add the perturbing potential to $\tfrac{4}{3}\pi a^3/r$, give r its value (53) in this term, put $r = a$ in the perturbing potential, and make the whole potential constant by equating to zero the coefficients of each harmonic term.

We will begin by putting $r/a = 1 - \tfrac{1}{6}\epsilon\delta^2 w_4/r^2$, and considering only the perturbing potentials (54) and (58). We have then, for the coefficient of $\delta^2 w_4/r^2$,

$$\tfrac{4}{3}\pi a^3 . \tfrac{1}{6}\epsilon - \tfrac{1}{8}\omega^2 a^2 - \tfrac{1}{12}\omega^2 a^2$$

Now, with the value of ϵ in (37),

$$\tfrac{4}{3}\pi a^3 . \tfrac{1}{6}\epsilon = \tfrac{5}{24}\omega^2 a^2, \quad \text{and} \quad \tfrac{5}{24} - \tfrac{1}{8} - \tfrac{1}{12} = 0$$

Hence the coefficient of $\delta^2 w_4/r^2$ vanishes, and the term ϵ in (53) has been properly chosen to satisfy the perturbing potentials (54) and (58). Following the similar process with the remaining terms of (53), and equating to zero the coefficient of $\delta^2 w_{k+2}$, we have, from (55′), (56), (57′),

$$\frac{2k+1}{2k-2}m_k - \frac{3m_k}{2k-2} - \tfrac{1}{10}\epsilon\left(\frac{A}{c}\right)^2 - \tfrac{3}{2}\left(\frac{a}{c}\right)^3 \sum_{i=2}^{i=\infty} \frac{k+i\,!}{i-2\,!\,k+2\,!}\frac{\Gamma^{i-1}}{i-1}M_i = 0$$

or

$$m_k = \tfrac{1}{10}\epsilon\left(\frac{A}{c}\right)^2 + \tfrac{3}{2}\left(\frac{a}{c}\right)^3 \sum_{i=2}^{i=\infty} \frac{k+i\,!}{i-2\,!\,k+2\,!}\frac{\Gamma^{i-1}}{i-1}M_i \quad \ldots\ldots\ldots(59)$$

By symmetry the condition that the spheroid A may be a level surface is

$$M_r = \tfrac{1}{10}\epsilon\left(\frac{a}{c}\right)^2 + \tfrac{3}{2}\left(\frac{A}{c}\right)^3 \sum_{i=2}^{i=\infty} \frac{r+i\,!}{i-2\,!\,r+2\,!}\frac{\gamma^{i-1}}{i-1}m_i \quad \ldots\ldots\ldots(60)$$

Multiply both sides of (60) by $\frac{3}{2}\left(\dfrac{a}{c}\right)^3 \dfrac{k+r!}{r-2!\,k+2!}\dfrac{\Gamma^{r-1}}{r-1}$, and perform $\displaystyle\sum_{r=2}^{r=\infty}$ on the whole, and substitute from (59), and we have

$$m_k - \tfrac{1}{10}\epsilon\left(\frac{A}{c}\right)^2 = \tfrac{1}{10}\epsilon\left(\frac{a}{c}\right)^2 \tfrac{3}{2}\left(\frac{a}{c}\right)^3 \sum_{r=2}^{r=\infty} \frac{k+r!}{r-2!\,k+2!}\frac{\Gamma^{r-1}}{r-1}$$

$$+ \left(\tfrac{3}{2}\right)^2 \left(\frac{a}{c}\right)^3 \left(\frac{A}{c}\right)^3 \sum_{r=2}^{r=\infty}\sum_{i=2}^{i=\infty} \frac{r+i!\,k+r!}{i-2!\,r+2!\,r-2!\,k+2!}\frac{\Gamma^{r-1}}{r-1}\frac{\gamma^{i-1}}{i-1}\,m_i \quad (61)$$

Now let us write

$$\left.\begin{aligned}
\{k,\ \Gamma\} &= \sum_{r=2}^{r=\infty} \frac{k+r!}{r-1!\,k+2!}\,\Gamma^{r-1}\\[2mm]
\boxed{k,\,i,\,\Gamma} &= \sum_{r=2}^{r=\infty} \frac{r+i!\,k+r!}{i-1!\,r+2!\,r-1!\,k+2!}\,\Gamma^{r-1}
\end{aligned}\right\} \quad\ldots\ldots(62)$$

so that (61) may be written

$$m_k = \tfrac{1}{10}\epsilon\left(\frac{A}{c}\right)^2 + \tfrac{1}{10}\epsilon\left(\frac{a}{c}\right)^2 \tfrac{3}{2}\left(\frac{a}{c}\right)^3 \{k,\ \Gamma\} + \left(\tfrac{3}{2}\right)^2\left(\frac{a}{c}\right)^3\left(\frac{A}{c}\right)^3 \sum_{i=2}^{i=\infty}\boxed{k,\,i,\,\Gamma}\,\gamma^{i-1}\,m_i$$

$$\ldots\ldots(63)$$

Next put $\qquad m_k = \tfrac{1}{10}\epsilon\left(\dfrac{A}{c}\right)^2 \mu_k, \qquad M_k = \tfrac{1}{10}\epsilon\left(\dfrac{a}{c}\right)^2 \mathrm{M}_k \ldots\ldots\ldots\ldots(64)$

and (63) becomes

$$\mu_k = 1 + \tfrac{3}{2}\left(\frac{a}{c}\right)^3\left(\frac{a}{A}\right)^2 \{k,\ \Gamma\} + \left(\tfrac{3}{2}\right)^2\left(\frac{a}{c}\right)^3\left(\frac{A}{c}\right)^3 \sum_{i=2}^{i=\infty}\boxed{k,\,i,\,\Gamma}\,\gamma^{i-1}\,\mu_i \quad\ldots(65)$$

We attribute to k in (65) all values from ∞ to 2, and thus find a series of equations for the μ's. A similar series of equations holds for the M's.

We now have to sum the series (62).

Consider the function

$$\frac{1}{k+2}\left[(1+\beta)^{k+2}-1\right] = \frac{1}{k+2}\left[(1-\gamma)^{-k-2}-1\right]$$

$$= \frac{1}{k+2}\left[\frac{k+2}{1!}\gamma + \frac{k+2\,.\,k+3}{2!}\gamma^2 + \ldots\right]$$

$$= \frac{1}{k+2}\sum_{r=2}^{r=\infty}\frac{k+r!}{k+1!\,r-1!}\gamma^{r-1} = \sum_{r=2}^{r=\infty}\frac{k+r!}{k+2!\,r-1!}\gamma^{r-1}$$

Hence $\qquad \{k,\ \gamma\} = \dfrac{1}{k+2}\left[(1+\beta)^{k+2}-1\right] \quad\ldots\ldots\ldots\ldots\ldots(66)$

Next $\qquad \dfrac{1}{(k+2)\gamma^3}\cdot\dfrac{1}{i-1!}\dfrac{d^{i-2}}{d\gamma^{i-2}}\{\gamma^{i+1}\left[(1+\beta)^{k+2}-1\right]\}$

$$= \frac{1}{\gamma^3\,.\,i-1!}\frac{d^{i-2}}{d\gamma^{i-2}}\sum_{r=2}^{r=\infty}\frac{k+r!}{k+2!\,r-1!}\gamma^{i+r}$$

$$= \frac{1}{\gamma^3}\sum_{r=2}^{r=\infty}\frac{k+r!\,i+r!}{i-1!\,k+2!\,r-1!\,r+2!}\gamma^{r+2}$$

Hence

$$\overline{|k, i, \gamma|} = \frac{1}{(k+2)\gamma^3} \cdot \frac{1}{i-1!} \frac{d^{i-2}}{d\gamma^{i-2}} \{\gamma^{i+1}[(1+\beta)^{k+2}-1]\} \quad \text{.........(67)}$$

The differential in (67) must now be evaluated. We have, by Leibnitz's theorem,

$$\frac{d^{i-2}}{d\gamma^{i-2}} \{\gamma^{i+1}[(1+\beta)^{k+2}-1]\}$$

$$= -\frac{d^{i-2}}{d\gamma^{i-2}}\gamma^{i+1} + \sum_{r=0}^{r=i-2} \frac{i-2!}{r!\,i-r-2!} \frac{d^r\gamma^{i+1}}{d\gamma^r} \frac{d^{i-r-2}}{d\gamma^{i-r-2}}(1-\gamma)^{-k-2}$$

$$= -\frac{i+1!}{3!}\gamma^3 + \sum_{r=0}^{r=i-2} \frac{i-2!}{r!\,i-r-2!} \frac{i+1!}{i-r+1!} \frac{i+k-r-1!}{k+1!} \frac{\gamma^{i-r+1}}{(1-\gamma)^{i+k-r}}$$

$$= -\frac{i+1!}{3!}\gamma^3 + \frac{\gamma^3}{(1-\gamma)^{k+2}} \sum_{r=0}^{r=i-2} \ldots \beta^{i-r-2}$$

Substituting in (67), we have

$$\overline{|k, i, \gamma|} =$$

$$\frac{i(i+1)}{6(k+2)}\left[-1 + (1+\beta)^{k+2} \sum_{r=0}^{r=i-2} 3!\,\frac{i-2!\,i+k-r-1!}{r!\,i-r-2!\,i-r+1!\,k+1!}\beta^{i-r-2}\right]$$

......(68)

The following Tables, computed from (66) and (68), give the values of $\{k, \gamma\}$ and $\overline{|k, i, \gamma|}$ as far as $k=5$, and $k=5$, $i=5$.

Table of $\{k, \gamma\}$.

$\{2, \gamma\} = \frac{1}{4}[(1+\beta)^4 - 1]$

$\{3, \gamma\} = \frac{1}{5}[(1+\beta)^5 - 1]$

$\{4, \gamma\} = \frac{1}{6}[(1+\beta)^6 - 1]$

$\{5, \gamma\} = \frac{1}{7}[(1+\beta)^7 - 1]$

Table of $\overline{|k, i, \gamma|}$.

$k=2, i=2;\ \frac{1}{4}[(1+\beta)^4 - 1]$

$k=2, i=3;\ \frac{1}{2}[(1+\beta)^5 - 1]$

$k=2, i=4;\ \frac{5}{6}[(1+\beta)^6 - 1]$

$k=2, i=5;\ \frac{5}{4}[(1+\beta)^7 - 1]$

$k=3, i=2;\ \frac{1}{5}[(1+\beta)^5 - 1]$

$k=3, i=3;\ \frac{2}{5}[(1+\beta)^5(1+\frac{5}{4}\beta) - 1]$

$k=3, i=4;\ \frac{2}{3}[(1+\beta)^6(1+\frac{3}{2}\beta) - 1]$

$k=3, i=5;\ [(1+\beta)^7(1+\frac{7}{4}\beta) - 1]$

$k=4, i=2;\ \frac{1}{6}[(1+\beta)^6 - 1]$

$k=4, i=3;\ \frac{1}{3}[(1+\beta)^6(1+\frac{3}{2}\beta) - 1]$

$k=4, i=4;\ \frac{5}{9}[(1+\beta)^6(1+3\beta+\frac{21}{10}\beta^2) - 1]$

$k=4, i=5;\ \frac{5}{6}[(1+\beta)^7(1+\frac{7}{2}\beta+\frac{14}{5}\beta^2) - 1]$

$k=5, i=2;\ \frac{1}{7}[(1+\beta)^7 - 1]$

$k=5, i=3;\ \frac{2}{7}[(1+\beta)^7(1+\frac{7}{4}\beta) - 1]$

$k=5, i=4;\ \frac{10}{21}[(1+\beta)^7(1+\frac{7}{2}\beta+\frac{14}{5}\beta^2) - 1]$

$k=5, i=5;\ \frac{5}{7}[(1+\beta)^7(1+\frac{21}{4}\beta+\frac{42}{5}\beta^2+\frac{21}{5}\beta^3) - 1]$

§ 6. *Determination of the Angular Velocity of the System.*

The angular velocity of the system must now be determined in such a way as to annul the outstanding potential of the first degree of harmonics.

Referring to origin o, we have from (35) $-\omega^2 dw_1$ directly from the rotation potential; the remaining terms are $u_1 + v_1$, since the sectorial harmonic term does not contribute anything.

Thus, taking u_1 from (32), and v_1 from (50), we get for the potential

$$-\omega^2 dw_1 + \frac{4\pi A^3}{3c}\frac{w_1}{c}\left\{1 + \tfrac{3}{2}\left(\frac{a}{c}\right)^3 \sum_{i=2}^{i=\infty} \frac{i+1}{i-1} \Gamma^{i-1} H_i \right.$$
$$\left. + \tfrac{1}{10}\epsilon\left(\frac{a}{c}\right)^2 \left[6\left(\frac{A}{a}\right)^2 + \tfrac{3}{2}\left(\frac{a}{c}\right)^3 \sum_{i=2}^{i=\infty} \frac{(i+1)^2(i+2)}{i-1} \Gamma^{i-1}\Lambda_i\right]\right\}$$

Equating this to zero,

$$\frac{3\omega^2 c^2}{4\pi} d = A^3\left\{1 + \tfrac{3}{2}\left(\frac{a}{c}\right)^3 \sum_{i=2}^{i=\infty} \frac{i+1}{i-1} \Gamma^{i-1} H_i \right.$$
$$\left. + \tfrac{1}{10}\epsilon\left(\frac{a}{c}\right)^2 \left[6\left(\frac{A}{a}\right)^2 + \tfrac{3}{2}\left(\frac{a}{c}\right)^3 \sum_{i=2}^{i=\infty} \frac{(i+1)^2(i+2)}{i-1} \Gamma^{i-1}\Lambda_i\right]\right\} \quad \text{......(69)}$$

And, by symmetry,

$$\frac{3\omega^2 c^2}{4\pi} D = a^3\left\{1 + \tfrac{3}{2}\left(\frac{A}{c}\right)^3 \sum_{i=2}^{i=\infty} \frac{i+1}{i-1} \gamma^{i-1} h_i \right.$$
$$\left. + \tfrac{1}{10}\epsilon\left(\frac{A}{c}\right)^2 \left[6\left(\frac{a}{A}\right)^2 + \tfrac{3}{2}\left(\frac{A}{c}\right)^3 \sum_{i=2}^{i=\infty} \frac{(i+1)^2(i+2)}{i-1} \gamma^{i-1}\lambda_i\right]\right\} \quad \text{......(70)}$$

Add (70) to (69), note that $d + D = c$, and $\frac{1}{10}\epsilon = 3\omega^2/32\pi$, and solve for ω^2, and we have

$$\frac{3\omega^2}{4\pi} = \left[\left(\frac{A}{c}\right)^3 + \left(\frac{a}{c}\right)^3\right] \times$$
$$\frac{1 + \tfrac{3}{2}\dfrac{A^3 a^3}{(A^3+a^3)c^3} \displaystyle\sum_{i=2}^{i=\infty} \frac{i+1}{i-1}(\Gamma^{i-1}H_i + \gamma^{i-1}h_i)}{1 - \tfrac{3}{4}\left[\left(\dfrac{A}{c}\right)^5 + \left(\dfrac{a}{c}\right)^5\right] - \tfrac{3}{16}\left(\dfrac{A}{c}\right)^3\left(\dfrac{a}{c}\right)^3 \displaystyle\sum_{i=2}^{i=\infty} \frac{(i+1)^2(i+2)}{i-1}(\gamma\Gamma^{i-1}\Lambda_i + \Gamma\gamma^{i-1}\lambda_i)} \quad \text{......(71)}$$

Now let $1 + K$ denote the factor by which $(A/c)^3 + (a/c)^3$ is multiplied in (71). Then, if the two masses were particles, K would be zero, and (71) would simply be the usual formula connecting masses, mean motion, and mean distance in a circular orbit. Hence $1 + K$ is an augmenting factor by which the value of the square of the angular velocity must be multiplied if it be derived from the law of the periodic time of two particles revolving about one another. K, in fact, gives the correction to Kepler's law for the non-sphericity of the masses.

This completes the solution of the problem, for we have determined the angular velocity in such a way as to justify the neglect of the harmonic terms of the first degree in §§ 2 and 4.

§ 7. *Solution of the Problem.*

We may now collect from the preceding paragraphs the complete solution of the problem.

In (38) and (53) we have found that there are terms in r/a as follows:—

$$\tfrac{1}{3}\epsilon\frac{w_2}{r^2}-\tfrac{1}{6}\epsilon\frac{\delta^2 w_4}{r^2}$$

Now $\qquad w_2 = z^2 - \tfrac{1}{2}x^2 - \tfrac{1}{2}y^2, \quad \text{and} \quad \delta^2 w_4 = 3(x^2-y^2)$

hence $\qquad w_2 - \tfrac{1}{2}\delta^2 w_4 = z^2 - 2x^2 + y^2 = r^2 - 3x^2$

and these terms are therefore equal to $\epsilon(\tfrac{1}{3} - x^2/r^2)$.

We note that $\epsilon = 15\omega^2/16\pi = \tfrac{5}{4}\omega^2/\tfrac{4}{3}\pi$, and that $\omega^2/\tfrac{4}{3}\pi$ is the ratio generally written m in works on the figure of the Earth. Then, from (17), (38), (53), the equations to the two surfaces are

$$\left.\begin{aligned}\frac{r}{a} &= 1+\epsilon\left(\tfrac{1}{3}-\frac{x^2}{r^2}\right)+\left(\frac{A}{a}\right)^3\sum_{i=2}^{i=\infty}\frac{2i+1}{2i-2}\left(\frac{a}{c}\right)^{i+1}\left\{(h_i+l_i)\frac{w_i}{r^i}-m_i\frac{\delta^2 w_{i+2}}{r^i}\right\}\\ \frac{R}{A} &= 1+\epsilon\left(\tfrac{1}{3}-\frac{X^2}{R^2}\right)+\left(\frac{a}{A}\right)^3\sum_{i=2}^{i=\infty}\frac{2i+1}{2i-2}\left(\frac{A}{c}\right)^{i+1}\left\{(H_i+L_i)\frac{W_i}{R^i}-M_i\frac{\delta^2 W_{i+2}}{R^i}\right\}\end{aligned}\right\} \quad \text{......(72)}$$

From (27), (49), (65), we see that $h_2, h_3 \dots h_i \dots, \lambda_2, \lambda_3 \dots \lambda_i \dots, \mu_2, \mu_3 \dots \mu_i \dots$, are to be found by solving the equations resulting from all values of k from 2 to infinity in the following:—

$$\left.\begin{aligned}h_k - 1 &= \tfrac{3}{2}\left(\frac{a}{c}\right)^3(k,\Gamma)+(\tfrac{3}{2})^2\left(\frac{a}{c}\right)^3\left(\frac{A}{c}\right)^3\sum_{i=2}^{i=\infty}[k,i,\Gamma]\frac{\gamma^{i-1}}{i-1}h_i\\ \lambda_k - 1 &= \frac{3}{(k+1)(k+2)}\left(\frac{a}{c}\right)^3\left(\frac{a}{A}\right)^3[k,2,\Gamma]\\ &\quad+\frac{1}{(k+1)(k+2)}(\tfrac{3}{2})^2\left(\frac{a}{c}\right)^3\left(\frac{A}{c}\right)^3\sum_{i=2}^{i=\infty}[k,i,\Gamma]\frac{(i+1)(i+2)}{i-1}\gamma^{i-1}\lambda_i\\ \mu_k - 1 &= \tfrac{3}{2}\left(\frac{a}{c}\right)^3\left(\frac{a}{A}\right)^2\{k,\Gamma\}+(\tfrac{3}{2})^2\left(\frac{a}{c}\right)^3\left(\frac{A}{c}\right)^3\sum_{i=2}^{i=\infty}\boxed{k,i,\Gamma}\,\gamma^{i-1}\mu_i\end{aligned}\right\} \quad \text{......(73)}$$

and symmetrical systems of equations for obtaining the H's, Λ's, and M's.

With the values found by the solution of these equations we then evaluate K by formula (71); and we have

$$\tfrac{4}{5}\epsilon = \frac{3\omega^2}{4\pi} = \left[\left(\frac{A}{c}\right)^3+\left(\frac{a}{c}\right)^3\right](1+K) \quad \text{..................(74)}$$

We are now enabled to find the l's and m's by the formulæ (48) and (64), viz.,

$$\left.\begin{aligned} l_k &= \tfrac{1}{10}\epsilon (k+1)(k+2)\left(\frac{A}{c}\right)^2 \lambda_k \\ m_k &= \tfrac{1}{10}\epsilon \left(\frac{A}{c}\right)^2 \mu_k \end{aligned}\right\} \quad \text{.....................(75)}$$

and the symmetrical forms give us the L's and M's.

Having thus evaluated all the auxiliary constants, (72) gives the solution of the problem.

It is well known that $\frac{5}{4} \times 3\omega^2/4\pi$ is the ellipticity of a single homogeneous mass of fluid rotating with angular velocity ω. Hence the first terms of (72) simply denote the ellipticity due to rotation in each of the masses, as if the other did not exist. Now the rigorous solution for the form of equilibrium of a rotating mass of fluid is an ellipsoid of revolution with eccentricity sin g, the value of g being given by the solution of

$$\frac{\omega^2}{2\pi} = \cot^3 \text{g} \, [(3 + \tan^2 \text{g}) \, \text{g} - 3 \tan \text{g}]^* \quad \text{...............(76)}$$

Hence it will undoubtedly be more correct to construct the surface, of which the equation is (72), by regarding the part of r under the symbol Σ as the correction to the radius-vector of an ellipsoid of revolution with eccentricity determined by (76), where $\omega^2/2\pi$ is found from (74).

§ 8. *Examples of the Solution.*

The principal object of the preceding investigation is to trace the forms of the two masses when they approach to close proximity; we shall thus be able to determine the forms when they are on the point of coalescing into a single mass, and shall finally obtain at least an approximate figure of the single mass. For this purpose we require to push the approximation by spherical harmonic analysis as far as it will bear. We shall below endeavour to estimate the degree of departure from correctness involved by the use of this analysis. The results will, therefore, be worked out numerically for such values of c/a as bring the two masses close together, and it will appear that the largest value of c/a assumed for numerical solution is such that the surfaces cross; in this case the reality will be a single mass of a shape which it will be possible to draw with tolerable accuracy.

The computations are facilitated if, instead of assuming c to be an exact multiple of a, we take c^2 a multiple of a^2; that is to say, we shall take $1/\gamma$ as an integer, and therefore $1/\beta$ also an integer.

We shall in the first instance suppose the two masses to be equal. In the following examples, then, we have $A = a$, $\Gamma = \gamma$, $B = \beta$, and the two masses assume the same shape.

* See, for example, Thomson and Tait's *Natural Philosophy* (3), § 771, with $f = \tan \text{g}$.

The computations will be carried through in detail in two cases, viz., when $\beta=\frac{1}{7}$, and when $\beta=\frac{1}{5}$. The results will also be given for $\beta=\frac{1}{6}$.

When $\beta=\frac{1}{7}$, $\gamma=\frac{1}{8}$, $c/a=2{\cdot}8284$, and when $\beta=\frac{1}{5}$, $\gamma=\frac{1}{6}$, $c/a=2{\cdot}449$. Thus the distances of the centres apart are $2\frac{4}{5}$ and $2\frac{2}{5}$ of the mean radius respectively. The numerical details of the two computations may be stated *pari passû*, and the numbers applying to $\beta=\frac{1}{5}$ will be distinguished by being printed in small type.

In the case of $\beta=\frac{1}{6}$, we have $\gamma=\frac{1}{7}$, $c/a=2{\cdot}6458$; but only the final result is given, without intermediate details.

The first step is to compute the values of the several series by means of the Tables in §§ 2 and 5.

The numerical results are as follows.

TABLE of (k, γ).

	$\beta=\frac{1}{7}$	$\beta=\frac{1}{5}$
$k=2$	·839	1·167
$k=3$	1·433	2·012
$k=4$	2·204	3·125
$k=5$	3·163	4·536

TABLE of $\{k, \gamma\}$.

	$\beta=\frac{1}{7}$	$\beta=\frac{1}{5}$
$k=2$	·177	·268
$k=3$	·190	·298
$k=4$	·205	·331
$k=5$	·221	·369

TABLE of $[k, i, \gamma]$.

	$\beta=\frac{1}{7}$	$\beta=\frac{1}{5}$
$k=2,\ i=2$	5·460	7·847
$k=2,\ i=3$	9·494	13·895
$k=2,\ i=4$	14·875	22·201
$k=2,\ i=5$	21·780	33·190
$k=3,\ i=2$	9·494	13·895
$k=3,\ i=3$	16·667	24·969
$k=3,\ i=4$	26·384	40·517
$k=3,\ i=5$	39·047	61·574
$k=4,\ i=2$	14·875	22·201
$k=4,\ i=3$	26·384	40·517
$k=4,\ i=4$	42·214	66·840
$k=4,\ i=5$	63·183	103·372
$k=5,\ i=2$	21·780	33·190
$k=5,\ i=3$	39·047	61·574
$k=5,\ i=4$	63·183	103·372
$k=5,\ i=5$	95·690	162·831

TABLE of $\boxed{k, i, \gamma}$.

	$\beta=\frac{1}{7}$	$\beta=\frac{1}{5}$
$k=2,\ i=2$	·177	·268
$k=2,\ i=3$	·475	·744
$k=2,\ i=4$	1·024	1·655
$k=2,\ i=5$	1·933	3·229
$k=3,\ i=2$	·190	·298
$k=3,\ i=3$	·519	·844
$k=3,\ i=4$	1·137	1·921
$k=3,\ i=5$	2·183	3·837
$k=4,\ i=2$	·205	·331
$k=4,\ i=3$	·569	·961
$k=4,\ i=4$	1·266	2·238
$k=4,\ i=5$	2·471	4·578
$k=5,\ i=2$	·221	·369
$k=5,\ i=3$	·624	1·096
$k=5,\ i=4$	1·412	2·616
$k=5,\ i=5$	2·803	5·479

With these values for the series, we have to compute the coefficients of the systems of simultaneous equations (73). The equations lend themselves more readily to solution if we consider $h_i - 1$, $\lambda_i - 1$, $\mu_i - 1$, as the unknowns instead of h_i, λ_i, μ_i. The results are given in the following equations.

The upper coefficients correspond to the case of $\beta = \frac{1}{7}$; the lower ones, printed in small type, to the case of $\beta = \frac{1}{5}$.

$$h_2 - 1 = \underset{\cdot 13516}{\cdot 05902} + \underset{\cdot 01362}{\cdot 00300}\,(h_2 - 1) + \underset{\cdot 00201}{\cdot 00033}\,(h_3 - 1) + \underset{\cdot 00036}{\cdot 00004}\,(h_4 - 1) + \underset{\cdot 00007}{\cdot 00001}\,(h_5 - 1) + \ldots$$

$$h_3 - 1 = \underset{\cdot 23385}{\cdot 10086} + \underset{\cdot 02412}{\cdot 00522} \quad + \underset{\cdot 00361}{\cdot 00057} \quad + \underset{\cdot 00065}{\cdot 00008} \quad + \underset{\cdot 00012}{\cdot 00001} \quad + \ldots$$

$$h_4 - 1 = \underset{\cdot 36467}{\cdot 15529} + \underset{\cdot 03854}{\cdot 00817} \quad + \underset{\cdot 00586}{\cdot 00091} \quad + \underset{\cdot 00108}{\cdot 00012} \quad + \underset{\cdot 00021}{\cdot 00002} \quad + \ldots$$

$$h_5 - 1 = \underset{\cdot 53150}{\cdot 22321} + \underset{\cdot 05762}{\cdot 01196} \quad + \underset{\cdot 00891}{\cdot 00134} \quad + \underset{\cdot 00165}{\cdot 00018} \quad + \underset{\cdot 00033}{\cdot 00003} \quad + \ldots$$

. . . .

$$\lambda_2 - 1 = \underset{\cdot 15158}{\cdot 06400} + \underset{\cdot 01362}{\cdot 00300}\,(\lambda_2 - 1) + \underset{\cdot 00335}{\cdot 00054}\,(\lambda_3 - 1) + \underset{\cdot 00089}{\cdot 00011}\,(\lambda_4 - 1) + \underset{\cdot 00023}{\cdot 00002}\,(\lambda_5 - 1) + \ldots$$

$$\lambda_3 - 1 = \underset{\cdot 16114}{\cdot 06677} + \underset{\cdot 01447}{\cdot 00313} \quad + \underset{\cdot 00361}{\cdot 00057} \quad + \underset{\cdot 00098}{\cdot 00011} \quad + \underset{\cdot 00026}{\cdot 00002} \quad + \ldots$$

$$\lambda_4 - 1 = \underset{\cdot 17174}{\cdot 06976} + \underset{\cdot 01542}{\cdot 00327} \quad + \underset{\cdot 00391}{\cdot 00060} \quad + \underset{\cdot 00108}{\cdot 00012} \quad + \underset{\cdot 00029}{\cdot 00002} \quad + \ldots$$

$$\lambda_5 - 1 = \underset{\cdot 18352}{\cdot 07297} + \underset{\cdot 01646}{\cdot 00342} \quad + \underset{\cdot 00424}{\cdot 00064} \quad + \underset{\cdot 00118}{\cdot 00013} \quad + \underset{\cdot 00033}{\cdot 00003} \quad + \ldots$$

. . . .

$$\mu_2 - 1 = \underset{\cdot 02818}{\cdot 01184} + \underset{\cdot 00047}{\cdot 00010}\,(\mu_2 - 1) + \underset{\cdot 00022}{\cdot 00003}\,(\mu_3 - 1) + \underset{\cdot 00008}{\cdot 00001}\,(\mu_4 - 1) + \underset{\cdot 00003}{\cdot 00000}\,(\mu_5 - 1) + \ldots$$

$$\mu_3 - 1 = \underset{\cdot 03127}{\cdot 01274} + \underset{\cdot 00052}{\cdot 00010} \quad + \underset{\cdot 00024}{\cdot 00004} \quad + \underset{\cdot 00009}{\cdot 00001} \quad + \underset{\cdot 00003}{\cdot 00000} \quad + \ldots$$

$$\mu_4 - 1 = \underset{\cdot 03478}{\cdot 01374} + \underset{\cdot 00057}{\cdot 00011} \quad + \underset{\cdot 00028}{\cdot 00004} \quad + \underset{\cdot 00011}{\cdot 00001} \quad + \underset{\cdot 00004}{\cdot 00000} \quad + \ldots$$

$$\mu_5 - 1 = \underset{\cdot 03879}{\cdot 01482} + \underset{\cdot 00064}{\cdot 00012} \quad + \underset{\cdot 00032}{\cdot 00004} \quad + \underset{\cdot 00013}{\cdot 00001} \quad + \underset{\cdot 00004}{\cdot 00000} \quad + \ldots$$

. . . .

The solutions of these equations are obviously found by an easy approximation; they are

$$h_2 = \underset{1\cdot 1377}{1\cdot 0593} \qquad \lambda_2 = \underset{1\cdot 1544}{1\cdot 0642} \qquad \mu_2 = \underset{1\cdot 0282}{1\cdot 0118}$$

$$h_3 = \underset{1\cdot 2382}{1\cdot 1012} \qquad \lambda_3 = \underset{1\cdot 1642}{1\cdot 0670} \qquad \mu_3 = \underset{1\cdot 0313}{1\cdot 0127}$$

$$h_4 = \underset{1\cdot 3719}{1\cdot 1559} \qquad \lambda_4 = \underset{1\cdot 1750}{1\cdot 0700} \qquad \mu_4 = \underset{1\cdot 0348}{1\cdot 0137}$$

$$h_5 = \underset{1\cdot 5424}{1\cdot 2241} \qquad \lambda_5 = \underset{1\cdot 1870}{1\cdot 0732} \qquad \mu_5 = \underset{1\cdot 0388}{1\cdot 0148}$$

the small figures corresponding, as before, to the case of $\beta = \frac{1}{5}$.

With these values of the h's and λ's, I find

$$2\Sigma\frac{i+1}{i-1}\gamma^{i-1}h_i=\underset{1\cdot3005}{\cdot8718};\quad 2\Sigma\frac{(i+1)^2(i+2)}{i-1}\gamma^i\lambda_i=\underset{2\cdot8542}{1\cdot3949};\quad \tfrac{3}{2}\left(\frac{a}{c}\right)^5=\underset{\cdot01701}{\cdot00829}$$

the summations, of course, stopping with $i=5$.

Applying these in (71), we have, when

$$\beta=\tfrac{1}{7},\quad 1+K=\frac{1+\cdot02891}{1-\cdot00880}=1\cdot0380$$

or, when

$$\beta=\tfrac{1}{5},\quad 1+K=\frac{1+\cdot0664}{1-\cdot0195}=1\cdot0877$$

whence

$$\tfrac{4}{5}\epsilon=\frac{3\omega^2}{4\pi}=\underset{\cdot13608}{\cdot08839}\times\underset{1\cdot0877}{1\cdot0380}=\underset{\cdot1481}{\cdot09175}$$

Thus the angular velocity of the system has been found.

Next we have

$$\tfrac{1}{10}\epsilon\left(\frac{a}{c}\right)^2=\underset{\cdot00309}{\cdot001434}$$

Introducing this into (48) and (64) with the previously found values of the λ's and μ's,

$$\left.\begin{aligned}
l_2&=\underset{\cdot0428}{\cdot0183}, & m_2&=\underset{\cdot0032}{\cdot00145}\\
l_3&=\underset{\cdot0720}{\cdot0306}, & m_3&=\underset{\cdot0032}{\cdot00145}\\
l_4&=\underset{\cdot1089}{\cdot0460}, & m_4&=\underset{\cdot0032}{\cdot00145}\\
l_5&=\underset{\cdot1541}{\cdot0646}, & m_5&=\underset{\cdot0032}{\cdot00145}
\end{aligned}\right\};\text{ and hence }\left\{\begin{aligned}
h_2+l_2&=\underset{1\cdot1806}{1\cdot0776}\\
h_3+l_3&=\underset{1\cdot3100}{1\cdot1318}\\
h_4+l_4&=\underset{1\cdot4808}{1\cdot2019}\\
h_5+l_5&=\underset{1\cdot6965}{1\cdot2887}
\end{aligned}\right.$$

By taking the differences of $h+l$, we may conclude that

$$h_6+l_6=\underset{1\cdot96}{1\cdot39}$$

and this sixth harmonic term will now be included.

It appears from the values of the m's that the harmonics of the type $\delta^2 w_{i+2}$ are practically negligible, excepting the term $\delta^2 w_4$, and that in that we may neglect the part depending on m_2.

Now, if r denotes the radius-vector due to the rotation, and δr the increase of radius-vector due to the mutual influence of the two masses, we have

$$\frac{\delta r}{a}=\underset{\cdot2008}{\cdot1191}\frac{w_2}{r^2}+\underset{\cdot0637}{\cdot0309}\frac{w_3}{r^3}+\underset{\cdot0252}{\cdot0100}\frac{w_4}{r^4}+\underset{\cdot0108}{\cdot0035}\frac{w_5}{r^5}+\underset{\cdot0048}{\cdot0013}\frac{w_6}{r^6}+\ldots \quad\ldots\ldots(77)$$

We next have to consider the values of r, the radius-vector of the ellipsoid, due to rotation.

We might compute from the spherical harmonic formula

$$\frac{r}{a} = 1 + \epsilon\left(\tfrac{1}{3} - \frac{x^2}{r^2}\right)$$

The results so computed will be compared with the others computed as shown below.

The following Table of the angular velocity and corresponding eccentricity e of the equilibrium ellipsoid of revolution is extracted from Thomson and Tait's *Natural Philosophy*, § 772:—

e	$\dfrac{\omega^2}{2\pi}$
·3	·0243
·4	·0436
·5	·0690
·6	·1007
·7	·1387
·8	·1816

From this we find by interpolation that, when $3\omega^2/4\pi = {\cdot}09175$, $e = {\cdot}472$; and, when $3\omega^2/4\pi = {\cdot}1481$, $e = {\cdot}594$.

These, then, are the eccentricities of the ellipsoids whose radius-vector is r in the two cases $\beta = \frac{1}{7}$, $\beta = \frac{1}{5}$.

The equations to the generating ellipses are

$$\frac{r}{a} = \frac{1 - {\cdot}0806}{1 - {\cdot}2228\cos^2\theta} \text{ for } \beta = \tfrac{1}{7}, \quad \text{and } \frac{r}{a} = \frac{1 - {\cdot}1353}{1 - {\cdot}3535\cos^2\theta} \text{ for } \beta = \tfrac{1}{5}$$

The following are the computed values of r/a for each 15° of θ, the latitude, the small figures written below appertaining to the case of $\beta = \frac{1}{5}$.

$\theta =$	0°	15°	30°	45°	60°	75°	90°
$\beta = \frac{1}{7}$: $\frac{r}{a} =$	1·0429,	1·0330,	1·0074,	·9753,	·9461,	·9264,	·9194
$\beta = \frac{1}{5}$:	1·075,	1·056,	1·009,	·953,	·906,	·875,	·865

Computing from the spherical harmonic formula, I find

$\beta = \frac{1}{7}$: $\frac{r}{a} =$	1·0382,	1·0305,	1·0096,	·9809,	·9522,	·9312,	·9235
$\beta = \frac{1}{5}$:	1·0616,	1·0490,	1·0154,	·9692,	·9230,	·8892,	·8768

The greatest discrepancy occurs when $\beta = \frac{1}{5}$ and $\theta = 90°$, and the difference between the two results is $\frac{1}{70}$ of either. It follows, therefore, that in drawing the figures it is not of much importance which results we take. But, as above remarked, the radius-vectors computed from the true ellipsoidal figure are the more correct.

The formula (77) for δr consists of a series of zonal harmonics. The pole of symmetry of these harmonics lies in the equator of the ellipsoid of revolution defined by r, and is that point of each mass which lies nearest to the other. Then, denoting by θ co-latitude estimated from this pole, I find that the numerical values of δr for each 15° of θ are as follows:—

$\theta =$	0°	15°	30°	45°	60°	75°	90°
$\beta = \frac{1}{7}: \frac{\delta r}{a} =$	+ ·165	+ ·141	+ ·084	+ ·019	− ·031	− ·055	− ·056
$\beta = \frac{1}{5}:$	+ ·280	+ ·257	+ ·142	+ ·024	− ·060	− ·094	− ·092

	105°	120°	135°	150°	165°	180°
$\beta = \frac{1}{7}$	− ·037	− ·004	+ ·032	+ ·065	+ ·088	+ ·096
$\beta = \frac{1}{5}$	− ·059	− ·002	+ ·055	+ ·106	+ ·143	+ ·159

These have to be combined with r, so as to give the radius-vectors of the mass of fluid along two sections, one perpendicular to the axis of rotation (which may be called the equatorial section), the other through the axis and the two centres (which may be called the section through the prime meridian). Taking the case of $\beta = \frac{1}{7}$, we add the successive values of δr to the equatorial value of r, viz., 1·043, and thus find the equally-spaced radius-vectors along the equatorial section. Next we add the successive values of δr to the corresponding values of r, and thus find the equally-spaced radius-vectors along the prime meridian. The results are as follows:—

$\theta =$	0°	15°	30°	45°	60°	75°	90°
$\beta = \frac{1}{7}:$ Equator, $\frac{r + \delta r}{a} =$	1·208	1·184	1·126	1·062	1·012	·988	·987
Pr. Merid. $=$	1·208	1·174	1·091	·994	·915	·871	·863

	105°	120°	135°	150°	165°	180°
Equator	1·006	1·039	1·075	1·108	1·131	1·139
Pr. Merid.	·889	·942	1·008	1·072	1·121	1·139

These results apply to the case of $\beta = \frac{1}{7}$; those for $\beta = \frac{1}{5}$ are found in the same way, and are given in the figures referred to below.

When $\beta = \frac{1}{7}$ the distance between the centres is given by $c/a = 2{\cdot}828$. I have also worked out the case of $\beta = \frac{1}{6}$, although none of the numerical details are given here.

In figs. 2, 3, 4, and 5 are exhibited the figures which result from some of these computations.

Figs. 2 and 3 refer to $\beta = \frac{1}{5}$, 4 and 5 to that of $\beta = \frac{1}{6}$, and the numerical values for $\beta = \frac{1}{7}$, given above, make it easy to draw a figure for $\beta = \frac{1}{7}$.

Since in these cases the masses are equal, the two halves of the figure are the images of one another. The numerical value of each radius-vector is entered on the figures; and other numerical data and explanations are given.

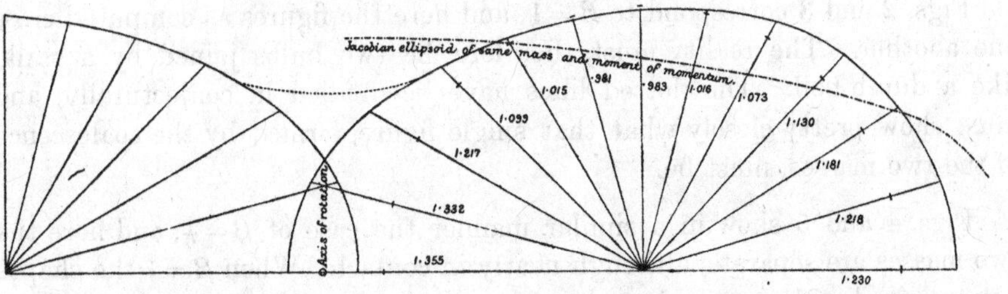

Fig. 2. Dumb-bell figure of equilibrium. Section perpendicular to axis of rotation.

[$A=a$; $c/a=2\cdot449$; $\beta=\frac{1}{5}$; $\gamma=\frac{1}{6}$; $\omega^2/4\pi=\cdot0494$; momentum$=(\frac{4}{3}\pi)^{\frac{3}{2}}\, b^5\times\cdot482$.]

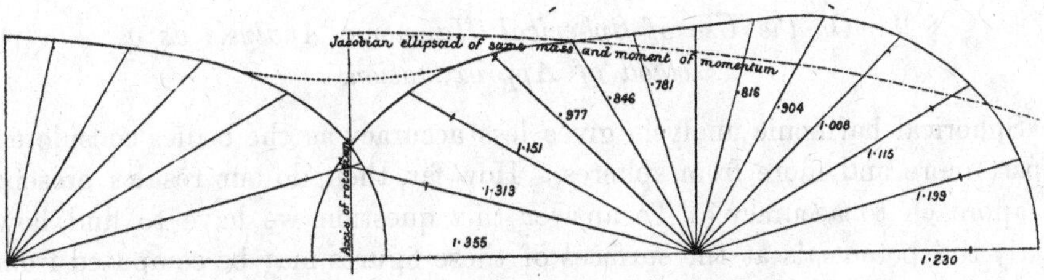

Fig. 3. Dumb-bell figure of equilibrium. Section through axis of rotation.

[$A=a$; $c/a=2\cdot449$; $\beta=\frac{1}{5}$; $\gamma=\frac{1}{6}$; $\omega^2/4\pi=\cdot0494$; momentum$=(\frac{4}{3}\pi)^{\frac{3}{2}}\, b^5\times\cdot482$.]

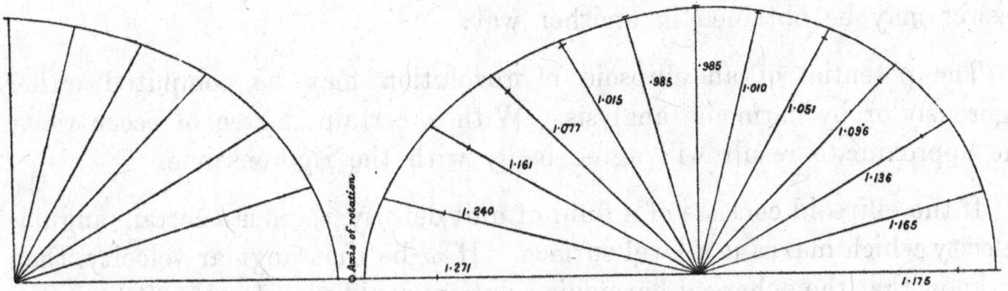

Fig. 4. Equal masses nearly in contact. Section perpendicular to axis of rotation.

[$A=a$; $c/a=2\cdot646$; $\beta=\frac{1}{6}$; $\gamma=\frac{1}{7}$; $\omega^2/4\pi=\cdot038$; momentum$=(\frac{4}{3}\pi)^{\frac{3}{2}}\, b^5\times\cdot472$.]

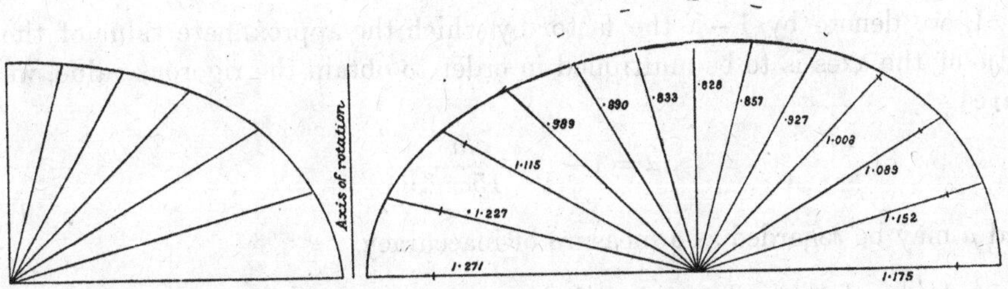

Fig. 5. Equal masses nearly in contact. Section through axis of rotation.

[$A=a$; $c/a=2\cdot646$; $\beta=\frac{1}{6}$; $\gamma=\frac{1}{7}$; $\omega^2/4\pi=\cdot038$; momentum$=(\frac{4}{3}\pi)^{\frac{3}{2}}\, b^5\times\cdot472$.]

Figs. 2 and 3 correspond to $\beta = \frac{1}{5}$, and here the figures as computed cross one another. The reality must, therefore, be two bulbs joined by a stalk, like a dumb-bell. The dotted lines have been filled in conjecturally, and must show pretty closely what that single figure, formed by the coalescence of the two masses, must be.

Figs. 4 and 5 show in a similar manner the case of $\beta = \frac{1}{6}$, and here the two masses are separate, although nearly in contact. When $\beta = \frac{1}{7}$ the shapes present similar characters, but are wider apart.

§ 9. *On the Use of Spherical Harmonic Analysis as a Method of Approximation.*

Spherical harmonic analysis gives less accuracy as the bodies considered depart more and more from spheres. How far, then, do our results present an approach to accuracy? To answer this question, we have to find how nearly the potentials at the surfaces of these figures may be computed from the spherical harmonic formulæ.

It would be laborious to make an accurate computation of the potential, and it fortunately appears to be unnecessary to do so, since a sufficient answer may be obtained in another way.

The potential of an ellipsoid of revolution may be computed either rigorously or by harmonic analysis. With a certain degree of eccentricity the approximate result will agree badly with the rigorous one.

If the ellipsoid consists of a fluid of unit density, there is a certain angular velocity which makes it a level surface. If ω be that angular velocity, then we know that the spherical harmonic solution would give $1 - 15\omega^2/16\pi$ as the ratio of the minor to the major axis. If then c, a, are the rigorous values of the minor and major semi-axes, the harmonic approximation is good if c/a does not differ much from $1 - 15\omega^2/16\pi$.

If we denote by $1 - \mu$ the factor by which the approximate value of the ratio of the axes is to be multiplied in order to obtain the rigorous value, we have

$$\mu = 1 - \frac{c/a}{1 - 15\omega^2/16\pi}$$

and μ may be regarded as a measure of inaccuracy.

A table of the values of $\omega^2/2\pi$, corresponding to various eccentricities $e = \sqrt{\{1 - (c/a)^2\}}$, is computed from the transcendental equation in Thomson and Tait's *Natural Philosophy*, § 772. From these I compute as follows:—

$e=\sqrt{\left(1-\frac{c^2}{a^2}\right)}$	$\frac{c}{a}$	$1-\frac{15\omega^2}{16\pi}$	Difference	$\frac{1}{\mu}$
·1	·9949	·9949	·0000	Large
·2	·9798	·9799	·0001	9799
·3	·9539	·9544	·0005	1909
·4	·9165	·9182	·0017	540
·5	·8660	·8705	·0045	193
·6	·8000	·8111	·0111	73
·7	·7141	·7399	·0258	29
·8	·6000	·6595	·0595	11·1
·9	·4359	·5869	·1510	3·9

The measures of inaccuracy corresponding to the values of e in the first column, or the values of c/a in the second, are the reciprocals of the numbers in the last column. We thus see that there is still a considerable degree of approximation when $e = ·8$, or when the ratio of the axes is 3 to 5, for the measure of inaccuracy is $\frac{1}{11}$; but for $e = ·9$ the approximation is bad.

Now the shapes of certain egg-like bodies have been computed by the spherical harmonic method, and it seems safe to assume that the approximation has given about the same degree of accuracy as would hold in the case of an ellipsoid of revolution whose minor axis bears to its major axis the same ratio as the shorter axis of the egg to the longer.

Turning now to our computation, and considering only the more elongated or meridional sections, we see that, when $\beta = \frac{1}{5}$, the longer axis is $1·355 + 1·230$ or $2·585$, while the shorter is $2(1 - ·227)$ or $1·546$. The ratio $1·546 : 2·585$ is ·6, which corresponds to the measure of inaccuracy 1/11·1. It might, however, be more legitimate to adopt two different measures, and at the pointed end of the egg to take the ratio $·773 : 1·355 = ·57$, which will correspond to a measure of inaccuracy about $\frac{1}{10}$; and at the blunt end to take the ratio $·773 : 1·230 = ·63$, which would correspond to a measure of inaccuracy $\frac{1}{12}$ or $\frac{1}{13}$.

In the case of $\beta = \frac{1}{5}$ the two masses cross one another, and the result has been used to give an approximate picture of the dumb-bell figure of equilibrium. We now see that even in this case there is a sufficient degree of approximation to give a very good idea of the accurate result.

In the case of the meridional section, where $\beta = \frac{1}{6}$, we have for the ratio of axes at the pointed end of the egg

$$\frac{1 - ·1717}{1·2712} = \frac{·8283}{1·2712} = ·65$$

and measure of inaccuracy about $\frac{1}{25}$; and at the blunt end

$$\frac{1 - ·1717}{1·1746} = \frac{·8283}{1·1746} = ·71$$

and measure of inaccuracy $\frac{1}{29}$.

In the case of $\beta = \frac{1}{7}$ the similar figures are, for the pointed end,

$$\frac{1 - {\cdot}137}{1{\cdot}208} = \frac{{\cdot}863}{1{\cdot}208} = {\cdot}72$$

and measure of inaccuracy about $\frac{1}{30}$; and for the blunt end

$$\frac{1 - {\cdot}137}{1{\cdot}139} = \frac{{\cdot}863}{1{\cdot}139} = {\cdot}76$$

and measure of inaccuracy perhaps about $\frac{1}{50}$.

It thus appears that as the bodies recede the accuracy increases with great rapidity, and in the two cases considered last it is hardly necessary, from a physical point of view, to consider greater accuracy than that attained.

It must be remarked, however, that this way of estimating the degree of inaccuracy must necessarily give much too unfavourable a view.

If we have a single mass of fluid departing considerably from the spherical form, it is clear that the potential computed on the hypothesis of a layer of surface density on the true sphere will come to depart largely from the potential at the surface of the fluid. If, however, we compute the potential of such a mass at points a little remote from the surface, the approximation will be much closer. Now, where there are two masses, as in our problem, the potential at the surface of either mass consists of two parts, one due to the mass itself, the other due to the other mass. As regards the first of these two parts, the above criterion is applicable, but as regards the second part it gives too unfavourable a view.

Now in the case of the single mass the deformative forces due to centrifugal force are considerably vitiated by computation at the spherical surface instead of the true surface, whilst in the case of the two masses the tide-generating forces are computed with greater accuracy than is shown by the criterion.

Under these circumstances it has appeared worth while to give another figure below, which, judged by the criterion, would be no approximation at all.

The reasons for giving this figure will be stated hereafter.

§ 10. *To find the Moment of Momentum of the System.*

Rotating figures of equilibrium are classified according to the amount of moment of momentum with which they are endued. It is, therefore, interesting to determine the moment of momentum of the systems now under consideration.

We must begin by finding the moments of inertia of the two masses. Let δI, δi, denote the moments of inertia of the shells of zero mass lying on the mean spheres of radii A, a.

Then
$$\delta i = \iint (y^2 + z^2)(r - a)\, a^2 d\varpi$$
where $d\varpi = \sin\theta\, d\theta\, d\phi$, and where the integral is taken throughout angular space.

Now
$$y^2 + z^2 = \tfrac{2}{3}a^2 + \tfrac{1}{3}a^2\left(\frac{w_2}{r^2} - \tfrac{1}{2}\frac{\delta^2 w_4}{r^2}\right)$$

and $r - a$ is the sum of a series of harmonics. Then, in consequence of the properties of harmonic functions, we need only consider the harmonics of the second degree in $r - a$, and

$$\delta i = \tfrac{1}{3}a^5 \iint \left\{\tfrac{1}{3}\epsilon\left[\left(\frac{w_2}{r^2}\right)^2 + \left(\tfrac{1}{2}\frac{\delta^2 w_4}{r^2}\right)^2\right]\right.$$
$$\left. + \tfrac{5}{2}\left(\frac{A}{c}\right)^3\left[(h_2 + l_2)\left(\frac{w_2}{r^2}\right)^2 + 2m_2\left(\tfrac{1}{2}\frac{\delta^2 w_4}{r^2}\right)^2\right]\right\} d\varpi$$

But
$$\iint\left(\frac{w_2}{r^2}\right)^2 d\varpi = \tfrac{4}{5}\pi, \quad \iint\left(\tfrac{1}{2}\frac{\delta^2 w_4}{r^2}\right)^2 d\varpi = \tfrac{4}{5}\pi \,.\, 3, \quad \tfrac{1}{3}\epsilon = \frac{5\omega^2}{16\pi}$$

and the moment of inertia of the mean sphere is $\frac{8}{15}\pi a^5$; hence, if we write

$$f = \frac{5\omega^2}{8\pi} + \tfrac{5}{2}\left(\frac{A}{c}\right)^3\left[\tfrac{1}{2}(h_2 + l_2) + 3m_2\right]$$
$$F = \frac{5\omega^2}{8\pi} + \tfrac{5}{2}\left(\frac{a}{c}\right)^3\left[\tfrac{1}{2}(H_2 + L_2) + 3M_2\right]$$

the moments of inertia, i and I, are given by
$$i = \tfrac{8}{15}\pi a^5(1 + f)$$
$$I = \tfrac{8}{15}\pi A^5(1 + F)$$

We already have in (71)
$$\frac{3\omega^2}{4\pi} = \left[\left(\frac{A}{c}\right)^3 + \left(\frac{a}{c}\right)^3\right](1 + K)$$

Hence the sum of the rotational momenta of the two masses is

$$(i + I)\,\omega = \tfrac{2}{5}(\tfrac{4}{3}\pi)^{\frac{3}{2}}\left[a^5(1 + f) + A^5(1 + F)\right]\left[\left(\frac{a}{c}\right)^3 + \left(\frac{A}{c}\right)^3\right]^{\frac{1}{2}}(1 + K)^{\frac{1}{2}}$$

The whole system revolves orbitally about the centre of inertia with an angular velocity ω: hence the orbital momentum is

$$\tfrac{4}{3}\pi\left[a^3\omega d^2 + A^3\omega D^2\right]$$

But
$$d = \frac{A^3 c}{a^3 + A^3}, \qquad D = \frac{a^3 c}{a^3 + A^3}$$

Hence the orbital momentum is

$$\frac{\tfrac{4}{3}\pi A^3 a^3}{A^3 + a^3}\,\omega c^2$$

and this is equal to
$$(\tfrac{4}{3}\pi)^{\frac{3}{2}}\frac{a^3 A^3 c^{\frac{1}{2}}}{(A^3 + a^3)^{\frac{1}{2}}}(1+K)^{\frac{1}{2}}$$

It will be convenient to refer the mass to the radius of a sphere of the same mass as the sum of the two.

Let b be the radius of such a sphere; then

$$b^3 = A^3 + a^3$$

Thus the whole moment of momentum is

$$(\tfrac{4}{3}\pi)^{\frac{3}{2}} b^5 \left\{\tfrac{2}{5}\left[\left(\frac{a}{b}\right)^5(1+f) + \left(\frac{A}{b}\right)^5(1+F)\right]\left(\frac{b}{c}\right)^{\frac{3}{2}} + \left(\frac{a}{b}\right)^3\left(\frac{A}{b}\right)^3\left(\frac{c}{b}\right)^{\frac{1}{2}}\right\}(1+K)^{\frac{1}{2}}$$

We shall therefore compute the coefficient of $(\tfrac{4}{3}\pi)^{\frac{3}{2}} b^5$.

Computing from this formula, I find the following values of the moment of momentum in the case where the masses are equal, when

$$\beta = \tfrac{1}{7}, \quad (\tfrac{4}{3}\pi)^{\frac{3}{2}} b^5 \times {\cdot}468$$
$$\beta = \tfrac{1}{6}, \quad \text{,,} \quad \times {\cdot}472$$
$$\beta = \tfrac{1}{5}, \quad \text{,,} \quad \times {\cdot}482$$

Now I find by a numerical investigation* that, if we imagine a mass of fluid equal to $\tfrac{4}{3}\pi b^3$ rotating in the form of a Jacobian ellipsoid of three unequal axes, then, when the momentum is $(\tfrac{4}{3}\pi)^{\frac{3}{2}} b^5 \times {\cdot}392$, the axes of the ellipsoid are $1{\cdot}898b$, $0{\cdot}8113b$, $0{\cdot}649b$; and when the momentum is $(\tfrac{4}{3}\pi)^{\frac{3}{2}} b^5 \times {\cdot}644$, the axes are $3{\cdot}136b$, $0{\cdot}586b$, $0{\cdot}545b$.

It seems probable, then, that the Jacobian ellipsoid of mass $\tfrac{4}{3}\pi b^3$ becomes unstable, at least as soon as when the moment of momentum is somewhere about $(\tfrac{4}{3}\pi)^{\frac{3}{2}} b^5 \times {\cdot}5$†.

It may be worth mentioning that the greatest moment of momentum for which the ellipsoid (of mass $\tfrac{4}{3}\pi b^3$) is stable, when it is a figure of revolution, is $(\tfrac{4}{3}\pi)^{\frac{3}{2}} b^5 \times {\cdot}30375$.

* *Roy. Soc. Proc.*, Vol. XLI., 1887, p. 319.

† [In § 7 of Paper 12 the factor here conjectured to be ·5 is proved to be ·389570.]

§ 11. *On the Conditions under which the two Masses may be close to one another.*

If at any point on the surface of either mass the sum of the tide-generating and centrifugal forces is greater than gravity, it is obvious that equilibrium cannot subsist. It is also clear that, if this condition is to be found anywhere, it will be at that point of the smaller mass which lies nearest to the larger mass. Hence, in order that the system may be a possible one, we must satisfy ourselves that at that point the gravity of the body itself exceeds the sum of the tide-generating and centrifugal forces.

To determine the limitations of size and proximity of the smaller of the two masses to a high degree of approximation would be very laborious, and we shall, therefore, content ourselves with a rough investigation, to be explained below.

We shall now find approximations for the shapes of the two masses and for their potentials.

The radius-vector of either mass and the potential may be expanded in powers of a/c and A/c, and a term involving c^n in the denominator will be referred to as being of the nth order.

Now the term of the highest order which can be included without the introduction of great complication is the 7th, and we shall content ourselves with that term.

The expressions for the various parts of the potential have been developed above, but it may be observed that the terms involving the first order of harmonics may be omitted, since they are subsequently annulled by a proper choice of the angular velocity.

From (22–i.) we have

$$\frac{4\pi a^2}{3}\frac{a}{r}+\frac{4\pi A^3}{3c}\tfrac{3}{2}\left[h_2\left(\frac{a}{c}\right)^2\left(\frac{a}{r}\right)^3\frac{w_2}{r^2}+\tfrac{1}{2}h_3\left(\frac{a}{c}\right)^3\left(\frac{a}{r}\right)^4\frac{w_3}{r^3}+\dots\right]$$

The last term in the development to the 7th order is that involving w_6. Then it is clear that we require h_2 correct to the 4th order, h_3 to the 3rd, and so on. But (25) shows us that the h's are equal to unity to the 4th order inclusive. Hence, in the above, all the h's may be treated as unity.

Again (22–ii.) when written in reference to the origin o affords other terms, in which all those included under $\Sigma\Sigma$ are of the 8th and higher orders, and negligible; and the rest (with omission of the first harmonic term) gives

$$\frac{4\pi A^3}{3c}\left[\left(\frac{a}{c}\right)^2\frac{w_2}{a^2}+\left(\frac{a}{c}\right)^3\frac{w_3}{a^3}+\dots\right]$$

Thus this first part of the potential is, to the 7th order inclusive,

$$\frac{4\pi a^2}{3}\cdot\frac{a}{r}+\frac{4\pi A^3}{3c}\sum_{k=2}^{k=6}\left(\frac{a}{c}\right)^k\left\{\frac{3}{2(k-1)}\left(\frac{a}{r}\right)^{k+1}+\left(\frac{r}{a}\right)^k\right\}\frac{w_k}{r^k}$$

Next, from the expression for Ω in § 3, we have a term in the potential due to rotation $+\frac{1}{3}\omega^2 r^2$. The remaining terms due to rotation will be taken up later.

From (71) we see that, to the 7th order inclusive,

$$\frac{3\omega^2}{4\pi} = \left(\frac{A}{c}\right)^3 + \left(\frac{a}{c}\right)^3$$

Hence ω^2 and ϵ are of the 3rd order; and from (48) and (64) it follows that the factors by which the l's and m's are derived from the λ's and μ's are of the 5th order. And, since the λ's and μ's only differ from unity in terms of the 5th order, it follows that the l's and m's are of the 5th order. Then (41) and (56) show us that all the terms in l and m are negligible.

The first set of terms due to rotation and to the corresponding deformation are given in (39) and (43), and together contribute

$$\tfrac{1}{6}\omega^2 a^2 \left[\tfrac{3}{2}\left(\frac{a}{r}\right)^3 + \left(\frac{r}{a}\right)^2\right] \frac{w_2}{r^2}$$

The second set of terms due to rotation, and to the corresponding deformation, are given in (54) and (58), and together contribute

$$-\tfrac{1}{6}\omega^2 a^2 \left[\tfrac{3}{2}\left(\frac{a}{r}\right)^3 + \left(\frac{r}{a}\right)^2\right] \tfrac{1}{2}\frac{\delta^2 w_4}{r^2}$$

Hence, to the 7th order inclusive, we have

$$V = \frac{4\pi a^2}{3}\frac{a}{r} + \frac{4\pi A^3}{3c}\sum_{k=2}^{k=6}\left(\frac{a}{c}\right)^k \left\{\frac{3}{2(k-1)}\left(\frac{a}{r}\right)^{k+1} + \left(\frac{r}{a}\right)^k\right\}\frac{w_k}{r^k}$$
$$+ \tfrac{1}{3}\omega^2 a^2 \left(\frac{r}{a}\right)^2 + \tfrac{1}{6}\omega^2 a^2 \left\{\tfrac{3}{2}\left(\frac{a}{r}\right)^3 + \left(\frac{r}{a}\right)^2\right\}\left(\frac{w_2 - \frac{1}{2}\delta^2 w_4}{r^2}\right) \quad \ldots(78)$$

Now the expression (72) for the radius-vector of the mass a to the same order of approximation gives us

$$\frac{r}{a} = 1 + \frac{5\omega^2}{16\pi}\left(\frac{w_2 - \frac{1}{2}\delta^2 w_4}{r^2}\right) + \left(\frac{A}{c}\right)^3 \sum_{k=2}^{k=6}\frac{2k+1}{2k-2}\left(\frac{a}{c}\right)^{k-2}\frac{w_k}{r^k}$$

and a similar expression for R/A.

To determine the inward force at the pole of the mass a, where it is nearest to the mass A, we must evaluate $-dV/dr$, and in the first term substitute the above expression for r, and in the remaining terms put $r=a$; also at this pole $w_2/r^2 = 1$, and $\delta^2 w_4 = 0$.

Then, differentiating (78),

$$-\frac{dV}{dr} = \frac{4\pi a}{3}\frac{a^2}{r^2} - \frac{4\pi a}{3}\left(\frac{A}{c}\right)^3 \sum_{k=2}^{k=6}\left(\frac{a}{c}\right)^{k-2}\left\{-\frac{3(k+1)}{2(k-1)} + k\right\}$$
$$-\tfrac{2}{3}\omega^2 a - \tfrac{1}{6}\omega^2 a^2\left\{-\frac{3\,.\,3}{2} + 2\right\} \quad \ldots(79)$$

But at the pole

$$\frac{4\pi a}{3}\frac{a^2}{r^2} = \frac{4\pi a}{3} - \tfrac{5}{6}\omega^2 a - \frac{4\pi a}{3}\left(\frac{A}{c}\right)^3 \sum_{k=2}^{k=6} \frac{2k+1}{k-1}\left(\frac{a}{c}\right)^{k-2}$$

Substituting this for the first term of (79), we have

$$-\frac{dV}{dr} = \frac{4\pi a}{3} - \tfrac{13}{12}\omega^2 a - \frac{4\pi a}{3}\left(\frac{A}{c}\right)^3 \sum_{k=2}^{k=6} (k+\tfrac{1}{2})\left(\frac{a}{c}\right)^{k-2}$$

But $$\omega^2 a = \frac{4\pi a}{3}\left[\left(\frac{A}{c}\right)^3 + \left(\frac{a}{c}\right)^3\right]$$

hence

$$-\frac{dV}{dr} = \frac{4\pi a}{3}\left[1 - \tfrac{13}{12}\left\{\left(\frac{A}{c}\right)^3 + \left(\frac{a}{c}\right)^3\right\} - \left(\frac{A}{c}\right)^3\left\{\tfrac{5}{2} + \tfrac{7}{2}\frac{a}{c} + \tfrac{9}{2}\left(\frac{a}{c}\right)^2 + \tfrac{11}{2}\left(\frac{a}{c}\right)^3 + \tfrac{13}{2}\left(\frac{a}{c}\right)^4\right\}\right]$$

$$= \frac{4\pi a}{3}\left[1 - \frac{43A^3 + 13a^3}{12c^3} - \left(\frac{A}{c}\right)^3\left\{\tfrac{7}{2}\frac{a}{c} + \tfrac{9}{2}\left(\frac{a}{c}\right)^2 + \tfrac{11}{2}\left(\frac{a}{c}\right)^3 + \tfrac{13}{2}\left(\frac{a}{c}\right)^4\right\}\right]$$

Thus the criterion of the possibility of equilibrium is that

$$C = 1 - \frac{43A^3 + 13a^3}{12c^3} - \left(\frac{A}{c}\right)^3\left\{\tfrac{7}{2}\frac{a}{c} + \tfrac{9}{2}\left(\frac{a}{c}\right)^2 + \tfrac{11}{2}\left(\frac{a}{c}\right)^3 + \tfrac{13}{2}\left(\frac{a}{c}\right)^4\right\} \quad \ldots(80)$$

should be positive.

But the radius-vectors of the poles are

$$\frac{r}{a} = 1 + \tfrac{5}{12}\left[\left(\frac{A}{c}\right)^3 + \left(\frac{a}{c}\right)^3\right] + \left(\frac{A}{c}\right)^3\left[\tfrac{5}{2} + \tfrac{7}{4}\left(\frac{a}{c}\right) + \tfrac{9}{6}\left(\frac{a}{c}\right)^2 + \tfrac{11}{8}\left(\frac{a}{c}\right)^3 + \tfrac{13}{10}\left(\frac{a}{c}\right)^4\right]$$

and, similarly,

$$\frac{R}{A} = 1 + \tfrac{5}{12}\left[\left(\frac{A}{c}\right)^3 + \left(\frac{a}{c}\right)^3\right] + \left(\frac{a}{c}\right)^3\left[\tfrac{5}{2} + \tfrac{7}{4}\left(\frac{A}{c}\right) + \tfrac{9}{6}\left(\frac{A}{c}\right)^2 + \tfrac{11}{8}\left(\frac{A}{c}\right)^3 + \tfrac{13}{10}\left(\frac{A}{c}\right)^4\right]$$

Therefore

$$r + R = a + A + \frac{5}{12c^3}[(7a+A)A^3 + (a+7A)a^3] + \frac{7}{4c^4}A^2a^2(A+a)$$
$$+ \frac{3}{c^5}A^3a^3 + \frac{11}{8c^6}A^3a^3(a+A) + \frac{13}{10c^7}A^3a^3(a^2+A^2)$$

Now the interval between the two masses is $c-(r+R)$; hence, if the two masses are just in contact,

$$c = a + A + \frac{5}{12c^3}[(7a+A)A^3 + (a+7A)a^3] + \frac{7}{4c^4}A^2a^2(A+a)$$
$$+ \frac{3}{c^5}A^3a^3 + \frac{11}{8c^6}A^3a^3(a+A) + \frac{13}{10c^7}A^3a^3(a^2+A^2) \ldots\ldots(81)$$

In order, then, to test whether equilibrium is still possible when the two masses are just in contact, it is necessary to determine c from (81); and then, substituting in (80), find whether C is positive or not.

The solution of such an equation as

$$c=\alpha+\frac{\beta}{c^3}+\frac{\gamma}{c^4}+\frac{\delta}{c^5}+\frac{\epsilon}{c^6}+\frac{\zeta}{c^7}$$

and the determination of

$$C=1-\frac{\mathrm{B}}{c^3}-\frac{\Gamma}{c^4}-\frac{\Delta}{c^5}-\frac{\mathrm{E}}{c^6}-\frac{\mathrm{Z}}{c^7}$$

can only be performed by trial and error.

Now suppose that the solution is $c_0+\delta c$, where δc is small; and that

$$c_1=\alpha+\frac{\beta}{c_0^{\,3}}+\frac{\gamma}{c_0^{\,4}}+\dots,\qquad C_1=1-\frac{\mathrm{B}}{c_0^{\,3}}-\frac{\Gamma}{c_0^{\,4}}-\dots$$

Then it is obvious that

$$\frac{\delta c}{c_0}=\frac{c_1-c_0}{c_0+\dfrac{3\beta}{c_0^{\,3}}+\dfrac{4\gamma}{c_0^{\,4}}+\dots}$$

and $$\delta C=\left(\frac{3\mathrm{B}}{c_0^{\,3}}+\frac{4\Gamma}{c_0^{\,4}}+\dots\right)\frac{\delta c}{c_0},\quad\text{and}\quad C=C_1+\delta C$$

It is not hard to find an approximate solution c_0 by trial and error, and the correct results may then be found thus.

Consider the case where the two bodies are equal to one another, and put $a=A=1$. The equations then become

$$c=2+\frac{20}{3c^3}+\frac{7}{2c^4}+\frac{3}{c^5}+\frac{11}{4c^6}+\frac{13}{5c^7}$$

$$C=1-\frac{14}{3c^3}-\frac{7}{2c^4}-\frac{9}{2c^5}-\frac{11}{2c^6}-\frac{13}{2c^7}$$

By trial and error we find $c=2{\cdot}535$, $C=+{\cdot}557$.

From this we conclude that equilibrium still subsists when the two masses are in contact.

When $a=A=1$, $c=2{\cdot}535$, we have

$$\gamma=(a/c)^2=\frac{1}{6{\cdot}43},\quad\text{and}\quad\beta=\gamma/(1-\gamma)=\frac{1}{5{\cdot}43}$$

Our figures showed that when $\beta=\frac{1}{6}$ the two masses were nearly in contact, and when $\beta=\frac{1}{5}$ they crossed.

This result is, therefore, in accordance with the figures.

Next pass to the case of an infinitesimal satellite, and suppose a infinitely small compared with A and c, and that $A=1$. The equations are

$$c=1+\frac{5}{12c^3}$$

$$C=1-\frac{43}{12c^3}$$

The solution of the first equation is $c = 1{\cdot}226$, and this value of c makes $C = -{\cdot}94$. Hence we conclude that an infinitesimal fluid satellite cannot revolve with its surface in contact with its planet.

C vanishes when $c = (\frac{43}{12})^{\frac{1}{2}} = 1{\cdot}89$. Hence it appears that the nearest approach of the infinitesimal satellite to the planet is 1·89 mean radii of the planet. The nature of the approximation adopted is, however, such that in reality the satellite must lie further from the planet than this, perhaps at two radii distance*. The satellite and planet of which we here speak are, of course, supposed to revolve as parts of a rigid body. Now, if for equal masses equilibrium still subsists when the two masses are in contact, whilst for infinitesimal mass of one equilibrium is impossible with the masses in contact, it follows that for some ratio of masses equilibrium can just subsist when they are in contact.

The question, therefore, remains to determine this limiting ratio of masses.

I find, then, that when $a = 1$, $A = 3{\cdot}4$, we have

$$c = 4{\cdot}4 + [2{\cdot}25684]\,c^{-3} + [1{\cdot}94945]\,c^{-4} + [2{\cdot}07156]\,c^{-5} + [2{\cdot}37619]\,c^{-6} + [2{\cdot}80737]\,c^{-7}$$

$$C = 1 - [2{\cdot}15205]\,c^{-3} - [2{\cdot}13850]\,c^{-4} - [2{\cdot}24765]\,c^{-5} - [2{\cdot}33480]\,c^{-6} - [2{\cdot}40735]\,c^{-7}$$

the numbers in [ ] being the logarithms of the coefficients.

The solution of this is $c = 5{\cdot}57$, which makes $C = -{\cdot}006$.

Again, when $a = 1$, $A = 3{\cdot}3$, we have

$$c = 4{\cdot}3 + [2{\cdot}21556]\,c^{-3} + [1{\cdot}91353]\,c^{-4} + [2{\cdot}03266]\,c^{-5} + [2{\cdot}32731]\,c^{-6} + [2{\cdot}74467]\,c^{-7}$$

$$C = 1 - [2{\cdot}11347]\,c^{-3} - [2{\cdot}09961]\,c^{-4} - [2{\cdot}20876]\,c^{-5} - [2{\cdot}29591]\,c^{-6} - [2{\cdot}36846]\,c^{-7}$$

the solution of which is $c = 5{\cdot}45$, which makes $C = +{\cdot}010$. Since $(3{\cdot}4)^3 = 39{\cdot}3$, and $(3{\cdot}3)^3 = 35{\cdot}9$, it follows that the ratio of the masses in the first case is $1 : 39{\cdot}3$, and in the second $1 : 35{\cdot}9$.

From this it appears that when the ratio of the masses is about 1 to 38 equilibrium is still just possible when the two masses touch.

It must be borne in mind, however, that the nature of the approximations adopted in this investigation is such that the results in this limiting case are only given very roughly, and it is certain that actually the limiting size of the smaller of the two masses must be greater than as thus computed.

* See Roche, *Montpellier Acad. Sci. Mém.*, Vol. I., 1847–1850, p. 243. [The accurate result is proved in Paper 15 to be 2·4553.]

We can only conclude that the limiting case occurs when the ratio of the masses is about 1 to 30, or the ratio of the radii about 1 to 3.

There is one other case which it is interesting to consider, namely, to find the limiting proximity of the Moon to the Earth, both bodies being treated as homogeneous fluids of the same density, revolving as a rigid body.

The case of Moon and Earth is well represented by $a = 1$, $A = 4{\cdot}333$; for this gives 1 to 81·35 as the ratio of the masses. With these values I find

$$C = 1 - [2{\cdot}46626]\, c^{-3} - [2{\cdot}45443]\, c^{-4} - [2{\cdot}56358]\, c^{-5} - [2{\cdot}65073]\, c^{-6} - [2{\cdot}72328]\, c^{-7}$$

and

$$r + R = 5{\cdot}333 + [2{\cdot}59898]\, c^{-3} + [2{\cdot}24358]\, c^{-4} + [2{\cdot}38748]\, c^{-5} + [2{\cdot}77563]\, c^{-6} + [3{\cdot}32042]\, c^{-7}$$

Now $c = 7{\cdot}0$ will be found to make C vanish, and, with this value of c, $c - (r + R) = {\cdot}414$.

If A be 4000 miles, $c = 6500$ miles, and $c - (r + R) = 380$ miles.

Thus, as far as this investigation goes, it appears that when the fluid Moon is on the point of breaking up from stress of tidal and centrifugal forces the distance between the centres of Moon and Earth is 6500 miles, and the shortest distance between the two surfaces is 380 miles.

This result must, however, from the nature of the approximation, be an underestimate of the distances.

The whole of the present section has been suggested by a pamphlet by Mr James Nolan* in which he criticises some of my previous papers. I have commented elsewhere on his criticisms†.

§ 12. *On the Case where the two Masses are unequal.*

The results of the previous section point to a very remarkable limitation to the possibility of approach of two masses of unequal size. It has, therefore, seemed worth while to consider this case numerically, and a case is therefore chosen which shall approach near to that which we know is the limit of possibility. I choose, therefore, $a = 1$, $A = 3$, which makes the ratio of the masses 1 to 27, and $c = 5{\cdot}3$, which brings the protuberances into close proximity.

The numerical details are omitted, but figs. 6 and 7 give the results, the numerical values of the radius-vectors being, as before, entered on the figure.

* *Darwin's Theory of the Genesis of the Moon.* Robertson: Melbourne, Sydney, Adelaide, and Brisbane, 1885.

† *Nature*, February 18 and July 29, 1886.

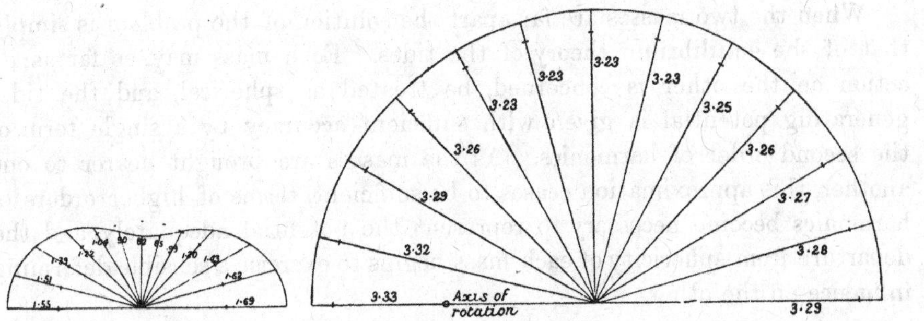

FIG. 6. Unequal masses. Section perpendicular to axis of rotation.

[$A/a=3$; $(A/a)^3=27$; $a/c=\cdot189$; $A/c=\cdot566$; $\gamma=\cdot036$; $\Gamma=\cdot321$; $\omega^2/4\pi=\cdot066$; momentum $=(\frac{4}{3}\pi)^{\frac{3}{2}}\, b^5\times\cdot29$.]

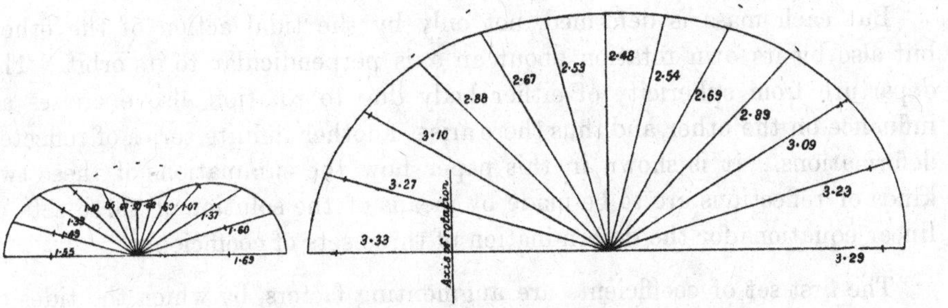

FIG. 7. Unequal masses. Section through axis of rotation.

[$A/a=3$; $(A/a)^3=27$; $a/c=\cdot189$; $A/c=\cdot566$; $\gamma=\cdot036$; $\Gamma=\cdot321$; $\omega^2/4\pi=\cdot066$; momentum $=(\frac{4}{3}\pi)^{\frac{3}{2}}\, b^5\times\cdot29$.]

The elongation of the smaller mass is so extreme that it is obvious that, strictly speaking, the spherical harmonic approximation must be considered to break down. Nevertheless, I conceive that these curious figures may be held to indicate the general nature of the true result.

It is remarkable that the smaller mass exhibits a marked furrowing round the middle. This seems to indicate that such a system tends to break up by the separation of the smaller mass into two parts.

§ 13. *Summary.*

The intention of this paper is, first, to investigate the forms which two masses of fluid assume when they revolve in close proximity about one another, without relative motion of their parts; and, secondly, to obtain a representation of the single form of equilibrium which must exist when the two masses approach so near to one another as just to coalesce into a single mass.

When the two masses are far apart the solution of the problem is simply that of the equilibrium theory of the tides. Each mass may, as far as its action on the other is concerned, be treated as spherical, and the tide-generating potential is given with sufficient accuracy by a single term of the second order of harmonics. As the masses are brought nearer to one another, this approximation ceases to be sufficient, terms of higher orders of harmonics become necessary to represent the potential adequately, and the departure from sphericity of each mass begins to exercise a sensible deforming influence on the other.

When the departure from sphericity of one body produces a sensible deformation in the other, that deformation in its turn reacts on the first, and thus the actual figure assumed by either mass may be regarded as a deformation due to the primitive influence of the other mass, on which is superposed the sum of an infinite series of reflected deformations.

But each mass is deformed, not only by the tidal action of the other, but also by its own rotation about an axis perpendicular to its orbit. The departure from sphericity of either body due to rotation also exercises an influence on the other, and thus there arises another infinite series of reflected deformations. It is shown in this paper how the summations of these two kinds of reflections are to be made by means of the solution of three sets of linear equations for the determination of three sets of coefficients.

The first set of coefficients are augmenting factors, by which the tides of each order of harmonics are to be raised above the value which they would have if the perturbing mass were spherical. It appears that, the higher the order of harmonics, the more do these factors exceed unity.

The second set of coefficients correspond to one part of the rotational effects. They appertain to terms of exactly the same form as the tidal terms, and in the final result the terms to which they apply become fused with the tidal terms. These terms are the zonal harmonics of the several orders with respect to the axis joining the centres of the two masses.

The third set of coefficients correspond to the remainder of the rotational effect, and they appertain to a different kind of deformation. These deformations are represented by sectorial harmonics involving $\cos 2\phi$, where ϕ is azimuth measured from the plane passing through the axis of rotation of the system and the centres of the two masses. That term of this set which is of the second order of harmonics, and which represents the ellipticity of either mass augmented by mutual influence, is the only term which is considerable, even when the two masses are very close together; but the existence of the other harmonic deformations of this class is interesting. We may say, then, that all the tides of either mass are augmented above the values which they would have if the other mass were spherical; that the ellipticity corresponding to rotation is augmented; and that the deformation due to rotation is no longer exactly elliptic-spheroidal.

The angular velocity of the system is found by the consideration that the repulsion due to centrifugal force between the two masses shall exactly balance the resultant attraction between them. If the masses were spherical, the square of the orbital angular velocity, multiplied by the cube of the distance between the centres, would be equal to the sum of the masses. When the masses are deformed, however, this law is no longer true, and the angular velocity has to be augmented by a factor a little greater than unity, which depends on the amounts of the deformations.

The theory here sketched is applied above numerically to several cases, and the results will be found in the preceding paragraphs. We shall first consider the cases where the two masses are equal to one another.

In the first example ($\beta = \frac{1}{7}$) solved numerically, the distance between the centres of the two masses is 2·83 times the mean radius of either of them. The two bodies are found to be elongated until they approach near to one another; but, as the character of the distortion is better illustrated in a subsequent case, the result is not given graphically. All the data, however are found which will enable the reader to draw the figure if he should wish to do so.

In the next example ($\beta = \frac{1}{6}$), with the masses still equal, the distance between the centres is reduced to 2·646 of the mean radius of either. The result of the solution is illustrated by two figures. In fig. 4, the section of the masses by a plane perpendicular to the axis of rotation is shown, and in fig. 5 we have the section by a plane passing through the axis and the centres of the two masses. On both figures are inscribed the values of the radii for each 15° of latitude in terms of the mean radius as unity, and the mean sphere, from which the distortion is computed, is marked by a short line on each radius. The elongation of the masses is, of course, considerably greater in the section through the axis than in the other section. Each mass is shaped somewhat like an egg, and the small ends face one another and come very nearly into contact.

In the headings to the figures, amongst other numerical data, are given the square of the angular velocity and the angular momentum of the system. The density of the fluid being unity, the angular velocity ω is given by the value of $\omega^2/4\pi$; this is the function of angular velocity which is usually given when reference is made to figures of equilibrium of rotating fluid, such as the revolutional or Jacobian ellipsoids of equilibrium. The moment of momentum of the system is given by reference to the angular velocity of a sphere, of the same mass as the sum of our two masses, rotating so as to have the same momentum. If, in fact, b be such a length that a sphere of fluid of that radius has the same mass as our system (so that $b^3 = a^3 + A^3$), then the moment of momentum is given by a number μ in the expression $(\frac{4}{3}\pi)^{\frac{3}{2}} b^5 \times \mu$. By this notation the angular velocity and moment of momentum are made comparable with the results given in a previous paper* on the Jacobian ellipsoid of equilibrium.

In figs. 4 and 5, $\omega^2/4\pi$ is ·038, and the momentum μ is ·472. On comparison with the Table of Jacobi's ellipsoids, we see that this corresponds with a considerably slower rotation than the 6th solution, and nearly the same moment of momentum.

In the next case the two masses are still closer $(\beta = \frac{1}{5})$, the distance between the centres being only 2·449 times either mean radius. The result is illustrated in figs. 2 and 3; the explanation of figs. 4 and 5 serves, *mutatis mutandis*, for these two figures also.

This case is interesting because the masses have approached so near to one another that they partially overlap. Two portions of matter cannot, of course, occupy the same space, and the continuity of figures of equilibrium leads us to believe that the reality must consist of a single mass of fluid. In figs. 2 and 3 conjectural dotted lines are drawn to show how it is probable that the overlapping of the two masses is replaced by a neck of fluid joining them. The figures as thus amended serve to give a good representation of the single dumb-bell shaped figure of equilibrium.

The angular velocity is here given by $\omega^2/4\pi = {\cdot}049$, and the moment of momentum by ·482. In the sixth entry of the Table of Solutions of Jacobi's problem we find $\omega^2/4\pi = {\cdot}0536$, and the moment of momentum $\mu = {\cdot}481$. This ellipsoid has, then, the same moment of momentum, and only about 4 per cent. more angular velocity, than our dumb-bell. It has seemed, therefore, worth while to mark (in chain-dot) on figs. 2 and 3 the outline of this Jacobian ellipsoid of the same mass as the dumb-bell. The actual vertex of the ellipsoid just falls outside the limits to which it was possible to extend the figure.

In the paper above referred to it is shown how the energy of the Jacobian ellipsoid is to be computed. If we denote the kinetic energy by $(\frac{4}{3}\pi)^2 b^5 \times \epsilon$, and the intrinsic energy by $(\frac{4}{3}\pi)^2 b^5 \times (i - 1)$†, then it appears that in the case of the ellipsoid drawn in these figures $\epsilon = {\cdot}0964$, $i = {\cdot}4808$, and the total energy $E = \epsilon + i = {\cdot}5772$.

Now in the case of our dumb-bell figure it appears, from calculations which are not reproduced, that $\epsilon = {\cdot}0925$, $i = {\cdot}4873$, and $E = {\cdot}5798$. Hence in the dumb-bell figure the kinetic energy is less, but the intrinsic energy is so much greater that the total energy is about a half per cent. greater. These

* *Roy. Soc. Proc.*, Vol. XLI., 1887, p. 319. [Paper 8, p. 119.]

† The intrinsic energy being negative, it is more convenient to tabulate i a positive quantity.

numbers are, of course, computed from the approximate formulæ, and must not be taken as rigorously correct for the dumb-bell figure of equilibrium.

With reference to a figure of transition from the Jacobian ellipsoid, Sir William Thomson has remarked*:—

"We have a most interesting gap between the unstable Jacobian ellipsoid, when too slender for stability, and the case of smallest moment of momentum consistent with stability in two equal detached portions. The consideration of how to fill up this gap with intermediate figures is a most attractive question, towards answering which we at present offer no contribution†."

Figs. 2 and 3 are intended to form such a contribution, but it is certain that the matter is far from being probed to the bottom.

M. Poincaré has made an admirable investigation of the forms of equilibrium of a single rotating mass of fluid, and has especially considered the stability of Jacobi's ellipsoid‡. He has shown, by a difficult analytical process, that when the ellipsoid is moderately elongated (he has not arrived at a numerical result§) instability sets in by a furrowing of the ellipsoid along a line which lies in a plane perpendicular to the longest axis. It is, however, extremely remarkable that the furrow is not symmetrical with respect to the two ends, and thus there appears to be a tendency to form a dumb-bell with unequal bulbs.

If M. Poincaré's result shall appear to be not only true, but to contain the whole truth concerning the mode in which instability of the ellipsoid supervenes, then there must be some other transitional form between the unsymmetrically furrowed Jacobian ellipsoid and the dumb-bell; except, perhaps, in the case where the two bulbs pass on to two masses of a definite ratio.

M. Poincaré's work seemed so important that this paper was kept back for a year, whilst I endeavoured to apply the principles, which he has pointed out, to the discussion of the stability of the two masses. The attempt, which [was originally] given in the Appendix, [was erroneous, and is now omitted.]

* Thomson and Tait's *Natural Philosophy*, 1883, § 778″ (*i*).

† In 778″ (*g*) he remarks that "a deviation from the ellipsoidal figure in the way of thinning it in the middle and thickening it towards the end would, with the same moment of momentum, give less energy." I conceive that the energy referred to throughout this paragraph is kinetic only, and we have seen that the kinetic energy is less for the dumb-bell than for the ellipsoid.

If we write U for a quantity proportional to the excess of kinetic above intrinsic energy, so that $U=\epsilon+(1-i)$, then figures of equilibrium are to be determined by making U stationary for variations of the parameters involved in it. This course is actually pursued in the Appendix below, the function (viii.) being, in fact, this U; and the variations of it, being made stationary, afford a controlling solution of the problem of this paper. The similar method may easily be applied to the case of Jacobi's ellipsoids. From this point of view the interesting function to tabulate is $\epsilon+(1-i)$, and we observe that in the case of the Jacobian ellipsoid referred to on p. 130 it is ·6052, and for the dumb-bell it is ·6156. Is not the energy referred to by Sir W. Thomson this function U?

‡ *Acta Mathematica*, VII., 3 and 4, 1885.

§ [The numerical result will be given in Paper 12 below.]

We must, therefore, leave this complex question in abeyance*.

Allusion has just been made to the imperfection of spherical harmonic analysis, and this brings us naturally to face the question whether that analysis may not have been pushed altogether too far in the computation of the figures of equilibrium under discussion. This question is considered in § 9, and a rough criterion of the limits of applicability of this analysis is there found. From this it appears that even in the cases of figs. 2 and 3 the result must present a fair approximation to correctness. The criterion, indeed, appears to be such as necessarily to give too unfavourable a view of the correctness of the result.

The rigorous method of discussing the stability of the system having failed, certain considerations are adduced in § 11 which bear on the conditions under which there is a form of equilibrium consisting of two fluid masses in close proximity. It appears that there cannot be such a form with the two masses just in contact, unless the smaller of the two masses exceeds in mass about one-thirtieth of the larger.

If we take into consideration the fact that the criterion of the applicability of harmonic analysis is too severe, it appears to be worth while to find to what results the analysis leads when two masses, one 27 times as great as the other, are brought close together. The numerical work of the calculation is omitted, since the numbers can only represent the true conclusion very roughly; but the result is illustrated graphically in figs. 6 and 7. These figures can only serve to give a general idea of the truth, but the form into which the smaller mass is thrown is so remarkable as to be worthy of attention. The deep furrow round the smaller mass, lying in a plane parallel to the axis of rotation, cannot be due merely to the imperfection of the solution; and it appears to point to the conclusion that there is a tendency for the smaller body to separate into two, just as we have seen the Jacobian ellipsoid become dumb-bell shaped and separate into two parts.

In this paper, indeed, we have sought to trace the process in the opposite direction, and to observe the coalescence of two masses into one. The investigation is complementary to, but far less perfect than, that of M. Poincaré, who describes the series of changes which he has been tracing in the following words:—

"Considérons une masse fluide homogène animée originairement d'un mouvement de rotation; imaginons que cette masse se contracte en se refroidissant lentement, mais de façon à rester toujours homogène. Supposons que le refroidissement soit assez lent et le frottement intérieur du fluide assez fort pour que le mouvement de rotation reste le même dans les diverses portions du fluide. Dans ces conditions le fluide tendra toujours à prendre

* [See Paper 15.]

une figure d'équilibre séculairement stable. Le moment de la quantité de mouvement restera d'ailleurs constant.

"Au début, la densité étant très faible, la figure de la masse est un ellipsoïde de révolution très peu différent d'une sphère. Le refroidissement aura d'abord pour effet d'augmenter l'aplatissement de l'ellipsoïde, qui restera cependant de révolution. Quand l'aplatissement sera devenu à peu près égal à $\frac{2}{5}$, l'ellipsoïde cessera d'être de révolution et deviendra un ellipsoïde de Jacobi. Le refroidissement continuant, la masse cessera d'être ellipsoïdale; elle deviendra dissymétrique par rapport au plan des *yz*, et elle affectera la forme représentée dans la figure, p. 347*.

"Comme nous l'avons fait observer à propos de cette figure, l'ellipsoïde semble se creuser légèrement dans sa partie moyenne, mais plus près de l'un des deux sommets du grand axe; la plus grande partie de la matière tend à se rapprocher de la forme sphérique, pendant que la plus petite partie sort de l'ellipsoïde par un des sommets du grand axe, comme si elle cherchait à se détacher de la masse principale.

"Il est difficile d'annoncer avec certitude ce qui arrivera ensuite si le refroidissement continue, mais il est permis de supposer que la masse ira en se creusant de plus en plus, puis en s'étranglant dans la partie moyenne, et finira par se partager en deux corps isolés.

"On pourrait être tenté de chercher dans ces considérations une confirmation ou une réfutation de l'hypothèse de Laplace, mais on ne doit pas oublier que les conditions sont ici très différentes, car notre masse est homogène, tandis que la nébuleuse de Laplace devait être très fortement condensée vers le centre†."

It was in the hope that the investigation might throw some light on the nebular hypothesis of Laplace and Kant that I first undertook the work. It must be admitted, however, that we do not obtain much help from the results. It is justly remarked by M. Poincaré that the conditions for the separation of a satellite from a nebula differ from those of his problem in the great concentration of density in the central body. But both his investigation and the considerations adduced here seem to show that, when a portion of the central body becomes detached through increasing angular velocity, the portion should bear a far larger ratio to the remainder than is observed in the satellites of our system as compared with their planets; and it is hardly probable that the heterogeneity of the central body can make so great a difference in the result as would be necessary if we are to make an application of these ideas. It appears then at present necessary to suppose that after the birth of a satellite, if it takes place at all in this way, a series of changes occur which are quite unknown.

* The furrowed ellipsoid of Jacobi.

† Poincaré, *Acta Mathematica*, VII., 1885, p. 379.

APPENDIX.

On the Energy and Stability of the System.

M. Poincaré has shown in his admirable memoir, referred to in the Summary, how the dynamical stability of a rotating fluid system in relative equilibrium depends on the energy. Certain coefficients in the expression for the energy, which he calls coefficients of stability, are there proved to afford the required criterion.

[An attempt was made in this Appendix to determine these coefficients of stability, but a mistake of principle was made and the attempt failed. The subject is treated in Paper 15 in this volume, and it will appear that all these figures of equilibrium are unstable. The instability arises from tidal friction, and as this cause is very slow in its operation, it seems possible that the figures drawn in the present paper might subsist for a considerable time.

My attempt to discuss this problem in this Appendix throws an interesting light on the methods by which the equations of the two masses were determined, and therefore a portion of the Appendix is now reproduced.]

The task before us is to determine the "exhaustion of potential energy" of the two masses in presence of one another as due to the deformation of each from the spherical figure by yielding to gravitation and to centrifugal force.

The work will be rendered simpler by the introduction of a new notation. Let us write, then, as the equations to two shapes, which are not necessarily together a figure of equilibrium :—

$$\left.\begin{aligned} \frac{r}{a} &= 1 + \sum_{k=2}^{k=\infty} \frac{2k+1}{2k-2}\left(\frac{A}{c}\right)^3\left(\frac{a}{c}\right)^{k-2}\left\{n_k\frac{w_k}{r^k} - p_k\frac{\delta^2 w_{k+2}}{r^k}\right\} \\ \frac{R}{A} &= 1 + \sum_{k=2}^{k=\infty} \frac{2k+1}{2k-2}\left(\frac{a}{c}\right)^3\left(\frac{A}{c}\right)^{k-2}\left\{N_k\frac{W_k}{R^k} - P_k\frac{\delta^2 W_{k+2}}{R^k}\right\} \end{aligned}\right\} \quad \text{......(i.)}$$

It will be observed that these equations have the same form as (72), but that the constants introduced are different from the h, l, m, ϵ, which were determined, so that the figures might be figures of equilibrium. At present we do not assume that (i.) do represent figures of equilibrium.

The energy lost may be divided into several parts :—

e_1, the energy lost by the mass a yielding from its spherical figure to the gravitation of the mean sphere a.

e_2, the exhaustion of mutual energy of that layer of matter on the mass a which constitutes its departure from sphericity.

e_3, the loss of energy due to the deformation of the mass a in presence of the mean sphere A.

E_1, E_2, E_3, the similar quantities for the mass A.

$(Ee)_4$, the loss of mutual energy of the two layers in presence of one another.

e_5, the loss of energy due to the deformation of the mass a in the presence of centrifugal force due to rotation ω.

E_5, the similar loss for A.

1st. e_1 is equal and opposite to the work required to raise each element of the layer on a through half its own height against the gravity due to the mean sphere a. This gravity is $\frac{4}{3}\pi a$. Co-latitude and longitude being denoted by θ, ϕ, let $d\varpi = \sin\theta d\theta d\phi$, an element of solid angle. In effecting the integrations, the properties of spherical harmonic functions are used without comment, viz.:—

$$\iint\left(\frac{w_k}{r^k}\right)^2 d\varpi = \frac{4\pi}{2k+1}, \quad \iint\left(\frac{\delta^2 w_{k+2}}{r^k}\right)^2 d\varpi = \frac{4\pi}{2k+1}\frac{k+2\,!}{2\,.\,k-2\,!}$$

$$\iint w_k\delta^2 w_{k+2} d\varpi = 0, \quad \iint w_i w_k d\varpi = 0, \quad \iint w_i \delta^2 w_{k+2} d\varpi = 0$$

Then, taking only a typical term of the first of (i.),

$$e_1 = -\iint \tfrac{4}{3}\pi a^5\,.\,\tfrac{1}{2}\,.\left[\frac{2k+1}{2k-2}\left(\frac{A}{c}\right)^3\left(\frac{a}{c}\right)^{k-2}\left\{n_k\frac{w_k}{r^k} - p_k\frac{\delta^2 w_{k+2}}{r^k}\right\}\right]^2 d\varpi$$

$$= -(\tfrac{4}{3}\pi)^2\,\tfrac{1}{2}a^5\frac{2k+1}{(2k-2)^2}\,3\left(\frac{A}{c}\right)^6\left(\frac{a}{c}\right)^{2k-4}\left[n_k^2 + \frac{k+2\,!}{2\,.\,k-2\,!}p_k^2\right]$$

whence, with all the terms, and remembering that $(a/c)^2 = \gamma$, $(A/c)^2 = \Gamma$,

$$e_1 = -\tfrac{1}{2}\left[(\tfrac{4}{3}\pi)^2\frac{A^3a^3}{c}\right]\tfrac{3}{2}\left(\frac{A}{c}\right)^3\sum_{k=2}^{k=\infty}\frac{2k+1}{2k-2}\frac{\gamma^{k-1}}{k-1}\left[n_k^2 + \frac{k+2\,!}{2\,.\,k-2\,!}p_k^2\right] \quad \text{...(ii.)}$$

The formula for E_1 may be written down by symmetry.

2nd. e_2, the exhaustion of mutual energy of the layer on itself, is half the potential of the layer at any element, multiplied by the mass of the element, and integrated over the whole sphere.

The potential of the layer is

$$\frac{4\pi A^3}{3c}\sum_{k=0}^{k=\infty}\frac{3}{2k-2}\left(\frac{a}{c}\right)^k\left(\frac{a}{r}\right)^{k+1}\left(n_k\frac{w_k}{r^k} - p_k\frac{\delta^2 w_{k+2}}{r^k}\right)$$

Then, at an element of the layer $r = a$, and taking a typical term only, we have

$$e_2 = \tfrac{1}{2}\,.\,\frac{4\pi A^3}{3c}\,.\,\frac{3(2k+1)}{(2k-2)^2}\left(\frac{a}{c}\right)^{2k-2}\left(\frac{A}{c}\right)^3 a^3\iint\left[n_k\left(\frac{w_k}{r^k}\right)^2 + p_k\left(\frac{\delta^2 w_{k+2}}{r^k}\right)^2\right]d\varpi$$

whence

$$e_2 = \tfrac{1}{2}\left[(\tfrac{4}{3}\pi)^2\frac{A^3a^3}{c}\right]\tfrac{3}{2}\left(\frac{A}{c}\right)^3\sum_{k=2}^{k=\infty}\frac{3}{2k-2}\frac{\gamma^{k-1}}{k-1}\left[n_k^2 + \frac{k+2\,!}{2\,.\,k-2\,!}p_k^2\right] \quad \text{...(iii.)}$$

The formula for E_2 may be written down by symmetry.

The addition of e_1 to e_2, and of E_1 to E_2, simplifies these expressions by cutting out the factor immediately following the Σ in either, and replacing it by unity.

3rd. e_3 is the loss of energy due to raising the layer on a in presence of the mean sphere A. We multiply the potential of the sphere A by the mass of the element on a, and integrate throughout angular space.

The potential of the sphere A, when transferred to the origin o, is

$$\frac{4\pi A^3}{3c}\sum_{k=0}^{k=\infty}\left(\frac{a}{c}\right)^k\left(\frac{r}{a}\right)^k\frac{w_k}{r^k}$$

Then, at an element of the layer $r=a$ and taking a typical term,

$$e_3=\frac{4\pi A^3}{3c}\frac{2k+1}{2k-2}\left(\frac{A}{c}\right)^3\left(\frac{a}{c}\right)^{2k-2}a^3\iint n_k\left(\frac{w_k}{r^k}\right)^2 d\varpi$$

whence

$$e_3=(\tfrac{4}{3}\pi)^2\frac{A^3a^3}{c}\,\tfrac{3}{2}\left(\frac{A}{c}\right)^3\sum_{k=2}^{k=\infty}\frac{\gamma^{k-1}}{k-1}n_k \quad\text{..................(iv.)}$$

The expression for E_3 may be written down by symmetry. On collecting results from (ii.), (iii.), and (iv.), we have

$$e_1+e_2+e_3=(\tfrac{4}{3}\pi)^2\frac{A^3a^3}{c}\cdot\tfrac{3}{2}\left(\frac{A}{c}\right)^3\sum_{k=2}^{k=\infty}\frac{\gamma^{k-1}}{k-1}\left\{n_k-\tfrac{1}{2}n_k^2-\frac{\overline{k+2}\,!}{4\,.\,\overline{k-2}\,!}p_k^2\right\}\ldots\text{(v.)}$$

and a similar expression for $E_1+E_2+E_3$.

4th. $(Ee)_4$ is the loss of energy of one layer in the presence of the other. We take the potential of the layer on A, multiply it by the mass of an element on a, and integrate.

The potential of the layer on A when transferred to the origin o, as in (22–ii.), is

$$\frac{4\pi A^3}{3c}\tfrac{3}{2}\left(\frac{a}{c}\right)^3\left\{\sum_{k=2}^{k=\infty}\sum_{i=2}^{i=\infty}\left[\frac{\overline{k+i}\,!}{k!\,i!}\frac{\Gamma^{i-1}}{i-1}N_i\left(\frac{a}{c}\right)^k\left(\frac{r}{a}\right)^k\frac{w_k}{r^k}\right.\right.$$
$$\left.\left.-\frac{\overline{k+i}\,!}{\overline{i-2}\,!\,\overline{k+2}\,!}\frac{\Gamma^{i-1}}{i-1}P_i\left(\frac{a}{c}\right)^k\left(\frac{r}{a}\right)^k\frac{\delta^2 w_{k+2}}{r^k}\right]\right\}$$

Introducing this into the integral, only taking a typical term, and neglecting those terms in the integral which must vanish, we get

$$(Ee)_4=\frac{4\pi A^3}{3c}\tfrac{3}{2}\left(\frac{a}{c}\right)^3\frac{2k+1}{2k-2}\left(\frac{A}{c}\right)^3\left(\frac{a}{c}\right)^{2k-2}a^3\frac{\Gamma^{i-1}}{i-1}\left\{\iint\frac{\overline{k+i}\,!}{k!\,i!}N_in_k\left(\frac{w_k}{r^k}\right)^2d\varpi\right.$$
$$\left.+\iint\frac{\overline{k+i}\,!}{\overline{i-2}\,!\,\overline{k+2}\,!}P_ip_k\left(\frac{\delta^2w_{k+2}}{r^k}\right)^2d\varpi\right\}$$

Effecting the integrations, and putting in the $\Sigma\Sigma$, we get

$$(Ee)_4=(\tfrac{4}{3}\pi)^2\frac{A^3a^3}{c}\cdot(\tfrac{3}{2})^2\left(\frac{a}{c}\right)^3\left(\frac{A}{c}\right)^3\sum_{k=2}^{k=\infty}\sum_{i=2}^{i=\infty}\frac{\Gamma^{i-1}}{i-1}\frac{\gamma^{k-1}}{k-1}$$
$$\times\left\{\frac{\overline{k+i}\,!}{k!\,i!}N_in_k+\frac{\overline{k+2}\,!}{2\,.\,\overline{k-2}\,!}\frac{\overline{k+i}\,!}{\overline{i-2}\,!\,\overline{k+2}\,!}P_ip_k\right\}\ldots\text{(vi.)}$$

This involves the two figures symmetrically.

5th. e_5 is the loss of energy in the yielding of the figure a to centrifugal force. To find it, we multiply the potential due to rotation by the mass of each element of the layer a, and integrate.

By (35) and (52) we know that the rotation potential is

$$\tfrac{1}{6}\omega^2 a^2 \left(\frac{r}{a}\right)^2 \left\{\frac{w_2}{r^2} - \tfrac{1}{2}\frac{\delta^2 w_4}{r^2}\right\}$$

As this only involves harmonics of the second order, we may neglect in the layer a all terms except those of the second order. Thus we get

$$\begin{aligned} e_5 &= \tfrac{1}{6}\omega^2 a^2 \tfrac{5}{2}\left(\frac{A}{c}\right)^3 a^3 \iint \left\{n_2\left(\frac{w_2}{r^2}\right)^2 + \tfrac{1}{2}p_2\left(\frac{\delta^2 w_4}{r^2}\right)^2\right\} d\varpi \\ &= 4\pi \cdot \tfrac{1}{12}\omega^2 a^5 \left(\frac{A}{c}\right)^3 \left\{n_2 + \frac{4\,!}{2.2.0\,!}\,p_2\right\} \\ &= (\tfrac{4}{3}\pi)^2 \frac{A^3 a^3}{c}\left\{\frac{3\omega^2}{16\pi}\left(\frac{a}{c}\right)^2 (n_2 + 6p_2)\right\} \quad \text{..........................(vii.)} \end{aligned}$$

The expression for E_5 may be written down by symmetry. Collecting results from (v.), (vi.), and (vii.), we get, for the whole exhaustion of energy,

$$\begin{aligned} E \div (\tfrac{4}{3}\pi)^2 \frac{A^3 a^3}{c} = \sum_{k=2}^{k=\infty} &\left[\tfrac{3}{2}\left(\frac{A}{c}\right)^3 \frac{\gamma^{k-1}}{k-1}\left\{n_k - \tfrac{1}{2}n_k^2 - \frac{k+2\,!}{4.\,k-2\,!}\,p_k^2\right\}\right. \\ &\left. + \tfrac{3}{2}\left(\frac{a}{c}\right)^3 \frac{\Gamma^{k-1}}{k-1}\left\{N_k - \tfrac{1}{2}N_k^2 - \frac{k+2\,!}{4.\,k-2\,!}\,P_k^2\right\}\right] \\ &+ (\tfrac{3}{2})^2\left(\frac{A}{c}\right)^3\left(\frac{a}{c}\right)^3 \sum_{k=2}^{k=\infty}\sum_{i=2}^{i=\infty} \frac{\Gamma^{i-1}}{i-1}\frac{\gamma^{k-1}}{k-1} \\ &\times\left\{\frac{k+i\,!}{k\,!\,i\,!}N_i n_k + \frac{k+2\,!}{2.\,k-2\,!}\frac{k+i\,!}{i-2\,!\,k+2\,!}P_i p_k\right\} \\ &+ \frac{3\omega^2}{16\pi}\left[\left(\frac{a}{c}\right)^2(n_2+6p_2) + \left(\frac{A}{c}\right)^2(N_2+6P_2)\right] \quad \text{......(viii.)} \end{aligned}$$

The expression is found without any assumption that the two masses are bounded by level surfaces, and therefore in equilibrium. But the condition for equilibrium is that the differential coefficients of E with respect to any one and all of the parameters n, p, N, P, shall vanish. If we equate to zero dV/dn_k, we get

$$1 - n_k + \tfrac{3}{2}\left(\frac{a}{c}\right)^3 \sum_{i=2}^{i=\infty} \frac{\Gamma^{i-1}}{i-1}\frac{k+i\,!}{k\,!\,i\,!}N_i = 0$$

If, however, $k = 2$, there is on the left-hand side an additional term

$$\frac{3\omega^2}{16\pi}\left(\frac{a}{c}\right)^2 \div \tfrac{3}{2}\left(\frac{A}{c}\right)^3 \frac{\gamma}{1} = \frac{\omega^2}{8\pi}\left(\frac{c}{A}\right)^3 = \tfrac{2}{15}\epsilon\left(\frac{c}{A}\right)^3$$

The equation of dV/dN_i to zero gives a similar equation.

If we equate to zero dV/dp_k, we get

$$-p_k + \tfrac{3}{2}\left(\frac{a}{c}\right)^3 \Sigma \frac{\Gamma^{i-1}}{i-1} \frac{k+i\,!}{i-2\,!\,k+2\,!} P_i = 0$$

If, however, $k = 2$, there is on the left-hand side an additional term

$$\frac{3\omega^2}{16\pi} 6\left(\frac{a}{c}\right)^2 \div \tfrac{3}{2}\left(\frac{A}{c}\right)^3 \frac{\gamma}{1} \frac{4\,!}{2\,!\,0\,!} = \frac{\omega^2}{8\pi}\left(\frac{c}{A}\right)^3 \frac{6}{4\,.\,3} = \tfrac{1}{15}\epsilon\left(\frac{c}{A}\right)^3$$

The equation of dV/dP_i to zero gives a similar equation.

Now, if we put $h_k + l_k$ for n_k, except when $k = 2$, and then put

$$h_2 + l_2 + \tfrac{2}{15}\epsilon\,(c/A)^3 = n_2$$

and similarly introduce the H's and L's; and if we put $p_k = m_k$, except when $k = 2$, and then put $p_2 = m_2 + \frac{1}{15}\epsilon\,(c/A)^3$, and similarly introduce the M's, it is easy to see that the equations (i.) to the two surfaces become the same as (72), and the equations of condition between n and N, and between p and P, become exactly those which we found by a different method above in (23), (44), and (59). The only difference is that the equations for h and l are fused together.

This, therefore, forms a valuable confirmation of the correctness of the long analysis employed for the determination of the forms of equilibrium.

The formula (viii.) also enables us to obtain the intrinsic energy of the system, that is to say, the exhaustion of energy of the concentration of the matter from a state of infinite dispersion to its actual shape, with its sign changed.

The last line of (viii.) depends on the yielding of the fluid to centrifugal force, and must be omitted from the exhaustion of energy.

Besides the rest of (viii.), we have in the exhaustion of energy of the system, the exhaustion of the two spheres and their mutual exhaustion.

It is clear, then, that the exhaustion of energy of the system, apart from that due to centrifugal force, is

$$\begin{aligned}
&(\tfrac{4}{3}\pi)^2 . \tfrac{3}{5}(a^5 + A^5) + (\tfrac{4}{3}\pi)^2 \frac{A^3 a^3}{c} \\
&+ (\tfrac{4}{3}\pi)^2 \frac{A^3 a^3}{c} \sum_{k=2}^{k=\infty} \left[\tfrac{3}{2}\left(\frac{A}{c}\right)^3 \frac{\gamma^{k-1}}{k-1}\left\{n_k - \tfrac{1}{2}n_k^2 - \frac{k+2\,!}{4\,.\,k-2\,!} p_k^2\right\}\right. \\
&\left. + \tfrac{3}{2}\left(\frac{a}{c}\right)^3 \frac{\Gamma^{k-1}}{k-1}\left\{N_k - \tfrac{1}{2}N_k^2 - \frac{k+2\,!}{4\,.\,k-2\,!} P_k^2\right\}\right] \\
&+ (\tfrac{4}{3}\pi)^2 \frac{A^3 a^3}{c} \sum_{k=2}^{k=\infty} \sum_{i=2}^{i=\infty} (\tfrac{3}{2})^2 \left(\frac{A}{c}\right)^3 \left(\frac{a}{c}\right)^3 \frac{\Gamma^{i-1}}{i-1} \frac{\gamma^{k-1}}{k-1} \\
&\times \left\{\frac{k+i\,!}{k\,!\,i\,!} N_i n_k + \frac{k+i\,!}{2\,.\,k-2\,!\,i-2\,!} P_i p_k\right\} \quad \text{..................(ix.)}
\end{aligned}$$

[This is the function called E in § 2 of Paper 15 below.

In order to apply the methods indicated in that paper we require the moment of inertia I.

Now

$$I = \tfrac{4}{3}\pi \left\{\tfrac{2}{5}(a^5 + A^5) + \frac{A^3 a^3 c^2}{A^3 + a^3} + \tfrac{1}{2}\frac{A^3 a^3}{c^3}\left[a^2 (n_2 + 6p_2) + A^2 (N_2 + 6P_2)\right]\right\}$$

The parameters which define the system are c, n_k, p_k, N_k, P_k.

The attempt to form the determinants, indicated in § 2 of Paper 15, leads to formulæ of inextricable complication, and thus the hope of obtaining the coefficients of stability must be abandoned. However the investigations of Paper 15 show how instability must first arise and prove the instability of these figures, as I have already said.

The remainder of this Appendix is omitted on account of the mistakes involved in it.]

10.

ELLIPSOIDAL HARMONIC ANALYSIS.

[*Philosophical Transactions of the Royal Society*, Vol. 197, A (1901), pp. 461—557.]

TABLE OF CONTENTS.

Introduction.

Lamé's functions or ellipsoidal harmonics have been successfully used in many investigations, but the form in which they have been presented has always been such as to render numerical calculation so difficult as to be practically impossible. The object of the present investigation is to remove this imperfection in the method. I believe that I have now reduced these functions to such a form that numerical results will be accessible, although by the nature of the case the arithmetic will necessarily remain tedious.

Throughout my work on ellipsoidal harmonics I have enjoyed the immense advantage of frequent discussions with Mr E. W. Hobson. He has helped me freely from his great store of knowledge, and beginning, as I did, in almost complete ignorance of the subject, I could hardly have brought my attempt to a successful issue without his advice. In many cases the help derived from him has been of immense value, even where it is not possible to indicate a specific point as due to him. In other cases he has put me in the way of giving succinct proofs of propositions which I had only proved by clumsy and tedious methods, or where I merely felt sure of the truth of a result without rigorous proof. In particular, I should have been quite unable to carry out the investigation of § 19, unless he had shown me how the needed series were to be determined.

My original object in attacking this problem was the hope of being thereby enabled to obtain exact numerical results with respect to M. Poincaré's pear-shaped figure of equilibrium of a mass of liquid in rotation. But I soon found that a partial investigation with one particular point in view was impracticable, and I was thus led on little by little to cover the whole field, in as far as it was necessary to do so for the purpose of practical application. This paper had then grown to such considerable dimensions that it seemed best that it should stand by itself, and that the discussion of the specific problem should be deferred.

A paper of this kind is hardly read even by the mathematician, unless he happens to be working at a cognate subject. It appears therefore to be useful to present a summary, which shall render it possible for the mathematical reader to understand the nature of the method and results, without having to pick it out from a long and complex train of analysis. Such a summary is given in Part III.

PART I.

FORMATION OF THE FUNCTIONS.

§ 1. *The Principles of Ellipsoidal Harmonic Analysis.*

The basis of this method of analysis is expounded in various works on the subject. I begin with a statement of results in my own notation.

If u_1^2, u_2^2, u_3^2 denote the three roots of the cubic

$$\frac{x^2}{a^2+u^2}+\frac{y^2}{b^2+u^2}+\frac{z^2}{c^2+u^2}=1$$

it may be proved that

$$x^2=\frac{(a^2+u_1^2)(a^2+u_2^2)(a^2+u_3^2)}{(b^2-a^2)(c^2-a^2)}$$

and y^2, z^2 may be written down by cyclical changes.

If for brevity we write

$$A_n^2=u_n^2+a^2,\quad B_n^2=u_n^2+b^2,\quad C_n^2=u_n^2+c^2,\quad (n=1,\,2,\,3)$$

Laplace's equation becomes

$$(u_2^2-u_3^2)\left(A_1B_1C_1\frac{d}{u_1du_1}\right)^2V_i+(u_3^2-u_1^2)\left(A_2B_2C_2\frac{d}{u_2du_2}\right)^2V_i$$
$$+(u_1^2-u_2^2)\left(A_3B_3C_3\frac{d}{u_3du_3}\right)^2V_i=0$$

The solution is

$$V_i=U_1U_2U_3$$

where U_1, U_2, U_3 are functions of u_1, u_2, u_3 respectively, and satisfy

$$\left(A_1B_1C_1\frac{d}{u_1du_1}\right)^2U_1=[i(i+1)u_1^2+\kappa^2]\,U_1$$

and two other equations with suffixes 2 and 3, involving the same κ, a constant, and the same i, a positive integer.

If a, b, c are in ascending order of magnitude we may suppose u_1^2 to lie between $-a^2$ and ∞, u_2^2 between $-c^2$ and $-b^2$, and u_3^2 between $-b^2$ and $-a^2$.

If s_1, s_2, s_3 denote the three orthogonal arcs formed by the intersections of the three orthogonal quadrics,

$$\left(\frac{ds_1}{u_1du_1}\right)^2=\frac{(u_2^2-u_1^2)(u_3^2-u_1^2)}{A_1^2B_1^2C_1^2}$$

and two other equations found by cyclical changes of suffixes.

§ 2. *Notation; limits of β so as to represent all Ellipsoids.*

I now change the notation, and let the three roots be defined thus:—

$$u_1^2 = k^2\nu^2$$

$$u_2^2 = k^2\mu^2$$

$$u_3^2 = k^2\frac{1-\beta\cos 2\phi}{1-\beta}$$

where ν ranges from ∞ to 0, μ between ± 1, ϕ between 0 and 2π.

Let the axes of the fundamental ellipsoid of reference be

$$a^2 = -k^2\frac{1+\beta}{1-\beta}$$

$$b^2 = -k^2$$

$$c^2 = 0$$

The ellipsoid defined by ν has its three axes a, b, c given by

$$a^2 = k^2\left(\nu^2 - \frac{1+\beta}{1-\beta}\right),\quad b^2 = k^2(\nu^2-1),\quad c^2 = k^2\nu^2,\quad (a<b<c)$$

This mode of defining the axis is such as to indicate the relationship to the prolate ellipsoid $a=b<c$. But another hypothesis may be made which will bring the axes into relationship with those of the oblate ellipsoid $a=b>c$; for if we take a new k, numerically equal to the old one but imaginary, and replace ν^2 by $-\zeta^2$, we have

$$a^2 = k^2\left(\zeta^2 + \frac{1+\beta}{1-\beta}\right),\quad b^2 = k^2(\zeta^2+1),\quad c^2 = k^2\zeta^2,\quad (a>b>c)$$

If β be made to range from 0 to ∞, all possible ellipsoids are comprised in either of these types. It will, however, now be shown that, by a proper choice of type, all ellipsoids may be included with the range of β from 0 to $\frac{1}{3}$.

Let us suppose the axes to be expressed in three forms, as follows:—

$$\begin{array}{llll} & (1) & (2) & (3) \\ a^2 = & k^2\left(\nu^2-\dfrac{1+\beta}{1-\beta}\right) & = k_1^2\zeta_1^2 & = k_2^2\nu_2^2 \\ b^2 = & k^2(\nu^2-1) & = k_1^2(\zeta_1^2+1) & = k_2^2\left(\nu_2^2-\dfrac{1+\beta_2}{1-\beta_2}\right) \\ c^2 = & k^2\nu^2 & = k_1^2\left(\zeta_1^2+\dfrac{1+\beta_1}{1-\beta_1}\right) & = k_2^2(\nu_2^2-1) \end{array}$$

Then we have
$$b^2 - a^2 = \frac{2k^2\beta}{1-\beta} = k_1^2 = -k_2^2\frac{1+\beta_2}{1-\beta_2}$$

$$c^2 - b^2 = k^2 = \frac{2k_1^2\beta_1}{1-\beta_1} = \frac{2k_2^2\beta_2}{1-\beta_2}$$

Therefore
$$\frac{b^2-a^2}{c^2-b^2}=\frac{2\beta}{1-\beta}=\frac{1-\beta_1}{2\beta_1}=-\frac{1+\beta_2}{2\beta_2}$$
$$\frac{c^2-a^2}{c^2-b^2}=\frac{1+\beta}{1-\beta}=\frac{1+\beta_1}{2\beta_1}=-\frac{1-\beta_2}{2\beta_2}$$

And
$$\frac{b^2-a^2}{2c^2-a^2-b^2}=\beta=\frac{1-\beta_1}{1+3\beta_1}=\frac{1+\beta_2}{1-3\beta_2}$$

Now let β increase from 0 to ∞.

As β passes from 0 to $\frac{1}{3}$, form (1) is appropriate.

As β passes from $\frac{1}{3}$ to 1, β_1 decreases from $\frac{1}{3}$ to 0, so that form (2) is appropriate.

Lastly, as β passes from 1 to ∞, β_2 increases from 0 to $\frac{1}{3}$, so that form (3) is appropriate.

But we might equally well have written forms (1) and (3) so as to involve ζ, and form (2) so as to involve ν, and it follows that all possible ellipsoids are comprised in the range of β from 0 to $\frac{1}{3}$, provided that the type be appropriately chosen.

The developments in this paper are made in powers of β. It will, therefore, be well to show that there is a class of ellipsoids, analogous to ellipsoids of revolution, which might form the basis of developments similar to those carried out below.

Ellipsoids of revolution are defined by the condition
$$a^2-c^2=b^2-c^2, \quad \text{or } a^2=b^2$$

In the class to which I refer
$$a^2-c^2=c^2-b^2, \quad \text{or } c^2=\tfrac{1}{2}(a^2+b^2)$$

Ellipsoids of this kind are given by $\beta=\beta_1=-\beta_2=\frac{1}{3}$; for in this case $b^2=\frac{1}{2}(a^2+c^2)$. They are also given by
$$\beta=\infty, \quad -\beta_1=\beta_2=\tfrac{1}{3}; \text{ for then } c^2=\tfrac{1}{2}(a^2+b^2)$$

Hence if we only allow β to range from 0 to $\frac{1}{3}$, $\beta=0$ corresponds with ellipsoids of revolution, to which spheroidal harmonic analysis is applicable; and $\beta=\frac{1}{3}$ corresponds with this new class for which the corresponding analysis has not yet been worked out.

We shall see below that the solid harmonic for this case where $\beta=\frac{1}{3}$ will be of the form $\mathrm{B}(\nu)\,\mathrm{B}(\mu)\,\mathrm{E}(\phi)$, where B and E satisfy the equations
$$(\nu^2+1)(\nu^2-1)\frac{d^2\mathrm{B}}{d\nu^2}+2\nu^3\frac{d\mathrm{B}}{d\nu}-i(i+1)\nu^2\mathrm{B}+s^2\mathrm{B}=0$$
$$\cos 2\phi\frac{d^2\mathrm{E}}{d\phi^2}-\sin 2\phi\frac{d\mathrm{E}}{d\phi}+i(i+1)\mathrm{E}\cos 2\phi-s^2\mathrm{E}=0$$

I am not clear whether or not it would be advisable to proceed *ab initio* from these equations, but at any rate I shall show hereafter how the B- and E-functions may be determined from the analysis of the present paper with any degree of accuracy desirable.

If it were proposed to use the functions corresponding to $\beta = \frac{1}{3}$ as a basis for the development of general ellipsoidal harmonics, we should have to assume

$$a^2 = k'^2 \nu'^2, \quad b^2 = k'^2 (\nu'^2 - 1), \quad c^2 = k'^2 \left(\nu'^2 - \frac{2}{1-\eta}\right)$$

or else
$$a^2 = k'^2 \zeta'^2, \quad b^2 = k'^2 (\zeta'^2 + 1), \quad c^2 = k'^2 \left(\zeta'^2 + \frac{2}{1-\eta}\right)$$

The developments would then proceed by powers of η.

In order to discover what is the greatest value of η which must be used so as to comprise all ellipsoids, when we proceed from both bases of development, a comparison must be made between this assumption and the previous one. Suppose in fact that

$$a^2 = k^2 \left(\nu^2 - \frac{1+\beta}{1-\beta}\right) = k'^2 \zeta'^2; \quad b^2 = k^2 (\nu^2 - 1) = k'^2 (\zeta'^2 + 1)$$

$$c^2 = k^2 \nu^2 = k'^2 \left(\zeta'^2 + \frac{2}{1-\eta}\right)$$

Then
$$b^2 - a^2 = \frac{2k^2\beta}{1-\beta} = k'^2; \quad c^2 - b^2 = k^2 = k'^2 \frac{1+\eta}{1-\eta}$$

and therefore
$$\frac{2\beta}{1-\beta} = \frac{1-\eta}{1+\eta}, \quad \text{or } \eta = \frac{1-3\beta}{1+\beta}$$

When η and β are both equally great, they must each equal the positive root of $\beta = \dfrac{1-3\beta}{1+\beta}$. This root is $\sqrt{5} - 2$ or ·236. Thus the greatest values will be

$$\beta = \eta = \frac{1}{\sqrt{5}+2} = \frac{1}{4{\cdot}236}$$

In this case $\eta^2 = \beta^2 = \frac{1}{18}$ very nearly, whereas when $\beta = \frac{1}{3}$, $\beta^2 = \frac{1}{9}$. Thus if the developments were to stop with β^2 we should double the accuracy of the result. However, I do not at present propose to carry out the process suggested.

§ 3. *The Differential Equations.*

We now put $u_1^2 = k^2\nu^2$, $u_2^2 = k^2\mu^2$, $u_3^2 = k^2\dfrac{1-\beta\cos 2\phi}{1-\beta}$

$$a^2 = -k^2\frac{1+\beta}{1-\beta}, \quad b^2 = -k^2, \quad c^2 = 0$$

and find from the formulæ of § 1,

$$\left.\begin{aligned}\frac{x^2}{k^2} &= -\frac{1-\beta}{1+\beta}\left(\nu^2 - \frac{1+\beta}{1-\beta}\right)\left(\mu^2 - \frac{1+\beta}{1-\beta}\right)\cos^2\phi \\ \frac{y^2}{k^2} &= -(\nu^2-1)(\mu^2-1)\sin^2\phi \\ \frac{z^2}{k^2} &= \nu^2\mu^2\frac{1-\beta\cos 2\phi}{1+\beta}\end{aligned}\right\}\quad\ldots\ldots\ldots\ldots(1)$$

It will be observed that y is independent of β, and that it has the same form as in spheroidal harmonic analysis when β vanishes. Since μ^2 is less than 1 and ν^2 greater than $\dfrac{1+\beta}{1-\beta}$, x and y are real.

In all the earlier portion of this paper I always write μ^2-1 and not $1-\mu^2$, so as to maintain perfect symmetry with respect to ν and μ.

We now have

$$A_1^2 = k^2\left(\nu^2 - \frac{1+\beta}{1-\beta}\right), \quad B_1^2 = k^2(\nu^2-1), \quad C_1^2 = k^2\nu^2$$

$$A_2^2 = k^2\left(\mu^2 - \frac{1+\beta}{1-\beta}\right), \quad B_2^2 = k^2(\mu^2-1), \quad C_2^2 = k^2\mu^2$$

$$A_3^2 = \frac{-2k^2\beta\cos^2\phi}{1-\beta}, \quad B_3^2 = \frac{2k^2\beta\sin^2\phi}{1-\beta}, \quad C_3^2 = k^2\frac{1-\beta\cos 2\phi}{1-\beta}$$

Let us denote the differential operators involved in our equations, thus:—

$$D_1 = (1-\beta)^{\frac{1}{2}}\frac{A_1B_1C_1}{ku_1}\frac{d}{du_1}, \quad D_2 = (1-\beta)^{\frac{1}{2}}\frac{A_2B_2C_2}{ku_2}\frac{d}{du_2}$$

$$D_3 = -\sqrt{-1}\,.\,(1-\beta)^{\frac{1}{2}}\frac{A_3B_3C_3}{ku_3}\frac{d}{du_3}$$

Then

$$\left.\begin{aligned}D_1 &= (1-\beta)^{\frac{1}{2}}\left(\nu^2 - \frac{1+\beta}{1-\beta}\right)^{\frac{1}{2}}(\nu^2-1)^{\frac{1}{2}}\frac{d}{d\nu} \\ D_2 &= (1-\beta)^{\frac{1}{2}}\left(\mu^2 - \frac{1+\beta}{1-\beta}\right)^{\frac{1}{2}}(\mu^2-1)^{\frac{1}{2}}\frac{d}{d\mu} \\ D_3 &= (1-\beta\cos 2\phi)^{\frac{1}{2}}\frac{d}{d\phi}\end{aligned}\right\}\quad\ldots\ldots\ldots\ldots(2)$$

$$A_1B_1C_1 \frac{d}{u_1 du_1} = \frac{kD_1}{(1-\beta)^{\frac{1}{2}}}, \quad A_2B_2C_2 \frac{d}{u_2 du_2} = \frac{kD_2}{(1-\beta)^{\frac{1}{2}}}$$

$$A_3B_3C_3 \frac{d}{u_3 du_3} = \sqrt{-1} \cdot \frac{kD_3}{(1-\beta)^{\frac{1}{2}}}$$

Hence our differential equations are

$$\frac{D_1{}^2 U_1}{1-\beta} = \left[i(i+1)\nu^2 + \frac{\kappa^2}{k^2}\right] U_1$$

a similar equation with suffix 2, and

$$\frac{-D_3{}^2 U_3}{1-\beta} = \left[i(i+1)\frac{1-\beta\cos 2\phi}{1-\beta} + \frac{\kappa^2}{k^2}\right] U_3$$

Let us replace κ^2 by another constant such that

$$\left[i(i+1)\nu^2 + \frac{\kappa^2}{k^2}\right](1-\beta) = i(i+1)\left[\nu^2(1-\beta) - 1\right] + s^2 - \sigma$$

so that*

$$\frac{\kappa^2}{k^2} = -\frac{i(i+1) - s^2 + \sigma}{1-\beta}$$

In this formula s is a constant integer and σ a constant to be determined.

Our equations are now

$$\left.\begin{aligned} &[D_1{}^2 - i(i+1)[\nu^2(1-\beta) - 1] - s^2 + \sigma]U_1 = 0 \\ &\text{a similar equation for } \mu \\ &\text{and} \quad [D_3{}^2 - i(i+1)\beta\cos 2\phi + s^2 - \sigma]U_3 = 0 \end{aligned}\right\} \quad \text{..........(3)}$$

Laplace's equation is

$$\left[\left(\mu^2 - \frac{1-\beta\cos 2\phi}{1-\beta}\right)D_1{}^2 + \left(\frac{1-\beta\cos 2\phi}{1-\beta} - \nu^2\right)D_2{}^2 - (\nu^2-\mu^2)D_3{}^2\right]U_1U_2U_3 = 0 \quad \text{......(4)}$$

Laplace's operator ∇^2 is equal to the differential operator in (4), divided by

$$-k^2(\nu^2-\mu^2)\left(\nu^2 - \frac{1-\beta\cos 2\phi}{1-\beta}\right)\left(\frac{1-\beta\cos 2\phi}{1-\beta} - \mu^2\right)$$

It is well known that in spheroidal harmonic analysis there are two kinds of functions of ν and μ which satisfy the differential equation, and they are usually denoted P_i^s, Q_i^s. The Q-functions of the variable μ have no significance, so that virtually there are P- and Q-functions of ν, but only P-functions of μ. The like is true in the present case, however, with the additional complication that each of the functions may assume one of two alternative forms. I adopt a parallel notation and write for U_1 and U_2 either $\mathfrak{P}_i^s$, $\mathfrak{Q}_i^s$, or P_i^s, Q_i^s, as the case may be. Since ν and μ enter in the first two equations

* The quantity which is here denoted by σ was written as $\beta\sigma$ in the original paper.

in exactly the same way, we need only consider one of them, and we may usually write simply (for example) $\mathfrak{P}_i^s$ where the full notation would be $\mathfrak{P}_i^s$ (ν or μ). In the early part of the investigation I shall only refer to the P-functions, and the Q-functions will be considered later.

In spheroidal harmonic analysis the third function is a cosine or sine of $s\phi$. So here also we find functions of two kinds associated with cosines and sines, which I shall denote $\mathfrak{C}_i^s$, $\mathfrak{S}_i^s$, C_i^s, S_i^s, the variable ϕ being understood.

Throughout the greater part of this paper the functions will be of degree denoted by i, and it seems useless to print the subscript i hundreds of times. I shall accordingly drop the subscript i except where it shall be necessary or advisable to retain it; for example, $\mathfrak{P}^s$ will be the abridged notation for $\mathfrak{P}_i^s(\nu)$.

The operators involved in the differential equations (3) will occur so frequently that an abridged notation seems justifiable. I therefore write

$$\left.\begin{aligned}\psi_s &= D_1^2 - i(i+1)[\nu^2(1-\beta)-1] - s^2 + \sigma \\ \chi_s &= D_3^2 - i(i+1)\beta\cos 2\phi + s^2 - \sigma \\ \text{where}\qquad D_1 &= (1-\beta)^{\frac{1}{2}}\left(\nu^2 - \tfrac{1+\beta}{1-\beta}\right)^{\frac{1}{2}}(\nu^2-1)^{\frac{1}{2}}\frac{d}{d\nu} \\ D_3 &= (1-\beta\cos 2\phi)^{\frac{1}{2}}\frac{d}{d\phi}\end{aligned}\right\} \quad \text{.............(5)}$$

The equations are then

$$\left.\begin{aligned}\psi_s(\mathfrak{P}^s \text{ or } \mathsf{P}^s) &= 0 \\ \chi_s(\mathfrak{C}^s \text{ or } \mathfrak{S}^s \text{ or } \mathsf{C}^s \text{ or } \mathsf{S}^s) &= 0\end{aligned}\right\} \quad \text{.....................(5)}$$

§ 4. *The Forms of the Functions.*

It is well known that the function U_i is a linear function of u_1 of degree i made up in one of the eight following ways:—

1. When i is even, a linear function of u_1^2 of degree $\frac{1}{2}i$.

2, 3, 4. When i is odd, a linear function of u_1^2 of degree $\frac{1}{2}(i-1)$, multiplied by A_1, or B_1, or C_1.

5, 6, 7. When i is even, a linear function of u_1^2 of degree $\frac{1}{2}(i-2)$, multiplied by B_1C_1, or C_1A_1, or A_1B_1.

8. When i is odd, a linear function of u_1^2 of degree $\frac{1}{2}(i-3)$, multiplied by $A_1B_1C_1$.

These eight classes might be conveniently specified by the initials O, A, B C, BC, CA, AB, ABC, but it is better to rearrange them according as they are associated with the evenness or oddness of i and s, and with the cosine or sine functions. This new grouping may be defined by a shorthand notation involving the initials E, O and C or S, which shall denote successively the evenness or oddness of i and s, and cosine or sine.

We shall see below that this arrangement is as follows:—

O or EEC; i even, s even, cosine.
AB or EES; i even, s even, sine.
A or OOC; i odd, s odd, cosine.
B or OOS; i odd, s odd, sine.
C or OEC; i odd, s even, cosine.
ABC or OES; i odd, s even, sine.
CA or EOC; i even, s odd, cosine.
CB or EOS; i even, s odd, sine.

Since the several functions are linear in u_1^2, they are in the new notation functions of ν^2 or μ^2, or of $\nu^2 - 1$ and $\mu^2 - 1$.

Hence $\mathfrak{P}^s(\nu)$ and $\mathsf{P}^s(\nu)$ involve linear functions of $\nu^2 - 1$ of various degrees multiplied by various factors; and the same is true of the functions of μ.

In the case of the third root the linear function of powers of $\cos 2\phi$ may be replaced by a series of cosines of even multiples of ϕ. Further, in forming the $\mathfrak{C}$, $\mathfrak{S}$, C, S functions we may regard A_3 as being $\cos\phi$, B_3 as $\sin\phi$, and C_3 as $(1 - \beta\cos 2\phi)^{\frac{1}{2}}$, since this only amounts to dropping constant factors which may be deemed to be included in the, as yet, undetermined coefficients of the several series.

I will now consider in detail the forms of the several P-functions of ν (those for μ following by symmetry), and at the same time indicate more precisely the nature of the notation adopted.

In the following series, indicated by Σ, the variable t is supposed *to proceed from the lower to the upper limit by* 2 *at a time.* The reader will be able to perceive the manner of the formation of the functions when he bears in mind that

$$A_1 = k\left(\nu^2 - \tfrac{1+\beta}{1-\beta}\right)^{\frac{1}{2}}, \quad B_1 = k(\nu^2 - 1)^{\frac{1}{2}}, \quad C_1 = k\nu$$

Type O or EEC; $\quad \mathfrak{P}^s = \sum_0^i \alpha_t (\nu^2 - 1)^{\frac{1}{2}t}$

Type AB or EES; $\quad \mathsf{P}^s = \sum_2^i \alpha_t (\nu^2 - 1)^{\frac{1}{2}(t-1)} \left(\nu^2 - \tfrac{1+\beta}{1-\beta}\right)^{\frac{1}{2}}$

Type A or OOC; $\quad \mathsf{P}^s = \sum_1^i \alpha_t (\nu^2 - 1)^{\frac{1}{2}(t-1)} \left(\nu^2 - \tfrac{1+\beta}{1-\beta}\right)^{\frac{1}{2}}$

Type B or OOS; $\mathfrak{P}^s = \sum_1^i \alpha_t (\nu^2 - 1)^{\frac{1}{2}t}$

Type C or OEC; $\mathfrak{P}^s = \sum_1^i \alpha_t \nu (\nu^2 - 1)^{\frac{1}{2}(t-1)}$

Type ABC or OES; $\mathsf{P}^s = \sum_3^i \alpha_t \nu (\nu^2 - 1)^{\frac{1}{2}(t-2)} \left(\nu^2 - \frac{1+\beta}{1-\beta}\right)^{\frac{1}{2}}$

Type CA or EOC; $\mathsf{P}^s = \sum_2^i \alpha_t \nu (\nu^2 - 1)^{\frac{1}{2}(t-2)} \left(\nu^2 - \frac{1+\beta}{1-\beta}\right)^{\frac{1}{2}}$

Type CB or EOS; $\mathfrak{P}^s = \sum_2^i \alpha_t \nu (\nu^2 - 1)^{\frac{1}{2}(t-1)}$

Observe that P is always associated with $\left(\nu^2 - \frac{1+\beta}{1-\beta}\right)^{\frac{1}{2}}$, and that, each form being repeated twice, there are two forms of function of each kind. Moreover, a cosine and a sine function are always associated with different kinds. It is obvious that the $\mathfrak{P}$-functions are expressible in terms of the ordinary P-functions of spherical harmonic analysis, and that if we take out the factor $\left(\frac{\nu^2 - \frac{1+\beta}{1-\beta}}{\nu^2 - 1}\right)^{\frac{1}{2}}$ the P-functions are similarly expressible. This factor will occur so frequently that I write

$$\Omega(\nu) = \left(\frac{\nu^2 - \frac{1+\beta}{1-\beta}}{\nu^2 - 1}\right)^{\frac{1}{2}}$$

and as elsewhere commonly put Ω to denote $\Omega(\nu)$.

We assume then the following forms for the functions:—

For EEC, OEC, OOS, EOS

$$\mathfrak{P}^s = q_s \mathrm{P}^s + \Sigma \beta^n q_{s-2n} \mathrm{P}^{s-2n} + \Sigma \beta^n q_{s+2n} \mathrm{P}^{s+2n}$$

For EES, OES, OOC, EOC

$$\mathsf{P}^s = \Omega \{q'_s \mathrm{P}^s + \Sigma \beta^n q'_{s-2n} \mathrm{P}^{s-2n} + \Sigma \beta^n q'_{s+2n} \mathrm{P}^{s+2n}\} \qquad \text{.........(6)}$$

In these series n proceeds by intervals of one at a time, beginning from a lower limit of unity. In both forms the upper limit of the first Σ is $\frac{1}{2}s$ or $\frac{1}{2}(s-1)$ according as s is even or odd; and the upper limit of the second Σ is $\frac{1}{2}(i-s)$ or $\frac{1}{2}(i-s-1)$ according as i and s agree or do not agree in evenness or in oddness.

The factor Ω contains $(\nu^2 - 1)^{\frac{1}{2}}$ in the denominator, but P^s does not become infinite when $\nu = \pm 1$, because when s is not zero P^s is divisible by

$(\nu^2-1)^{\frac{1}{2}}$ and we shall see that q'_0 is zero*. When s is zero there is no function of the P type.

It may be noted that the limits of the series are such that neither q nor q' can ever have a negative suffix.

We shall ultimately make q_s and q'_s equal to unity, and this will be justifiable because there must be one arbitrary constant.

We have now to consider the forms of the cosine and sine functions. They may be derived at once from the preceding results, for we have only to read $(\nu^2-1)^{\frac{1}{2}t}$ as $\cos t\phi$ where t is even; $(\nu^2-1)^{\frac{1}{2}}$ as $\sin\phi$, $\left(\nu^2-\frac{1+\beta}{1-\beta}\right)^{\frac{1}{2}}$ as $\cos\phi$, and ν as $(1-\beta\cos 2\phi)^{\frac{1}{2}}$.

The factor $(1-\beta\cos 2\phi)^{\frac{1}{2}}$ will occur frequently, and I write

$$\Phi(\phi)=(1-\beta\cos 2\phi)^{\frac{1}{2}}$$

and as elsewhere I commonly write Φ to denote $\Phi(\phi)$.

The following are the results:—

$$\text{Type O or EEC;}\qquad \mathfrak{C}^s=\sum_0^i \gamma_t\cos t\phi$$

$$\text{Type AB or EES;}\qquad \mathfrak{S}^s=\sum_2^i \gamma_t\sin t\phi$$

It is clear that we may equally well regard the lower limit in the latter as zero.

Type A, or OOC; each term is of type

$$\cos(t-1)\phi\cos\phi \quad\text{or}\quad \cos(t-2)\phi+\cos t\phi$$

Hence
$$\mathfrak{C}^s=\sum_1^i \gamma_t\cos t\phi$$

Type B, or OOS; since we now have $\cos(t-1)\phi\sin\phi$,

$$\mathfrak{S}^s=\sum_1^i \gamma_t\sin t\phi$$

$$\text{Type C, or OEC;}\qquad \mathsf{C}^s=\Phi\sum_1^i \gamma_t\cos(t-1)\phi$$

Type ABC, or OES; each term is of type $\Phi\cos(t-1)\phi\sin\phi\cos\phi$, which gives $[\sin(t+1)\phi-\sin(t-3)\phi]\,\Phi$. Hence

$$\mathsf{S}^s=\Phi\sum_3^i \gamma_t\sin(t-1)\phi$$

* This also follows from the fact that the series for P^s begins with $\Omega a_2(\nu^2-1)$ in the case of EES, and with $\Omega a_3\nu(\nu^2-1)$ in the case of OES. Thus in the former case there is no term Ωa_0 and in the latter no term $\Omega a_1\nu$.

It is clear that we may equally well regard the lower limit as unity.

Type CA, or EOC; each term is of type $\Phi \cos(t-2)\phi \cos\phi$. Hence

$$\mathbf{C}^s = \Phi \sum_2^i \gamma_t \cos(t-1)\phi$$

Type CB, or EOS; each term is of type $\Phi \cos(t-2)\phi \sin\phi$. Hence

$$\mathbf{S}^s = \Phi \sum_2^i \gamma_t \sin(t-1)\phi$$

When i and s agree as to evenness or oddness we have the forms independent of Φ, when they differ in this respect the factor Φ occurs.

Therefore (in alternative form) for EEC, EES, OOC, OOS

$$\begin{cases}\mathfrak{C}^s \\ \mathfrak{S}^s\end{cases} = p_s \begin{cases}\cos \\ \sin\end{cases} s\phi + \Sigma\beta^n p_{s-2n} \begin{cases}\cos \\ \sin\end{cases} (s-2n)\phi + \Sigma\beta^n p_{s+2n} \begin{cases}\cos \\ \sin\end{cases} (s+2n)\phi$$

and for OEC, OES, EOC, EOS

$$\begin{cases}\mathbf{C}^s \\ \mathbf{S}^s\end{cases} = \Phi \left[p'_s \begin{cases}\cos \\ \sin\end{cases} s\phi + \Sigma\beta^n p'_{s-2n} \begin{cases}\cos \\ \sin\end{cases} (s-2n)\phi + \Sigma\beta^n p'_{s+2n} \begin{cases}\cos \\ \sin\end{cases} (s+2n)\phi \right] \quad \ldots\ldots(7)$$

In these series n proceeds by intervals of one at a time, beginning with unity. In both forms the upper limit of the first Σ is $\frac{1}{2}s$ or $\frac{1}{2}(s-1)$ according as s is even or odd. In the first form the upper limit of the second Σ is $\frac{1}{2}(i-s)$, and in the second form it is $\frac{1}{2}(i-s-1)$.

We shall ultimately put p_s and p'_s, which may be regarded as arbitrary constants, equal to unity.

§ 5. *Preparation for determination of the Functions.*

In order to determine the coefficients q, q', p, p' and σ, we have to substitute these assumed forms in the differential equations.

Where the functions involve Ω and Φ as factors, the forms already given for the differential equations are perhaps the most convenient, but in the other cases a reduction seems desirable.

By considering the forms of D_1 and D_3 in (3) it is easy to show that

$$\psi_s = \left[(\nu^2-1)\frac{d}{d\nu}\right]^2 - i(i+1)(\nu^2-1) - s^2$$
$$- \beta\left[(\nu^2-1)(\nu^2+1)\frac{d^2}{d\nu^2} + 2\nu^3\frac{d}{d\nu} - i(i+1)\nu^2 - \frac{\sigma}{\beta}\right] \quad \ldots(8)$$

$$\chi_s = \frac{d^2}{d\phi^2} + s^2 - \beta\left[\cos 2\phi \frac{d^2}{d\phi^2} - \sin 2\phi \frac{d}{d\phi} + i(i+1)\cos 2\phi + \frac{\sigma}{\beta}\right] \quad \ldots(9)$$

By making β vanish we reduce these operators to the forms appropriate to spheroidal harmonic analysis. By making β infinite we obtain the differential equations specified in § 2 as appropriate to ellipsoids of the class $c^2 = \frac{1}{2}(a^2 + b^2)$.

It is now necessary to perform the operation ψ_s on typical terms P^t and ΩP^t, and χ_s on typical terms $\begin{Bmatrix}\cos\\ \sin\end{Bmatrix} t\phi$ and $\Phi \begin{Bmatrix}\cos\\ \sin\end{Bmatrix} t\phi$.

(α) *To find* $\psi_s(P^t)$.

The form (8) for ψ_s is here convenient.

It is clear that

$$\left\{\left[(\nu^2-1)\frac{d}{d\nu}\right]^2 - i(i+1)(\nu^2-1) - s^2\right\} P^t = (t^2 - s^2) P^t$$

because P^t is the solution of the differential equation found by erasing the term $-s^2 P^t$ from each side.

Again we have from the same differential equation

$$(\nu^2-1)\frac{d^2}{d\nu^2} P^t = -2\nu \frac{dP^t}{d\nu} + i(i+1)P^t + \frac{t^2}{\nu^2-1} P^t$$

It may be noted in passing that this is equally true when the subject of operation is Q^t, the function of the other form.

Therefore

$$\left[(\nu^2-1)(\nu^2+1)\frac{d^2}{d\nu^2} + 2\nu^3 \frac{d}{d\nu} - i(i+1)\nu^2 - \frac{\sigma}{\beta}\right] P^t$$

$$= \left[-2\nu\frac{d}{d\nu} + i(i+1) + t^2\frac{\nu^2+1}{\nu^2-1} - \frac{\sigma}{\beta}\right] P^t$$

Hence

$$\psi_s(P^t) = (t^2 - s^2) P^t - \beta\left\{-2\nu\frac{d}{d\nu} + i(i+1) + t^2\frac{\nu^2+1}{\nu^2-1} - \frac{\sigma}{\beta}\right\} P^t$$

We have now to eliminate $\nu \dfrac{dP^t}{d\nu}$ and $\dfrac{\nu^2+1}{\nu^2-1} P^t$.

It is known that $$P = \frac{1}{2^i . i!}\left(\frac{d}{d\nu}\right)^i (\nu^2-1)^i$$

and $$P^t = (\nu^2-1)^{\frac{1}{2}t}\left(\frac{d}{d\nu}\right)^t P$$

The differential equation satisfied by P^t involves t in the form t^2. Hence $(\nu^2-1)^{-\frac{1}{2}t}\left(\dfrac{d}{d\nu}\right)^{-t} P$ can only differ from P^t by a constant factor. In order to find that factor suppose ν to be infinitely large;

then $$P = \frac{2i!}{2^i (i!)^2}\nu^i$$

and
$$\mathrm{P}^t = \frac{2i\,!}{2^i\,i\,!} \cdot \frac{\nu^i}{i-t\,!}$$

Also
$$(\nu^2-1)^{-\frac{1}{2}t}\left(\frac{d}{d\nu}\right)^{-t}\mathrm{P} = \nu^{-t}\frac{2i\,!}{2^i\,(i\,!)^2}\cdot\frac{i\,!}{i+t\,!}\,\nu^{i+t} = \frac{2i\,!}{2^i i\,!}\cdot\frac{\nu^i}{i+t\,!}$$

Therefore the factor is $\dfrac{i+t\,!}{i-t\,!}$, and

$$\mathrm{P}^t = (\nu^2-1)^{\frac{1}{2}t}\left(\frac{d}{d\nu}\right)^t \mathrm{P} = \frac{i+t\,!}{i-t\,!}(\nu^2-1)^{-\frac{1}{2}t}\left(\frac{d}{d\nu}\right)^{-t}\mathrm{P}$$

It will be convenient to pause here and obtain the corresponding formulæ for the Q-functions. Various writers have adopted various conventions as to the factors involved in these functions. I write

$$\mathrm{Q} = \mathrm{P}\int_\nu^\infty \frac{d\nu}{(\nu^2-1)(\mathrm{P})^2}$$

and
$$\mathrm{Q}^t = (\nu^2-1)^{\frac{1}{2}t}\left(\frac{d}{d\nu}\right)^t \mathrm{Q}$$

As in the case of P^t we may change the sign of t, if we introduce a constant factor, and this may be found by making ν infinitely great. In that case it is easy to show that

$$\mathrm{Q} = \frac{2^i\,(i\,!)^2}{2i+1\,!}\cdot\frac{1}{\nu^{i+1}}$$

By performing $\left(\dfrac{d}{d\nu}\right)^t$ and $\left(\dfrac{d}{d\nu}\right)^{-t}$ on Q it follows that the constant factor is the same as before, and that the alternative forms for Q^t are exactly the same as for P^t.

Hence the transformations which follow for the P-functions are equally applicable to the Q-functions.

If we differentiate P^t in its two forms we find

$$\frac{d\mathrm{P}^t}{d\nu} = t\nu\,(\nu^2-1)^{\frac{1}{2}(t-2)}\left(\frac{d}{d\nu}\right)^t\mathrm{P} + (\nu^2-1)^{\frac{1}{2}t}\left(\frac{d}{d\nu}\right)^{t+1}\mathrm{P} = \frac{t\nu}{\nu^2-1}\,\mathrm{P}^t + \frac{\mathrm{P}^{t+1}}{(\nu^2-1)^{\frac{1}{2}}}$$

And

$$\frac{d\mathrm{P}^t}{d\nu} = \frac{i+t\,!}{i-t\,!}\left\{-t\nu\,(\nu^2-1)^{-\frac{1}{2}(t+2)}\left(\frac{d}{d\nu}\right)^{-t}\mathrm{P} + (\nu^2-1)^{-\frac{1}{2}t}\left(\frac{d}{d\nu}\right)^{-(t-1)}\mathrm{P}\right\}$$

$$= -\frac{t\nu}{\nu^2-1}\,\mathrm{P}^t + \frac{i+t\,!}{i-t\,!}\cdot\frac{i-t+1\,!}{i+t-1\,!}\,\frac{\mathrm{P}^{t-1}}{(\nu^2-1)^{\frac{1}{2}}}$$

I now write

$$\{i,\,t\} = (i+t)\,(i-t+1) = i\,(i+1) - t\,(t-1)$$

It is clear that

$$\{i,\,-t\} = \{i,\,t+1\}, \quad \text{and } \{i,\,0\} = \{i,\,1\} = i\,(i+1)$$

Now since $\dfrac{i+t!}{i-t!}\dfrac{i-t+1!}{i+t-1!}=\{i,t\}$, by taking the sum and difference of the two forms of $\dfrac{dP^t}{d\nu}$, we have

$$\left.\begin{aligned}\frac{dP^t}{d\nu}&=\frac{1}{2(\nu^2-1)^{\frac{1}{2}}}\left[P^{t+1}+\{i,t\}P^{t-1}\right]\\ \frac{\nu P^t}{(\nu^2-1)^{\frac{1}{2}}}&=\tfrac{1}{2}\left[-\frac{1}{t}P^{t+1}+\frac{\{i,t\}}{t}P^{t-1}\right]\end{aligned}\right\}\quad\ldots\ldots\ldots\ldots(10)$$

It is easy to verify, by means of the relationship $P^{-t}=\dfrac{i-t!}{i+t!}P^t$, that these equations are true when t is negative. They are also true when $t=0$, although the second equation then becomes nugatory.

Multiply the first of (10) by ν and the second by $\dfrac{\nu}{(\nu^2-1)^{\frac{1}{2}}}$, and apply them a second time.

Then since
$$\frac{\{i,t+1\}}{t+1}-\frac{\{i,t\}}{t-1}=-\frac{2i(i+1)}{t^2-1}$$

$$\frac{\{i,t+1\}}{t(t+1)}+\frac{\{i,t\}}{t(t-1)}=2\left[\frac{i(i+1)}{t^2-1}-1\right]$$

$$\frac{\nu^2+1}{\nu^2-1}=\frac{2\nu^2}{\nu^2-1}-1$$

$$\left.\begin{aligned}2\nu\frac{dP^t}{d\nu}&=\tfrac{1}{2}\left[-\frac{P^{t+2}}{t+1}-\frac{2i(i+1)}{t^2-1}P^t+\frac{\{i,t\}\{i,t-1\}}{t-1}P^{t-2}\right]\\ \frac{\nu^2+1}{\nu^2-1}P^t&=\tfrac{1}{2}\left[\frac{P^{t+2}}{t(t+1)}-\frac{2i(i+1)}{t^2-1}P^t+\frac{\{i,t\}\{i,t-1\}}{t(t-1)}P^{t-2}\right]\end{aligned}\right\}\ldots(11)$$

These equations are always true although for $t=\pm 1$ and 0 they become nugatory.

Then
$$-2\nu\frac{dP^t}{d\nu}+t^2\frac{\nu^2+1}{\nu^2-1}P^t=\tfrac{1}{2}\left[P^{t+2}+\{i,t\}\{i,t-1\}P^{t-2}\right]-i(i+1)P^t$$

Hence
$$\psi_s(P^t)=-\tfrac{1}{2}\beta\left[\frac{2(s^2-t^2)}{\beta}P^t+P^{t+2}-2\frac{\sigma}{\beta}P^t+\{i,t\}\{i,t-1\}P^{t-2}\right]\ldots(12)$$

(β) *To find* $\psi_s(\Omega P^t)$.

It is now best to use ψ_s in the form (5), where D_1 is defined by (2).

Now $\quad D_1(\Omega P^t)=\dfrac{1}{(1-\beta)^{\frac{1}{2}}}\left\{\left[(\nu^2-1)(1-\beta)-2\beta\right]\dfrac{dP^t}{d\nu}+\dfrac{2\beta\nu}{\nu^2-1}P^t\right\}$

and

$$D_1^2(\Omega P^t) = (\nu^2 - 1)\,\Omega\left\{[(\nu^2-1)(1-\beta) - 2\beta]\frac{d^2P^t}{d\nu^2}\right.$$

$$\left. + \left(1 - \beta + \frac{\beta}{\nu^2-1}\right)2\nu\frac{dP^t}{d\nu} - \frac{2\beta P^t}{\nu^2-1}\left(\frac{\nu^2+1}{\nu^2-1}\right)\right\}$$

The latter terms of ψ_s contribute

$$\Omega\,\{-i(i+1)[(\nu^2-1)(1-\beta)-\beta]P^t - (s^2-\sigma)P^t\}$$

Therefore

$$\psi_s(\Omega P^t) = \Omega\left\{(\nu^2-1)[(\nu^2-1)(1-\beta)-2\beta]\frac{d^2P^t}{d\nu^2}\right.$$

$$+ [(\nu^2-1)(1-\beta)+\beta]\,2\nu\frac{dP^t}{d\nu} - i(i+1)(1-\beta)(\nu^2-1)P^t$$

$$\left. - s^2P^t + \beta i(i+1)P^t + \sigma P^t - 2\beta P^t\frac{\nu^2+1}{\nu^2-1}\right\}$$

But $$(\nu^2-1)\frac{d^2P^t}{d\nu^2} = -2\nu\frac{dP^t}{d\nu} + i(i+1)P^t + \frac{t^2}{\nu^2-1}P^t$$

and we find on reduction that

$$\psi_s(\Omega P^t) = \Omega P^t(t^2-s^2) - \beta\Omega\left[-6\nu\frac{dP^t}{d\nu} + i(i+1)P^t\right.$$

$$\left. + \frac{\nu^2+1}{\nu^2-1}(t^2+2)P^t - \frac{\sigma}{\beta}P^t\right]$$

On substituting for $\frac{\nu dP^t}{d\nu}$ and $\frac{\nu^2+1}{\nu^2-1}P^t$ their values, we have

$$\psi_s(\Omega P^t) = -\tfrac{1}{2}\beta\Omega\left[\frac{2(s^2-t^2)}{\beta}P^t + \frac{t+2}{t}P^{t+2} - 2\frac{\sigma}{\beta}P^t\right.$$

$$\left. + \frac{t-2}{t}\{i,t\}\{i,t-1\}P^{t-2}\right] \quad \text{...(13)}$$

(γ) *To find* $\chi_s\left(\begin{Bmatrix}\cos\\ \sin\end{Bmatrix} t\phi\right)$.

In this case the most convenient form for χ_s is that in (9), and we easily find

$$\chi_s\left(\begin{Bmatrix}\cos\\ \sin\end{Bmatrix} t\phi\right) = -\tfrac{1}{2}\beta\left[-\frac{2(s^2-t^2)}{\beta}\begin{Bmatrix}\cos\\ \sin\end{Bmatrix} t\phi + \{i,t+1\}\begin{Bmatrix}\cos\\ \sin\end{Bmatrix}(t+2)\phi\right.$$

$$\left. + 2\frac{\sigma}{\beta}\begin{Bmatrix}\cos\\ \sin\end{Bmatrix} t\phi + \{i,t\}\begin{Bmatrix}\cos\\ \sin\end{Bmatrix}(t-2)\phi\right] \quad \text{...(14)}$$

(δ) *To find* $\chi_s\left(\Phi\begin{Bmatrix}\cos\\ \sin\end{Bmatrix}t\phi\right)$.

I now use the form χ_s as defined in (5), where D_3 is given in (2), so that $D_3 = (1-\beta\cos 2\phi)^{\frac{1}{2}}\dfrac{d}{d\phi} = \Phi\dfrac{d}{d\phi}$.

We have

$$D_3\left(\Phi\begin{Bmatrix}\cos\\ \sin\end{Bmatrix}t\phi\right) = \mp t\begin{Bmatrix}\sin\\ \cos\end{Bmatrix}t\phi \pm \tfrac{1}{2}\beta(t+1)\begin{Bmatrix}\sin\\ \cos\end{Bmatrix}(t+2)\phi \pm \tfrac{1}{2}\beta(t-1)\begin{Bmatrix}\sin\\ \cos\end{Bmatrix}(t-2)\phi$$

and

$$D_3^2\left(\Phi\begin{Bmatrix}\cos\\ \sin\end{Bmatrix}t\phi\right) = \Phi\left[-t^2\begin{Bmatrix}\cos\\ \sin\end{Bmatrix}t\phi + \tfrac{1}{2}\beta(t+1)(t+2)\begin{Bmatrix}\cos\\ \sin\end{Bmatrix}(t+2)\phi + \tfrac{1}{2}\beta(t-1)(t-2)\begin{Bmatrix}\cos\\ \sin\end{Bmatrix}(t-2)\phi\right]$$

The latter terms of χ_s contribute

$$\Phi\left\{(s^2-\sigma)\begin{Bmatrix}\cos\\ \sin\end{Bmatrix}t\phi - \tfrac{1}{2}\beta i(i+1)\left[\begin{Bmatrix}\cos\\ \sin\end{Bmatrix}(t+2)\phi + \begin{Bmatrix}\cos\\ \sin\end{Bmatrix}(t-2)\phi\right]\right\}$$

Therefore

$$\chi_s\left[\Phi\begin{Bmatrix}\cos\\ \sin\end{Bmatrix}t\phi\right] = -\tfrac{1}{2}\Phi\left[-2(s^2-t^2)\begin{Bmatrix}\cos\\ \sin\end{Bmatrix}t\phi + \beta\{i, t+2\}\begin{Bmatrix}\cos\\ \sin\end{Bmatrix}(t+2)\phi + 2\sigma\begin{Bmatrix}\cos\\ \sin\end{Bmatrix}t\phi + \{i, t-1\}\begin{Bmatrix}\cos\\ \sin\end{Bmatrix}(t-2)\phi\right] \quad \text{...(15)}$$

§ 6. *Determination of the Coefficients in the Functions.*

In this section I use successively the four results (12) (13) (14) (15) obtained in the last section under the headings (α), (β), (γ), (δ).

(α) $$\mathfrak{P}^s = q_s P^s + \Sigma\beta^n q_{s-2n} P^{s-2n} + \Sigma\beta^n q_{s+2n} P^{s+2n}$$

The limits of the first Σ are 1 to $\frac{1}{2}s$ or $\frac{1}{2}(s-1)$, and of the second 1 to $\frac{1}{2}(i-s)$ or $\frac{1}{2}(i-s-1)$.

Applying the operation ψ_s to $\mathfrak{P}^s$ and equating $-\dfrac{2}{\beta}\psi_s(\mathfrak{P}^s)$ to zero, we have

$$\begin{aligned}&\Sigma 8n(s-n)\beta^{n-1}q_{s-2n}P^{s-2n} - \Sigma 8n(s+n)\beta^{n-1}q_{s+2n}P^{s+2n}\\ &+ q_s\left[P^{s+2} - 2\frac{\sigma}{\beta}P^s + \{i, s\}\{i, s-1\}P^{s-2}\right]\\ &+ \Sigma\beta^n q_{s-2n}\left[P^{s-2n+2} - 2\frac{\sigma}{\beta}P^{s-2n} + \{i, s-2n\}\{i, s-2n-1\}P^{s-2n-2}\right]\\ &+ \Sigma\beta^n q_{s+2n}\left[P^{s+2n+2} - 2\frac{\sigma}{\beta}P^{s+2n} + \{i, s+2n\}\{i, s+2n-1\}P^{s+2n-2}\right] = 0\end{aligned}$$

The coefficients of the P's must vanish separately. This gives from the coefficients of P^{s-2n} and P^{s+2n} the following:—

$$2[4n(s-n)-\sigma]\,q_{s-2n}+\beta^2 q_{s-2n-2}+\{i,\,s-2n+2\}\{i,\,s-2n+1\}\,q_{s-2n+2}=0$$

$$-2[4n(s+n)+\sigma]\,q_{s+2n}+q_{s+2n-2}+\beta^2\{i,\,s+2n+2\}\{i,\,s+2n+1\}\,q_{s+2n+2}=0$$

These equations may be written in the form

$$\left.\begin{aligned}\frac{2q_{s-2n}}{q_{s-2n+2}}&=\frac{-\{i,\,s-2n+2\}\{i,\,s-2n+1\}}{4n(s-n)-\sigma+\frac{1}{4}\beta^2\left(\dfrac{2q_{s-2n-2}}{q_{s-2n}}\right)}\\ \frac{2q_{s+2n}}{q_{s+2n-2}}&=\frac{1}{4n(s+n)+\sigma-\frac{1}{4}\beta^2\{i,\,s+2n+2\}\{i,\,s+2n+1\}\left(\dfrac{2q_{s+2n+2}}{q_{s+2n}}\right)}\end{aligned}\right\}\quad(16)$$

Whence by continued application, the continued fractions

$$\left.\begin{aligned}\frac{2q_{s-2n}}{q_{s-2n+2}}&=\frac{-\{i,\,s-2n+2\}\{i,\,s-2n+1\}}{4n(s-n)-\sigma-}\;\frac{\frac{1}{4}\beta^2\{i,\,s-2n\}\{i,\,s-2n-1\}}{4(n+1)(s-n-1)-\sigma-\ldots}\\ &\qquad\ldots\frac{-\frac{1}{4}\beta^2\{i,\,s-2n-2r+2\}\{i,\,s-2n-2r+1\}}{4(n+r)(s-n-r)+\frac{1}{4}\beta^2\left(\dfrac{2q_{s-2n-2r-2}}{q_{s-2n-2r}}\right)}\\ \frac{2q_{s+2n}}{q_{s+2n-2}}&=\frac{1}{4n(s+n)+\sigma-}\;\frac{\frac{1}{4}\beta^2\{i,\,s+2n+2\}\{i,\,s+2n+1\}}{4(n+1)(s+n+1)+\sigma-\ldots}\\ &\qquad\ldots\frac{-\frac{1}{4}\beta^2\{i,\,s+2n+2r\}\{i,\,s+2n+2r-1\}}{4(n+r)(s+n+r)-\frac{1}{4}\beta^2\{i,\,s+2n+2r+2\}\{i,\,s+2n+2r+1\}\left(\dfrac{2q_{s+2n+2r+2}}{q_{s+2n+2r}}\right)}\end{aligned}\right\}\quad\ldots\ldots(16)$$

We must now consider what I may call the middle of the series, which corresponds with $n=0$. In this case each of the Σ's contributes one term and the q_s term gives another. The result is

$$-2\sigma q_s+\beta^2 q_{s-2}+\beta^2 q_{s+2}\{i,\,s+2\}\{i,\,s+1\}=0$$

or

$$\sigma=\tfrac{1}{4}\beta^2\left(\frac{2q_{s-2}}{q_s}\right)+\tfrac{1}{4}\beta^2\{i,\,s+2\}\{i,\,s+1\}\left(\frac{2q_{s+2}}{q_s}\right)$$

Since $2q_{s-2}/q_s$ and $2q_{s+2}/q_s$ are expressible as continued fractions, we have an equation for σ, if the continued fractions terminate.

We shall now consider those terminations.

First, suppose that s is even, corresponding to types EEC, OEC.

The first continued fraction depends only on the first Σ. The condition to be satisfied is

$$2s^2\beta^{\frac{1}{2}(s-2)}q_0 P + 2(s^2-4)\beta^{\frac{1}{2}(s-4)}q_2 P^2 + \dots$$
$$+\beta^{\frac{1}{2}s}q_0\left[P^2 - 2\frac{\sigma}{\beta}P + \{i, 0\}\{i, -1\}P^{-2}\right]$$
$$+\beta^{\frac{1}{2}(s-2)}q_2\left[P^4 - 2\frac{\sigma}{\beta}P^2 + \{i, 2\}\{i, 1\}P\right]$$
$$+\beta^{\frac{1}{2}(s-4)}q_4\left[P^6 + 2\frac{\sigma}{\beta}P^4 + \{i, 4\}\{i, 3\}P^2\right] + \dots = 0$$

Since $\{i, 0\}\{i, -1\}P^{-2} = P^2$, we have, by equating to zero the coefficients of P and P^2, results which may be written

$$\frac{2q_0}{q_2} = \frac{-\{i, 2\}\{i, 1\}}{s^2-\sigma}, \quad \frac{2q_2}{q_4} = \frac{-\{i, 4\}\{i, 3\}}{s^2 - 4 - \sigma + \frac{1}{2}\beta^2\left(\frac{2q_0}{q_2}\right)}$$

Hence the q's disappear from the first continued fraction, which terminates with

$$\frac{-\frac{1}{2}\beta^2\{i, 2\}\{i, 1\}}{s^2-\sigma}$$

In this last term the $\frac{1}{4}\beta^2$ which prevails elsewhere is replaced by $\frac{1}{2}\beta^2$.

Observe that when $s=2$ the first continued fraction is replaced by a simple fraction, so that the equation for σ becomes

$$\sigma = \frac{-\frac{1}{2}\beta^2\{i, 2\}\{i, 1\}}{4-\sigma} + \frac{1}{4}\beta^2\{i, 4\}\{i, 3\}\left(\frac{2q_4}{q_2}\right)$$

Secondly, suppose that s is odd, corresponding to the types OOS, EOS.

The condition to be satisfied is now

$$2(s^2-1)\beta^{\frac{1}{2}(s-3)}q_1 P^1 + 2(s^2-9)\beta^{\frac{1}{2}(s-5)}q_3 P^3 + \dots$$
$$+\beta^{\frac{1}{2}(s-1)}q_1\left[P^3 - 2\frac{\sigma}{\beta}P^1 + \{i, 1\}\{i, 0\}P^{-1}\right]$$
$$+\beta^{\frac{1}{2}(s-3)}q_3\left[P^5 - 2\frac{\sigma}{\beta}P^3 + \{i, 3\}\{i, 2\}P^1\right]$$
$$+\beta^{\frac{1}{2}(s-5)}q_5\left[P^7 - 2\frac{\sigma}{\beta}P^5 + \{i, 5\}\{i, 4\}P^3\right] + \dots = 0$$

Now $\{i, 1\}\{i, 0\}P^{-1} = i(i+1)\dfrac{i+1!}{i-1!}P^{-1} = i(i+1)P^1$, and if we equate to zero the coefficients of P^1 and P^3 we obtain results which may be written

$$\frac{2q_1}{q_3} = \frac{-\{i, 3\}\{i, 2\}}{s^2 - 1 - \sigma + \frac{1}{2}\beta i(i+1)}$$
$$\frac{2q_3}{q_5} = \frac{-\{i, 5\}\{i, 4\}}{s^2 - 9 - \sigma + \frac{1}{4}\beta^2\left(\frac{2q_1}{q_3}\right)}$$

Thus the q's again disappear, and the first continued fraction ends with

$$\frac{-\tfrac{1}{4}\beta^2\{i, 3\}\{i, 2\}}{s^2-1-\sigma+\tfrac{1}{2}\beta i(i+1)}$$

Observe that when $s=3$, the continued fraction reduces to a simple fraction, and the equation for σ becomes

$$\sigma = \frac{-\tfrac{1}{4}\beta^2\{i, 3\}\{i, 2\}}{8-\sigma+\tfrac{1}{2}\beta i(i+1)} + \tfrac{1}{4}\beta^2\{i, 5\}\{i, 4\}\left(\frac{2q_5}{q_3}\right)$$

The case of $s=1$ must be considered separately.

We have next to consider the termination of the second fraction, which depends only on the second Σ.

First, when i and s are either both even or both odd, the types are EEC and OOS, and the limits are $\tfrac{1}{2}(i-s)$ to 1. The condition to be satisfied is

$$-2(i^2-s^2)\beta^{\frac{1}{2}(i-s-2)}q_i\mathrm{P}^i - 2[(i-2)^2-s^2]\beta^{\frac{1}{2}(i-s-4)}q_{i-2}\mathrm{P}^{i-2} - \dots$$

$$+\beta^{\frac{1}{2}(i-s)}q_i\left[\mathrm{P}^{i+2} - 2\frac{\sigma}{\beta}\mathrm{P}^i + \{i, i\}\{i, i-1\}\mathrm{P}^{i-2}\right]$$

$$+\beta^{\frac{1}{2}(i-s-2)}q_{i-2}\left[\mathrm{P}^i - 2\frac{\sigma}{\beta}\mathrm{P}^{i-2} + \{i, i-2\}\{i, i-3\}\mathrm{P}^{i-4}\right]$$

$$+\beta^{\frac{1}{2}(i-s-4)}q_{i-4}\left[\mathrm{P}^{i-2} - 2\frac{\sigma}{\beta}\mathrm{P}^{i-4} + \{i, i-4\}\{i, i-5\}\mathrm{P}^{i-6}\right] + \dots = 0$$

Now P^{i+2} is zero, and equating the coefficients of P^i and P^{i-2} to zero we obtain results which may be written

$$\frac{2q_i}{q_{i-2}} = \frac{1}{i^2-s^2+\sigma}$$

$$\frac{2q_{i-2}}{q_{i-4}} = \frac{1}{(i-2)^2-s^2+\sigma-\tfrac{1}{4}\beta^2\{i, i\}\{i, i-1\}\left(\dfrac{2q_i}{q_{i-2}}\right)}$$

Hence this continued fraction ends with

$$\frac{-\tfrac{1}{4}\beta^2\{i, i\}\{i, i-1\}}{i^2-s^2+\sigma}$$

Secondly, when i and s differ as to evenness or oddness, the types are OEC and EOS, and the limits are $\tfrac{1}{2}(i-s-1)$ to 1. The same investigation applies again when i is changed into $i-1$.

Hence the continued fraction ends with

$$\frac{-\tfrac{1}{4}\beta^2\{i, i-1\}\{i, i-2\}}{(i-1)^2-s^2+\sigma}$$

The cases of $s=0$, $s=1$ must be considered by themselves.

When $s=0$, the types are EEC and OEC. The "middle" of the series is now also an end, and the condition is

$$-8q_2\text{P}^2 - 8.2^2\beta q_4\text{P}^4 - \ldots + q_0\left[\text{P}^2 - 2\frac{\sigma}{\beta}\text{P} + \{i, 0\}\{i, -1\}\text{P}^{-2}\right]$$
$$+\beta q_2\left[\text{P}^4 - 2\frac{\sigma}{\beta}\text{P}^2 + \{i, 2\}\{i, 1\}\text{P}\right]$$
$$+\beta^2 q_4\left[\text{P}^6 - 2\frac{\sigma}{\beta}\text{P}^4 + \{i, 4\}\{i, 3\}\text{P}^2\right] + \ldots = 0$$

Writing P^2 for $\{i, 0\}\{i, -1\}\text{P}^{-2}$ and equating the coefficients of P and P^2 to zero, we have

$$\sigma = \tfrac{1}{2}\beta^2\{i, 1\}\{i, 2\}\left(\frac{q_2}{q_0}\right)$$

$$\frac{q_2}{q_0} = \frac{1}{4 + \sigma - \frac{1}{4}\beta^2\{i, 3\}\{i, 4\}\left(\dfrac{2q_4}{q_2}\right)}$$

Therefore $\sigma = \dfrac{\frac{1}{2}\beta^2\{i, 1\}\{i, 2\}}{4.1^2 + \sigma -}\;\dfrac{\frac{1}{4}\beta^2\{i, 3\}\{i, 4\}}{4.2^2 + \sigma -}\;\dfrac{-\frac{1}{4}\beta^2\{i, 5\}\{i, 6\}}{4.3^2 + \sigma - \ldots}$

ending with $\dfrac{-\frac{1}{4}\beta^2\{i, i\}\{i, i-1\}}{i^2 + \sigma}$ for EEC

and with $\dfrac{-\frac{1}{4}\beta^2\{i, i-1\}\{i, i-2\}}{(i-1)^2 + \sigma}$ for OEC

Next when $s=1$ the types are OOS, EOS; the "middle" is again an end, and the condition is

$$-8.1.2q_3\text{P}^3 - 8.2.3\beta q_5\text{P}^5 - \ldots + q_1\left[\text{P}^3 - 2\frac{\sigma}{\beta}\text{P}^1 + \{i, 1\}\{i, 0\}\text{P}^{-1}\right]$$
$$+\beta q_3\left[\text{P}^5 - 2\frac{\sigma}{\beta}\text{P}^3 + \{i, 3\}\{i, 2\}\text{P}^1\right]$$
$$+\beta^2 q_5\left[\text{P}^7 - 2\frac{\sigma}{\beta}\text{P}^5 + \{i, 5\}\{i, 4\}\text{P}^3\right] + \ldots = 0$$

Writing $i(i+1)\text{P}^1$ for $\{i, 1\}\{i, 0\}\text{P}^{-1}$ and equating to zero the coefficients of P^1 and P^3, we have

$$\sigma - \tfrac{1}{2}\beta i(i+1) = \tfrac{1}{4}\beta^2\{i, 3\}\{i, 2\}\left(\frac{2q_3}{q_1}\right)$$

$$\frac{2q_3}{q_1} = \frac{1}{4.1.2 + \sigma - \frac{1}{4}\beta^2\{i, 5\}\{i, 4\}\left(\dfrac{2q_5}{q_3}\right)}$$

Therefore

$$\sigma - \tfrac{1}{2}\beta i(i+1) = \frac{\frac{1}{4}\beta^2\{i, 3\}\{i, 2\}}{4.1.2 + \sigma -}\;\frac{\frac{1}{4}\beta^2\{i, 5\}\{i, 4\}}{4.2.3 + \sigma -}\;\frac{\frac{1}{4}\beta^2\{i, 7\}\{i, 6\}}{4.3.4 + \sigma - \ldots}$$

ending with $\dfrac{-\frac{1}{4}\beta^2 \{i, i\} \{i, i-1\}}{i^2+\sigma}$ for OOS

and with $\dfrac{-\frac{1}{4}\beta^2 \{i, i-1\} \{i, i-2\}}{(i-1)^2+\sigma}$ for EOS

(β) We have next to consider the other form of P-function for types EES, OES, OOC, EOC, namely,

$$\mathsf{P}^s = \Omega\, [q_s' \mathrm{P}^s + \Sigma\beta^n q'_{s-2n} \mathrm{P}^{s-2n} + \Sigma\beta^n q'_{s+2n} \mathrm{P}^{s+2n}]$$

where
$$\Omega = \left(\frac{\nu^2 - \frac{1+\beta}{1-\beta}}{\nu^2 - 1}\right)^{\frac{1}{2}}$$

Let us write $q'_{s\pm 2n} = (s \pm 2n)\, q_{s\pm 2n}$. The q's are not now the actual coefficients of any P-function, but we shall see that they are determinable by almost the same relationships as those already found, and therefore the notation is convenient.

We now have

$$\mathsf{P}^s = \Omega\, [q_s s \mathrm{P}^s + \Sigma\beta^n q_{s-2n} (s-2n)\, \mathrm{P}^{s-2n} + \Sigma\beta^n q_{s+2n} (s+2n)\, \mathrm{P}^{s+2n}]$$

Applying the operation ψ_s to P^s and equating $-\dfrac{2}{\Omega\beta}\psi_s(\mathsf{P}^s)$ to zero, we have

$$\begin{aligned}
&\Sigma 8n (s-n)(s-2n)\, q_{s-2n} \mathrm{P}^{s-2n}\, \Sigma 8n (s+n)(s+2n)\, q_{s+2n} \mathrm{P}^{s+2n} \\
&\quad + q_s \left[(s+2)\, \mathrm{P}^{s+2} - 2\frac{\sigma}{\beta} s \mathrm{P}^s + \{i, s\} \{i, s-1\} (s-2)\, \mathrm{P}^{s-2}\right] \\
&\quad + \Sigma\beta^n q_{s-2n} \Big[(s-2n+2)\, \mathrm{P}^{s-2n+2} - 2\frac{\sigma}{\beta}(s-2n)\, \mathrm{P}^{s-2n} \\
&\qquad\qquad + \{i, s-2n\} \{i, s-2n-1\} (s-2n-2)\, \mathrm{P}^{s-2n-2}\Big] \\
&\quad + \Sigma\beta^n q_{s+2n} \Big[(s+2n+2)\, \mathrm{P}^{s+2n+2} - 2\frac{\sigma}{\beta}(s+2n)\, \mathrm{P}^{s+2n} \\
&\qquad\qquad + \{i, s+2n\} \{i, s+2n-1\} (s+2n-2)\, \mathrm{P}^{s+2n-2}\Big] = 0
\end{aligned}$$

This is the same equation as before, if we replace $t\mathrm{P}^t$ by P^t. As we may equate coefficients of $t\mathrm{P}^t$ to zero (instead of coefficients of P^t), we obtain the same equations for the q's as before.

A certain change must, however, be noted with respect to the beginning of the first series, which determines the end of the first continued fraction.

We previously wrote

P^2 for $\{i, 0\} \{i, -1\} \mathrm{P}^{-2}$ and $i(i+1)\, \mathrm{P}^1$ for $\{i, 1\} \{i, 0\} \mathrm{P}^{-1}$

But the corresponding terms will now be

$$\{i, 0\}\{i, -1\}(-2)\,\mathrm{P}^{-2} \quad \text{and} \quad \{i, 1\}\{i, 0\}(-1)\,\mathrm{P}^{-1}$$

and these are equal to $-(2\mathrm{P}^2)$ and $-(1 \,.\, \mathrm{P}^1)$.

Hence it follows that when s is even (EES, OES)

$$\frac{2q_2}{q_0} = \frac{-\{i, 2\}\{i, 1\}}{s^2 - \sigma}, \qquad \frac{2q_2}{q_4} = \frac{-\{i, 4\}\{i, 3\}}{s^2 - 4 - \sigma}$$

The q_0 term has disappeared from the latter of these, and thus the continued fraction is independent of q_0. This is correct, since whatever value (short of infinity) q_0 may have q_0', being equal to $0q_0$, vanishes. Hence the continued fraction is docked of one term and ends with

$$\frac{-\frac{1}{4}\beta^2\{i, 4\}\{i, 3\}}{s^2 - 4 - \sigma}$$

It is important to note the deficiency of one term in the fraction, since it indicates that when $s = 2$ the first continued fraction entirely disappears.

When $s = 0$ there is no function of the P form, so the question of interpretation does not arise.

When s is odd (OOC, EOC) the only change is that $i(i+1)$ enters with the opposite sign, so that the first fraction ends with

$$\frac{-\frac{1}{4}\beta^2\{i, 3\}\{i, 2\}}{s^2 - 1 - \sigma - \frac{1}{2}\beta i(i+1)}$$

When $s = 1$, we have $\sigma + \frac{1}{2}\beta i(i+1)$ equal to the same fraction as before.

When the q's are determined we have $q_t' = tq_t$. But it is desired that in the case (α) q_s should be unity, and that in the case (β) q_s' should be unity. This condition will be satisfied in the present case if we determine all the q's, put q_s equal to unity, and finally take

$$q'_{s\pm 2n} = \frac{s \pm 2n}{s}\, q_{s\pm 2n}$$

Thus in both (α) and (β) we put q_s equal to unity, and in (β) determine the q''s by the above equation.

(γ) We now have to consider the cosine and sine functions.

For EEC, EES, OOC, OOS

$$\left\{\begin{matrix}\mathfrak{C}^s \\ \mathfrak{S}^s\end{matrix}\right. = p_s \left\{\begin{matrix}\cos \\ \sin\end{matrix}\right. s\phi + \Sigma\beta^n p_{s-2n}\left\{\begin{matrix}\cos \\ \sin\end{matrix}\right. (s-2n)\,\phi + \Sigma\beta^n p_{s+2n}\left\{\begin{matrix}\cos \\ \sin\end{matrix}\right. (s+2n)\,\phi$$

The first Σ has limits $\frac{1}{2}s$ or $\frac{1}{2}(s-1)$ to 1, the second $\frac{1}{2}(i-s)$ to 1.

Apply the operation χ_s and equate $-\frac{2}{\beta}\chi_s\left(\begin{Bmatrix}\mathfrak{C}^s\\ \mathfrak{S}^s\end{Bmatrix}\right)$ to zero; then

$$-\Sigma 8n(s-n)\beta^{n-1}p_{s-2n}\begin{Bmatrix}\cos\\ \sin\end{Bmatrix}(s-2n)\phi+\Sigma 8n(s+n)\beta^{n-1}p_{s+2n}\begin{Bmatrix}\cos\\ \sin\end{Bmatrix}(s+2n)\phi$$

$$+p_s\left[\{i, s+1\}\begin{Bmatrix}\cos\\ \sin\end{Bmatrix}(s+2)\phi+2\frac{\sigma}{\beta}\begin{Bmatrix}\cos\\ \sin\end{Bmatrix}s\phi+\{i, s\}\begin{Bmatrix}\cos\\ \sin\end{Bmatrix}(s-2)\phi\right]$$

$$+\Sigma\beta^n p_{s-2n}\left[\{i, s-2n+1\}\begin{Bmatrix}\cos\\ \sin\end{Bmatrix}(s-2n+2)\phi+2\frac{\sigma}{\beta}\begin{Bmatrix}\cos\\ \sin\end{Bmatrix}(s-2n)\phi\right.$$

$$\left.+\{i, s-2n\}\begin{Bmatrix}\cos\\ \sin\end{Bmatrix}(s-2n-2)\phi\right]$$

$$+\Sigma\beta^n p_{s+2n}\left[\{i, s+2n+1\}\begin{Bmatrix}\cos\\ \sin\end{Bmatrix}(s+2n+2)\phi+2\frac{\sigma}{\beta}\begin{Bmatrix}\cos\\ \sin\end{Bmatrix}(s+2n)\phi\right.$$

$$\left.+\{i, s+2n\}\begin{Bmatrix}\cos\\ \sin\end{Bmatrix}(s+2n-2)\phi\right]=0$$

If we equate to zero the coefficients of $\begin{Bmatrix}\cos\\ \sin\end{Bmatrix}(s\pm 2n)\phi$, we find

$$\frac{2p_{s-2n}}{p_{s-2n+2}}=\frac{\{i, s-2n+2\}}{4n(s-n)-\sigma-\frac{1}{4}\beta^2\{i, s-2n-1\}\left(\dfrac{2p_{s-2n-2}}{p_{s-2n}}\right)}$$

$$\frac{2p_{s+2n}}{p_{s+2n-2}}=\frac{-\{i, s+2n-1\}}{4n(s+n)+\sigma+\frac{1}{4}\beta^2\{i, s+2n+2\}\left(\dfrac{2p_{s+2n+2}}{p_{s+2n}}\right)}$$

These will, as before, lead to continued fractions, and by elimination of the p's to an equation for σ. The equation will agree with our former result, for it can of course make no difference from which equation we determine σ*. It follows then by comparison with the previous result (16) that

$$\frac{p_{s-2n}}{p_{s-2n+2}}=-\frac{1}{\{i, s-2n+1\}}\frac{q_{s-2n}}{q_{s-2n+2}}$$

$$\frac{p_{s+2n}}{p_{s+2n-2}}=-\{i, s+2n-1\}\frac{q_{s+2n}}{q_{s+2n-2}}$$

Hence when the q's are found, the p's follow at once.

(δ) For OEC, OES, EOC, EOS

$$\begin{Bmatrix}\mathbf{C}^s\\ \mathbf{S}^s\end{Bmatrix}=\Phi\left[p_s'\begin{Bmatrix}\cos\\ \sin\end{Bmatrix}s\phi+\Sigma\beta^n p'_{s-2n}\begin{Bmatrix}\cos\\ \sin\end{Bmatrix}(s-2n)\phi+\Sigma\beta^n p'_{s+2n}\begin{Bmatrix}\cos\\ \sin\end{Bmatrix}(s+2n)\phi\right]$$

where $\Phi=(1-\beta\cos 2\phi)^{\frac{1}{2}}$.

The limits of the first Σ are $\frac{1}{2}s$ or $\frac{1}{2}(s-1)$ to 1, of the second $\frac{1}{2}(i-s-1)$ to 1.

* I have of course verified that this is so.

Proceeding exactly as before we find

$$\frac{2p'_{s-2n}}{p'_{s-2n+2}} = \frac{\{i, s-2n+1\}}{4n(s-n) - \sigma - \frac{1}{4}\beta^2 \{i, s-2n\} \left(\frac{2p'_{s-2n-2}}{p'_{s-2n}}\right)}$$

$$\frac{2p'_{s+2n}}{p'_{s+2n-2}} = \frac{-\{i, s+2n\}}{4n(s+n) + \sigma + \frac{1}{4}\beta^2 \{i, s+2n+1\} \left(\frac{2p'_{s+2n+2}}{p'_{s+2n}}\right)}$$

By comparison with (16) we see that

$$\frac{p'_{s-2n}}{p'_{s-2n+2}} = -\frac{1}{\{i, s-2n+2\}} \frac{q_{s-2n}}{q_{s-2n+2}}$$

$$\frac{p'_{s+2n}}{p'_{s+2n-2}} = -\{i, s+2n\} \frac{q_{s+2n}}{q_{s+2n-2}}$$

Therefore when the q's are found, the p's follow at once.

We may now summarise our results, as follows:—

In the general case where s is neither 0 nor 1, σ is the root which nearly vanishes of the equation

$$\sigma = \frac{-\frac{1}{4}\beta^2 \{i, s\} \{i, s-1\}}{4.1(s-1) - \sigma -} \; \frac{\frac{1}{4}\beta^2 \{i, s-2\} \{i, s-3\}}{4.2(s-2) - \sigma - \ldots}$$

$$+ \frac{\frac{1}{4}\beta^2 \{i, s+1\} \{i, s+2\}}{4.1(s+1) + \sigma -} \; \frac{\frac{1}{4}\beta^2 \{i, s+3\} \{i, s+4\}}{4.2(s+2) + \sigma - \ldots}$$

The continued fractions terminate variously for the various types of function. The end of the first continued fraction is as follows:—

For EEC $\dfrac{-\frac{1}{2}\beta^2 \{i, 1\} \{i, 2\}}{s^2 - \sigma}$; and when $s = 2$ this is the whole fraction.

For EES $\dfrac{-\frac{1}{4}\beta^2 \{i, 3\} \{i, 4\}}{s^2 - 4 - \sigma}$; and when $s = 2$ the fraction disappears.

For OOC $\dfrac{-\frac{1}{4}\beta^2 \{i, 2\} \{i, 3\}}{s^2 - 1 - \sigma - \frac{1}{2}\beta i(i+1)}$; and when $s = 3$ this is the whole fraction.

For OOS $\dfrac{-\frac{1}{4}\beta^2 \{i, 2\} \{i, 3\}}{s^2 - 1 - \sigma + \frac{1}{2}\beta i(i+1)}$; and when $s = 3$ this is the whole fraction.

For OEC $\dfrac{-\frac{1}{2}\beta^2 \{i, 1\} \{i, 2\}}{s^2 - \sigma}$; and when $s = 2$ this is the whole fraction.

For OES $\dfrac{-\frac{1}{4}\beta^2 \{i, 3\} \{i, 4\}}{s^2 - 4 - \sigma}$; and when $s = 2$ the fraction disappears.

For EOC $\dfrac{-\frac{1}{4}\beta^2 \{i, 2\} \{i, 3\}}{s^2 - 1 - \sigma - \frac{1}{2}\beta i(i+1)}$; and when $s = 3$ this is the whole fraction.

For EOS $\dfrac{-\frac{1}{4}\beta^2 \{i, 2\} \{i, 3\}}{s^2 - 1 - \sigma + \frac{1}{2}\beta i(i+1)}$; and when $s = 3$ this is the whole fraction.

For the first four of these types, viz., EEC, EES, OOC, OOS, the second continued fraction ends with

$\dfrac{-\frac{1}{4}\beta^2 \{i, i\} \{i, i-1\}}{i^2 - s^2 + \sigma}$; and when $s = i$ the second fraction disappears.

For the last four, viz., OEC, OES, EOC, EOS, it ends with

$\dfrac{-\frac{1}{4}\beta^2 \{i, i-1\} \{i, i-2\}}{(i-1)^2 - s^2 + \sigma}$; and when $s = i - 1$ the second fraction disappears*.

When $s = 0$, the equation becomes

$$\sigma = \frac{\frac{1}{2}\beta^2 \{i, 1\} \{i, 2\}}{4 \,.\, 1^2 + \sigma -} \quad \frac{\frac{1}{4}\beta^2 \{i, 3\} \{i, 4\}}{4 \,.\, 2^2 + \sigma - \ldots}$$

ending when i is even (EEC) with

$$\frac{-\frac{1}{4}\beta^2 \{i, i\} \{i, i-1\}}{i^2 + \sigma}$$

and when i is odd (OEC) with

$$\frac{-\frac{1}{4}\beta^2 \{i, i-1\} \{i, i-2\}}{(i-1)^2 + \sigma}$$

When $s = 1$ the equation has two forms, which may, however, be written together. If the upper sign refers to cosines (OOC, EOC) and the lower to sines (OOS, EOS), the equations are:—

$$\sigma \pm \tfrac{1}{2}\beta i (i+1) = \frac{\frac{1}{4}\beta^2 \{i, 2\} \{i, 3\}}{4 \,.\, 1 \,.\, 2 + \sigma -} \quad \frac{\frac{1}{4}\beta^2 \{i, 4\} \{i, 5\}}{4 \,.\, 2 \,.\, 3 + \sigma - \ldots}$$

ending when i is even (EOC, EOS) with

$$\frac{-\frac{1}{4}\beta^2 \{i, i-1\} \{i, i-2\}}{(i-1)^2 - 1 + \sigma}$$

and when i is odd (OOC, OOS) with

$$\frac{-\frac{1}{4}\beta^2 \{i, i\} \{i, i-1\}}{i^2 - 1 + \sigma}$$

It might appear at first sight that a difficulty will arise in the interpretation of these results when i is small, for the numbers in the denominators of the fractions increase, and yet it is possible that the number at the end should be smaller than that at the beginning; thus apparently the fraction ends before it begins. But this difficulty does not really arise, because in such cases the numerator will always be found to vanish, and thus the whole fraction disappears. For example, in the last case specified, if $s = 1$, $i = 2$ the denominators, according to the formula, begin with $8 + \sigma$ and end with $0 + \sigma$; but the fraction has for numerator $\{2, 2\} \{2, 3\}$, which vanishes.

* [There was a mis-statement about the two cases, $s = i$ or $i - 1$, in the original paper.]

Reviewing these statements we see that, without attributing to i any specific value, there are sixteen cases in which one of the two continued fractions disappears, and sixteen other cases in which one of them is reduced to a simple fraction.

[In order to distinguish the values of σ corresponding to any harmonic we may append to that symbol the affixes s, i as in the case of harmonic functions, and write it as σ_i^s. It will not however be necessary to adopt a distinctive symbol to distinguish between the cosine and sine functions, not because that distinction is immaterial, but because the two kinds of functions may be considered apart.

Suppose in the first instance that we are considering the cosine functions of even rank, so that the types are EEC, OEC. The equation for σ_i^s is, as above,—

$$\sigma_i^s = \frac{-\frac{1}{4}\beta^2 \{i, s\} \{i, s-1\}}{4.1(s-1) - \sigma_i^s - \dots} + \frac{\frac{1}{4}\beta^2 \{i, s+1\} \{i, s+2\}}{4.1(s+1) + \sigma_i^s - \dots}$$

Put $\sigma_i^s = x - (s+2)^2 + s^2 = x - 4(s+1)$. The successive denominators of the first continued fraction are $4n(s-n) - \sigma_i^s$, and of the second $4n(s+n) + \sigma_i^s$, with $n = 1, 2, 3$ &c. On effecting the transformation these become respectively $4(n+1)(s-n+1) - x$ and $4(n-1)(s+n+1) + x$. Hence the equation becomes

$$x - 4(s+1) = \frac{-\frac{1}{4}\beta^2 \{i, s\} \{i, s-1\}}{4.2s - x -} \; \frac{\frac{1}{4}\beta^2 \{i, s-2\} \{i, s-3\}}{4.3(s-1) - x -} \; \frac{\frac{1}{4}\beta^2 \{i, s-4\} \{i, s-5\}}{4.4(s-2) - x - \dots}$$
$$+ \frac{\frac{1}{4}\beta^2 \{i, s+1\} \{i, s+2\}}{x -} \; \frac{\frac{1}{4}\beta^2 \{i, s+3\} \{i, s+4\}}{4.1(s+3) + x -} \; \frac{\frac{1}{4}\beta^2 \{i, s+4\} \{i, s+5\}}{4.2(s+4) + x - \dots}$$

Denote the first of these continued fractions by A, and that portion of the second continued fraction, which is found by omitting its first term, by B; then this equation may be written

$$x - 4(s+1) = A + \frac{\frac{1}{4}\beta^2 \{i, s+1\} \{i, s+2\}}{x + B}$$

whence $$[x - 4(s+1) - A][x + B] = \tfrac{1}{4}\beta^2 \{i, s+1\} \{i, s+2\} \quad \dots\dots\dots(a)$$

Now from the general formula the equation for σ_i^{s+2} is

$$\sigma_i^{s+2} = \frac{-\frac{1}{4}\beta^2 \{i, s+2\} \{i, s+1\}}{4.1(s+1) - \sigma_i^{s+2} + A} - B$$

Therefore

$$[\sigma_i^{s+2} - 4(s+1) - A][\sigma_i^{s+2} + B] = \tfrac{1}{4}\beta^2 \{i, s+1\} \{i, s+2\} \quad \dots\dots(b)$$

On comparing (a) and (b) we see that $x = \sigma_i^{s+2}$. Hence it follows that two of the roots of the s equation for σ are σ_i^s and $\sigma_i^{s+2} - (s+2)^2 + s^2$. Extending the same argument from the $s+2$ equation to the $s+4$ equation, and thence to the $s+6$ equation and so on, we see that the required roots of all these successive equations may be determined from the s equation and may be written

$$\sigma_i^s, \quad \sigma_i^{s+2}-(s+2)^2+s^2, \quad \sigma_i^{s+4}-(s+4)^2+s^2, \quad \sigma_i^{s+6}-(s+6)^2+s^2$$

If we have begun this operation with the equation for $s=0$, we see that all the roots of the equation for $s=0$ are

$$\sigma_i, \quad \sigma_i^2-2^2, \quad \sigma_i^4-4^2, \quad \sigma_i^6-6^2, \ldots$$

When i is even the largest even value of s is i, and when i is odd it is $i-1$. It follows that when i is even the equation for σ for $s=0$ is of order $\frac{1}{2}i+1$, and when i is odd it is of order $\frac{1}{2}(i+1)$, and in both cases all the roots are real. The roots are approximately $0, -2^2, -4^2 \ldots -i^2$ or $-(i-1)^2$ and the corrections to these roots are respectively $\sigma_i, \sigma_i^2, \sigma_i^4 \ldots$.

Hence we can find all the values of σ from the equation for $s=0$ corresponding to cosine functions of even rank.

There is nothing in this argument, excepting in the last stage, which compels us to apply it only to cosine functions of even rank. It is equally applicable to all the others, let us consider therefore the sine functions of even rank EES, OES. These are distinguished from the cosine functions by the fact that there is no function corresponding to $s=0$, but they begin with $s=2$. Hence we see that the roots of the equation, corresponding to $s=2$,

$$\sigma = \frac{\frac{1}{4}\beta^2\{i,3\}\{i,4\}}{4.1.3+\sigma-}\;\frac{\frac{1}{4}\beta^2\{i,5\}\{i,6\}}{4.2.4+\sigma-\ldots}$$

are $$\sigma_i^2, \quad \sigma_i^4-4^2+2^2, \quad \sigma_i^6-6^2+2^2, \quad \sigma_i^8-8^2+2^2$$

When i is even there are $\frac{1}{2}i$ such roots, and when i is odd $\frac{1}{2}(i-1)$. This equation is of one degree lower order than the corresponding equation for the cosine-harmonics. These cosine and sine functions of even rank together account for $\frac{1}{2}i+1+\frac{1}{2}i$, or $i+1$ functions when i is even; and for $\frac{1}{2}(i+1)+\frac{1}{2}(i-1)$ or i functions when i is odd.

The remaining functions may be considered all together, and we see that the roots of the equation corresponding to $s=1$, namely

$$\sigma \pm \tfrac{1}{2}\beta i(i+1) = \frac{\frac{1}{4}\beta^2\{i,2\}\{i,3\}}{4.1.2+\sigma-}\;\frac{\frac{1}{4}\beta^2\{i,4\}\{i,5\}}{4.2.3+\sigma-\ldots}$$

are $$\sigma_i^1, \quad \sigma_i^3-2^2+1^2, \quad \sigma_i^5-4^2+1^2, \quad \sigma_i^7-6^2+1^2\ldots$$

The succession ends with $\sigma_i^{i-1}-(i-2)^2+1^2$ when i is even, and with $\sigma_i^i-(i-1)^2+1^2$ when i is odd.

When i is even the number of these roots is $\frac{1}{2}i$, when i is odd it is $\frac{1}{2}(i+1)$.

If we take the upper sign of β on the left the equation corresponds to the cosine functions EOC, OOC; with the lower sign to the sine functions EOS, OOS. There are thus two equations of the same algebraic form and one is derived from the other by changing the sign of β. Together they account for i functions when i is even and for $i+1$ functions when i is odd.

Thus we see that in order to find all the σ's corresponding to the $2i+1$ functions of order i, it is necessary to solve four equations.

The following schedule gives a list of the four equations:—

i even		i odd		
Type of Function	Order of equation	Type of Function	Order of equation	Specification of equation to be solved
EEC	$\frac{1}{2}i+1$	OEC	$\frac{1}{2}(i+1)$	$s=0$, cosine function
EES	$\frac{1}{2}i$	OES	$\frac{1}{2}(i-1)$	$s=2$, sine function
EOC	$\frac{1}{2}i$	OOC	$\frac{1}{2}(i+1)$	$s=1$, cosine function
EOS	$\frac{1}{2}i$	OOS	$\frac{1}{2}(i+1)$	$s=1$, sine function

For harmonics of the third order there are one quadratic, and three equations of the first degree, and of the latter two have the same algebraic form. Hence all the harmonics of the third order are determinable algebraically.

For harmonics of the fourth order, there are one cubic and three quadratics, and all these harmonics but three are determinable algebraically.

I shall determine the expressions for the harmonics of the third order in Paper 12, and shall derive the values of σ in the way indicated. At the same time the corresponding equations for σ for the harmonics of the fourth order will be found and solved, so far as is possible.]

When σ has been determined we find the q's by the formulæ—

$$\frac{2q_{s-2n}}{q_{s-2n+2}} = \frac{-\{i, s-2n+2\}\{i, s-2n+1\}}{4n(s-n)-\sigma-} \; \frac{\frac{1}{4}\beta^2\{i, s-2n\}\{i, s-2n-1\}}{4(n+1)(s-n-1)-\sigma-\dots}$$

$$\frac{2q_{s+2n}}{q_{s+2n-2}} = \frac{1}{4n(s+n)+\sigma-} \; \frac{\frac{1}{4}\beta^2\{i, s+2n+2\}\{i, s+2n+1\}}{4(n+1)(s+n+1)+\sigma-\dots}$$

The terminations of the continued fractions are as specified above in the equation for σ.

By forming continued products of ratios of successive q's, we can find all the q's as multiples of q_s, and $q_s=1$.

In the cases EEC, OEC, OOS, EOS, these are the required coefficients for $\mathfrak{P}^s$.

In the cases EES, OES, OOC, EOC we put $q'_{s\pm 2n} = \frac{s\pm 2n}{s} q_{s\pm 2n}$, and thus find the coefficients for P^s.

The coefficients for $\mathfrak{C}$, $\mathfrak{S}$ in EEC, EES, OOC, OOS are determined by

$$\frac{p_{s-2n}}{p_{s-2n+2}} = -\frac{1}{\{i,\, s-2n+1\}} \frac{q_{s-2n}}{q_{s-2n+2}}$$

$$\frac{p_{s+2n}}{p_{s+2n-2}} = -\{i,\, s+2n-1\} \frac{q_{s+2n}}{q_{s+2n-2}}$$

The coefficients for C, S in OEC, OES, EOC, EOS are determined by

$$\frac{p'_{s-2n}}{p'_{s-2n+2}} = -\frac{1}{\{i,\, s-2n+2\}} \frac{q_{s-2n}}{q_{s-2n+2}}$$

$$\frac{p'_{s+2n}}{p'_{s+2n-2}} = -\{i,\, s+2n\} \frac{q_{s+2n}}{q_{s+2n-2}}$$

It follows that if we put $q_s = 1$ and $p_s = 1$

$$p_{s-2n} = (-)^n \frac{1}{\{i,\, s-2n+1\}\{i,\, s-2n+3\} \ldots \{i,\, s-1\}} q_{s-2n}$$

$$p_{s+2n} = (-)^n \{i,\, s+2n-1\}\{i,\, s+2n-3\} \ldots \{i,\, s+1\}\, q_{s+2n}$$

$$p'_{s-2n} = (-)^n \frac{1}{\{i,\, s-2n+2\}\{i,\, s-2n+4\} \ldots \{i,\, s\}} q_{s-2n}$$

$$p'_{s+2n} = (-)^n \{i,\, s+2n\}\{i,\, s+2n-2\} \ldots \{i,\, s+2\}\, q_{s+2n}$$

When $s=0$, q_2/q_0 is equal to that which would be given by the general formula for $\dfrac{2q_{s+2n}}{q_{s+2n-2}}$ when we put in it $n=1$, $s=0$. Hence it follows that the q's for $s=0$ have double the values given by the general formula.

If we change the sign of s, the two continued fractions in the equation for σ are simply interchanged. Hence σ is unchanged when s changes sign. Also, since $\{i, t\}$ is equal to $\{-i-1, t\}$, σ is unchanged when $-i-1$ is written for i. A consideration of the forms of the q's and p's shows that $q_{-s+2k}\mathrm{P}^{-s+2k}$ is equal to $\dfrac{i-s\,!}{i+s\,!} q_{s-2k}\mathrm{P}^{s-2k}$, and therefore

$$\begin{Bmatrix} \mathfrak{P}_i^s \\ \mathsf{P}_i^s \end{Bmatrix} = \frac{i+s\,!}{i-s\,!} \begin{Bmatrix} \mathfrak{P}_i^{-s} \\ \mathsf{P}_i^{-s} \end{Bmatrix}$$

$$\begin{Bmatrix} \mathfrak{P}_i^s \\ \mathsf{P}_i^s \end{Bmatrix} = \begin{Bmatrix} \mathfrak{P}^s_{-i-1} \\ \mathsf{P}^s_{-i-1} \end{Bmatrix}$$

§ 7. *Rigorous determination of the Functions of the second order.*

If a numerical value be attributed to β it is obviously possible to obtain the rigorous expressions for the several functions. Thus, if β were $\frac{1}{3}$ we could determine the harmonics of the ellipsoids of the class $c^2 = \frac{1}{2}(a^2+b^2)$. In order to show how our formulæ lead to the required result I will determine the five functions corresponding to $i=2$. The case of $i=3$ will be considered in Paper 12.

When $s=0$ the equation for determining σ is

$$\sigma = \frac{\frac{1}{2}\beta^2 \{2, 1\} \{2, 2\}}{4+\sigma} = \frac{12\beta^2}{4+\sigma}$$

It appears from § 6 that the two roots of this equation are σ for $s=0$, and $\sigma - 2^2$ for $s=2$.

Therefore for $s=0$, $\sigma = -2 + 2(1+3\beta^2)^{\frac{1}{2}}$, or writing $B^2 = 1+3\beta^2$ for brevity, $\sigma = 2(B-1)$, and for $s=2$, $\sigma = -2(B-1)$.

Again when $s=1$, the equation for σ is $\sigma \pm \frac{1}{2}\beta i(i+1) = 0$ with the upper sign for the type EEC, and the lower for the type EES. Since in the present case $i=2$, $\sigma = \mp 3\beta$. Now for $s=0$, we have on putting $q_0 = 1$, and remembering that the value of q_2 is twice that given by the general formula,

$$q_2 = \frac{1}{4+\sigma} = \frac{B-1}{6\beta^2}$$

Therefore $$\mathfrak{P}_2 = \mathrm{P}_2 + \frac{B-1}{6\beta} \mathrm{P}_2^2 \quad\ldots\ldots(17)$$

where $$\mathrm{P}_2 = \tfrac{3}{2}\nu^2 - \tfrac{1}{2}, \quad \mathrm{P}_2^2 = 3(\nu^2 - 1)$$

The coefficient of the cosine function is given by

$$p_2 = -\{2, 1\} q_2 = -6q_2$$

Therefore $$\mathfrak{C}_2 = 1 - \frac{B-1}{\beta} \cos 2\phi \quad\ldots\ldots(18)$$

$s=1$, cosine; EOC type.

Here $\sigma = -3\beta$. But the continued fraction is not required.

$$\mathsf{P}_2^1 = \Omega\,[q_1' \mathrm{P}_2^1], \text{ and } q_1' = 1$$

Therefore $$\mathsf{P}_2^1 = \sqrt{\frac{\nu^2 - \frac{1+\beta}{1-\beta}}{\nu^2 - 1}}\, \mathrm{P}_2^1 = 3\nu\left(\nu^2 - \tfrac{1+\beta}{1-\beta}\right)^{\frac{1}{2}} \quad\ldots\ldots(19)$$

Clearly $$\mathsf{C}_2^1 = \cos\phi\,(1 - \beta\cos 2\phi)^{\frac{1}{2}} \quad\ldots\ldots(20)$$

$s=1$, sine; EOS type.

Here $\sigma = 3\beta$. But the continued fraction is not required.

Putting $q_1' = 1$,

$$\mathfrak{P}_2^1 = q_1' \mathrm{P}_1^1 = \mathrm{P}_1^1 = 3\nu(\nu^2 - 1)^{\frac{1}{2}} \quad\ldots\ldots(21)$$

$$\mathsf{S}_2^1 = \sin\phi\,(1 - \beta\cos 2\phi)^{\frac{1}{2}} \quad\ldots\ldots(22)$$

$s=2$, cosine; EEC type.

Here $\sigma = 2(1-B)$.

Putting $q_2 = 1$,

$$q_0 = -\tfrac{1}{2}\frac{\{2, 2\}\{2, 1\}}{4-\sigma} = \frac{-2(B-1)}{\beta^2}$$

Therefore
$$\mathfrak{P}_2^2 = \frac{-2(B-1)}{\beta} P_2 + P_2^2 \text{ ....(23)}$$
where
$$P_2 = \tfrac{3}{2}\nu^2 - \tfrac{1}{2}$$
$$P_2^2 = 3(\nu^2 - 1)$$
Then
$$p_0 = -\frac{1}{\{2, 1\}} q_0 = -\tfrac{1}{6} q_0$$
and
$$\mathfrak{C}_2^2 = \frac{B-1}{3\beta} + \cos 2\phi \text{ ....(24)}$$

$s = 2$, sine; EES type.

Both fractions disappear and σ vanishes, but is not needed for determining the functions. Noting that $q_0' = 0$, and $q_2' = 1$,

$$P_2^2 = \Omega\,[q_2' P_2^2] = 3\left(\nu^2 - \tfrac{1+\beta}{1-\beta}\right)^{\frac{1}{2}} (\nu^2 - 1)^{\frac{1}{2}} \text{ ....(25)}$$
$$\mathfrak{S}_2^2 = \sin 2\phi \text{ ....(26)}$$

We can write down the functions of μ by symmetry, and the products of the three functions give rigorously the five solid harmonic solutions of Laplace's equation of the second degree.

§ 8. *Approximate Form of the Functions.*

It is clear that the first approximation to σ is zero, and that the second approximation, in the general case, is

$$\sigma = -\tfrac{1}{16}\beta^2 \frac{\{i, s\}\{i, s-1\}}{s-1} + \tfrac{1}{16}\beta^2 \frac{\{i, s+1\}\{i, s+2\}}{s+1}$$
$$= \tfrac{1}{8}\beta^2 \left[3s^2 - 2i(i+1) - \frac{i^2(i+1)^2}{s^2-1}\right]$$

If this expression were inserted in $\frac{q_{s\pm 2}}{q_s}$ we should obtain $q_{s\pm 2}$ correct to β^2. But since the next approximation would only introduce β^4, it follows that $q_{s\pm 2}$ would be correct to β^3 inclusive. Now $q_{s\pm 2}$ enters in the functions with a factor β, and therefore this approximation would give results correct to β^4 inclusive. Since the similar operation could be applied with equal ease in all the cases in which the continued fractions assume special forms, it follows that this degree of accuracy is very easily attainable. However, the forms of the coefficients would be rather complicated, and it would render the subsequent algebra so tedious that I do not propose at present to carry the approximation beyond β^2.

It now suffices to put $\sigma = 0$ in the denominators of all the continued fractions, whereby the coefficients are determined, except in the cases of $s = 1$, $s = 3$, where we put $\sigma = \pm\tfrac{1}{2}i(i+1)$.

In the general case we have

$$\left.\begin{aligned}
q_{s-2} &= -\tfrac{1}{8}\frac{\{i, s\}\{i, s-1\}}{s-1}, & q_{s+2} &= \frac{1}{8(s+1)} \\
q_{s-4} &= \frac{\{i, s\}\{i, s-1\}\{i, s-2\}\{i, s-3\}}{128(s-1)(s-2)}, & q_{s+4} &= \frac{1}{128(s+1)(s+2)} \\
q'_{s-2} &= \frac{s-2}{s}q_{s-2}, & q'_{s+2} &= \frac{s+2}{s}q_{s+2} \\
q'_{s-4} &= \frac{s-4}{s}q_{s-4}, & q'_{s+4} &= \frac{s+4}{s}q_{s+4} \\
p_{s-2} &= \frac{\{i, s\}}{8(s-1)}, & p_{s+2} &= \frac{-\{i, s+1\}}{8(s+1)} \\
p_{s-4} &= \frac{\{i, s\}\{i, s-2\}}{128(s-1)(s-2)}, & p_{s+4} &= \frac{\{i, s+1\}\{i, s+3\}}{128(s+1)(s+2)} \\
p'_{s-2} &= \frac{\{i, s-1\}}{8(s-1)}, & p'_{s+2} &= \frac{-\{i, s+2\}}{8(s+1)} \\
p'_{s-4} &= \frac{\{i, s-1\}\{i, s-3\}}{128(s-1)(s-2)}, & p'_{s+4} &= \frac{\{i, s+2\}\{i, s+4\}}{128(s+1)(s+2)}
\end{aligned}\right\} \quad (27)$$

When $s=0$, we double the results given by the general formula and find

$$q_2 = \tfrac{1}{4},\quad q_4 = \tfrac{1}{128},\quad p_2 = -\tfrac{1}{4}\{i, 1\},\quad p_4 = \tfrac{1}{128}\{i, 1\}\{i, 3\}$$

There are no

$$q_2',\quad q_4',\quad \text{and}\quad p_2' = -\tfrac{1}{4}\{i, 2\},\quad p_4' = \tfrac{1}{128}\{i, 2\}\{i, 4\} \quad \text{......(28)}$$

When $s=1$,

$$q_3 = \frac{1}{16 \mp \beta i(i+1)} = \tfrac{1}{16}[1 \pm \tfrac{1}{16}\beta i(i+1)]$$

with upper sign for cosines (EOC, OOC) and lower sign for sines (OOS, EOS).

$$q_5 = \frac{1}{128.2.3} = \tfrac{1}{768} \text{ for all cases}$$

But for OOS, EOS we use the $\mathfrak{P}$ form, and for EOC, OOC the P form; and for the latter $\dfrac{s+2}{s} = 3$, $\dfrac{s+4}{s} = 5$.

Therefore for OOS, EOS (sines)

$$q_3 = \tfrac{1}{16}[1 - \tfrac{1}{16}\beta i(i+1)],\quad q_5 = \tfrac{1}{768}$$

and for EOC, OOC (cosines)

$$q_3' = \tfrac{3}{16}[1 + \tfrac{1}{16}\beta i(i+1)],\quad q_5' = \tfrac{5}{768} \quad \text{......(29)}$$

For OOC, OOS, with upper sign for cosine and lower sign for sine,

$$p_3 = -\tfrac{1}{16}\{i, 2\}[1 \pm \tfrac{1}{16}\beta i(i+1)],\quad p_5 = \tfrac{1}{768}\{i, 2\}\{i, 4\} \quad \text{......(29)}$$

For EOC, EOS, with upper sign for cosine and lower sign for sine,

$$p_3' = -\tfrac{1}{16}\{i, 3\}[1 \pm \tfrac{1}{16}\beta i(i+1)], \quad p_5' = \tfrac{1}{768}\{i, 3\}\{i, 5\} \quad \ldots\ldots(29)$$

When $s=2$ the coefficients may be derived from the general formula.

When $s=3$

$$q_1 = \frac{-\frac{1}{2}\{i, 3\}\{i, 2\}}{8 \mp \frac{1}{2}\beta i(i+1)} = -\tfrac{1}{16}\{i, 2\}\{i, 3\}[1 \pm \tfrac{1}{16}\beta i(i+1)]$$

the upper sign applying to cosines (OOC, EOC) the lower to sines (OOS, EOS);

$$q_5 = \tfrac{1}{32}, \quad q_7 = \tfrac{1}{2560}$$

But for OOS, EOS the $\mathfrak{P}$ form applies, and for OOC, EOC the P form applies.

Also with $\quad s=3, \quad \dfrac{s-2}{s} = \tfrac{1}{3}, \quad \dfrac{s+2}{s} = \tfrac{5}{3}, \quad \dfrac{s+4}{s} = \tfrac{7}{3}$

Therefore for OOS, EOS

$$q_1 = -\tfrac{1}{16}\{i, 2\}\{i, 3\}[1 - \tfrac{1}{16}\beta i(i+1)], \quad q_5 = \tfrac{1}{32}, \quad q_7 = \tfrac{1}{2560}$$

For OOC, EOC

$$q_1' = -\tfrac{1}{48}\{i, 2\}\{i, 3\}[1 + \tfrac{1}{16}\beta i(i+1)], \quad q_5' = \tfrac{5}{96}, \quad q_7' = \tfrac{7}{7680}$$

For OOC, OOS, with upper sign for cosine and lower for sine,

$$p_1 = \tfrac{1}{16}\{i, 3\}[1 \pm \tfrac{1}{16}\beta i(i+1)], \quad p_5 = -\tfrac{1}{32}\{i, 4\}, \quad p_7 = \tfrac{1}{2560}\{i, 4\}\{i, 6\} \quad \ldots(30)$$

For EOC, EOS, with upper sign for cosine and lower for sine.

$$p_1' = \tfrac{1}{16}\{i, 2\}[1 \pm \tfrac{1}{16}\beta i(i+1)], \quad p_5' = -\tfrac{1}{32}\{i, 5\}, \quad p_7' = \tfrac{1}{2560}\{i, 5\}\{i, 7\}$$

It will save much trouble to note that if we were to admit negative suffixes to the q's, the general formula would give us the term $\beta^2 q_{-1} \mathrm{P}^{-1}$, where

$$q_{-1} = \frac{\{i, 3\}\{i, 2\}\{i, 1\}\{i, 0\}}{128 \cdot 2 \cdot 1}$$

Thus this term is $\dfrac{1}{(16)^2}\beta^2 i(i+1) \cdot \{i, 3\}\{i, 2\}\mathrm{P}^1$. But this is exactly that part of the term in (30) which arises from $\beta q_1 \mathrm{P}^1$, but which is not included in the general formula.

Similarly the general formula gives for q'_{-1}, p_{-1}, p'_{-1} those parts of the terms arising from q_1', p_1, p_1' which are not included in the general formula.

It follows that in much of the subsequent work we need not devote special consideration to the case of $s=3$.

§ 9. *Factors of Transformation between the two forms of P-function and of C- or S-function.*

The rigorous expressions $\mathfrak{P}^s$ and P^s always differ from one another, but approximately they are the same up to a certain power of β, provided that s is greater than a certain quantity.

Since $\Omega = \left(\dfrac{\nu^2 - \frac{1+\beta}{1-\beta}}{\nu^2-1}\right)^{\frac{1}{2}} = \left(1 + \dfrac{2\beta}{(\nu^2-1)(1-\beta)}\right)^{\frac{1}{2}}$, it is legitimate to develop Ω in powers of $1/(\nu^2-1)$ up to a certain power, say t, provided that it is to be multiplied by a function involving at least $(\nu^2-1)^t$ as a factor; for this condition insures that there shall be no infinite terms when $\nu = \pm 1$. At present, I limit the development to β^2, so that

$$\Omega = 1 - \frac{\beta+\beta^2}{\nu^2-1} - \frac{\frac{1}{2}\beta^2}{(\nu^2-1)^2}$$

Therefore

$$\mathsf{P}^s = \left(1 - \frac{\beta+\beta^2}{\nu^2-1} - \frac{\frac{1}{2}\beta^2}{(\nu^2-1)^2}\right)\mathrm{P}^s + \left(1 - \frac{\beta}{\nu^2-1}\right)(\beta q'_{s-2}\mathrm{P}^{s-2} + \beta q'_{s+2}\mathrm{P}^{s+2}) + \beta^2 q'_{s-4}\mathrm{P}^{s-4} + \beta^2 q'_{s+4}\mathrm{P}^{s+4}$$

It is obvious on inspection that we cannot rely on this development if s is less than 4.

If then s is equal to, or greater than 4, this value of P^s, when properly developed, to the adopted order of approximation can only differ from $\mathfrak{P}^s$ by a constant factor, say C_i^s or shortly C^s; so that

$$\mathfrak{P}^s = C^s\mathsf{P}^s \quad\text{.............................}(31)$$

and we have to determine the constant C^s.

We might develop the above expression for P^s completely and compare it with $\mathfrak{P}^s$, but this is unnecessary since the comparison of a single term suffices.

I now write

$$\Sigma_i^s = \frac{i(i+1)}{s^2-1} \quad\text{.............................}(32)$$

or shortly Σ. This notation is introduced because this function occurs very frequently hereafter.

We have seen in (11) (slightly modified) that

$$\frac{\mathrm{P}^s}{\nu^2-1} = \frac{1}{4}\left\{\frac{\mathrm{P}^{s+2}}{s(s+1)} - 2(\Sigma+1)\mathrm{P}^s + \frac{\{i, s\}\{i, s-1\}}{s(s-1)}\mathrm{P}^{s-2}\right\}$$

We may write this

$$\frac{\mathrm{P}^s}{\nu^2-1} = \alpha_s\mathrm{P}^{s+2} + \beta_s\mathrm{P}^{s+1} + \gamma_s\mathrm{P}^{s-2}$$

where $\alpha_s = \dfrac{1}{4s(s+1)}$, $\beta_s = -\frac{1}{2}(\Sigma+1)$, $\gamma_s = \dfrac{\{i, s\}\{i, s-1\}}{4s(s-1)}$

Then
$$\frac{P^s}{(\nu^2-1)^2} = \alpha_s(\alpha_{s+2}P^{s+4} + \beta_{s+2}P^{s+2} + \gamma_{s+2}P^s)$$
$$+ \beta_s(\alpha_s P^{s+2} + \beta_s P^s + \gamma_s P^{s-2})$$
$$+ \gamma_s(\alpha_{s-2}P^s + \beta_{s-2}P^{s-2} + \gamma_{s-2}P^{s-4})$$

Therefore the coefficient of P^s is $\alpha_s\gamma_{s+2} + (\beta_s)^2 + \alpha_{s-2}\gamma_s$, or

$$\tfrac{1}{16}\left[\frac{\{i, s+2\}\{i, s+1\}}{s(s+1)^2(s+2)} + 4(\Sigma+1)^2 + \frac{\{i, s\}\{i, s-1\}}{s(s-1)^2(s-2)}\right]$$

I now introduce a further abridgement and write

$$\Upsilon_i^s = \frac{(i-1)i(i+1)(i+2)}{s^2-4} \quad \text{......................(32)}$$

or shortly Υ.

Then, after reduction, I find

$$\frac{P^s}{(\nu^2-1)^2} = \tfrac{1}{8}[-\Sigma^2 s^2 + \Sigma^2 + 4\Sigma + 3 + \Upsilon]\,P^s + \dots$$

Accordingly the coefficient of P^s in P^s is

$$1 + \tfrac{1}{2}\beta(\Sigma+1) + \tfrac{1}{2}\beta^2(\Sigma+1) - \tfrac{1}{16}\beta^2[-\Sigma^2 s^2 + \Sigma^2 + 4\Sigma + 3 + \Upsilon]$$
$$-\tfrac{1}{4}\beta^2\left[\frac{q'_{s-2}}{(s-1)(s-2)} + q'_{s+2}\frac{\{i, s+1\}\{i, s+2\}}{(s+1)(s+2)}\right]$$

But $q'_{s-2} = -\dfrac{(s-2)\{i, s\}\{i, s-1\}}{8s(s-1)}$, $q'_{s+2} = \dfrac{s+2}{8s(s+1)}$, and the last term in the above expression will be found to be equal to $+\tfrac{1}{8}\beta^2(\Sigma^2-1)$. Thus the coefficient of P^s in the development of P^s is

$$1 + \tfrac{1}{2}\beta(\Sigma+1) + \tfrac{1}{16}\beta^2(\Sigma^2 s^2 + \Sigma^2 + 4\Sigma + 3 - \Upsilon)$$

but the same coefficient in $\mathfrak{P}^s$ is unity.

Therefore

$$\left.\begin{aligned}
\frac{1}{C_i^s} &= 1 + \tfrac{1}{2}\beta(\Sigma+1) + \tfrac{1}{16}\beta^2(\Sigma^2 s^2 + \Sigma^2 + 4\Sigma + 3 - \Upsilon)\\
C_i^s &= 1 - \tfrac{1}{2}\beta(\Sigma+1) + \tfrac{1}{16}\beta^2(-\Sigma^2 s^2 + 3\Sigma^2 + 4\Sigma + 1 + \Upsilon)\\
\frac{1}{(C_i^s)^2} &= 1 + \beta(\Sigma+1) + \tfrac{1}{8}\beta^2(\Sigma^2 s^2 + 3\Sigma^2 + 8\Sigma + 5 - \Upsilon)\\
(C_i^s)^2 &= 1 - \beta(\Sigma+1) + \tfrac{1}{8}\beta^2(-\Sigma^2 s^2 + 5\Sigma^2 + 8\Sigma + 3 + \Upsilon)
\end{aligned}\right\} \quad \text{...(33)}$$

The squares of this constant and of its reciprocal are given because they will be needed at a later stage.

We next consider the cosine and sine functions.

$$\begin{Bmatrix}\mathsf{C}^s\\ \mathsf{S}^s\end{Bmatrix} = \Phi\left[\begin{Bmatrix}\cos\\ \sin\end{Bmatrix} s\phi + \beta p'_{s-2}\begin{Bmatrix}\cos\\ \sin\end{Bmatrix}(s-2)\phi + \dots\right]$$

As far as β^2

$$\Phi = (1 - \beta \cos 2\phi)^{\frac{1}{2}} = 1 - \tfrac{1}{2}\beta \cos 2\phi - \tfrac{1}{16}\beta^2 (1 + \cos 4\phi)$$

Therefore

$$\begin{cases}\mathsf{C}^s \\ \mathsf{S}^s\end{cases} = [1 - \tfrac{1}{2}\beta \cos 2\phi - \tfrac{1}{16}\beta^2 (1 + \cos 4\phi)] \begin{cases}\cos \\ \sin\end{cases} s\phi$$

$$+ \beta \left[p'_{s-2} \begin{cases}\cos \\ \sin\end{cases} (s-2)\phi + p'_{s+2} \begin{cases}\cos \\ \sin\end{cases} (s+2)\phi \right] [1 - \tfrac{1}{2}\beta \cos 2\phi]$$

$$+ \beta^2 p'_{s-4} \begin{cases}\cos \\ \sin\end{cases} (s-4)\phi + \beta^2 p'_{s+4} \begin{cases}\cos \\ \sin\end{cases} (s+4)\phi$$

This expression, when developed, must lead to $\mathfrak{C}^s$ or $\mathfrak{S}^s$ multiplied by a constant factor.

Let

$$\begin{cases}\mathfrak{C}^s \\ \mathfrak{S}^s\end{cases} = \mathrm{D}_i^s \begin{cases}\mathsf{C}^s \\ \mathsf{S}^s\end{cases} \quad \text{.................................(34)}$$

Then D_i^s or D^s may be found by considering only the coefficient of $\begin{cases}\cos \\ \sin\end{cases} s\phi$.

Hence

$$\frac{1}{\mathrm{D}^s} = 1 - \tfrac{1}{16}\beta^2 - \tfrac{1}{4}\beta^2 p'_{s-2} - \tfrac{1}{4}\beta^2 p'_{s+2}$$

But

$$p'_{s-2} = \frac{\{i, s-1\}}{8(s-1)}, \qquad p'_{s+2} = \frac{-\{i, s+2\}}{8(s+1)}$$

and

$$p'_{s-2} + p'_{s+2} = \tfrac{1}{4}(\Sigma + 2)$$

Therefore

$$\left.\begin{aligned}\frac{1}{\mathrm{D}_i^s} &= 1 - \tfrac{1}{16}\beta^2 (\Sigma + 3) \\ \frac{1}{(\mathrm{D}_i^s)^2} &= 1 - \tfrac{1}{8}\beta^2 (\Sigma + 3)\end{aligned}\right\} \quad \text{..........................(35)}$$

The reciprocals may clearly be written down at once.

There are no factors by which $\mathfrak{P}^3$, $\mathfrak{P}^2$, $\mathfrak{P}^1$ can be converted into P^3, P^2, P^1; but this is not true of the cosine and sine functions.

In the case of $s = 3$, it will be found that the general formula holds good for the factor whereby $\mathfrak{C}^3$, $\mathfrak{S}^3$ are convertible into C^3, S^3.

When $s = 2$,

$$\begin{cases}\mathsf{C}^2 \\ \mathsf{S}^2\end{cases} = [1 - \tfrac{1}{2}\beta \cos 2\phi - \tfrac{1}{16}\beta^2 (1 + \cos 4\phi)] \begin{cases}\cos \\ \sin\end{cases} 2\phi + \beta p_0' (1 - \tfrac{1}{2}\beta \cos 2\phi) \begin{cases}1 \\ 0\end{cases}$$

$$+ \beta p_4' (1 - \tfrac{1}{2}\beta \cos 2\phi) \begin{cases}\cos \\ \sin\end{cases} 4\phi + \beta^2 p_6' \begin{cases}\cos \\ \sin\end{cases} 6\phi$$

Then

$$\mathsf{C}^2 = [1 - (\tfrac{3}{32} + \tfrac{1}{2}p_0' + \tfrac{1}{4}p_4')\beta^2] \cos 2\phi + \dots$$

$$\mathsf{S}^2 = [1 - (\tfrac{1}{32} + \tfrac{1}{4}p_4')\beta^2] \sin 2\phi + \dots$$

But $p_0' = \frac{1}{8}\{i, 1\}, \quad p_4' = -\frac{1}{24}\{i, 4\}, \quad p_6' = \frac{1}{1536}\{i, 4\}\{i, 6\}$

and
$$\tfrac{3}{32} + \tfrac{1}{2}p_0' + \tfrac{1}{4}p_4' = \tfrac{1}{32}(5\Sigma + 7)$$
$$\tfrac{1}{32} + \tfrac{1}{4}p_4' = -\tfrac{1}{32}(\Sigma - 5), \text{ where } \Sigma = \frac{i(i+1)}{2^2 - 1} = \tfrac{1}{3}i(i+1)$$

Therefore the factors are
$$\left.\begin{aligned} \frac{1}{D_i^2}(\cos) &= 1 - \tfrac{1}{32}\beta^2(5\Sigma^2 + 7) \\ \frac{1}{D_i^2}(\sin) &= 1 + \tfrac{1}{32}\beta^2(\Sigma^2 - 5) \end{aligned}\right\} \quad \ldots\ldots\ldots\ldots\ldots\ldots\ldots(36)$$

It is easy to verify that the other coefficients of $\mathfrak{C}^2$ and $\mathfrak{S}^2$ are in fact reproduced.

The notation adopted here and below for distinguishing the cosine and sine factors is perhaps rather clumsy, but I have not thought it worth while to take distinctive symbols for the factors in these cases, because they will not be of frequent occurrence.

When $s = 1$,
$$\begin{aligned} \begin{Bmatrix} C^1 \\ S^1 \end{Bmatrix} &= [1 - \tfrac{1}{2}\beta\cos 2\phi - \tfrac{1}{16}\beta^2(1 + \cos 4\phi)] \begin{Bmatrix} \cos \\ \sin \end{Bmatrix} \phi + \beta p_3'(1 - \tfrac{1}{2}\beta\cos 2\phi) \begin{Bmatrix} \cos \\ \sin \end{Bmatrix} 3\phi \\ &\qquad + \beta^2 p_5' \begin{Bmatrix} \cos \\ \sin \end{Bmatrix} 5\phi \\ &= [1 \mp \tfrac{1}{4}\beta - \tfrac{1}{4}\beta^2(p_3' + \tfrac{1}{4})] \begin{Bmatrix} \cos \\ \sin \end{Bmatrix} \phi + \ldots \end{aligned}$$

This must be equal to $\frac{1}{D_i^1}\begin{Bmatrix} \mathfrak{C}^1 \\ \mathfrak{S}^1 \end{Bmatrix}$.

Now, with upper sign for cosine and lower for sine,
$$p_3' = -\tfrac{1}{16}\{i, 3\}[1 \pm \tfrac{1}{16}\beta i(i+1)], \quad p_5' = \tfrac{1}{768}\{i, 3\}\{i, 5\}$$

Substituting for p_3' its values, we find with the upper sign
$$\frac{1}{D_i^1}(\cos) = 1 - \tfrac{1}{4}\beta - \tfrac{1}{4}\beta^2(p_3' + \tfrac{1}{4}) = 1 - \tfrac{1}{4}\beta + \tfrac{1}{64}\beta^2[i(i+1) - 10]$$

And with the lower sign $\qquad$ (37)
$$\frac{1}{D_i^1}(\sin) = 1 + \tfrac{1}{4}\beta - \tfrac{1}{4}\beta^2(p_3' + \tfrac{1}{4}) = 1 + \tfrac{1}{4}\beta + \tfrac{1}{64}\beta^2[i(i+1) - 10]$$

It follows that
$$\left.\begin{aligned} D_i^1(\cos) &= 1 + \tfrac{1}{4}\beta - \tfrac{1}{64}\beta^2[i(i+1) - 14] \\ D_i^1(\sin) &= 1 - \tfrac{1}{4}\beta - \tfrac{1}{64}\beta^2[i(i+1) - 14] \\ \left[\frac{1}{D_i^1}(\cos)\right]^2 &= 1 - \tfrac{1}{2}\beta + \tfrac{1}{32}\beta^2[i(i+1) - 8] \\ [D_i^1(\cos)]^2 &= 1 + \tfrac{1}{2}\beta - \tfrac{1}{32}\beta^2[i(i+1) - 16] \\ \left[\frac{1}{D_i^1}(\sin)\right]^2 &= 1 + \tfrac{1}{2}\beta + \tfrac{1}{32}\beta^2[i(i+1) - 8] \\ [D_i^1(\sin)]^2 &= 1 - \tfrac{1}{2}\beta - \tfrac{1}{32}\beta^2[i(i+1) - 16] \end{aligned}\right\} \quad \ldots\ldots\ldots\ldots(37)$$

We cannot in the present case use Σ_i^1 as an abridged notation, because it is infinite as involving $s^2 - 1$ in the denominator.

It is easy to verify that the other coefficients of $\mathfrak{C}^1$ and $\mathfrak{S}^1$ are, in fact reproduced in the transformation.

Lastly when $s = 0$, we have only cosine functions. As before

$$\mathsf{C} = 1 - \tfrac{1}{2}\beta \cos 2\phi - \tfrac{1}{16}\beta^2 (1 + \cos 4\phi) + \beta p_2' (1 - \tfrac{1}{2}\beta \cos 2\phi) \cos 2\phi + \beta^2 p_4' \cos 4\phi$$

$$= 1 - \tfrac{1}{16}\beta^2 - \tfrac{1}{4}\beta^2 p_2' + \ldots$$

This must be equal to $\dfrac{\mathfrak{C}}{\mathrm{D}_i}$, and therefore $\dfrac{1}{\mathrm{D}_i} = 1 - \tfrac{1}{16}\beta^2 - \tfrac{1}{4}\beta^2 p_2'$.

Now $$p_2' = -\tfrac{1}{4}\{i, 2\}, \quad p_4' = \tfrac{1}{128}\{i, 2\}\{i, 4\}$$

Hence $$\frac{1}{\mathrm{D}_i} = 1 + \tfrac{1}{16}\beta^2 [i(i+1) - 3] \qquad (38)$$

Since in this case $\Sigma_i = \dfrac{i(i+1)}{-1}$,

$$\frac{1}{\mathrm{D}_i} = 1 - \tfrac{1}{16}\beta^2 [\Sigma_i + 3] \qquad (38)$$

Thus the general formula again holds good.

It is easy to verify that the other coefficients of $\mathfrak{C}$ are in fact reproduced.

The principal use of the transforming factors, determined in this section, is that it will enable us to avoid some tedious analysis hereafter.

§ 10. *The Functions of the Second Kind.*

[I found in subsequent investigations that the amount of accuracy attainable by developing these functions of the second kind was disappointing; nevertheless I proceed to show how approximate forms may be obtained for them.]

The second continued fraction of § 6 terminates because

$$\{i, s + 2n + 2\}\{i, s + 2n + 1\} = 0$$

when $n = \frac{1}{2}(i - s)$ or $\frac{1}{2}(i - s - 1)$, since one of the two factors then assumes the form $\{i, i + 1\}$.

Hence it follows that the equation for determining σ is the same as before; but we cannot on that account assume that the q coefficients vanish when their suffixes are greater than i.

In considering the P-functions it was immaterial whether or not we regarded them as vanishing, because P^t vanishes if t is greater than i.

But the Q-functions do not vanish in this case, and therefore we must postulate the existence of q's with suffix greater than i.

In fact, whilst we have as before, when i and s are both odd or both even,

$$\frac{2q_i}{q_{i-2}} = \frac{1}{i^2 - s^2 + \sigma}$$

we also have

$$\frac{2q_{i+2}}{q_i} = \frac{1}{(i+2)^2 - s^2 + \sigma - \frac{1}{4}\beta^2 \{i, i+4\} \{i, i+3\} \left(\frac{2q_{i+4}}{q_{i+2}}\right)}$$

and similarly a fraction for $\frac{2q_{i+4}}{q_{i+2}}$, and so forth.

It follows therefore that while the q's with suffixes less or equal to i depend on finite continued fractions, those with suffixes greater than i depend on infinite continued fractions.

It thus appears that while the first series in the expression for $\mathfrak{Q}_i^s$ or for $\mathbf{Q}_i^s$ has limits 1 to $\frac{1}{2}s$ or $\frac{1}{2}(s-1)$, as before, the limits of the second series are 1 to ∞.

Thus we have found an expansion for this class of functions in powers of β.

In the limited case in which the coefficients have been actually evaluated, namely, where the development is only carried as far as the squares of β, we have

$$q_{s+2} = \frac{1}{8(s+1)}, \qquad q_{s+4} = \frac{1}{128(s+1)(s+2)}$$

$$q'_{s+2} = \frac{s+2}{8s(s+1)}, \qquad q'_{s+4} = \frac{s+4}{128s(s+1)(s+2)}$$

These coefficients do not vanish when $s+2$ or $s+4$ are greater than i, and this confirms the conclusion already arrived at.

In spherical harmonic analysis there is no occasion to consider the value of Q_i^s when s is greater than i, and the values are therefore not familiar. I will therefore now determine them.

It is known* that

$$Q_i = \frac{2^i (i!)^2}{2i+1!} \left[\frac{1}{\nu^{i+1}} + \frac{i+2!}{2 . 1! \, i!} \cdot \frac{1}{(2i+3)\, \nu^{i+3}} + \frac{i+4!}{2^2 . 2! \, i! \, (2i+3)(2i+5)\, \nu^{i+5}} + \dots\right]$$

* Bryan, *Camb. Phil. Soc. Proc.*, Vol. VI., 1888, p. 293.

Therefore differentiating

$$\left(\frac{d}{d\nu}\right)^{i+1} Q_i = (-)^{i+1}\, 2^i \,.\, i!\left[\frac{1}{\nu^{2i+2}} + \frac{i+1}{1!}\,\frac{1}{\nu^{2i+4}} + \frac{(i+1)(i+2)}{2!}\cdot\frac{1}{\nu^{2i+6}} + \dots\right]$$

$$= (-)^{i+1}\frac{2^i \,.\, i!}{(\nu^2-1)^{i+1}}$$

And

$$\left(\frac{d}{d\nu}\right)^{i+2} Q_i = (-)^i \frac{2^{i+1}\,.\,\overline{i+1}!}{(\nu^2-1)^{i+2}}\,\nu$$

$$\left(\frac{d}{d\nu}\right)^{i+3} Q_i = (-)^i\, 2^{i+1}\,.\,\overline{i+1}!\left[\frac{-2(i+2)\nu^2}{(\nu^2-1)^{i+3}} + \frac{1}{(\nu^2-1)^{i+2}}\right]$$

$$= (-)^{i+1}\left[\frac{2^{i+2}\,.\,\overline{i+2}!}{(\nu^2-1)^{i+3}} + \frac{2^{i+1}(2i+3)\,.\,\overline{i+1}!}{(\nu^2-1)^{i+2}}\right]$$

$$\left(\frac{d}{d\nu}\right)^{i+4} Q_i = (-)^i\left[\frac{2^{i+3}\,.\,\overline{i+3}!}{(\nu^2-1)^{i+4}} + \frac{2^{i+2}(2i+3)\,.\,\overline{i+2}!}{(\nu^2-1)^{i+3}}\right]\nu$$

But $\qquad Q^t = (\nu^2-1)^{\frac{1}{2}t}\left(\frac{d}{d\nu}\right)^t Q_i$

therefore

$$\left.\begin{aligned}
Q_i^{i+1} &= (-)^{i+1}\frac{2^i\, i!}{(\nu^2-1)^{\frac{1}{2}(i+1)}}\\
Q_i^{i+2} &= (-)^i \frac{2^{i+1}\,.\,\overline{i+1}!\,\nu}{(\nu^2-1)^{\frac{1}{2}(i+2)}}\\
Q_i^{i+3} &= (-)^{i+1}\frac{2^{i+1}\,.\,\overline{i+1}!}{(\nu^2-1)^{\frac{1}{2}(i+3)}}\left[2i+4+(2i+3)(\nu^2-1)\right]\\
Q_i^{i+4} &= (-)^i \frac{2^{i+2}\,.\,\overline{i+2}!\,\nu}{(\nu^2-1)^{\frac{1}{2}(i+4)}}\left[2i+6+(2i+3)(\nu^2-1)\right]
\end{aligned}\right\}\quad\dots\dots(39)$$

These are all the functions which can be needed for the expression as far as β^2 of $\mathfrak{Q}_i^s$ or of $\mathbf{Q}_i^s$ when s is less or equal to i. If s is equal to i, we shall have terms $\beta^2 q_{s+4} Q^{s+4}$ or $\Omega\beta^2 q'_{s+4} Q^{s+4}$, and these are the furthest.

But it is well known that there is another expression for these functions of the second kind.

The differential equation is

$$\left[(\nu^2-1)^2\frac{d^2}{d\nu^2} + 2\nu(\nu^2-1)\frac{d}{d\nu} - i(i+1)(\nu^2-1) - s^2\right]\mathfrak{Q}_i^s$$
$$-\beta\left[(\nu^2-1)(\nu^2+1)\frac{d^2}{d\nu^2} + 2\nu^3\frac{d}{d\nu} - i(i+1)\nu^2 - \frac{\sigma}{\beta}\right]\mathfrak{Q}_i^s = 0$$

where $\mathfrak{Q}_i^s$ may be interpreted as meaning also $\mathbf{Q}_i^s$.

Let us assume that $\qquad \mathfrak{Q}^s = \mathfrak{P}^s \int_\nu^\infty V\,d\nu$

is a solution, where $\mathfrak{Q}^s$, $\mathfrak{P}^s$ may be interpreted as meaning also $\mathbf{Q}^s$, $\mathbf{P}^s$.

Then since $\mathfrak{P}^s$ is a solution of the differential equation, we have

$$(\nu^2-1)^2\left[2\mathrm{V}\frac{d\mathfrak{P}^s}{d\nu}+\mathfrak{P}^s\frac{d\mathrm{V}}{d\nu}\right]+2\nu(\nu^2-1)\mathfrak{P}^s\mathrm{V}$$
$$-\beta\left[(\nu^2-1)(\nu^2+1)\left(2\mathrm{V}\frac{d\mathfrak{P}^s}{d\nu}+\mathfrak{P}^s\frac{d\mathrm{V}}{d\nu}\right)+2\nu^3\mathfrak{P}^s\mathrm{V}\right]=0$$

This is easily reducible to

$$\frac{d}{d\nu}\log\left[\mathrm{V}(\mathfrak{P}^s)^2(\nu^2-1)^{\frac{1}{2}}\left(\nu^2-\tfrac{1+\beta}{1-\beta}\right)^{\frac{1}{2}}\right]=0$$

whence $\mathrm{V}=\dfrac{\mathfrak{E}_i^s}{(\mathfrak{P}^s)^2(\nu^2-1)^{\frac{1}{2}}\left(\nu^2-\frac{1+\beta}{1-\beta}\right)^{\frac{1}{2}}}$, where $\mathfrak{E}_i^s$ is a constant.

Hence

$$\left.\begin{aligned}\mathfrak{Q}_i^s &= \mathfrak{E}_i^s\mathfrak{P}_i^s\int_\nu^\infty\frac{d\nu}{(\mathfrak{P}_i^s)^2(\nu^2-1)^{\frac{1}{2}}\left(\nu^2-\frac{1+\beta}{1-\beta}\right)^{\frac{1}{2}}}\\ \mathsf{Q}_i^s &= \mathsf{E}_i^s\,\mathsf{P}_i^s\int_\nu^\infty\frac{d\nu}{(\mathsf{P}_i^s)^2(\nu^2-1)^{\frac{1}{2}}\left(\nu^2-\frac{1+\beta}{1-\beta}\right)^{\frac{1}{2}}}\end{aligned}\right\}\quad\text{..........(40)}$$

The general solution of the differential equation must be

$$\alpha\mathfrak{P}_i^s+\gamma\mathfrak{Q}_i^s$$

and we have already found both $\mathfrak{P}^s$ and $\mathfrak{Q}^s$. Hence the two $\mathfrak{Q}^s$'s must be different expressions for the same thing, for the form of $\mathfrak{Q}^s$ as a series negatives the hypothesis that it involves $\mathfrak{P}^s$ in the form $\gamma_1\mathfrak{P}^s+\gamma_2\mathfrak{Q}^s$.

Having then two forms of $\mathfrak{Q}^s$ or of Q^s, it remains to evaluate the coefficients $\mathfrak{E}_i^s$, E_i^s, which are involved in the equations (40). In order to do this it will suffice to consider the case where ν is very great, so that

$$\mathrm{P}^t=\frac{2i\,!}{2^i\,i\,!}\cdot\frac{\nu^i}{i-t\,!},\qquad \mathrm{Q}^t=(-)^t\frac{2^i\,.\,i\,!}{2i+1\,!}\,\frac{i+t\,!}{\nu^{i+1}}$$

As far as concerns the first term in the series

$$\mathfrak{P}^s=\frac{2i\,!}{2^i\,i\,!}\frac{\nu^i}{i-s\,!}\left[1+\beta q_{s-2}\frac{i-s\,!}{i-s+2\,!}+\beta q_{s+2}\frac{i-s\,!}{i-s-2\,!}\right.$$
$$\left.+\beta^2 q_{s-4}\frac{i-s\,!}{i-s+4\,!}+\beta^2 q_{s+4}\frac{i-s\,!}{i-s-4\,!}\right]$$

$$\mathfrak{Q}^s=(-)^s\frac{2^i\,.\,i\,!}{2i+1\,!}\,\frac{i+s\,!}{\nu^{i+1}}\left[1+\beta q_{s-2}\frac{i+s-2\,!}{i+s\,!}+\beta q_{s+2}\frac{i+s+2\,!}{i+s\,!}\right.$$
$$\left.+\beta^2 q_{s-4}\frac{i+s-4\,!}{i+s\,!}+\beta^2 q_{s+4}\frac{i+s+4\,!}{i+s\,!}\right]$$

It will be observed that if s is equal to i or $i-1$ the terms in $\mathfrak{P}^s$ in q_{s+2} and q_{s+4} disappear; and if s is equal to $i-2$ or $i-3$ that in q_{s+4} disappears.

This agrees, as it should do, with the vanishing of P^{s+2} and P^{s+4} when the order is greater than the degree.

If we write $\mathfrak{P}^s = \alpha\nu^i$ and $\mathfrak{Q}^s = \dfrac{\gamma}{\nu^{i+1}}$, the first of our equations (40) becomes, when ν is very large,

$$\frac{\gamma}{\nu^{i+1}} = \mathfrak{E}^s \alpha\nu^i \int_\nu^\infty \frac{d\nu}{\nu^2(\alpha\nu^i)^2} = \mathfrak{E}^s \frac{\nu^i}{\alpha}\int_\nu^\infty \frac{d\nu}{\nu^{2i+2}}$$

$$= \frac{\mathfrak{E}^s}{\alpha} \cdot \frac{1}{\nu^{i+1}} \cdot \frac{1}{2i+1}$$

Therefore $\mathfrak{E}^s = (2i+1)\,\alpha\gamma$, and since the α, γ in the case of the P^s, Q^s only differ from these in the accenting of the q's we have

$$\mathfrak{E}^s = (-)^s \frac{i+s\,!}{i-s\,!}\left[1 + \beta q_{s-2}\frac{i-s\,!}{i-s+2\,!} + \ldots\right]\left[1 + \beta q_{s-2}\frac{i+s-2\,!}{i+s\,!} + \ldots\right]$$

$\mathsf{E}^s =$ the same with accented q's.

Effecting the multiplication of the series

$$\begin{aligned}\mathfrak{E}^s = (-)^s \frac{i+s\,!}{i-s\,!}\Bigg[1 + \beta\Bigg(&q_{s-2}\frac{i-s\,!}{i-s+2\,!} + q_{s-2}\frac{i+s-2\,!}{i+s\,!} \\ &+ q_{s+2}\frac{i-s\,!}{i-s-2\,!} + q_{s+2}\frac{i+s+2\,!}{i+s\,!}\Bigg) \\ + \beta^2\Bigg(&q_{s-4}\frac{i-s\,!}{i-s+4\,!} + q_{s-4}\frac{i+s-4\,!}{i+s\,!} + q_{s+4}\frac{i-s\,!}{i-s-4\,!} \\ &+ q_{s+4}\frac{i+s+4\,!}{i+s\,!} + q_{s-2}q_{s-2}\frac{i-s\,!\,i+s-2\,!}{i+s\,!\,i-s+2\,!} \\ &+ q_{s+2}q_{s+2}\frac{i-s\,!\,i+s+2\,!}{i+s\,!\,i-s-2\,!} + q_{s-2}q_{s+2}\frac{i-s\,!\,i+s+2\,!}{i+s\,!\,i-s+2\,!} \\ &+ q_{s+2}q_{s-2}\frac{i-s\,!\,i+s-2\,!}{i+s\,!\,i-s-2\,!}\Bigg)\Bigg]\end{aligned}$$

$\mathsf{E}^s =$ the same with accented q's.

If we substitute for the q's their values, the coefficient of β inside [ ] in the expression for $\mathfrak{E}^s$ is

$$\tfrac{1}{8}\left[-\frac{(i+s)(i+s-1)}{s-1} - \frac{(i-s+1)(i-s+2)}{s-1} + \frac{(i-s)(i-s-1)}{s+1} + \frac{(i+s+1)(i+s+2)}{s+1}\right]$$

In the expression for E^s the first pair of these terms are multiplied by $\dfrac{s-2}{s}$, and the second pair by $\dfrac{s+2}{s}$.

The coefficient of β^2 in the expression for $\mathfrak{E}^s$ is

$$\frac{1}{128}\left[\frac{(i+s)(i+s-1)(i+s-2)(i+s-3)}{(s-1)(s-2)}\right.$$

$$+\frac{(i-s+1)(i-s+2)(i-s+3)(i-s+4)}{(s-1)(s-2)}$$

$$+\frac{(i-s)(i-s-1)(i-s-2)(i-s-3)}{(s+1)(s+2)}$$

$$+\frac{(i+s+1)(i+s+2)(i+s+3)(i+s+4)}{(s+1)(s+2)}$$

$$+\frac{2(i+s)(i+s-1)(i-s+1)(i-s+2)}{(s-1)^2}$$

$$+\frac{2(i-s)(i-s-1)(i+s+1)(i+s+2)}{(s+1)^2}$$

$$-\frac{2(i+s-1)(i+s)(i+s+1)(i+s+2)}{(s-1)(s+1)}$$

$$\left.-\frac{2(i-s-1)(i-s)(i-s+1)(i-s+2)}{(s-1)(s+1)}\right]$$

In the expression for E^s the first pair of these terms are multiplied by $\frac{s-4}{s}$; the second pair by $\frac{s+4}{s}$; the first of the third pair by $\left(\frac{s-2}{s}\right)^2$, and the second by $\left(\frac{s+2}{s}\right)^2$; and the last pair by $\frac{s^2-4}{s^2}$.

The reduction of terms such as these will occur frequently hereafter, and I will therefore say a word on the most convenient way of carrying it out. It is obvious that the coefficient of β may be arranged in the form

$$Ai(i+1)+B(2i+1)+C$$

The coefficient A is equal to the coefficient of i^2 in the original expression, and if we put $i=0$ we have $B+C$, and with $i=-1$, $-B+C$. Hence A, B, C may be easily determined.

Again the coefficient of β^2 may be arranged in the form

$$Ai^2(i+1)^2+B(2i+1)i(i+1)+Ci(i+1)+D(2i+1)+E$$

This may be written

$$Ai^4+2(A+B)i^3+(A+3B+C)i^2+(B+C+D)i+D+E$$

It is easy to pick out the coefficients of i^4, i^3, i^2, and we thus obtain A, B, C. Then putting i successively equal to 0 and -1 we have $D+E$ and $-D+E$.

In order to express the results succinctly I use as before the notation

$$\Sigma_i^s = \frac{i(i+1)}{s^2-1}, \quad \Upsilon_i^s = \frac{(i-1)i(i+1)(i+2)}{s^2-4}$$

and I usually omit the superscript and subscript s and i.

Proceeding in this way I find

$$\left.\begin{aligned} \mathfrak{E}_i^s &= (-)^s \frac{i+s!}{i-s!}\{1 - \tfrac{1}{2}\beta(\Sigma - 1) \\ &\quad + \tfrac{1}{32}\beta^2[-s^2(\Sigma^2 + 2\Sigma - 1) + 3(\Sigma^2 - 2\Sigma + 2) + 2\Upsilon]\} \\ \mathsf{E}_i^s &= (-)^s \frac{i+s!}{i-s!}\{1 + \tfrac{1}{2}\beta(\Sigma + 3) \\ &\quad + \tfrac{1}{32}\beta^2[s^2(3\Sigma^2 - 2\Sigma + 1) - (\Sigma^2 - 26\Sigma - 42) - 2\Upsilon]\} \end{aligned}\right\} \quad (41)$$

These results may be verified, for if we multiply $\mathfrak{E}^s$ by $\dfrac{1}{(\mathsf{C}^s)^2}$, as given in (33), we ought to find E^s; and this is so.

The formulæ apparently fail when $s = 0, 1, 2, 3$; but when $s = 3$ they still hold good because, as remarked above, the general formula for $s = 3$ gives correct results when properly interpreted. Thus it only remains to consider $s = 0, 1, 2$.

When $s = 2$ the coefficients of β remain as in (41). In the coefficients of β^2

$$q_{s-4} = 0, \quad q_{s-2} = -\tfrac{1}{8}\{i, 2\}\{i, 1\}, \quad q_{s+2} = \frac{1}{8 \cdot 3}, \quad q_{s+4} = \frac{1}{128 \cdot 3 \cdot 4}$$

$$q'_{s-4} = 0, \quad q'_{s-2} = 0, \quad q'_{s+2} = \frac{1}{4 \cdot 3}, \quad q'_{s+4} = \frac{1}{128 \cdot 4}$$

In the expression for $\mathfrak{E}^2$ the coefficient of β^2 inside the bracket is

$$\begin{aligned} \frac{1}{128 \cdot 3^2 \cdot 4}[&3(i-2)(i-3)(i-4)(i-5) + 3(i+3)(i+4)(i+5)(i+6) \\ &+ 72(i-1)i(i+1)(i+2) + 8(i-2)(i-3)(i+3)(i+4) \\ &- 24(i+1)(i+2)(i+3)(i+4) - 24(i-3)(i-2)(i-1)i] \end{aligned} \quad (42)$$

Effecting the reduction and writing Σ for $\frac{1}{3}i(i+1)$, we find

$$\mathfrak{E}_i^2 = \frac{i+2!}{i-2!}\{1 - \tfrac{1}{2}\beta(\Sigma - 1) + \tfrac{1}{256}\beta^2(19\Sigma^2 - 130\Sigma + 80)\} \quad \text{......(43)}$$

The coefficient of β^2 for E^2 may be got from (42) thus:—Multiply the first and second terms by 3, erase the third, fifth, and sixth terms, and multiply the fourth term by 4.

Effecting the reduction we find

$$\mathsf{E}_i^2 = \frac{i+2!}{i-2!}\{1 + \tfrac{1}{2}\beta(\Sigma + 3) + \tfrac{1}{256}\beta^2(25\Sigma^2 + 186\Sigma + 368)\} \quad \text{...(44)}$$

Observe that there is no factor by which $\mathfrak{P}^2$ can be converted into P^2, so that this case cannot be verified like the general one.

When $s = 1$ we have

$$q_{s-4} = 0, \quad q_{s-2} = 0, \quad q_{s+2} = \tfrac{1}{16}[1 - \tfrac{1}{16}\beta i(i+1)], \quad q_{s+4} = \tfrac{1}{768}$$
$$q'_{s-4} = 0, \quad q'_{s-2} = 0, \quad q'_{s+2} = \tfrac{3}{16}[1 + \tfrac{1}{16}\beta i(i+1)], \quad q'_{s+4} = \tfrac{5}{768}$$

The terms in βq_{s+2} and $\beta q'_{s+2}$ now contribute to the terms in β^2.

For $\mathfrak{E}^1$ the term in β inside the bracket is

$$\frac{1}{8.2}[(i-1)(i-2)+(i+2)(i+3)] = \tfrac{1}{8}[i(i+1)+4]$$

The term in β^2, of which the first portion is carried over from the term in β, is

$$-\tfrac{1}{256}i(i+1)[(i-1)(i-2)+(i+2)(i+3)]$$
$$+\frac{1}{128.2.3}[(i-1)(i-2)(i-3)(i-4)+(i+2)(i+3)(i+4)(i+5)$$
$$+3(i-1)(i-2)(i+2)(i+3)]$$

This is equal to $-\tfrac{1}{768}[i^2(i+1)^2 - 56i(i+1) - 180]$

As we cannot now use the abridged notation with Σ_i^1, which is infinite, I write

$$\left.\begin{aligned} &j = i(i+1) \\ \text{Thus}\qquad &\mathfrak{E}_i^1 = -\frac{i+1!}{i-1!}[1 + \tfrac{1}{8}\beta(j+4) - \tfrac{1}{768}\beta^2(j^2 - 56j - 180)] \end{aligned}\right\} \ldots\ldots(45)$$

For E_i^1 the coefficient of β is three times as great as before, and the coefficient of β^2 is

$$\tfrac{3}{128}i(i+1)[i(i+1)+4] + \frac{1}{128.2.3}[5(i-1)(i-2)(i-3)(i-4)$$
$$+5(i+2)(i+3)(i+4)(i+5) + 27(i-1)(i-2)(i+2)(i+3)]$$

On effecting the reduction I find

$$\mathsf{E}_i^1 = -\frac{i+1!}{i-1!}[1 + \tfrac{3}{8}\beta(j+4) + \tfrac{1}{768}\beta^2(55j^2 + 376j + 1044)] \ldots(46)$$

When $s = 0$ we have only $\mathfrak{E}_i$ to determine. Here

$$q_{s-4} = q_{s-2} = 0, \quad q_{s+2} = \tfrac{1}{4}, \quad q_{s+4} = \tfrac{1}{128}$$

The term in β is $\tfrac{1}{4}[i(i-1)+(i+1)(i+2)] = \tfrac{1}{2}(j+1)$

That in β^2 is

$$\tfrac{1}{128}[i(i-1)(i-2)(i-3)+(i+1)(i+2)(i+3)(i+4)$$
$$+8i(i-1)(i+1)(i+2)] = \tfrac{1}{64}(5j^2+14j+12)$$

Therefore

$$\left.\begin{aligned}\mathfrak{E}_i &= 1+\tfrac{1}{2}\beta(j+1)+\tfrac{1}{64}\beta^2(5j^2+14j+12)\\ &= 1-\tfrac{1}{2}\beta(\Sigma-1)+\tfrac{1}{64}\beta^2(5\Sigma^2-14\Sigma+12)\end{aligned}\right\}\text{.............}(47)$$

since $\Sigma_i = -i(i+1) = -j$

Collecting results from (41), (43), (44), (45), (46), and (47),

$$\left.\begin{aligned}
(s>2)\ \mathfrak{E}_i^s &= (-)^s\frac{i+s!}{i-s!}\{1-\tfrac{1}{2}\beta(\Sigma-1)\\
&\qquad+\tfrac{1}{32}\beta^2[-s^2(\Sigma^2+2\Sigma-1)+3(\Sigma^2-2\Sigma+2)+2\Upsilon]\}\\
\mathfrak{E}_i^2 &= \frac{i+2!}{i-2!}\{1-\tfrac{1}{2}\beta(\Sigma-1)+\tfrac{1}{256}\beta^2[19\Sigma^2-130\Sigma+80]\}\\
\mathfrak{E}_i^1 &= -\frac{i+1!}{i-1!}\{1+\tfrac{1}{8}\beta(j+4)-\tfrac{1}{768}\beta^2(j^2-56j-180)\}\\
\mathfrak{E}_i &= 1+\tfrac{1}{2}\beta(j+1)+\tfrac{1}{64}\beta^2(5j^2+14j+12)\\
(s>2)\ \mathsf{E}_i^s &= (-)^s\frac{i+s!}{i-s!}\{1+\tfrac{1}{2}\beta(\Sigma+3)\\
&\qquad+\tfrac{1}{32}\beta^2[s^2(3\Sigma^2-2\Sigma+1)-(\Sigma^2-26\Sigma-42)-2\Upsilon]\}\\
\mathsf{E}_i^2 &= \frac{i+2!}{i-2!}\{1+\tfrac{1}{2}\beta(\Sigma+3)+\tfrac{1}{256}\beta^2[25\Sigma^2+186\Sigma+368]\}\\
\mathsf{E}_i^1 &= -\frac{i+1!}{i-1!}\{1+\tfrac{3}{8}\beta(j+4)+\tfrac{1}{768}\beta^2(55j^2+376j+1044)\}\\
\text{where}\quad \Sigma &= \frac{i(i+1)}{s^2-1},\quad \Upsilon=\frac{(i-1)i(i+1)(i+2)}{s^2-4},\quad j=i(i+1)
\end{aligned}\right\}(48)$$

PART II.

Application of Ellipsoidal Harmonic Analysis.

§ 11. *The Potential of an harmonic deformation of an Ellipsoid.*

A solid harmonic, or solution of Laplace's equation, is the product of two P-functions of ν and of μ respectively, and of a cosine or sine function of ϕ. A surface harmonic is a P-function of μ multiplied by a cosine or sine function of ϕ.

We found

$$\mathfrak{P}^s(\nu) = \mathrm{P}^s(\nu) + \Sigma\beta^n q_{s-2n}\mathrm{P}^{s-2n}(\nu) + \Sigma\beta^n q_{s+2n}\mathrm{P}^{s+2n}(\nu)$$

where $\mathrm{P}^t(\nu) = \dfrac{(\nu^2-1)^{\frac{1}{2}t}}{2^i \,.\, i!}\left(\dfrac{d}{d\nu}\right)^{i+t}(\nu^2-1)^i$; and a similar formula held for $\mathsf{P}^s(\nu)$.

Hitherto we have supposed $\mathrm{P}^t(\mu)$ to have exactly the same form as $\mathrm{P}^t(\nu)$. But since μ is less than unity this introduces an imaginary factor when t is odd, and makes the succession of P's alternately positive and negative when t is even. As this is practically inconvenient I now define

$$\mathrm{P}^t(\mu) = \frac{(1-\mu^2)^{\frac{1}{2}t}}{2^i i!}\left(\frac{d}{d\mu}\right)^{i+t}(\mu^2-1)^i$$

and then retaining the former meaning for the q coefficients, we give the following definition—

$$\mathfrak{P}^s(\mu) = \mathrm{P}^s(\mu) + \Sigma(-)^n \beta^n q_{s-2n} \mathrm{P}^{s-2n}(\mu) + \Sigma(-)^n \beta^n q_{s+2n} \mathrm{P}^{s+2n}(\mu)$$

with a similar formula for $\mathrm{P}^s(\mu)$.

Thus we need only remark that in the functions of μ the q's corresponding to odd powers of β enter with the opposite sign from that which holds in the functions of ν, and the whole of our preceding results are true with this definition of $\mathrm{P}^t(\mu)$.

If ν_0 defines the ellipsoid to which the surface harmonic applies, we require the expression for the perpendicular p on the tangent plane at ν_0, μ, ϕ, and that for an element of area of the surface of the ellipsoid at the same point.

By the usual formula

$$\frac{k^2}{p^2} = \frac{x^2}{k^2\left(\nu_0^2 - \frac{1+\beta}{1-\beta}\right)^2} + \frac{y^2}{k^2(\nu_0^2-1)^2} + \frac{z^2}{k^2\nu_0^4}$$

$$= -\frac{(1-\beta)\left(\mu^2 - \frac{1+\beta}{1-\beta}\right)}{(1+\beta)\left(\nu_0^2 - \frac{1+\beta}{1-\beta}\right)}\cos^2\phi - \frac{\mu^2-1}{\nu_0^2-1}\sin^2\phi + \frac{\mu^2(1-\beta\cos 2\phi)}{\nu_0^2(1+\beta)}$$

$$= \frac{(\nu_0^2-\mu^2)\left(\nu_0^2 - \frac{1-\beta\cos 2\phi}{1-\beta}\right)}{\nu_0^2(\nu_0^2-1)\left(\nu_0^2 - \frac{1+\beta}{1-\beta}\right)} \quad \text{.......................................(49)}$$

Let dn, dm, df be the three elements of the orthogonal arcs corresponding to variations of ν, μ, ϕ respectively.

Then by the formula at the end of § 1,

$$\left.\begin{aligned}\left(\frac{dn}{k^2\nu_0 d\nu_0}\right)^2 &= \frac{(\nu_0^2-\mu^2)\left(\nu_0^2 - \frac{1-\beta\cos 2\phi}{1-\beta}\right)}{k^2\nu_0^2(\nu_0^2-1)\left(\nu_0^2 - \frac{1+\beta}{1-\beta}\right)} = \frac{1}{p^2} \\ \left(\frac{dm}{k^2 d\mu}\right)^2 &= \frac{(\nu_0^2-\mu^2)\left(\frac{1-\beta\cos 2\phi}{1-\beta} - \mu^2\right)}{k^2(1-\mu^2)\left(\frac{1+\beta}{1-\beta} - \mu^2\right)}\end{aligned}\right\} \quad \text{.............(50)}$$

$$\left(\frac{df}{\frac{k\beta\sin 2\phi}{1-\beta}d\phi}\right)^2 = \frac{\left(\nu_0^2 - \frac{1-\beta\cos 2\phi}{1-\beta}\right)\left(\frac{1-\beta\cos 2\phi}{1-\beta} - \mu^2\right)}{\frac{\beta^2\sin^2 2\phi}{(1-\beta)^3}(1-\beta\cos 2\phi)}$$

Therefore

$$\left(\frac{df}{kd\phi}\right)^2 = \frac{\left(\nu_0^2 - \frac{1-\beta\cos 2\phi}{1-\beta}\right)\left(\frac{1-\beta\cos 2\phi}{1-\beta} - \mu^2\right)(1-\beta)}{1-\beta\cos 2\phi} \quad \text{.....(50)}$$

and

$$\left.\begin{aligned}\left(\frac{p\,dm\,df}{k^3 d\mu d\phi}\right)^2 &= \nu_0^2(\nu_0^2-1)\left(\nu_0^2-\tfrac{1+\beta}{1-\beta}\right)(1-\beta)\frac{\left(\frac{1-\beta\cos 2\phi}{1-\beta}-\mu^2\right)^2}{(1-\beta\cos 2\phi)(1-\mu^2)\left(\frac{1+\beta}{1-\beta}-\mu^2\right)}\\ \frac{d}{dn} &= \frac{p}{k^2}\cdot\frac{d}{\nu_0 d\nu_0}\end{aligned}\right\} \quad \text{.........(50)}$$

Two functions, written in alternative form,

$$A\mathfrak{P}_i^s\left(\begin{Bmatrix}\nu\\ \nu_0\end{Bmatrix}\right)\mathfrak{Q}_i^s\left(\begin{Bmatrix}\nu_0\\ \nu\end{Bmatrix}\right)\mathfrak{P}_i^s(\mu)\,\mathfrak{C}_i^s(\phi)$$

are solutions of Laplace's equation, and together form a function V continuous at the surface of the ellipsoid $\nu=\nu_0$. Reading the upper line we have a function always finite inside the ellipsoid, and reading the lower line one always finite outside. Hence V is the potential of a layer of surface density on the ellipsoid ν_0, and by Poisson's equation that density is equal to

$$-\frac{1}{4\pi}\left[\frac{dV}{dn}\,(\text{outside})-\frac{dV}{dn}\,(\text{inside})\right]$$

Let the surface density, which it is our object to find, be

$$p\mathfrak{P}_i^s(\mu)\,\mathfrak{C}_i^s(\phi)\,.\,\rho$$

a surface harmonic multiplied by the perpendicular on to the tangent plane and by a quantity ρ.

Then since $\dfrac{d}{dn}=\dfrac{p}{k^2}\dfrac{d}{\nu_0 d\nu_0}$

$$\rho=-\frac{A}{4\pi k^2\nu_0}\left[\mathfrak{P}_i^s(\nu_0)\frac{d}{d\nu_0}\mathfrak{Q}_i^s(\nu_0)-\mathfrak{Q}_i^s(\nu_0)\frac{d}{d\nu_0}\mathfrak{P}_i^s(\nu_0)\right]$$

But $$\mathfrak{Q}_i^s(\nu_0)=\mathfrak{E}_i^s\,\mathfrak{P}_i^s(\nu_0)\int_{\nu_0}^{\infty}\frac{d\nu}{[\mathfrak{P}_i^s(\nu)]^2(\nu^2-1)^{\frac12}\left(\nu^2-\frac{1+\beta}{1-\beta}\right)^{\frac12}}$$

Differentiating this logarithmically we find

$$\rho=\frac{A\mathfrak{E}_i^s}{4\pi k^2\nu_0(\nu_0^2-1)^{\frac12}\left(\nu_0^2-\frac{1+\beta}{1-\beta}\right)^{\frac12}},\ \text{a constant}$$

Hence surface density $p\mathfrak{P}_i^s(\mu)\,\mathfrak{C}_i^s(\phi)\,.\,\rho$, where ρ is constant, gives rise to potential

$$\begin{Bmatrix}\text{inside}\\ \text{outside}\end{Bmatrix}\ \frac{4\pi k^2\rho\nu_0}{\mathfrak{E}_i^s}(\nu_0^2-1)^{\frac12}\left(\nu_0^2-\tfrac{1+\beta}{1-\beta}\right)^{\frac12}\mathfrak{P}_i^s\left(\begin{Bmatrix}\nu\\ \nu_0\end{Bmatrix}\right)\mathfrak{Q}_i^s\left(\begin{Bmatrix}\nu_0\\ \nu\end{Bmatrix}\right)\mathfrak{P}_i^s(\mu)\,\mathfrak{C}_i^s(\phi) \quad \text{.........(51)}$$

The same investigation holds good with $\mathfrak{S}_i^s(\phi)$, or with P, Q, C, S in place of the corresponding letters above.

Imagine that the surface of a homogeneous ellipsoid of density ρ, defined by ν_0, receives a normal displacement δn, such that

$$\delta n = p \,.\, \epsilon \mathfrak{P}_i^s(\mu)\, \mathfrak{C}_i^s(\phi)$$

Then the equivalent surface density is $p\,.\,\epsilon\rho \mathfrak{P}_i^s(\mu)\, \mathfrak{C}_i^s(\phi)$, and we can at once write down the expressions for the internal and external potentials by means of (51).

If x_0, y_0, z_0 be the co-ordinates of a point on the surface, it is clear that the co-ordinates of the corresponding point on the deformed surface are

$$x = x_0\left(1 + \frac{p^2 \epsilon \mathfrak{P}_i^s(\mu)\, \mathfrak{C}_i^s(\phi)}{k^2\left(\nu_0^2 - \frac{1+\beta}{1-\beta}\right)}\right), \qquad y = y_0\left(1 + \frac{p^2 \epsilon \mathfrak{P}_i^s(\mu)\, \mathfrak{C}_i^s(\phi)}{k^2(\nu_0^2 - 1)}\right)$$

$$z = z_0\left(1 + \frac{p^2 \epsilon \mathfrak{P}_i^s(\mu)\, \mathfrak{C}_i^s(\phi)}{k^2\nu^2}\right)$$

Hence the equation to the deformed surface is

$$\frac{x^2}{k^2\left(\nu_0^2 - \frac{1+\beta}{1-\beta}\right)} + \frac{y^2}{k^2(\nu_0^2 - 1)} + \frac{z^2}{k^2\nu_0^2} = 1 + 2\epsilon \mathfrak{P}_i^s(\mu)\, \mathfrak{C}_i^s(\phi) \quad \text{......(52)}$$

or since

$$\frac{x^2}{k^2\left(\nu^2 - \frac{1+\beta}{1-\beta}\right)} + \frac{y^2}{k^2(\nu^2 - 1)} + \frac{z^2}{k^2\nu^2} = 1$$

it may be written

$$(\nu^2 - \nu_0^2)\frac{k^2}{p^2} = 2\epsilon \mathfrak{P}_i^s(\mu)\, \mathfrak{C}_i^s(\phi)$$

If we substitute for $\frac{k^2}{p^2}$ its value from (49), this may be written in the form

$$\frac{(\nu^2 - \nu_0^2)(\mu^2 - \nu_0^2)\left(\frac{1-\beta\cos 2\phi}{1-\beta} - \nu_0^2\right)}{\nu_0^2(\nu_0^2 - 1)\left(\nu_0^2 - \frac{1+\beta}{1-\beta}\right)} = 2\epsilon \mathfrak{P}_i^s(\mu)\, \mathfrak{C}_i^s(\phi) \quad \text{........(52)}$$

This is the equation in elliptic co-ordinates to the deformed surface, but in actual computation the form involving rectangular co-ordinates might perhaps be more convenient.

§ 12. *The Potential of a homogeneous solid Ellipsoid.*

It is well known that the potential of a solid ellipsoid externally is equal to that of a "focaloid" shell of the same mass coincident with its external surface.

If ρ' be the density of the shell defined by ν_0 and $\nu_0 + \delta\nu$, we have

$$\tfrac{4}{3}\pi k^3 \rho' \left[\left(\nu_0^2 + 2\nu_0\delta\nu - \tfrac{1+\beta}{1-\beta}\right)^{\frac{1}{2}}(\nu_0^2 + 2\nu_0\delta\nu - 1)^{\frac{1}{2}}(\nu_0^2 + 2\nu_0\delta\nu)^{\frac{1}{2}} - \left(\nu_0^2 - \tfrac{1+\beta}{1-\beta}\right)^{\frac{1}{2}}(\nu_0^2 - 1)^{\frac{1}{2}}\nu_0\right] = \tfrac{4}{3}\pi k^3 \rho \left(\nu_0^2 - \tfrac{1+\beta}{1-\beta}\right)^{\frac{1}{2}}(\nu_0^2 - 1)^{\frac{1}{2}}\nu_0$$

Therefore $$\rho' \nu_0 \delta\nu \left(\frac{1}{\nu_0^2 - \frac{1+\beta}{1-\beta}} + \frac{1}{\nu_0^2 - 1} + \frac{1}{\nu_0^2}\right) = \rho$$

or $$\rho' \nu_0 \delta\nu = \frac{\rho\nu_0^2 (\nu_0^2 - 1)\left(\nu_0^2 - \frac{1+\beta}{1-\beta}\right)}{3\nu_0^4 - \frac{4\nu_0^2}{1-\beta} + \frac{1+\beta}{1-\beta}} \quad \text{......................(53)}$$

If δn be the thickness of the shell at the point where p is the perpendicular on the tangent plane,

$$\delta n = \nu_0 \delta\nu \cdot \frac{k^2}{p}$$

If we multiply both sides of (53) by $\frac{k^2}{p}$, we see that the surface density of the focaloid shell is

$$p \cdot \frac{\rho\nu_0^2 (\nu_0^2 - 1)\left(\nu_0^2 - \frac{1+\beta}{1-\beta}\right)}{3\nu_0^4 - \frac{4\nu_0^2}{1-\beta} + \frac{1+\beta}{1-\beta}} \cdot \frac{k^2}{p^2}$$

If therefore we can express $\frac{k^2}{p^2}$ in the form of surface harmonics, it will be easy to write down the external potential of the ellipsoid by means of the formula (51).

Before doing this I will, however, take one other step.

It is easy to see that

$$3\nu_0^4 - \frac{4\nu_0^2}{1-\beta} + \frac{1+\beta}{1-\beta} = 3\left(\nu_0^2 - \frac{2+B}{3(1-\beta)}\right)\left(\nu_0^2 - \frac{2-B}{3(1-\beta)}\right) \quad \text{......(54)}$$

where for brevity $B = (1+3\beta^2)^{\frac{1}{2}}$.

Now on referring to § 7, (17) and (23), we see that

$$\mathfrak{P}_2(\nu) = \mathrm{P}_2(\nu) + \frac{B-1}{6\beta}\mathrm{P}_2^2(\nu), \qquad \mathfrak{P}_2^2(\nu) = -\frac{2(B-1)}{\beta}\mathrm{P}_2(\nu) + \mathrm{P}_2^2(\nu)$$

where $$\mathrm{P}_2(\nu) = \tfrac{3}{2}\nu^2 - \tfrac{1}{2}, \qquad \mathrm{P}_2^2(\nu) = 3(\nu^2 - 1)$$

If then we put $$\mathfrak{P}_2(\nu) = \alpha\nu^2 + \gamma$$

$$\mathfrak{P}_2^2(\nu) = -\alpha'\nu^2 - \gamma', \qquad \text{or} \qquad \mathfrak{P}_2^2(\mu) = \alpha'\mu^2 + \gamma'$$

it is clear that $$\alpha = \frac{B-1+3\beta}{2\beta}, \qquad \gamma = \frac{-B+1-\beta}{2\beta}$$

$$\alpha' = \frac{3(B-1-\beta)}{\beta}, \qquad \gamma' = \frac{-B+1+3\beta}{\beta}$$

and $$\frac{\gamma}{\alpha} = \frac{B-2}{3(1-\beta)}, \qquad \frac{\gamma'}{\alpha'} = \frac{-B-2}{3(1-\beta)}$$

It is obvious then that our expression (54) is equal to $-\frac{3}{\alpha\alpha'}\mathfrak{P}_2(\nu_0)\mathfrak{P}_2^2(\nu_0)$.

Then since $-\frac{3}{\alpha\alpha'} = \frac{1+B}{3(1-\beta)}$, we have the surface density of the focaloid given by

$$p \cdot \frac{3(1+\beta)}{1+B} \cdot \frac{\rho\nu_0^2(\nu_0^2-1)\left(\nu_0^2-\frac{1+\beta}{1-\beta}\right)}{\mathfrak{P}_2(\nu_0)\,\mathfrak{P}_2^2(\nu_0)} \cdot \left(\frac{k}{p}\right)^2$$

where
$$\frac{k^2}{p^2} = \frac{x_0^2}{k^2\left(\nu_0^2-\frac{1+\beta}{1-\beta}\right)^2} + \frac{y_0^2}{k^2(\nu_0^2-1)^2} + \frac{z_0^2}{k^2\nu_0^4}$$

But since
$$\frac{x_0^2}{k^2\left(\nu_0^2-\frac{1+\beta}{1-\beta}\right)} + \frac{y_0^2}{k^2(\nu_0^2-1)} + \frac{z_0^2}{k^2\nu_0^2} = 1 \qquad (55)$$

we have
$$\frac{k^2}{p^2} = \frac{1}{\nu_0^2} + \frac{\frac{1+\beta}{1-\beta}\cdot x_0^2}{k^2\nu_0^2\left(\nu_0^2-\frac{1+\beta}{1-\beta}\right)^2} + \frac{y_0^2}{k^2\nu_0^2(\nu_0^2-1)^2} \qquad (56)$$

With the object of writing this function in surface harmonics, and besides to enable us to express a rotation potential in similar form, we have to reduce x^2, y^2, z^2 in the required manner.

I now drop the suffix zero, since we are not concerned with any particular ellipsoid.

Referring again to § 7, (18) and (24), we have

$$\mathfrak{C}_2(\phi) = 1 - \frac{B-1}{\beta}\cos 2\phi, \qquad \mathfrak{C}_2^2(\phi) = \frac{B-1}{3\beta} + \cos 2\phi$$

If then we put
$$\epsilon = \frac{2(1-B)}{\beta}, \qquad \zeta = \frac{B-1+\beta}{\beta}$$

$$\epsilon' = 2, \qquad \zeta' = \frac{B-1-3\beta}{3\beta}$$

we may write $\mathfrak{C}_2(\phi) = \epsilon\cos^2\phi + \zeta$, $\quad \mathfrak{C}_2^2(\phi) = \epsilon'\cos^2\phi + \zeta'$

Let us assume, if possible,

$$\frac{-x^2}{k^2\left(\frac{1-\beta}{1+\beta}\right)\left(\nu^2-\frac{1+\beta}{1-\beta}\right)} = F\mathfrak{P}_2(\mu)\,\mathfrak{C}_2(\phi) + G\mathfrak{P}_2^2(\mu)\,\mathfrak{C}_2^2(\phi) + H$$

or

$$\left(\mu^2-\frac{1+\beta}{1-\beta}\right)\cos^2\phi = F(\alpha\mu^2+\gamma)(\epsilon\cos^2\phi+\zeta) + G(\alpha'\mu^2+\gamma')(\epsilon'\cos^2\phi+\zeta') + H$$

From which it follows that

$$F\alpha\zeta + G\alpha'\zeta' = 0, \qquad F\gamma\zeta + G\gamma'\zeta' + H = 0$$

$$F\alpha\epsilon + G\alpha'\epsilon' = 1, \qquad F\gamma\epsilon + G\gamma'\epsilon' = -\frac{1+\beta}{1-\beta}$$

These equations give

$$F = \frac{1}{\alpha\zeta}\cdot\frac{1}{\epsilon/\zeta - \epsilon'/\zeta'}, \qquad G = -\frac{1}{\alpha'\zeta'}\cdot\frac{1}{\epsilon/\zeta-\epsilon'/\zeta'}, \qquad H = -\frac{\gamma/\alpha - \gamma'/\alpha'}{\epsilon/\zeta - \epsilon'/\zeta'}$$

and the condition

$$(1-\beta)\left(\frac{\gamma\epsilon}{\alpha\zeta}-\frac{\gamma'\epsilon'}{\alpha'\zeta'}\right)+(1+\beta)\left(\frac{\epsilon}{\zeta}-\frac{\epsilon'}{\zeta'}\right)=0$$

Now $\qquad \dfrac{\gamma\epsilon}{\alpha\zeta}=\dfrac{1+\beta-B}{1-\beta}, \qquad \dfrac{\gamma'\epsilon'}{\alpha'\zeta'}=\dfrac{1+\beta+B}{1-\beta}$

$$\frac{\epsilon}{\zeta}=\frac{-1-3\beta+B}{1+\beta}, \qquad \frac{\epsilon'}{\zeta'}=\frac{-1-3\beta-B}{1+\beta}$$

Since these values satisfy the condition amongst the coefficients, the assumed form for x^2 is justifiable.

I find then

$$F=\frac{1+B}{2B}\cdot\frac{B-2\beta}{3(1-\beta)}, \qquad G=-\frac{1+B}{4B}\cdot\frac{B+2\beta}{3(1-\beta)}, \qquad H=-\frac{1+\beta}{3(1-\beta)}$$

Whence

$$\frac{3x^2}{k^2\left(\nu^2-\frac{1+\beta}{1-\beta}\right)}=-\frac{1+B}{2B}\cdot\frac{B-2\beta}{1+\beta}\mathfrak{P}_2(\mu)\mathfrak{C}_2(\phi)$$
$$+\frac{1+B}{4B}\cdot\frac{B+2\beta}{1+\beta}\mathfrak{P}_2^2(\mu)\mathfrak{C}_2^2(\phi)+1 \;\ldots\ldots\ldots\ldots(57)$$

This is the required expression for x^2 in surface harmonics.

Next assume

$$\frac{-y^2}{k^2(\nu^2-1)}=F_1\mathfrak{P}_2(\mu)\mathfrak{C}_2(\phi)+G_1\mathfrak{P}_2^2(\mu)\mathfrak{C}_2^2(\phi)+H_1$$

If we put $\quad \mathfrak{C}_2(\phi)=\epsilon_1\sin^2\phi+\zeta_1, \qquad \mathfrak{C}_2^2(\phi)=\epsilon_1'\sin^2\phi+\zeta_1'$

we have $\quad \epsilon_1=\dfrac{2(B-1)}{\beta}, \qquad \zeta_1=\dfrac{-B+1+\beta}{\beta}$

$$\epsilon_1'=-2, \qquad \zeta_1'=\frac{B-1+3\beta}{3\beta}$$

and

$$(\mu^2-1)\sin^2\phi=F_1(\alpha\mu^2+\gamma)(\epsilon_1\sin^2\phi+\zeta_1)+G_1(\alpha'\mu^2+\gamma')(\epsilon_1'\sin^2\phi+\zeta_1')+H_1$$

Whence F_1, G_1, H_1 have the same forms as before, and the condition to be satisfied by the coefficients is

$$\frac{\gamma\epsilon_1}{\alpha\zeta_1}-\frac{\gamma'\epsilon_1'}{\alpha'\zeta_1'}+\frac{\epsilon_1}{\zeta_1}-\frac{\epsilon_1'}{\zeta_1'}=0$$

It will be found that the condition is satisfied, and that

$$F_1=\frac{1+B}{6B}, \qquad G_1=\frac{1+B}{12B}, \qquad H_1=-\tfrac{1}{3}$$

Hence

$$\frac{3y^2}{k^2(\nu^2-1)}=-\frac{1+B}{2B}\mathfrak{P}_2(\mu)\mathfrak{C}_2(\phi)-\frac{1+B}{4B}\mathfrak{P}_2^2(\mu)\mathfrak{C}_2^2(\phi)+1 \ldots\ldots(58)$$

It follows from (55) that

$$\frac{3z^2}{k^2\nu^2} = \frac{1+B}{2B}\left(\frac{B-2\beta}{1+\beta}+1\right)\mathfrak{P}_2(\mu)\mathfrak{C}_2(\phi) - \frac{1+B}{4B}\left(\frac{B+2\beta}{1+\beta}-1\right)\mathfrak{P}_2^2(\mu)\mathfrak{C}_2^2(\phi)+1 \quad \text{.........(59)}$$

Whence

$$\frac{3}{k^2}(y^2+z^2) = \frac{1+B}{2B}\left(1+\nu^2\frac{B-2\beta}{1+\beta}\right)\mathfrak{P}_2(\mu)\,\mathfrak{C}_2(\phi)$$

$$+\frac{1+B}{4B}\left(1-\nu^2\frac{B+2\beta}{1+\beta}\right)\mathfrak{P}_2(\mu)\,\mathfrak{C}_2(\phi)+2\nu^2-1 \quad \text{............(60)}$$

This is needed to express the rotation potential $\frac{1}{2}\omega^2(y^2+z^2)$. If we add $\frac{3x^2}{k^2}$ to this we have

$$\frac{3}{k^2}(x^2+y^2+z^2) = \frac{1+B}{2B}\cdot\frac{B+1-3\beta}{1-\beta}\mathfrak{P}_2(\mu)\,\mathfrak{C}_2(\phi)$$

$$-\frac{1+B}{4B}\cdot\frac{B-1+3\beta}{1-\beta}\mathfrak{P}_2^2(\mu)\,\mathfrak{C}_2^2(\phi)+3\nu^2-\frac{2}{1-\beta} \quad \text{.........(61)}$$

This expression will be needed hereafter.

Returning now to the formation of the expression for k^2/p^2, I find

$$\frac{3k^2}{p^2} = \frac{1}{\nu^2(\nu^2-1)\left(\nu^2-\frac{1+\beta}{1-\beta}\right)}\left[-\frac{1+B}{2B}\cdot\frac{\nu^2(B+1-3\beta)-(B+1-\beta)}{1-\beta}\mathfrak{P}_2(\mu)\,\mathfrak{C}_2(\phi)\right.$$

$$+\frac{1+B}{4B}\cdot\frac{\nu^2(B-1+3\beta)-(B-1+\beta)}{1-\beta}\mathfrak{P}_2^2(\mu)\,\mathfrak{C}_2^2(\phi)$$

$$\left.+3\nu^4-\frac{4\nu^2}{1-\beta}+\frac{1+\beta}{1-\beta}\right]$$

On considering the forms of the functions $\mathfrak{P}_2(\nu)$, $\mathfrak{P}_2^2(\nu)$, it is found that this result may be written thus:

$$\frac{k^2}{p^2} = \frac{\mathfrak{P}_2(\nu)\,\mathfrak{P}_2^2(\nu)}{3\nu^2(\nu^2-1)\left(\nu^2-\frac{1+\beta}{1-\beta}\right)}\cdot\frac{1+B}{3(1-\beta)}$$

$$\times\left[-\frac{1+B}{2B}\frac{\mathfrak{P}_2(\mu)\,\mathfrak{C}_2(\phi)}{\mathfrak{P}_2(\nu)}+\frac{3\beta}{2B}\frac{\mathfrak{P}_2^2(\mu)\,\mathfrak{C}_2^2(\phi)}{\mathfrak{P}_2^2(\nu)}+1\right]$$

Therefore, writing $\mathfrak{P}_0(\mu)\,\mathfrak{C}_0(\phi)$ for unity, the surface density of the focaloid shell, for which $\nu=\nu_0$, is

$$p\rho\left[-\frac{1+B}{6B}\frac{\mathfrak{P}_2(\mu)\,\mathfrak{C}_2(\phi)}{\mathfrak{P}_2(\nu_0)}+\frac{\beta}{2B}\frac{\mathfrak{P}_2^2(\mu)\,\mathfrak{C}_2^2(\phi)}{\mathfrak{P}_2^2(\nu_0)}+\tfrac{1}{3}\mathfrak{P}_0(\mu)\mathfrak{C}_0(\phi)\right] \quad \text{...(62)}$$

By means of (51), we now at once write down the external potential of the ellipsoid. It is

$$\mathrm{V} = \frac{\mathrm{M}_0}{k}\left[-\frac{1+B}{2B}\frac{\mathfrak{Q}_2(\nu)}{\mathfrak{E}_2}\mathfrak{P}_2(\mu)\,\mathfrak{C}_2(\phi)+\frac{3\beta}{2B}\frac{\mathfrak{Q}_2^2(\nu)}{\mathfrak{E}_2^2}\mathfrak{P}_2^2(\mu)\,\mathfrak{C}_2^2(\phi)+\frac{\mathfrak{Q}_0(\nu)}{\mathfrak{E}_0}\right] \quad \text{.........(63)}$$

In this expression M_0 denotes the mass of the ellipsoid, and the $\mathfrak{C}$'s are merely coefficients determined approximately in § 10.

In order to find the potential internally, let

$$r^2 = x^2 + y^2 + z^2$$

and, as suggested by the form of (61), let

$$\frac{r_0^2}{k^2} = \frac{1+B}{6B} \cdot \frac{B+1-3\beta}{1-\beta} \frac{\mathfrak{P}_2(\nu)}{\mathfrak{P}_2(\nu_0)} \mathfrak{P}_2(\mu)\, \mathfrak{C}_2(\phi)$$
$$- \frac{1+B}{12B} \cdot \frac{B-1+3\beta}{1-\beta} \frac{\mathfrak{P}_2^2(\nu)}{\mathfrak{P}_2^2(\nu_0)} \mathfrak{P}_2^2(\mu)\, \mathfrak{C}_2^2(\phi) + \nu_0^2 - \frac{2}{3(1-\beta)}$$

Then r_0^2 is a solution of Laplace's equation throughout the interior of the ellipsoid, and at the surface, where $\nu = \nu_0$, it is equal to $x^2 + y^2 + z^2$.

Now consider the function

$$V = -\tfrac{2}{3}\pi\rho (r^2 - r_0^2) + \frac{M_0}{k}\left[-\frac{1+B}{2B} \frac{\mathfrak{Q}_2(\nu_0)}{\mathfrak{C}_2 \mathfrak{P}_2(\nu_0)} \mathfrak{P}_2(\nu)\, \mathfrak{P}_2(\mu)\, \mathfrak{C}_2(\phi) \right.$$
$$\left. + \frac{3\beta}{2B} \frac{\mathfrak{Q}_2^2(\nu_0)}{\mathfrak{C}_2^2 \mathfrak{P}_2^2(\nu_0)} \mathfrak{P}_2^2(\nu)\, \mathfrak{P}_2^2(\mu)\, \mathfrak{C}_2^2(\phi) + \frac{\mathfrak{Q}_0(\nu_0)}{\mathfrak{C}_0} \right] \quad \text{.........(64)}$$

The whole of it, excepting the term in r^2, is a solution of Laplace's equation for space inside the ellipsoid, and the term in r^2 gives $\nabla^2 V = -4\pi\rho$. Also at the surface, where $\nu = \nu_0$, this expression agrees with (63). Hence we have found the potential of the ellipsoid internally.

The potential at an internal point does not lend itself to expression in elliptic co-ordinates, but it may be given another form which is perhaps more convenient.

In our present notation the well-known formula is

$$V = \tfrac{3}{2}\frac{M_0}{k}\int_{\nu_0}^{\infty}\left(1 - \frac{x^2}{k^2\left(\nu^2 - \frac{1+\beta}{1-\beta}\right)} - \frac{y^2}{k^2(\nu^2-1)} - \frac{z^2}{k^2\nu^2}\right)\frac{d\nu}{\left(\nu^2 - \frac{1+\beta}{1-\beta}\right)^{\frac{1}{2}}(\nu^2-1)^{\frac{1}{2}}}$$

Since $\mathfrak{P}_0(\nu) = 1$, $\mathsf{P}_1^1(\nu) = \left(\nu^2 - \frac{1+\beta}{1-\beta}\right)^{\frac{1}{2}}$, $\mathfrak{P}_1^1(\nu) = (\nu^2-1)^{\frac{1}{2}}$, $\mathfrak{P}_1(\nu) = \nu$, the integrals may be expressed in terms of the Q-functions, and we have (omitting the divisors $\mathfrak{C}$ and E for brevity)

$$V = \tfrac{3}{2}\frac{M_0}{k}\left(\frac{\mathfrak{Q}_0(\nu_0)}{\mathfrak{P}_0(\nu_0)} - \frac{x^2}{k^2}\frac{\mathsf{Q}_1^1(\nu_0)}{\mathsf{P}_1^1(\nu_0)} - \frac{y^2}{k^2}\frac{\mathfrak{Q}_1^1(\nu_0)}{\mathfrak{P}_1^1(\nu_0)} - \frac{z^2}{k^2}\frac{\mathfrak{Q}_1(\nu_0)}{\mathfrak{P}_1(\nu_0)}\right) \quad \text{..... (65)}$$

In this we may substitute the expressions for x^2, y^2, z^2 found above.

It may be worth noting that

$$\frac{\mathsf{Q}_1^1(\nu)}{\mathsf{P}_1^1(\nu)} + \frac{\mathfrak{Q}_1^1(\nu)}{\mathfrak{P}_1^1(\nu)} + \frac{\mathfrak{Q}_1(\nu)}{\mathfrak{P}_1(\nu)} = \frac{1}{\nu(\nu^2-1)^{\frac{1}{2}}\left(\nu^2 - \frac{1+\beta}{1-\beta}\right)^{\frac{1}{2}}}$$

Also

$$\mathsf{P}_1^1(\nu)\,\mathsf{Q}_1^1(\nu) + \mathfrak{P}_1^1(\nu)\,\mathfrak{Q}_1^1(\nu) + \mathfrak{P}_1(\nu)\,\mathfrak{Q}_1(\nu) + \mathfrak{P}_0(\nu)\,\mathfrak{Q}_0(\nu) = 0$$

This last follows from the fact that if a, b, c are the axes of the ellipsoid, and if Ψ denotes the function $\int_0^\infty \frac{du}{ABC}$ (proportional to our $\mathfrak{Q}_0(\nu_0)$), Ψ is a homogeneous function of degree -1 in a, b, c, and therefore

$$a\frac{d\Psi}{da} + b\frac{d\Psi}{db} + c\frac{d\Psi}{dc} = -\Psi$$

§ 13. *Preparation for the Integration of the square of a surface harmonic over the Ellipsoid.*

If it is intended to express any function in harmonics, it is necessary to know the integrals over the surface of the ellipsoid of the squares of surface harmonics multiplied by the perpendicular on the tangent plane.

The surface harmonic has one of the eight forms

$$V_i^s = [\mathfrak{P}_i^s(\mu) \text{ or } \mathsf{P}_i^s(\mu)] \times \begin{bmatrix} \mathfrak{C}_i^s(\phi) & \text{or} & \mathsf{C}_i^s(\phi) \\ \mathfrak{S}_i^s(\phi) & & \mathsf{S}_i^s(\phi) \end{bmatrix}$$

and the P-functions are expressible in terms of the P's where

$$P_i^s(\mu) = \frac{(1-\mu^2)^{\frac{1}{2}s}}{2^i \cdot i\,!}\left(\frac{d}{d\mu}\right)^{i+s}(\mu^2-1)^i$$

I shall in this portion of the investigation frequently write $\mu = \sin\theta$, and shall omit the μ or θ or ϕ in the P-, C-, S-functions. Also I may very generally omit the subscript i, as elsewhere.

If $d\sigma$ denotes the element of surface of the ellipsoid, and

$$M = k^3\nu(\nu^2-1)^{\frac{1}{2}}\left(\nu^2 - \tfrac{1+\beta}{1-\beta}\right)^{\frac{1}{2}}$$

so that $\frac{4}{3}\pi M$ is the volume of the ellipsoid, we have, by (50) of § 11,

$$\frac{pd\sigma}{d\theta\, d\phi} = \frac{M(1-\beta)^{\frac{1}{2}}}{(1-\beta\cos 2\phi)^{\frac{1}{2}}} \frac{\left(\cos^2\theta + \frac{\beta}{1-\beta} - \frac{\beta\cos 2\phi}{1-\beta}\right)}{\left(\frac{1+\beta}{1-\beta} - \sin^2\theta\right)^{\frac{1}{2}}}$$

Then

$$\int p(V_i^s)^2 d\sigma = M(1-\beta)^{\frac{1}{2}}\iint \frac{\cos^2\theta + \frac{\beta}{1-\beta} - \frac{\beta\cos 2\phi}{1-\beta}}{(1-\beta\cos 2\phi)^{\frac{1}{2}}\left(\frac{1+\beta}{1-\beta} - \sin^2\theta\right)^{\frac{1}{2}}}(V_i^s)^2\, d\theta\, d\phi$$

where the limits of θ are $\frac{1}{2}\pi$ to $-\frac{1}{2}\pi$, and of ϕ are 2π to 0.

It will be legitimate to develop $pd\sigma$ in powers of $\sec^2\theta$ up to any given power, provided $(V_i^s)^2$ involves as a factor such a power of $\cos^2\theta$ that the whole function to be integrated does not become infinite at the poles where $\theta = \pm\frac{1}{2}\pi$.

I shall at present limit the developments to the square of β.

We know that P^s is of the same form as $\mathfrak{P}^s$, but with the additional factor

$$\left(\frac{\frac{1+\beta}{1-\beta}-\sin^2\theta}{\cos^2\theta}\right)^{\frac{1}{2}}$$

Suppose then that

$$\Pi_0+\beta\Pi_1+\beta^2\Pi_2=(\mathfrak{P}^s)^2 \quad\text{or}\quad \frac{\cos^2\theta}{\frac{1+\beta}{1-\beta}-\sin^2\theta}(\mathsf{P}^s)^2$$

and let

$$\gamma=1-\cos 2\phi$$

Then we put

$$F_1=\frac{\cos^2\theta+\gamma(\beta+\beta^2)}{\left(\frac{1+\beta}{1-\beta}-\sin^2\theta\right)^{\frac{1}{2}}}(\Pi_0+\beta\Pi_1+\beta^2\Pi_2)$$

and

$$F_2=\frac{\frac{1+\beta}{1-\beta}-\sin^2\theta}{\cos^2\theta}F_1$$

Now suppose that K^2, a function independent of θ, denotes one of the four

$$\frac{(\mathfrak{C}^s)^2 \text{ or } (\mathfrak{S}^s)^2 \text{ or } (\mathsf{C}^s)^2 \text{ or } (\mathsf{S}^s)^2}{(1-\beta\cos 2\phi)^{\frac{1}{2}}}$$

Then in the cases involving $\mathfrak{P}$-functions and P-functions respectively, we have in alternative form—

$$\text{for }\begin{Bmatrix}\mathfrak{P}^s\\ \mathsf{P}^s\end{Bmatrix},\quad \int p\,(\mathrm{V}_i^s)^2\,d\sigma=\mathrm{M}(1-\beta)^{\frac{1}{2}}\iint K^2\begin{Bmatrix}F_1\\ F_2\end{Bmatrix}d\theta\,d\phi$$

If it be supposed that the development in powers of $\sec^2\theta$ is justifiable

$$F_1=\cos\theta\left[1+\frac{\gamma(\beta+\beta^2)}{\cos^2\theta}\right]\left[1-\frac{\beta+\beta^2}{\cos^2\theta}+\frac{3}{2}\frac{\beta^2}{\cos^4\theta}\right][\Pi_0+\beta\Pi_1+\beta^2\Pi_2]$$

$$=\Pi_0\cos\theta+\beta\left[\frac{\Pi_0(\gamma-1)}{\cos\theta}+\Pi_1\cos\theta\right]$$

$$+\beta^2\left[\frac{\Pi_0(\gamma-1)}{\cos\theta}+\frac{\Pi_0(\frac{3}{2}-\gamma)}{\cos^3\theta}+\frac{\Pi_1(\gamma-1)}{\cos\theta}+\Pi_2\cos\theta\right]$$

And F_2 has a similar form, save that $\gamma+1$ replaces $\gamma-1$, and $\gamma-\frac{1}{2}$ replaces $\frac{3}{2}-\gamma$.

It is clear that unless Π_0 is divisible by $\cos^3\theta$ and Π_1 by $\cos\theta$, $\int F_1 d\theta$ and $\int F_2 d\theta$ will have infinite elements at the poles, and the development is not legitimate.

Since $\mathsf{P}^s=\frac{\cos^s\theta}{2^i\,.\,i\,!}\frac{d^{i+s}}{d\mu}(\mu^2-1)^i$, it follows that the power of $\cos\theta$ by which P^s is divisible increases as s increases.

Let us consider the case of $s=2$.

Then $\Pi_0 + \beta\Pi_1 + \beta^2\Pi_2 = [P^2 - \beta q_0 P - \beta q_4 P^4 + \beta^2 q_6 P^6]^2$

$$= (P^2)^2 - 2\beta(q_0 PP^2 + q_4 P^2 P^4) + \beta^2[2q_6 P^2 P^6 + (q_0 P)^2 + (q_4 P^4)^2 + 2q_0 q_4 PP^4]$$

(or the same with accented q's for the other case).

From this it is clear that Π_0 is divisible by $\cos^4\theta$ and Π_1 by $\cos^2\theta$, and the method of development is legitimate when $s = 2$, but it is not so when $s = 0$ and $s = 1$.

The investigation then separates into the general case, and the cases $s = 0$, $s = 1$.

§ 14. *Integration in the general case.*

We have

$$\mathfrak{P}^s = P^s - \beta q_{s-2} P^{s-2} - \beta q_{s+2} P^{s+2} + \beta^2 q_{s-4} P^{s-4} + \beta^2 q_{s+4} P^{s+4}$$

and

$$(\mathfrak{P}^s)^2 = (P^s)^2 - 2\beta(q_{s-2} P^s P^{s-2} + q_{s+2} P^s P^{s+2}) + 2\beta^2(q_{s-4} P^s P^{s-4} + q_{s+4} P^s P^{s+4}) + \beta^2[(q_{s-2} P^{s-2})^2 + (q_{s+2} P^{s+2})^2 + 2q_{s-2} q_{s+2} P^{s-2} P^{s+2}]$$

Also $\left[P^s\left(\dfrac{\cos^2\theta}{\frac{1+\beta}{1-\beta} - \sin^2\theta}\right)^{\frac{1}{2}}\right]^2$ has the same form with accented q's, so that it will be merely necessary to accent the q's to obtain the second case.

We have then

$$\Pi_0 = (P^s)^2, \quad \Pi_1 = -2(q_{s-2} P^s P^{s-2} + q_{s+2} P^s P^{s+2})$$

$$\Pi_2 = 2(q_{s-4} P^s P^{s-4} + q_{s+4} P^s P^{s+4}) + (q_{s-2} P^{s-2})^2 + (q_{s+2} P^{s+2})^2 + 2q_{s-2} q_{s+2} P^{s-2} P^{s+2}$$

Then since $\cos\theta\, d\theta = d\mu$,

$$\begin{aligned}\int_{-\frac{1}{2}\pi}^{\frac{1}{2}\pi} F_1 d\theta = &\int_{-1}^{+1} (P^s)^2 d\mu + \beta\int_{-1}^{+1} \frac{(\gamma-1)(P^s)^2}{1-\mu^2} d\mu \\ &- 2\beta\int_{-1}^{+1} (q_{s-2} P^s P^{s-2} + q_{s+2} P^s P^{s+2}) d\mu \\ &+ \beta^2\int_{-1}^{+1} \frac{(\gamma-1)(P^s)^2}{1-\mu^2} d\mu + \beta^2\int_{-1}^{+1} \frac{(\frac{3}{2}-\gamma)(P^s)^2}{(1-\mu^2)^2} d\mu \\ &- 2\beta^2\int_{-1}^{+1} \frac{(\gamma-1)(q_{s-2} P^s P^{s-2} + q_{s+2} P^s P^{s+2})}{1-\mu^2} d\mu \\ &+ \beta^2\int_{-1}^{+1} [2q_{s-4} P^s P^{s-4} + 2q_{s+4} P^s P^{s+4} + (q_{s-2} P^{s-2})^2 \\ &\quad + (q_{s+2} P^{s+2})^2 + 2q_{s-2} q_{s+2} P^{s-2} P^{s+2}]\, d\mu \quad \ldots\ldots(66)\end{aligned}$$

And $\int F_2 d\theta$ has the same form, but with accented q's, and with $\gamma+1$ replacing $\gamma-1$, and $\gamma-\frac{1}{2}$ replacing $\frac{3}{2}-\gamma$.

It is now necessary to evaluate the several definite integrals involved in this expression.

It is well known that

$$\int_{-1}^{+1} [P_i^s(\mu)]^2 d\mu = \frac{2}{2i+1} \cdot \frac{i+s!}{i-s!}$$

It is easy to see that it is possible to express P^{s+2k} in the form

$$P_i^{s+2k} = AP_i^s + BP^s_{i-2} + CP^s_{i-4} + \ldots$$

where A, B, C ... do not involve μ.

The value of A may be found by considering only the highest power of μ on each side of the identity.

Now $$P_i^{s+2k} = \frac{(1-\mu^2)^{\frac{1}{2}s+k}}{2^i . i!} \left(\frac{d}{d\mu}\right)^{i+s+2k} (\mu^2-1)^i$$

$$= (-)^{\frac{1}{2}s+k} \frac{2i!}{2^i . i! \, i-s-2k!} \mu^i + \ldots$$

and $$P^s = (-)^{\frac{1}{2}s} \frac{2i!}{2^i . i! \, i-s!} \mu^i + \ldots$$

Therefore $$A = (-)^k \frac{i-s!}{i-s-2k!}$$

Then, since the integral of the product of two P's of different orders vanishes, we have

$$\int_{-1}^{+1} P^s P^{s+2k} d\mu = (-)^k \frac{i-s!}{i-s-2k!} \int_{-1}^{+1} (P^s)^2 d\mu = (-)^k \frac{2}{2i+1} \cdot \frac{i+s!}{i-s-2k!}$$

We will next consider $\int \frac{P^s P^{s+2k}}{1-\mu^2} d\mu$, where k is not zero.

The differential equation gives

$$\frac{d}{d\mu}\left[(1-\mu^2)\frac{dP^{s+2k}}{d\mu}\right] + i(i+1) P^{s+2k} - \frac{(s+2k)^2}{1-\mu^2} P^{s+2k} = 0$$

$$\frac{d}{d\mu}\left[(1-\mu^2)\frac{dP^s}{d\mu}\right] + i(i+1) P^s - \frac{s^2}{1-\mu^2} P^s = 0$$

Multiply the first of these by P^s and the second by P^{s+2k} and subtract, and we have

$$\frac{4k(s+k) P^s P^{s+2k}}{1-\mu^2} = P^{s+2k} \frac{d}{d\mu}\left[(1-\mu^2)\frac{dP^s}{d\mu}\right] - P^s \frac{d}{d\mu}\left[(1-\mu^2)\frac{dP^{s+2k}}{d\mu}\right]$$

Therefore*

$$4k(s+k)\int_{-1}^{+1}\frac{P^sP^{s+2k}}{1-\mu^2}d\mu=(1-\mu^2)\left[P^{s+2k}\frac{dP^s}{d\mu}-P^s\frac{dP^{s+2k}}{d\mu}\right], \text{ between limits } \pm 1$$
$$=0$$

Again since by (11)

$$\frac{P^{s+2k}}{1-\mu^2}=AP^{s+2k+2}+BP^{s+2k}+CP^{s+2k-2}$$

it follows that $\int_{-1}^{+1}\frac{P^sP^{s+2k}}{(1-\mu^2)^2}d\mu=0$, unless $k=0$ or 1

It remains to find the integrals of $\frac{(P^s)^2}{1-\mu^2}$, $\frac{(P^s)^2}{(1-\mu^2)^2}$, and $\frac{P^sP^{s+2}}{(1-\mu^2)^2}$.

We have seen in (11) (transformed to accord with our present definition of P^s) that

$$\frac{P^s}{1-\mu^2}=\frac{1}{4s(s+1)}P^{s+2}+\tfrac{1}{2}\left[\frac{i(i+1)}{s^2-1}+1\right]P^s+\frac{\{i,s\}\{i,s-1\}}{4s(s-1)}P^{s-2}$$

Hence

$$\int\frac{(P^s)^2}{1-\mu^2}d\mu=\frac{1}{4s(s+1)}\int P^sP^{s+2}d\mu+\tfrac{1}{2}\left[\frac{i(i+1)}{s^2-1}+1\right]\int(P^s)^2d\mu$$
$$+\frac{\{i,s\}\{i,s-1\}}{4s(s-1)}\int P^sP^{s-2}d\mu$$

$$\int\frac{(P^s)^2}{(1-\mu^2)^2}d\mu=\frac{1}{4s(s+1)}\int\frac{P^sP^{s+2}}{1-\mu^2}d\mu+\tfrac{1}{2}\left[\frac{i(i+1)}{s^2-1}+1\right]\int\frac{(P^s)^2}{1-\mu^2}d\mu$$
$$+\frac{\{i,s\}\{i,s-1\}}{4s(s-1)}\int\frac{P^sP^{s-2}}{1-\mu^2}d\mu$$

$$\int\frac{P^sP^{s+2}}{(1-\mu^2)^2}d\mu=\frac{1}{4s(s+1)}\int\frac{(P^{s+2})^2}{1-\mu^2}d\mu+\tfrac{1}{2}\left[\frac{i(i+1)}{s^2-1}+1\right]\int\frac{P^sP^{s+2}}{1-\mu^2}d\mu$$
$$+\frac{\{i,s\}\{i,s-1\}}{4s(s-1)}\int\frac{P^{s-2}P^{s+2}}{1-\mu^2}d\mu$$

The first of these involves integrals already determined; on introducing them on the right and reducing we find the result to be $\frac{1}{s}\cdot\frac{i+s\,!}{i-s\,!}$.

The first and last terms of the second integral vanish, and the integral is clearly $\tfrac{1}{2}\left[\frac{i(i+1)}{s^2-1}+1\right]\frac{1}{s}\frac{i+s\,!}{i-s\,!}$.

The second and third terms of the third integral vanish, and the whole is clearly $\frac{1}{4s(s+1)(s+2)}\frac{i+s+2\,!}{i-s-2\,!}$.

* I owe this method of finding these last two integrals to Mr Hobson.

Collecting results we have

$$\left.\begin{aligned}
&\int_{-1}^{+1} (P^s)^2\, d\mu \quad = \frac{2}{2i+1}\,\frac{i+s\,!}{i-s\,!}\\
&\int_{-1}^{+1} P^s P^{s+2k}\, d\mu = (-)^k \frac{2}{2i+1}\,\frac{i+s\,!}{i-s-2k\,!}\\
&\int_{-1}^{+1} \frac{P^s P^{s+2k}}{1-\mu^2}\, d\mu = 0\\
&\int_{-1}^{+1} \frac{(P^s)^2}{1-\mu^2}\, d\mu \quad = \frac{1}{s}\,\frac{i+s\,!}{i-s\,!}\\
&\int_{-1}^{+1} \frac{(P^s)^2}{(1-\mu^2)^2}\, d\mu = \tfrac{1}{2}\left[\frac{i(i+1)}{s^2-1}+1\right]\frac{1}{s}\,\frac{i+s\,!}{i-s\,!}\\
&\int_{-1}^{+1} \frac{P^s P^{s+2}}{(1-\mu^2)^2}\, d\mu = \frac{1}{4s(s+1)(s+2)}\,\frac{i+s+2\,!}{i-s-2\,!}\\
&\int_{-1}^{+1} \frac{P^s P^{s-2}}{(1-\mu^2)^2}\, d\mu = \frac{1}{4s(s-1)(s-2)}\,\frac{i+s\,!}{i-s\,!}
\end{aligned}\right\} \quad \text{............(67)}$$

Then by means of (66) and (67)

$$\begin{aligned}
\int F_1\, d\theta = {} & \frac{2}{2i+1}\,\frac{i+s\,!}{i-s\,!} + \beta(\gamma-1)\,\frac{1}{s}\,\frac{i+s\,!}{i-s\,!}\\
& + 2\beta\left(q_{s-2}\,\frac{2}{2i+1}\,\frac{i+s-2\,!}{i-s\,!} + q_{s+2}\,\frac{2}{2i+1}\,\frac{i+s\,!}{i-s-2\,!}\right)\\
& + \beta^2(\gamma-1)\,\frac{1}{s}\,\frac{i+s\,!}{i-s\,!} + \beta^2\left(\tfrac{3}{2}-\gamma\right)\tfrac{1}{2}\left[\frac{i(i+1)}{s^2-1}+1\right]\frac{1}{s}\,\frac{i+s\,!}{i-s\,!}\\
& + \frac{2\beta^2}{2i+1}\left[2q_{s-4}\,\frac{i+s-4\,!}{i-s\,!} + 2q_{s+4}\,\frac{i+s\,!}{i-s-4\,!} + (q_{s-2})^2\,\frac{i+s-2\,!}{i-s+2\,!}\right.\\
& \qquad \left. + (q_{s+2})^2\,\frac{i+s+2\,!}{i-s-2\,!} + q_{s-2}q_{s+2}\,\frac{i+s-2\,!}{i-s-2\,!}\right]
\end{aligned}$$

Therefore

$$\begin{aligned}
\int F_1\, d\theta = \frac{2}{2i+1}\,\frac{i+s\,!}{i-s\,!}\Bigg[1 & + \frac{2\beta q_{s-2}}{(i+s)(i+s-1)} + 2\beta q_{s+2}(i-s)(i-s-1)\\
& + \frac{2\beta^2 q_{s-4}}{(i+s)(i+s-1)(i+s-2)(i+s-3)}\\
& + 2\beta^2 q_{s+4}(i-s)(i-s-1)(i-s-2)(i-s-3)\\
& + \frac{\beta^2 (q_{s-2})^2}{(i-s+1)(i-s+2)(i+s)(i+s-1)}\\
& + \beta^2 (q_{s+2})^2 (i+s+1)(i+s+2)(i-s)(i-s-1)\\
& + 2\beta^2 q_{s-2} q_{s+2}\,\frac{(i-s)(i-s-1)}{(i+s)(i+s-1)}\Bigg]\\
+ \beta\,\frac{1}{s}\,\frac{i+s\,!}{i-s\,!}\left\{(1+\beta)(\gamma-1) + \beta\left(\tfrac{3}{2}-\gamma\right)\tfrac{1}{2}\left[\frac{i(i+1)}{s^2-1}+1\right]\right\} & \quad \text{...(68)}
\end{aligned}$$

Also $\int F_2 d\theta$ has a similar form with accented q's, $\gamma+1$ for $\gamma-1$, and $\gamma-\frac{1}{2}$ for $\frac{3}{2}-\gamma$.

When we substitute for γ its value $1-\cos 2\phi$, and write as before

$$\Sigma = \frac{i(i+1)}{s^2-1}$$

the last term in $\int F_1 d\theta$ becomes

$$+\beta \frac{1}{s}\frac{i+s!}{i-s!}\{\tfrac{1}{4}\beta(\Sigma+1) - \cos 2\phi[1-\tfrac{1}{2}\beta(\Sigma-1)]\} \quad \text{.........(68)}$$

Also the last term in $\int F_2 d\theta$ becomes

$$+\beta \frac{1}{s}\frac{i+s!}{i-s!}\{2[1+\tfrac{1}{8}\beta(\Sigma+9)] - \cos 2\phi[1+\tfrac{1}{2}\beta(\Sigma+3)]\} \quad \text{...(68)}$$

But it will appear later that we only need the parts of these terms which involve $\cos 2\phi$ developed as far as the first power of β; hence in both cases we may write the latter term inside { } simply as $-\cos 2\phi$.

Our general formulæ for the q coefficients apply for all values of s down to $s=3$, inclusive, although the result for $s=3$ needs proper interpretation. Hence the present result applies down to $s=3$, inclusive.

I have just re-defined Σ, and I remind the reader that

$$\Upsilon = \frac{(i-1)i(i+1)(i+2)}{s^2-4}$$

Then if in (68) we introduce for the q's their values, we find that the coefficient of the term in β is

$$\frac{2q_{s-2}}{(i+s)(i+s-1)} + q_{s+2}(i-s)(i-s-1) = -\tfrac{1}{2}(\Sigma-1)$$

The coefficient of the term in β^2 is

$$\tfrac{1}{64}\left\{\frac{(i-s+1)(i-s+2)(i-s+3)(i-s+4)}{(s-1)(s-2)}\right.$$
$$+\frac{(i-s)(i-s-1)(i-s-2)(i-s-3)}{(s+1)(s+2)}$$
$$+\frac{(i+s)(i+s-1)(i-s+1)(i-s+2)}{(s-1)^2}$$
$$+\frac{(i+s+1)(i+s+2)(i-s)(i-s-1)}{(s+1)^2}$$
$$\left.-2\frac{(i-s+1)(i-s+2)(i-s)(i-s-1)}{(s^2-1)}\right\}$$

If this be reduced by a process similar to that employed in § 10, we find

$$\int F_1 d\theta = \frac{2}{2i+1}\frac{i+s!}{i-s!}\{1-\tfrac{1}{2}\beta(\Sigma-1)+\tfrac{1}{32}\beta^2[3\Sigma^2-6\Sigma+6-s^2(\Sigma^2+2\Sigma-1)+2\Upsilon]\}$$
$$+\beta\frac{1}{s}\frac{i+s!}{i-s!}\{\tfrac{1}{4}\beta(\Sigma+1)-\cos 2\phi\} \quad \ldots(69)$$

We know that P^s is derivable from $\mathfrak{P}^s$ by multiplication by $1/\mathrm{C}_i^s$, and we have found in (33), § 9,

$$\frac{1}{(\mathrm{C}_i^s)^2} = 1+\beta(\Sigma+1)+\tfrac{1}{8}\beta^2[3\Sigma^2+8\Sigma+5+s^2\Sigma^2-\Upsilon]$$

Hence multiplying (69) by $\frac{1}{(\mathrm{C}_i^s)^2}$ we have

$$\int F_2 d\theta = \frac{2}{2i+1}\frac{i+s!}{i-s!}\{1+\tfrac{1}{2}\beta(\Sigma+3)+\tfrac{1}{32}\beta^2[-\Sigma^2+26\Sigma+42$$
$$+s^2(3\Sigma^2-2\Sigma+1)-2\Upsilon]\}+\beta\frac{1}{s}\frac{i+s!}{i-s!}\{\tfrac{1}{4}\beta(\Sigma+1)-\cos 2\phi\}\ldots(70)$$

I have also obtained this result by direct development. It may be thought surprising that the last term is now the same in both formulæ, notwithstanding the difference in the earlier stages, but if the reader will go through the analysis he will see how this has been brought about. The formulæ (69) and (70) also hold true when $s=3$ (as I have verified), notwithstanding the fact that P^3 is not to be derived from $\mathfrak{P}^3$ by a factor.

The next step is the integration with respect to ϕ.

We have

$$\left\{\begin{matrix}\mathfrak{C}^s\\ \mathfrak{S}^s\end{matrix}\right. = \left\{\begin{matrix}\cos\\ \sin\end{matrix}\right. s\phi+\beta\left[p_{s-2}\left\{\begin{matrix}\cos\\ \sin\end{matrix}\right.(s-2)\phi+p_{s+2}\left\{\begin{matrix}\cos\\ \sin\end{matrix}\right.(s+2)\phi\right]$$
$$+\beta^2\left[p_{s-4}\left\{\begin{matrix}\cos\\ \sin\end{matrix}\right.(s-4)\phi+p_{s+4}\left\{\begin{matrix}\cos\\ \sin\end{matrix}\right.(s+4)\phi\right]$$

Therefore

$$\left\{\begin{matrix}(\mathfrak{C}_i^s)^2\\ (\mathfrak{S}_i^s)^2\end{matrix}\right. = \tfrac{1}{2}\pm\tfrac{1}{2}\cos 2s\phi+\beta[(p_{s-2}+p_{s+2})\cos 2\phi\pm p_{s-2}\cos 2(s-1)\phi$$
$$\pm p_{s+2}\cos 2(s+1)\phi]+\beta^2[\tfrac{1}{2}(p_{s-2})^2+\tfrac{1}{2}(p_{s+2})^2+(p_{s-4}+p_{s+4}$$
$$+p_{s-2}p_{s+2})\cos 4\phi\pm p_{s-2}p_{s+2}\cos 2s\phi\pm(p_{s-4}+\tfrac{1}{2}(p_{s-2})^2)\cos 2(s-2)\phi$$
$$\pm(p_{s+4}+\tfrac{1}{2}(p_{s+2})^2)\cos 2(s+2)\phi]$$

Also $\frac{(\mathsf{C}_i^s)^2 \text{ or } (\mathsf{S}_i^s)^2}{1-\beta\cos 2\phi}$ have the same forms with accented p's.

Accordingly, with unaccented p's, we have to multiply this expression by $(1-\beta\cos 2\phi)^{-\frac{1}{2}}$, and with accented p's we multiply by $(1-\beta\cos 2\phi)^{\frac{1}{2}}$, and we shall then have the functions denoted above by K^2.

The function K^2 has to be multiplied by a function of the form $\mathrm{A}+\mathrm{B}\beta\cos 2\phi$, and integrated from $\phi=2\pi$ to 0. It follows that the only terms in K^2 which will not vanish are those independent of ϕ and those in $\cos 2\phi$; moreover, the latter terms are only required as far as the first power of β.

Now
$$(1-\beta\cos 2\phi)^{-\frac{1}{2}} = 1 + \tfrac{1}{2}\beta\cos 2\phi + \tfrac{3}{16}\beta^2(1+\cos 4\phi)$$
$$(1-\beta\cos 2\phi)^{\frac{1}{2}} = 1 - \tfrac{1}{2}\beta\cos 2\phi - \tfrac{1}{16}\beta^2(1+\cos 4\phi)$$

Then omitting terms which will vanish on integration

$$\frac{(\mathfrak{C}_i^s \text{ or } \mathfrak{S}_i^s)^2}{(1-\beta\cos 2\phi)^{\frac{1}{2}}} = \tfrac{1}{2}\left\{1 + \beta^2\left[(p_{s-2})^2 + (p_{s+2})^2 + \tfrac{1}{2}p_{s-2} + \tfrac{1}{2}p_{s+2} + \tfrac{3}{16}\right]\right\} + \beta\left(p_{s-2} + p_{s+2} + \tfrac{1}{4}\right)\cos 2\phi$$

$$\frac{(\mathsf{C}_i^s \text{ or } \mathsf{S}_i^s)^2}{(1-\beta\cos 2\phi)^{\frac{1}{2}}} = \tfrac{1}{2}\left\{1 + \beta^2\left[(p'_{s-2})^2 + (p'_{s+2})^2 - \tfrac{1}{2}p'_{s-2} - \tfrac{1}{2}p'_{s+2} - \tfrac{1}{16}\right]\right\} + \beta\left(p'_{s-2} + p'_{s+2} - \tfrac{1}{4}\right)\cos 2\phi$$

However, the latter formula is not needed except for verification, because it will be derivable from the former by multiplication by $\dfrac{1}{(\mathrm{D}_i^s)^2}$.

Now if we substitute for the p's their values as given in (27), § 8, we find

$$\frac{(\mathfrak{C}_i^s \text{ or } \mathfrak{S}_i^s)^2}{(1-\beta\cos 2\phi)^{\frac{1}{2}}} = \tfrac{1}{2}\left\{1 + \tfrac{1}{32}\beta^2\left[\Sigma^2 + 4\Sigma + 6 + s^2(\Sigma^2 - 2\Sigma + 1)\right]\right\} + \tfrac{1}{4}\beta(\Sigma+1)\cos 2\phi$$

And multiplying by $\dfrac{1}{(\mathrm{D}_i^s)^2}$ or $1 - \tfrac{1}{8}\beta^2(\Sigma+3)$, or developing directly

$$\frac{(\mathsf{C}_i^s \text{ or } \mathsf{S}_i^s)^2}{(1-\beta\cos 2\phi)^{\frac{1}{2}}} = \tfrac{1}{2}\left\{1 + \tfrac{1}{32}\beta^2\left[\Sigma^2 - 6 + s^2(\Sigma^2 - 2\Sigma + 1)\right]\right\} + \tfrac{1}{4}\beta(\Sigma+1)\cos 2\phi$$

These represent the K^2 of our integrals.

Then

$$\int p\left(\mathfrak{P}_i^s\begin{Bmatrix}\mathfrak{C}_i^s\\ \mathfrak{S}_i^s\end{Bmatrix}\right)^2 d\sigma = \mathrm{M}(1-\beta)^{\frac{1}{2}}\iint F_1\,\frac{(\mathfrak{C}^s \text{ or } \mathfrak{S}^s)^2}{(1-\beta\cos 2\phi)^{\frac{1}{2}}}\,d\theta\, d\phi$$

$$\begin{aligned}
&= \frac{2\pi\mathrm{M}(1-\beta)^{\frac{1}{2}}}{2i+1}\cdot\frac{i+s!}{i-s!}\Big\{1 - \tfrac{1}{2}\beta(\Sigma-1) \\
&\qquad + \tfrac{1}{32}\beta^2\left[3\Sigma^2 - 6\Sigma + 6 - s^2(\Sigma^2 + 2\Sigma - 1) + 2\Upsilon\right]\Big\} \\
&\qquad \times\left\{1 + \tfrac{1}{32}\beta^2\left[\Sigma^2 + 4\Sigma + 6 + s^2(\Sigma^2 - 2\Sigma + 1)\right]\right\} \\
&\quad + \tfrac{1}{4}\pi\mathrm{M}(1-\beta)^{\frac{1}{2}}\left[\beta^2\frac{1}{s}\frac{i+s!}{i-s!}(\Sigma+1) - \beta^2\frac{1}{s}\frac{i+s!}{i-s!}(\Sigma+1)\right] \\
&= \frac{2\pi\mathrm{M}}{2i+1}\frac{i+s!}{i-s!}(1-\beta)^{\frac{1}{2}}\Big\{1 - \tfrac{1}{2}\beta(\Sigma-1) \\
&\qquad + \tfrac{1}{16}\beta^2\left[2\Sigma^2 - \Sigma + 6 - s^2(2\Sigma - 1) + \Upsilon\right]\Big\} \\
&= \frac{2\pi\mathrm{M}}{2i+1}\frac{i+s!}{i-s!}\left\{1 - \tfrac{1}{2}\beta\Sigma + \tfrac{1}{16}\beta^2\left[2\Sigma^2 + 3\Sigma - s^2(2\Sigma - 1) + \Upsilon\right]\right\}
\end{aligned} \qquad (71)$$

The following results may be obtained either by direct development, or by multiplication by either or both the factors $\frac{1}{(C_i^s)^2}$ and $\frac{1}{(D_i^s)^2}$. The former converts $\mathfrak{P}$ into P, the latter $\mathfrak{C}$ or $\mathfrak{S}$ into C or S.

$$\left.\begin{aligned}
\int p\left(\mathfrak{P}_i^s \begin{Bmatrix} \mathsf{C}_i^s \\ \mathsf{S}_i^s \end{Bmatrix}\right)^2 d\sigma &= \frac{2\pi M}{2i+1}\frac{i+s\,!}{i-s\,!}\{1-\tfrac{1}{2}\beta\Sigma \\
&\qquad + \tfrac{1}{16}\beta^2\left[2\Sigma^2+\Sigma-6-s^2(2\Sigma-1)+\Upsilon\right]\} \\
\int p\left(\mathsf{P}_i^s \begin{Bmatrix} \mathfrak{C}_i^s \\ \mathfrak{S}_i^s \end{Bmatrix}\right)^2 d\sigma &= \quad\ldots\ldots\ldots \quad \{1+\tfrac{1}{2}\beta(\Sigma+2) \\
&\qquad + \tfrac{1}{16}\beta^2\left[11\Sigma+10+s^2(2\Sigma^2-2\Sigma+1)-\Upsilon\right]\} \\
\int p\left(\mathsf{P}_i^s \begin{Bmatrix} \mathsf{C}_i^s \\ \mathsf{S}_i^s \end{Bmatrix}\right)^2 d\sigma &= \quad\ldots\ldots\ldots \quad \{1+\tfrac{1}{2}\beta(\Sigma+2) \\
&\qquad + \tfrac{1}{16}\beta^2\left[9\Sigma+4+s^2(2\Sigma^2-2\Sigma+1)-\Upsilon\right]\}
\end{aligned}\right\} \quad \ldots\ldots\ldots(71)$$

§ 15. *Integration in the case of* $s=2$.

Although the development in powers of $\sec^2\theta$ is still legitimate in this case, yet the formulæ found in the last section fail because Υ contains s^2-4 in the denominator. Moreover since $\mathfrak{P}^2$ is not convertible into P^2 by a factor each case must be considered separately.

We now have $q_{s-4}=0$, $q'_{s-4}=0$, and therefore from (68)

$$\begin{aligned}
\int F_1 d\theta = \frac{2}{2i+1}\frac{i+2\,!}{i-2\,!}\Big\{1 &+ 2\beta\left[\frac{q_0}{(i+1)(i+2)}+q_4(i-2)(i-3)\right] \\
&+ \beta^2\Big[2q_6(i-2)(i-3)(i-4)(i-5)+\frac{(q_0)^2}{(i-1)i(i+1)(i+2)} \\
&+ (q_4)^2(i+3)(i+4)(i-2)(i-3)+2q_0q_4\frac{(i-2)(i-3)}{(i+1)(i+2)}\Big]\Big\} \\
&+ \tfrac{1}{2}\beta\frac{i+2\,!}{i-2\,!}\{\tfrac{1}{4}\beta(\Sigma+1)-\cos 2\phi\}
\end{aligned}$$

$\int F_2 d\theta$ is equal to the same with accented q's, and the last term equal to

$$\tfrac{1}{2}\beta\frac{i+2\,!}{i-2\,!}\{2+\tfrac{1}{4}\beta(\Sigma+9)-\cos 2\phi\}$$

We now have

$q_0 = -\frac{1}{8}\{i, 1\}\{i, 2\}$, $q_4 = \frac{1}{24}$, $q_6 = \frac{1}{1536}$, associated with a cosine function.

$q_0' = 0$, $q_4' = \frac{1}{12}$, $q_6' = \frac{1}{512}$, associated with a sine function.

It is well to note that these values are given by the general formula, because this consideration shows that much of the previous reductions is still applicable.

Effecting the reductions I find

$$\int F_1 d\theta = \frac{2}{2i+1}\frac{i+2\,!}{i-2\,!}\{1 - \tfrac{1}{2}\beta(\Sigma - 1) + \tfrac{1}{64}\beta^2[\tfrac{19}{4}\Sigma^2 - \tfrac{65}{2}\Sigma + 20$$

$$+ 2(2i+1)(3\Sigma - 1)]\} + \tfrac{1}{2}\beta\frac{i+2\,!}{i-2\,!}\{\tfrac{1}{4}\beta(\Sigma+1) - \cos 2\phi\}$$

$$= \frac{2}{2i+1}\frac{i+2\,!}{i-2\,!}\{1 - \tfrac{1}{2}\beta(\Sigma - 1) + \tfrac{1}{256}\beta^2(19\Sigma^2 - 130\Sigma + 80)\}$$

$$+ \tfrac{1}{2}\beta\frac{i+2\,!}{i-2\,!}\{\tfrac{1}{8}\beta(5\Sigma + 1) - \cos 2\phi\}$$

This integral will be associated with $\mathfrak{C}_i^2$ and $\mathbf{C}_i^2$, and in the present case $\Sigma = \frac{1}{3}i(i+1)$.

In the same way

$$\int F_2 d\theta = \frac{2}{2i+1}\frac{i+2\,!}{i-2\,!}\{1 + \tfrac{1}{2}\beta[\Sigma + 3 - (2i+1)] + \tfrac{1}{64}\beta^2[\tfrac{25}{4}\Sigma^2 + \tfrac{93}{2}\Sigma + 92$$

$$- 6(2i+1)(\Sigma+5)]\} + \tfrac{1}{2}\beta\frac{i+2\,!}{i-2\,!}\{2 + \tfrac{1}{4}\beta(\Sigma+9) - \cos 2\phi\}$$

$$= \frac{2}{2i+1}\frac{i+2\,!}{i-2\,!}\{1 + \tfrac{1}{2}\beta(\Sigma+3) + \tfrac{1}{256}\beta^2(25\Sigma^2 + 186\Sigma + 368)\}$$

$$+ \tfrac{1}{2}\beta\frac{i+2\,!}{i-2\,!}\{-\tfrac{1}{8}\beta(\Sigma - 3) - \cos 2\phi\}$$

This will be associated with $\mathfrak{S}_i^2$ and $\mathbf{S}_i^2$.

Now turning to the cosine and sine functions, we find that they must be treated apart, but the integral involving $\mathbf{C}_i^2$ may be derived from that in $\mathfrak{C}_i^2$ by the factor $\left[\frac{1}{\mathrm{D}_i^2}(\cos)\right]^2$; and similarly $\mathbf{S}_i^2$ from $\mathfrak{S}_i^2$ by the factor $\left[\frac{1}{\mathrm{D}_i^2}(\sin)\right]^2$. These factors were evaluated in (36), § 9.

We now have $p_{s-4} = 0$, $p'_{s-4} = 0$; also for the sine function $p_{s-2} = p_0 = 0$. Then

$$(\mathfrak{C}_i^2)^2 = \tfrac{1}{2} + \tfrac{1}{2}\cos 4\phi + \beta[(2p_0 + p_4)\cos 2\phi + p_4 \cos 6\phi]$$

$$+ \beta^2[(p_0)^2 + \tfrac{1}{2}(p_4)^2 + (p_6 + p_0p_4)\cos 4\phi + p_0p_4\cos 4\phi$$

$$+ (p_6 + \tfrac{1}{2}(p_4)^2)\cos 8\phi]$$

$$(\mathfrak{S}_i^2)^2 = \tfrac{1}{2} - \tfrac{1}{2}\cos 4\phi + \beta[p_4\cos 2\phi - p_4\cos 6\phi]$$

$$+ \beta^2[\tfrac{1}{2}(p_4)^2 + p_6\cos 4\phi - (p_6 + \tfrac{1}{2}(p_4)^2)\cos 8\phi]$$

Then as far as material

$$\frac{(\mathfrak{C}_i^2)^2}{(1-\beta\cos 2\phi)^{\frac{1}{2}}} = \tfrac{1}{2}\{1 + \beta^2[(p_4)^2 + 2(p_0)^2 + p_0 + \tfrac{1}{2}p_4 + \tfrac{9}{32}]\}$$

$$+ \beta(p_4 + 2p_0 + \tfrac{3}{8})\cos 2\phi$$

$$\frac{(\mathfrak{S}_i^2)^2}{(1-\beta\cos 2\phi)^{\frac{1}{2}}} = \tfrac{1}{2}\{1 + \beta^2[(p_4)^2 + \tfrac{1}{2}p_4 + \tfrac{3}{32}]\} + \beta(p_4 + \tfrac{1}{8})\cos 2\phi$$

Now $p_0 = \frac{1}{8}\{i, 2\}$, $p_4 = -\frac{1}{24}\{i, 3\}$, whence

$$\frac{(\mathfrak{C}_i^2)^2}{(1-\beta\cos 2\phi)^{\frac{1}{2}}} = \tfrac{1}{2}\{1 + \tfrac{1}{64}\beta^2(19\Sigma^2 - 8\Sigma + 22)\} + \tfrac{1}{8}\beta(5\Sigma + 1)\cos 2\phi$$

$$\frac{(\mathfrak{S}_i^2)^2}{(1-\beta\cos 2\phi)^{\frac{1}{2}}} = \tfrac{1}{2}\{1 + \tfrac{1}{64}\beta^2(\Sigma^2 - 8\Sigma + 18)\} - \tfrac{1}{8}\beta(\Sigma - 3)\cos 2\phi$$

We now multiply these by $\int F_1 d\theta$ and $\int F_2 d\theta$ respectively, and the last terms disappear as before. I remark that the disappearance of the terms which do not involve the factor $1/(2i+1)$ affords an excellent test of the correctness of the laborious reductions throughout all this part of the work.

Then we have

$$\int p(\mathfrak{P}_i^2\,\mathfrak{C}_i^2)^2\,d\sigma = \frac{2\pi M}{2i+1}\frac{i+2!}{i-2!}(1-\beta)^{\frac{1}{2}}[1 - \tfrac{1}{2}\beta(\Sigma - 1) + \tfrac{1}{256}\beta^2(19\Sigma^2 - 130\Sigma + 80)][1 + \tfrac{1}{64}\beta^2(19\Sigma^2 - 8\Sigma + 22)]$$

$$= \frac{2\pi M}{2i+1}\frac{i+2!}{i-2!}\{1 - \tfrac{1}{2}\beta\Sigma + \tfrac{1}{256}\beta^2(95\Sigma^2 - 98\Sigma + 72)\} \qquad \text{.........(72)}$$

If we multiply this by $\left[\frac{1}{D_i^2}(\cos)\right]^2$ or $1 - \frac{1}{16}\beta^2(5\Sigma + 7)$, we obtain the result when C_i^2 replaces $\mathfrak{C}_i^2$; the only change is that the last term inside { } now becomes $+\frac{1}{256}\beta^2(95\Sigma^2 - 178\Sigma - 40)$.

Again

$$\int p(\mathsf{P}_i^2\,\mathfrak{S}_i^2)^2\,d\sigma = \frac{2\pi M}{2i+1}\frac{i+2!}{i-2!}(1-\beta)^{\frac{1}{2}}[1 + \tfrac{1}{2}\beta(\Sigma + 3) + \tfrac{1}{256}\beta^2(25\Sigma^2 + 186\Sigma + 368)][1 + \tfrac{1}{64}\beta^2(\Sigma^2 - 8\Sigma + 18)]$$

$$= \frac{2\pi M}{2i+1}\frac{i+2!}{i-2!}[1 + \tfrac{1}{2}\beta(\Sigma + 2) + \tfrac{1}{256}\beta^2(29\Sigma^2 + 90\Sigma + 216)] \qquad \text{.........(72)}$$

If we multiply this by $\left[\frac{1}{D_i^2}(\sin)\right]^2$ or $1 + \frac{1}{16}\beta^2(\Sigma + 5)$, we obtain the result when S_i^2 replaces $\mathfrak{S}_i^2$; the only change is that the last term inside { } now becomes $+\frac{1}{256}\beta^2(29\Sigma^2 + 106\Sigma + 136)$.

This terminates the integrals, which can be completely determined by this method of developing in powers of $\sec^2\theta$.

§ 16. *Portion of the Integration in the case of $s=1$.*

The preceding method may be used for finding the four integrals

$$\int p \begin{Bmatrix} (\mathfrak{S}_i^1)^2 \\ (\mathsf{S}_i^1)^2 \end{Bmatrix} [(\mathfrak{P}_i^1)^2 - (\mathrm{P}_i^1)^2]\, d\sigma$$

and

$$\int p \begin{Bmatrix} (\mathfrak{C}_i^1)^2 \\ (\mathsf{C}_i^1)^2 \end{Bmatrix} \left[(\mathsf{P}_i^1)^2 - \left(\mathrm{P}_i^1 \sqrt{\frac{\frac{1+\beta}{1-\beta} - \sin^2\theta}{\cos^2\theta}} \right)^2 \right] d\theta$$

There will then remain four integrals of the type $\int p\, (\mathfrak{S}_i^1)^2 (\mathrm{P}_i^1)^2\, d\sigma$ to evaluate.

The first pair of our integrals are clearly to be treated by putting $\Pi_0 = 0$, $q_{s-2} = q_{-1} = 0$, $q_{s-4} = q_{-3} = 0$, and then determining $\int \mathrm{F}_1 d\theta$. The condition for the second pair only differs in the accentuation of the q's which vanish, and in the use of $\int \mathrm{F}_2 d\theta$.

The vanishing of Π_0 makes

$$\mathrm{F}_1 = \beta \Pi_1 \cos\theta + \beta^2 \left(\frac{\Pi_1(\gamma - 1)}{\cos\theta} + \Pi_2 \cos\theta \right)$$

$$\mathrm{F}_2 = \beta \Pi_1 \cos\theta + \beta^2 \left(\frac{\Pi_1(\gamma + 1)}{\cos\theta} + \Pi_2 \cos\theta \right)$$

In the first of these

$$\Pi_1 = -\, 2q_3 \mathrm{P}^1 \mathrm{P}^3, \quad \Pi_2 = 2q_5 \mathrm{P}^1 \mathrm{P}^5 + (q_3)^2\, (\mathrm{P}^3)^2$$

and in the second the form is the same with accented q's.

Also since $\int \frac{\mathrm{P}^1 \mathrm{P}^3}{1-\mu^2}\, d\mu = 0$, we have $\int \frac{\Pi_1 d\theta}{\cos\theta} = 0$.

Hence

$$\int \mathrm{F}_1 d\theta = -\, 2\beta q_3 \int \mathrm{P}^1 \mathrm{P}^3\, d\mu + 2\beta^2 \int [q_5 \mathrm{P}^1 \mathrm{P}^5 + \tfrac{1}{2} (q_3)^2 (\mathrm{P}^3)^2]\, d\mu$$

$$= \frac{2}{2i+1} \frac{i+1\,!}{i-1\,!} [2\beta q_3 (i-1)(i-2) + 2\beta^2 q_5 (i-1)(i-2)(i-3)(i-4) + \beta^2 (q_3)^2 (i+2)(i+3)(i-1)(i-2)]$$

and $\int \mathrm{F}_2 d\theta$ is the same with accented q's.

It is only necessary to pursue the cases $\int p\, (\mathfrak{S}_i^1)^2\, \mathrm{F}_1 d\sigma$ and $\int p\, (\mathfrak{C}_i^1)^2\, \mathrm{F}_2 d\sigma$, since the other pair of integrals may be determined by means of multiplication by the appropriate factors, determined in § 9.

Now for $\mathfrak{P}_i^1\, \mathfrak{S}_i^1$, associated with F_1,

$$q_3 = \tfrac{1}{16}[1 - \tfrac{1}{16}\beta i(i+1)], \quad q_5 = \tfrac{1}{768}$$

and for $\mathsf{P}_i^1\, \mathfrak{C}_i^1$, associated with F_2,

$$q_3' = \tfrac{3}{16}[1 + \tfrac{1}{16}\beta i(i+1)], \quad q_5' = \tfrac{5}{768}$$

Therefore

$$\int F_1 d\theta = \frac{2}{2i+1}\frac{i+1\,!}{i-1\,!}\{\tfrac{1}{8}\beta(i-1)(i-2) + \tfrac{1}{128}\beta^2[\tfrac{1}{3}(i-1)(i-2)(i-3)(i-4)$$
$$+ \tfrac{1}{2}(i+2)(i+3)(i-1)(i-2) - (i-1)(i-2)i(i+1)]\}$$

$$\int F_2 d\theta = \frac{2}{2i+1}\frac{i+1\,!}{i-1\,!}\{\tfrac{3}{8}\beta(i-1)(i-2) + \tfrac{1}{128}\beta^2[\tfrac{5}{3}(i-1)(i-2)(i-3)(i-4)$$
$$+ \tfrac{9}{2}(i+2)(i+3)(i-1)(i-2) + 3(i-1)(i-2)i(i+1)]\}$$

In the present case we cannot use Σ as an abridgement, since it is infinite; I therefore now write

$$j = i(i+1)$$

Effecting the reductions we have

$$\int F_1 d\theta = \frac{2}{2i+1}\frac{i+1\,!}{i-1\,!}\{\tfrac{1}{8}\beta[j - 2(2i+1) + 4] + \tfrac{1}{128}\beta^2[-\tfrac{1}{6}j^2 + \tfrac{28}{3}j + 30 - 16(2i+1)]\}$$

$$\int F_2 d\theta = \frac{2}{2i+1}\frac{i+1\,!}{i-1\,!}\{\tfrac{3}{8}\beta[j - 2(2i+1) + 4]$$
$$+ \tfrac{1}{128}\beta^2[\tfrac{55}{6}j^2 + \tfrac{188}{3}j + 174 - 16(2i+1)(j+5)]\}$$

The former of these is associated with $\mathfrak{S}$, the latter with $\mathfrak{C}$.

In the cosine and sine functions we have

$$p_{s-2} = p_{-1} = 0, \qquad p_{s-4} = p_{-3} = 0,$$

and

$$(\mathfrak{S}_i^1)^2 = \tfrac{1}{2} - \tfrac{1}{2}\cos 2\phi + \beta(p_3 \cos 2\phi - p_3 \cos 4\phi)$$
$$+ \beta^2[\tfrac{1}{2}(p_3)^2 + p_5 \cos 4\phi - (p_5 + \tfrac{1}{2}(p_3)^2)\cos 6\phi]$$

$$(\mathfrak{C}_i^1)^2 = \tfrac{1}{2} + \tfrac{1}{2}\cos 2\phi + \beta(p_3 \cos 2\phi + p_3 \cos 4\phi)$$
$$+ \beta^2[\tfrac{1}{2}(p_3)^2 + p_5 \cos 4\phi + (p_5 + \tfrac{1}{2}(p_3)^2)\cos 6\phi]$$

As far as material, we then have

$$\frac{(\mathfrak{S}_i^1)^2}{(1-\beta\cos 2\phi)^{\frac{1}{2}}} = \tfrac{1}{2}\{1 - \tfrac{1}{4}\beta + \beta^2[(p_3)^2 + \tfrac{1}{2}p_3 + \tfrac{3}{16}]\}$$
$$+ (-\tfrac{1}{2} + \tfrac{1}{4}\beta + \beta p_3)\cos 2\phi$$
$$= \tfrac{1}{2}\{1 - \tfrac{1}{4}\beta + \beta^2[(p_3)^2 + \tfrac{1}{2}p_3 + \tfrac{3}{16}]\}\{1 - (1 - 2\beta p_3 - \tfrac{1}{4}\beta)\cos 2\phi\}$$

$$\frac{(\mathfrak{C}_i^1)^2}{(1-\beta\cos 2\phi)^{\frac{1}{2}}} = \tfrac{1}{2}\{1 + \tfrac{1}{4}\beta + \beta^2[(p_3)^2 + \tfrac{1}{2}p_3 + \tfrac{3}{16}]\} + (\tfrac{1}{2} + \tfrac{1}{4}\beta + \beta p_3)\cos 2\phi$$
$$= \tfrac{1}{2}\{1 + \tfrac{1}{4}\beta + \beta^2[(p_3)^2 + \tfrac{1}{2}p_3 + \tfrac{3}{16}]\}\{1 + (1 + 2\beta p_3 + \tfrac{1}{4}\beta)\cos 2\phi\}$$

Now both for sines and cosines, to the order necessary for our present purpose, $p_3 = -\frac{1}{16}\{i, 2\}$. Therefore, introducing j for $i(i+1)$,

$$\left.\begin{aligned}\frac{(\mathfrak{S}_i^1)^2}{(1-\beta\cos 2\phi)^{\frac{1}{2}}} &= \tfrac{1}{2}\{1-\tfrac{1}{4}\beta+\tfrac{1}{256}\beta^2(j^2-12j+68)\}\{1-[1+\tfrac{1}{8}\beta j-\tfrac{1}{2}\beta]\cos 2\phi\}\\ \frac{(\mathfrak{C}_i^1)^2}{(1-\beta\cos 2\phi)^{\frac{1}{2}}} &= \tfrac{1}{2}\{1+\tfrac{1}{4}\beta+\tfrac{1}{256}\beta^2(j^2-12j+68)\}\{1+[1-\tfrac{1}{8}\beta j+\tfrac{1}{2}\beta]\cos 2\phi\}\end{aligned}\right\} \quad \text{.........(73)}$$

Observe that $\int F_1 d\theta$ and $\int F_2 d\theta$ do not involve $\cos 2\phi$, and are of the first order in β. Hence, as far as material for the *present* portion of the work,

$$\frac{(\mathfrak{S}_i^1)^2}{(1-\beta\cos 2\phi)^{\frac{1}{2}}} = \tfrac{1}{2}(1-\tfrac{1}{4}\beta), \qquad \frac{(\mathfrak{C}_i^1)^2}{(1-\beta\cos 2\phi)^{\frac{1}{2}}} = \tfrac{1}{2}(1+\tfrac{1}{4}\beta) \text{...(73)}$$

Also, to the first order, from (37)

$$\left[\frac{1}{D_i^1}(\sin)\right]^2 = 1+\tfrac{1}{2}\beta, \qquad \left[\frac{1}{D_i^1}(\cos)\right]^2 = 1-\tfrac{1}{2}\beta$$

Therefore as far as necessary

$$\frac{(\mathsf{S}_i^1)^2}{(1-\beta\cos 2\phi)^{\frac{1}{2}}} = \tfrac{1}{2}(1+\tfrac{1}{4}\beta), \qquad \frac{(\mathsf{C}_i^1)^2}{(1-\beta\cos 2\phi)^{\frac{1}{2}}} = \tfrac{1}{2}(1-\tfrac{1}{4}\beta)$$

Hence

$$\begin{aligned}\int p(\mathfrak{S}_i^1)^2[(\mathfrak{P}_i^1)^2-(P_i^1)^2]\,d\sigma &= \frac{2\pi M}{2i+1}\frac{\overline{i+1}\,!}{\overline{i-1}\,!}(1-\beta)^{\frac{1}{2}}(1-\tfrac{1}{4}\beta)\{\tfrac{1}{8}\beta[j-2(2i+1)+4]\\ &\qquad + \tfrac{1}{128}\beta^2[-\tfrac{1}{6}j^2+\tfrac{28}{3}j+30-16(2i+1)]\}\\ &= \frac{2\pi M}{2i+1}\frac{\overline{i+1}\,!}{\overline{i-1}\,!}\{\tfrac{1}{8}\beta(j+4)-\tfrac{1}{768}\beta^2(j^2+16j+108)\}\\ &\qquad + \pi M\frac{\overline{i+1}\,!}{\overline{i-1}\,!}(-\tfrac{1}{2}\beta+\tfrac{1}{8}\beta^2) \quad \text{............(74)}\end{aligned}$$

For S_i^1 we have only to replace the factor $1-\frac{1}{4}\beta$ by $1+\frac{1}{4}\beta$, and find

$$\begin{aligned}\int p(\mathsf{S}_i^1)^2[(\mathfrak{P}_i^1)^2-(P_i^1)^2]\,d\sigma &= \frac{2\pi M}{2i+1}\frac{\overline{i+1}\,!}{\overline{i-1}\,!}\{\tfrac{1}{8}\beta(j+4)-\tfrac{1}{768}\beta^2(j^2-32j-84)\}\\ &\qquad + \pi M\frac{\overline{i+1}\,!}{\overline{i-1}\,!}(-\tfrac{1}{2}\beta-\tfrac{1}{8}\beta^2) \quad \text{..............(74)}\end{aligned}$$

Again, omitting intermediate steps,

$$\begin{aligned}&\int p(\mathfrak{C}_i^1)^2\left[(\mathsf{P}_i^1)^2-\left(P_i^1\sqrt{\frac{\frac{1+\beta}{1-\beta}-\sin^2\theta}{\cos^2\theta}}\right)^2\right]d\sigma\\ &\qquad = \frac{2\pi M}{2i+1}\frac{\overline{i+1}\,!}{\overline{i-1}\,!}\{\tfrac{3}{8}\beta(j+4)+\tfrac{1}{768}\beta^2(55j^2+304j+756)\}\\ &\qquad\qquad - \pi M\frac{\overline{i+1}\,!}{\overline{i-1}\,!}[\tfrac{3}{2}\beta+\tfrac{1}{8}\beta^2(2j+7)] \quad \text{.........(74)}\end{aligned}$$

$$\int p\,(\mathsf{C}_i^1)^2\left[(\mathsf{P}_i^1)^2-\left(\mathsf{P}_i^1\sqrt{\frac{\frac{1+\beta}{1-\beta}-\sin^2\theta}{\cos^2\theta}}\right)^2\right]d\sigma$$

$$=\frac{2\pi\mathrm{M}}{2i+1}\frac{i+1\,!}{i-1\,!}\{\tfrac{3}{8}\beta(j+4)+\tfrac{1}{768}\beta^2(55j^2+160j+180)\}$$

$$-\pi\mathrm{M}\frac{i+1\,!}{i-1\,!}[\tfrac{3}{2}\beta+\tfrac{1}{8}\beta^2(2j+1)]\text{ ....(74)}$$

§ 17. *Portion of the Integration in the case of $s=0$.*

We are to find $\int p\begin{Bmatrix}(\mathfrak{C}_i)^2\\(\mathsf{C}_i)^2\end{Bmatrix}[(\mathfrak{P}_i)^2-(\mathrm{P}_i)^2]\,d\sigma$, leaving two integrals of the type $\int p\,(\mathfrak{C}_i)^2(\mathrm{P}_i)^2\,d\sigma$ to be determined subsequently.

It is only necessary to consider $\mathfrak{C}_i$, since the other case is determinable from it by multiplication to $\frac{1}{(\mathrm{D}_i)^2}$, as found in (38) of § 9.

Following the procedure of the case where $s=1$, we have

$$\int \mathrm{F}_1 d\theta=-2\beta q_2\int \mathrm{PP}^2 d\mu+2\beta^2\int[q_4\mathrm{PP}^4+\tfrac{1}{2}(q_2)^2(\mathrm{P}^2)^2]\,d\mu$$

$$=\frac{2}{2i+1}[2\beta q_2 i(i-1)+2\beta^2 q_4 i(i-1)(i-2)(i-3)$$
$$+\beta^2(q_2)^2(i+1)(i+2)i(i-1)]$$

Then since $q_2=\frac{1}{4}$, $q_4=\frac{1}{128}$,

$$\int \mathrm{F}_1 d\theta=\frac{2}{2i+1}\{\tfrac{1}{2}\beta[j+1-(2i+1)]+\tfrac{1}{64}\beta^2[5j^2+14j+12-4(2i+1)(j+3)]\}$$

Now $(\mathfrak{C}_i)^2=1+2\beta p_2\cos 2\phi+\beta^2[(2p_4+\frac{1}{2}(p_2)^2)\cos 4\phi+\frac{1}{2}(p_2)^2]$
and as far as material

$$\frac{(\mathfrak{C}_i)^2}{(1-\beta\cos 2\phi)^{\frac{1}{2}}}=1+\beta^2[\tfrac{1}{2}p_2+\tfrac{1}{2}(p_2)^2+\tfrac{3}{16}]+\beta(\tfrac{1}{2}+2p_2)\cos 2\phi$$
$$=\{1+\tfrac{1}{32}\beta^2(j^2-4j+6)\}\{1-\tfrac{1}{2}\beta(j-1)\cos 2\phi\}\text{...(75)}$$

since $p_2=-\frac{1}{4}i(i+1)=-\frac{1}{4}j$.

At present we only require this to the first power of β, and since $\int \mathrm{F}_1 d\theta$ does not contain $\cos 2\phi$, the expression (75) as far as at present needed is simply unity.

Again, by (38) of § 9,

$$\frac{1}{(\mathrm{D}_i)^2}=1+\tfrac{1}{8}\beta^2(j-3)$$

therefore by multiplication

$$\frac{(\mathsf{C}_i)^2}{(1-\beta\cos 2\phi)^{\frac{1}{2}}}=\{1+\tfrac{1}{32}\beta^2(j^2-6)\}\{1-\tfrac{1}{2}\beta(j-1)\cos 2\phi\}\text{ ......(76)}$$

This is also unity to the order at present needed.

Hence

$$\int p \begin{Bmatrix}(\mathfrak{C}_i)^2 \\ (\mathsf{C}_i)^2\end{Bmatrix} [(\mathfrak{P}_i)^2 - (\mathrm{P}_i)^2]\, d\sigma = \frac{4\pi}{2i+1} \mathrm{M} (1-\beta)^{\frac{1}{2}} \{\tfrac{1}{2}\beta [j+1-(2i+1)]$$
$$+ \tfrac{1}{64}\beta^2 [5j^2 + 14j + 12 - 4(2i+1)(j+3)]\}$$
$$= \frac{4\pi \mathrm{M}}{2i+1} \{\tfrac{1}{2}\beta (j+1) + \tfrac{1}{64}\beta^2 (5j^2 - 2j - 4)\}$$
$$- \pi \mathrm{M} \{2\beta + \tfrac{1}{4}\beta^2 (j-1)\} \quad \text{......(77)}$$

§ 18. *Preparation for the Integrations when $s=1$ and 0.*

We have now to evaluate the three integrals

$$\left.\begin{aligned} \mathrm{L} &= \int p (\mathfrak{S}_i^1 \mathrm{P}_i^1)^2\, d\sigma \\ \mathrm{M} &= \int p \left(\mathfrak{C}_i^1 \frac{\mathrm{P}_i^1}{\cos\theta}\right)^2 \left(\tfrac{1+\beta}{1-\beta} - \sin^2\theta\right) d\sigma \\ \mathrm{N} &= \int p (\mathfrak{C}_i \mathrm{P}_i)^2\, d\sigma \end{aligned}\right\} \quad \text{................(78)}$$

and from these to determine three others when S, C replace $\mathfrak{S}$, $\mathfrak{C}$.

We have

$$\frac{p\,d\sigma}{d\theta\, d\phi} = \left\{\frac{M(1-\beta)^{\frac{1}{2}} \left(\frac{1-\beta}{1+\beta}\right)^{\frac{1}{2}}}{\left(1 - \frac{1-\beta}{1+\beta}\sin^2\theta\right)^{\frac{1}{2}}}\right\} \left\{\frac{\cos^2\theta + \frac{\beta(1-\cos 2\phi)}{1-\beta}}{(1-\beta\cos 2\phi)^{\frac{1}{2}}}\right\}$$

$$\frac{p\left(\frac{1+\beta}{1-\beta} - \sin^2\theta\right) d\sigma}{d\theta\, d\phi} = \left\{M(1-\beta)^{\frac{1}{2}} \left(\tfrac{1+\beta}{1-\beta}\right)^{\frac{1}{2}} \left(1 - \tfrac{1-\beta}{1+\beta}\sin^2\theta\right)^{\frac{1}{2}}\right\} \left\{\frac{\cos^2\theta + \frac{\beta(1-\cos 2\phi)}{1-\beta}}{(1-\beta\cos 2\phi)^{\frac{1}{2}}}\right\}$$

It is the second factor which alone involves ϕ, and as I shall now first integrate with respect to ϕ, the first factor may be dropped for the moment, and the second factor multiplied by the squares of the cosine or sine functions. Since the integration is from $\phi = 2\pi$ to 0, those terms which vanish on integration may be dropped.

For brevity write

$$j = i(i+1)$$
$$j_0 = \tfrac{1}{32}(j^2 - 4j + 6)$$
$$j_1 = \tfrac{1}{256}(j^2 - 12j + 68)$$

Then we have seen in (73) and (75) that

$$\frac{(\mathfrak{S}_i^1)^2}{(1-\beta\cos 2\phi)^{\frac{1}{2}}} = \tfrac{1}{2}(1 - \tfrac{1}{4}\beta + \beta^2 j_1)\{1 - [1 + \tfrac{1}{8}\beta(j-4)]\cos 2\phi\}$$

$$\frac{(\mathfrak{C}_i^1)^2}{(1-\beta\cos 2\phi)^{\frac{1}{2}}} = \tfrac{1}{2}(1 + \tfrac{1}{4}\beta + \beta^2 j_1)\{1 + [1 - \tfrac{1}{8}\beta(j-4)]\cos 2\phi\}$$

$$\frac{(\mathfrak{C}_i)^2}{(1-\beta\cos 2\phi)^{\frac{1}{2}}} = (1 + \beta^2 j_0)\{1 - \tfrac{1}{2}\beta(j-1)\cos 2\phi\}$$

Therefore

$$\left.\begin{aligned}
&\int_0^{2\pi} \frac{\cos^2\theta + \frac{\beta(1-\cos 2\phi)}{1-\beta}}{(1-\beta\cos 2\phi)^{\frac{1}{2}}} \cdot (\mathfrak{S}_i^1)^2\, d\phi = \pi\,(1-\tfrac{1}{4}\beta+\beta^2 j_1)\,\{\cos^2\theta + \tfrac{3}{2}\beta + \tfrac{1}{16}\beta^2 (j+20)\} \\
&\int_0^{2\pi} \ldots\ldots\ldots\ldots\ldots\ldots\ (\mathfrak{C}_i^1)^2\, d\phi = \pi\,(1+\tfrac{1}{4}\beta+\beta^2 j_1)\,\{\cos^2\theta + \tfrac{1}{2}\beta + \tfrac{1}{16}\beta^2 (j+4)\} \\
&\int_0^{2\pi} \ldots\ldots\ldots\ldots\ldots\ldots\ (\mathfrak{C}_i)^2\, d\phi = 2\pi\,(1+\beta^2 j_0)\,\{\cos^2\theta + \beta + \tfrac{1}{4}\beta^2 (j+3)\}
\end{aligned}\right\} \quad \ldots\ldots(78)$$

Now pick out the parts of $pd\sigma$ and of these integrals (78) which are independent of θ, and write

$$\left.\begin{aligned}
F &= \pi M (1-\beta)^{\frac{1}{2}} \left(\frac{1-\beta}{1+\beta}\right)^{\frac{1}{2}} (1-\tfrac{1}{4}\beta+\beta^2 j_1) \\
&= \pi M \{1 - \tfrac{7}{4}\beta + \tfrac{1}{256}\beta^2 (j^2 - 12j + 388)\} \\
G &= \pi M (1-\beta)^{\frac{1}{2}} \left(\frac{1+\beta}{1-\beta}\right)^{\frac{1}{2}} (1+\tfrac{1}{4}\beta+\beta^2 j_1) \\
&= \pi M \{1 + \tfrac{3}{4}\beta + \tfrac{1}{256}\beta^2 (j^2 - 12j + 68)\} \\
H &= 2\pi M (1-\beta)^{\frac{1}{2}} \left(\frac{1-\beta}{1+\beta}\right)^{\frac{1}{2}} (1+\beta^2 j_0) \\
&= 2\pi M \{1 - \tfrac{3}{2}\beta + \tfrac{1}{32}\beta^2 (j^2 - 4j + 34)\}
\end{aligned}\right\} \quad \ldots\ldots\ldots(79)$$

$$\text{Also write}\quad \left.\begin{aligned}
f &= \tfrac{3}{2}(1+\tfrac{1}{24}\beta j + \tfrac{5}{6}\beta) \\
g &= \tfrac{1}{2}(1+\tfrac{1}{8}\beta j + \tfrac{1}{2}\beta) \\
h &= 1 + \tfrac{1}{4}\beta j + \tfrac{3}{4}\beta \\
\kappa^2 &= 1 - \kappa'^2 = \frac{1-\beta}{1+\beta},\ \text{so that } \kappa'^2 = 2\beta + 2\beta^2 + \ldots
\end{aligned}\right\} \quad \ldots\ldots\ldots(80)$$

Lastly, in accordance with the usual notation for elliptic integrals, write

$$\Delta^2 = 1 - \frac{1-\beta}{1+\beta}\sin^2\theta = 1 - \kappa^2 \sin^2\theta \quad \ldots\ldots\ldots\ldots\ldots\ldots(80)$$

Then we have

$$\left.\begin{aligned}
L &= F \int_{-\frac{1}{2}\pi}^{\frac{1}{2}\pi} \frac{\cos^2\theta + \beta f}{\Delta} (P_i^1)^2\, d\theta \\
M &= G \int_{-\frac{1}{2}\pi}^{\frac{1}{2}\pi} (\cos^2\theta + \beta g) \,.\, \Delta \,.\, (P_i^1)^2\, d\theta \\
N &= H \int_{-\frac{1}{2}\pi}^{\frac{1}{2}\pi} \frac{\cos^2\theta + \beta h}{\Delta} (P_i)^2\, d\theta
\end{aligned}\right\} \quad \ldots\ldots\ldots\ldots(81)$$

The next step is to express the squares of the P's in a series of powers of $\cos^2\theta$.

When $P_i^s(\mu) = \frac{(1-\mu^2)^{\frac{1}{2}s}}{2^i . i!}\left(\frac{d}{d\mu}\right)^{i+s}(\mu^2-1)^i$, it is known that

$$P_i[\mu\mu' + (1-\mu^2)^{\frac{1}{2}}(1-\mu'^2)^{\frac{1}{2}}\cos\phi] = P_i(\mu)P_i(\mu') + 2\sum_{s=1}^{s=i}\frac{i-s!}{i+s!}P_i^s(\mu)P_i^s(\mu')\cos s\phi$$

By putting $\mu=\mu'$ we see that $2\frac{i-s!}{i+s!}(P_i^s(\mu))^2$ is the coefficient of $\cos s\phi$ in the expansion of $P_i[1-(1-\mu^2)\,2\sin^2\frac{1}{2}\phi]$. By Taylor's theorem this last is equal to

$$\sum_{r=0}^{r=i}(-)^r\frac{(1-\mu^2)^r}{r!}(2\sin^2\tfrac{1}{2}\phi)^2\left[\frac{d^r}{d\mu^r}P_i(\mu),\quad \mu=1\right]$$

Now

$$\left(\frac{d}{d\mu}\right)^r P_i = \frac{1}{2^i i!}\left(\frac{d}{d\mu}\right)^{i+r}(\mu^2-1)^i$$

$$= \frac{1}{2^r r!}\frac{i+r!}{i-r!}[1 + \text{terms involving powers of } \mu^2-1]$$

$$= \frac{1}{2^r r!}\frac{i+r!}{i-r!}, \text{ when } \mu=1$$

Also
$$\sin^{2r}\tfrac{1}{2}\phi = \left(\frac{e^{\frac{1}{2}\phi\sqrt{-1}} - e^{-\frac{1}{2}\phi\sqrt{-1}}}{2\sqrt{-1}}\right)^{2r}$$

$$= \sum_{t=0}^{t=r}(-)^{r-t}\frac{2r!}{2^{2r}.\,2r-t!\,t!}e^{(r-t)\phi\sqrt{-1}}$$

On putting $r-t=s$, we see that the coefficient of $\cos s\phi$ in $\sin^{2r}\frac{1}{2}\phi$ is

$$\frac{(-)^s}{2^{2r-1}}\cdot\frac{2r!}{r-s!\,r+s!}$$

Hence we have*

$$2\frac{i-s!}{i+s!}[P_i^s(\mu)]^2 = 2\sum_{r=s}^{r=i}(-)^{r+s}\frac{2r!}{2^{2r}\,r-s!\,(r!)^2\,r+s!}\frac{i+r!}{i-r!}\cos^{2r}\theta \ldots(82)$$

Now suppose

$$(P_i^1)^2 = \sum_1^i \gamma_{2r-2}\cos^{2r}\theta$$

$$(P_i)^2 = \sum_0^i \alpha_{2r}\cos^{2r}\theta$$

Then clearly

$$\gamma_{2r-2} = (-)^{r+1}\frac{2r!}{2^{2r}\,r-1!\,(r!)^2\,r+1!}\cdot\frac{i+r!}{i-r!}\cdot\frac{i+1!}{i-1!}$$

$$\alpha_{2r} = (-)^r\frac{2r!}{2^{2r}(r!)^4}\frac{i+r!}{i-r!}$$

$\ldots\ldots\ldots(82)$

* Mr Hobson kindly gave me this proof when I had shown him the series which I believed to hold true.

Therefore

$$\left.\begin{aligned}\frac{\mathrm{L}}{\mathrm{F}} &= \sum_{r=1}^{r=i} \gamma_{2r-2} \int_{-\frac{1}{2}\pi}^{\frac{1}{2}\pi} \frac{\cos^{2r+2}\theta + \beta f \cos^{2r}\theta}{\Delta}\, d\theta \\ \frac{\mathrm{M}}{\mathrm{G}} &= \sum_{r=1}^{r=i} \gamma_{2r-2} \int_{-\frac{1}{2}\pi}^{\frac{1}{2}\pi} (\cos^{2r}\theta + \beta g \cos^{2r-2}\theta)\, \Delta\, d\theta \\ \frac{\mathrm{N}}{\mathrm{H}} &= \sum_{r=0}^{r=i} \alpha_{2r} \int_{-\frac{1}{2}\pi}^{\frac{1}{2}\pi} \frac{\cos^{2r+2}\theta + \beta h \cos^{2r}\theta}{\Delta}\, d\theta\end{aligned}\right\} \quad \text{.........(83)}$$

The evaluation of these integrals depends on two integrals only, namely, $\int \frac{\cos^{2n}\theta}{\Delta}\, d\theta$ and $\int \cos^{2n}\theta \,.\, \Delta d\theta$, and these will be considered in the next section.

§ 19. *Evaluation of the Integrals* $\int \frac{\cos^{2n}\theta}{\Delta}\, d\theta$ *and* $\int \cos^{2n}\theta \Delta d\theta$.

I will denote these integrals D and E respectively, and I propose to find their values in series proceeding by powers of κ'^2.

The usual notation is adopted where $\Pi(x)$ is such a function that it is equal to $x\Pi(x-1)$; accordingly when x is a positive integer $\Pi(x) = x\,!$.

Since κ^2 is less than unity

$$\frac{1}{\Delta} = \sum_0^\infty \frac{\frac{1}{2}\cdot\frac{3}{2}\cdots\frac{2r-1}{2}}{r\,!}\kappa^{2r}\sin^{2r}\theta$$

and since $$\int_{-\frac{1}{2}\pi}^{\frac{1}{2}\pi} \cos^{2n}\theta \sin^{2r}\theta d\theta = \pi \frac{\frac{1}{2}\cdot\frac{3}{2}\cdots\frac{2r-1}{2}\cdot\frac{1}{2}\cdot\frac{3}{2}\cdots\frac{2n-1}{2}}{n+r\,!}$$

$$\mathrm{D} = \int \frac{\cos^{2n}\theta}{\Delta}\, d\theta = \pi \frac{\frac{1}{2}\cdot\frac{3}{2}\cdots\frac{2n-1}{2}}{n\,!} \sum_0^\infty \frac{\left(\frac{1}{2}\cdot\frac{3}{2}\cdots\frac{2r-1}{2}\right)^2 \kappa^2}{(n+1)(n+2)\dots(n+r)\,r\,!}$$

or, with the usual notation for hypergeometric series,

$$\mathrm{D} = \pi \frac{2n\,!}{2^{2n}(n\,!)^2} F(\tfrac{1}{2}, \tfrac{1}{2}, n+1, \kappa^2)$$

This series is of no service, since it proceeds by powers of κ^2, which in our case is nearly unity. It is required then to transform the series into one proceeding by powers of κ'^2.

It is known that, if $\kappa^2 + \kappa'^2 = 1$,

$$F(a, b, c, \kappa^2) = \frac{\Pi(c-a-b-1)\,\Pi(c-1)}{\Pi(c-a-1)\,\Pi(c-b-1)} F(a, b, 1+a+b-c, \kappa'^2)$$

$$+ \kappa'^{2(c-a-b)} \frac{\Pi(a+b-c-1)\,\Pi(c-1)}{\Pi(a-1)\,\Pi(b-1)} F(c-a, c-b, c-a-b+1, \kappa'^2)^*$$

* I have to thank Mr Hobson for giving me this formula, and for showing me the procedure whereby it can be made effective.

If we apply this theorem with $a=b=\frac{1}{2}$, $c=n+1$, the first F becomes $F(\frac{1}{2}, \frac{1}{2}, 1-n, \kappa'^2)$, whose nth and all subsequent terms involve zero factors in the denominators. Also the coefficient of the second F involves $\Pi(-n-1)$, which has an infinite factor. Hence the formula leads to an indeterminate result. Let us therefore put $c=n+1+\epsilon$, and proceed to the limit when $\epsilon=0$.

We have then

$$D = \text{Limit}\ \pi \frac{2n!}{2^{2n}(n!)^2}\left\{\frac{\Pi(n-1+\epsilon)\,\Pi(n+\epsilon)}{[\Pi(n-\frac{1}{2}+\epsilon)]^2} F(\tfrac{1}{2}, \tfrac{1}{2}, 1-n-\epsilon, \kappa'^2)\right.$$

$$\left.+\,\kappa'^{2(n+\epsilon)}\frac{\Pi(-n-1-\epsilon)\,\Pi(n+\epsilon)}{[\Pi(-\frac{1}{2})]^2} F(n+\tfrac{1}{2}+\epsilon, n+\tfrac{1}{2}+\epsilon, n+1+\epsilon, \kappa'^2)\right\}$$

Now $\Pi(\epsilon) = 1+\epsilon\Pi'(0)$, $\quad \Pi(-\frac{1}{2}+\epsilon) = \Pi(-\frac{1}{2})\left(1+\epsilon\dfrac{\Pi'(-\frac{1}{2})}{\Pi(-\frac{1}{2})}\right)$

Therefore, when ϵ is very small,

$$\frac{\Pi(n-1+\epsilon)}{\Pi(n-1)} = 1+\epsilon\left(\frac{1}{n-1}+\frac{1}{n-2}+\dots+\tfrac{1}{2}+1\right)+\epsilon\Pi'(0)$$

$$= 1+\epsilon\left(\Pi'(0)+\sum_1^n\frac{1}{t}-\frac{1}{n}\right)$$

$$\frac{\Pi(n+\epsilon)}{\Pi(n)} = 1+\epsilon\left(\Pi'(0)+\sum_1^n\frac{1}{t}\right)$$

$$\frac{\Pi(n-\frac{1}{2}+\epsilon)}{\Pi(n-\frac{1}{2})} = 1+2\epsilon\left(\frac{1}{2n-1}+\frac{1}{2n-3}+\dots+\tfrac{1}{3}+1\right)+\epsilon\frac{\Pi'(-\frac{1}{2})}{\Pi(-\frac{1}{2})}$$

$$= 1+\epsilon\left(\frac{\Pi'(-\frac{1}{2})}{\Pi(-\frac{1}{2})}+2\sum_1^n\frac{1}{2t-1}\right)$$

Hence for the coefficient of the first series we have

$$\frac{\Pi(n-1+\epsilon)\,\Pi(n+\epsilon)}{[\Pi(n-\frac{1}{2}+\epsilon)]^2}$$

$$= \frac{\Pi(n-1)\,\Pi(n)}{\Pi(n-\frac{1}{2})}\left\{1+\epsilon\left(2\Pi'(0)+2\sum_1^n\frac{1}{t}-\frac{1}{n}-2\frac{\Pi'(-\frac{1}{2})}{\Pi(-\frac{1}{2})}-4\sum_1^n\frac{1}{2t-1}\right)\right\}$$

But $\qquad \Pi(-\frac{1}{2}) = \pi^{\frac{1}{2}}, \qquad \Pi'(0)-\dfrac{\Pi'(-\frac{1}{2})}{\Pi(-\frac{1}{2})} = \log_e 4$*

$$[\Pi(n-\tfrac{1}{2})]^2 = \pi\left(\frac{2n!}{2^{2n}\,n!}\right)^2, \quad \Pi(n)\,\Pi(n-1) = n!\,n-1!$$

Therefore

$$\pi\frac{2n!}{2^{2n}(n!)^2}\,\frac{\Pi(n-1+\epsilon)\,\Pi(n+\epsilon)}{[\Pi(n-\frac{1}{2}+\epsilon)]^2}$$

$$= 2^{2n}\frac{n!\,n-1!}{2n!}\left\{1+\epsilon\left(2\log 4-\frac{1}{n}+2\sum_1^n\frac{1}{t}-4\sum_1^n\frac{1}{2t-1}\right)\right\}$$

* Proved by differentiating the known formula $\Pi(x-1)\,\Pi(x-\frac{1}{2}) = \Pi(2x-1)\,.\,\dfrac{(4\pi)^{\frac{1}{2}}}{4^x}$, and putting $x=\frac{1}{2}$.

This is true from $n=\infty$ to 1, but in the case of $n=0$ we have

$$\Pi(-1+\epsilon)=\frac{1}{\epsilon}\Pi(\epsilon)=\frac{1}{\epsilon}+\Pi'(0)$$

so that in that case $$\pi\frac{\Pi(-1+\epsilon)\Pi(\epsilon)}{[\Pi(-\frac{1}{2}+\epsilon)]^2}=\frac{1}{\epsilon}+2\log 4$$

Now consider the coefficient of the second series.

We have $$\kappa'^{2n+2\epsilon}=\kappa'^{2n}(1+2\epsilon\log_e\kappa')$$

and since $$\Pi(-x)\Pi(x-1)=\frac{\pi}{\sin\pi x}$$

$$\Pi(-n-1-\epsilon)\Pi(n+\epsilon)=\frac{\pi}{\sin(n+1+\epsilon)\pi}=\frac{(-)^{n+1}}{\epsilon}\pi,\qquad [\Pi(-\tfrac{1}{2})]^2=\pi$$

Therefore the coefficient of the second series is

$$(-)^{n+1}\frac{2n\,!}{2^{2n}(n\,!)^2}\frac{\kappa'^{2n}}{\epsilon}(1+2\epsilon\log\kappa')$$

and

$$D=\frac{2^{2n}n\,!\,n-1\,!}{2n\,!}\left[1+\epsilon\left(2\log 4-\frac{1}{n}+2\sum_1^n\frac{1}{t}-4\sum_1^n\frac{1}{2t-1}\right)\right]F(\tfrac{1}{2},\tfrac{1}{2},1-n-\epsilon,\kappa'^2)$$

$$+(-)^{n+1}\frac{2n\,!}{2^{2n}(n\,!)^2}\kappa'^{2n}\frac{1}{\epsilon}(1+2\epsilon\log\kappa')F(n+\tfrac{1}{2}+\epsilon,n+\tfrac{1}{2}+\epsilon,n+1+\epsilon,\kappa'^2)$$

The case of $n=0$ is an exception, for the coefficient of the first F has the part inside [ ] replaced by $\frac{1}{\epsilon}(1+2\epsilon\log 4)$.

It remains to consider these two F series.

$$F(\tfrac{1}{2},\tfrac{1}{2},1-n-\epsilon,\kappa'^2)=\sum_0^\infty\frac{[1.3\ldots(2r-1)]^2\kappa'^{2r}}{2^{2r}(1-n-\epsilon)(2-n-\epsilon)\ldots(r-n-\epsilon).r\,!}$$

$$=\sum_0^{n-1}(-)^r\frac{[1.3\ldots(2r-1)]^2\kappa'^{2r}}{2^{2r}(n-1+\epsilon)(n-2+\epsilon)\ldots(n-r+\epsilon)r\,!}$$

$$+(-)^n\sum_n^\infty\frac{[1.3\ldots(2r-1)]^2\kappa'^{2r}}{2^{2r}(n-1+\epsilon)\ldots(1+\epsilon)\epsilon(1-\epsilon)\ldots(r-n-\epsilon).r\,!}$$

When $r<n$

$$\frac{1}{(n-1+\epsilon)(n-2+\epsilon)\ldots(n-r+\epsilon)}=\frac{1}{(n-1)(n-2)\ldots(n-r)}\left(1-\epsilon\sum_{n-r}^{n-1}\frac{1}{t}\right)$$

When $r>n$, put $r=n+s$, and

$$\frac{1}{(n-1+\epsilon)(n-2+\epsilon)\ldots(1+\epsilon)\epsilon(1-\epsilon)\ldots(s-\epsilon)}$$

$$=\frac{1}{n-1\,!\,s\,!}\cdot\frac{1}{\epsilon}\left[1-\epsilon\sum_1^n\frac{1}{t}+\epsilon\sum_1^s\frac{1}{r}+\frac{\epsilon}{n}\right]$$

Also when $r = n + s$

$$\kappa'^{2r}\frac{[1.3\dots(2r-1)]^2}{2^{2r}.r!} = \frac{(2n!)^2}{2^{4n}(n!)^3}\kappa'^{2n}\frac{[(2n+1)(2n+3)\dots(2n+2s-1)]^2}{2^{2s}(n+1)(n+2)\dots(n+s)}\kappa'^{2s}$$

Thus

$$F(\tfrac{1}{2}, \tfrac{1}{2}, 1-n-\epsilon, \kappa'^2) = \sum_{r=0}^{r=n-1}(-)^r\frac{[1.3\dots(2r-1)]^2}{2^{2r}(n-1)\dots(n-r)r!}\left(1-\epsilon\sum_{n-r}^{n-1}\frac{1}{t}\right)\kappa'^{2r}$$

$$+(-)^n\frac{(2n!)^2}{2^{4n}(n!)^4}n\kappa'^{2n}\sum_{s=0}^{s=\infty}\frac{[(2n+1)\dots(2n+2s-1)]^2}{2^{2s}(n+1)\dots(n+s)s!}$$

$$\times\frac{1}{\epsilon}\left(1-\epsilon\sum_1^n\frac{1}{t}+\frac{\epsilon}{n}+\epsilon\sum_1^s\frac{1}{t}\right)\kappa'^{2s}$$

It follows that we may write the first term of D as follows:—

$$(-)^n\frac{2n!}{2^{2n}(n!)^2}\frac{\kappa'^{2n}}{\epsilon}F(n+\tfrac{1}{2}, n+\tfrac{1}{2}, n+1, \kappa'^2)$$

$$+\frac{2^{2n}n!\,n-1!}{2n!}\sum_0^{n-1}\frac{(-)^r[1.3\dots(2r-1)]^2}{2^{2r}(n-1)\dots(n-r).r!}\kappa'^{2r}$$

$$+(-)^n\frac{2n!}{2^{2n}(n!)^2}\kappa'^{2n}\sum_0^\infty\frac{[(2n+1)\dots(2n+2s-1)]^2}{2^{2s}(n+1)\dots(n+s).s!}$$

$$\times\left[\sum_1^n\frac{1}{t}+\sum_1^s\frac{1}{t}-4\sum_1^n\frac{1}{2t-1}+2\log 4\right]\kappa'^{2s}$$

The first of these terms becomes infinite when $\epsilon = 0$.

Turning to the second F we have

$$F(n+\tfrac{1}{2}+\epsilon, n+\tfrac{1}{2}+\epsilon, n+1+\epsilon, \kappa'^2)$$

$$=\sum_0^\infty\frac{[(2n+1+2\epsilon)(2n+3+2\epsilon)\dots(2n+2s-1+2\epsilon)]^2}{2^{2s}(n+1+\epsilon)(n+2+\epsilon)\dots(n+s+\epsilon).s!}\kappa'^{2s}$$

$$=\sum_0^\infty\frac{[(2n+1)(2n+3)\dots(2n+2s-1)]^2}{2^{2s}(n+1)(n+2)\dots(n+s).s!}\left[1+4\epsilon\sum_{n+1}^{n+s}\frac{1}{2t-1}-\epsilon\sum_{n+1}^{n+s}\frac{1}{t}\right]\kappa'^{2s}$$

Thus the second term of D is

$$(-)^{n+1}\frac{2n!}{2^{2n}(n!)^2}\frac{\kappa'^{2n}}{\epsilon}F(n+\tfrac{1}{2}, n+\tfrac{1}{2}, n+1, \kappa'^2)$$

$$+(-)^{n+1}\frac{2n!}{2^{2n}(n!)^2}\kappa'^{2n}\sum_0^\infty\frac{[(2n+1)\dots(2n+2s-1)]^2}{2^{2s}(n+1)\dots(n+s)s!}$$

$$\times\left[4\sum_{n+1}^{n+s}\frac{1}{2t-1}-\sum_{n+1}^{n+s}\frac{1}{t}+2\log\kappa'\right]\kappa'^{2s}$$

The first term of this becomes infinite when $\epsilon = 0$, but it is equal and opposite to the infinite term in the first part of D, and they annihilate one another.

Hence

$$D=\frac{2^{2n}\,n\,!\,\overline{n-1}\,!}{2n\,!}\sum_{0}^{n-1}\frac{(-)^r\,[1\,.\,3\ldots(2r-1)]^2}{2^{2r}\,(n-1)\ldots(n-r)\,r\,!}\,\kappa'^{2r}$$

$$+(-)^n\,\frac{2n\,!}{2^{2n}\,(n\,!)^2}\,\kappa'^{2n}\sum_{0}^{\infty}\frac{[(2n+1)\ldots(2n+2s-1)]^2}{2^{2s}\,(n+1)\ldots(n+s)\,.\,s\,!}$$

$$\times\left[2\log\frac{4}{\kappa'}+\sum_{1}^{n+s}\frac{1}{t}+\sum_{1}^{s}\frac{1}{t}-4\sum_{1}^{n+s}\frac{1}{2t-1}\right]\kappa'^{2s}$$

On examining the case of $n=0$ we find that this formula also embraces it, provided we interpret $\sum_{1}^{0}$ as meaning zero.

The coefficient in the last term admits of some simplification, for

$$\sum_{1}^{n+s}\frac{1}{t}+\sum_{1}^{s}\frac{1}{t}-4\sum_{1}^{n+s}\frac{1}{2t-1}=-\left[2\sum_{1}^{n+s}\frac{1}{t\,(2t-1)}+\sum_{t=1}^{t=n}\frac{1}{t+s}\right]$$

We thus conclude that D or

$$\int_{-\frac{1}{2}\pi}^{\frac{1}{2}\pi}\frac{\cos^{2n}\theta}{\Delta}\,d\theta$$

$$=\frac{2^{2n}\,n\,!\,\overline{n-1}\,!}{2n\,!}\left[1-\frac{1^2}{2^2\,(n-1)\,1\,!}\,\kappa'^2+\frac{1^2\,.\,3^2}{2^4\,(n-1)\,(n-2)\,2\,!}\,\kappa'^4-\ldots\text{ to } n \text{ terms}\right]$$

$$+(-)^n\,\frac{2n\,!}{2^{2n}\,(n\,!)^2}\,\kappa'^{2n}\left[\quad\left(2\log\frac{4}{\kappa'}-\sum_{1}^{n}\frac{1}{t}-2\sum_{1}^{n}\frac{1}{t\,(2t-1)}\right)\right.$$

$$+\frac{(2n+1)^2}{2^2\,(n+1)\,1\,!}\left(2\log\frac{4}{\kappa'}-\sum_{2}^{n+1}\frac{1}{t}-2\sum_{1}^{n+1}\frac{1}{t\,(2t-1)}\right)\kappa'^2$$

$$\left.+\frac{(2n+1)^2\,(2n+3)^2}{2^4\,(n+1)\,(n+2)\,2\,!}\left(2\log\frac{4}{\kappa'}-\sum_{3}^{n+2}\frac{1}{t}-2\sum_{1}^{n+2}\frac{1}{t\,(2t-1)}\right)\kappa'^4+\ldots\right]\quad\ldots(84)$$

The second integral E may be found as follows:—

$$E_n=\int\cos^{2n}\theta\,\Delta\,d\theta=\int\cos^{2n}\theta\,[\kappa'^2+(1-\kappa'^2)\cos^2\theta]\,\frac{d\theta}{\Delta}$$

$$=\kappa'^2 D_n+(1-\kappa'^2)\,D_{n+1}\quad\ldots\ldots\ldots\ldots(85)$$

From this I find E or

$$\int_{-\frac{1}{2}\pi}^{\frac{1}{2}\pi}\cos^{2n}\theta\,.\,\Delta\,d\theta$$

$$=\frac{2^{2n+1}\,(n\,!)^2}{2n+1\,!}+\frac{2^{2n-1}\,n\,!\,\overline{n-1}\,!}{2n+1\,!}\sum_{0}^{n-1}(-)^r\,\frac{1\,.\,3^2\ldots(2r-1)^2\,(2r+1)}{2^{2r}\,(n-1)\,(n-2)\ldots(n-r)\,.\,\overline{r+1}\,!}\,\kappa'^{2r+2}$$

$$+\frac{(-)^n\,2n\,!}{2^{2n+1}\,n\,!\,\overline{n+1}\,!}\,\kappa'^{2n}\sum_{1}^{\infty}\frac{(2n+1)\,(2n+3)^2\ldots(2n+2s-3)^2\,(2n+2s-1)}{2^{2s-2}\,(n+2)\,(n+3)\ldots(n+s)\,.\,\overline{s-1}\,!}$$

$$\times\left[2\log\frac{4}{\kappa'}+\sum_{1}^{n+s}\frac{1}{t}+\sum_{1}^{s-1}\frac{1}{t}-4\sum_{1}^{n+s-1}\frac{1}{2t-1}-\frac{2}{2n+2s-1}\right]\kappa'^{2s}$$

This is applicable also to the case of $n=0$, provided that $\sum\limits_0^{n-1}$ is interpreted as zero. In the particular case in hand I find, however, that it is shorter not to use this general formula, but to carry out the transformation (85) in the particular cases where the result is needed.

§ 20. *Reduction of preceding integrals; disappearance of logarithmic terms.*

In the application of the integrals of the last section, we are to put $\kappa'^2 = 1 - \dfrac{1-\beta}{1+\beta}$, and only to develop as far as β^2.

Then to the proposed order $\kappa'^2 = 2\beta(1-\beta)$, $\kappa'^4 = 4\beta^2$.

Also $$2\log\frac{4}{\kappa'} = \log\frac{8}{\beta} + \log(1+\beta) = \log\frac{8}{\beta} + \beta - \tfrac{1}{2}\beta^2$$

It will now facilitate future developments to adopt an abridged notation. I write then

$$f(n) = \frac{2^{2n}\, n!\, \overline{n-1}!}{2n!}$$

and observe that $f(n+1) = \dfrac{2n}{2n+1} f(n)$, and $f(1) = 2$, $f(2) = \tfrac{4}{3}$.

Since κ'^2 is of the first order in β, only the first series in the D integral (84) enters when n is greater than 2. In that case

$$\mathrm{D} = f(n)\left[1 - \frac{\beta-\beta^2}{2(n-1)} + \frac{9\beta^2}{8(n-1)(n-2)}\right]$$

$$= f(n)\left[1 - \frac{\beta}{2(n-1)} + \frac{(4n+1)\beta^2}{8(n-1)(n-2)}\right] \quad \ldots\ldots\ldots\ldots(86)$$

This result may be obtained very shortly without reference to the general formula; for when n is greater than 2

$$\mathrm{D} = \left(\frac{1+\beta}{1-\beta}\right)^{\frac{1}{2}} \int_{-\frac{1}{2}\pi}^{\frac{1}{2}\pi} \frac{\cos^{2n}\theta\, d\theta}{(\cos^2\theta + 2\beta + 2\beta^2)^{\frac{1}{2}}}$$

$$= \left(\frac{1+\beta}{1-\beta}\right)^{\frac{1}{2}} \int_{-\frac{1}{2}\pi}^{\frac{1}{2}\pi} \cos^{2n-1}\theta \left[1 - \frac{\beta+\beta^2}{\cos^2\theta} + \frac{3\beta^2}{2\cos^4\theta}\right] d\theta$$

The integral of an odd power of $\cos\theta$ is easily determined, and it will be found that the result (86) is obtained. It is, however, clear that if n is not greater than 2 the development in powers of $\sec^2\theta$ is not legitimate.

When n is not greater than 2 the formula (84) of the last section is necessary, and we find

$$\left.\begin{aligned}\int\frac{\cos^4\theta}{\Delta}d\theta &= \tfrac{3}{2}\beta^2\log\frac{8}{\beta}+f(2)\left[1-\tfrac{1}{2}\beta-\tfrac{61}{16}\beta^2\right]\\ \int\frac{\cos^2\theta}{\Delta}d\theta &= -(\beta+\tfrac{5}{4}\beta^2)\log\frac{8}{\beta}+f(1)\left[1+\tfrac{3}{2}\beta+\tfrac{19}{16}\beta^2\right]\\ \int\frac{d\theta}{\Delta} &= (1+\tfrac{1}{2}\beta+\tfrac{1}{16}\beta^2)\log\frac{8}{\beta}-\tfrac{5}{16}\beta^2\end{aligned}\right\}\quad\text{......(87)}$$

We have now to find the second integral E, and this may be done more easily than by reference to the general formula of the last section.

We have

$$\begin{aligned}E=\int\cos^{2n}\theta\Delta d\theta &= (1-\kappa'^2)\int\frac{\cos^{2n+2}\theta}{\Delta}d\theta+\kappa'^2\int\frac{\cos^{2n}\theta}{\Delta}d\theta\\ &= (1-2\beta+2\beta^2)\int\frac{\cos^{2n+2}\theta}{\Delta}d\theta+2\beta(1-\beta)\int\frac{\cos^{2n}\theta}{\Delta}d\theta\end{aligned}$$

It will be observed that even when n is 2 the general formula (86) gives the D integral as far as the first power of β. Hence in finding E we may use that general formula except when $n=0, 1$.

Then since $f(n)=\dfrac{2n+1}{2n}f(n+1)$, when n is greater than 1,

$$\begin{aligned}E &= f(n+1)\left[(1-2\beta+2\beta^2)\left(1-\frac{\beta}{2n}+\frac{(4n+5)\beta^2}{8n(n-1)}\right)\right.\\ &\qquad\left.+\frac{2n+1}{n}\beta(1-\beta)\left(1-\frac{\beta}{2(n-1)}\right)\right]\\ &= f(n+1)\left[1+\frac{\beta}{2n}-\frac{(4n-1)}{8n(n-1)}\beta^2\right]\quad\text{......(88)}\end{aligned}$$

But when $n=1$,

$$\begin{aligned}E &= (1-2\beta+2\beta^2)\,.\,\tfrac{3}{2}\beta^2\log\frac{8}{\beta}-2\beta(1-\beta)\,.\,\beta(1+\tfrac{5}{4}\beta)\log\frac{8}{\beta}\\ &\qquad+f(2)\left[(1-2\beta+2\beta^2)(1-\tfrac{1}{2}\beta-\tfrac{61}{16}\beta^2)+3\beta(1-\beta)(1+\tfrac{3}{2}\beta)\right]\\ &= -\tfrac{1}{2}\beta^2\log\frac{8}{\beta}+f(2)\left[1+\tfrac{1}{2}\beta+\tfrac{11}{16}\beta^2\right]\quad\text{......(89)}\end{aligned}$$

And when $n=0$,

$$\begin{aligned}E &= -(1-2\beta+2\beta^2)\beta(1+\tfrac{5}{4}\beta)\log\frac{8}{\beta}+2\beta(1-\beta)(1+\tfrac{1}{2}\beta)\log\frac{8}{\beta}\\ &\qquad+f(1)\left[(1-2\beta+2\beta^2)(1+\tfrac{3}{2}\beta+\tfrac{19}{16}\beta^2)-2\beta(1-\beta)\,.\,\tfrac{5}{16}\beta^2\right]\\ &= \beta(1-\tfrac{1}{4}\beta)\log\frac{8}{\beta}+f(1)\left[1-\tfrac{1}{2}\beta+\tfrac{3}{16}\beta^2\right]\quad\text{......(90)}\end{aligned}$$

I now wish to show that, in the use to which these integrals are to be put, the logarithmic terms disappear.

The following is a table of these integrals collected from (87), (89), (90), in as far only as they involve logarithms:—

$$\int\frac{d\theta}{\Delta}=(1+\tfrac{1}{2}\beta+\tfrac{1}{16}\beta^2)\log\frac{8}{\beta},\qquad \int\Delta\,d\theta=\beta(1-\tfrac{1}{4}\beta)\log\frac{8}{\beta}$$

$$\int\frac{\cos^2\theta}{\Delta}\,d\theta=-\beta(1+\tfrac{5}{4}\beta)\log\frac{8}{\beta},\qquad \int\cos^2\theta\Delta\,d\theta=-\tfrac{1}{2}\beta^2\log\frac{8}{\beta}$$

$$\int\frac{\cos^4\theta}{\Delta}\,d\theta=\tfrac{3}{2}\beta^2\log\frac{8}{\beta}$$

Then the formulæ (83) for L, M, N, in so far only as is at present material, are

$$\frac{L}{F}=\int\left[\frac{\gamma_0}{\Delta}(\cos^4\theta+\beta f\cos^2\theta)+\frac{\gamma_2}{\Delta}\beta f\cos^4\theta\right]d\theta$$

$$\frac{M}{G}=\int[\gamma_0\Delta(\cos^2\theta+\beta g)+\gamma_2\Delta\beta g\cos^2\theta]\,d\theta$$

$$\frac{N}{H}=\int\left[\frac{\alpha_0}{\Delta}(\cos^2\theta+\beta h)+\frac{\alpha_2}{\Delta}(\cos^4\theta+\beta h\cos^2\theta)+\frac{\alpha_4}{\Delta}\beta h\cos^4\theta\right]d\theta$$

On using the integrals and only retaining squares of β, we find

$$\frac{L}{F}=\beta^2\gamma_0(\tfrac{3}{2}-f)\log\frac{8}{\beta}$$

$$\frac{M}{G}=\beta^2\gamma_0(g-\tfrac{1}{2})\log\frac{8}{\beta}$$

$$\frac{N}{H}=\{\beta\alpha_0[-(1+\tfrac{5}{4}\beta)+h(1+\tfrac{1}{2}\beta)]+\beta^2\alpha_2(\tfrac{3}{2}-h)\}\log\frac{8}{\beta}$$

But by definition of f and g in (80) and of the α's in (82), to the order zero of small quantities,

$$f=\tfrac{3}{2},\quad g=\tfrac{1}{2},\quad h=1+\tfrac{1}{4}\beta j+\tfrac{3}{4}\beta,\quad \alpha_0=1,\quad \alpha_2=-\tfrac{1}{2}i(i+1)=-\tfrac{1}{2}j$$

Thus the logarithmic terms entirely disappear, and henceforth may be dropped.

Thus, *as far as material*, we have the following table of integrals:—

$$\left.\begin{aligned}
&\int\frac{d\theta}{\Delta}=-\tfrac{5}{16}\beta^2,\qquad \int\frac{\cos^2\theta}{\Delta}\,d\theta=f(1)[1+\tfrac{3}{2}\beta+\tfrac{19}{16}\beta^2]\\
&\int\frac{\cos^4\theta}{\Delta}\,d\theta=f(2)[1-\tfrac{1}{2}\beta-\tfrac{61}{16}\beta^2]\\
&\int\frac{\cos^{2n}\theta}{\Delta}\,d\theta=f(n)\left[1-\frac{\beta}{2(n-1)}+\frac{(4n+1)\beta^2}{8(n-1)(n-2)}\right],\ n>2\\
&\int\Delta\,d\theta=f(1)[1-\tfrac{1}{2}\beta+\tfrac{3}{16}\beta^2]\\
&\int\cos^2\theta\Delta\,d\theta=f(2)[1+\tfrac{1}{2}\beta+\tfrac{11}{16}\beta^2]\\
&\int\cos^{2n}\theta\Delta\,d\theta=f(n+1)\left[1+\frac{\beta}{2n}-\frac{(4n-1)\beta^2}{8n(n-1)}\right],\ n>1
\end{aligned}\right\}\quad\ldots\ldots(91)$$

Before using these for the determination of L, M, N, it is well to obtain one other result.

We have seen in (82) that

$$(\mathrm{P}^s)^2 = \frac{\overline{i+s}\,!}{\overline{i-s}\,!}\sum_s^i (-)^{r+s} \frac{\overline{2r}\,!}{2^{2r}\,\overline{r+s}\,!\,(r\,!)^2\,\overline{r-s}\,!}\cos^{2r}\theta$$

Therefore

$$\int_{-1}^{+1}(\mathrm{P}^s)^2\,d\mu = \frac{\overline{i+s}\,!}{\overline{i-s}\,!}\sum_s^i (-)^{r+s} f(r+1)\frac{\overline{2r}\,!}{2^{2r}\,\overline{r+s}\,!\,(r\,!)^2\,\overline{r-s}\,!}$$

But this integral is equal to $\dfrac{2}{2i+1}\dfrac{\overline{i+s}\,!}{\overline{i-s}\,!}$; therefore

$$\sum_s^i (-)^{r+s} f(r+1)\frac{\overline{2r}\,!}{2^{2r}\,\overline{r+s}\,!\,(r\,!)^2\,\overline{r-s}\,!} = \frac{2}{2i+1}$$

Putting $s=1$ and 0, and comparing with the values of α_{2r}, γ_{2r-2} in (82), we have

$$\left.\begin{aligned}\sum_0^i \alpha_{2r} f(r+1) &= \frac{2}{2i+1}\\ \sum_1^i \gamma_{2r-2} f(r+1) &= \frac{2}{2i+1}\cdot\frac{\overline{i+1}\,!}{\overline{i-1}\,!}\end{aligned}\right\}\quad\ldots\ldots\ldots\ldots\ldots\ldots(92)$$

§ 21. *Integrals of the squares of harmonics when $s=1$ and $s=0$.*

In (83) we have

$$\frac{\mathrm{L}}{\mathrm{F}} = \sum_1^i \int_{-\frac{1}{2}\pi}^{\frac{1}{2}\pi}\frac{\gamma_{2r-2}}{\Delta}(\cos^{2r+2}\theta + \beta f\cos^{2r}\theta)\,d\theta$$

Therefore, noting that $f(r) = \dfrac{2r+1}{2r}f(r+1)$, and using the integrals (91),

$$\frac{\mathrm{L}}{\mathrm{F}} = \sum_2^i \gamma_{2r-2} f(r+1)\left[1 - \frac{\beta}{2r} + \frac{(4r+5)\beta^2}{8r(r-1)} + \frac{2r+1}{2r}\beta\left(1 - \frac{\beta}{2(r-1)}\right)f\right]$$
$$+\gamma_0 f(2)\left[1 - \tfrac{1}{2}\beta - \tfrac{61}{16}\beta^2 + \tfrac{3}{2}\beta\left(1 + \tfrac{3}{2}\beta\right)f\right]$$

Substituting for f (which the reader must not confuse with the functional f in use here) its value (80), the term of order zero is $\sum_1^i \gamma_{2r-2} f(r+1)$. By (92) this is equal to $\dfrac{2}{2i+1}\cdot\dfrac{\overline{i+1}\,!}{\overline{i-1}\,!}$.

The term of the first order in β is

$$\beta\sum_2^i \gamma_{2r-2} f(r+1)\left(-\frac{1}{2r} + \frac{3(2r+1)}{4r}\right) + \beta\gamma_0 f(2)\left(-\tfrac{1}{2} + \tfrac{9}{4}\right)$$

which may be reduced to the form

$$\beta\sum_1^i \gamma_{2r-2} f(r+1) + \tfrac{1}{2}\beta\sum_1^i \gamma_{2r-2} f(r)$$

and is equal to $$\frac{2\beta}{2i+1}\frac{i+1!}{i-1!} + \tfrac{1}{2}\beta\sum_1^i \gamma_{2r-2} f(r)$$

The term of the second order in β is

$$\beta^2\sum_2^i \gamma_{2r-2} f(r+1)\left[\frac{4r+5}{8r(r-1)} + \frac{3(2r+1)}{4r}\left(-\frac{1}{2(r-1)} + \tfrac{1}{24}j + \tfrac{5}{6}\right)\right]$$
$$+ \beta^2\gamma_0 f(2)\left[-\tfrac{61}{16} + \tfrac{9}{4}\left(\tfrac{3}{2} + \tfrac{1}{24}j + \tfrac{5}{6}\right)\right]$$

This may be reduced to the form

$$\beta^2\left[\tfrac{1}{2}\sum_1^i \gamma_{2r-2} f(r+1) + \tfrac{1}{16}(j+12)\sum_1^i \gamma_{2r-2} f(r) - \tfrac{1}{4}\gamma_0\right]$$

of which the first term is $\tfrac{1}{2}\beta^2 . \dfrac{2}{2i+1}\dfrac{i+1!}{i-1!}$.

Therefore

$$\frac{L}{F} = (1+\beta+\tfrac{1}{2}\beta^2)\frac{2}{2i+1}\frac{i+1!}{i-1!} + \tfrac{1}{2}\beta\left[1+\tfrac{1}{8}\beta(j+12)\right]\sum_1^i \gamma_{2r-2} f(r) - \tfrac{1}{4}\beta^2\gamma_0$$

Now $$\gamma_{2r-2} f(r) = (-)^{r+1}\frac{i+1!}{i-1!}\frac{1}{r+1!\,r!}\frac{i+r!}{i-r!}$$

and $$\sum_1^i \frac{(-)^{r+1}}{r+1!\,r!}\frac{i+r!}{i-r!} = \frac{(i+1)i}{1!\,2} - \frac{(i+1)(i+2)i(i-1)}{2!\,2.3} + \dots$$
$$= 1 - F(i+1, -i, 2, 1)$$

It is known that

$$F(a, b, c, 1) = \frac{\Pi(c-1)\,\Pi(c-a-b-1)}{\Pi(c-a-1)\,\Pi(c-b-1)}\text{*} \quad \dots\dots\dots\dots(93)$$

Then since $\Pi(-i)$ contains an infinite factor

$$F(i+1, -i, 2, 1) = \frac{\Pi(1)\,\Pi(0)}{\Pi(-i)\,\Pi(i+1)} = 0$$

therefore $$\sum_1^i \gamma_{2r-2} f(r) = \frac{i+1!}{i-1!} \quad \dots\dots\dots\dots(93)$$

Also $$\gamma_0 = \tfrac{1}{4}i^2(i+1)^2 = \tfrac{1}{4}j\frac{i+1!}{i-1!}$$

Hence $$\frac{L}{F} = \frac{2}{2i+1}\frac{i+1!}{i-1!}(1+\beta+\tfrac{1}{2}\beta^2) + \tfrac{1}{2}\beta\frac{i+1!}{i-1!}(1+\tfrac{3}{2}\beta)$$

Introducing the value of F as defined in (79), we have L or

$$\int p(\mathfrak{P}_i^1 P_i^1)^2\, d\sigma = \frac{2\pi M}{2i+1}\frac{i+1!}{i-1!}\left[1 - \tfrac{3}{4}\beta + \tfrac{1}{256}\beta^2(j^2 - 12j + 68)\right]$$
$$+ \pi M\frac{i+1!}{i-1!}\left[\tfrac{1}{2}\beta - \tfrac{1}{8}\beta^2\right]$$

* I have again to thank Mr Hobson for this formula, which is due to Gauss.

We have in (74) obtained $\int p\,(\mathfrak{S}_i^1)^2\,[(\mathfrak{P}_i^1)^2-(\mathrm{P}_i^1)^2]\,d\sigma$, and if it be added to our last result we see that the term which does not involve the factor $1/(2i+1)$ is annihilated, and

$$\int p\,(\mathfrak{P}_i^1\mathfrak{S}_i^1)^2\,d\sigma=\frac{2\pi\mathrm{M}}{2i+1}\frac{i+1!}{i-1!}[1+\tfrac{1}{8}\beta(j-2)+\tfrac{1}{384}\beta^2(j^2-26j+48)]\ldots(94)$$

Now from (37) the square of the factor for converting $\mathfrak{S}^1$ into S^1 is

$$\left[\frac{1}{\mathrm{D}_i^1}(\sin)\right]^2=1+\tfrac{1}{2}\beta+\tfrac{1}{32}\beta^2(j-8)$$

Therefore

$$\int p\,(\mathfrak{P}_i^1\mathrm{S}_i^1)^2\,d\sigma=\frac{2\pi\mathrm{M}}{2i+1}\cdot\frac{i+1!}{i-1!}[1+\tfrac{1}{8}\beta(j+2)+\tfrac{1}{384}\beta^2(j^2+10j-96)]\ldots(94)$$

These are two of the required integrals.

Next we have from (83)

$$\frac{\mathrm{M}}{\mathrm{G}}=\sum_1^i\int\gamma_{2r-2}(\cos^{2r}\theta+\beta g\cos^{2r-2}\theta)\,\Delta\,d\theta$$

Noting as before that $f(r)=\dfrac{2r+1}{2r}f(r+1)$, and using the integrals (91),

$$\frac{\mathrm{M}}{\mathrm{G}}=\sum_2^i\gamma_{2r-2}f(r+1)\left[1+\frac{\beta}{2r}-\frac{(4r-1)\beta^2}{8(r-1)r}+\frac{2r+1}{2r}\beta g\left(1+\frac{\beta}{2(r-1)}\right)\right]$$
$$+\gamma_0 f(2)[1+\tfrac{1}{2}\beta+\tfrac{11}{16}\beta^2+\tfrac{3}{2}\beta g(1-\tfrac{1}{2}\beta)]$$

Substituting for g its value from (80), I find the term of order zero to be $\sum_1^i\gamma_{2r-2}f(r+1)$, or $\dfrac{2}{2i+1}\cdot\dfrac{i+1!}{i-1!}$.

The term of the first order is

$$\beta\sum_2^i\gamma_{2r-2}f(r+1)\left[\frac{1}{2r}+\frac{2r+1}{4r}\right]+\beta\gamma_0 f(2)(\tfrac{3}{4}+\tfrac{1}{2})$$

This may be reduced to the form $-\beta\sum_1^i\gamma_{2r-2}f(r+1)+\tfrac{3}{2}\beta\sum_1^i\gamma_{2r-2}f(r)$; which by (92) and (93) becomes $-\dfrac{2\beta}{2i+1}\dfrac{i+1!}{i-1!}+\tfrac{3}{2}\beta\dfrac{i+1!}{i-1!}$.

The term of the second order is

$$\beta^2\sum_2^i\gamma_{2r-2}f(r+1)\left[-\frac{(4r-1)}{8r(r-1)}+\frac{2r+1}{4r}\left(\frac{1}{2(r-1)}+\tfrac{1}{8}j+\tfrac{1}{2}\right)\right]$$
$$+\beta^2\gamma_0 f(2)(\tfrac{11}{16}+\tfrac{3}{32}j)$$

This is reducible to

$$\tfrac{1}{2}\beta^2\sum_1^i\gamma_{2r-2}f(r+1)+\tfrac{1}{16}\beta^2(j-4)\sum_1^i\gamma_{2r-2}f(r)+\tfrac{3}{4}\beta^2\gamma_0$$

which becomes

$$\tfrac{1}{2}\beta^2 \frac{2}{2i+1} \frac{i+1!}{i-1!} + \tfrac{1}{16}\beta^2 (j-4) \frac{i+1!}{i-1!} + \tfrac{3}{16}\beta^2 \frac{i+1!}{i-1!} j$$

Therefore

$$\frac{M}{G} = \frac{2}{2i+1} \frac{i+1!}{i-1!} (1 - \beta + \tfrac{1}{2}\beta^2) + \frac{i+1!}{i-1!} [\tfrac{3}{2}\beta + \tfrac{1}{4}\beta^2 (j-1)]$$

Introducing for G its value (79), we find M or

$$\int_{-\frac{1}{2}\pi}^{\frac{1}{2}\pi} p \left(\mathfrak{C}_i^1 \frac{P_i^1}{\cos\theta}\right)^2 \left(\tfrac{1+\beta}{1-\beta} - \sin^2\theta\right) d\sigma$$

$$= \frac{2\pi M}{2i+1} \frac{i+1!}{i-1!} [1 - \tfrac{1}{4}\beta + \tfrac{1}{256}\beta^2 (j^2 - 12j + 4)] + \pi M \frac{i+1!}{i-1!} [\tfrac{3}{2}\beta + \tfrac{1}{8}\beta^2 (2j+7)]$$

But in (74) we have $\int p (\mathfrak{C}_i^1)^2 \left[(P_i^1)^2 - \left(P_i^1 \sqrt{\frac{\frac{1+\beta}{1-\beta} - \sin^2\theta}{\cos^2\theta}}\right)^2\right] d\sigma$. If this be added to the result just found the term which has not $1/(2i+1)$ as a factor is annihilated, and

$$\int_{-\frac{1}{2}\pi}^{\frac{1}{2}\pi} p (P_i^1 \mathfrak{C}_i^1)^2 d\sigma$$

$$= \frac{2\pi M}{2i+1} \frac{i+1!}{i-1!} [1 + \tfrac{1}{8}\beta (3j+10) + \tfrac{1}{384}\beta^2 (29j^2 + 134j + 384)] \quad \ldots(95)$$

Now from (37) the square of the factor for converting $\mathfrak{C}^1$ into C^1 is

$$\left[\frac{1}{D_i^1}(\cos)\right]^2 = 1 - \tfrac{1}{2}\beta + \tfrac{1}{32}\beta^2 (j-8)$$

Therefore

$$\int_{-\frac{1}{2}\pi}^{\frac{1}{2}\pi} p (P_i^1 C_i^1)^2 d\sigma = \frac{2\pi M}{2i+1} \frac{i+1!}{i-1!} [1 + \tfrac{3}{8}\beta (j+2) + \tfrac{1}{384}\beta^2 (29j^2 + 74j + 48)]$$

$$\ldots\ldots(95)$$

These last two complete the solution for $s=1$.

Next we have from (83)

$$\frac{N}{H} = \sum_0^i \int \frac{\alpha_{2r}}{\Delta} (\cos^{2r+2}\theta + \beta h \cos^{2r}\theta)\, d\theta$$

Proceeding as before,

$$\frac{N}{H} = \sum_2^i \alpha_{2r} f(r+1) \left[1 - \frac{\beta}{2r} + \frac{4r+5}{8r(r-1)} \beta^2 + \frac{2r+1}{2r} \beta h \left(1 - \frac{\beta}{2(r-1)}\right)\right]$$

$$+ \alpha_2 f(2) [1 - \tfrac{1}{2}\beta - \tfrac{61}{16}\beta^2 + \tfrac{3}{2}\beta h (1 + \tfrac{3}{2}\beta)]$$

$$+ \alpha_0 f(1) [1 + \tfrac{3}{2}\beta + \tfrac{19}{16}\beta^2] + \alpha_0 \beta h (-\tfrac{5}{16}\beta^2)$$

Substituting for h its value (80), we find that the term of order zero is $\sum_0^i \alpha_{2r} f(r+1)$, and by (92) this is equal to $\frac{2}{2i+1}$.

The term of the first order is

$$\beta\sum_2^i \alpha_{2r} f(r+1)\left[-\frac{1}{2r}+\frac{2r+1}{2r}\right]+\tfrac{4}{3}\beta\alpha_2+3\beta\alpha_0$$

which may be written in the form

$$\beta\sum_0^i \alpha_{2r} f(r+1)+\beta\alpha_0, \text{ and is equal to } \frac{2\beta}{2i+1}+\beta$$

The term of the second order is

$$\beta^2\sum_2^i \alpha_{2r} f(r+1)\left[\frac{4r+5}{8r(r-1)}-\frac{2r+1}{4r(r-1)}+\frac{2r+1}{8r}(j+3)\right]$$
$$+\beta^2\alpha_2\left(\tfrac{1}{2}j-\tfrac{7}{12}\right)+\tfrac{19}{8}\beta^2\alpha_0$$

which is equal to

$$\tfrac{3}{8}\beta^2\sum_2^i \alpha_{2r}\frac{f(r+1)}{r(r-1)}+\tfrac{1}{4}\beta^2(j+3)\sum_2^i \alpha_{2r} f(r)+\beta^2\alpha_2\left(\tfrac{1}{2}j-\tfrac{7}{12}\right)+\tfrac{19}{8}\beta^2\alpha_0$$

Now $$\tfrac{3}{8}\frac{f(r+1)}{r(r-1)}=\tfrac{1}{2}f(r+1)-\tfrac{3}{4}f(r)+\tfrac{1}{4}\frac{r}{r-1}f(r)$$

Hence the term may be written

$$\tfrac{1}{2}\beta^2\sum_2^i \alpha_{2r} f(r+1)+\tfrac{1}{4}\beta^2\sum_1^{i-1}\alpha_{2r+2}\frac{r+1}{r}f(r+1)+\tfrac{1}{4}\beta^2 j\sum_2^i \alpha_{2r} f(r)$$
$$+\beta^2\alpha_2\left(\tfrac{1}{2}j-\tfrac{7}{12}\right)+\tfrac{19}{8}\beta^2\alpha_0$$

But $$\alpha_{2r} f(r)=(-)^r\frac{1}{(r!)^2 r}\frac{i+r!}{i-r!}$$

And

$$\frac{r+1}{r}\alpha_{2r+2} f(r+1)=-(-)^r\frac{1}{(r!)^2(r+1)^2 r}\left[i(i+1)-r(r+1)\right]\frac{i+r!}{i-r!}$$
$$=\left[-\frac{(-)^r}{(r!)^2 r}\frac{j}{(r+1)^2}+\frac{(-)^r}{(r!)^2(r+1)}\right]\frac{i+r!}{i-r!}$$

In the preceding formula the sum of this last function had limits $i-1$ to 1, but as we now see that it vanishes when $r=i$, the upper limit may be changed to i.

It follows that the terms of the second order are

$$\tfrac{1}{2}\beta^2\sum_0^i \alpha_{2r} f(r+1)-\tfrac{1}{2}\beta^2\alpha_2 f(2)-\tfrac{1}{2}\beta^2\alpha_0 f(1)$$
$$+\tfrac{1}{4}\beta^2\sum_1^i\left[-\frac{(-)^r}{(r!)^2 r}\frac{j}{(r+1)^2}+\frac{(-)^r}{(r!)^2(r+1)}\right]\frac{i+r!}{i-r!}+\tfrac{1}{4}\beta^2 j\sum_1^i\frac{(-)^r}{r(r!)^2}\frac{i+r!}{i-r!}$$
$$-\tfrac{1}{4}\beta^2 j\alpha_2 f(1)+\beta^2\alpha_2\left(\tfrac{1}{2}j-\tfrac{7}{12}\right)+\tfrac{19}{8}\beta^2\alpha_0$$

The term in α_2 in this expression will be found to be $-\tfrac{5}{4}\alpha_2$. That in α_0 will be found to be $\tfrac{11}{8}\alpha_0$. Then since $\alpha_2=-\tfrac{1}{2}j$, $\alpha_0=1$, these terms are together $\tfrac{1}{8}\beta^2(5j+11)$.

The whole may then be written

$$\tfrac{1}{2}\beta^2\sum_0^i \alpha_{2r} f(r+1) + \tfrac{1}{4}\beta^2 (j+1)\sum_1^i \frac{(-)^r}{r!\,r+1!}\frac{i+r!}{i-r!}$$

$$+\tfrac{1}{4}\beta^2 j\sum_1^i \frac{(-)^r}{(r+1!)^2}\frac{i+r!}{i-r!} + \tfrac{1}{8}\beta^2(5j+11)$$

Now

$$\sum_1^i \frac{(-)^r}{r!\,r+1!}\frac{i+r!}{i-r!} = -\frac{(i+1)\,i}{1!\,2} + \frac{(i+1)(i+2)\,i\,(i-1)}{2!\,2.3} - \dots$$

$$= F(i+1, -i, 2, 1) - 1 = -1$$

$$\sum_1^i \frac{(-)^r}{(r+1)^2}\frac{i+r!}{i-r!} = -\frac{i(i+1)}{2!\,2!} + \frac{(i+1)(i+2)\,i\,(i-1)}{3!\,3!} - \dots$$

$$= -\frac{1}{i(i+1)}\left[\frac{i(i+1)(-i-1)(-i)}{2!\,1.2} + \frac{i(i+1)(i+2)(-i-1)(-i)(-i+1)}{3!\,1.2.3} + \dots\right]$$

$$= -\frac{1}{i(i+1)}\{F(i, -i-1, 1, 1) - 1 + i(i+1)\}$$

$$= -\frac{1}{j}F(i, -i-1, 1, 1) + \frac{1}{j} - 1$$

$$= \frac{1}{j} - 1$$

The last result follows from the fact that in accordance with (93) the sum of the hypergeometric series has an infinite factor in the denominator, and vanishes.

Then since by (92) $\sum_0^i \alpha_{2r} f(r+1) = \dfrac{2}{2i+1}$, the terms of the second order are found to be

$$\tfrac{1}{2}\beta^2\frac{2}{2i+1} + \tfrac{1}{8}\beta^2 j + \tfrac{11}{8}\beta^2$$

Hence, collecting terms,

$$\frac{\mathrm{N}}{\mathrm{H}} = (1+\beta+\tfrac{1}{2}\beta^2)\frac{2}{2i+1} + \beta + \tfrac{1}{8}\beta^2(j+11)$$

Substituting for H its value (79), we have N or

$$\int_{-\frac{1}{2}\pi}^{\frac{1}{2}\pi} p\,(\mathfrak{C}_i \mathrm{P}_i)^2\,d\sigma = \frac{4\pi \mathrm{M}}{2i+1}\left[1 - \tfrac{1}{2}\beta + \tfrac{1}{32}\beta^2(j^2 - 4j + 2)\right]$$

$$+ 2\pi\mathrm{M}\left[\beta + \tfrac{1}{8}\beta^2(j-1)\right]$$

But we have already found in (77) the value of $\int p\,(\mathfrak{C}_i)^2\,[(\mathfrak{P}_i)^2 - (\mathrm{P}_i)^2]\,d\sigma$, and on adding it to the last result the term independent of $1/(2i+1)$ disappears, and we have

$$\int_{-\frac{1}{2}\pi}^{\frac{1}{2}\pi} p\,(\mathfrak{P}_i \mathfrak{C}_i)^2\,d\sigma = \frac{4\pi M}{2i+1}\left[1 + \tfrac{1}{2}\beta j + \tfrac{1}{64}\beta^2(7j^2 - 10j)\right] \quad \ldots\ldots(96)$$

The square of the factor whereby $\mathfrak{C}_i$ is converted into C_i was found in (38), namely,

$$\frac{1}{(\mathrm{D}_i)^2} = 1 + \tfrac{1}{8}\beta^2(j-3)$$

Hence

$$\int_{-\frac{1}{2}\pi}^{\frac{1}{2}\pi} p\,(\mathfrak{P}_i \mathsf{C}_i)^2\,d\sigma = \frac{4\pi M}{2i+1}\left[1 + \tfrac{1}{2}\beta j + \tfrac{1}{64}\beta^2(7j^2 - 2j - 24)\right] \quad \ldots(96)$$

These are the last of the required integrals.

§ 22. *Table of Integrals of squares of harmonics*.*

In this section the results obtained in (71), (72), (94), (95), and (96) are collected.

* After having completed the evaluation of all these integrals, I found that they may be evaluated very shortly by means of the factors $\mathfrak{E}$ and E of (48), § 10.

I find that for all values of s (writing the eight forms in a single formula),

$$\int p \begin{Bmatrix} (\mathfrak{P}_i^s)^2 \\ (\mathsf{P}_i^s)^2 \end{Bmatrix} \times \begin{matrix} (\mathfrak{C}_i^s)^2 \text{ or } (\mathsf{C}_i^s)^2 \\ (\mathfrak{S}_i^s)^2 \text{ or } (\mathsf{S}_i^s)^2 \end{matrix}\, d\sigma = \frac{4\pi M}{2i+1}(1-\beta)^{\frac{1}{2}} \cdot \begin{Bmatrix} \mathfrak{E}_i^s \\ \mathsf{E}_i^s \end{Bmatrix} \times \left\{\text{const. part of } \frac{(\mathfrak{C}_i^s \text{ or } \mathsf{C}_i^s \text{ or } \mathfrak{S}_i^s \text{ or } \mathsf{S}_i^s)^2}{(1-\beta\cos 2\phi)^{\frac{1}{2}}}\right\}$$

I leave the reader to verify that this is so.

Unfortunately I have hitherto been unable to prove the truth of this except by the laborious method in the text. I do not therefore know whether the result remains true for higher degrees of approximation, although I suspect it does so. If it should be true, it would be very easy to compute the integrals when higher powers of β are included.

It may be worth mentioning that the variables are separable in the integrals. Thus, when $\mathfrak{P}_i^s \mathfrak{C}_i^s$ denotes any one of the eight forms,

$$\frac{1}{M(1-\beta)^{\frac{1}{2}}}\int p\,[\mathfrak{P}_i^s \mathfrak{C}_i^s]^2\,d\sigma = \frac{1}{\sqrt{(1+\beta)}}\int_{-\frac{1}{2}\pi}^{\frac{1}{2}\pi} \frac{(\mathfrak{P}_i^s)^2\,d\theta}{\Delta}\int_0^{2\pi}(1-\beta\cos 2\phi)^{\frac{1}{2}}(\mathfrak{C}_i^s)^2\,d\phi$$
$$- \sqrt{\left(\frac{1-\beta}{1+\beta}\right)}\int_{-\frac{1}{2}\pi}^{\frac{1}{2}\pi}\frac{\sin^2\theta\,(\mathfrak{P}_i^s)^2}{\Delta}\,d\theta\int_0^{2\pi}\frac{(\mathfrak{C}_i^s)^2}{(1-\beta\cos 2\phi)^{\frac{1}{2}}}\,d\phi$$

The ϕ integrals present no difficulty, but with regard to the others we are met by the impossibility of expanding in powers of $\sec^2\theta$ for the lower orders. It would be a great step in the right direction, if it could be proved that all the terms which do not involve the factor $\frac{1}{2i+1}$ necessarily vanish.

It may be well to remind the reader that $M = k^3\nu(\nu^2-1)^{\frac{1}{2}}\left(\nu^2 - \frac{1+\beta}{1-\beta}\right)^{\frac{1}{2}}$

$$\Sigma = \frac{i(i+1)}{s^2-1}$$

$$\Upsilon = \frac{(i-1)i(i+1)(i+2)}{s^2-4}$$

$$j = i(i+1)$$

First when $s > 2$.

Types $\begin{cases}\text{EEC}\\ \text{OOS}\end{cases}$ $\int p\left(\mathfrak{P}_i^s(\mu)\begin{cases}\mathfrak{C}_i^s\\ \mathfrak{S}_i^s\end{cases}(\phi)\right)^2 d\sigma$

$$= \frac{2\pi M}{2i+1}\frac{i+s!}{i-s!}\{1 - \tfrac{1}{2}\beta\Sigma + \tfrac{1}{16}\beta^2[2\Sigma^2 + 3\Sigma - s^2(2\Sigma-1) + \Upsilon]\}$$

Types $\begin{cases}\text{OEC}\\ \text{EOS}\end{cases}$ $\int p\left(\mathfrak{P}_i^s(\mu)\begin{cases}\mathsf{C}_i^s\\ \mathsf{S}_i^s\end{cases}(\phi)\right)^2 d\sigma$

$$= \frac{2\pi M}{2i+1}\frac{i+s!}{i-s!}\{1 - \tfrac{1}{2}\beta\Sigma + \tfrac{1}{16}\beta^2[2\Sigma^2 + \Sigma - 6 - s^2(2\Sigma-1) + \Upsilon]\}$$

Types $\begin{cases}\text{OOC}\\ \text{EES}\end{cases}$ $\int p\left(\mathsf{P}_i^s(\mu)\begin{cases}\mathfrak{C}_i^s\\ \mathfrak{S}_i^s\end{cases}(\phi)\right)^2 d\sigma$

$$= \frac{2\pi M}{2i+1}\frac{i+s!}{i-s!}\{1 + \tfrac{1}{2}\beta(\Sigma+2) + \tfrac{1}{16}\beta^2[11\Sigma + 10 + s^2(2\Sigma^2 - 2\Sigma + 1) - \Upsilon]\}$$

Types $\begin{cases}\text{EOC}\\ \text{OES}\end{cases}$ $\int p\left(\mathsf{P}_i^s(\mu)\begin{cases}\mathsf{C}_i^s\\ \mathsf{S}_i^s\end{cases}(\phi)\right)^2 d\sigma$

$$= \frac{2\pi M}{2i+1}\frac{i+s!}{i-s!}\{1 + \tfrac{1}{2}\beta(\Sigma+2) + \tfrac{1}{16}\beta^2[9\Sigma + 4 + s^2(2\Sigma^2 - 2\Sigma + 1) - \Upsilon]\}$$

Secondly, when $s = 2$, $\Sigma = \frac{1}{3}i(i+1)$.

Type EEC $\int p[\mathfrak{P}_i^2(\mu)\,\mathfrak{C}_i^2(\phi)]^2 d\sigma$

$$= \frac{2\pi M}{2i+1}\frac{i+2!}{i-2!}\{1 - \tfrac{1}{2}\beta\Sigma + \tfrac{1}{256}\beta^2(95\Sigma^2 - 98\Sigma + 72)\}$$

Type OEC $\int p[\mathfrak{P}_i^2(\mu)\,\mathsf{C}_i^2(\phi)]^2 d\sigma$

$$= \frac{2\pi M}{2i+1}\frac{i+2!}{i-2!}\{1 - \tfrac{1}{2}\beta\Sigma + \tfrac{1}{256}\beta^2(95\Sigma^2 - 178\Sigma - 40)\}$$

Type EES $\int p[\mathsf{P}_i^2(\mu)\,\mathfrak{S}_i^2(\phi)]^2 d\sigma$

$$= \frac{2\pi M}{2i+1}\frac{i+2!}{i-2!}\{1 + \tfrac{1}{2}\beta(\Sigma+2) + \tfrac{1}{256}\beta^2(29\Sigma^2 + 90\Sigma + 216)\}$$

Type OES $\int p[\mathsf{P}_i^2(\mu)\,\mathsf{S}_i^2(\phi)]^2 d\sigma$

$$= \frac{2\pi M}{2i+1}\frac{i+2!}{i-2!}\{1 + \tfrac{1}{2}\beta(\Sigma+2) + \tfrac{1}{256}\beta^2(29\Sigma^2 + 106\Sigma + 136)\}$$

Thirdly, when $s=1$, Σ is infinite and we must use $j=i(i+1)$.

Type OOS $\int p\,[\mathfrak{P}_i^1(\mu)\,\mathfrak{S}_i^1(\phi)]^2\,d\sigma$

$$=\frac{2\pi M}{2i+1}\frac{i+1!}{i-1!}\{1+\tfrac{1}{8}\beta(j-2)+\tfrac{1}{384}\beta^2(j^2-26j+48)\}$$

Type EOS $\int p\,[\mathfrak{P}_i^1(\mu)\,\mathsf{S}_i^1(\phi)]^2\,d\sigma$

$$=\frac{2\pi M}{2i+1}\frac{i+1!}{i-1!}\{1+\tfrac{1}{8}\beta(j+2)+\tfrac{1}{384}\beta^2(j^2+10j-96)\}$$

Type OOC $\int p\,[\mathsf{P}_i^1(\mu)\,\mathfrak{C}_i^1(\phi)]^2\,d\sigma$

$$=\frac{2\pi M}{2i+1}\frac{i+1!}{i-1!}\{1+\tfrac{1}{8}\beta(3j+10)+\tfrac{1}{384}\beta^2(29j^2+134j+384)\}$$

Type EOC $\int p\,[\mathsf{P}_i^1(\mu)\,\mathsf{C}_i^1(\phi)]^2\,d\sigma$

$$=\frac{2\pi M}{2i+1}\frac{i+1!}{i-1!}\{1+\tfrac{3}{8}\beta(j+2)+\tfrac{1}{384}\beta^2(29j^2+74j+48)\}$$

Lastly, when $s=0$; $\Sigma=-i(i+1)=-j$. There are only two types—

Type EEC $\int p\,[\mathfrak{P}_i(\mu)\,\mathfrak{C}_i(\phi)]^2\,d\sigma=\frac{4\pi M}{2i+1}[1+\tfrac{1}{2}\beta j+\tfrac{1}{64}\beta^2(7j^2-10j)]$

Type OEC $\int p\,[\mathfrak{P}_i(\mu)\,\mathsf{C}_i(\phi)]^2\,d\sigma=\frac{4\pi M}{2i+1}[1+\tfrac{1}{2}\beta j+\tfrac{1}{64}\beta^2(7j^2-2j-24)]$

PART III.

SUMMARY.

The symmetrical form in which Lamé presented the three functions whose product is a solid ellipsoidal harmonic is such as to render purely analytical investigations both elegant and convenient. But it seemed to me that facility for computation might be gained by the surrender of symmetry, and I have acted on this idea in the preceding paper.

Spheroidal analysis has been successfully employed where the ellipsoid is one of revolution, and it therefore seemed advisable to make that method the point of departure for the treatment of ellipsoids with three unequal axes. In spheroidal harmonics we start with a fundamental prolate ellipsoid of revolution, with imaginary semi-axes $k\sqrt{-1}$, $k\sqrt{-1}$, 0. The position of a point is then defined by three co-ordinates; the first of these, ν, is such that its reciprocal is the eccentricity of a meridional section of an ellipsoid confocal with the fundamental ellipsoid and passing through the point. Since that eccentricity diminishes as we recede from the origin, ν plays the part of a reciprocal to the radius vector. The second co-ordinate, μ, is the cosine of the auxiliary angle in the meridional ellipse measured from the axis of symmetry. It therefore plays the part of sine of latitude. The third co-ordinate is simply the longitude ϕ. The three co-ordinates may then be described as the radial, latitudinal, and longitudinal co-ordinates. The parameter k defines the absolute scale on which the figure is drawn.

It is equally possible to start with a fundamental oblate ellipsoid with real axes $k, k, 0$. We should then take the first co-ordinate, ζ, as such that $\zeta^2 = -\nu^2$. All that follows would then be equally applicable; but, in order not to complicate the statement by continual reference to alternative forms, I shall adhere to the first form as a standard.

In this paper a closely parallel notation is adopted for the ellipsoid of three unequal axes. The squares of semi-axes of the fundamental ellipsoid are taken to be $-k^2\frac{1+\beta}{1-\beta}$, $-k^2$, 0, and the three co-ordinates are still ν, μ, ϕ. Although their geometrical meanings are now by no means so simple, they may still be described as radial, latitudinal, and longitudinal co-ordinates. As before, we might equally well start with a fundamental ellipsoid whose squares of semi-axes are $k^2\frac{1+\beta}{1-\beta}$, k^2, 0, and replace ν^2 by ζ^2, where $\zeta^2 = -\nu^2$. All possible ellipsoids are comprised in either of these types by making β vary from zero to infinity. But it is shown in § 2 that, by a proper choice of type, all possible ellipsoids are comprised in a range of β from zero to one-third. When β is zero we have the spheroids for which harmonic analysis already exists; and when $\beta = \frac{1}{3}$ the ellipsoid is such that the mean axis is the square root of mean square of the extreme axes. The harmonic analysis for this class of ellipsoid has not been yet worked out, but the method of this paper would render it possible to do so. We may then regard β as essentially less than $\frac{1}{3}$, and may conveniently make developments in powers of β.

In spheroidal analysis, for space internal to an ellipsoid ν_0, two of the three functions are the same P-functions that occur in spherical analysis; one P being a function of ν, the other of μ. The third function is a cosine or sine of a multiple of the longitude ϕ. In external space the P-function of ν is replaced by a Q-function, being a solution of the differential equation of the second kind.

The like is true in ellipsoidal analysis, and we have P- and Q-functions of ν for internal and external space, a P-function of μ, and a cosine- or sine-function of ϕ. I will now for a time set aside the Q-functions and consider them later.

There are eight cases to consider (§ 4); these are determined by the evenness or oddness of the degree i and of the order s of the harmonic, and by the alternative of whether they correspond with a cosine- or sine-function of ϕ. I indicate these eight types by the initials E, O, C, or S—for example,

EOS means the type in which i is even, s is odd, and that there is association with a sine-function.

It appears that the new P-functions fall into two forms. The first form, which I write $\mathfrak{P}_i^s$, is found to be expressible in a finite series in terms of the $P_i^{s\pm 2k}$, where the P's are the ordinary functions of spherical analysis. The terms in this series are arranged in powers of β, so that the coefficient of $P_i^{s\pm 2k}$ has β^k as part of its coefficient. The second form, which I write P_i^s, is such that $\sqrt{\dfrac{\nu^2-1}{\nu^2-\frac{1+\beta}{1-\beta}}}\,\mathsf{P}_i^s(\nu)$ or $\sqrt{\dfrac{1-\mu^2}{\frac{1+\beta}{1-\beta}-\mu^2}}\,\mathsf{P}_i^s(\mu)$ is expressible by a series of the same kind as that for $\mathfrak{P}_i^s$. Amongst the eight types four involve $\mathfrak{P}$-functions and four P-functions; and if for given s a $\mathfrak{P}_i^s$-function is associated with a cosine-function, the corresponding P_i^s is associated with a sine-function, and *vice versâ*.

Lastly, a $\mathfrak{P}$-function of ν is always associated with a $\mathfrak{P}$-function of μ; and the like is true of the P's.

Again, the cosine- and sine-functions fall into two forms. In the first form s and i are either both even or both odd, and the function, which I write $\mathfrak{C}_i^s$ or $\mathfrak{S}_i^s$, is expressed by a series of terms consisting of a coefficient multiplied by β^k cos or $\sin(s\pm 2k)\phi$. In the second form s and i differ as to evenness and oddness, and the function, written C_i^s or S_i^s, is expressed by a similar series multiplied by $(1-\beta\cos 2\phi)^{\frac{1}{2}}$.

The combination of the two forms of P-function with the four forms of cosine- and sine-function gives the eight types of solid harmonic.

Corresponding to the two forms of P-function there are two forms of Q-function, such that $\mathfrak{Q}_i^s$ and $\mathsf{Q}_i^s\sqrt{\dfrac{\nu^2-1}{\nu^2-\frac{1+\beta}{1-\beta}}}$ are expansible in a series of ordinary Q-functions; but whereas the series for $\mathfrak{P}_i^s$ and P_i^s are terminable, because P_i^s vanishes when s is greater than i, this is not the case with the Q-functions. In fact the series for the Q-functions begins with Q_i or Q_i^1, and the order of the Q's increases by two at a time up to s when we have the principal or central term; it then goes on increasing up to $s=i$ or $i-1$, and on to infinity. [Unfortunately it appears that the approximation to these functions is found not to be very close, unless β is very small.]

In spherical and spheroidal analysis the differential equation satisfied by P_i^s involves the integer s, whereby the order is specified. So here also the differential equations, satisfied by $\mathfrak{P}_i^s$ or P_i^s and by $\mathfrak{C}_i^s$, $\mathfrak{S}_i^s$, or C_i^s, S_i^s, involve a constant, but it is no longer an integer. It seemed convenient to assume $s^2-\sigma$ as the form for this constant, where s is the known integer specifying the order of harmonic, and σ remains to be determined from the differential equations.

When the assumed forms for the P-function and for the cosine- and sine-functions are substituted in the differential equations, it is found (§ 6) that, in order to satisfy the equations, σ must be equal to the difference between two finite-continued fractions, each of which involves σ. We thus have an equation for σ, and the required root is that which vanishes when β vanishes.

For the harmonics of degrees 0, 1, 2, 3, and for all orders, [and for certain harmonics of the 4th order], σ may be found rigorously in algebraic form, but for higher degrees the equation can only be solved approximately, unless β should have a definite numerical value.

When σ has been determined, either rigorously or approximately, the successive coefficients of the series are determinable in such a way that the ratio of each coefficient to the preceding one is expressed by a continued fraction, which is, in fact, a portion of one of the two fractions involved in the equation for σ.

Throughout the rest of the paper the greater part of the work is carried out with approximate forms, and, although it would be easy to attain to greater accuracy, I have thought it sufficient, in the first instance, to stop at β^2. With this limitation the coefficients of the series assume simple forms (§ 8), and we have thus definite, if approximate, expressions for all the functions which can occur in ellipsoidal analysis.

In rigorous expressions, $\mathfrak{P}_i^s$ and P_i^s are essentially different from one another, but in approximate forms, when s is greater than a certain integer dependent on the degree of approximation, the two are the same thing in different shapes, except as to a constant factor. I have, therefore, in § 9 determined up to squares of β the factors whereby P_i^s is convertible into $\mathfrak{P}_i^s$, and C_i^s or S_i^s into $\mathfrak{C}_i^s$ or $\mathfrak{S}_i^s$. With the degree of approximation adopted there is no factor for converting the P's when $s=3, 2, 1$. Similarly, down to $s=3$ inclusive, the same factor serves for converting C_i^s into $\mathfrak{C}_i^s$ and S_i^s into $\mathfrak{S}_i^s$. But for $s=2, 1, 0$ one form is needed for changing C into $\mathfrak{C}$, and another for changing S into $\mathfrak{S}$. It may be well to note that there is no sine-function when s is zero.

The use of these factors does much to facilitate the laborious reductions involved in the whole investigation.

It is well known that the Q-functions are expressible in terms of the P-functions by means of a definite integral. Hence $\mathfrak{Q}_i^s$ and Q_i^s must have a second form, which can only differ from the other by a constant factor. The factors connecting the two forms are determined in § 10.

The second part of the paper is devoted to applications of the harmonic method. In § 11 the perpendicular from the centre on to the tangent plane to an ellipsoid ν_0, and the area of an element of surface of the ellipsoid, are found in terms of the co-ordinates μ, ϕ, and the constant ν_0.

It is easy to form a function, continuous at the surface ν_0, which shall be a solid harmonic both for external and for internal space. Poisson's equation then enables us to determine the surface density of which this continuous function is the potential, and it is found to be a surface harmonic of μ, ϕ multiplied by the perpendicular on to the tangent plane. This application of Poisson's equation involves the use of the Q-function in its integral form. Accordingly, if the serial form for the Q-function is adopted as a standard, the expression for the potential of a layer of surface density involves the use of the factor for conversion between the two forms of Q-function.

This result may obviously be employed to determine the potential of an harmonic deformation of a solid ellipsoid.

The potential of the solid ellipsoid itself may be found by the consideration that it is externally equal to that of a focaloid shell of the same mass. It appears that in order to express the equivalent surface density in surface harmonics, it is only necessary to express the reciprocal of the square of the perpendicular on the tangent plane in that form. This result is attained by expressing x^2, y^2, z^2 in surface harmonics. When this is done, an application of the preceding theorem enables us to write down the external potential of the solid ellipsoid at once. In § 12 the external potential of the solid ellipsoid is expressed rigorously in terms of solid harmonics of degrees zero and 2.

Since x^2, y^2, z^2 have been found in surface harmonics, we can also write down a rotation-potential about any one of the three axes in the same form.

The internal potential of a solid ellipsoid does not lend itself well to elliptic co-ordinates, but expressions for it are given in § 12.

If it be desired to express any arbitrary function of μ, ϕ in surface harmonics, it is necessary to know the integrals, over the surface of the ellipsoid of the squares of the several surface harmonics, each multiplied by the perpendicular on to the tangent plane. The rest of the paper is devoted to the evaluation of these integrals. No attempt is made to carry the developments beyond β^2, although the methods employed would render it possible to do so.

When s is greater than unity, it appears that it is legitimate to develop the function to be integrated in powers of $1/(1-\mu^2)$; and when this is done, the integration, although laborious, does not present any great difficulty.

But when s is either 1 or 0, the method of development breaks down, because it would give rise to infinite elements in the integrals at the poles where μ^2 is unity. However, portions of the integrals in these cases can still be found by the former method of development. As to the residues which cannot be so treated, it appears that they depend on integrals of the forms

$$\int_{-\frac{1}{2}\pi}^{\frac{1}{2}\pi} \frac{\cos^{2n}\theta\, d\theta}{(1-\kappa^2\sin^2\theta)^{\frac{1}{2}}} \text{ and } \int_{-\frac{1}{2}\pi}^{\frac{1}{2}\pi} \cos^{2n}\theta\,(1-\kappa^2\sin^2\theta)^{\frac{1}{2}}\, d\theta$$

where κ^2 is nearly equal to unity.

Development of the square-roots in powers of κ^2 is useless on account of the slow convergence, and it is required to find series which proceed by powers of κ'^2, where $\kappa'^2 = 1 - \kappa^2$.

By a somewhat difficult investigation, in respect to which I owe my special thanks to Mr Hobson, the needed series are found (§ 19).

It appears that portions of the two integrals involve logarithms which become infinite when κ' vanishes. Since, in the application of these integrals, the vanishing of κ' implies the vanishing of β, we appear to be met by a difficulty. It is known that in spheroidal analysis no such terms appear, and we may feel confident that they cannot really exist in ellipsoidal analysis. In § 20 it is proved that the logarithmic terms do as a fact disappear. The residues of the integrals in the cases $s=1$, 0 are thus found, and added to the previous portions to form the complete results.

The second part of the paper ends (§ 22) with a list of the integrals of the squares of the surface harmonics for all values of s, as far as the squares of β.

Finally, an appendix below contains a table of all the functions as far as $i=5$, $s=5$. It is probable that for the higher values of s the results would only be applicable when β is very small.

Appendix 1.

Table of the P- and Q-Functions.

$i=0$ (EEC)
$$\left\{\begin{matrix}\mathfrak{P}_0\\ \mathfrak{Q}_0\end{matrix}\right. = \left\{\begin{matrix}\mathrm{P}_0\\ \mathrm{Q}_0\end{matrix}\right. + \tfrac{1}{4}\beta\left\{\begin{matrix}0\\ \mathrm{Q}_0{}^2\end{matrix}\right. + \tfrac{1}{128}\beta^2\left\{\begin{matrix}0\\ \mathrm{Q}_0{}^4\end{matrix}\right.$$

$i=1$ (OEC)
$$\left\{\begin{matrix}\mathfrak{P}_1\\ \mathfrak{Q}_1\end{matrix}\right. = \left\{\begin{matrix}\mathrm{P}_1\\ \mathrm{Q}_1\end{matrix}\right. + \tfrac{1}{4}\beta\left\{\begin{matrix}0\\ \mathrm{Q}_1{}^2\end{matrix}\right. + \tfrac{1}{128}\beta^2\left\{\begin{matrix}0\\ \mathrm{Q}_1{}^4\end{matrix}\right.$$

(OOC)
$$\left\{\begin{matrix}\mathsf{P}_1{}^1\\ \mathsf{Q}_1{}^1\end{matrix}\right. = \Omega\left[\left\{\begin{matrix}\mathrm{P}_1{}^1\\ \mathrm{Q}_1{}^1\end{matrix}\right. + \tfrac{3}{16}\beta\left(1+\tfrac{1}{8}\beta\right)\left\{\begin{matrix}0\\ \mathrm{Q}_1{}^3\end{matrix}\right. + \tfrac{5}{768}\beta^2\left\{\begin{matrix}0\\ \mathrm{Q}_1{}^5\end{matrix}\right.\right]$$

(OOS)
$$\left\{\begin{matrix}\mathfrak{P}_1{}^1\\ \mathfrak{Q}_1{}^1\end{matrix}\right. = \left\{\begin{matrix}\mathrm{P}_1{}^1\\ \mathrm{Q}_1{}^1\end{matrix}\right. + \tfrac{1}{16}\beta\left(1-\tfrac{1}{8}\beta\right)\left\{\begin{matrix}0\\ \mathrm{Q}_1{}^3\end{matrix}\right. + \tfrac{1}{768}\beta^2\left\{\begin{matrix}0\\ \mathrm{Q}_1{}^5\end{matrix}\right.$$

$i=2$ (EEC)
$$\left\{\begin{matrix}\mathfrak{P}_2\\ \mathfrak{Q}_2\end{matrix}\right. = \left\{\begin{matrix}\mathrm{P}_2\\ \mathrm{Q}_2\end{matrix}\right. + \tfrac{1}{4}\beta\left\{\begin{matrix}\mathrm{P}_2{}^2\\ \mathrm{Q}_2{}^2\end{matrix}\right. + \tfrac{1}{128}\beta^2\left\{\begin{matrix}0\\ \mathrm{Q}_2{}^4\end{matrix}\right.$$

(EOC)
$$\left\{\begin{matrix}\mathsf{P}_2{}^1\\ \mathsf{Q}_2{}^1\end{matrix}\right. = \Omega\left[\left\{\begin{matrix}\mathrm{P}_2{}^1\\ \mathrm{Q}_2{}^1\end{matrix}\right. + \tfrac{3}{16}\beta\left(1+\tfrac{3}{8}\beta\right)\left\{\begin{matrix}0\\ \mathrm{Q}_2{}^3\end{matrix}\right. + \tfrac{5}{768}\beta^2\left\{\begin{matrix}0\\ \mathrm{Q}_2{}^5\end{matrix}\right.\right]$$

(EOS)
$$\left\{\begin{matrix}\mathfrak{P}_2{}^1\\ \mathfrak{Q}_2{}^1\end{matrix}\right. = \left\{\begin{matrix}\mathrm{P}_2{}^1\\ \mathrm{Q}_2{}^1\end{matrix}\right. + \tfrac{1}{16}\beta\left(1-\tfrac{3}{8}\beta\right)\left\{\begin{matrix}0\\ \mathrm{Q}_2{}^3\end{matrix}\right. + \tfrac{1}{768}\beta^2\left\{\begin{matrix}0\\ \mathrm{Q}_2{}^5\end{matrix}\right.$$

(EEC) $$\left\{\begin{matrix}\mathfrak{P}_2^2\\ \mathfrak{Q}_2^2\end{matrix}\right. = -3\beta\left\{\begin{matrix}P_2\\ Q_2\end{matrix}\right. + \left\{\begin{matrix}P_2^2\\ Q_2^2\end{matrix}\right. + \tfrac{1}{24}\beta\left\{\begin{matrix}0\\ Q_2^4\end{matrix}\right. + \tfrac{1}{1536}\beta^2\left\{\begin{matrix}0\\ Q_2^6\end{matrix}\right.$$

(EES) $$\left\{\begin{matrix}\mathsf{P}_2^2\\ \mathsf{Q}_2^2\end{matrix}\right. = \Omega\left[\left\{\begin{matrix}P_2^2\\ Q_2^2\end{matrix}\right. + \tfrac{1}{12}\beta\left\{\begin{matrix}0\\ Q_2^4\end{matrix}\right. + \tfrac{1}{512}\beta^2\left\{\begin{matrix}0\\ Q_2^6\end{matrix}\right.\right]$$

$i=3$ (OEC) $$\left\{\begin{matrix}\mathfrak{P}_3\\ \mathfrak{Q}_3\end{matrix}\right. = \left\{\begin{matrix}P_3\\ Q_3\end{matrix}\right. + \tfrac{1}{4}\beta\left\{\begin{matrix}P_3^2\\ Q_3^2\end{matrix}\right. + \tfrac{1}{128}\beta^2\left\{\begin{matrix}0\\ Q_3^4\end{matrix}\right.$$

(OOC) $$\left\{\begin{matrix}\mathsf{P}_3^1\\ \mathsf{Q}_3^1\end{matrix}\right. = \Omega\left[\left\{\begin{matrix}P_3^1\\ Q_3^1\end{matrix}\right. + \tfrac{3}{16}\beta\left(1+\tfrac{3}{4}\beta\right)\left\{\begin{matrix}P_3^3\\ Q_3^3\end{matrix}\right. + \tfrac{5}{768}\beta^2\left\{\begin{matrix}0\\ Q_3^5\end{matrix}\right.\right]$$

(OOS) $$\left\{\begin{matrix}\mathfrak{P}_3^1\\ \mathfrak{Q}_3^1\end{matrix}\right. = \left\{\begin{matrix}P_3^1\\ Q_3^1\end{matrix}\right. + \tfrac{1}{16}\beta\left(1-\tfrac{3}{4}\beta\right)\left\{\begin{matrix}P_3^3\\ Q_3^3\end{matrix}\right. + \tfrac{1}{768}\beta^2\left\{\begin{matrix}0\\ Q_3^5\end{matrix}\right.$$

(OEC) $$\left\{\begin{matrix}\mathfrak{P}_3^2\\ \mathfrak{Q}_3^2\end{matrix}\right. = -15\beta\left\{\begin{matrix}P_3\\ Q_3\end{matrix}\right. + \left\{\begin{matrix}P_3^2\\ Q_3^2\end{matrix}\right. + \tfrac{1}{24}\beta\left\{\begin{matrix}0\\ Q_3^4\end{matrix}\right. + \tfrac{1}{1536}\beta^2\left\{\begin{matrix}0\\ Q_3^6\end{matrix}\right.$$

(OES) $$\left\{\begin{matrix}\mathsf{P}_3^2\\ \mathsf{Q}_3^2\end{matrix}\right. = \Omega\left[\left\{\begin{matrix}P_3^2\\ Q_3^2\end{matrix}\right. + \tfrac{1}{12}\beta\left\{\begin{matrix}0\\ Q_3^4\end{matrix}\right. + \tfrac{1}{512}\beta^2\left\{\begin{matrix}0\\ Q_3^6\end{matrix}\right.\right]$$

(OOC) $$\left\{\begin{matrix}\mathsf{P}_3^3\\ \mathsf{Q}_3^3\end{matrix}\right. = \Omega\left[-\tfrac{5}{4}\beta\left(1+\tfrac{3}{4}\beta\right)\left\{\begin{matrix}P_3^1\\ Q_3^1\end{matrix}\right. + \left\{\begin{matrix}P_3^3\\ Q_3^3\end{matrix}\right. + \tfrac{5}{96}\beta\left\{\begin{matrix}0\\ Q_3^5\end{matrix}\right. + \tfrac{7}{7680}\beta^2\left\{\begin{matrix}0\\ Q_3^7\end{matrix}\right.\right]$$

(OOS) $$\left\{\begin{matrix}\mathfrak{P}_3^3\\ \mathfrak{Q}_3^3\end{matrix}\right. = -\tfrac{15}{4}\beta\left(1-\tfrac{3}{4}\beta\right)\left\{\begin{matrix}P_3^1\\ Q_3^1\end{matrix}\right. + \left\{\begin{matrix}P_3^3\\ Q_3^3\end{matrix}\right. + \tfrac{1}{32}\beta\left\{\begin{matrix}0\\ Q_3^5\end{matrix}\right. + \tfrac{1}{2560}\beta^2\left\{\begin{matrix}0\\ Q_3^7\end{matrix}\right.$$

$i=4$ (EEC) $$\left\{\begin{matrix}\mathfrak{P}_4\\ \mathfrak{Q}_4\end{matrix}\right. = \left\{\begin{matrix}P_4\\ Q_4\end{matrix}\right. + \tfrac{1}{4}\beta\left\{\begin{matrix}P_4^2\\ Q_4^2\end{matrix}\right. + \tfrac{1}{128}\beta^2\left\{\begin{matrix}P_4^4\\ Q_4^4\end{matrix}\right.$$

(EOC) $$\left\{\begin{matrix}\mathsf{P}_4^1\\ \mathsf{Q}_4^1\end{matrix}\right. = \Omega\left[\left\{\begin{matrix}P_4^1\\ Q_4^1\end{matrix}\right. + \tfrac{3}{16}\beta\left(1+\tfrac{5}{4}\beta\right)\left\{\begin{matrix}P_4^3\\ Q_4^3\end{matrix}\right. + \tfrac{5}{768}\beta^2\left\{\begin{matrix}0\\ Q_4^5\end{matrix}\right.\right]$$

(EOS) $$\left\{\begin{matrix}\mathfrak{P}_4^1\\ \mathfrak{Q}_4^1\end{matrix}\right. = \left\{\begin{matrix}P_4^1\\ Q_4^1\end{matrix}\right. + \tfrac{1}{16}\beta\left(1-\tfrac{5}{4}\beta\right)\left\{\begin{matrix}P_4^3\\ Q_4^3\end{matrix}\right. + \tfrac{1}{768}\beta^2\left\{\begin{matrix}0\\ Q_4^5\end{matrix}\right.$$

(EEC) $$\left\{\begin{matrix}\mathfrak{P}_4^2\\ \mathfrak{Q}_4^2\end{matrix}\right. = -45\beta\left\{\begin{matrix}P_4\\ Q_4\end{matrix}\right. + \left\{\begin{matrix}P_4^2\\ Q_4^2\end{matrix}\right. + \tfrac{1}{24}\beta\left\{\begin{matrix}P_4^4\\ Q_4^4\end{matrix}\right. + \tfrac{1}{1536}\beta^2\left\{\begin{matrix}0\\ Q_4^6\end{matrix}\right.$$

(EES) $$\left\{\begin{matrix}\mathsf{P}_4^2\\ \mathsf{Q}_4^2\end{matrix}\right. = \Omega\left[\left\{\begin{matrix}P_4^2\\ Q_4^2\end{matrix}\right. + \tfrac{1}{12}\beta\left\{\begin{matrix}P_4^4\\ Q_4^4\end{matrix}\right. + \tfrac{1}{512}\beta^2\left\{\begin{matrix}0\\ Q_4^6\end{matrix}\right.\right]$$

(EOC) $$\left\{\begin{matrix}\mathsf{P}_4^3\\ \mathsf{Q}_4^3\end{matrix}\right. = \Omega\left[-\tfrac{21}{4}\beta\left(1+\tfrac{5}{4}\beta\right)\left\{\begin{matrix}P_4^1\\ Q_4^1\end{matrix}\right. + \left\{\begin{matrix}P_4^3\\ Q_4^3\end{matrix}\right. + \tfrac{5}{96}\beta\left\{\begin{matrix}0\\ Q_4^5\end{matrix}\right. + \tfrac{7}{7680}\beta^2\left\{\begin{matrix}0\\ Q_4^7\end{matrix}\right.\right]$$

(EOS) $$\left\{\begin{matrix}\mathfrak{P}_4^3\\ \mathfrak{Q}_4^3\end{matrix}\right. = -\tfrac{63}{4}\left(1-\tfrac{5}{4}\beta\right)\left\{\begin{matrix}P_4^1\\ Q_4^1\end{matrix}\right. + \left\{\begin{matrix}P_4^3\\ Q_4^3\end{matrix}\right. + \tfrac{1}{32}\beta\left\{\begin{matrix}0\\ Q_4^5\end{matrix}\right. + \tfrac{1}{2560}\beta^2\left\{\begin{matrix}0\\ Q_4^7\end{matrix}\right.$$

(EEC) $$\left\{\begin{matrix}\mathfrak{P}_4^4\\ \mathfrak{Q}_4^4\end{matrix}\right. = \tfrac{105}{2}\beta^2\left\{\begin{matrix}P_4\\ Q_4\end{matrix}\right. - \tfrac{14}{3}\beta\left\{\begin{matrix}P_4^2\\ Q_4^2\end{matrix}\right. + \left\{\begin{matrix}P_4^4\\ Q_4^4\end{matrix}\right. + \tfrac{1}{40}\beta\left\{\begin{matrix}0\\ Q_4^6\end{matrix}\right. + \tfrac{1}{3840}\beta^2\left\{\begin{matrix}0\\ Q_4^8\end{matrix}\right.$$

(EES) $$\left\{\begin{matrix}\mathsf{P}_4^4\\ \mathsf{Q}_4^4\end{matrix}\right. = \Omega\left[-\tfrac{7}{3}\beta\left\{\begin{matrix}P_4^2\\ Q_4^2\end{matrix}\right. + \left\{\begin{matrix}P_4^4\\ Q_4^4\end{matrix}\right. + \tfrac{3}{80}\beta\left\{\begin{matrix}0\\ Q_4^6\end{matrix}\right. + \tfrac{1}{1920}\beta^2\left\{\begin{matrix}0\\ Q_4^8\end{matrix}\right.\right]$$

$$i=5 \quad (\text{OEC}) \quad \begin{cases}\mathfrak{P}_5\\ \mathfrak{Q}_5\end{cases} = \begin{cases}P_5\\ Q_5\end{cases} + \tfrac{1}{4}\beta\begin{cases}P_5^2\\ Q_5^2\end{cases} + \tfrac{1}{128}\beta^2\begin{cases}P_5^4\\ Q_5^4\end{cases}$$

$$(\text{OOC}) \quad \begin{cases}\mathsf{P}_5^1\\ \mathsf{Q}_5^1\end{cases} = \Omega\left[\begin{cases}P_5^1\\ Q_5^1\end{cases} + \tfrac{3}{16}\beta\left(1+\tfrac{15}{8}\beta\right)\begin{cases}P_5^3\\ Q_5^3\end{cases} + \tfrac{5}{768}\beta^2\begin{cases}P_5^5\\ Q_5^5\end{cases}\right]$$

$$(\text{OOS}) \quad \begin{cases}\mathfrak{P}_5^1\\ \mathfrak{Q}_5^1\end{cases} = \begin{cases}P_5^1\\ Q_5^1\end{cases} + \tfrac{1}{16}\beta\left(1-\tfrac{15}{8}\beta\right)\begin{cases}P_5^3\\ Q_5^3\end{cases} + \tfrac{1}{768}\beta^2\begin{cases}P_5^5\\ Q_5^5\end{cases}$$

$$(\text{OEC}) \quad \begin{cases}\mathfrak{P}_5^2\\ \mathfrak{Q}_5^2\end{cases} = -105\beta\begin{cases}P_5\\ Q_5\end{cases} + \begin{cases}P_5^2\\ Q_5^2\end{cases} + \tfrac{1}{24}\beta\begin{cases}P_5^4\\ Q_5^4\end{cases} + \tfrac{1}{1536}\beta^2\begin{cases}0\\ Q_5^6\end{cases}$$

$$(\text{OES}) \quad \begin{cases}\mathsf{P}_5^2\\ \mathsf{Q}_5^2\end{cases} = \Omega\left[\begin{cases}P_5^2\\ Q_5^2\end{cases} + \tfrac{1}{12}\beta\begin{cases}P_5^4\\ Q_5^4\end{cases} + \tfrac{1}{512}\beta^2\begin{cases}0\\ Q_5^6\end{cases}\right]$$

$$(\text{OOC}) \quad \begin{cases}\mathsf{P}_5^3\\ \mathsf{Q}_5^3\end{cases} = \Omega\left[-14\beta\left(1+\tfrac{15}{8}\beta\right)\begin{cases}P_5^1\\ Q_5^1\end{cases} + \begin{cases}P_5^3\\ Q_5^3\end{cases} + \tfrac{5}{96}\beta\begin{cases}P_5^5\\ Q_5^5\end{cases} + \tfrac{7}{7680}\beta^2\begin{cases}0\\ Q_5^7\end{cases}\right]$$

$$(\text{OOS}) \quad \begin{cases}\mathfrak{P}_5^3\\ \mathfrak{Q}_5^3\end{cases} = -42\beta\left(1-\tfrac{15}{8}\beta\right)\begin{cases}P_5^1\\ Q_5^1\end{cases} + \begin{cases}P_5^3\\ Q_5^3\end{cases} + \tfrac{1}{32}\beta\begin{cases}P_5^5\\ Q_5^5\end{cases} + \tfrac{1}{2560}\beta^2\begin{cases}0\\ Q_5^7\end{cases}$$

$$(\text{OEC}) \quad \begin{cases}\mathfrak{P}_5^4\\ \mathfrak{Q}_5^4\end{cases} = \tfrac{945}{2}\beta^2\begin{cases}P_5\\ Q_5\end{cases} - 18\beta\begin{cases}P_5^2\\ Q_5^2\end{cases} + \begin{cases}P_5^4\\ Q_5^4\end{cases} + \tfrac{1}{40}\beta\begin{cases}0\\ Q_5^6\end{cases} + \tfrac{1}{3840}\beta^2\begin{cases}0\\ Q_5^8\end{cases}$$

$$(\text{OES}) \quad \begin{cases}\mathsf{P}_5^4\\ \mathsf{Q}_5^4\end{cases} = \Omega\left[-9\beta\begin{cases}P_5^2\\ Q_5^2\end{cases} + \begin{cases}P_5^4\\ Q_5^4\end{cases} + \tfrac{3}{80}\beta\begin{cases}0\\ Q_5^6\end{cases} + \tfrac{1}{1920}\beta^2\begin{cases}0\\ Q_5^8\end{cases}\right]$$

$$(\text{OOC}) \quad \begin{cases}\mathsf{P}_5^5\\ \mathsf{Q}_5^5\end{cases} = \Omega\left[\tfrac{63}{4}\beta^2\begin{cases}P_5^1\\ Q_5^1\end{cases} - \tfrac{27}{8}\beta\begin{cases}P_5^3\\ Q_5^3\end{cases} + \begin{cases}P_5^5\\ Q_5^5\end{cases} + \tfrac{7}{240}\beta\begin{cases}0\\ Q_5^7\end{cases} + \tfrac{3}{8960}\beta^2\begin{cases}0\\ Q_5^9\end{cases}\right]$$

$$(\text{OOS}) \quad \begin{cases}\mathfrak{P}_5^5\\ \mathfrak{Q}_5^5\end{cases} = \tfrac{315}{4}\beta^2\begin{cases}P_5^1\\ Q_5^1\end{cases} - \tfrac{45}{8}\beta\begin{cases}P_5^3\\ Q_5^3\end{cases} + \begin{cases}P_5^5\\ Q_5^5\end{cases} + \tfrac{1}{48}\beta\begin{cases}0\\ Q_5^7\end{cases} + \tfrac{1}{5376}\beta^2\begin{cases}0\\ Q_5^9\end{cases}$$

Note that in this table P_i^s denotes $\dfrac{(\nu^2-1)^{\frac{1}{2}s}}{2^i\, i\,!}\left(\dfrac{d}{d\nu}\right)^{i+s}(\nu^2-1)^i$, and Ω is $\left(\dfrac{\nu^2-\frac{1+\beta}{1-\beta}}{\nu^2-1}\right)^{\frac{1}{2}}$.

If the variable is μ, and if accordingly the factor $(\nu^2-1)^{\frac{1}{2}s}$ in P_i^s is replaced by $(1-\mu^2)^{\frac{1}{2}s}$, the signs of all the terms which have β as coefficient must be changed. Ω has still the same meaning, but must be written in the form $\left(\dfrac{\frac{1+\beta}{1-\beta}-\mu^2}{1-\mu^2}\right)^{\frac{1}{2}}$.

Table of the Cosine and Sine Functions.

$i=0$ (EEC) $\mathfrak{C}_0 = 1$

$i=1$ (OEC) $\mathsf{C}_1 = \Phi$

(OOC)
(OOS) $\left\{\begin{matrix}\mathfrak{C}_1^1\\ \mathfrak{S}_1^1\end{matrix}\right. = \left\{\begin{matrix}\cos\\ \sin\end{matrix}\right.\phi$

$i=2$ (EEC) $\mathfrak{C}_2 = 1 - \tfrac{3}{2}\beta\cos 2\phi$

(EOC)
(EOS) $\left\{\begin{matrix}\mathsf{C}_2^1\\ \mathsf{S}_2^1\end{matrix}\right. = \Phi\left(\left\{\begin{matrix}\cos\\ \sin\end{matrix}\right.\phi\right)$

(EEC)
(EES) $\left\{\begin{matrix}\mathfrak{C}_2^2\\ \mathfrak{S}_2^2\end{matrix}\right. = \tfrac{1}{2}\beta\left\{\begin{matrix}1\\ 0\end{matrix}\right. + \left\{\begin{matrix}\cos\\ \sin\end{matrix}\right. 2\phi$

$i=3$ (OEC) $\mathsf{C}_3 = \Phi\left[1 - \tfrac{5}{2}\beta\cos 2\phi\right]$

(OOC)
(OOS) $\left\{\begin{matrix}\mathfrak{C}_3^1\\ \mathfrak{S}_3^1\end{matrix}\right. = \left\{\begin{matrix}\cos\\ \sin\end{matrix}\right.\phi - \tfrac{5}{8}\beta\left(1 \pm \tfrac{3}{4}\beta\right)\left\{\begin{matrix}\cos\\ \sin\end{matrix}\right. 3\phi$

(OEC)
(OES) $\left\{\begin{matrix}\mathsf{C}_3^2\\ \mathsf{S}_3^2\end{matrix}\right. = \Phi\left[\tfrac{3}{2}\beta\left\{\begin{matrix}1\\ 0\end{matrix}\right. + \left\{\begin{matrix}\cos\\ \sin\end{matrix}\right. 2\phi\right]$

(OOC)
(OOS) $\left\{\begin{matrix}\mathfrak{C}_3^3\\ \mathfrak{S}_3^3\end{matrix}\right. = \tfrac{3}{8}\beta\left(1 \pm \tfrac{3}{4}\beta\right)\left\{\begin{matrix}\cos\\ \sin\end{matrix}\right.\phi + \left\{\begin{matrix}\cos\\ \sin\end{matrix}\right. 3\phi$

$i=4$ (EEC) $\mathfrak{C}_4 = 1 - 5\beta\cos 2\phi + \tfrac{35}{16}\beta^2\cos 4\phi$

(EOC)
(EOS) $\left\{\begin{matrix}\mathsf{C}_4^1\\ \mathsf{S}_4^1\end{matrix}\right. = \Phi\left[\left\{\begin{matrix}\cos\\ \sin\end{matrix}\right.\phi - \tfrac{7}{8}\beta\left(1 \pm \tfrac{5}{4}\beta\right)\left\{\begin{matrix}\cos\\ \sin\end{matrix}\right. 3\phi\right]$

(EEC)
(EES) $\left\{\begin{matrix}\mathfrak{C}_4^2\\ \mathfrak{S}_4^2\end{matrix}\right. = \tfrac{9}{4}\beta\left\{\begin{matrix}1\\ 0\end{matrix}\right. + \left\{\begin{matrix}\cos\\ \sin\end{matrix}\right. 2\phi - \tfrac{7}{12}\beta\left\{\begin{matrix}\cos\\ \sin\end{matrix}\right. 4\phi$

(EOC)
(EOS) $\left\{\begin{matrix}\mathsf{C}_4^3\\ \mathsf{S}_4^3\end{matrix}\right. = \Phi\left[\tfrac{9}{8}\beta\left(1 \pm \tfrac{5}{4}\beta\right)\left\{\begin{matrix}\cos\\ \sin\end{matrix}\right.\phi + \left\{\begin{matrix}\cos\\ \sin\end{matrix}\right. 3\phi\right]$

(EEC)
(EES) $\left\{\begin{matrix}\mathfrak{C}_4^4\\ \mathfrak{S}_4^4\end{matrix}\right. = \tfrac{3}{16}\beta^2\left\{\begin{matrix}1\\ 0\end{matrix}\right. + \tfrac{1}{3}\beta\left\{\begin{matrix}\cos\\ \sin\end{matrix}\right. 2\phi + \left\{\begin{matrix}\cos\\ \sin\end{matrix}\right. 4\phi$

$i=5$ (OEC) $\mathsf{C}_5 = \Phi\left[1 - 7\beta\cos 2\phi + \tfrac{63}{16}\beta^2\cos 4\phi\right]$

(OOC)
(OOS) $\left\{\begin{matrix}\mathfrak{C}_5^1\\ \mathfrak{S}_5^1\end{matrix}\right. = \left\{\begin{matrix}\cos\\ \sin\end{matrix}\right.\phi - \tfrac{7}{4}\beta\left(1 \pm \tfrac{15}{8}\beta\right)\left\{\begin{matrix}\cos\\ \sin\end{matrix}\right. 3\phi + \tfrac{21}{32}\beta^2\left\{\begin{matrix}\cos\\ \sin\end{matrix}\right. 5\phi$

(OEC)
(OES) $\left\{\begin{matrix}\mathsf{C}_5^2\\ \mathsf{S}_5^2\end{matrix}\right. = \Phi\left[\tfrac{15}{4}\beta\left\{\begin{matrix}1\\ 0\end{matrix}\right. + \left\{\begin{matrix}\cos\\ \sin\end{matrix}\right. 2\phi - \tfrac{3}{4}\beta\left\{\begin{matrix}\cos\\ \sin\end{matrix}\right. 4\phi\right]$

(OOC) (OOS) $\begin{cases}\mathfrak{C}_5^3 \\ \mathfrak{S}_5^3\end{cases} = \frac{3}{2}\beta\,(1 \pm \frac{15}{8}\beta)\begin{cases}\cos \\ \sin\end{cases}\phi + \begin{cases}\cos \\ \sin\end{cases} 3\phi - \frac{9}{16}\beta\begin{cases}\cos \\ \sin\end{cases} 5\phi$

(OEC) (OES) $\begin{cases}\mathrm{C}_5^4 \\ \mathrm{S}_5^4\end{cases} = \Phi\left[\frac{15}{16}\beta^2\begin{cases}1 \\ 0\end{cases} + \beta\begin{cases}\cos \\ \sin\end{cases} 2\phi + \begin{cases}\cos \\ \sin\end{cases} 4\phi\right]$

(OOC) (OOS) $\begin{cases}\mathfrak{C}_5^5 \\ \mathfrak{S}_5^5\end{cases} = \frac{5}{32}\beta^2\begin{cases}\cos \\ \sin\end{cases}\phi + \frac{5}{16}\beta\begin{cases}\cos \\ \sin\end{cases} 3\phi + \begin{cases}\cos \\ \sin\end{cases} 5\phi$

Note that in this table

$$\Phi = (1 - \beta \cos 2\phi)^{\frac{1}{2}}$$

A table of $\mathrm{P}(\nu)$ and $\mathrm{Q}(\nu)$ up to $i = 5$, $s = 5$ is contained in Professor Bryan's paper (*Proc. Camb. Phil. Soc.*, vol. VI., 1888, p. 297). The functions, there tabulated as $\mathrm{T}_n^s(\nu)$ and $\mathrm{U}_n^s(\nu)$, in the notation here adopted would be $\mathrm{P}_n^s(\nu)$ (with the factor $(\nu^2 - 1)^{\frac{1}{2}s}$) and $(-)^s \frac{i-s\,!}{i+s\,!}\mathrm{Q}_n^s(\nu)$.

The formula for $\mathrm{Q}_i^s(\nu)$, where s is greater than i, is given in § 10 above.

APPENDIX 2.

On the Symmetry of the Cosine and Sine-functions with the P-*functions.*

In writing this paper I failed to notice that the symmetry between the P-functions and the cosine and sine-functions is not destroyed, but is only masked, in the approximate expressions for the harmonic functions.

For example, we have

$$\mathfrak{P}_3(\mu) = P_3(\mu) - \tfrac{1}{4}\beta P_3^2(\mu)$$

therefore, in consequence of the symmetry which subsists, we ought to find

$$\mathrm{C}_3(\phi) = P_3\left[\left(\frac{1 - \beta\cos 2\phi}{1 - \beta}\right)^{\frac{1}{2}}\right] - \tfrac{1}{4}\beta P_3^2\left[\left(\frac{1 - \beta\cos 2\phi}{1 - \beta}\right)^{\frac{1}{2}}\right]$$

Now

$$P_3\left[\left(\frac{1 - \beta\cos 2\phi}{1 - \beta}\right)^{\frac{1}{2}}\right] = \frac{(1 - \beta\cos 2\phi)^{\frac{1}{2}}}{(1 - \beta)^{\frac{3}{2}}}(1 + \tfrac{3}{2}\beta - \tfrac{5}{2}\beta\cos 2\phi)$$

$$P_3^2\left[\left(\frac{1 - \beta\cos 2\phi}{1 - \beta}\right)^{\frac{1}{2}}\right] = -\,15\,\frac{(1 - \beta\cos 2\phi)^{\frac{1}{2}}}{(1 - \beta)^{\frac{3}{2}}}\,\beta\,(1 - \cos 2\phi)$$

Whence

$$\mathrm{C}_3(\phi) = \frac{(1 + \frac{3}{2}\beta + \frac{15}{4}\beta^2)}{(1 - \beta)^{\frac{3}{2}}}(1 - \beta\cos 2\phi)^{\frac{1}{2}}\,(1 - \tfrac{5}{2}\beta\cos 2\phi)$$

This only differs by a constant factor from the expression which arises from the approximate formulae.

It would be possible then to have only one type of function, viz. $\mathfrak{P}$ or P, and to express all the cosine and sine-functions by means of the appropriate one of them. This would be found to be equivalent to expressing the latter functions in terms of powers of $\sin\phi$. For the purposes of practical application I do not think this would be so convenient as the use of cosines and sines of multiples of ϕ, and the advantage of using only one type of function would not compensate for the loss of convenience in the result. Accordingly I do not think it worth while to undertake the very laborious task of revising all the analysis from this point of view.

11.

ON THE PEAR-SHAPED FIGURE OF EQUILIBRIUM OF A ROTATING MASS OF LIQUID.

[*Philosophical Transactions of the Royal Society*, Vol. 198, A (1901), pp. 301—331.]

INTRODUCTION.

THIS is the sequel to a previous paper on "Ellipsoidal Harmonic Analysis" [Paper 10]. I here make use of the methods of that paper, and for brevity shall refer to it as "Harmonics."

The sections 1 to 4 are preparatory, and might have been included in "Harmonics," but seem more appropriate here. Section 5 is an independent investigation of so much of M. Poincaré's celebrated memoir on rotating liquid* as relates to the immediate object in view.

It is not necessary to say more here, since I give a short summary in the last section.

§ 1. *The Harmonics of the Third Degree.*

[In "Harmonics" § 6, p. 211, it is shown how σ corresponding to each harmonic may be found independently of all the other harmonics, but it is further shown in an addition made for the present volume how all the values may be found from only four equations. In this paper as originally published each value was found independently, but I shall now find the required values more shortly.

In the case of third harmonics the equation for determining σ for $s=0$ (cosine functions), is

$$\sigma = \frac{\frac{1}{2}\beta^2 \{3, 1\} \{3, 2\}}{4 \cdot 1^2 + \sigma} = \frac{60\beta^2}{4+\sigma}$$

* *Acta Mathematica*, Vol. VII., 1885.

The solutions of this quadratic are

$$\sigma = -2 \pm 2\sqrt{(1+15\beta^2)}$$

Now we know (p. 213, "Harmonics") that the roots are σ_3 and $\sigma_3^2 - 2^2$.

If therefore we write $(B_1)^2 = 1 + 15\beta^2$

$$\text{For } s=0, \text{ type OEC, } \sigma = -2(1-B_1)$$

$$s=2, \text{ type OEC, } \sigma = \ \ 2(1-B_1)$$

The equation for σ for the sine function, when $s=2$, is $\sigma = 0$.

Thus for $s=2$, type OES, $\sigma = 0$

The equation for σ for the cosine function, when $s=1$, is

$$\sigma + 6\beta = \frac{\frac{1}{4}\beta^2\{3, 2\}\{3, 3\}}{4 \,.\, 1 \,.\, 2 + \sigma} = \frac{15\beta^2}{8+\sigma}$$

The solutions of this quadratic are

$$-4 - 3\beta \pm 4\sqrt{[1 - \tfrac{3}{2}\beta(1-\beta)]}$$

We know that the roots are σ_3^1 and $\sigma_3^3 - 3^2 + 1^2$.

If therefore we write $(B_2)^2 = 1 - \frac{3}{2}\beta(1-\beta)$

$$\text{For } s=1, \text{ type OOC, } \sigma = -4 - 3\beta + 4B_2$$

$$s=3, \text{ type OOC, } \sigma = \ \ 4 - 3\beta - 4B_2$$

For the corresponding sine functions the equation has the same algebraic form, but the sign of β is changed, and if we write

$$(B_3)^2 = 1 + \tfrac{3}{2}\beta(1+\beta)$$

$$\text{For } s=1, \text{ type OOS, } \sigma = -4 + 3\beta + 4B_3$$

$$s=3, \text{ type OOS, } \sigma = \ \ 4 + 3\beta - 4B_3$$

We thus have the values of σ corresponding to the seven functions of the third order.]

$s=0$; type OEC, and $\mathfrak{P}_3(\nu) = q_0 P_3(\nu) + \beta q_2 P_3^2(\nu)$, with $q_0 = 1$.

We have

$$\sigma = 2(B_1 - 1)$$

where

$$(B_1)^2 = 1 + 15\beta^2$$

Also

$$\frac{q_2}{q_0} = \frac{1}{4+\sigma} = \frac{1}{2(B_1+1)}, \text{ with } q_0 = 1$$

Then since

$$P_3(\nu) = \tfrac{5}{2}\nu^3 - \tfrac{3}{2}\nu, \quad P_3^2 = 15\nu(\nu^2 - 1)$$

we find

$$\mathfrak{P}_3(\nu) = \frac{1}{2\beta}(B_1 - 1 + 5\beta)\,\nu\left(\nu^2 - \frac{4 - B_1}{5(1-\beta)}\right) \quad \text{..............(1)}$$

$s=1$; type OOC, and

$$\mathsf{P}_3^1(\nu) = \sqrt{\frac{\nu^2 - \frac{1+\beta}{1-\beta}}{\nu^2 - 1}}\,[q_1' P_3^1(\nu) + \beta q_3' P_3^3(\nu)], \text{ with } q_1' = 1$$

We have $$\sigma = 4(B_2 - 1 - \tfrac{3}{4}\beta)$$
where $$(B_2)^2 = 1 - \tfrac{3}{2}\beta(1-\beta)$$
Also
$$\frac{2q_3}{q_1} = \frac{1}{8+\sigma} = \frac{1}{4(B_2 + 1 - \frac{3}{4}\beta)}, \text{ and } \frac{q_3'}{q_1'} = 3\frac{q_3}{q_1}, \text{ with } q_1' = 1$$

Then since $P_3^1(\nu) = \frac{3}{2}(5\nu^2 - 1)(\nu^2-1)^{\frac{1}{2}}$, $P_3^3(\nu) = 15(\nu^2-1)(\nu^2-1)^{\frac{1}{2}}$, we find
$$\mathsf{P}_3^1(\nu) = \frac{6}{\beta}(B_2 - 1 + 2\beta)\left(\nu^2 - \frac{1+\beta}{1-\beta}\right)^{\frac{1}{2}}\left(\nu^2 - \frac{3-\beta-2B_2}{5(1-\beta)}\right) \quad \ldots\ldots(2)$$

$s=1$, type OOS and $\mathfrak{P}_3^1(\nu) = q_1 P_3^1(\nu) + \beta q_3 P_3^3(\nu)$, with $q_1 = 1$.

We have $$\sigma = 4(B_3 - 1 + \tfrac{3}{4}\beta)$$
where $$(B_3)^2 = 1 + \tfrac{3}{2}\beta(1+\beta)$$
$$\frac{2q_3}{q_1} = \frac{1}{4(B_2 + 1 + \frac{3}{4}\beta)}, \text{ with } q_1 = 1$$

On substitution we find
$$\mathfrak{P}_3^1(\nu) = \frac{2}{\beta}(B_3 - 1 + 3\beta)(\nu^2-1)^{\frac{1}{2}}\left(\nu^2 - \frac{3+\beta-2B_3}{5(1-\beta)}\right) \quad \ldots\ldots\ldots(3)$$

$s=2$; type OEC, $\mathfrak{P}_3^2(\nu) = \beta q_0 P_3(\nu) + q_2 P_3^2(\nu)$, with $q_2 = 1$.

We have $$\sigma = -2(B_1 - 1)$$
Also $$\frac{2q_0}{q_2} = \frac{-\{3,1\}\{3,2\}}{4-\sigma} = \frac{-4(B_1-1)}{\beta^2}, \text{ with } q_2 = 1$$

With the known values of P_3 and of P_3^2, we find
$$\mathfrak{P}_3^2(\nu) = \frac{5}{\beta}(1 + 3\beta - B_1)\,\nu\left(\nu^2 - \frac{4+B_1}{5(1-\beta)}\right) \quad \ldots\ldots\ldots\ldots\ldots(4)$$

A comparison with (1) for $s=0$ shows that the last factors in each only differ in the sign of B_1.

$s=2$; type OES, $\mathsf{P}_3^2(\nu) = \sqrt{\dfrac{\nu^2 - \frac{1+\beta}{1-\beta}}{\nu^2-1}}\,.\,P_3^2(\nu)$.

We have $\sigma = 0$, and $P_3^2(\nu) = 15\nu(\nu^2-1)$,
$$\mathsf{P}_3^2(\nu) = 15\nu(\nu^2-1)^{\frac{1}{2}}\left(\nu^2 - \frac{1+\beta}{1-\beta}\right)^{\frac{1}{2}} \quad \ldots\ldots\ldots\ldots\ldots\ldots\ldots(5)$$

$s=3$; type OOC,
$$\mathsf{P}_3^3(\nu) = \sqrt{\frac{\nu^2 - \frac{1+\beta}{1-\beta}}{\nu^2-1}}\,[\beta q_1' P_3^1(\nu) + q_3' P_3^3(\nu)], \text{ with } q_3' = 1$$

We have $$\sigma = 4(1 - \tfrac{3}{4}\beta - B_2)$$

Also
$$\frac{2q_1}{q_3} = \frac{-\{3,2\}\{3,3\}}{8-\sigma-6\beta} = -\frac{16}{\beta^2}(B_2 - 1 + \tfrac{3}{4}\beta)$$

and
$$\frac{q_1'}{q_3'} = \frac{q_1}{3q_3}, \text{ with } q_3' = 1$$

Whence on substitution
$$\mathsf{P}_3^3(\nu) = \frac{20}{\beta}(1-B_2)\left(\nu^2 - \frac{1+\beta}{1-\beta}\right)^{\frac{1}{2}}\left(\nu^2 - \frac{3-\beta+2B_2}{5(1-\beta)}\right) \text{.........} (6)$$

$s=3$; OOS, $\mathfrak{P}_3^3(\nu) = \beta q_1 P_3^1(\nu) + q_3 P_3^3(\nu)$, with $q_3 = 1$.

We have
$$\sigma = 4(1 + \tfrac{3}{4}\beta - B_3)$$
$$\frac{2q_1}{q_3} = -\frac{16}{\beta^2}(B_3 - 1 - \tfrac{3}{4}\beta), \text{ with } q_3 = 1$$

Whence on substitution
$$\mathfrak{P}_3^3(\nu) = \frac{60}{\beta}(1+\beta-B_3)(\nu^2-1)^{\frac{1}{2}}\left(\nu^2 - \frac{3+\beta+2B_3}{5(1-\beta)}\right) \text{.........} (7)$$

The forms of the corresponding functions of μ are the same, except that $(1-\mu^2)^{\frac{1}{2}}$ and $\left(\frac{1+\beta}{1-\beta} - \mu^2\right)^{\frac{1}{2}}$ replace the corresponding factors.

I have not determined the cosine- and sine-functions, because they may be written down at once from the results already obtained. The three roots of the fundamental cubic are ν^2, μ^2, and $\frac{1-\beta\cos 2\phi}{1-\beta}$. Hence we have only to replace ν^2 by this last function in the seven formulæ (1)—(7) in order to obtain functions *proportional* to the seven cosine- and sine-functions. If the definition of the latter functions is to agree with that given in "Harmonics," the factors must be determined appropriately, but the question as to the value of the factor will not arise here.

§ 2. *Change of Notation.*

It will be convenient, with a view to future work, to change the notation, and I desire to adopt a notation which shall not only agree in the main with that used in "Harmonics," but shall also facilitate reference to a previous paper on Jacobi's ellipsoid [Paper 8, p. 119].

I write
$$\kappa^2 = \frac{1-\beta}{1+\beta}, \quad \kappa'^2 = 1 - \kappa^2$$

I have in general written the current co-ordinates ν, μ, ϕ, and the ellipsoid of reference ν_0, so that the squares of the semi-axes are
$$k^2\left(\nu_0^2 - \frac{1+\beta}{1-\beta}\right), \quad k^2(\nu_0^2 - 1), \quad k^2\nu_0^2$$

I now propose to write as the squares of three semi-axes of the ellipsoid of reference

$$c^2 \cos^2 \gamma, \quad c^2 (1 - \kappa^2 \sin^2 \gamma), \quad c^2$$

Comparing these two we see that

$$k = c\kappa \sin \gamma, \quad \text{and } \nu_0 = \frac{1}{\kappa \sin \gamma}$$

For the current co-ordinates I retain ϕ and write

$$\nu = \frac{1}{\kappa \sin \psi}, \quad \mu = \sin \theta$$

The three roots of the fundamental cubic are therefore

$$\nu^2 = \frac{1}{\kappa^2 \sin^2 \psi}, \quad \mu^2 = \sin^2 \theta, \quad \frac{1 - \beta \cos 2\phi}{1 - \beta} = \frac{1}{\kappa^2} (1 - \kappa'^2 \cos^2 \phi)$$

The rectangular co-ordinates x, y, z are now expressible as follows:—

$$\left. \begin{aligned} x &= \frac{c \sin \gamma}{\sin \psi} . \cos \psi \, (1 - \kappa^2 \sin^2 \theta)^{\frac{1}{2}} \cos \phi \\ y &= \frac{c \sin \gamma}{\sin \psi} . (1 - \kappa^2 \sin^2 \psi)^{\frac{1}{2}} \cos \theta \sin \phi \\ z &= \frac{c \sin \gamma}{\sin \psi} . \sin \theta \, (1 - \kappa'^2 \cos^2 \phi)^{\frac{1}{2}} \end{aligned} \right\} \quad \text{..................(8)}$$

These give $\quad \dfrac{x^2}{\cos^2 \psi} + \dfrac{y^2}{1 - \kappa^2 \sin^2 \psi} + z^2 = \dfrac{c^2 \sin^2 \gamma}{\sin^2 \psi}$

At the surface $\psi = \gamma$, and we have

$$\frac{x^2}{\cos^2 \gamma} + \frac{y^2}{1 - \kappa^2 \sin^2 \gamma} + z^2 = c^2$$

In the formulæ for the third harmonics, in every case but one, and in two out of the five harmonics of the second degree, there occurs a factor of the form (ν^2 − constant); in each such case I write that constant in the form q^2/κ^2, and $q'^2 = 1 - q^2$. Thus q will have a different value for each harmonic.

It has been already remarked that for most purposes it is immaterial by what constants the several functions are multiplied. Although it would be easy to determine the constant in each case so as to make the function agree with its value as defined in "Harmonics," yet I shall not take that course, and shall omit factors as being in most cases redundant.

For the sake of completeness I will give the first and second harmonics in the new notation, as well as the third.

Since the harmonics of the first degree are expressed by

$$\mathfrak{P}_1(\nu) = \nu, \quad \mathsf{P}_1^1(\nu) = \left(\nu^2 - \frac{1 + \beta}{1 - \beta}\right)^{\frac{1}{2}}, \quad \mathfrak{P}_1^1(\nu) = (\nu^2 - 1)^{\frac{1}{2}}$$

it is clear that in the new notation

$$\left.\begin{array}{lll} \mathfrak{P}_1(\nu) = \dfrac{1}{\sin\psi}, & \mathsf{P}_1^1(\nu) = \cot\psi, & \mathfrak{P}_1^1(\nu) = \dfrac{(1-\kappa^2\sin^2\psi)^{\frac{1}{2}}}{\sin\psi} \\ \mathfrak{P}_1(\mu) = \sin\theta, & \mathsf{P}_1^1(\mu) = (1-\kappa^2\sin^2\theta)^{\frac{1}{2}}, & \mathfrak{P}_1^1(\mu) = \cos\theta \\ \mathsf{C}_1(\phi) = (1-\kappa'^2\cos^2\phi)^{\frac{1}{2}}, & \mathfrak{C}_1^1(\phi) = \cos\phi, & \mathfrak{S}_1^1(\phi) = \sin\phi \end{array}\right\} \quad \ldots\ldots(9)$$

It appears from § 12 of "Harmonics" that

$$\mathfrak{P}_2(\nu) = \nu^2 + \frac{\gamma}{\alpha}, \qquad \mathfrak{P}_2^2(\nu) = \nu^2 + \frac{\gamma'}{\alpha'}$$

where

$$\frac{\gamma}{\alpha} = \frac{B-2}{3(1-\beta)}, \qquad \frac{\gamma'}{\alpha'} = -\frac{B+2}{3(1-\beta)}, \qquad \text{and} \quad B^2 = 1 + 3\beta^2$$

In accordance with the notation suggested above, let

$$\frac{q^2}{\kappa^2} = \frac{2 \mp B}{3(1-\beta)}$$

Then substituting $\dfrac{1-\kappa^2}{1+\kappa^2}$ for β, we find

$$q^2 = \tfrac{1}{3}[1 + \kappa^2 \mp (1-\kappa^2\kappa'^2)^{\frac{1}{2}}]$$

and for both cases

$$\kappa^2 = q^2\frac{2-3q^2}{1-2q^2}$$

Hence

$$\left.\begin{array}{l} \mathfrak{P}_2(\nu) \text{ and } \mathfrak{P}_2^2(\nu) = \dfrac{1-q^2\sin^2\psi}{\sin^2\psi} \\ \mathfrak{P}_2(\mu) \text{ and } \mathfrak{P}_2^2(\mu) = -\left(1 - \dfrac{\kappa^2}{q^2}\sin^2\theta\right) \\ \mathfrak{C}_2(\phi) \text{ and } \mathfrak{C}_2^2(\phi) = 1 - \dfrac{\kappa'^2}{q'^2}\cos^2\phi \end{array}\right\} \quad \ldots\ldots\ldots\ldots(10)$$

where $\kappa^2 = q^2\dfrac{2-3q^2}{1-2q^2}$, and $q^2 = \frac{1}{3}[1+\kappa^2 \mp (1-\kappa^2\kappa'^2)^{\frac{1}{2}}]$, with upper sign for the first and the lower sign for the second.

It appears from (19) and (20), § 7, of "Harmonics" that

$$\left.\begin{array}{l} \mathsf{P}_2^1(\nu) = \dfrac{\cos\psi}{\sin^2\psi} \\ \mathsf{P}_2^1(\mu) = \sin\theta\,(1-\kappa^2\sin^2\theta)^{\frac{1}{2}} \\ \mathsf{C}_2^1(\phi) = \cos\phi\,(1-\kappa'^2\cos^2\phi)^{\frac{1}{2}} \end{array}\right\} \quad \ldots\ldots\ldots\ldots\ldots\ldots(11)$$

and from (21) and (22) that

$$\left.\begin{array}{l} \mathfrak{P}_2^1(\nu) = \dfrac{(1-\kappa^2\sin^2\psi)^{\frac{1}{2}}}{\sin^2\psi} \\ \mathfrak{P}_2^1(\mu) = \sin\theta\cos\theta \\ \mathsf{S}_2^1(\phi) = \sin\phi\,(1-\kappa'^2\cos^2\phi)^{\frac{1}{2}} \end{array}\right\} \quad \ldots\ldots\ldots\ldots\ldots\ldots(12)$$

Lastly, from (25) and (26)

$$\left.\begin{aligned} \mathsf{P}_2^2(\nu) &= \frac{\cos\psi\,(1-\kappa^2\sin^2\psi)^{\frac{1}{2}}}{\sin^2\psi} \\ \mathsf{P}_2^2(\mu) &= \cos\theta\,(1-\kappa^2\sin^2\theta)^{\frac{1}{2}} \\ \mathfrak{S}_2^2(\phi) &= \sin\phi\cos\phi \end{aligned}\right\} \quad \text{.....................(13)}$$

Turning to the harmonics of the third degree, we found that in the two cases where the type is OEC,

$$\mathfrak{P}_3(\nu) \text{ and } \mathfrak{P}_3^2(\nu) = \nu\left(\nu^2 - \frac{4 \mp B_1}{5(1-\beta)}\right)$$

If we put

$$\frac{q^2}{\kappa^2} = \frac{4 \mp B_1}{5(1-\beta)}$$

we find

$$q^2 = \tfrac{2}{5}\left[1 + \kappa^2 \mp (1 - \tfrac{7}{4}\kappa^2 + \kappa^4)^{\frac{1}{2}}\right]$$

and

$$\kappa^2 = q^2\,\frac{4-5q^2}{3-4q^2}$$

Therefore, with the above alternative form for q^2,

$$\left.\begin{aligned} \mathfrak{P}_3(\nu) \text{ and } \mathfrak{P}_3^2(\nu) &= \frac{1-q^2\sin^2\psi}{\sin^3\psi} \\ \mathfrak{P}_3(\mu) \text{ and } \mathfrak{P}_3^2(\mu) &= -\sin\theta\left(1 - \frac{\kappa^2}{q^2}\sin^2\theta\right) \\ \mathsf{C}_3(\phi) \text{ and } \mathsf{C}_3^2(\phi) &= \left(1 - \frac{\kappa'^2}{q'^2}\cos^2\phi\right)(1-\kappa'^2\cos^2\phi)^{\frac{1}{2}} \end{aligned}\right\} \quad \text{......(14)}$$

Again in the two cases where the type is OOC we found

$$\mathsf{P}_3^1(\nu) \text{ and } \mathsf{P}_3^3(\nu) = \left(\nu^2 - \frac{3-\beta \mp 2B_2}{5(1-\beta)}\right)\left(\nu^2 - \frac{1+\beta}{1-\beta}\right)^{\frac{1}{2}}$$

Putting

$$\frac{q^2}{\kappa^2} = \frac{3-\beta \mp 2B_2}{5(1-\beta)}$$

we find

$$q^2 = \tfrac{1}{5}\left(1 + 2\kappa^2 \mp (1 - \kappa^2 + 4\kappa^4)^{\frac{1}{2}}\right)$$

and

$$\kappa^2 = q^2\,\frac{2-5q^2}{1-4q^2}$$

Therefore, with the above alternative form for q^2,

$$\left.\begin{aligned} \mathsf{P}_3^1(\nu) \text{ and } \mathsf{P}_3^3(\nu) &= \frac{\cos\psi\,(1-q^2\sin^2\psi)}{\sin^3\psi} \\ \mathsf{P}_3^1(\mu) \text{ and } \mathsf{P}_3^3(\mu) &= -(1-\kappa^2\sin^2\theta)^{\frac{1}{2}}\left(1 - \frac{\kappa^2}{q^2}\sin^2\theta\right) \\ \mathfrak{C}_3^1(\phi) \text{ and } \mathfrak{C}_3^3(\phi) &= \cos\phi\left(1 - \frac{\kappa'^2}{q'^2}\cos^2\phi\right) \end{aligned}\right\} \quad \text{......(15)}$$

In the two cases where the type is OOS we found

$$\mathfrak{P}_3^1(\nu) \text{ and } \mathfrak{P}_3^3(\nu) = \left(\nu^2 - \frac{3+\beta \mp 2B_3}{5(1-\beta)}\right)(\nu^2-1)^{\frac{1}{2}}$$

Putting
$$\frac{q^2}{\kappa^2} = \frac{3+\beta \mp 2B_3}{5(1-\beta)}$$

we find
$$q^2 = \tfrac{1}{5}\left(2+\kappa^2 \mp (4-\kappa^2\kappa'^2)^{\frac{1}{2}}\right)$$

and
$$\kappa^2 = q^2\,\frac{4-5q^2}{1-2q^2}$$

Therefore, with the above alternative form for q^2,

$$\left.\begin{aligned} \mathfrak{P}_3^1(\nu) \text{ and } \mathfrak{P}_3^3(\nu) &= \frac{(1-\kappa^2\sin^2\psi)^{\frac{1}{2}}(1-q^2\sin^2\psi)}{\sin^3\psi} \\ \mathfrak{P}_3^1(\mu) \text{ and } \mathfrak{P}_3^3(\mu) &= -\cos\theta\left(1-\frac{\kappa^2}{q^2}\sin^2\theta\right) \\ \mathfrak{S}_3^1(\phi) \text{ and } \mathfrak{S}_3^3(\phi) &= \sin\phi\left(1-\frac{\kappa'^2}{q'^2}\cos^2\phi\right) \end{aligned}\right\} \quad\ldots\ldots\ldots(16)$$

The seventh of these harmonics, which is of type OES, stands by itself. We had

$$\mathsf{P}_3^2(\nu) = \nu(\nu^2-1)^{\frac{1}{2}}\left(\nu^2 - \frac{1+\beta}{1-\beta}\right)^{\frac{1}{2}}$$

This gives in the new notation

$$\left.\begin{aligned} \mathsf{P}_3^2(\nu) &= \frac{\cos\psi\,(1-\kappa^2\sin^2\psi)^{\frac{1}{2}}}{\sin^3\psi} \\ \mathsf{P}_3^2(\mu) &= \sin\theta\cos\theta\,(1-\kappa^2\sin^2\theta)^{\frac{1}{2}} \\ \mathsf{S}_3^2(\phi) &= \sin\phi\cos\phi\,(1-\kappa'^2\cos^2\phi)^{\frac{1}{2}} \end{aligned}\right\} \quad\ldots\ldots\ldots\ldots\ldots(17)$$

The formulæ (9) to (17) give the fifteen sets of three functions constituting the fifteen harmonic functions of the first three degrees. It would be easy, although somewhat tedious, to find the coefficient by which each function is to be multiplied so that its definition might agree with that of the previous paper*.

* [Although the fourth harmonics are not required in this paper I may mention that I have obtained the values of σ in the same way with the following results.

The equation corresponding to $i=4$, $s=0$, where the type is EEC, may be written

$$\sigma^3 + 20\sigma^2 + (64 - 208\beta^2)\,\sigma - 2880\beta^2 = 0$$

The roots of this are σ_4, $\sigma_4{}^2 - 2^2$, $\sigma_4{}^4 - 4^2$.

If we write $\sigma = x - \frac{20}{3}$, the cubic equation becomes

$$x^3 - \tfrac{208}{3}(1+3\beta^2) + \tfrac{4480}{27}(1-9\beta^2) = 0$$

If α be the smallest angle such that

$$\cos 3\alpha = \frac{35}{13^{\frac{3}{2}}}\,\frac{1-9\beta^2}{(1+3\beta^2)^{\frac{3}{2}}}$$

§ 3. *Expressions for the Solid Harmonics in Rectangular Coordinates.*

The three roots of the original cubic equation were $\nu^2, \mu^2, \dfrac{1-\beta\cos 2\phi}{1-\beta}$, and in the new notation the three roots of

$$\frac{x^2}{\omega^2 - 1/\kappa^2} + \frac{y^2}{\omega^2 - 1} + \frac{z^2}{\omega^2} = c^2\kappa^2 \sin^2\gamma \text{ are } \frac{1}{\kappa^2 \sin^2\psi},\ \sin^2\theta,\ \frac{1-\kappa'^2\cos^2\phi}{\kappa^2}$$

Hence it follows that we have the identity

$$\frac{x^2}{\omega^2 - 1/\kappa^2} + \frac{y^2}{\omega^2 - 1} + \frac{z^2}{\omega^2} - c^2\kappa^2\sin^2\gamma$$

$$= c^2\kappa^2\sin^2\gamma \frac{\left(\frac{1}{\kappa^2\sin^2\psi} - \omega^2\right)\left(\sin^2\theta - \omega^2\right)\left(\frac{1-\kappa'^2\cos^2\phi}{\kappa^2} - \omega^2\right)}{(1/\kappa^2 - \omega^2)(1-\omega^2)\,\omega^2}$$

Putting $\omega^2 = \dfrac{q^2}{\kappa^2}$,

$$\frac{x^2}{q'^2} + \frac{y^2}{\kappa^2 - q^2} - \frac{z^2}{q^2} + c^2\sin^2\gamma$$

$$= \frac{c^2\sin^2\gamma}{(\kappa^2 - q^2)\sin^2\psi}(1 - q^2\sin^2\psi)\left(1 - \frac{\kappa^2}{q^2}\sin^2\theta\right)\left(1 - \frac{\kappa'^2}{q'^2}\cos^2\phi\right)$$

---

the three roots of the equation in x are $-p\cos\alpha$, $p\cos(\alpha \mp 60°)$, where

$$p = \tfrac{8}{3}\, 13^{\frac{1}{2}}\, (1 + 3\beta^2)^{\frac{1}{2}}$$

By adding $\frac{20}{3}$ to each of these roots we get the roots of the equation in σ, and by proper choice of these roots we obtain the values of σ_4, σ_4^2, σ_4^4, corresponding to the three cosine harmonics $\mathfrak{P}_4^s$, with $s=0, 2, 4$.

However when $\beta = \frac{1}{3}$, which as we have seen applies to the class of ellipsoids defined by $c^2 = \frac{1}{2}(a^2+b^2)$, the cubic is soluble algebraically, and $x=0$ or $\pm\frac{8}{3}\sqrt{13}$.

Whence

$$\text{for } s=0,\ \sigma = \tfrac{4}{3}(2\sqrt{13} - 5)$$
$$\text{for } s=2,\ \sigma = -\tfrac{8}{3}$$
$$\text{for } s=4,\ \sigma = \tfrac{4}{3}(7 - 2\sqrt{13})$$

For the sine harmonics of even rank we proceed from the equation for the case of $s=2$. The equation is

$$\sigma^2 + 12\sigma = 28\beta^2$$

and its two roots are σ_4^2, $\sigma_4^4 - 4^2 + 2^2$. Hence we find for sine harmonics P_4^s, with $s=2, 4$,

$$\sigma_4^2 = -6 + 6\sqrt{(1+\tfrac{7}{9}\beta^2)}$$
$$\sigma_4^4 = \quad 6 - 6\sqrt{(1+\tfrac{7}{9}\beta^2)}$$

For the cosine harmonics of odd rank we proceed from the equation for $s=1$; it is

$$\sigma^2 + (8 + 10\sigma) = -80\beta + 63\beta^2$$

The roots are σ_4^1 and $\sigma_4^3 - 3^2 + 1^2$. Hence we find for cosine harmonics P_4^s, with $s=1, 3$,

$$\sigma_4^1 = -4 - 5\beta + 4\sqrt{(1 - \tfrac{5}{2}\beta + \tfrac{11}{2}\beta^2)}$$
$$\sigma_4^3 = \quad 4 - 5\beta - 4\sqrt{(1 - \tfrac{5}{2}\beta + \tfrac{11}{2}\beta^2)}$$

By changing the sign of β, we have for the sine harmonics $\mathfrak{P}_4^s$, with $s=1, 3$,

$$\sigma_4^1 = -4 + 5\beta + 4\sqrt{(1 + \tfrac{5}{2}\beta + \tfrac{11}{2}\beta^2)}$$
$$\sigma_4^3 = \quad 4 - 5\beta - 4\sqrt{(1 + \tfrac{5}{2}\beta + \tfrac{11}{2}\beta^2)}]$$

This expression, together with those for x, y, z in (8), enables us to write down the results at once. As before, I drop the several factors as being redundant for most purposes.

From (9)

$$\mathfrak{P}_1(\nu)\,\mathfrak{P}_1(\mu)\,\mathsf{C}_1(\phi)=z,\quad \mathsf{P}_1^1(\nu)\,\mathsf{P}_1^1(\mu)\,\mathfrak{C}_1^1(\phi)=x,\quad \mathfrak{P}_1^1(\nu)\,\mathfrak{P}_1^1(\mu)\,\mathfrak{S}_1^1(\phi)=y \quad\ldots\ldots(18)$$

From (10)

$\mathfrak{P}_2(\nu)\,\mathfrak{P}_2(\mu)\,\mathfrak{C}_2(\phi)$ and $\mathfrak{P}_2^2(\nu)\,\mathfrak{P}_2^2(\mu)\,\mathfrak{C}_2^2(\phi)$

$$=q^2x^2+\frac{q^2q'^2}{\kappa^2-q^2}y^2-q'^2z^2+c^2q^2q'^2\sin^2\gamma\ \ldots\ldots(19)$$

where $\quad q^2=\tfrac{1}{3}[1+\kappa^2\mp(1-\kappa^2\kappa'^2)^{\frac{1}{2}}]$, and $\kappa^2=q^2\dfrac{2-3q^2}{1-2q^2}$

so that $\quad \dfrac{q^2q'^2}{\kappa^2-q^2}=1-2q^2$

From (11), (12), and (13)

$$\mathsf{P}_2^1(\nu)\,\mathsf{P}_2^1(\mu)\,\mathsf{C}_2^1(\phi)=xz,\quad \mathfrak{P}_2^1(\nu)\,\mathfrak{P}_2^1(\mu)\,\mathsf{S}_2^1(\phi)=yz,\quad \mathsf{P}_2^2(\nu)\,\mathsf{P}_2^2(\mu)\,\mathfrak{S}_2^2(\phi)=xy \quad\ldots\ldots(20)$$

From (14)

$\mathfrak{P}_3(\nu)\,\mathfrak{P}_3(\mu)\,\mathsf{C}_3(\phi)$ and $\mathfrak{P}_3^2(\nu)\,\mathfrak{P}_3^2(\mu)\,\mathsf{C}_3^2(\phi)$

$$=z\left(q^2x^2+\frac{q^2q'^2}{\kappa^2-q^2}y^2-q'^2z^2+c^2q^2q'^2\sin^2\gamma\right)\ldots\ldots(21)$$

where $\quad q^2=\tfrac{2}{5}[1+\kappa^2\mp(1-\tfrac{7}{4}\kappa^2+\kappa^4)^{\frac{1}{2}}]$, and $\kappa^2=q^2\dfrac{4-5q^2}{3-4q^2}$

so that $\quad \dfrac{q^2q'^2}{\kappa^2-q^2}=3-4q^2$

From (15)

$\mathsf{P}_3^1(\nu)\,\mathsf{P}_3^1(\mu)\,\mathfrak{C}_3^1(\phi)$ and $\mathsf{P}_3^3(\nu)\,\mathsf{P}_3^3(\mu)\,\mathfrak{C}_3^3(\phi)$

$$=x\left(q^2x^2+\frac{q^2q'^2}{\kappa^2-q^2}y^2-q'^2z^2+c^2q^2q'^2\sin^2\gamma\right)\ldots\ldots(22)$$

where $\quad q^2=\tfrac{1}{5}(1+2\kappa^2\mp(1-\kappa^2+4\kappa^4)^{\frac{1}{2}})$, and $\kappa^2=q^2\dfrac{2-5q^2}{1-4q^2}$

so that $\quad \dfrac{q^2q'^2}{\kappa^2-q^2}=1-4q^2$

From (16)

$\mathfrak{P}_3^1(\nu)\,\mathfrak{P}_3^1(\mu)\,\mathfrak{S}_3^1(\phi)$ and $\mathfrak{P}_3^3(\nu)\,\mathfrak{P}_3^3(\mu)\,\mathfrak{S}_3^3(\phi)$

$$=y\left(q^2x^2+\frac{q^2q'^2}{\kappa^2-q^2}y^2-q'^2z^2+c^2q^2q'^2\sin^2\gamma\right)\ldots\ldots(23)$$

where $\quad q^2=\tfrac{1}{5}(2+\kappa^2\mp(4-\kappa^2\kappa'^2)^{\frac{1}{2}})$ and $\kappa^2=q^2\dfrac{4-5q^2}{1-2q^2}$

so that $\quad \dfrac{q^2q'^2}{\kappa^2-q^2}=\tfrac{1}{3}(1-2q^2)$

Lastly, from (17),

$$\mathsf{P}_3^2(\nu)\,\mathsf{P}_3^2(\mu)\,\mathsf{S}_3^2(\phi) = xyz \qquad\qquad (24)$$

It is easy to verify that each of these expressions satisfies Laplace's equation.

§ 4. *The Expression for the Q-functions in Elliptic Integrals.*

In this paper I drop the factors $\mathfrak{E}$ and E which were found to be necessary when the Q-functions were expressed in series.

We make the following definition:—

$$\mathfrak{P}_i^s(\nu_0)\,\mathfrak{Q}_i^s(\nu_0) = [\mathfrak{P}_i^s(\nu_0)]^2 \int_{\nu_0}^{\infty} \frac{d\nu}{[\mathfrak{P}_i^s(\nu)^2](\nu^2-1)^{\frac{1}{2}}\left(\nu^2-\frac{1+\beta}{1-\beta}\right)^{\frac{1}{2}}}$$

and a similar formula holds for $\mathsf{P}_i^s\,\mathsf{Q}_i^s$.

It is clear that $\mathfrak{P}_i^s$ may be multiplied by any constant factor without changing the result; hence we may use the forms which have been found in §§ 2, 3.

The notation must now be changed.

We have $\nu = \dfrac{1}{\kappa \sin\psi}$ and $\nu_0 = \dfrac{1}{\kappa\sin\gamma}$. Therefore, when ψ is adopted as variable, the limits are γ to 0, and the sign of the whole is changed.

But
$$d\nu = -\frac{\cos\psi}{\kappa\sin^2\psi}\,d\psi$$

and
$$(\nu^2-1)^{\frac{1}{2}}\left(\nu^2-\frac{1+\beta}{1-\beta}\right)^{\frac{1}{2}} = \frac{\cos\psi\,(1-\kappa^2\sin^2\psi)^{\frac{1}{2}}}{\kappa^2\sin^2\psi}$$

Therefore
$$\int_{\nu_0}^{\infty}\frac{d\nu}{(\nu^2-1)^{\frac{1}{2}}\left(\nu^2-\frac{1+\beta}{1-\beta}\right)^{\frac{1}{2}}} = \kappa\int_0^{\gamma}\frac{d\psi}{(1-\kappa^2\sin^2\psi)^{\frac{1}{2}}}$$

In accordance with the usage in elliptic integrals, I write

$$\Delta^2 = 1-\kappa^2\sin^2\psi$$

under the integral sign, or $1-\kappa^2\sin^2\gamma$ outside the integral.

I shall also for brevity write

$$\Delta_1^2 = 1-q^2\sin^2\psi$$

under the integral, or $1-q^2\sin^2\gamma$ outside the integral.

We have then

$$\mathfrak{P}_i^s(\nu_0)\,\mathfrak{Q}_i^s(\nu_0) = \kappa\,[\mathfrak{P}_i^s(\nu_0)]^2\int_0^{\gamma}\frac{d\psi}{[\mathfrak{P}_i^s(\nu)]^2\,\Delta}$$

I apply this formula successively to the several functions, as given in (9) to (17), and introduce the abridged notation just defined, but I do not reiterate the special meanings to be attached to the symbol q in each case.

Since $\mathfrak{P}_0(\nu) = 1$, we have (dropping the now unnecessary suffix 0 to ν),

$$\left.\begin{aligned}
&\mathfrak{P}_0(\nu)\,\mathfrak{Q}_0(\nu) = \kappa\int_0^\gamma \frac{d\psi}{\Delta}\\
&\mathfrak{P}_1(\nu)\,\mathfrak{Q}_1(\nu) = \frac{\kappa}{\sin^2\gamma}\int_0^\gamma \frac{\sin^2\psi\, d\psi}{\Delta}\\
&\mathsf{P}_1^1(\nu)\,\mathsf{Q}_1^1(\nu) = \kappa\cot^2\gamma\int_0^\gamma \frac{\tan^2\psi}{\Delta}\,d\psi\\
&\mathfrak{P}_1^1(\nu)\,\mathfrak{Q}_1^1(\nu) = \frac{\kappa\Delta^2}{\sin^2\gamma}\int_0^\gamma \frac{\sin^2\psi}{\Delta^3}\,d\psi\\
&\mathfrak{P}_2(\nu)\,\mathfrak{Q}_2(\nu) \text{ and } \mathfrak{P}_2^2(\nu)\,\mathfrak{Q}_2^2(\nu) = \frac{\kappa\Delta_1^4}{\sin^4\gamma}\int_0^\gamma \frac{\sin^4\psi}{\Delta_1^4\Delta}\,d\psi\\
&\mathsf{P}_2^1(\nu)\,\mathsf{Q}_2^1(\nu) = \frac{\kappa\cos^2\gamma}{\sin^4\gamma}\int_0^\gamma \frac{\sin^4\psi}{\cos^2\psi\Delta}\,d\psi\\
&\mathfrak{P}_2^1(\nu)\,\mathfrak{Q}_2^1(\nu) = \frac{\kappa\Delta^2}{\sin^4\gamma}\int_0^\gamma \frac{\sin^4\psi}{\Delta^3}\,d\psi\\
&\mathsf{P}_2^2(\nu)\,\mathsf{Q}_2^2(\nu) = \frac{\kappa\cos^2\gamma\Delta^2}{\sin^4\gamma}\int_0^\gamma \frac{\sin^4\psi}{\cos^2\psi\Delta^3}\,d\psi\\
&\mathfrak{P}_3(\nu)\,\mathfrak{Q}_3(\nu) \text{ and } \mathfrak{P}_3^2(\nu)\,\mathfrak{Q}_3^2(\nu) = \frac{\kappa\Delta_1^4}{\sin^6\gamma}\int_0^\gamma \frac{\sin^6\psi}{\Delta_1^4\Delta}\,d\psi\\
&\mathsf{P}_3^1(\nu)\,\mathsf{Q}_3^1(\nu) \text{ and } \mathsf{P}_3^3(\nu)\,\mathsf{Q}_3^3(\nu) = \frac{\kappa\cos^2\gamma\Delta_1^4}{\sin^6\gamma}\int_0^\gamma \frac{\sin^6\psi}{\cos^2\psi\Delta_1^4\Delta}\,d\psi\\
&\mathfrak{P}_3^1(\nu)\,\mathfrak{Q}_3^1(\nu) \text{ and } \mathfrak{P}_3^3(\nu)\,\mathfrak{Q}_3^3(\nu) = \frac{\kappa\Delta_1^4\Delta^2}{\sin^6\gamma}\int_0^\gamma \frac{\sin^6\psi}{\Delta_1^4\Delta^3}\,d\psi\\
&\mathsf{P}_3^2(\nu)\,\mathsf{Q}_3^2(\nu) = \frac{\kappa\cos^2\gamma\Delta^2}{\sin^6\gamma}\int_0^\gamma \frac{\sin^6\psi}{\cos^2\psi\Delta^3}\,d\psi
\end{aligned}\right\}\quad\ldots(25)$$

All these integrals are expressible in terms of the elliptic integrals

$$F = \int_0^\gamma \frac{d\psi}{\Delta},\qquad E = \int_0^\gamma \Delta\, d\psi,\qquad \Pi = \int_0^\gamma \frac{d\psi}{\Delta_1^2\Delta}$$

It will, however, be found that in fact the coefficient of Π vanishes in every case.

The cases of $i = 0$ and $i = 1$ are very simple, and we have

$$\mathfrak{P}_0\,\mathfrak{Q}_0 = \kappa F$$

$$\mathfrak{P}_1\,\mathfrak{Q}_1 = \frac{\kappa}{\sin^2\gamma}\left(\frac{1}{\kappa^2}F - \frac{1}{\kappa^2}E\right)$$

$$\mathsf{P}_1^1\,\mathsf{Q}_1^1 = \kappa\cot^2\gamma\left(\frac{1}{\kappa'^2}\Delta\tan\gamma - \frac{1}{\kappa'^2}E\right)$$

$$\mathfrak{P}_1^1\,\mathfrak{Q}_1^1 = \frac{\kappa\Delta^2}{\sin^2\gamma}\left(-\frac{1}{\kappa^2}F + \frac{1}{\kappa^2\kappa'^2}E - \frac{\sin\gamma\cos\gamma}{\kappa'^2\Delta}\right)$$

It is possible by direct differentiation to verify the following results, although the verification will be found pretty tedious.

$$\int \frac{\sin^4\psi}{\Delta_1^4\Delta}\,d\psi = \frac{(2-3q^2)\,q^2 - \kappa^2(1-2q^2)}{2q^4q'^2(\kappa^2-q^2)}\,\Pi + \frac{2q'^2-1}{2q^4q'^2}\,F - \frac{1}{2q^2q'^2(\kappa^2-q^2)}\,E + \frac{\Delta\sin\psi\cos\psi}{2q'^2(\kappa^2-q^2)\,\Delta_1^2}$$

$$\int \frac{\sin^4\psi}{\cos^2\psi\,\Delta}\,d\psi = -\frac{1}{\kappa^2}F + \frac{1-2\kappa^2}{\kappa^2\kappa'^2}E + \frac{1}{\kappa'^2}\Delta\tan\psi$$

$$\int \frac{\sin^4\psi}{\Delta^3}\,d\psi = -\frac{2}{\kappa^4}F + \frac{1+\kappa'^2}{\kappa^4\kappa'^2}E - \frac{\sin\psi\cos\psi}{\kappa^2\kappa'^2\Delta}$$

$$\int \frac{\sin^4\psi}{\cos^2\psi\,\Delta^3}\,d\psi = \frac{1}{\kappa^2\kappa'^2}F - \frac{1+\kappa^2}{\kappa^2\kappa'^4}E + \frac{\tan\psi}{\kappa'^4\Delta}[2-(1+\kappa^2)\sin^2\psi]$$

These are all the integrals needed for the harmonics of the second degree. In the case of the first we have

$$\kappa^2 = q^2\,\frac{2-3q^2}{1-2q^2}$$

Thus the coefficient of Π vanishes and the results are

$$\mathfrak{P}_2(\nu)\,\mathfrak{Q}_2(\nu) \text{ and } \mathfrak{P}_2^2(\nu)\,\mathfrak{Q}_2^2(\nu) = \frac{\kappa\Delta_1^4}{\sin^4\gamma}\left[\frac{1-2q^2}{2q^4q'^2}F - \frac{1-2q^2}{2q^4q'^4}E + \frac{(1-2q^2)\,\Delta\sin\gamma\cos\gamma}{2q^2q'^4\Delta_1^2}\right]$$

$$\mathsf{P}_2^1(\nu)\,\mathsf{Q}_2^1(\nu) = \frac{\kappa\cos^2\gamma}{\sin^4\gamma}\left[-\frac{1}{\kappa^2}F + \frac{1-2\kappa^2}{\kappa^2\kappa'^2}E + \frac{1}{\kappa'^2}\Delta\tan\gamma\right]$$

$$\mathfrak{P}_2^1(\nu)\,\mathfrak{Q}_2^1(\nu) = \frac{\kappa\Delta^2}{\sin^4\gamma}\left[-\frac{2}{\kappa^4}F + \frac{1+\kappa'^2}{\kappa^4\kappa'^2}E - \frac{\sin\gamma\cos\gamma}{\kappa^2\kappa'^2\Delta}\right]$$

$$\mathsf{P}_2^2(\nu)\,\mathsf{Q}_2^2(\nu) = \frac{\kappa\cos^2\gamma\,\Delta^2}{\sin^4\gamma}\left[\frac{1}{\kappa^2\kappa'^2}F - \frac{1+\kappa^2}{\kappa^2\kappa'^4}E + \frac{\tan\gamma\,(2-(1+\kappa^2)\sin^2\gamma)}{\kappa'^4\Delta}\right]$$

In the first of these

$$q^2 = \tfrac{1}{3}[1+\kappa^2 \mp (1-\kappa^2\kappa'^2)^{\frac{1}{2}}] \quad\text{and}\quad \kappa^2 = q^2\,\frac{2-3q^2}{1-2q^2}$$

The following integrals may also be verified by differentiation:

$$\int \frac{\sin^6\psi}{\Delta_1^4\Delta}\,d\psi = \frac{q^2(4-5q^2)-\kappa^2(3-4q^2)}{2q^6q'^2(\kappa^2-q^2)}\,\Pi + \frac{2q^2q'^2+\kappa^2(3-4q^2)}{2\kappa^2q^6q'^2}\,F - \frac{\kappa^2(3-2q^2)-2q^2q'^2}{2\kappa^2q^4q'^2(\kappa^2-q^2)}\,E + \frac{\Delta\sin\psi\cos\psi}{2q^2q'^2(\kappa^2-q^2)\,\Delta_1^2} \quad\text{......(26)}$$

$$\int \frac{\sin^6\psi}{\cos^2\psi\,\Delta_1^4\Delta}\,d\psi = \frac{\kappa^2(1-4q^2)-q^2(2-5q^2)}{2q^4q'^4(\kappa^2-q^2)}\,\Pi + \frac{4q^2-1}{2q^4q'^4}\,F + \frac{1+2q^4-\kappa^2(2q^2+1)}{2\kappa'^2q^2q'^4(\kappa^2-q^2)}\,E + \frac{\Delta\tan\psi}{\kappa'^2q'^4} - \frac{\Delta\sin\psi\cos\psi}{2q'^4(\kappa^2-q^2)\,\Delta_1^2} \quad\text{......(27)}$$

$$\int \frac{\sin^6 \psi}{\Delta_1^4 \Delta^3} d\psi = \frac{\kappa^2 (1-2q^2) - q^2 (4-5q^2)}{2q^4 q'^2 (\kappa^2 - q^2)^2} \Pi + \frac{2q^2 q'^2 - \kappa^2 (1-2q^2)}{2\kappa^2 q^4 q'^2 (\kappa^2 - q^2)} F$$

$$+ \frac{2q^2 q'^2 + \kappa^2 \kappa'^2}{2\kappa^2 \kappa'^2 q^2 q'^2 (\kappa^2 - q^2)^2} E - \frac{\sin\psi \cos\psi}{\kappa'^2 (\kappa^2 - q^2)^2 \Delta} - \frac{\Delta \sin\psi \cos\psi}{2q'^2 (\kappa^2 - q^2)^2 \Delta_1^2} \quad \ldots\ldots(28)$$

$$\int \frac{\sin^6 \psi}{\cos^2 \psi \Delta^3} d\psi = \frac{2-\kappa^2}{\kappa^4 \kappa'^2} F - \frac{2(1-\kappa^2\kappa'^2)}{\kappa^4 \kappa'^4} E + \frac{\sin\psi\cos\psi}{\kappa^2 \kappa'^4 \Delta} + \frac{\Delta \tan\psi}{\kappa'^4} \quad \ldots\ldots(29)$$

Now in (26) we have to put

$$\kappa^2 = q^2 \frac{4-5q^2}{3-4q^2}$$

in (27)
$$\kappa^2 = q^2 \frac{2-5q^2}{1-4q^2}$$

and in (28)
$$\kappa^2 = q^2 \frac{4-5q^2}{1-2q^2}$$

Introducing these values, and taking the integrals between the limits γ and 0, we find:

$$\mathfrak{P}_3 \mathfrak{Q}_3 \text{ and } \mathfrak{P}_3^2 \mathfrak{Q}_3^2 = \frac{\kappa \Delta_1^4}{\sin^6 \gamma} \left\{ \frac{7q'^2 - 1}{2\kappa^2 q^4 q'^2} F - \frac{2q'^4 + 5q'^2 - 1}{2\kappa^2 q^4 q'^4} E \right.$$
$$\left. + \frac{(4q'^2 - 1)}{2q^4 q'^4} \frac{\Delta \sin\gamma \cos\gamma}{\Delta_1^2} \right\} \quad \ldots\ldots(30)$$

$$\mathsf{P}_3^1 \mathsf{Q}_3^1 \text{ and } \mathsf{P}_3^3 \mathsf{Q}_3^3 = \frac{\kappa \cos^2\gamma \Delta_1^4}{\sin^6\gamma} \left\{ \frac{4q^2 - 1}{2q^4 q'^4} F + \frac{1 - 5q^2 - 2q^4}{2\kappa'^2 q^4 q'^4} E \right.$$
$$\left. - \left(\frac{1 - 7q^2 - (1 - 5q^2 - 2q^4)\sin^2\gamma}{2\kappa'^2 q^2 q'^4} \right) \frac{\Delta \tan\gamma}{\Delta_1^2} \right\} \quad \ldots\ldots(31)$$

$$\mathfrak{P}_3^1 \mathfrak{Q}_3^1 \text{ and } \mathfrak{P}_3^3 \mathfrak{Q}_3^3 = \frac{\kappa \Delta_1^4 \Delta^2}{\sin^6\gamma} \left\{ - \frac{(1-2q^2)(2-3q^2)}{6\kappa^2 q^4 q'^4} F + \frac{2 - 11q^2 q'^2}{6\kappa^2 \kappa'^2 q^4 q'^4} E \right.$$
$$\left. - \left(\frac{1 - 5q^2 + 6q^4 - q^2(2 - 11q^2 q'^2)\sin^2\gamma}{6\kappa'^2 q^4 q'^4} \right) \frac{\sin\gamma\cos\gamma}{\Delta\Delta_1^2} \right\} \quad \ldots\ldots(32)$$

$$\mathsf{P}_3^2 \mathsf{Q}_3^2 = \frac{\kappa \cos^2\gamma \Delta^2}{\sin^6\gamma} \left\{ \frac{1 + \kappa'^2}{\kappa^4 \kappa'^2} F - \frac{2(1 - \kappa^2\kappa'^2)}{\kappa^4 \kappa'^4} E \right.$$
$$\left. + \left(\frac{1 + \kappa^2 - (1 + \kappa^4)\sin^2\gamma}{\kappa^2 \kappa'^4} \right) \frac{\tan\gamma}{\Delta} \right\} \quad \ldots\ldots(33)$$

In (30) $\quad q^2 = \frac{2}{5}[1 + \kappa^2 \mp (1 - \frac{7}{4}\kappa^2 + \kappa^4)^{\frac{1}{2}}], \quad \kappa^2 = q^2 \dfrac{4-5q^2}{3-4q^2}$

In (31) $\quad q^2 = \frac{1}{5}[1 + 2\kappa^2 \mp (1 - \kappa^2 + 4\kappa^4)^{\frac{1}{2}}], \quad \kappa^2 = q^2 \dfrac{2-5q^2}{1-4q^2}$

In (32) $\quad q^2 = \frac{1}{5}[2 + \kappa^2 \mp (4 - \kappa^2\kappa'^2)^{\frac{1}{2}}], \quad \kappa^2 = q^2 \dfrac{4-5q^2}{1-2q^2}$

[The expressions for the functions of the second kind will be generalised so as to be applicable to any order and rank in § 19 of Paper 12, p. 369.]

§ 5. *Bifurcation of Jacobi's Ellipsoid.*

If a mass of liquid be rotating like a rigid body about an axis, x, with uniform angular velocity ω, the determination of the figure of equilibrium may be treated as a statical problem, if the mass be subjected to a potential $\frac{1}{2}\omega^2(y^2+z^2)$.

The energy lost in the concentration of a body from a condition of infinite dispersion is equal to the potential of the body in its final configuration at the position of each molecule, multiplied by the mass of the molecule and summed throughout the body. In the proposed system, as rendered a statical one, it is necessary to add $\frac{1}{2}\omega^2(y^2+z^2)$ to the gravitation potential before making the summation. If A denotes the moment of inertia of the body about x, this latter portion of the sum is $\frac{1}{2}A\omega^2$, and is therefore the kinetic energy of the system.

If dm_1, dm_2 denote any pair of molecules and D_{12} the distance between them, and E the energy lost, we have

$$E = \tfrac{1}{2}\int \frac{dm_1 dm_2}{D_{12}} + \tfrac{1}{2}A\omega^2$$

If the system had been considered as a dynamical one, the expression for the energy of the system, say U, would have resembled that for E, but the former of these terms would have presented itself with a negative sign.

It is clear that the variation of $\frac{1}{2}A\omega^2$, when the moment of momentum is kept constant, is equal and opposite to the variation of the same function when the angular velocity is kept constant.

The condition for a figure of equilibrium is that U shall be stationary for constant moment of momentum, or E stationary for constant ω, in both cases subject to the condition of constancy of volume. The variations in question lead to identical results, and I shall proceed from the variation of E.

If $$\Psi = \int_0^\infty \frac{du}{(u+a^2)^{\frac{1}{2}}(u+b^2)^{\frac{1}{2}}(u+c^2)^{\frac{1}{2}}}$$

the internal potential of an ellipsoid of mass M and semi-axes a, b, c is

$$\tfrac{3}{4}M\left[\Psi + \frac{x^2}{a}\frac{d\Psi}{da} + \frac{y^2}{b}\frac{d\Psi}{db} + \frac{z^2}{c}\frac{d\Psi}{dc}\right]$$

Hence $$\tfrac{1}{2}\int \frac{dm_1 dm_2}{D_{12}} = \tfrac{3}{8}M\int_0^\infty \left[\Psi + \frac{x^2}{a}\frac{d\Psi}{da} + \dots\right] dm$$

Now if A, B, C denote the principal moments of inertia of the ellipsoid about x, y, z,

$$\int x^2 dm = \tfrac{1}{2}(C+B-A) = \tfrac{1}{5}Ma^2$$

and similar formulæ hold for the two other axes.

Therefore

$$\tfrac{1}{2}\int \frac{dm_1 dm_2}{D_{12}} = \tfrac{3}{8}M^2\left[\Psi + \tfrac{1}{5}\left(a\frac{d\Psi}{da} + b\frac{d\Psi}{db} + c\frac{d\Psi}{dc}\right)\right]$$

But since Ψ is a homogeneous function of degree -1 in a, b, c, the sum of the three differential terms is equal to $-\Psi$. Hence this expression is equal to $\tfrac{3}{10}M^2\Psi$.

Since

$$\tfrac{1}{2}A\omega^2 = \tfrac{1}{10}M(b^2+c^2)\omega^2$$

we have

$$E = \tfrac{3}{10}M^2\left[\Psi + \frac{b^2+c^2}{3M}\omega^2\right]$$

If E be varied, whilst abc and ω are constant, it is stationary if

$$\frac{d\Psi}{da}\delta a + \left(\frac{d\Psi}{db} + \frac{2b}{3M}\omega^2\right)\delta b + \left(\frac{d\Psi}{dc} + \frac{2c}{3M}\omega^2\right)\delta c = 0$$

$$\frac{\delta a}{a} + \frac{\delta b}{b} + \frac{\delta c}{c} = 0$$

Eliminating $\delta a, \delta b, \delta c$ we have the well-known conditions for Jacobi's ellipsoid

$$\left.\begin{aligned} \frac{2\omega^2 b^2}{3M} &= a\frac{d\Psi}{da} - b\frac{d\Psi}{db} \\ \frac{2\omega^2 c^2}{3M} &= a\frac{d\Psi}{da} - c\frac{d\Psi}{dc} \\ \frac{1}{b^2}\left(a\frac{d\Psi}{da} - b\frac{d\Psi}{db}\right) &= \frac{1}{c^2}\left(a\frac{d\Psi}{da} - c\frac{d\Psi}{dc}\right) \end{aligned}\right\} \quad \text{..............(34)}$$

If we add together the first two of these, and avail ourselves of the property that Ψ is homogeneous of degree -1, we easily prove that the stationary value of E is

$$E = \tfrac{9}{20}M^2\left[\Psi + a\frac{d\Psi}{da}\right]$$

Since the potential of the ellipsoid must satisfy Poisson's equation

$$\frac{d\Psi}{ada} + \frac{d\Psi}{bdb} + \frac{d\Psi}{cdc} = -\frac{2}{abc}$$

Also

$$a\frac{d\Psi}{da} + b\frac{d\Psi}{db} + c\frac{d\Psi}{dc} = -\Psi$$

By means of these and two out of the three equations (34), we may eliminate the differentials of Ψ, and writing ρ for the density find

$$\frac{\omega^2}{2\pi\rho} = \frac{\Psi abc\left(\frac{1}{a^2} + \frac{1}{b^2} + \frac{1}{c^2}\right) - 6}{(b^2+c^2)\left(\frac{1}{a^2} + \frac{1}{b^2} + \frac{1}{c^2}\right) - 6} \quad \text{..................(35)}$$

I do not happen to have seen this form for the angular velocity of Jacobi's ellipsoid in any book.

It is easy also to show that the stationary value of E may be written

$$E = \tfrac{9}{20}M^2 \frac{\left[(b^2+c^2)\left(\frac{1}{a^2}+\frac{1}{b^2}+\frac{1}{c^2}\right)-4\right]\Psi - 2\,\frac{b^2+c^2}{abc}}{(b^2+c^2)\left(\frac{1}{a^2}+\frac{1}{b^2}+\frac{1}{c^2}\right)-6}$$

We may now express the potential, say V, of the system entirely in terms of Ψ and $a\dfrac{d\Psi}{da}$, for

$$V = \tfrac{3}{4}M\left[\Psi + \frac{x^2}{a}\frac{d\Psi}{da} + \frac{y^2}{b^2}\left(a\frac{d\Psi}{da} - \frac{2\omega^2 b^2}{3M}\right) + \frac{z^2}{c^2}\left(a\frac{d\Psi}{da} - \frac{2\omega^2 c^2}{3M}\right)\right] + \tfrac{1}{2}\omega^2(y^2+z^2)$$

$$= \tfrac{3}{4}M\left[\Psi + a\frac{d\Psi}{da}\left(\frac{x^2}{a^2}+\frac{y^2}{b^2}+\frac{z^2}{c^2}\right)\right]$$

We thus verify that V is constant over the surface of the ellipsoid.

Let g denote the value of gravity at the surface. Then if dn be an element of the outward normal, $g = -\dfrac{dV}{dn}$. Since

$$\frac{dx}{dn} = \frac{px}{a^2}, \quad \frac{dy}{dn} = \frac{py}{b^2}, \quad \frac{dz}{dn} = \frac{pz}{c^2}, \text{ where } \frac{1}{p^2} = \frac{x^2}{a^4} + \frac{y^2}{b^4} + \frac{z^2}{c^4}$$

we have
$$g = -\tfrac{3}{2}Ma\frac{d\Psi}{da}\,p\left(\frac{x^2}{a^4}+\frac{y^2}{b^4}+\frac{z^2}{c^4}\right) = -\tfrac{3}{2}\frac{M}{p}\,a\frac{d\Psi}{da}$$

Now change the notation and write

$$a^2 = k^2\left(\nu_0^2 - \frac{1+\beta}{1-\beta}\right), \quad b^2 = k^2(\nu_0^2 - 1), \quad c^2 = k^2\nu_0^2$$

$$u = k^2(\nu^2 - \nu_0^2)$$

Then
$$\Psi = \frac{2}{k}\int_{\nu_0}^{\infty} \frac{d\nu}{\left(\nu^2 - \frac{1+\beta}{1-\beta}\right)^{\frac{1}{2}}(\nu^2-1)^{\frac{1}{2}}}$$

$$a\frac{d\Psi}{da} = -\frac{2}{k}\left(\nu_0^2 - \frac{1+\beta}{1-\beta}\right)\int_{\nu_0}^{\infty} \frac{d\nu}{\left(\nu^2 - \frac{1+\beta}{1-\beta}\right)^{\frac{3}{2}}(\nu^2-1)^{\frac{1}{2}}}$$

Now
$$\mathfrak{P}_0(\nu) = 1, \quad \mathsf{P}_1^1(\nu) = \left(\nu^2 - \frac{1+\beta}{1-\beta}\right)^{\frac{1}{2}}$$

and
$$\mathfrak{P}_i^s(\nu_0)\,\mathfrak{Q}_i^s(\nu_0) = [\mathfrak{P}_i^s(\nu_0)]^2\int_{\nu_0}^{\infty} \frac{d\nu}{[\mathfrak{P}_i^s]^2\left(\nu^2 - \frac{1+\beta}{1-\beta}\right)^{\frac{1}{2}}(\nu^2-1)^{\frac{1}{2}}}$$

so that

$$\left.\begin{aligned} \Psi &= \frac{2}{k}\mathfrak{P}_0(\nu_0)\,\mathfrak{Q}_0(\nu_0) \\ a\frac{d\Psi}{da} &= -\frac{2}{k}\mathsf{P}_1{}^1(\nu_0)\,\mathsf{Q}_1{}^1(\nu_0) \\ g &= \frac{3M}{pk}\mathsf{P}_1{}^1(\nu_0)\,\mathsf{Q}_1{}^1(\nu_0) \end{aligned}\right\} \quad\text{......................(36)}$$

and

We may note in passing that the condition for a Jacobian ellipsoid (the last equation of (34)) is reducible [as shown in (22) of Paper 8, p. 123] to the form

$$\frac{\kappa\Delta^2}{\sin^4\gamma}\int_0^\gamma \frac{\sin^4\psi}{\Delta^3}\,d\psi = \kappa\cot^2\gamma\int_0^\gamma \frac{\tan^2\psi}{\Delta}\,d\psi$$

On examining the series of functions given in (25), we see that it may be written

$$\mathfrak{P}_2{}^1(\nu_0)\,\mathfrak{Q}_2{}^1(\nu_0) = \mathsf{P}_1{}^1(\nu_0)\,\mathsf{Q}_1{}^1(\nu_0)$$

This agrees with M. Poincaré's equation (1) on p. 341 of his memoir.

We will now suppose that the body, instead of being an ellipsoid, is an ellipsoidal harmonic deformation of an ellipsoid, which is itself a figure of equilibrium for rotation ω.

The addition to E will consist of three parts; first that due to the mutual energy of the layer of deformation; secondly that due to the ellipsoid and the layer; thirdly that due to the change in the moment of inertia.

If a subscript l denotes integration throughout the space occupied by the layer, U the potential of the ellipsoid, and dv an element of volume,

$$\delta E = \tfrac{1}{2}\int_l \frac{dm_1\,dm_2}{D_{12}} + \int_l U\rho\,dv + \tfrac{1}{2}\omega^2\int_l (y^2+z^2)\,\rho\,dv$$

If ζ denotes the thickness of the layer standing on the element $d\sigma$, the first of these terms is $\tfrac{1}{2}\rho^2\iint\frac{\zeta_1\zeta_2\,d\sigma_1\,d\sigma_2}{D_{12}}$.

The value of $U+\tfrac{1}{2}\omega^2(y^2+z^2)$ throughout the layer is equal to $V_0 - g\zeta'$, where V_0 is the constant value of $U+\tfrac{1}{2}\omega^2(y^2+z^2)$ over the surface of the ellipsoid, and ζ' is the distance measured along the normal to the element $d\zeta'\,d\sigma$ of volume.

Hence

$$\int_l U\rho\,dv + \tfrac{1}{2}\omega^2\int_l (y^2+z^2)\,\rho\,dv = \iint_0^\zeta \rho\,(V_0 - g\zeta')\,d\zeta'\,d\sigma$$

Since V_0 is constant and the total mass of the layer is zero, this is equal to $-\tfrac{1}{2}\rho\int g\zeta^2\,d\sigma$.

It follows that

$$\delta E = \tfrac{1}{2}\rho^2\iint\frac{\zeta_1\zeta_2\,d\sigma_1\,d\sigma_2}{D_{12}} - \tfrac{1}{2}\rho\int g\zeta^2\,d\sigma$$

The axes of the ellipsoid have been chosen so as to make our original E stationary, and the further condition to be satisfied is that δE shall be stationary.

Let us suppose that $\qquad \zeta = pe\mathfrak{P}_i^s(\mu)\,\mathfrak{C}_i^s(\phi)$

which expression shall be deemed to include any one of the other types of harmonic.

Then it is shown in (51) of "Harmonics" that the potential of this layer at the surface of the ellipsoid is

$$\frac{3M}{k}e\mathfrak{P}_i^s(\nu_0)\,\mathfrak{Q}_i^s(\nu_0)\,\mathfrak{P}_i^s(\mu)\,\mathfrak{C}_i^s(\phi)$$

Since the mass of an element is $pe\rho\,\mathfrak{P}_i^s(\mu)\,\mathfrak{C}_i^s(\phi)\,d\sigma$, we have

$$\tfrac{1}{2}\rho^2\int\frac{\zeta_1\zeta_2 d\sigma_1 d\sigma_2}{D_{12}} = \tfrac{3}{2}\frac{M\rho}{k}e^2\mathfrak{P}_i^s(\nu_0)\,\mathfrak{Q}_i^s(\nu_0)\int[\mathfrak{P}_i^s(\mu)\,\mathfrak{C}_i^s(\phi)]^2 p\,d\sigma$$

With the value of g found in (36)

$$\tfrac{1}{2}\rho\int g\zeta^2\,d\sigma = \tfrac{3}{2}\frac{M\rho}{k}e^2\mathsf{P}_1^1(\nu_0)\,\mathsf{Q}_1^1(\nu_0)\int[\mathfrak{P}_i^s(\mu)\,\mathfrak{C}_i^s(\phi)]^2 p\,d\sigma$$

Hence

$$\delta E = -\tfrac{3}{2}\frac{M\rho}{k}e^2\mathsf{P}_1^1(\nu_0)\,\mathsf{Q}_1^1(\nu_0)\left[1 - \frac{\mathfrak{P}_i^s(\nu_0)\,\mathfrak{Q}_i^s(\nu_0)}{\mathsf{P}_1^1(\nu_0)\,\mathsf{Q}_1^1(\nu_0)}\right]\int[\mathfrak{P}_i^s(\mu)\,\mathfrak{C}_i^s(\phi)]^2 p\,d\sigma$$

In order that the new figure may be one of equilibrium, this expression must be stationary for variations of e. It follows that we must either have $e = 0$, which leads back to Jacobi's ellipsoid, or else

$$1 - \frac{\mathfrak{P}_i^s(\nu_0)\,\mathfrak{Q}_i^s(\nu_0)}{\mathsf{P}_1^1(\nu_0)\,\mathsf{Q}_1^1(\nu_0)} = 0$$

This last condition is what M. Poincaré calls the vanishing of a coefficient of stability*. It shows that if ν_0 and β satisfy not only the condition for the Jacobian ellipsoid, namely, $\mathfrak{P}_2^1(\nu_0)\,\mathfrak{Q}_2^1(\nu_0) = \mathsf{P}_1^1(\nu_0)\,\mathsf{Q}_1^1(\nu_0)$, but also this equation, we have arrived at a figure which belongs at the same time to two series, and there is a bifurcation at this point. The form of the figure is found by attributing to e any arbitrary but small value.

* *Acta Math.*, Vol. VII., 1885, p. 321. The factors $\frac{1}{3}$ and $1/2n+1$ (or $1/(2i+1)$, if i is the degree of the harmonic) which occur in his form of the condition are included in my functions.

§ 6. *The Properties of the Successive Coefficients of Stability.*

Corresponding to each harmonic deformation of the ellipsoid, there is a coefficient of stability of one of the two forms

$$1-\frac{\mathfrak{P}_i^s(\nu_0)\,\mathfrak{Q}_i^s(\nu_0)}{P_1^1(\nu_0)\,Q_1^1(\nu_0)} \quad \text{or} \quad 1-\frac{P_i^s(\nu_0)\,Q_i^s(\nu_0)}{P_1^1(\nu_0)\,Q_1^1(\nu_0)}$$

These coefficients may be written $\mathfrak{K}_i^s$ or K_i^s according to an easily intelligible notation. The Jacobian ellipsoid is defined by ν_0, and the question arises as to the possibility of the vanishing of the several $\mathfrak{K}$'s as ν_0 gradually diminishes from infinity, that is to say, as the ellipsoid lengthens.

An harmonic of the first order merely denotes a shift of the centre of inertia along one of the three axes; one of the second order denotes a change of ellipticity of the ellipsoid. Since we must keep the centre of inertia at the origin, and since the ellipticity is determined by the consideration that the ellipsoid is a Jacobian, these harmonics need not be considered, and we may begin with those of the third order.

I shall not attempt to follow M. Poincaré in his masterly discussion of the properties of the coefficients of stability*, but will merely restate in my own notation the principal conclusions at which he has arrived.

1st. The equation

$$P_1^1(\nu)\,Q_1^1(\nu)-\mathfrak{P}_i^s(\nu)\,\mathfrak{Q}_i^s(\nu) \text{ or } P_i^s(\nu)\,Q_i^s(\nu)=0, \quad (i>2)$$

is not satisfied by any value of ν between 1 and infinity, if $\mathfrak{P}_i^s$ or P_i^s is divisible by $\left(\nu^2-\frac{1+\beta}{1-\beta}\right)^{\frac{1}{2}}$. It appears from the forms of the functions as given in § 4 of "Harmonics" that the P functions are so divisible. These functions appertain to the types EES, OOC, OES, EOC, and therefore the ellipsoid cannot bifurcate into deformations of these types.

2nd. The equation has no solution if $\mathfrak{P}_i^s$ is divisible by $(\nu^2-1)^{\frac{1}{2}}$. We again see from § 4 of "Harmonics" that $\mathfrak{P}_i^s$ is so divisible if it is of the types OOS, EOS. Hence the ellipsoid cannot bifurcate into these types. The only types remaining are EEC, OEC.

3rd. The equation has no solution if any of the roots of $\mathfrak{P}_i^s(\nu)=0$ lie outside the limits $+1$ to -1. The only $\mathfrak{P}_i^s$ of the types EEC, OEC which has all its roots inside the limits $+1$ to -1 is the zonal harmonic for which $s=0$.

Hence the ellipsoid can only bifurcate into a zonal harmonic.

* Sections 10 and 12 of his memoir. I have to thank him for saving me from making a serious mistake in this portion of my work.

4th. The equation

$$\mathsf{P}_1^1\,\mathsf{Q}_1^1 - \mathfrak{P}_i\,\mathfrak{Q}_i = 0 \quad (i > 2)$$

must have a solution between 1 and infinity for all values of i.

It follows from these four propositions that the Jacobian ellipsoid is stable for all deformations except the zonal ones, and that as it lengthens it must at successive stages bifurcate into each and all the zonal deformations.

5th. As the ellipsoid lengthens, the first coefficient of stability to vanish is that of the third zonal harmonic. This stage is the end of the stability of the Jacobian ellipsoids, and there is almost certainly exchange of stability with the pear-shaped figure defined by this harmonic*.

6th. It has not been rigorously proved that there is only one solution of the equation $\mathfrak{K}_i = 0$ even in the case where $i = 3$, but M. Poincaré believes that this is almost certainly the case.

7th. The functions

$$\left.\begin{matrix}\mathfrak{P}_i^s(\nu_0)\\ \text{or}\\ \mathsf{P}_i^s(\nu_0)\end{matrix}\right\} \times \left.\begin{matrix}\mathfrak{P}_i^t(\nu)\\ \text{or}\\ \mathsf{P}_i^t(\nu)\end{matrix}\right\} - \left.\begin{matrix}\mathfrak{P}_i^s(\nu)\\ \text{or}\\ \mathsf{P}_i^s(\nu)\end{matrix}\right\} \times \left.\begin{matrix}\mathfrak{P}_i^t(\nu_0)\\ \text{or}\\ \mathsf{P}_i^t(\nu_0)\end{matrix}\right\}$$

have always the same sign as ν increases from ν_0 to infinity, provided that s and t are both greater than zero, and i greater than 2.

The seventh of the preceding propositions renders it easy to determine the relative magnitudes of all the $\mathfrak{K}$'s belonging to a single degree i.

In what follows I may take the symbols $\mathfrak{P}$, $\mathfrak{Q}$ as including also P, Q.

Now
$$\mathfrak{K}_i^s > = < \mathfrak{K}_i^t$$
as,
$$\mathfrak{P}_i^s(\nu_0)\,\mathfrak{Q}_i^s(\nu_0) - \mathfrak{P}_i^t(\nu_0)\,\mathfrak{Q}_i^t(\nu_0) < = > 0$$

If we express the $\mathfrak{Q}$'s in terms of integrals this becomes

$$\int_{\nu_0}^{\infty} \frac{[\mathfrak{P}_i^s(\nu_0)\,\mathfrak{P}_i^t(\nu)]^2 - [\mathfrak{P}_i^s(\nu)\,\mathfrak{P}_i^t(\nu_0)]^2}{[\mathfrak{P}_i^s(\nu)\,\mathfrak{P}_i^t(\nu)]^2\,(\nu^2-1)^{\frac{1}{2}}\left(\nu^2 - \frac{1+\beta}{1-\beta}\right)^{\frac{1}{2}}}\,d\nu < = > 0$$

The seventh proposition shows that when s and t are greater than zero, and i is greater than 2, all the elements of the integral have the same sign. Hence the question is whether

$$\frac{\mathfrak{P}_i^s(\nu_0)}{\mathfrak{P}_i^s(\nu)} < = > \frac{\mathfrak{P}_i^t(\nu_0)}{\mathfrak{P}_i^t(\nu)}$$

Therefore we have to arrange all the $\dfrac{\mathfrak{P}_i^s(\nu_0)}{\mathfrak{P}_i^s(\nu)}$ in descending order of magnitude, and shall thereby obtain the non-zonal $\mathfrak{K}$'s in ascending order.

* [See Paper 12.]

I wish first to show that these coefficients may to a great extent be sorted by considering the inequality

$$\frac{P_i^s(\nu_0)}{P_i^s(\nu)} <=> \frac{P_i^t(\nu_0)}{P_i^t(\nu)} \quad (s=1, 2, 3 \ldots, i;\ t=1, 2, 3 \ldots, i)$$

Suppose, if possible, that whereas, for the ellipsoids defined by β, ν, ν_0,

$$\frac{\mathfrak{P}_i^s(\nu_0)}{\mathfrak{P}_i^s(\nu)} < \frac{\mathfrak{P}_i^t(\nu_0)}{\mathfrak{P}_i^t(\nu)}, \text{ yet } \frac{P_i^s(\nu_0)}{P_i^s(\nu)} > \frac{P_i^t(\nu_0)}{P_i^t(\nu)}$$

Then there must be some value of β for which

$$\mathfrak{P}_i^s(\nu_0)\,\mathfrak{P}_i^t(\nu) = \mathfrak{P}_i^s(\nu)\,\mathfrak{P}_i^t(\nu_0)$$

for all values of ν greater than ν_0.

It is almost obvious that there is no one value of β which renders this equation possible; but consider for example the case of $s=2$, $t=0$.

Now

$$\mathfrak{P}_3^2(\nu) = -\beta q_0 P_3(\nu) + P_3^2(\nu), \quad \mathfrak{P}_3(\nu) = P_3(\nu) + \beta q_2 P_3^2(\nu)$$

If we substitute this in the equation we find

$$P_3^2(\nu_0)\,P_3(\nu) = P_3^2(\nu)\,P_3(\nu_0)$$

This can only be satisfied by $\nu = \nu_0$, and hence the hypothesis is negatived. Similarly the assumption of other values of s and t leads to an impossibility.

Thus we may consider the P functions in place of the $\mathfrak{P}$ functions.

Consider the inequality

$$\frac{P_i^s(\nu_0)}{P_i^s(\nu)} >=< \frac{P_i^{s+1}(\nu_0)}{P_i^{s+1}(\nu)}, \quad \text{for } s=1, 2 \ldots, i-1$$

If the inequality is determined for any value of ν, it is determined for all values. Now when ν is very large

$$P_i^s(\nu) = \frac{2i!}{2^i\, i!\, \overline{i-s}!}\nu^i, \qquad P_i^{s+1}(\nu) = \frac{2i!}{2^i\, i!\, \overline{i-s-1}!}\nu^i$$

Hence our inequality becomes

$$(i-s)\,P_i^s(\nu_0) >=< P_i^{s+1}(\nu_0)$$

This inequality is of the same kind for all values of ν_0. Now $P_i^s(\nu_0)$ involves the factor $(\nu_0^2-1)^{\frac{1}{2}s}$ and $P_i^{s+1}(\nu_0)$ involves $(\nu_0^2-1)^{\frac{1}{2}(s+1)}$. Putting therefore $\nu_0^2 = 1+\epsilon$, the left-hand side involves $\epsilon^{\frac{1}{2}s}$ and the right $\epsilon^{\frac{1}{2}(s+1)}$. It follows that unless s is equal to i the left-hand side is greater than the right; but s is necessarily equal to $i-1$ at greatest.

Therefore
$$\frac{P_i^s(\nu_0)}{P_i^s(\nu)} > \frac{P_i^{s+1}(\nu_0)}{P_i^{s+1}(\nu)}$$

Hence K's with smaller s are less than those with greater s.

It remains to discriminate between the two sorts of P-functions which occur in ellipsoidal harmonic analysis; that is to say we must determine

$$\frac{\mathfrak{P}_i^s(\nu_0)}{\mathfrak{P}_i^s(\nu)} >=< \frac{\mathsf{P}_i^s(\nu_0)}{\mathsf{P}_i^s(\nu)}$$

Since the β of "Harmonics" is equal to $\kappa'^2/(2-\kappa'^2)$ in the present notation, when β and κ' are small, we have by the formulæ of that paper

$$\mathfrak{P}_i^s(\nu) = \mathrm{P}_i^s(\nu) + \tfrac{1}{2}\kappa'^2 q_{s+2}\mathrm{P}_i^{s+2}(\nu) + \tfrac{1}{2}\kappa'^2 q_{s-2}\mathrm{P}_i^{s-2}(\nu) + \dots$$

$$\mathsf{P}_i^s(\nu) = \frac{(\nu^2 - 1/\kappa^2)^{\frac{1}{2}}}{(\nu^2-1)^{\frac{1}{2}}}\left[\mathrm{P}_i^s(\nu) + \frac{s+2}{2s}\kappa'^2 q_{s+2}\mathrm{P}_i^{s+2}(\nu) + \frac{s-2}{2s}\kappa'^2 q_{s-2}\mathrm{P}_i^{s-2}(\nu) + \dots\right]$$

When ν is very great and κ' very small $\mathfrak{P}_i^s = \mathsf{P}_i^s$, so it suffices to determine the inequality

$$\mathfrak{P}_i^s(\nu_0) >=< \mathsf{P}_i^s(\nu_0)$$

and this may be considered for any value of ν_0 greater than unity. By taking ν_0 very large and κ' very small the inequality becomes

$$(\nu^2-1)^{\frac{1}{2}} >=< \left(\nu_0{}^2 - \frac{1}{\kappa^2}\right)^{\frac{1}{2}}$$

or

$$1 >=< \kappa$$

But $\kappa < 1$, hence the first sign holds true and

$$\frac{\mathfrak{P}_i^s(\nu_0)}{\mathfrak{P}_i^s(\nu)} > \frac{\mathsf{P}_i^s(\nu_0)}{\mathsf{P}_i^s(\nu)}$$

whence

$$\mathfrak{K}_i^s < \mathsf{K}_i^s$$

Thus it follows that for order i

$$\mathfrak{K}_i^1 < \mathsf{K}_i^1 < \mathfrak{K}_i^2 < \mathsf{K}_i^2 \dots < \mathfrak{K}_i^i < \mathsf{K}_i^i$$

The order of magnitude of these coefficients is therefore completely determined.

As confirmatory of the correctness of this result it may be mentioned that I find that when $\gamma = 69° \, 50'$ and $\kappa = \sin 73° \, 56'$,

$$\mathfrak{K}_3^1 = \cdot 1765,\ \mathsf{K}_3^1 = \cdot 2990,\ \mathfrak{K}_3^2 = \cdot 4467,\ \mathsf{K}_3^2 = \cdot 4550,\ \mathfrak{K}_3^3 = \cdot 5719,\ \mathsf{K}_3^3 = \cdot 5876$$

When $\gamma = 75°$ and $\kappa = \sin 81° \, 4'$ (another Jacobian ellipsoid) the numbers run ·130, ·224, ·460, ·465, ·604, ·614.

We see that for the harmonics of higher order the ellipsoid is more stable than it was and for those of lower order less stable.

§ 7. *The critical Jacobian Ellipsoid.*

From a number of preliminary calculations I saw reason to believe that the critical ellipsoid would be found within the region comprised between $\gamma = 69° 48'$ and $69° 50'$, and $\sin^{-1}\kappa = 73° 52'$ and $73° 56'$.

If we write

$$f(\gamma, \sin^{-1}\kappa) = \frac{E}{\kappa'^2}\left(1 + \frac{\kappa^4 \sin^2\gamma \cos^2\gamma}{1 - \kappa^2 \sin^2\gamma}\right) - (2F - E) - \frac{\kappa^2 \sin\gamma \cos\gamma\,(1 + \kappa^2 \sin^2\gamma)}{\kappa'^2 (1 - \kappa^2 \sin^2\gamma)^{\frac{1}{2}}}$$

where the amplitudes of E and F are γ and their moduli κ, the existence of the Jacobian ellipsoid is determined by

$$f(\gamma, \sin^{-1}\kappa) = 0^*$$

The coefficient of stability is

$$\mathfrak{K}_3(\gamma, \sin^{-1}\kappa) = 1 - \frac{\mathfrak{P}_3(\nu_0)\,\mathfrak{Q}_3(\nu_0)}{P_1^1(\nu_0)\,Q_1^1(\nu_0)}$$

The formulæ for computing $\mathfrak{K}_3$ are given in § 4.

The values of E and F are from Legendre's tables.

Now I find

$$f(69° 48', 73° 52') = +\cdot000191\,; \qquad f(69° 50', 73° 52') = +\cdot001319$$
$$f(69° 48', 73° 56') = -\cdot001186\,; \qquad f(69° 50', 73° 56') = -\cdot000031$$
$$\mathfrak{K}_3(69° 48', 73° 52') = +\cdot001058\,; \qquad \mathfrak{K}_3(69° 50', 73° 52') = -\cdot000885$$
$$\mathfrak{K}_3(69° 48', 73° 56') = +\cdot000655\,; \qquad \mathfrak{K}_3(69° 50', 73° 56') = -\cdot000765$$

By interpolation we get the following results:—

The Jacobian ellipsoid is given by

$$(\gamma - 69° 48') - \cdot59642\,(\sin^{-1}\kappa - 73° 52') + \cdot33091 = 0$$

The vanishing of the coefficient of stability is given by

$$(\gamma - 69° 48') + \cdot041625\,(\sin^{-1}\kappa - 73° 52') - 1\cdot0890 = 0$$

In these equations the minute of arc is the unit.

Solving them I find

$$\gamma = 69° 48'\cdot997 = 69° 49'\cdot0$$
$$\sin^{-1}\kappa = 73° 54'\cdot225 = 73° 54'\cdot2$$

With these values I find that the three axes a, b, c, where $abc = \mathrm{a}^3$ are

$$\frac{a}{\mathrm{a}} = \cdot650659$$

$$\frac{b}{\mathrm{a}} = \cdot814975$$

$$\frac{c}{\mathrm{a}} = 1\cdot885827$$

The last place of decimals in these is certainly doubtful.

* See [Paper 8, equation (16), p. 122] where the formula is reduced to a form convenient for computation.

The formula for ω^2 is given in (35).

Now

$$\Psi = \frac{2}{k} \mathfrak{P}_0(\nu_0)\,\mathfrak{Q}_0(\nu_0), \qquad k = c\kappa \sin\gamma, \qquad \mathfrak{P}_0(\nu_0)\,\mathfrak{Q}_0(\nu_0) = \kappa F$$

Then since $a = c\cos\gamma$, $b = c\Delta$,

$$\frac{\omega^2}{2\pi\rho} = \frac{2F\Delta\cot\gamma - \frac{6}{1+\Delta^{-2}+\sec^2\gamma}}{1+\Delta^2 - \frac{6}{1+\Delta^{-2}+\sec^2\gamma}}$$

In this formula, F, γ, Δ must correspond with values interpolated amongst those used in obtaining the solution.

From this I find

$$\frac{\omega^2}{2\pi\rho} = \cdot 1419990 = \cdot 14200$$

In the paper on the Jacobian ellipsoid referred to above the moment of momentum is tabulated by means of μ, where the moment of momentum is $(\frac{4}{3}\pi\rho)^{\frac{3}{2}} a^5 \mu$. The formula for μ is given in (31) [p. 127] of that paper, and, modified to suit the present notation, is

$$\mu = \frac{3^{\frac{1}{2}}}{5}(\Delta\cos\gamma)^{-\frac{2}{3}}(1+\Delta^2)\left(\frac{\omega^2}{4\pi\rho}\right)^{\frac{1}{2}}$$

For the critical ellipsoid I find $\mu = \cdot 389570$.

A table of the numerical values for a number of Jacobian ellipsoids is given on p. 130, but I now give as a supplement to that table the various data for the critical Jacobian ellipsoid as follows:—

The critical Jacobian Ellipsoid.

$\gamma = 69^\circ\ 49'$, $\sin^{-1}\kappa = 73^\circ\ 54'$, $\cos^{-1}\Delta = 64^\circ\ 24'$, $a/a = \cdot 65066$, $b/a = \cdot 81498$, $c/a = 1\cdot 88583$, $\omega^2/2\pi\rho = \cdot 14200$, $\mu = \cdot 38957$

In order to determine the question as to whether or not it is possible that $\mathfrak{K}_3 = 0$ should have another solution than that found in this section, I have computed the value of this coefficient for the Jacobian ellipsoid $\gamma = 75^\circ$, $\kappa = \sin 81^\circ\ 4'\cdot 4$, and find it to be $-6\cdot 627$. From the manner in which the numbers in the computation present themselves, it is obvious that for more elongated ellipsoids $\mathfrak{K}_3$ will always remain negative, and will become numerically greater. I have therefore not thought it necessary to seek for an algebraic proof that there is no second root of the equation.

Very long Jacobian ellipsoids tend to become figures of revolution, and the coefficients of stability tend to assume the forms

$$1 - \frac{P_i(\nu)\,Q_i(\nu)}{P_1^1(\nu)\,Q_1^1(\nu)}$$

The forms of these functions are well known, and I think that fair approximations to the incidences of the successive figures of bifurcation might be derived from the vanishing of this expression.

For example

$$\mathrm{P}_1^1(\nu)\,\mathrm{Q}_1^1(\nu) = \tfrac{1}{2}\left[\nu - (\nu^2-1)\log\left(\frac{\nu+1}{\nu-1}\right)^{\frac{1}{2}}\right]$$

$$\mathrm{P}_4(\nu)\,\mathrm{Q}_4(\nu) = \tfrac{1}{64}\left[(35\nu^4-30\nu^2+3)\log\left(\frac{\nu+1}{\nu-1}\right)^{\frac{1}{2}} - \tfrac{5}{3}\nu(21\nu^2-11)(35\nu^4-30\nu^2+3)\right]$$

I have not, however, attempted to solve the equation found by equating these two expressions to one another.

Even when $i=3$ and $\gamma = 69°\,49'$ (the critical Jacobian) this rough approximation makes the coefficient of stability very small, but it is to be admitted that $\mathrm{P}_1^1\mathrm{Q}_1^1$ and $\mathrm{P}_3\mathrm{Q}_3$ differ very sensibly from $\mathsf{P}_1^1(\nu)\,\mathsf{Q}_1^1(\nu)$ and $\mathfrak{P}_3(\nu)\,\mathfrak{Q}_3(\nu)$, although in such a way that the errors compensate one another.

§ 8. *The pear-shaped Figure of Equilibrium.*

By (21) the normal displacement δn for the third zonal harmonic deformation may be written

$$\delta n = e\,\frac{z\,[q'^2z^2 - q^2x^2 - (3-4q^2)\,y^2 - c^2q^2q'^2\sin^2\gamma]}{c^2q'^2(1-q^2\sin^2\gamma)\,(x^2/\cos^4\gamma + y^2/\Delta^4 + z^2)^{\frac{1}{2}}}$$

subject to the condition

$$\frac{x^2}{\cos^2\gamma} + \frac{y^2}{\Delta^2} + z^2 = c^2$$

The expression has been arranged so that when $x=y=0$, $z=c$, we have $\delta n = e$. Hence $+e$ and $-e$ are the normal displacements at the stalk and blunt end of the pear respectively.

In the section $y=0$, this may be written

$$\delta n = \frac{e\cos\gamma}{q'^2}\cdot\frac{z\,(z^2-c^2q^2)}{c^2\,(c^2-z^2\sin^2\gamma)^{\frac{1}{2}}}$$

The nodal points are given by $\dfrac{z}{c} = \pm q = \pm\,{\cdot}758056$.

In the section $x=0$, since $\kappa^2 = q^2\dfrac{4-5q^2}{3-4q^2}$, it may be written

$$\delta n = e\,\frac{\Delta\,(4-5q^2)}{q'^2}\cdot\frac{z\,(\kappa^2z^2-c^2q^2)}{c^2\kappa^2\,(c^2-\kappa^2z^2\sin^2\gamma)^{\frac{1}{2}}}$$

The nodal points are given by $\dfrac{z}{c} = \pm\dfrac{q}{\kappa} = \pm\,{\cdot}788986$.

The section $z=0$ is obviously another nodal line for all sections.

By means of these formulæ it is easy to compute the normal displacements from the surface of the critical Jacobian.

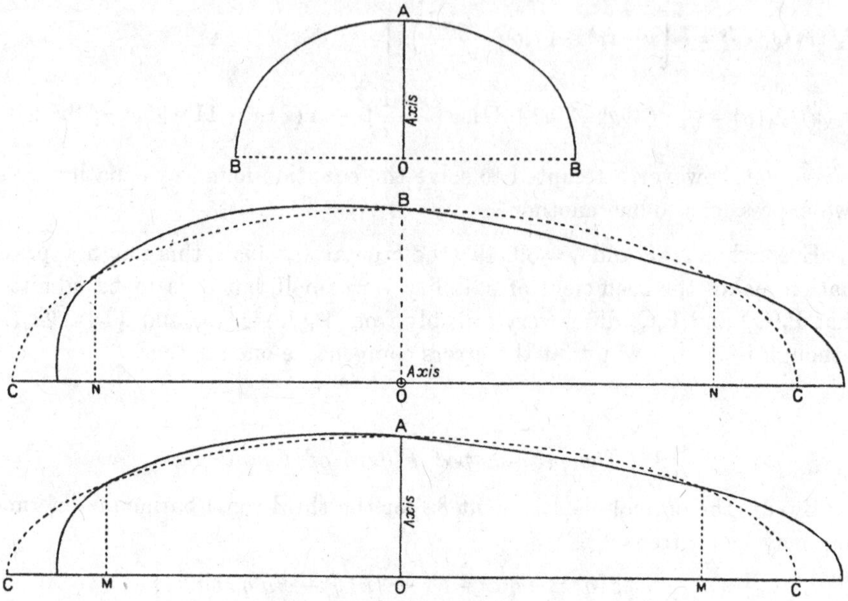

Pear-shaped figure of equilibrium.

OA $= \cdot 65066$, OB $= \cdot 81498$, OC $= 1\cdot 88583$; $\frac{\omega^2}{2\pi\rho} = \cdot 14200$, $\frac{\text{OM}}{\text{OC}} = \cdot 75806$, $\frac{\text{ON}}{\text{OC}} = \cdot 78899$.

The figure above showing the three sections $x=0$, $y=0$, $z=0$, is drawn from these formulæ, the dotted line being the critical Jacobian and the firm line the pear. The scale of the normal displacements is, of course, arbitrary.

Comparison with M. Poincaré's sketch shows that the figure is considerably longer than he supposed.

In this first approximation the positions of the nodal lines are independent of the magnitude of e, and they lie so near the ends that it is impossible to construct an exaggerated figure, for if we do so the blunt end acquires a dimple, which is absurd. It might have been hoped that such an exaggeration would afford us some idea of the mode of development of the pear.

M. Schwarzschild has remarked* that it is not absolutely certain that the principle of exchange of stability holds with reference to this figure, and that we cannot feel absolutely certain that the pear is stable unless we can prove that the moment of momentum is greater than in the critical Jacobian.

* "Die Poincarésche Theorie des Gleichgewichts," *Annalen der K. Sternwarte, München*, Bd. III.

With reference to this objection, M. Poincaré writes to me as follows:—

"Faisons croître le moment de rotation, que j'appellerai M. Deux hypothèses sont possibles.

"Ou bien pour $M < M_0$ (the moment of momentum of the Jacobian), nous aurons une seule figure, *stable*, à savoir l'ellipsoïde de Jacobi, et pour $M > M_0$ trois figures, une instable, l'ellipsoïde, et deux stables (d'ailleurs égales entre elles), les deux figures pyriformes.

"Ou bien pour $M < M_0$, nous aurons trois figures d'équilibre, deux pyriformes instables, une stable, l'ellipsoïde, et pour $M > M_0$ une seule figure instable, l'ellipsoïde—auquel cas la masse fluide devrait se dissoudre par un cataclysme subit.

"Il y a donc à vérifier si pour les figures pyriformes, $M >$ ou $< M_0$."

It seems very improbable that the latter can be the case; but this opinion is not a proof*.

Since ω^2 is stationary for the initial pear, a small change in the angular velocity will certainly produce a great change in the figure of the pear. If this investigation has, in fact, its counterpart in the genesis of satellites and planets, it seems clear that the birth of a new body, although not cataclysmal, is rapid.

§ 9. *Summary.*

It is possible by the methods explained in my previous paper on "Harmonics" to form rigorous expressions for the ellipsoidal harmonics of the third degree. Accordingly in § 1 I proceed to form those functions. In § 2 the notation is changed with a view to convenience in subsequent work, and for the sake of completeness the harmonics of the first and second degrees are also given. In § 3 the corresponding solid harmonics are expressed in rectangular coordinates x, y, z. In § 4 I find the Q-functions, the harmonic functions of the second kind, and express the results in terms of the elliptic integrals E and F. It appears that both the P- and Q-functions of the third degree of harmonics occur in three pairs which have the same algebraic forms, and that in each pair one of them only differs from the other in the value of a certain parameter. There is, lastly, a seventh function which stands by itself; this last corresponds to the solid harmonic xyz.

In § 5 the equations for Jacobi's ellipsoid are determined by the consideration that the energy must be stationary, and the superficial value of gravity is found in terms of the appropriate P- and Q-functions. I then proceed to find the additional terms in the energy when the mass of fluid is

* [It is denied by M. Liapounoff as will be seen in Paper 12.]

subject to an ellipsoidal harmonic deformation. This section is a paraphrase of M. Poincaré's work, but the notation and manner of presentation are somewhat different. The additional terms in the energy are shown to involve a certain coefficient, which is called by M. Poincaré a coefficient of stability. It is clear that when any coefficient vanishes we are at a point of bifurcation, and the particular Jacobian ellipsoid for which it vanishes is also a member of another series of figures of equilibrium.

In § 6 the principal properties of these coefficients, as established by M. Poincaré, are enumerated. He has shown that the ellipsoid can bifurcate only into figures defined by zonal harmonics; that it must do so for all degrees, and that the first bifurcation occurs with the third zonal harmonic. The order of magnitude of the coefficients of the several orders and of the same degree is determined. A numerical result seems to indicate that as the ellipsoid lengthens, it becomes more stable as regards deformations of the third degree and of higher orders, and less stable as regards the lower orders of the same degree.

In § 7 the numerical solution of the vanishing of the coefficient corresponding to the third zonal harmonic is found, and it is shown that the critical ellipsoid has its three axes proportional to ·65066, ·81498, 1·88583, and that the square of the angular velocity is given by $\omega^2/2\pi\rho = \cdot 14200$. The nature of the formula for the third zonal coefficient of stability seems to show that it can only vanish once—a point which it appears that M. Poincaré found himself unable to prove rigorously.

A suggestion is made for the approximate determination of the bifurcations into the successive zonal deformations, but no numerical results are given.

In § 8 the nature of the pear-shaped figure is determined numerically, and the reader may refer to the figure above, where it is delineated. It will be seen to be longer than was shown in M. Poincaré's conjectural sketch.

If, as M. Poincaré suggests, the bifurcation into the pear-shaped body leads onward stably and continuously to a planet attended by a satellite, the bifurcation into the fourth zonal harmonic probably leads unstably to a planet with a satellite on each side, that into the fifth to a planet with two satellites on one side and one on the other, and so on.

The pear-shaped bodies are almost certainly stable, but a rigorous and conclusive proof is wanting until the angular velocity and moment of momentum corresponding to a given pear are determined. Further approximation [which is carried out in Paper 12], is required to determine the stability.

12.

THE STABILITY OF THE PEAR-SHAPED FIGURE OF EQUILIBRIUM OF A ROTATING MASS OF LIQUID.

[*Philosophical Transactions of the Royal Society*, Vol. 200, A (1902), pp. 251—314, with which is incorporated "Further consideration of the stability &c.," *ibid.* Vol. 208, A (1908), pp. 1—19, and "Note on the stability of Jacobi's ellipsoid," *Proceedings of the Royal Society*, Vol. 82, A (1909), pp. 188—9.]

TABLE OF CONTENTS.

Introduction.

By aid of the methods of a paper on "Ellipsoidal Harmonic Analysis" [Paper 10] I here resume the subject of the immediately preceding paper [Paper 11]. These papers will be referred to hereafter by the abridged titles of "Harmonics" and "The Pear-shaped Figure."

At the end of the latter of these it was stated that the stability of the figure could not be proved definitely without approximation of a higher order of accuracy. After some correspondence with M. Poincaré during the course of my work, I made an attempt to carry out this further approximation, but found that the expression for a certain portion of the energy entirely foiled me. Meanwhile he had turned his attention to the subject, and he has shown (*Phil. Trans.*, A, Vol. 198, pp. 333—373) by a method of the greatest ingenuity and skill how the problem may be solved. He has not, however, pursued the arduous task of converting his analytical results into numbers, so that he left the question as to the stability of the pear still unanswered.

M. Poincaré was so kind as to allow me to detain his manuscript on its way to the Royal Society for two or three days, and I devoted that time almost entirely to understanding the method of his attack on the key of the position—namely, the method of double layers, expounded in my own language in § 9 below. Being thus furnished with the means, I was able to resume my attempt under favourable conditions, and this paper is the result.

The substance of the analysis of this paper is, of course, essentially the same as his, but the arrangement and notation are so different that the two present but little superficial resemblance. This difference arises partly from the fact that I desired to use my own notation for the ellipsoidal harmonics, and partly because during the time that I was working at the analysis his paper was still unprinted and therefore inaccessible to me. But it is, perhaps, well that the two investigations of so complicated a subject should be nearly independent of one another.

It is rather unfortunate that I did not feel myself sufficiently expert in the use of the methods of Weierstrass and Schwarz to evaluate the elliptic integrals after the methods suggested by M. Poincaré, but every exertion has been taken to insure correctness in the arithmetical results, on which the proof of stability depends. My choice of antiquated methods of computation leaves the way open for some one else to verify the conclusions by wholly independent and more elegant calculations. It is highly desirable that such a verification should be made.

As the body of this paper will hardly be studied by any one unless they should be actually working at the subject, I give a summary at the end. Even the mathematician who desires to study the subject in detail may find it advantageous to read the summary before looking at the analytical investigation.

PART I.

ANALYTICAL INVESTIGATION.

§ 1. *Method of Procedure.*

The pear-shaped figure is a deformation of the critical Jacobian ellipsoid, and to the first order of small quantities it is expressed by the third zonal harmonic with respect to the longest axis of the ellipsoid. In the higher approximation a number of other harmonic terms will arise, and the coefficients of these new terms will be of the second order of small quantities. The mass of an harmonic inequality vanishes only to the first order, and it can no longer be assumed that the centre of inertia of the pear coincides with the centre of the ellipsoid.

In order to define the pear, I describe an ellipsoid similar to and concentric with the original critical Jacobian; this new ellipsoid is taken to be sufficiently large to enclose the whole of the pear. It is clearly itself a critical Jacobian, and I adopt it as the ellipsoid of reference, and call it J. I call the region between J and the pear R. The pear may then be defined by density $+\rho$ throughout J, and density $-\rho$ throughout R.

If k is the parameter which defines J, its axes are expressed in the notation of "Harmonics" by $k\nu_0$, $k(\nu_0^2-1)^{\frac{1}{2}}$, $k\left(\nu_0^2-\frac{1+\beta}{1-\beta}\right)^{\frac{1}{2}}$; or in the notation of the "Pear-shaped Figure" by $k/\sin\beta$, $k\cos\beta/\sin\beta$, $k\cos\gamma/\sin\beta$, where $\sin\beta=\kappa\sin\gamma$.

Now let S_i^s denote any surface harmonic, so that S_i^s is the same thing as $[\mathfrak{P}_i^s(\mu)$ or $\mathsf{P}_i^s(\mu)]\times[\mathfrak{C}_i^s(\phi)$ or $\mathsf{C}_i^s(\phi)]$. The third zonal harmonic deformation will then be eS_3 or $e\mathfrak{P}_3(\mu)\,\mathsf{C}_3(\phi)$, where e is of the first order of small quantities. On account of the symmetry of the figure, the new terms cannot involve the sine functions $\mathfrak{S}$ or S, and moreover, the rank s must necessarily be even.

Suppose that the new terms are expressed by $\Sigma f_i^s S_i^s$ for all values of i from 1 to infinity, and with s equal to 0, 2, 4 ... i or $i-1$. Then all the f's are of order e^2, excepting f_3 which is zero.

[The ellipsoidal coordinates of the point x, y, z are ν, μ, ϕ. I propose to retain μ and ϕ as two of the coordinates, but to replace ν by a new coordinate τ such that

$$\tau = -\frac{k^2}{2p_0^2}(\nu^2 - \nu_0^2) \quad \text{.................................}(1)$$

where p_0 is the perpendicular from the origin on to the tangent plane to the ellipsoid ν_0 at the point ν_0, μ, ϕ.

The equation

$$\frac{x^2}{k^2(\nu_0^2 - 1/\kappa^2)} + \frac{y^2}{k^2(\nu_0^2 - 1)} + \frac{z^2}{k^2\nu_0^2} = 1 - 2c$$

where c is a constant, represents an ellipsoid similar to and concentric with J.

Suppose that from the point x, y, z or ν, μ, ϕ we draw the curve defined by $\mu = \text{const.}$, $\phi = \text{const.}$ and follow it until we reach J at the point x_0, y_0, z_0.

Then we have

$$x^2 = x_0^2 \frac{\nu^2 - 1/\kappa^2}{\nu_0^2 - 1/\kappa^2} = x_0^2\left[1 - \frac{2p_0^2\tau}{k^2(\nu_0^2 - 1/\kappa^2)}\right]$$

and similarly

$$y^2 = y_0^2\left[1 - \frac{2p_0\tau}{k^2(\nu_0^2 - 1)}\right], \quad z^2 = z_0^2\left[1 - \frac{2p_0\tau}{k^2\nu_0^2}\right]$$

Substituting these values of x, y, z in the equation to the ellipsoid similar to J we have

$$\frac{x_0^2}{k^2(\nu_0^2 - 1/\kappa^2)} + \frac{y_0^2}{k^2(\nu_0^2 - 1)} + \frac{z_0^2}{k^2\nu_0^2}$$
$$- \frac{2p_0^2\tau}{k^2}\left[\frac{x_0^2}{k^2(\nu_0^2 - 1/\kappa^2)^2} + \frac{y_0^2}{k^2(\nu_0^2 - 1)^2} + \frac{z_0^2}{k^2\nu_0^4}\right] = 1 - 2c$$

whence
$$1 - 2\tau = 1 - 2c$$

Hence $\tau = c$, a constant, defines an interior ellipsoid similar to and concentric with J.]

The equation to the pear may now be written

$$\tau = c - eS_3 - \sum_1^\infty f_i^s S_i^s$$

The only condition which is imposed on c is that it shall be great enough to make τ always positive.

In order to solve our problem it is necessary to determine the energy lost in the process of concentration from a condition of infinite dispersion into the final configuration. This involves the use of the formula for the gravity of J, inclusive of rotation. It is well known that this formula is simple for the inside of J and more complicated for the outside. Since the whole region R lies inside J there is no necessity in the present case to use the more complicated formula.

The final expression for the lost energy cannot involve the size of J, the exterior ellipsoid of reference, and therefore the arbitrary constant c must ultimately disappear. It is therefore legitimate to make c zero from the beginning.

It is clear that we might with equal justice have discussed the problem by means of an ellipsoid which should lie entirely inside the pear, the region between the pear and the ellipsoid would then have been filled with positive density, and the formula for external gravity would have been needed. The same argument as before would then have justified our putting the constant c equal to zero.

We thus arrive at the same conclusion as does M. Poincaré, namely, that it is immaterial whether the formula for external or internal gravity be used.

I now revert to my first hypothesis of the enveloping ellipsoid, but put c equal to zero from the first. In order, however, to afford clearness to our conceptions, I shall continue to discuss the problem as though c were not zero and as though J enclosed the whole pear. With this explanation, we may write the equation to the pear in the form

$$\tau = -eS_3 - \sum_1^\infty f_i^s S_i^s \quad \text{.............................(2)}$$

§ 2. *The Lost Energy of the System.*

If the negative density in R is transported along tubes formed by a family of orthogonal curves and deposited as surface density on J, we may refer to such a condensation as $-C$. I do not suppose the condensation actually effected, but imagine the surface of J to be coated with equal and opposite condensations $+C$ and $-C$.

The system of masses forming the pear may then be considered as being as follows:—

Density $+\rho$ throughout J, say $+J$.

Negative condensation on J, say $-C$.

Positive condensation $+C$ on J and negative volume density $-\rho$ throughout R. This last forms a double system of zero mass, say D, and $D = C - R$.

Let V_j, V_r be the potentials of $+J$ and $+R$, and V_{j-r} the potential of the pear.

An element of volume being written dv, let $\int_j dv$, $\int_r dv$, $\int_{j-r} dv$, denote integrations throughout J, R and the pear respectively.

Let d be the distance along the z axis from the centre of the ellipsoid as origin to the centre of inertia of the pear; let ω be the angular velocity of the critical Jacobian about the axis x, so that $\omega^2/2\pi\rho = \cdot 14200$; and let $\omega^2 + \delta\omega^2$ be the square of the angular velocity of the pear. Lastly, let M be the mass of the pear.

Then the lost energy E is given by

$$E = \tfrac{1}{2}\int_{j-r} V_{j-r}\rho dv + \tfrac{1}{2}(\omega^2 + \delta\omega^2)\int_{j-r} [y^2 + (z-d)^2]\rho dv$$

Now $\int_{j-r} z\rho dv = Md$, so that $\int_{j-r} (-2zd + d^2)\rho dv = -Md^2$

Again, since

$$V_{j-r} = V_j - V_r, \quad \int_{j-r} = \int_j - \int_r, \quad \int_j V_r\rho dv = \int_r V_j\rho dv$$

we have $$\tfrac{1}{2}\int_{j-r} V_{j-r}\rho dv = \tfrac{1}{2}\int_j V_j\rho dv - \int_r V_j\rho dv + \tfrac{1}{2}\int_r V_r\rho dv$$

Also

$$\tfrac{1}{2}(\omega^2 + \delta\omega^2)\int_{j-r} [y^2 + (z-d)^2]\rho dv = \tfrac{1}{2}\omega^2\int_j (y^2+z^2)\rho dv - \tfrac{1}{2}\omega^2\int_r (y^2+z^2)\rho dv$$
$$+ \tfrac{1}{2}\delta\omega^2\int_{j-r} (y^2+z^2)\rho dv - \tfrac{1}{2}(\omega^2+\delta\omega^2) Md^2$$

Hence

$$E = \tfrac{1}{2}\int_j [V_j + \omega^2(y^2+z^2)]\rho dv - \int_r [V_j + \tfrac{1}{2}\omega^2(y^2+z^2)]\rho dv + \tfrac{1}{2}\int_r V_r\rho dv$$
$$+ \tfrac{1}{2}\delta\omega^2\int_{j-r} (y^2+z^2)\rho dv - \tfrac{1}{2}(\omega^2+\delta\omega^2) Md^2$$

As the several terms will be considered separately, it will be convenient to have an abridged notation to specify them. I may denote the lost energy of J, inclusive of rotation, by $\tfrac{1}{2}JJ$; the mutual lost energy of J (inclusive of rotation) and of the region R, considered as filled with positive density, by JR; the lost energy of the region R by $\tfrac{1}{2}RR$.

The moment of inertia of the pear is A, and it is equal to $A_j - A_r$, the moment of inertia of J less that of R.

Then

$$E = \tfrac{1}{2}JJ - JR + \tfrac{1}{2}RR + \tfrac{1}{2}(A_j - A_r)\delta\omega^2 - \tfrac{1}{2}(\omega^2 + \delta\omega^2) Md^2$$

where $$\tfrac{1}{2}JJ = \tfrac{1}{2}\int_j [V_j + \omega^2(y^2+z^2)]\rho dv$$

$$JR = \int_r [V_j + \tfrac{1}{2}\omega^2(y^2+z^2)]\rho dv$$

$$A_j = \int_j (y^2+z^2)\rho dv, \quad A_r = \int_r (y^2+z^2)\rho dv$$

and $$\tfrac{1}{2}RR = \int_r V_r\rho dv$$

If $\frac{1}{2}DD$ denotes the lost energy of the double system described above, we clearly have

$$\tfrac{1}{2}RR = \tfrac{1}{2}(C-R)(C-R) + CR - \tfrac{1}{2}CC = \tfrac{1}{2}DD + CR - \tfrac{1}{2}CC$$

We require to evaluate E to the fourth order; now d is at least of the second order and d^2 of the fourth order; hence $d^2 . \delta\omega^2$ is at least of the fifth order and negligible.

Hence, finally, to the required degree of approximation

$$E = \tfrac{1}{2}JJ - JR + CR - \tfrac{1}{2}CC + \tfrac{1}{2}DD + \tfrac{1}{2}(A_j - A_r)\,\delta\omega^2 - \tfrac{1}{2}Md^2\omega^2 \ldots (3)$$

It will appear below that d is not even of the second order, so that the last term will, in fact, entirely disappear, although we cannot see at the present stage that this will be so.

§ 3. *Expression for the Element of Volume.*

The parameter β of "Harmonics" is connected with κ of the "Pear-shaped Figure" by the equations

$$\frac{1-\beta}{1+\beta} = \kappa^2, \quad \beta = \frac{1-\kappa^2}{1+\kappa^2} = \frac{\kappa'^2}{1+\kappa^2}, \quad \frac{2\beta}{1-\beta} = \frac{\kappa'^2}{\kappa^2}, \quad \frac{2\beta}{1+\beta} = \kappa'^2$$

There will, I think, be no confusion if I also use β in a second sense, defining it by the equations

$$\sin\beta = \kappa \sin\gamma, \quad \cos^2\beta = 1 - \kappa^2 \sin^2\gamma$$

It has already been remarked above that the squares of the semi-axes of J are

$$k^2\nu_0^2 = \frac{k^2}{\sin^2\beta}, \quad k^2(\nu_0^2 - 1) = \frac{k^2\cos^2\beta}{\sin^2\beta}, \quad k^2\left(\nu_0^2 - \frac{1+\beta}{1-\beta}\right) = \frac{k^2\cos^2\gamma}{\sin^2\beta}$$

The mass of J is then $\frac{4}{3}\pi\rho k^3 \dfrac{\cos\beta\cos\gamma}{\sin^3\beta}$.

I now take the mass M of the pear to be

$$M = \tfrac{4}{3}\pi\rho k_0^3 \frac{\cos\beta\cos\gamma}{\sin^3\beta}$$

Thus k_0 is a constant which specifies the volume of liquid in the pear, and the mass of J is $M(k/k_0)^3$.

It will be convenient to introduce certain new symbols, namely,

$$\begin{aligned} \Delta^2 &= 1 - \kappa^2\sin^2\theta, & \Gamma^2 &= 1 - \kappa'^2\cos^2\phi \\ \Delta_1^2 &= 1 - \kappa^2\sin^2\gamma\sin^2\theta, & \Gamma_1^2 &= \cos^2\gamma + \kappa'^2\sin^2\gamma\cos^2\phi \\ &= \cos^2\beta + \kappa^2\sin^2\gamma\cos^2\theta, & &= \cos^2\beta - \kappa'^2\sin^2\gamma\sin^2\phi \end{aligned}$$

where $\sin\theta$ is the μ of "Harmonics."

It will be observed that Γ_1 is the same function of $-\kappa'^2\sin^2\phi$ that Δ_1 is of $\kappa^2\cos^2\theta$.

The roots of the fundamental cubic were ν^2, μ^2, and $\frac{1-\beta\cos 2\phi}{1-\beta}$, and in the new notation they are ν^2, $\sin^2\theta$, $\frac{1-\kappa'^2\cos^2\phi}{\kappa^2}$ or $\frac{\Gamma^2}{\kappa^2}$.

Since $\nu_0^2=\frac{1}{\sin^2\beta}$, we now have

$$\nu_0^2-\mu^2=\frac{\Delta_1^2}{\sin^2\beta},\qquad \nu_0^2-\frac{1-\beta\cos 2\phi}{1-\beta}=\frac{\Gamma_1^2}{\sin^2\beta}$$

The expression for p_0, the perpendicular from the centre on to the tangent plane at θ, ϕ, is given in (49) of "Harmonics," namely,

$$\frac{p_0^2}{k^2}=\frac{\nu_0^2(\nu_0^2-1)\left(\nu_0^2-\frac{1+\beta}{1-\beta}\right)}{(\nu_0^2-\mu^2)\left(\nu_0^2-\frac{1-\beta\cos 2\phi}{1-\beta}\right)}=\frac{\cos^2\beta\cos^2\gamma}{\sin^2\beta}\,\frac{1}{\Delta_1^2\Gamma_1^2}\quad\text{.........(4)}$$

Also by (50) of "Harmonics" the element of surface $d\sigma$ of the ellipsoid is given by

$$\frac{p_0 d\sigma}{d\theta d\phi}=k^3\nu_0(\nu_0^2-1)^{\frac{1}{2}}\left(\nu_0^2-\frac{1+\beta}{1-\beta}\right)^{\frac{1}{2}}\frac{\frac{1-\beta\cos 2\phi}{1-\beta}-\mu^2}{\left(\frac{1-\beta\cos 2\phi}{1-\beta}\right)^{\frac{1}{2}}\left(\frac{1+\beta}{1-\beta}-\mu^2\right)^{\frac{1}{2}}}$$

Passing to the new notation this may be written

$$\frac{p_0 d\sigma}{d\theta d\phi}=\frac{3M}{4\pi\rho}\left(\frac{k}{k_0}\right)^3\frac{1-\kappa^2\sin^2\theta-\kappa'^2\cos^2\phi}{\Delta\Gamma}=\frac{3M}{4\pi\rho}\left(\frac{k}{k_0}\right)^3\cdot\frac{\Delta_1^2\Gamma_1^2}{\Delta\Gamma\sin^2\gamma}\left(\frac{1}{\Gamma_1^2}-\frac{1}{\Delta_1^2}\right)$$

The new independent variable τ is to replace ν; it was defined in (1) by

$$\tau=\frac{k^2}{2p_0^2}(\nu_0^2-\nu^2)$$

and in accordance with (2) the equation to the surface of the pear is

$$\tau=-eS_3-\sum_1^\infty f_i^s S_i^s$$

From (4) $\quad \nu^2=\nu_0^2-\frac{2p_0^2}{k^2}\tau=\frac{1}{\sin^2\beta}\left(1-\frac{2\tau\cos^2\beta\cos^2\gamma}{\Delta_1^2\Gamma_1^2}\right)$

For brevity I now write

$$\tau_1=\frac{2\tau\cos^2\beta\cos^2\gamma}{\Delta_1^2\Gamma_1^2}$$

so that

$$\nu^2=\frac{1}{\sin^2\beta}(1-\tau_1),\qquad \nu^2-1=\frac{\cos^2\beta}{\sin^2\beta}(1-\tau_1\sec^2\beta),$$

$$\nu^2-\frac{1+\beta}{1-\beta}=\frac{\cos^2\gamma}{\sin^2\beta}(1-\tau_1\sec^2\gamma)$$

$$\nu^2-\mu^2=\frac{\Delta_1^2}{\sin^2\beta}\left(1-\frac{\tau_1}{\Delta_1^2}\right),\qquad \nu^2-\frac{1-\beta\cos 2\phi}{1-\beta}=\frac{\Gamma_1^2}{\sin^2\beta}\left(1-\frac{\tau_1}{\Gamma_1^2}\right)$$

$$\frac{1-\beta\cos 2\phi}{1-\beta}-\mu^2=\frac{1-\kappa^2\sin^2\theta-\kappa'^2\cos^2\phi}{\kappa^2}=\frac{\Delta_1^2\Gamma_1^2}{\sin^2\beta}\left(\frac{1}{\Gamma_1^2}-\frac{1}{\Delta_1^2}\right)$$

Therefore

$$\frac{(\nu^2-\mu^2)\left(\nu^2-\frac{1-\beta\cos 2\phi}{1-\beta}\right)}{\nu(\nu^2-1)^{\frac{1}{2}}\left(\nu^2-\frac{1+\beta}{1-\beta}\right)^{\frac{1}{2}}}$$

$$=\frac{\Delta_1^2\Gamma_1^2}{\sin\beta\cos\beta\cos\gamma}\cdot\frac{\left(1-\frac{\tau_1}{\Delta_1^2}\right)\left(1-\frac{\tau_1}{\Gamma_1^2}\right)}{(1-\tau_1)^{\frac{1}{2}}(1-\tau_1\sec^2\beta)^{\frac{1}{2}}(1-\tau_1\sec^2\gamma)^{\frac{1}{2}}}$$

If we write

$$G=\tfrac{1}{2}(1+\sec^2\beta+\sec^2\gamma)$$

$$H=\tfrac{3}{8}(1+\sec^4\beta+\sec^4\gamma)+\tfrac{1}{4}(\sec^2\beta+\sec^2\gamma+\sec^2\beta\sec^2\gamma)$$

this expression, when expanded as far as τ_1^2, becomes

$$\frac{\Delta_1^2\Gamma_1^2}{\sin\beta\cos\beta\cos\gamma}\left[1-\tau_1\left(\frac{1}{\Delta_1^2}+\frac{1}{\Gamma_1^2}-G\right)-\tau_1^2\left\{-\frac{1}{\Delta_1^2\Gamma_1^2}+G\left(\frac{1}{\Delta_1^2}+\frac{1}{\Gamma_1^2}\right)-H\right\}\right]$$

The arcs of the three orthogonal curves were denoted dn, dm, df in "Harmonics," where dn was the outward normal. The element of volume dv is $dndmdf$, and the limits of integration are 0 to τ; but if we take the limits as being τ to 0, dv must be taken as being $-dndmdf$.

The equations (50) of "Harmonics" give

$$\frac{dv}{-\cos\theta\,\nu d\nu d\theta d\phi}=k^3\frac{(\nu^2-\mu^2)\left(\nu^2-\frac{1-\beta\cos 2\phi}{1-\beta}\right)\left(\frac{1-\beta\cos 2\phi}{1-\beta}-\mu^2\right)}{\nu(\nu^2-1)^{\frac{1}{2}}\left(\nu^2-\frac{1+\beta}{1-\beta}\right)^{\frac{1}{2}}\cos\theta\left(\frac{1+\beta}{1-\beta}-\mu^2\right)^{\frac{1}{2}}\left(\frac{1-\beta\cos 2\phi}{1-\beta}\right)^{\frac{1}{2}}}$$

But

$$-\nu d\nu=\frac{\cos^2\beta\cos^2\gamma}{\sin^2\beta}\cdot\frac{d\tau}{\Delta_1^2\Gamma_1^2}$$

and therefore

$$\frac{dv}{d\tau d\theta d\phi}=k^3\frac{\cos\beta\cos\gamma}{\sin^3\beta\sin^2\gamma}\frac{\Delta_1^2\Gamma_1^2}{\Delta\Gamma}\left(\frac{1}{\Gamma_1^2}-\frac{1}{\Delta_1^2}\right)\left[1-\tau_1\left(\frac{1}{\Delta_1^2}+\frac{1}{\Gamma_1^2}-G\right)\right.$$
$$\left.-\tau_1^2\left\{-\frac{1}{\Delta_1^2\Gamma_1^2}+G\left(\frac{1}{\Delta_1^2}+\frac{1}{\Gamma_1^2}\right)-H\right\}\right]$$

On comparing this with the expression for $p_0 d\sigma$, we see that

$$\frac{dv}{d\tau}=p_0 d\sigma\left[1-\tau_1\left(\frac{1}{\Delta_1^2}+\frac{1}{\Gamma_1^2}-G\right)-\tau_1^2\left\{-\frac{1}{\Delta_1^2\Gamma_1^2}+G\left(\frac{1}{\Delta_1^2}+\frac{1}{\Gamma_1^2}\right)-H\right\}\right]\quad(5)$$

Another form, which will be more generally useful, is found by substituting for τ_1 its value; it is

$$\frac{dv}{d\tau d\theta d\phi}=\frac{3M}{4\pi\rho}\left(\frac{k}{k_0}\right)^3\left\{\frac{\Delta_1^2-\Gamma_1^2}{\sin^2\gamma}-2\tau\frac{\cos^2\beta\cos^2\gamma}{\sin^2\gamma}\left[\frac{1}{\Gamma_1^4}-\frac{1}{\Delta_1^4}-G\left(\frac{1}{\Gamma_1^2}-\frac{1}{\Delta_1^2}\right)\right]\right.$$
$$-4\tau^2\frac{\cos^4\beta\cos^4\gamma}{\sin^2\gamma}\left[-\left(\frac{1}{\Gamma_1^6\Delta_1^4}-\frac{1}{\Gamma_1^4\Delta_1^6}\right)+G\left(\frac{1}{\Gamma_1^6\Delta_1^2}-\frac{1}{\Gamma_1^2\Delta_1^6}\right)\right.$$
$$\left.\left.-H\left(\frac{1}{\Gamma_1^4\Delta_1^2}-\frac{1}{\Gamma_1^2\Delta_1^4}\right)\right]\right\}\frac{1}{\Delta\Gamma}$$

In order to express this more succinctly let

$$\left.\begin{aligned}
\Phi &= \frac{6}{\pi}\frac{\Delta_1^2-\Gamma_1^2}{\Delta\Gamma\sin^2\gamma}\\
\Psi &= \frac{6}{\pi}\frac{\cos^2\beta\cos^2\gamma}{\sin^2\gamma}\left[\frac{1}{\Gamma_1^4}-\frac{1}{\Delta_1^4}-G\left(\frac{1}{\Gamma_1^2}-\frac{1}{\Delta_1^2}\right)\right]\frac{1}{\Delta\Gamma}\\
\Omega &= \frac{6}{\pi}\frac{\cos^4\beta\cos^4\gamma}{\sin^2\gamma}\left[-\left(\frac{1}{\Gamma_1^6\Delta_1^4}-\frac{1}{\Gamma_1^4\Delta_1^6}\right)+G\left(\frac{1}{\Gamma_1^6\Delta_1^2}-\frac{1}{\Gamma_1^2\Delta_1^6}\right)\right.\\
&\qquad\left.-H\left(\frac{1}{\Gamma_1^4\Delta_1^2}-\frac{1}{\Gamma_1^2\Delta_1^4}\right)\right]\frac{1}{\Delta\Gamma}
\end{aligned}\right\}\ldots(6)$$

Other forms of these functions are as follows:—

$$\left.\begin{aligned}
\Phi &= \frac{6}{\pi}\frac{\kappa^2\cos^2\theta+\kappa'^2\sin^2\phi}{\Delta\Gamma}\\
\Psi &= \frac{6}{\pi}\cos^2\beta\cos^2\gamma\,(\kappa^2\cos^2\theta+\kappa'^2\sin^2\phi)\left[\frac{1}{\Gamma_1^2}+\frac{1}{\Delta_1^2}-G\right]\frac{1}{\Delta_1^2\Delta\Gamma_1^2\Gamma}\\
\Omega &= \frac{6}{\pi}\cos^4\beta\cos^4\gamma\,(\kappa^2\cos^2\theta+\kappa'^2\sin^2\phi)\\
&\qquad\times\left[-\frac{1}{\Gamma_1^2\Delta_1^2}+G\left(\frac{1}{\Gamma_1^2}+\frac{1}{\Delta_1^2}\right)-H\right]\frac{1}{\Delta_1^4\Delta\Gamma_1^4\Gamma}
\end{aligned}\right\}\ldots(6)$$

We note that

$$\left.\begin{aligned}
\frac{p_0\,d\sigma}{d\theta\,d\phi} &= \tfrac{1}{8}\frac{M}{\rho}\left(\frac{k}{k_0}\right)^3\Phi\\
\Psi &= \cos^2\beta\cos^2\gamma\left(\frac{1}{\Gamma_1^2}+\frac{1}{\Delta_1^2}-G\right)\frac{\Phi}{\Delta_1^2\Gamma_1^2}
\end{aligned}\right\}\ldots\ldots\ldots\ldots\ldots(6)$$

Then $$\frac{dv}{d\tau\,d\theta\,d\phi}=\tfrac{1}{8}\frac{M}{\rho}\left(\frac{k}{k_0}\right)^3[\Phi-2\tau\Psi-4\tau^2\Omega]\ \ldots\ldots\ldots\ldots(6)$$

The surface $\tau = constant$ is an ellipsoid similar to J with squares of semi-axes reduced in the proportion $1-2\tau$ to unity. Therefore the volume enclosed between the two ellipsoids is

$$\int dv = \frac{M}{\rho}\left(\frac{k}{k_0}\right)^3[1-(1-2\tau)^{\frac{3}{2}}] = \frac{M}{\rho}\left(\frac{k}{k_0}\right)^3[3\tau-\tfrac{3}{2}\tau^2-\tfrac{1}{2}\tau^3]$$

But taking the limits of θ and ϕ as $\frac{1}{2}\pi$ to 0, so that we integrate through one octant and multiply the result by 8, we have another expression for the same thing, namely,

$$\int dv = \frac{M}{\rho}\left(\frac{k}{k_0}\right)^3\iint[\Phi\tau-\Psi\tau^2-\tfrac{4}{3}\Omega\tau^3]\,d\theta\,d\phi$$

Therefore equating coefficients of powers of τ in the two expressions,

$$\int_0^{\frac{1}{2}\pi}\int_0^{\frac{1}{2}\pi}\Phi\,d\theta\,d\phi=3,\quad \int_0^{\frac{1}{2}\pi}\int_0^{\frac{1}{2}\pi}\Psi\,d\theta\,d\phi=\tfrac{3}{2},\quad \int_0^{\frac{1}{2}\pi}\int_0^{\frac{1}{2}\pi}\Omega\,d\theta\,d\phi=\tfrac{3}{8}\ \ldots(7)$$

The first of these will be of use hereafter, and all three afford formulæ of verification in the numerical work.

§ 4. *Determination of k; Definition of Symbols for Integrals.*

The pear being defined by $\tau = -eS_3 - \sum_1^\infty f_i^s S_i^s$, with all the f's of order e^2, excepting f_3 which is zero, we have at the surface of the pear to the fourth order

$$\tau^2 = e^2 (S_3)^2 + 2\Sigma e f_i^s S_3 S_i^s + (\Sigma f_i^s S_i^s)^2$$
$$\tau^3 = -e^3 (S_3)^3 - 3\Sigma e^2 f_i^s (S_3)^2 S_i^s$$
$$\tau^4 = e^4 (S_3)^4$$

In all the integrations which follow, and especially in the present instance in the determination of the volume of the region R, it is important to note that Φ, Ψ, Ω are even functions of the angular co-ordinates, and that therefore the integral of any odd function of those co-ordinates multiplied by any of these functions will vanish. When the odd functions are omitted we may integrate throughout the octant defined by the limits $\frac{1}{2}\pi$ to 0 for θ and ϕ, and multiply the result by 8.

Then, only retaining terms as far as τ^3, we may in finding the volume R take

$$\tau = -\Sigma f_i^s S_i^s, \ i \text{ only even}$$
$$\tau^2 = e^2 (S_3)^2 + 2\Sigma e f_i^s S_3 S_i^s, \ i \text{ only odd}$$
$$\tau^3 = 0$$

To the cubes of small quantities we have, therefore,

$$\int_r dv = \frac{M}{\rho}\left(\frac{k}{k_0}\right)^3 \int_0^{\frac{1}{2}\pi}\int_0^{\frac{1}{2}\pi} [-\Phi\Sigma f_i^s S_i^s - \Psi\{e^2 (S_3)^2 + 2\Sigma e f_i^s S_3 S_i^s\}]\, d\theta\, d\phi$$

The first term vanishes because S_i^s is a surface harmonic and $\Phi d\theta d\phi$ is proportional to $p_0 d\sigma$.

Thus we are left with

$$\int_r dv = -\frac{M}{\rho}\left(\frac{k}{k_0}\right)^3 \int_0^{\frac{1}{2}\pi}\int_0^{\frac{1}{2}\pi} \Psi [e^2 (S_3)^2 + 2\Sigma e f_i^s S_3 S_i^s]\, d\theta\, d\phi$$

I now introduce symbols for certain integrals, and in order to bring all the definitions together I also define several others which will only occur later.

Let

$$\left.\begin{aligned} \phi_i^s &= \int_0^{\frac{1}{2}\pi}\int_0^{\frac{1}{2}\pi} \Phi (S_i^s)^2 d\theta\, d\phi \\ \omega_i^s &= \int_0^{\frac{1}{2}\pi}\int_0^{\frac{1}{2}\pi} \Psi (S_3)^2 S_i^s d\theta\, d\phi \\ \rho_i^s &= \frac{\cos^2\beta\cos^2\gamma}{\sin\beta}\int_0^{\frac{1}{2}\pi}\int_0^{\frac{1}{2}\pi} \frac{\Phi}{\Delta_1^2\Gamma_1^2} (S_3)^2 S_i^s d\theta\, d\phi \end{aligned}\right\} \quad \text{............}(8)$$

All these integrals vanish unless i is even. For immediate use I also introduce

$$\psi_i^s = \int_0^{\frac{1}{2}\pi}\int_0^{\frac{1}{2}\pi} \Psi S_3 S_i^s \, d\theta \, d\phi$$

The ψ integrals vanish unless i is odd, but it will appear later that they are not actually required.

I further write

$$\left.\begin{aligned}
\sigma_2 &= \int_0^{\frac{1}{2}\pi}\int_0^{\frac{1}{2}\pi} \Psi (S_3)^2 d\theta \, d\phi, \qquad \zeta_4 = \int_0^{\frac{1}{2}\pi}\int_0^{\frac{1}{2}\pi} \Omega (S_3)^4 d\theta \, d\phi \\
\sigma_4 &= \frac{6}{\pi}\frac{\cos^3\beta\cos^3\gamma\sin\beta}{\sin^2\gamma}\int_0^{\frac{1}{2}\pi}\int_0^{\frac{1}{2}\pi}\left[\tfrac{5}{2}\left(\frac{1}{\Delta_1^2\Gamma_1^6} - \frac{1}{\Delta_1^6\Gamma_1^2}\right)\right. \\
&\qquad\qquad \left. - 3G\left(\frac{1}{\Delta_1^2\Gamma_1^4} - \frac{1}{\Delta_1^4\Gamma_1^2}\right)\right]\frac{(S_3)^4}{\Delta\Gamma}\, d\theta \, d\phi \\
&= \frac{6}{\pi}\cos^3\beta\cos^3\gamma\sin\beta\int_0^{\frac{1}{2}\pi}\int_0^{\frac{1}{2}\pi}(\kappa^2\cos^2\theta + \kappa'^2\sin^2\phi) \\
&\qquad\qquad \times\left[\tfrac{5}{2}\left(\frac{1}{\Delta_1^4\Gamma_1^6} + \frac{1}{\Delta_1^6\Gamma_1^4}\right) - \frac{3G}{\Delta_1^4\Gamma_1^4}\right]\frac{(S_3)^4}{\Delta\Gamma}\, d\theta \, d\phi
\end{aligned}\right\} \quad \text{...(8)}$$

With this notation we have at once to cubes of small quantities,

$$\int_r \rho \, dv = -M\left(\frac{k}{k_0}\right)^3 [e^2\sigma_2 + 2\Sigma e f_i^s \psi_i^s] \quad \text{..................(9)}$$

But before using this I will obtain another integral to the fourth order. It is

$$\int_r \tau\rho \, dv = M\left(\frac{k}{k_0}\right)^3 \int_0^{\frac{1}{2}\pi}\int_0^{\frac{1}{2}\pi} \{\tfrac{1}{2}\Phi [e^2 (S_3)^2 + 2\Sigma e f_i^s S_3 S_i^s + (\Sigma f_i^s S_i^s)^2]$$
$$+ \tfrac{2}{3}\Psi [e^3 (S_3)^3 + 3\Sigma e^2 f_i^s (S_3)^2 S_i^s] - \Omega e^4 (S_3)^4\} \, d\theta \, d\phi$$

Omitting terms which vanish, amongst which are integrals of the type $\Phi S_i^s S_j^t$, we have

$$\int_r \tau\rho \, dv = \tfrac{1}{2}M\left(\frac{k}{k_0}\right)^3 \{e^2\phi_3 + \Sigma (f_i^s)^2 \phi_i^s + 4\Sigma e^2 f_i^s \omega_i^s - 2e^4\zeta_4\} \text{......(10)}$$

Returning now to the determination of the mass of $+R$, and observing that the mass of the pear is equal to that of $J - R$, we have

$$M = M\left(\frac{k}{k_0}\right)^3 [1 + e^2\sigma_2 + 2\Sigma e f_i^s \psi_i^s]$$

Therefore

$$\left(\frac{k_0}{k}\right)^3 = 1 + e^2\sigma_2 + 2\Sigma e f_i^s \psi_i^s + e^4\delta$$

A term $e^4\delta$ of the fourth order has been introduced, but it will appear that it is unnecessary to evaluate it.

There will be frequent occasion to express k^5 in terms of k_0^5. Now

$$\left(\frac{k}{k_0}\right)^5 = 1 - \frac{5}{3}\left[e^2\sigma_2 + 2\Sigma e f_i^s \psi_i^s + e^4\delta - \frac{8}{9}e^4(\sigma_2)^2\right]$$

But this will only be needed explicitly as far as e^2, and to that order

$$\left(\frac{k}{k_0}\right)^5 = 1 - \frac{5}{3}e^2\sigma_2 \text{..........................} (11)$$

It is, however, necessary to determine $\frac{3}{2}\left(\frac{k}{k_0}\right)^2 - \frac{3}{5}\left(\frac{k}{k_0}\right)^5$ to the fourth order.

Now
$$\frac{3}{2}\left(\frac{k}{k_0}\right)^2 = \frac{3}{2}\left\{1 - \frac{2}{3}\left[e^2\sigma_2 + 2\Sigma e f_i^s \psi_i^s + e^4\delta - \frac{5}{6}e^4(\sigma_2)^2\right]\right\}$$

$$\frac{3}{5}\left(\frac{k}{k_0}\right)^5 = \frac{3}{5}\left\{1 - \frac{5}{3}\left[e^2\sigma_2 + 2\Sigma e f_i^s \psi_i^s + e^4\delta - \frac{8}{9}e^4(\sigma_2)^2\right]\right\}$$

Hence to the fourth order

$$\frac{3}{2}\left(\frac{k}{k_0}\right)^2 - \frac{3}{5}\left(\frac{k}{k_0}\right)^5 = \frac{9}{10} - \frac{1}{2}e^4(\sigma_2)^2 \text{ ....} (11)$$

It will be observed that the ψ integrals and δ have both disappeared.

§ 5. *The Energies $\frac{1}{2}JJ$ and JR.*

If a_1, b_1, c_1 are the semi-axes of a Jacobian ellipsoid of mass M_1 and angular velocity ω, its lost energy, inclusive of rotation, is

$$\frac{3}{10}M_1^2\left[\Psi + \frac{b_1^2 + c_1^2}{3M_1}\omega^2\right]$$

where Ψ is the usual auxiliary function.

The equations to be satisfied by the ellipsoid afford expressions for $\omega^2 b_1^2$ and $\omega^2 c_1^2$ in terms of differentials of Ψ. If these expressions are added together, ω^2 may be eliminated, and the expression becomes

$$\frac{9}{20}M_1^2\left[\Psi + a_1\frac{d\Psi}{da_1}\right]$$

In reverting to the notation adopted here, I remark that $\mathfrak{P}_i^s$, $\mathfrak{Q}_i^s$ will be used to denote those functions when the variable is ν_0, and the variable will only be inserted explicitly when it has any other value.

In the present case M_1, the mass of the Jacobian ellipsoid, is $M(k/k_0)^3$, and it was shown in the "Pear-shaped Figure" that

$$\Psi = \frac{2}{k}\mathfrak{P}_0\mathfrak{Q}_0, \quad a_1\frac{d\Psi}{da_1} = -\frac{2}{k}\mathsf{P}_1^1\mathsf{Q}_1^1$$

Hence
$$\frac{1}{2}JJ = \frac{9}{10}\frac{M^2}{k_0}\left(\frac{k}{k_0}\right)^5\left[\mathfrak{P}_0\mathfrak{Q}_0 - \mathsf{P}_1^1\mathsf{Q}_1^1\right] \text{........................} (12)$$

It was shown in the same paper that the internal potential of the Jacobian inclusive of rotation, is

$$\tfrac{3}{4}M_1\left\{\Psi + a_1\frac{d\Psi}{da_1}\left(\frac{x^2}{a_1^2}+\frac{y^2}{b_1^2}+\frac{z^2}{c_1^2}\right)\right\}$$

Therefore in the present case

$$V_j + \tfrac{1}{2}\omega^2(y^2+z^2) = \tfrac{3}{2}\frac{M}{k_0}\left(\frac{k}{k_0}\right)^2\left\{\mathfrak{P}_0\mathfrak{Q}_0 - \frac{1}{k^2}\mathsf{P}_1{}^1\mathsf{Q}_1{}^1\sin^2\beta\,(x^2\sec^2\gamma + y^2\sec^2\beta + z^2)\right\}$$

But the equation to an inequality on the ellipsoid defined by τ is in our new notation

$$\sin^2\beta\,(x^2\sec^2\gamma + y^2\sec^2\beta + z^2) = k^2(1-2\tau)$$

therefore

$$V_j + \tfrac{1}{2}\omega^2(y^2+z^2) = \tfrac{3}{2}\frac{M}{k_0}\left(\frac{k}{k_0}\right)^2\{(\mathfrak{P}_0\mathfrak{Q}_0 - \mathsf{P}_1{}^1\mathsf{Q}_1{}^1) + 2\tau\mathsf{P}_1{}^1\mathsf{Q}_1{}^1\}$$

Let us divide this potential into two parts, say U', U'', of which the first is constant and the second a constant multiplied by τ. Also let $(JR)'$, $(JR)''$ be the two corresponding portions of the energy JR.

In order to find $(JR)'$ we have simply to multiply U' by the mass of R considered as consisting of positive density. The volume of R is the excess of the volume of J above that of the pear; hence the mass of R is $M\left[\left(\frac{k}{k_0}\right)^3 - 1\right]$.

Therefore $$(JR)' = \tfrac{3}{2}\frac{M^2}{k_0}\left[\left(\frac{k}{k_0}\right)^5 - \left(\frac{k}{k_0}\right)^2\right](\mathfrak{P}_0\mathfrak{Q}_0 - \mathsf{P}_1{}^1\mathsf{Q}_1{}^1)$$

Subtracting this from $\frac{1}{2}JJ$ as given in (12),

$$\tfrac{1}{2}JJ - (JR)' = \frac{M^2}{k_0}(\mathfrak{P}_0\mathfrak{Q}_0 - \mathsf{P}_1{}^1\mathsf{Q}_1{}^1)\left[\tfrac{3}{2}\left(\frac{k}{k_0}\right)^2 - \tfrac{3}{5}\left(\frac{k}{k_0}\right)^5\right]$$

But the latter factor was found in (11) as equal to $\frac{9}{10} - \frac{1}{2}e^4(\sigma_2)^2$. The term $\frac{9}{10}$ only contributes a constant to the whole energy and may therefore be dropped.

Accordingly

$$\tfrac{1}{2}JJ - (JR)' = \tfrac{3}{2}\frac{M^2}{k_0}\{-(\mathfrak{P}_0\mathfrak{Q}_0 - \mathsf{P}_1{}^1\mathsf{Q}_1{}^1)\tfrac{1}{3}e^4(\sigma_2)^2\} \quad\text{..........(13)}$$

For the other portion $(JR)''$ we have

$$U'' = 3\frac{M}{k_0}\left(\frac{k}{k_0}\right)^2\tau\mathsf{P}_1{}^1\mathsf{Q}_1{}^1$$

Then by means of (10)

$$(JR)'' = \int_r U''\rho\,dv = 3\frac{M}{k_0}\left(\frac{k}{k_0}\right)^2\mathsf{P}_1{}^1\mathsf{Q}_1{}^1\int_r \tau\rho\,dv$$

$$= \tfrac{3}{2}\frac{M^2}{k_0}\left(\frac{k}{k_0}\right)^5\mathsf{P}_1{}^1\mathsf{Q}_1{}^1\{e^2\phi_3 + \Sigma(f_i{}^s)^2\phi_i{}^s + 4\Sigma e^2 f_i{}^s\omega_i{}^s - 2e^4\zeta_4\} \quad\text{...(14)}$$

In the terms of the fourth order we may put $(k/k_0)^5$ equal to unity. Therefore combining (13) and (14)

$$\tfrac{1}{2}JJ - JR = -\tfrac{3}{2}\frac{M^2}{k_0}\left(\frac{k}{k_0}\right)^5 e^2 \mathsf{P}_1{}^1\mathsf{Q}_1{}^1\phi_3$$

$$+\tfrac{3}{2}\frac{M^2}{k_0}\{\tfrac{1}{3}e^4[-(3\mathfrak{P}_0\mathfrak{Q}_0 - \mathsf{P}_1{}^1\mathsf{Q}_1{}^1)(\sigma_2)^2 + 6\mathsf{P}_1{}^1\mathsf{Q}_1{}^1\zeta_4]$$

$$-4\mathsf{P}_1{}^1\mathsf{Q}_1{}^1\Sigma e^2 f_i^s\omega_i^s - \mathsf{P}_1{}^1\mathsf{Q}_1{}^1\Sigma(f_i^s)^2\phi_i^s\} \ldots\ldots (15)$$

§ 6. *Surface Density of Condensation C; Energy CR.*

The region R being filled with positive volume density ρ, is concentrated along orthogonal tubes on to J, and there gives surface density δ.

To the first order, by (5),

$$\frac{dv}{d\tau} = p_0 d\sigma\left[1 - 2\tau\frac{\cos^2\beta\cos^2\gamma}{\Delta_1{}^2\Gamma_1{}^2}\left(\frac{1}{\Delta_1{}^2} + \frac{1}{\Gamma_1{}^2} - G\right)\right]$$

Integrating with respect to τ from the pear to J, we have as far as squares of small quantities

$$\delta = -p_0\rho\left[eS_3 + \Sigma f_i^s S_i + e^2\frac{\cos^2\beta\cos^2\gamma}{\Delta_1{}^2\Gamma_1{}^2}\left(\frac{1}{\Delta_1{}^2} + \frac{1}{\Gamma_1{}^2} - G\right)(S_3)^2\right]$$

It is now necessary to express δ/p_0 in surface harmonics. The first two terms are already in the required form; for the remainder let

$$\sum_0^\infty \eta_i^s S_i^s = \frac{\cos^2\beta\cos^2\gamma}{\Delta_1{}^2\Gamma_1{}^2}\left(\frac{1}{\Delta_1{}^2} + \frac{1}{\Gamma_1{}^2} - G\right)(S_3)^2$$

Multiplying both sides by $S_i^s\Phi\, d\theta\, d\phi$ and integrating, we have

$$\eta_i^s\phi_i^s = \cos^2\beta\cos^2\gamma\iint\frac{\Phi}{\Delta_1{}^2\Gamma_1{}^2}\left(\frac{1}{\Delta_1{}^2} + \frac{1}{\Gamma_1{}^2} - G\right)(S_3)^2 S_i^s\, d\theta\, d\phi$$

$$= \iint\Psi(S_3)^2 S_i^s\, d\theta\, d\phi = \omega_i^s$$

Therefore $\eta_i^s = \omega_i^s/\phi_i^s$, and vanishes unless i is even.

When $i = 0$, $\eta_0 = \dfrac{\omega_0}{\phi_0}$; and since by (7) $\phi_0 = 3$, and $\omega_0 = \iint\Psi(S_3)^2 d\theta\, d\phi = \sigma_2$, we have $\eta_0 = \tfrac{1}{3}\sigma_2$.

Hence we have

$$\delta = -p_0\rho\left[eS_3 + \tfrac{1}{3}e^2\sigma_2 + \sum_1^\infty\left(e^2\frac{\omega_i^s}{\phi_i^s} + f_i^s\right)S_i^s\right]$$

This is expressed in surface harmonics, the middle term being of order zero.

By (51) of "Harmonics" the internal potential of δ is

$$V_c = -3\frac{M}{k_0}\left(\frac{k}{k_0}\right)^2 \left\{e\mathfrak{P}_3(\nu)\,\mathfrak{Q}_3(\nu_0)\,S_3 + \tfrac{1}{3}e^2\sigma_2\mathfrak{P}_0\mathfrak{Q}_0 + \Sigma\left(e^2\frac{\omega_i^s}{\phi_i^s} + f_i^s\right)\mathfrak{P}_i^s(\nu)\,\mathfrak{Q}_i^s(\nu_0)\,S_i^s\right\}$$

We have $\mathfrak{P}_i^s(\nu) = \mathfrak{P}_i^s - \dfrac{\nu_0^2 - \nu^2}{2\nu_0}\dfrac{d\mathfrak{P}_i^s}{d\nu_0} = \mathfrak{P}_i^s - \tau\dfrac{\cos^2\beta\cos^2\gamma}{\sin\beta}\dfrac{1}{\Delta_1^2\Gamma_1^2}\dfrac{d\mathfrak{P}_i^s}{d\nu_0}$.

But before proceeding to use this I will introduce a new abridgment, and let

$$\left.\begin{aligned}\mathfrak{A}_i^s &= \mathfrak{P}_i^s\mathfrak{Q}_i^s\\ \mathfrak{B}_i^s &= \mathfrak{Q}_i^s\frac{d\mathfrak{P}_i^s}{d\nu_0}\end{aligned}\right\}\qquad\qquad(16)$$

Then
$$\mathfrak{P}_i^s(\nu)\,\mathfrak{Q}_i^s(\nu_0) = \mathfrak{A}_i^s - \tau\frac{\cos^2\beta\cos^2\gamma}{\sin\beta}\frac{\mathfrak{B}_i^s}{\Delta_1^2\Gamma_1^2}$$

and

$$V_c = -3\frac{M}{k_0}\left(\frac{k}{k_0}\right)^2\left\{e\mathfrak{A}_3S_3 + \tfrac{1}{3}e^2\sigma_2\mathfrak{A}_0 + \Sigma\left(e^2\frac{\omega_i^s}{\phi_i^s} + f_i^s\right)\mathfrak{A}_i^sS_i^s\right\}$$
$$+ 3\frac{M}{k_0}\left(\frac{k}{k_0}\right)^2\tau\frac{\cos^2\beta\cos^2\gamma}{\sin\beta}\left\{e\mathfrak{B}_3\frac{S_3}{\Delta_1^2\Gamma_1^2} + \Sigma\left(e^2\frac{\omega_i^s}{\phi_i^s} + f_i^s\right)\mathfrak{B}_i^s\frac{S_i^s}{\Delta_1^2\Gamma_1^2}\right\}$$

In order to find the energy CR we multiply V_c by the element of mass

$$\rho\,dv = \tfrac{1}{8}M\left(\frac{k}{k_0}\right)^3[\Phi - 2\tau\Psi]\,d\tau\,d\theta\,d\phi$$

and integrate throughout R.

Now

$$\frac{V_c\rho\,dv}{d\tau\,d\theta\,d\phi} = -\tfrac{3}{8}\frac{M^2}{k_0}\left(\frac{k}{k_0}\right)^5\left\{e\mathfrak{A}_3\Phi S_3 + \tfrac{1}{3}e^2\sigma_2\mathfrak{A}_0\Phi + \Sigma\left(e^2\frac{\omega_i^s}{\phi_i^s} + f_i^s\right)\mathfrak{A}_i^s\Phi S_i^s\right\}$$
$$+ \tfrac{6}{8}\frac{M^2}{k_0}\left(\frac{k}{k_0}\right)^5\tau\left\{e\mathfrak{A}_3\Psi S_3 + \tfrac{1}{3}e^2\sigma_2\mathfrak{A}_0\Psi + \Sigma\left(e^2\frac{\omega_i^s}{\phi_i^s} + f_i^s\right)\mathfrak{A}_i^s\Psi S_i^s\right\}$$
$$+ \tfrac{3}{8}\frac{M^2}{k_0}\left(\frac{k}{k_0}\right)^5\tau\frac{\cos^2\beta\cos^2\gamma}{\sin\beta}\left\{e\mathfrak{B}_3\frac{\Phi S_3}{\Delta_1^2\Gamma_1^2} + \Sigma\left(e^2\frac{\omega_i^s}{\phi_i^s} + f_i^s\right)\mathfrak{B}_i^s\frac{\Phi S_i^s}{\Delta_1^2\Gamma_1^2}\right\}$$

Let us integrate these three lines separately.

First integral

$$= 3\frac{M^2}{k_0}\left(\frac{k}{k_0}\right)^5\iint\left\{e\mathfrak{A}_3\Phi S_3 + \tfrac{1}{3}e^2\sigma_2\mathfrak{A}_0\Phi + \Sigma\left(e^2\frac{\omega_i^s}{\phi_i^s} + f_i^s\right)\mathfrak{A}_i^s\Phi S_i^s\right\}\times\{eS_3 + \Sigma f_i^sS_i^s\}\,d\theta\,d\phi$$

$$= 3\frac{M^2}{k_0}\left(\frac{k}{k_0}\right)^5\left\{e^2\mathfrak{A}_3\phi_3 + \Sigma\left[e^2f_i^s\frac{\omega_i^s}{\phi_i^s} + (f_i^s)^2\right]\mathfrak{A}_i^s\phi_i^s\right\}$$

Second integral

$$= 3\frac{M^2}{k_0}\left(\frac{k}{k_0}\right)^5 \iint \left\{e\mathfrak{A}_3 \Psi S_3 + \tfrac{1}{3}e^2\sigma_2\mathfrak{A}_0\Psi + \Sigma\left(e^2\frac{\omega_i^s}{\phi_i^s} + f_i^s\right)\mathfrak{A}_i^s\Psi S_i^s\right\}$$

$$\times \{e^2(S_3)^2 + 2\Sigma e f_i^s S_3 S_i^s\}\, d\theta d\phi$$

$$= 3\frac{M^2}{k_0}\left(\frac{k}{k_0}\right)^5 \left\{\tfrac{1}{3}e^4\mathfrak{A}_0(\sigma_2)^2 + \Sigma\left[e^4\frac{(\omega_i^s)^2}{\phi_i^s} + e^2 f_i^s\omega_i^s\right]\mathfrak{A}_i^s + 2\Sigma e^2 f_i^s\mathfrak{A}_3\omega_i^s\right\}$$

Third integral

$$= \tfrac{3}{2}\frac{M^2}{k_0}\left(\frac{k}{k_0}\right)^5 \frac{\cos^2\beta\cos^2\gamma}{\sin\beta}\iint\left\{e\mathfrak{B}_3\frac{\Phi S_3}{\Delta_1^2\Gamma_1^2} + \Sigma\left(e^2\frac{\omega_i^s}{\phi_i^s} + f_i^s\right)\mathfrak{B}_i^s\frac{\Phi S_i^s}{\Delta_1^2\Gamma_1^2}\right\}$$

$$\times \{e^2(S_3)^2 + 2\Sigma e f_i^s S_3 S_i^s\}\, d\theta\, d\phi$$

$$= \tfrac{3}{2}\frac{M^2}{k_0}\left(\frac{k}{k_0}\right)^5 \left\{\Sigma\left(e^4\frac{\omega_i^s}{\phi_i^s} + e^2 f_i^s\right)\rho_i^s\mathfrak{B}_i^s + 2\Sigma e^2 f_i^s\mathfrak{B}_3\rho_i^s\right\}$$

All the terms, excepting the first of the first integral, are of the fourth order, and in them we may put $(k/k_0)^5$ equal to unity.

Therefore

$$CR = 3\frac{M^2}{k_0}\left(\frac{k}{k_0}\right)^5 e^2\mathfrak{A}_3\phi_3$$

$$+ \tfrac{3}{2}\frac{M^2}{k_0}\left\{e^4\left[\tfrac{2}{3}\mathfrak{A}_0(\sigma_2)^2 + \Sigma\left(2\mathfrak{A}_i^s\omega_i^s + \mathfrak{B}_i^s\rho_i^s\right)\frac{\omega_i^s}{\phi_i^s}\right]\right.$$

$$\left. + \Sigma e^2 f_i^s\left[4\left(\mathfrak{A}_i^s + \mathfrak{A}_3\right)\omega_i^s + \left(\mathfrak{B}_i^s + 2\mathfrak{B}_3\right)\rho_i^s\right] + 2\Sigma\left(f_i^s\right)^2\mathfrak{A}_i^s\phi_i^s\right\} \ldots (17)$$

§ 7. *The Energy $\tfrac{1}{2}CC$; Result for $\tfrac{1}{2}JJ - JR + CR - \tfrac{1}{2}CC$.*

From the last section it appears that the potential of C at the surface, where $\tau = 0$, is

$$V_0 = -3\frac{M}{k_0}\left(\frac{k}{k_0}\right)^2\left\{e\mathfrak{A}_3 S_3 + \tfrac{1}{3}e^2\sigma_2\mathfrak{A}_0 + \Sigma\left(e^2\frac{\omega_i^s}{\phi_i^s} + f_i^s\right)\mathfrak{A}_i^s S_i^s\right\}$$

For the mass of an element of the surface density we have

$$\rho_0\delta d\sigma = -\tfrac{1}{3}M\left(\frac{k}{k_0}\right)^3\Phi\left\{eS_3 + \tfrac{1}{3}e^2\sigma_2 + \Sigma\left(e^2\frac{\omega_i^s}{\phi_i^s} + f_i^s\right)S_i^s\right\} d\theta d\phi$$

These are to be multiplied together and half the product is to be integrated. Then bearing in mind that $\iint\Phi\, d\theta d\phi = 3$, we have

$$\tfrac{1}{2}CC = \tfrac{3}{2}\frac{M}{k_0}\left(\frac{k}{k_0}\right)^5\left\{e^2\mathfrak{A}_3\phi_3 + \tfrac{1}{3}e^4(\sigma_2)^2\mathfrak{A}_0 + \Sigma\left(e^2\frac{\omega_i^s}{\phi_i^s} + f_i^s\right)^2\mathfrak{A}_i^s\phi_i^s\right\}$$

In the terms of the fourth order we put $(k/k_0)^5$ equal to unity; thus

$$\tfrac{1}{2}CC = \tfrac{3}{2}\frac{M}{k_0}\left(\frac{k}{k_0}\right)^5 e^2\mathfrak{A}_3\phi_3 + \tfrac{3}{2}\frac{M}{k_0}\left\{e^4\left[\tfrac{1}{3}(\sigma_2)^2\mathfrak{A}_0 + \Sigma\mathfrak{A}_i^s\frac{(\omega_i^s)^2}{\phi_i^s}\right] + 2\Sigma e^2 f_i^s\omega_i^s\mathfrak{A}_i^s + \Sigma(f_i^s)^2\mathfrak{A}_i^s\phi_i^s\right\}$$

Combining this with (17)

$$CR - \tfrac{1}{2}CC = \tfrac{3}{2}\frac{M}{k_0}\left(\frac{k}{k_0}\right)^5 e^2\mathfrak{A}_3\phi_3 + \tfrac{3}{2}\frac{M}{k_0}\left\{e^4\left[\tfrac{1}{3}\mathfrak{A}_0(\sigma_2)^2 + \Sigma(\mathfrak{A}_i^s\omega_i^s + \mathfrak{B}_i^s\rho_i^s)\frac{\omega_i^s}{\phi_i^s}\right] + \Sigma e^2 f_i^s[2(\mathfrak{A}_i^s + 2\mathfrak{A}_3)\omega_i^s + (\mathfrak{B}_i^s + 2\mathfrak{B}_3)\rho_i^s] + \Sigma(f_i^s)^2\mathfrak{A}_i^s\phi_i^s\right\} \quad \text{...(18)}$$

We are in a position to collect together all the results obtained up to this point. Now $\frac{1}{2}JJ - JR$, as given in (15), contains $\mathsf{P}_1{}^1\mathsf{Q}_1{}^1$, $\mathfrak{P}_0\mathfrak{Q}_0$; the latter of these is what is now written $\mathfrak{A}_0$, and since the ellipsoid is critical $\mathsf{P}_1{}^1\mathsf{Q}_1{}^1 = \mathfrak{P}_3\mathfrak{Q}_3 = \mathfrak{A}_3$.

Collecting terms we find that the terms of the second order disappear, and that

$$\tfrac{1}{2}JJ - JR + CR - \tfrac{1}{2}CC = \tfrac{3}{2}\frac{M^2}{k_0}\left\{e^4\left[\mathfrak{A}_3\{\tfrac{1}{3}(\sigma_2)^2 + 2\zeta_4\} + \Sigma(\mathfrak{A}_i^s\omega_i^s + \mathfrak{B}_i^s\rho_i^s)\frac{\omega_i^s}{\phi_i^s}\right] + \Sigma e^2 f_i^s[2\mathfrak{A}_i^s\omega_i^s + (\mathfrak{B}_i^s + 2\mathfrak{B}_3)\rho_i^s] - \Sigma(f_i^s)^2(\mathfrak{A}_3 - \mathfrak{A}_i^s)\phi_i^s\right\} \quad \text{...(19)}$$

The reader will recognise that the last term involves the coefficient of stability for the deformation S_i^s. It is important to note that if S_i^s is of odd order there is no term with coefficient $e^2 f_i^s$.

§ 8. *The Term* $-\frac{1}{2}Md^2\omega^2$.

In the Jacobian ellipsoid

$$\frac{b_1^2 + c_1^2}{3M_1}\omega^2 = \Psi + \tfrac{3}{2}a_1\frac{d\Psi}{da_1}$$

In the present notation this is

$$\frac{k^2\omega^2}{3M}\left(\frac{1+\cos^2\beta}{\sin^2\beta}\right) = \frac{2}{k}(\mathfrak{P}_0\mathfrak{Q}_0 - \tfrac{3}{2}\mathsf{P}_1{}^1\mathsf{Q}_1{}^1) = \frac{2}{k}(\mathfrak{A}_0 - \tfrac{3}{2}\mathfrak{A}_3)$$

Hence

$$-\tfrac{1}{2}Md^2\omega^2 = -\tfrac{1}{2}(Md)^2\frac{\omega^2}{M} = -\frac{3\sin^2\beta}{1+\cos^2\beta}(\mathfrak{A}_0 - \tfrac{3}{2}\mathfrak{A}_3)\frac{(Md)^2}{k^3}$$

I now make the following definition

$$S_1 = \sin\theta(1 - \kappa'^2\cos^2\phi)^{\frac{1}{2}}$$

so that

$$z = kS_1$$

Then
$$Md = \int_{j-r} z\rho\,dv = \int_j z\rho\,dv - \int_r z\rho\,dv = -\int_r z\rho\,dv$$
$$= -M\left(\frac{k}{k_0}\right)^3 k \iiint [\Phi - 2\tau\Psi]\, S_1\, d\tau\, d\theta\, d\phi$$
$$= M\left(\frac{k}{k_0}\right)^3 k \iint [\Phi\,(eS_3 + \Sigma f_i^s S_i^s) + \Psi e^2 (S_3)^2]\, S_1\, d\theta\, d\phi$$
$$= M\left(\frac{k}{k_0}\right)^3 k f_1 \phi_1$$

Therefore to the required order
$$-\tfrac{1}{2} M d^2 \omega^2 = -\frac{3M^2}{k_0}\,\frac{\sin^2\beta}{1+\cos^2\beta}\,(\mathfrak{A}_0 - \tfrac{3}{2}\mathfrak{A}_3)\,(f_1\phi_1)^2 \quad \text{..........(20)}$$

We again note that this term in the energy does not introduce any term with a coefficient $e^2 f_1$. Hence thus far the whole energy for harmonic deformations of odd order is of the form $\mathrm{L}e^4 + \mathrm{M}\,(f_i^s)^2$.

§ 9. *Double Layers.*

It remains to determine the value of $\frac{1}{2}DD$ in the energy, and for this purpose we must consider double layers, according to the ingenious method devised by M. Poincaré.

Let a closed surface S be intersected at every point by a member of a family of curves, and let α be the angle between the curve and the outward normal at any point. At every point of S measure along the curve an infinitesimal arc τ, and let τ be a function of the two co-ordinates which determine position on S. The extremities of these arcs define a second surface S', and every element of area $d\sigma$ of S has its corresponding element $d\sigma'$ on S'. Suppose that S is coated with surface density δ, and that S' is coated with surface density $-\delta'$, where $\delta\, d\sigma = \delta' d\sigma'$. The system SS' may then be called a double layer, and its total mass is zero. We are to discuss the potential of such a system.

Let $U(+)$ and $U(-)$ be the external and internal potentials of density δ on S, and U_0 their common value at a point P of S. At P take a system of rectangular axes, n being along the outward normal, and s and t mutually at right angles in the tangent plane.

In the neighbourhood of P
$$U(+) = U_0 + n\frac{dU}{dn}(+) + s\frac{dU}{ds}(+) + t\frac{dU}{dt}(+) \ldots$$
$$U(-) = U_0 + n\frac{dU}{dn}(-) + s\frac{dU}{ds}(-) + t\frac{dU}{dt}(-) \ldots$$

In the first of these n is necessarily positive, in the second negative.

Now $\frac{dU}{ds}(+) = \frac{dU}{ds}(-) = \frac{dU}{ds}$; and the like holds for the differentials with respect to t.

Also by Poisson's equation

$$\frac{dU}{dn}(-) - \frac{dU}{dn}(+) = 4\pi\delta$$

Let PP' be one of the family of curves whereby the double layer is defined, and let P' lie on S', so that PP' is τ. By the definition of α the normal elevation of S' above S is $\tau \cos \alpha$.

Let v, v' be the potentials of the double layer at P and at P'.

The potential of S' at P' differs infinitely little in magnitude, but is of the opposite sign from that of S at P; it is therefore $-U_0$. The point P' lies on the positive side of S at a point whose co-ordinates may be taken to be

$$n = \tau \cos \alpha, \quad s = \tau \sin \alpha, \quad t = 0$$

Therefore the potential of S at P' is

$$U_0 + \tau \cos \alpha \frac{dU}{dn}(+) + \tau \sin \alpha \frac{dU}{ds}$$

Therefore
$$v' = \tau \cos \alpha \frac{dU}{dn}(+) + \tau \sin \alpha \frac{dU}{ds}$$

Again the potential of S at P is U_0, and since P lies on the negative side of S' and has co-ordinates relatively to the n, s, t axes at P' given by

$$n = -\tau \cos \alpha, \quad s = -\tau \sin \alpha, \quad t = 0$$

since further the density on S' is negative, we have

$$v = \tau \cos \alpha \frac{dU}{dn}(-) + \tau \sin \alpha \frac{dU}{ds}$$

Therefore

$$v - v' = \tau \cos \alpha \left[\frac{dU}{dn}(-) - \frac{dU}{dn}(+)\right] = 4\pi\tau\delta \cos \alpha$$

The differential with respect to n of the potential of S falls abruptly by $4\pi\delta$ as we cross S normally from the negative to the positive side; and the differential of the potential of S' rises abruptly by the same amount as we pass on across S'. It follows that dv/dn on the inside of S is continuous with its value on the outside of S'.

The surface S to which this theorem is to be applied is a slightly deformed ellipsoid, and the curves are the intersection of the two quadrics confocal with the ellipsoid which is deformed. The curves start normally to the ellipsoid, and where they meet S the angle α will be proportional to the deformation whereby S is derived from the ellipsoid. It follows that $\cos\alpha$ will only differ from unity by a term proportional to the square of the deformation, and as it is only necessary to retain terms of the order of the first power of the deformation, we may treat $\cos\alpha$ as unity.

We thus have the result $\quad v - v' = 4\pi\tau\delta$

Suppose the curve PP' produced both ways, and that M_0, M_1 are two points on it, either both on the same side or on opposite sides of the double layer.

Let M_0M_1 be equal to ζ, let ζ be measured in the same direction as n, and let ζ be a small quantity whose first power is to be retained in the results.

Let v_0, v_1 be the potentials of the double layer at M_0 and M_1 respectively.

When ζ does not cut the layer we have

$$v_0 - v_1 = -\zeta\frac{dv}{dn}$$

and when it does cut the layer

$$v_0 - v_1 = 4\pi\tau\delta - \zeta\frac{dv}{dn}$$

In the application which I shall make of this result the surface S' will actually be inside S. Then v_0 will denote the potential at any point not lying in the infinitely small space between S and S', and v_1 is the potential at a point more towards the inside of the ellipsoid by a distance ζ; δ is the surface density on the external surface S and τ is measured inwards. If then we still choose to measure n outwards, as I shall do, our formula becomes

$$v_0 - v_1 - \zeta\frac{dv}{dn} = 4\pi\tau\delta \text{ or } 0$$

according as ζ does or does not cut the double layer.

It may be well to remark that v being proportional to $\tau\delta$, $\zeta dv/dn$ is small compared with $4\pi\tau\delta$. It is also important to notice that the term $4\pi\tau\delta$ is independent of the form of the surface, and that dv/dn will be the same to the first order of small quantities for a slightly deformed ellipsoid as for the ellipsoid itself.

We have now to apply these results to our problem.

The position of a point in the region R may be defined by the distance measured inwards from J along one of the curves orthogonal to J. The surface of the pear as defined in this way is given by ϵ, a function of θ and ϕ. Any point on a curve may then be defined by $s\epsilon$, where s is a proper fraction. If s is the same at every point the surface s is a deformed ellipsoid; $s=1$ gives the pear and $s=0$ the ellipsoid J.

If $d\sigma$ is an element of area of J, the corresponding element on the surface s will be $(1-\lambda\epsilon s)\,d\sigma$. The value of λ will be determined hereafter, and it is only necessary to remark that it is positive because the areas must decrease as we travel inwards.

Let s and $s+ds$ be two adjacent surfaces; then the mass of negative density enclosed between them in the tube of which $(1-\lambda\epsilon s)\,d\sigma$ and $\{1-\lambda\epsilon(s+ds)\}\,d\sigma$ are the ends is $-\rho\epsilon(1-\lambda\epsilon s)\,d\sigma ds$. If this element of mass be regarded as surface density on s, that surface density is clearly $-\rho\epsilon ds$. If the same element of mass were carried along the orthogonal tube and deposited as surface density on J, that surface density would be $-\rho\epsilon(1-\lambda\epsilon s)$. The sum for all values of s of all such transportals would constitute the condensation $-C$ already considered.

The double system D consists of the volume density $-\rho$ in R, and the positive condensation $+C$ on J, the total mass being zero.

Let z, a proper fraction, define a surface between J and the pear. Consider one of the orthogonal curves, and let V_0 be the potential of D at the point P where the curve leaves J and V_z the potential at the point Q where it cuts z. Then I require to find $V_0 - V_z$.

Since s denotes a surface intermediate between J and the pear, $\frac{d}{ds}(V_0 - V_z)\,ds$ is the excess of the potential at P above that of Q of surface density $-\rho\epsilon ds$ on s and surface density $+\rho\epsilon(1-\lambda\epsilon s)\,ds$ on J. Such a system is a double layer, but there is a finite distance between the two surfaces, and the form of $\frac{d}{ds}(V_0 - V_z)$ will clearly be different according as z is greater or less than s.

The arc ϵs may be equally divided by a large number of surfaces, and we may take t to define any one of them. Now we may clothe each intermediate surface t with equal and opposite surface densities $\pm\rho\epsilon[1-\lambda\epsilon(s-t)]\,dt$.

The density $+\rho\epsilon[1-\lambda\epsilon(s-t)]dt$ on t, together with $-\rho\epsilon[1-\lambda\epsilon(s-t-dt)]dt$ on $t+dt$, constitute an infinitesimal double layer; and since the positive density on each t surface may be coupled with the negative density on the next interior surface, the finite double layer may be built up from a number of infinitesimal double layers. Hence $\frac{d^2}{dsdt}(V_0 - V_z)\,dtdt$ is the excess of the potential at P above that at Q of an infinitesimal double layer of thickness ϵdt, and with surface density $\rho\epsilon[1-\lambda\epsilon(s-t)]\,dt$ on its exterior surface.

We may now apply the result $v_0 - v_1 - \zeta\frac{dv}{dn} = 4\pi\delta\tau$ or 0, according as ζ does or does not cut the double layer, and it is clear that

$$\frac{d^2}{dsdt}\left[V_0 - V_z - \epsilon z\frac{dV}{dn}\right] = 4\pi\rho\epsilon^2[1-\lambda\epsilon(s-t)] \text{ or } 0$$

according as z is greater or less than t.

In the next place, we must integrate this from $t=s$ to $t=0$, and the result will have two forms.

First, suppose $z>s$; then for all the values of t, $z>t$, and the first alternative holds good. Therefore

$$\frac{d}{ds}\left(V_0 - V_z - \epsilon z\frac{dV}{dn}\right) = 4\pi\rho\epsilon^2\left[s - \tfrac{1}{2}\lambda\epsilon s^2\right]$$

Secondly, suppose $z<s$; then from $t=s$ to $t=z$, $z<t$ and the second alternative holds, while from $t=z$ to $t=0$, $z>t$ and the first holds. Therefore

$$\frac{d}{ds}\left(V_0 - V_z - \epsilon z\frac{dV}{dn}\right) = 4\pi\rho\epsilon^2\left[z - \lambda\epsilon\left(sz - \tfrac{1}{2}z^2\right)\right]$$

We have now to integrate again from $s=1$ to $s=0$.

From $s=1$ to $s=z$, $z<s$ and the second form is applicable; from $s=z$ to $s=0$, $z>s$ and the first form applies.

Therefore

$$\begin{aligned} V_0 - V_z - \epsilon z\frac{dV}{dn} &= 4\pi\rho\epsilon^2\int_z^1\left[z - \lambda\epsilon\left(sz - \tfrac{1}{2}z^2\right)\right]ds + 4\pi\rho\epsilon^2\int_0^z\left[s - \tfrac{1}{2}\lambda\epsilon s^2\right]ds \\ &= 4\pi\rho\epsilon^2\left\{z(1-z) - \lambda\epsilon\left[\tfrac{1}{2}z(1-z^2) - \tfrac{1}{2}z^2(1-z)\right] + \tfrac{1}{2}z^2 - \tfrac{1}{6}\lambda\epsilon z^3\right\} \\ &= 2\pi\rho\epsilon^2\left\{2z - z^2 - \lambda\epsilon\left(z - z^2 + \tfrac{1}{3}z^3\right)\right\} \end{aligned}$$

Finally, we have to multiply $-\frac{1}{2}(V_0 - V_z)$ by an element of negative mass at the point defined by z and integrate throughout R. The physical meaning of this integral will be considered subsequently.

We have already seen that such an element of mass is given by

$$-\rho\,dv = -\rho\epsilon(1-\lambda\epsilon z)\,d\sigma\,dz$$

and the limits of integration are $z=1$ to $z=0$.

Therefore

$$\begin{aligned} &\tfrac{1}{2}\int(V_0 - V_z)\rho\,dv \\ &= \pi\rho^2\iint\epsilon^3(1-\lambda\epsilon z)\left\{2z - z^2 - \lambda\epsilon\left(z - z^2 + \tfrac{1}{3}z^3\right)\right\}dz\,d\sigma + \tfrac{1}{2}\rho\iint\epsilon^2 z(1-\lambda\epsilon z)\frac{dV}{dn}\,dz\,d\sigma \end{aligned}$$

In this expression we neglect terms of the order ϵ^5 and note that $\epsilon^3 z^2\dfrac{dV}{dn}$ is of that order.

Thus

$$\begin{aligned} \tfrac{1}{2}\int(V_0 - V_z)\rho\,dv &= \pi\rho^2\iint\epsilon^3\left[2z - z^2 - \lambda\epsilon\left(z + z^2 - \tfrac{2}{3}z^3\right)\right]dz\,d\sigma \\ &\qquad + \tfrac{1}{2}\rho\iint\epsilon^2 z\frac{dV}{dn}\,dz\,d\sigma\ (z=1\text{ to }0) \\ &= \tfrac{2}{3}\pi\rho^2\int\epsilon^3(1-\lambda\epsilon)\,d\sigma + \tfrac{1}{4}\rho\int\epsilon^2\frac{dV}{dn}\,d\sigma \end{aligned}$$

the integrals being taken all over the surface of the ellipsoid.

We must now consider the meaning of the integral $\frac{1}{2}\int(V_0 - V_z)\rho dv$.

Let P be a point on J and Q a point in R on the same orthogonal curve.

Let $-U$ be the potential at Q of the density $-\rho$ throughout R, and $-U_0$ its value at P.

Let δ be the surface density of the positive concentration on J, W its potential at Q, and W_0 its value at P.

The lost energy of the double system consisting of $-\rho$ throughout R, and δ on J is

$$\frac{1}{2}\int U\rho dv + \frac{1}{2}\int W_0 \delta d\sigma - \frac{1}{2}\int U_0 \delta d\sigma - \frac{1}{2}\int W\rho dv$$

This is equal to $$\frac{1}{2}\int(U - W)\rho dv - \frac{1}{2}\int(U_0 - W_0)\delta d\sigma$$

Consider the triple integral $\iiint(U_0 - W_0)\rho dv$. Here $dv = \epsilon(1 - \lambda\epsilon s)d\sigma ds$; also $U_0 - W_0$ is not a function of s, and the limits of s are 1 to zero. Therefore

$$\iiint(U_0 - W_0)\rho dv = \iint(U_0 - W_0)\left[\int_0^1 \epsilon(1 - \lambda\epsilon s)\rho ds\right]d\sigma$$

But $\int_0^1 \epsilon(1 - \lambda\epsilon s)\rho ds$ is equal to δ the surface density of concentration. Therefore

$$\iint[U_0 - W_0]\delta d\sigma = \iiint(U_0 - W_0)\rho dv$$

We may now revert to the Gaussian notation with single integral sign, and we see that the lost energy of the system is

$$\frac{1}{2}\int[(W_0 - U_0) - (W - U)]\rho dv$$

But $W - U$ is the potential of the double system at Q, and is therefore V_z; and $W_0 - U_0$ is the potential of the double system at P, and is therefore V_0.

Accordingly the lost energy

$$\frac{1}{2}DD = \frac{1}{2}\int(V_0 - V_z)\rho dv$$

$$= \frac{2}{3}\pi\rho^2\int(\epsilon^3 - \lambda\epsilon^4)d\sigma + \frac{1}{4}\rho\int\epsilon^2\frac{dV}{dn}d\sigma \quad \text{......(21)}$$

§ 10. *Determination of ϵ and λ.*

ϵ is the arc of the orthogonal curve from J to the pear.

The arc of outward normal is connected with p and our variable τ by the equation

$$-dn = -\frac{k^2}{p}\nu d\nu = \frac{p_0^2}{p} d\tau$$

It follows that

$$\epsilon = p_0 \int \frac{p_0}{p} d\tau, \text{ integrated from } J \text{ to the pear}$$

By (50) of "Harmonics," with the notation of § 3 of this paper

$$\frac{k}{p} = \frac{(\nu^2-\mu^2)^{\frac{1}{2}}\left(\nu^2 - \frac{1-\beta\cos 2\phi}{1-\beta}\right)^{\frac{1}{2}}}{\nu(\nu^2-1)^{\frac{1}{2}}\left(\nu^2-\frac{1+\beta}{1-\beta}\right)^{\frac{1}{2}}} = \frac{\sin\beta}{\cos\beta\cos\gamma} \frac{\Delta_1\Gamma_1\left(1-\frac{\tau_1}{\Delta_1^2}\right)^{\frac{1}{2}}\left(1-\frac{\tau_1}{\Gamma_1^2}\right)^{\frac{1}{2}}}{(1-\tau_1)^{\frac{1}{2}}(1-\tau_1\sec^2\beta)^{\frac{1}{2}}(1-\tau_1\sec^2\gamma)^{\frac{1}{2}}}$$

Therefore

$$\frac{p_0}{p} = \frac{\left(1-\frac{\tau_1}{\Delta_1^2}\right)^{\frac{1}{2}}\left(1-\frac{\tau_1}{\Gamma_1^2}\right)^{\frac{1}{2}}}{(1-\tau_1)^{\frac{1}{2}}(1-\tau_1\sec^2\beta)^{\frac{1}{2}}(1-\tau_1\sec^2\gamma)^{\frac{1}{2}}} = 1 - \tfrac{1}{2}\tau_1\left(\frac{1}{\Delta_1^2}+\frac{1}{\Gamma_1^2}-2G\right)$$

$$= 1 - \tau\frac{\cos^2\beta\cos^2\gamma}{\Delta_1^2\Gamma_1^2}\left(\frac{1}{\Delta_1^2}+\frac{1}{\Gamma_1^2}-2G\right)$$

Integrating this from J to the pear

$$\epsilon = -p_0\left[eS_3 + \sum_1^\infty f_i^s S_i^s + \tfrac{1}{2}e^2\frac{\cos^2\beta\cos^2\gamma}{\Delta_1^2\Gamma_1^2}\left(\frac{1}{\Delta_1^2}+\frac{1}{\Gamma_1^2}-2G\right)(S_3)^2\right] \quad \text{...(22)}$$

We have, moreover, by the formula before integration

$$-dn = p_0\left[1-\tau\frac{\cos^2\beta\cos^2\gamma}{\Delta_1^2\Gamma_1^2}\left(\frac{1}{\Delta_1^2}+\frac{1}{\Gamma_1^2}-2G\right)\right]d\tau$$

Also to the order zero $-n = p_0\tau$.

Since $-n$ is what was denoted in § 9 by ϵs, the element of volume is $-(1+\lambda n)\,dn\,d\sigma$, and this is equal to

$$p_0[1-\lambda p_0\tau]\left[1-\tau\frac{\cos^2\beta\cos^2\gamma}{\Delta_1^2\Gamma_1^2}\left(\frac{1}{\Delta_1^2}+\frac{1}{\Gamma_1^2}-2G\right)\right]d\sigma\,d\tau$$

or

$$p_0\left[1-\tau\left\{\lambda p_0+\frac{\cos^2\beta\cos^2\gamma}{\Delta_1^2\Gamma_1^2}\left(\frac{1}{\Delta_1^2}+\frac{1}{\Gamma_1^2}-2G\right)\right\}\right]d\sigma\,d\tau$$

But by (5) the element of volume is

$$p_0\left[1-\frac{\tau\cos^2\beta\cos^2\gamma}{\Delta_1^2\Gamma_1^2}\left(\frac{2}{\Delta_1^2}+\frac{2}{\Gamma_1^2}-2G\right)\right]d\sigma\,d\tau$$

Equating coefficients of τ in the two expressions we find

$$\lambda = \frac{\cos^2\beta\cos^2\gamma}{p_0\Delta_1^2\Gamma_1^2}\left(\frac{1}{\Delta_1^2}+\frac{1}{\Gamma_1^2}\right) \quad \text{......................(23)}$$

§ 11. *The Energy* $\frac{2}{3}\pi\rho^2\int\epsilon^3(1-\lambda\epsilon)\,d\sigma$.

From (22) and (23) we have

$$\epsilon^3 = -p_0^3\left[e^3(S_3)^3 + 3\Sigma e^2 f_i^s(S_3)^2 S_i^s + \tfrac{3}{2}e^4\frac{\cos^2\beta\cos^2\gamma}{\Delta_1^2\Gamma_1^2}\left(\frac{1}{\Delta_1^2}+\frac{1}{\Gamma_1^2}-2G\right)(S_3)^4\right]$$

$$\lambda\epsilon^4 = p_0^3 e^4\frac{\cos^2\beta\cos^2\gamma}{\Delta_1^2\Gamma_1^2}\left(\frac{1}{\Delta_1^2}+\frac{1}{\Gamma_1^2}\right)(S_3)^4$$

So that

$$\epsilon^3(1-\lambda\epsilon) = -p_0^3\left[e^3(S_3)^3 + 3\Sigma e^2 f_i^s(S_3)^2 S_i^s + e^4\frac{\cos^2\beta\cos^2\gamma}{\Delta_1^2\Gamma_1^2}\left\{\tfrac{5}{2}\left(\frac{1}{\Delta_1^2}+\frac{1}{\Gamma_1^2}\right)-3G\right\}(S_3)^4\right]$$

Again from (6)

$$\tfrac{2}{3}\pi\rho^2 p_0^3 d\sigma = \tfrac{2}{3}\pi\rho^2\,.\,\tfrac{1}{8}\frac{M}{\rho}\left(\frac{k}{k_0}\right)^3\Phi\frac{k^2\cos^2\beta\cos^2\gamma}{\sin^2\beta}\frac{d\theta\,d\phi}{\Delta_1^2\Gamma_1^2}$$

$$= \tfrac{1}{16}\frac{M^2}{k_0}\left(\frac{k}{k_0}\right)^5\cos\beta\cos\gamma\sin\beta\,\frac{\Phi}{\Delta_1^2\Gamma_1^2}\,d\theta\,d\phi$$

Therefore

$$\tfrac{2}{3}\pi\rho^2\epsilon^3(1-\lambda\epsilon)\,d\sigma = -\tfrac{1}{16}\frac{M^2}{k_0}\left(\frac{k}{k_0}\right)^5\cos\beta\cos\gamma\sin\beta\left[e^3\frac{\Phi(S_3)^3}{\Delta_1^2\Gamma_1^2}\right.$$

$$\left. + 3\Sigma e^2 f_i^s\frac{\Phi(S_3)^2 S_i^s}{\Delta_1^2\Gamma_1^2} + e^4\frac{\cos^2\beta\cos^2\gamma}{\Delta_1^4\Gamma_1^4}\Phi\left\{\tfrac{5}{2}\left(\frac{1}{\Delta_1^2}+\frac{1}{\Gamma_1^2}\right)-3G\right\}(S_3)^4\right]d\theta\,d\phi$$

When this is integrated we may put $(k/k_0)^5$ equal to unity. In the integral the first term vanishes, and the second term gives

$$-\tfrac{3}{2}\frac{M^2}{k_0}\frac{\sin^2\beta}{\cos\beta\cos\gamma}\Sigma e^2 f_i^s\rho_i^s$$

In the third term we substitute for Φ its value and have

$$-\frac{3}{\pi}\frac{M^2}{k_0}\frac{\cos^3\beta\cos^3\gamma\sin\beta}{\sin^2\gamma}e^4\int_0^{\frac{1}{2}\pi}\int_0^{\frac{1}{2}\pi}\frac{1}{\Delta_1^2\Gamma_1^2}\left(\frac{1}{\Gamma_1^2}-\frac{1}{\Delta_1^2}\right)$$

$$\times\left\{\tfrac{5}{2}\left(\frac{1}{\Gamma_1^2}+\frac{1}{\Delta_1^2}\right)-3G\right\}(S_3)^4\frac{d\theta\,d\phi}{\Delta\Gamma}$$

which is equal to

$$-\tfrac{1}{2}\frac{M^2}{k_0}e^4\frac{6}{\pi}\frac{\cos^3\beta\cos^3\gamma\sin\beta}{\sin^2\gamma}\int_0^{\frac{1}{2}\pi}\int_0^{\frac{1}{2}\pi}\left\{\tfrac{5}{2}\left(\frac{1}{\Gamma_1^6\Delta_1^2}-\frac{1}{\Gamma_1^2\Delta_1^6}\right)\right.$$

$$\left.-3G\left(\frac{1}{\Gamma_1^4\Delta_1^2}-\frac{1}{\Gamma_1^2\Delta_1^4}\right)\right\}(S_3)^4\frac{d\theta\,d\phi}{\Delta\Gamma}$$

By the definition (8) this is equal to $-\frac{1}{2}\frac{M^2}{k_0}e^4\sigma_4$.

Hence the required term in the energy is

$$\tfrac{3}{2}\frac{M^2}{k_0}\left[-\frac{\sin^2\beta}{\cos\beta\cos\gamma}\Sigma e^2 f_i^s\rho_i^s - \tfrac{1}{3}e^4\sigma_4\right]\text{..................(24)}$$

§ 12. *The Energy* $\frac{1}{4}\rho\int \epsilon^2 \frac{dV}{dn}\,d\sigma$.

It is first necessary to determine dV/dn.

Suppose that the ellipsoid J is coated with surface density δ, and that a second surface is drawn inside J at an infinitesimal distance τ, and coated with negative surface density $-\delta'$, so that the two form a double layer. Then $\tau\delta$ being a function of the two angular co-ordinates on the ellipsoid may be expanded in surface harmonics; suppose then that

$$\tau\delta = \sum_0^\infty h_i^s S_i^s$$

Consider the two functions

$$V_e = \Sigma 4\pi h_i^s (\nu_0^2 - 1)^{\frac{1}{2}} \left(\nu_0^2 - \frac{1+\beta}{1-\beta}\right)^{\frac{1}{2}} \frac{d\mathfrak{P}_i^s(\nu_0)}{d\nu_0} \mathfrak{Q}_i^s(\nu) S_i^s, \text{ for external space}$$

$$V_i = \ldots\ldots\ldots\ldots\ldots\ldots\ldots\ldots\ldots \mathfrak{P}_i^s(\nu) \frac{d\mathfrak{Q}_i^s(\nu_0)}{d\nu_0} S_i^s, \text{ for internal space}$$

Since these functions are solid harmonics, the matter of which V_e and V_i are the potentials is entirely confined to the surface of the ellipsoid, and since they are not continuous with one another, the ellipsoid must be a double layer.

Now
$$\mathfrak{Q}_i^s(\nu) = \mathfrak{P}_i^s(\nu) \int_\nu^\infty \frac{d\nu}{[\mathfrak{P}_i^s(\nu)]^2 (\nu^2-1)^{\frac{1}{2}} \left(\nu^2 - \frac{1+\beta}{1-\beta}\right)^{\frac{1}{2}}}$$

and therefore

$$\mathfrak{Q}_i^s(\nu_0) \frac{d\mathfrak{P}_i^s(\nu_0)}{d\nu_0} - \mathfrak{P}_i^s(\nu_0) \frac{d\mathfrak{Q}_i^s(\nu_0)}{d\nu_0} = \frac{1}{(\nu_0-1)^{\frac{1}{2}} \left(\nu_0^2 - \frac{1+\beta}{1-\beta}\right)^{\frac{1}{2}}}$$

Hence at the surface of the ellipsoid

$$V_e - V_i = \Sigma 4\pi h_i^s S_i^s = 4\pi\tau\delta$$

But this is the law found in § 9 for the change of potential in crossing a double layer, and hence V_e, V_i are the external and internal potentials of the double layer $\tau\delta$.

Since
$$\frac{d}{dn} = \frac{p}{k^2}\frac{d}{\nu d\nu}$$

$$\frac{dV_e}{dn} = \frac{dV_i}{dn} = \frac{dV}{dn} = \sum_0^\infty \frac{4\pi p_0}{k^2 \nu_0} (\nu_0^2-1)^{\frac{1}{2}} \left(\nu_0^2 - \frac{1+\beta}{1-\beta}\right)^{\frac{1}{2}} h_i^s \frac{d\mathfrak{P}_i^s}{d\nu_0}\frac{d\mathfrak{Q}_i^s}{d\nu_0} S_i^s \ldots (25)$$

This result will hold good to the first order of small quantities if the surface be a slightly deformed ellipsoid, such as was the surface defined by t in § 9.

In the elementary double layer t the density was $\rho\epsilon[1 - \lambda\epsilon(s-t)]\,dt$, and the thickness was ϵdt, so that the thickness multiplied by the density was

$\rho\epsilon^2 [1 - \lambda\epsilon (s-t)]\, dt dt$. Since, however, we only need this to the first order, we may take it as $\rho\epsilon^2 dt dt$. It will now be convenient to change the meaning of h_i^s to some extent, and to write

$$\epsilon^2 = \sum_0^\infty h_i^s S_i^s$$

Thus for the elementary double layer we have

$$\tau\delta = \rho\, dt dt \sum_0^\infty h_i^s S_i^s$$

It follows that in applying the formula (25) to determine $\frac{dV}{dn}$ for the double system D, we may say that

$$\frac{d^2}{ds\,dt}\frac{dV}{dn} = \sum_0^\infty \frac{4\pi p_0 \rho}{k^2 \nu_0} (\nu_0^2 - 1)^{\frac{1}{2}} \left(\nu_0^2 - \frac{1+\beta}{1-\beta}\right)^{\frac{1}{2}} h_i^s \frac{d\mathfrak{P}_i^s}{d\nu_0} \frac{d\mathfrak{Q}_i^s}{d\nu_0} S_i^s$$

Since the right-hand side does not contain t, we have only to consider the integral

$$\int_0^1 \int_0^s ds\,dt = \int_0^1 s\,ds = \tfrac{1}{2}$$

Thus, for the system D,

$$\frac{dV}{dn} = \sum_0^\infty \frac{2\pi p_0 \rho}{k^2 \nu_0} (\nu_0^2 - 1)^{\frac{1}{2}} \left(\nu_0^2 - \frac{1+\beta}{1-\beta}\right)^{\frac{1}{2}} h_i^s \frac{d\mathfrak{P}_i^s}{d\nu_0} \frac{d\mathfrak{Q}_i^s}{d\nu_0} S_i^s \quad \text{......(26)}$$

This result may also be obtained as follows:—To the first order we may concentrate the negative density in the region R on a surface bisecting that region. We may then consider the positive concentration C on J, and the negative concentration on the bisecting surface as an infinitesimal double layer of thickness $\frac{1}{2}\epsilon$. We have seen that the surface density $+C$ is $-p\rho e S_3$, and that $\epsilon = -peS_3$ (in both cases to the first order only). Thus the density δ of $+C$ is $\rho\epsilon$, and the thickness τ of our layer is $\frac{1}{2}\epsilon$; the product therefore $\tau\delta$ is $\frac{1}{2}\rho\epsilon^2$.

Hence $\tau\delta = \frac{1}{2}\rho\epsilon^2 = \frac{1}{2}\rho \sum_0^\infty h_i^s S_i^s$, and thus we arrive at the same result as before.

I now introduce an abridged notation analogous to that used previously, and write

$$\mathfrak{D}_i^s = \frac{d\mathfrak{P}_i^s}{d\nu_0} \frac{d\mathfrak{Q}_i^s}{d\nu_0}$$

We then have by (26)

$$\frac{dV}{dn} = \sum_0^\infty \frac{2\pi p_0 \rho \cos\beta \cos\gamma}{k^2 \sin\beta} h_i^s \mathfrak{D}_i^s S_i^s \quad \text{..................(26)}$$

where

$$\epsilon^2 = \sum_0^\infty h_i^s S_i^s$$

By (22) to the first order

$$\epsilon^2 = e^2 p_0^2 (S_3)^2 = e^2 k^2 \frac{\cos^2\beta \cos^2\gamma}{\sin^2\beta} \frac{(S_3)^2}{\Delta_1^2 \Gamma_1^2}$$

Assume then
$$\frac{\cos^2\beta\cos^2\gamma}{\sin\beta}\frac{(S_3)^2}{\Delta_1^2\Gamma_1^2} = \sum_0^\infty \zeta_i^s S_i^s$$

Multiplying by ΦS_i^s and integrating, we have

$$\rho_i^s = \zeta_i^s \phi_i^s$$

Hence $\epsilon^2 = \dfrac{e^2k^2}{\sin\beta}\sum_0^\infty \dfrac{\rho_i^s}{\phi_i^s} S_i^s$, and therefore $h_i^s = \dfrac{e^2k^2}{\sin\beta}\dfrac{\rho_i^s}{\phi_i^s}$

Substituting in (26)

$$\frac{dV}{dn} = \sum_0^\infty 2\pi p_0 \rho e^2 \frac{\cos\beta\cos\gamma}{\sin^2\beta}\frac{\rho_i^s}{\phi_i^s}\mathfrak{D}_i^s S_i^s$$

$$= \sum_0^\infty \tfrac{3}{2}\frac{M}{k^3} e^2 p_0 \sin\beta \frac{\rho_i^s}{\phi_i^s}\mathfrak{D}_i^s S_i^s$$

Now
$$\epsilon^2 = \frac{e^2k^2}{\sin\beta}\sum_0^\infty \frac{\rho_i^s}{\phi_i^s} S_i^s$$

and
$$\epsilon^2\frac{dV}{dn} = \tfrac{3}{2}\frac{M}{k_0} e^4 p_0 \left(\sum_0^\infty \frac{\rho_i^s}{\phi_i^s}\mathfrak{D}_i^s S_i^s\right)\left(\sum_0^\infty \frac{\rho_i^s}{\phi_i^s} S_i^s\right)$$

Since on integration the terms involving products of unlike harmonics will disappear, we have, as far as material,

$$\epsilon^2\frac{dV}{dn} = \tfrac{3}{2}\frac{M}{k_0} e^4 p_0 \sum_0^\infty \left(\frac{\rho_i^s}{\phi_i^s}\right)^2 \mathfrak{D}_i^s (S_i^s)^2$$

Now
$$\tfrac{1}{4}\rho p_0 d\sigma = \tfrac{1}{32} M \left(\frac{k}{k_0}\right)^3 \Phi\, d\theta\, d\phi$$

Since the term which is being determined is of the fourth order in e, we may put $k/k_0 = 1$, and we have

$$\tfrac{1}{4}\rho\int \epsilon^2 \frac{dV}{dn} d\sigma = \tfrac{3}{8}\frac{M^2}{k_0} e^4 \sum_0^\infty \int_0^{\frac{1}{2}\pi}\int_0^{\frac{1}{2}\pi}\left(\frac{\rho_i^s}{\phi_i^s}\right)^2 \mathfrak{D}_i^s \Phi (S_i^s)^2 d\theta\, d\phi$$

$$= \tfrac{3}{8}\frac{M^2}{k_0} e^4 \sum_0^\infty \frac{(\rho_i^s)^2}{\phi_i^s}\mathfrak{D}_i^s \quad\text{.................................(27)}$$

Since $\mathfrak{P}_0(\nu) = 1$, $\dfrac{d}{d\nu}\mathfrak{P}_0(\nu) = 0$ and $\mathfrak{D}_0 = 0$, the term in Σ corresponding to $i = 0$ vanishes.

§ 13. *Terms in the Energy Depending on the Moment of Inertia.*

We have to determine A_r, the moment of inertia of the region R considered as filled with positive density.

In order to obtain this result, we must express $y^2 + z^2$ in terms of surface harmonics. This was done in § 12 of "Harmonics," but as a different definition of S_2 and S_2^2 was adopted there from that which I shall use here, it is easier to proceed *ab initio.*

Let
$$D^2 = 1 - \kappa^2\kappa'^2$$
and
$$(q_0)^2 = \tfrac{1}{3}(1 + \kappa^2 - D), \quad (q_2)^2 = \tfrac{1}{3}(1 + \kappa^2 + D)$$

For both the suffixes 0 and 2, we have $q^2 + q'^2 = 1$, and
$$\kappa^2 = q^2\frac{2 - 3q^2}{1 - 2q^2}, \quad \kappa'^2 = q'^2\frac{1 - 3q^2}{1 - 2q^2}, \quad \kappa^2 - q^2 = q'^2 - \kappa'^2 = \frac{q^2q'^2}{1 - 2q^2}$$

In accordance with equation (10) of "The Pear-shaped Figure" I define the harmonics as follows:—
$$S_2 = (\kappa^2\sin^2\theta - q_0^2)(q_0'^2 - \kappa'^2\cos^2\phi)$$
$$S_2^2 = (\kappa^2\sin^2\theta - q_2^2)(q_2'^2 - \kappa'^2\cos^2\phi)$$

Now $\quad y^2 = k^2(\nu^2 - 1)\cos^2\theta\sin^2\phi, \quad z^2 = k^2\nu^2\sin^2\theta(1 - \kappa'^2\cos^2\phi)$

and
$$\nu^2 = \frac{1 - \tau_1}{\sin^2\beta}, \text{ where } \tau_1 = \frac{2\tau\cos^2\beta\cos^2\gamma}{\Delta_1^2\Gamma_1^2}$$

Thus
$$\frac{\sin^2\beta}{k^2}(y^2 + z^2) = \cos^2\beta - \tau_1 + \sin^2\beta\sin^2\theta - (\cos^2\beta - \tau_1)\cos^2\phi$$
$$+ (\cos^2\beta - \kappa'^2 - \kappa^2\tau_1)\sin^2\theta\cos^2\phi$$

Let us assume, as we know to be justifiable,
$$\frac{\sin^2\beta}{k^2}(y^2 + z^2) = \mathrm{A}S_2 + \mathrm{B}S_2^2 + \mathrm{C}$$
$$= -[\mathrm{A}q_0^2q_0'^2 + \mathrm{B}q_2^2q_2'^2 - \mathrm{C}] + [\mathrm{A}q_0'^2 + \mathrm{B}q_2'^2]\kappa^2\sin^2\theta$$
$$+ [\mathrm{A}q_0^2 + \mathrm{B}q_2^2]\kappa'^2\cos^2\phi - [\mathrm{A} + \mathrm{B}]\kappa^2\kappa'^2\sin^2\theta\cos^2\phi$$

If we equate the coefficients of $\sin^2\theta$ and $\cos^2\phi$ in these two expressions, we have
$$\mathrm{A}q_0'^2 + \mathrm{B}q_2'^2 = \frac{\sin^2\beta}{\kappa^2}, \quad \mathrm{A}q_0^2 + \mathrm{B}q_2^2 = \frac{\tau_1 - \cos^2\beta}{\kappa'^2}$$

The solution of these equations may be written
$$\mathrm{A} = \frac{1}{2Dq_0^2}\left(1 - \frac{\cos^2\beta}{D + \kappa^2}\right) - \frac{\tau_1}{2Dq_0'^2}, \quad \mathrm{B} = -\frac{1}{2Dq_2^2}\left(1 + \frac{\cos^2\beta}{D - \kappa^2}\right) + \frac{\tau_1}{2Dq_2'^2}$$

The simplest way of finding C is to put $\sin^2\theta = \frac{q_0^2}{\kappa^2}$, $\cos^2\phi = \frac{q_2'^2}{\kappa'^2}$, so that $S_2 = S_2^2 = 0$; we thus find

$$C = \tfrac{1}{3}(1+\cos^2\beta) - \tfrac{2}{3}\tau_1$$

Now for brevity write

$$L = \frac{\sin\beta}{4Dq_0^2\cos\beta\cos\gamma}\left(1 - \frac{\cos^2\beta}{D+\kappa^2}\right), \qquad M = \frac{\sin\beta}{4Dq_2^2\cos\beta\cos\gamma}\left(1 + \frac{\cos^2\beta}{D-\kappa^2}\right)$$

We then have

$$A = 2\frac{\cos\beta\cos\gamma}{\sin\beta}L - \frac{\tau_1}{2Dq_0'^2}, \qquad B = -2\frac{\cos\beta\cos\gamma}{\sin\beta}M + \frac{\tau_1}{2Dq_2'^2}$$

Hence, substituting for τ_1 its value,

$$\frac{y^2+z^2}{k^2} = 2\frac{\cos\beta\cos\gamma}{\sin^3\beta}(LS_2 - MS_2^2) + \frac{1+\cos^2\beta}{3\sin^2\beta} - \tau\frac{\cos^2\beta\cos^2\gamma}{D\sin^2\beta}\left(\frac{1}{q_0'^2}\frac{S_2}{\Delta_1^2\Gamma_1^2} - \frac{1}{q_2'^2}\frac{S_2^2}{\Delta_1^2\Gamma_1^2} + \tfrac{4}{3}\frac{D}{\Delta_1^2\Gamma_1^2}\right)$$

Now
$$\frac{\rho\, dv}{d\tau\, d\theta\, d\phi} = \tfrac{1}{8}M\left(\frac{k}{k_0}\right)^3(\Phi - 2\tau\Psi)$$

Therefore

$$\frac{(y^2+z^2)\rho\, dv}{d\tau\, d\theta\, d\phi} = \tfrac{1}{8}Mk_0^2\left(\frac{k}{k_0}\right)^5\left\{2\frac{\cos\beta\cos\gamma}{\sin^3\beta}(L\Phi S_2 - M\Phi S_2^2) + \frac{1+\cos^2\beta}{3\sin^2\beta}\Phi - 4\tau\frac{\cos\beta\cos\gamma}{\sin^3\beta}(L\Psi S_2 - M\Psi S_2^2) - 2\tau\frac{1+\cos^2\beta}{3\sin^2\beta}\Psi - \tau\frac{\cos^2\beta\cos^2\gamma}{D\sin^2\beta}\left(\frac{1}{q_0'^2}\frac{\Phi S_2}{\Delta_1^2\Gamma_1^2} - \frac{1}{q_2'^2}\frac{\Phi S_2^2}{\Delta_1^2\Gamma_1^2} + \tfrac{4}{3}D\frac{\Phi}{\Delta_1^2\Gamma_1^2}\right)\right\}$$

When we integrate throughout the region R the limits of τ are $-eS_3 - \Sigma f_i^s S_i^s$ to zero.

Accordingly

$$\begin{aligned} A_r = -Mk_0^2\left(\frac{k}{k_0}\right)^5 \iint\Bigg\{&\left[2\frac{\cos\beta\cos\gamma}{\sin^3\beta}(L\Phi S_2 - M\Phi S_2^2) + \frac{1+\cos^2\beta}{3\sin^2\beta}\Phi\right] \times\left[eS_3 + \sum_2^\infty f_i^s S_i^s\right] \\ &+ \left[2\frac{\cos\beta\cos\gamma}{\sin^3\beta}(L\Psi S_2 - M\Psi S_2^2) + \frac{1+\cos^2\beta}{3\sin^2\beta}\Psi\right] e^2(S_3)^2 \\ &+ \frac{\cos^2\beta\cos^2\gamma}{2D\sin^2\beta}\left[\frac{1}{q_0'^2}\frac{\Phi S_2}{\Delta_1^2\Gamma_1^2} - \frac{1}{q_2'^2}\frac{\Phi S_2^2}{\Delta_1^2\Gamma_1^2} + \tfrac{4}{3}D\frac{\Phi}{\Delta_1^2\Gamma_1^2}\right] e^2(S_3)^2\Bigg\}\, d\theta\, d\phi \\ = -Mk_0^2\left(\frac{k}{k_0}\right)^5\Bigg\{&\frac{2\cos\beta\cos\gamma}{\sin^3\beta}(Lf_2\phi_2 - Mf_2^2\phi_2^2) + 2e^2\frac{\cos\beta\cos\gamma}{\sin^3\beta}(L\omega_2 - M\omega_2^2) \\ &+ e^2\frac{1+\cos^2\beta}{3\sin^2\beta}\sigma_2 + \frac{e^2}{2D\sin\beta}\left(\frac{\rho_2}{q_0'^2} - \frac{\rho_2^2}{q_2'^2} + \tfrac{4}{3}D\rho_0\right)\Bigg\} \end{aligned}$$

The moment of inertia of the ellipsoid J is

$$A_j = \tfrac{1}{5} M \left(\frac{k}{k_0}\right)^3 k^2 \frac{1+\cos^2\beta}{\sin^2\beta} = Mk_0^2 \left[\frac{1+\cos^2\beta}{5\sin^2\beta} - e^2 \frac{1+\cos^2\beta}{3\sin^2\beta}\sigma_2\right]$$

Also $$Mk_0^2 = \frac{M^2}{k_0} \cdot \frac{3\sin^3\beta}{4\pi\rho\cos\beta\cos\gamma} = \frac{3M^2}{2k_0} \cdot \frac{1}{2\pi\rho} \cdot \frac{\sin^3\beta}{\cos\beta\cos\gamma}$$

Lastly, to the required order we may put $(k/k_0)^5$ equal to unity in the expression for A_r.

Then

$$\tfrac{1}{2}(A_j - A_r)\,\delta\omega^2 = \frac{3M^2}{2k_0}\frac{\delta\omega^2}{2\pi\rho}\left\{\frac{(1+\cos^2\beta)\sin\beta}{10\cos\beta\cos\gamma} + L(f_2\phi_2 + e^2\omega_2)\right.$$
$$\left. - M(f_2^2\phi_2^2 + e^2\omega_2^2) + \frac{e^2}{4D}\frac{\sin^2\beta}{\cos\beta\cos\gamma}\left(\frac{\rho_2}{q_0'^2} - \frac{\rho_2^2}{q_2'^2} + \tfrac{4}{3}D\rho_0\right)\right\} \quad \text{...(28)}$$

This completes the expression for the lost energy E of the system, which may now be collected from (19), (20), (24), (27), and (28).

§ 14. *The Lost Energy of the System; Solution of the Problem.*

If the several contributions to the energy be examined, it will be seen that if i, the order of harmonics in $f_i^s S_i^s$, is odd, there is no term with coefficient $e^2 f_i^s$ in E; this follows from the fact that the ω and ρ integrals vanish for the odd harmonics. Hence, as far as concerns the odd harmonics, E involves f_i^s only in the form $(f_i^s)^2$. The condition that the pear shall be a level surface is that E shall be stationary for variations of the f's and of e. It follows that when i is odd f_i^s is zero. We may therefore drop all the odd harmonics, inclusive of f_1, and it is clear that the term $-\tfrac{1}{2}Md^2\omega^2$ in E (given in (20)) vanishes to our order of approximation.

For the sake of brevity, I adopt a single symbol for each of the coefficients of the several kinds of terms in E. Therefore let

$$A_0 = \mathfrak{A}_3[\tfrac{1}{3}(\sigma_2)^2 + 2\zeta_4] + \sum_2^\infty (\mathfrak{A}_i^s\omega_i^s + \mathfrak{B}_i^s\rho_i^s)\frac{\omega_i^s}{\phi_i^s} - \tfrac{1}{3}\sigma_4 + \tfrac{1}{4}\sum_2^\infty \frac{(\rho_i^s)^2}{\phi_i^s}\mathfrak{D}_i^s$$

$$2B_i^s = 2\mathfrak{A}_i^s\omega_i^s + (\mathfrak{B}_i^s + 2\mathfrak{B}_3)\rho_i^s - \frac{\sin^2\beta}{\cos\beta\cos\gamma}\rho_i^s$$

$$C_i^s = (\mathfrak{A}_3 - \mathfrak{A}_i^s)\phi_i^s$$

$$\mathfrak{a} = \frac{(1+\cos^2\beta)\sin\beta}{10\cos\beta\cos\gamma}$$

$$\mathfrak{b} = L\omega_2 - M\omega_2^2 + \frac{\sin^2\beta}{4D\cos\beta\cos\gamma}\left(\frac{\rho_2}{q_0'^2} - \frac{\rho_2^2}{q_2'^2} + \tfrac{4}{3}D\rho_0\right)$$

$$\mathfrak{c} = L\phi_2$$

$$\mathfrak{d} = M\phi_2^2, \text{ where } \mathfrak{A}_i^s = \mathfrak{P}_i^s\mathfrak{Q}_i^s,\ \mathfrak{B}_i^s = \mathfrak{Q}_i^s\frac{d\mathfrak{P}_i^s}{d\nu_0},\ \mathfrak{D}_i^s = \frac{d\mathfrak{P}_i^s}{d\nu_0}\frac{d\mathfrak{Q}_i^s}{d\nu_0}$$

With this notation

$$E = \tfrac{3}{2}\frac{M^2}{k_0}\left\{A_0 e^4 + 2\sum_2^\infty B_i^s e^2 f_i^s - \sum_2^\infty C_i^s (f_i^s)^2 + \frac{\delta\omega^2}{2\pi\rho}(\mathfrak{a} + \mathfrak{b}e^2 + \mathfrak{c}f_2 - \mathfrak{d}f_2^2)\right\}$$

Let us now make E stationary for variations of e and f_i^s.

First, by the variation of any f_i^s excepting f_2 and f_2^2, we have

$$f_i^s = \frac{B_i^s}{C_i^s} e^2 \quad \text{..............................(29)}$$

On eliminating all these f_i^s, we have

$$E = \tfrac{3}{2}\frac{M^2}{k_0}\left\{\left(A_0 + \sum_4^\infty \frac{(B_i^s)^2}{C_i^s}\right) e^4 + 2B_2 e^2 f_2 + 2B_2^2 e^2 f_2^2 - C_2 (f_2)^2 - C_2^2 (f_2^2)^2 + \frac{\delta\omega^2}{2\pi\rho}(\mathfrak{a} + \mathfrak{b}e^2 + \mathfrak{c}f_2 - \mathfrak{d}f_2^2)\right\}$$

By the variations of f_2, f_2^2, and e^2, we have

$$B_2 e^2 - C_2 f_2 + \frac{\delta\omega^2}{4\pi\rho}\mathfrak{c} = 0, \quad B_2^2 e^2 - C_2^2 f_2^2 - \frac{\delta\omega^2}{4\pi\rho}\mathfrak{d} = 0$$

$$\left(A_0 + \sum_4^\infty \frac{(B_i^s)^2}{C_i^s}\right) e^2 + B_2 f_2 + B_2^2 f_2^2 + \frac{\delta\omega^2}{4\pi\rho}\mathfrak{b} = 0$$

But from the first two of these equations

$$B_2 f_2 = \frac{(B_2)^2}{C_2} e^2 + \frac{\delta\omega^2}{4\pi\rho}\frac{B_2\mathfrak{c}}{C_2}, \quad B_2^2 f_2^2 = \frac{(B_2^2)^2}{C_2^2} e^2 - \frac{\delta\omega^2}{4\pi\rho}\frac{B_2^2\mathfrak{d}}{C_2^2}$$

Therefore

$$\left(A_0 + \sum_2^\infty \frac{(B_i^s)^2}{C_i^s}\right) e^2 + \frac{\delta\omega^2}{4\pi\rho}\left(\mathfrak{b} + \frac{B_2\mathfrak{c}}{C_2} - \frac{B_2^2\mathfrak{d}}{C_2^2}\right) = 0 \quad \text{...........(30)}$$

When $\delta\omega^2$ has been found, f_2 and f_2^2 are determined from

$$\left.\begin{aligned} f_2 &= \frac{B_2}{C_2} e^2 + \frac{\delta\omega^2}{4\pi\rho}\frac{\mathfrak{c}}{C_2} \\ f_2^2 &= \frac{B_2^2}{C_2^2} e^2 - \frac{\delta\omega^2}{4\pi\rho}\frac{\mathfrak{d}}{C_2^2} \end{aligned}\right\} \quad \text{..........................(31)}$$

A consideration of these formulæ shows that it is immaterial what definition is adopted for any one of the harmonics, provided, of course, that the same definition is maintained throughout.

In order to evaluate A_0, we must eliminate $\mathfrak{D}_i^s$.

Since $\quad \mathfrak{A}_i^s = \mathfrak{P}_i^s \mathfrak{Q}_i^s, \quad \mathfrak{B}_i^s = \mathfrak{Q}_i^s \dfrac{d\mathfrak{P}_i^s}{d\nu_0}, \quad \mathfrak{D}_i^s = \dfrac{d\mathfrak{P}_i^s}{d\nu_0}\dfrac{d\mathfrak{Q}_i^s}{d\nu_0}$

and $\quad \mathfrak{Q}_i^s \dfrac{d\mathfrak{P}_i^s}{d\nu_0} - \mathfrak{P}_i^s \dfrac{d\mathfrak{Q}_i^s}{d\nu_0} = \dfrac{\sin^2\beta}{\cos\beta\cos\gamma}$

we see that $\quad \mathfrak{D}_i^s = \left(\mathfrak{B}_i^s - \dfrac{\sin^2\beta}{\cos\beta\cos\gamma}\right)\dfrac{\mathfrak{B}_i^s}{\mathfrak{A}_i^s}$

Hence

$$\mathfrak{A}_i^s \frac{(\omega_i^s)^2}{\phi_i^s} + \mathfrak{B}_i^s \frac{\omega_i^s \rho_i^s}{\phi_i^s} + \tfrac{1}{4} \frac{\mathfrak{D}_i^s (\rho_i^s)^2}{\phi_i^s} = \frac{1}{\mathfrak{A}_i^s \phi_i^s} (\mathfrak{A}_i^s \omega_i^s + \tfrac{1}{2}\mathfrak{B}_i^s \rho_i^s)^2$$

$$- \tfrac{1}{4} \frac{\sin^2 \beta}{\cos \beta \cos \gamma} \frac{\mathfrak{B}_i^s}{\mathfrak{A}_i^s} \frac{(\rho_i^s)^2}{\phi_i^s} \quad \text{...(32)}$$

If for brevity we denote this last expression by $[i, s]$, we have

$$\left.\begin{aligned} A_0 &= \mathfrak{A}_3 [\tfrac{1}{3} (\sigma_2)^2 + 2\zeta_4] - \tfrac{1}{3}\sigma_4 + \sum_2^\infty [i, s] \\ B_i^s &= (\mathfrak{A}_i^s \omega_i^s + \tfrac{1}{2}\mathfrak{B}_i^s \rho_i^s) + \left(\mathfrak{B}_3 - \frac{\sin^2 \beta}{2 \cos \beta \cos \gamma}\right) \rho_i^s \\ C_i^s &= (\mathfrak{A}_3 - \mathfrak{A}_i^s)\, \phi_i^s \end{aligned}\right\} \quad \text{..........(32)}$$

We have now the complete analytical expressions necessary for the solution of the problem.

PART II.

Numerical Calculation.

§ 15. *The Determination of Certain Integrals*.*

The integrals ω_i^s, ρ_i^s, ϕ_i^s depend on certain others, namely

$$\left.\begin{aligned} {}^{2p}\Lambda_{2m}^{2n} &= \int_0^{\frac{1}{2}\pi} \frac{\sin^{2p} \theta \cos^{2n} \theta}{\Delta_1^{2m} \Delta} d\theta \\ \Omega_{2m}^{2n} &= \int_0^{\frac{1}{2}\pi} \frac{\sin^{2n} \phi}{\Gamma_1^{2m} \Gamma} d\phi \end{aligned}\right\} \quad \text{.....................(34)}$$

However we shall see that the integral ${}^{2p}\Lambda_{2m}^{2n}$ may easily be made to depend on ${}^{0}\Lambda_{2m}^{2n}$, and therefore we shall only consider this latter form for the present.

Since $\Delta_1^2 = 1 - \kappa^2 \sin^2 \gamma \sin^2 \theta = \cos^2 \beta + \sin^2 \beta \cos^2 \theta$, where $\sin \beta = \kappa \sin \gamma$

and $\Gamma_1^2 = \cos^2 \gamma + \kappa'^2 \sin^2 \gamma \cos^2 \phi = \cos^2 \beta - \kappa'^2 \sin^2 \gamma \sin^2 \phi$

we have

$$\left.\begin{aligned} {}^{0}\Lambda_{2m}^{2n} &= \operatorname{cosec}^2 \beta\, {}^{0}\Lambda_{2m-2}^{2n-2} - \cot^2 \beta\, {}^{0}\Lambda_{2m}^{2n-2} \\ \Omega_{2m}^{2n} &= \frac{\cos^2 \beta}{\kappa'^2 \sin^2 \gamma} \Omega_{2m}^{2n-2} - \frac{1}{\kappa'^2 \sin^2 \gamma} \Omega_{2m-2}^{2n-2} \end{aligned}\right\} \quad \text{..............(35)}$$

* [This section has been rewritten, and in order to retain the old numbering of the equations some few numbers are missing.]

I now write
$$F=\int_0^{\frac{1}{2}\pi}\frac{d\theta}{\Delta},\qquad E=\int_0^{\frac{1}{2}\pi}\Delta\,d\theta$$
$$F'=\int_0^{\frac{1}{2}\pi}\frac{d\phi}{\Gamma},\qquad E'=\int_0^{\frac{1}{2}\pi}\Gamma\,d\phi$$
$$F(\gamma)=\int_0^{\gamma}\frac{d\theta}{\Delta},\qquad E(\gamma)=\int_0^{\gamma}\Delta\,d\theta$$

It will be found from Legendre's tables that for $\gamma = 69^\circ\ 49'{\cdot}0$, $\kappa = \sin 73^\circ\ 54'{\cdot}2$
$$\left.\begin{aligned}\log F &= {\cdot}4317642, & \log E &= \ {\cdot}0355145\\ \log F' &= {\cdot}2047610, & \log E' &= \ {\cdot}1875655\\ \log F(\gamma) &= {\cdot}2117987, & \log E(\gamma) &= 9{\cdot}9856045\end{aligned}\right\}\quad\ldots\ldots\ldots\ldots(36)$$

If we use the fact that $d\Delta = -\dfrac{\kappa^2\sin\theta\cos\theta\,d\theta}{\Delta}$, and integrate $\dfrac{\cos^{2n-2}\theta\sin^2\theta}{\Delta}$ by parts we find
$$\left.\begin{aligned}{}^0\Lambda_0^{2n} &= \frac{2n-2}{2n-1}\left(\frac{\kappa^2-\kappa'^2}{\kappa^2}\right){}^0\Lambda_0^{2n-2}+\frac{2n-3}{2n-1}\frac{\kappa'^2}{\kappa^2}{}^0\Lambda_0^{2n-4}\\ \Omega_0^{2n} &= -\frac{2n-2}{2n-1}\left(\frac{\kappa^2-\kappa'^2}{\kappa'^2}\right)\Omega_0^{2n-2}+\frac{2n-3}{2n-1}\frac{\kappa^2}{\kappa'^2}\Omega_0^{2n-4}\end{aligned}\right\}\quad\ldots\ldots\ldots(37)$$

Now write
$$G=\tfrac{1}{2}(1+\sec^2\beta+\sec^2\gamma),\quad H'=\tfrac{1}{2}(\sec^2\beta+\sec^2\gamma+\sec^2\beta\sec^2\gamma)$$

The values of β and γ are $64^\circ\ 23'{\cdot}712$, $69^\circ\ 49'{\cdot}0$; whence $\log G = {\cdot}8679015$, $\log H' = 1{\cdot}4678555$. Also we require $\log H = 1{\cdot}7182664$ (see § 3, p. 323).

By differentiation it may be proved that
$$\sin^2\beta\sin^2\gamma\frac{d}{d\theta}\frac{\Delta\sin\theta\cos\theta}{\Delta_1^{2n}}=\frac{2n\cos^2\beta\cos^2\gamma}{\Delta_1^{2n+2}\Delta}-\frac{2(2n-1)\,G\cos^2\beta\cos^2\gamma}{\Delta_1^{2n}\Delta}$$
$$+\frac{2(2n-2)\,H'\cos^2\beta\cos^2\gamma}{\Delta_1^{2n-2}\Delta}-\frac{2n-3}{\Delta_1^{2n-4}\Delta}$$

Whence by integration
$${}^0\Lambda_{2n+2}^0=\frac{2n-1}{n}G\,{}^0\Lambda_{2n}^0-\frac{2n-2}{n}H'\,{}^0\Lambda_{2n-2}^0+\frac{2n-3}{2n}\sec^2\beta\sec^2\gamma\,{}^0\Lambda_{2n-4}^0\ \ldots(38)$$

On writing $\tan\gamma\sqrt{-1}$ for $\sin\gamma$, we find that exactly the same formula holds good for the Ω's.

To apply this to the determination of ${}^0\Lambda_0^2$, Ω_0^2, we note that
$${}^0\Lambda_{-2}^0=\cos^2\gamma F+\sin^2\gamma E,\quad \Omega_{-2}^0=F'-\sin^2\gamma E'\ \ldots\ldots\ldots\ldots(39)$$

Also
$${}^0\Lambda_0^2=\frac{1}{\kappa^2}E-\frac{\kappa'^2}{\kappa^2}F,\qquad \Omega_0^2=\frac{1}{\kappa'^2}E'-\frac{\kappa^2}{\kappa'^2}F'\ \ldots\ldots\ldots\ldots(40)$$

From the formulæ given in Cayley's *Elliptic Integrals* it appears that

$$\left.\begin{aligned} {}^0\Lambda_2^0 &= F + \frac{\sin\gamma}{\cos\beta\cos\gamma}\left[FE(\gamma) - EF(\gamma)\right] \\ \Omega_2^0 &= F' + \frac{\sin\gamma}{\cos\beta\cos\gamma}\left[F'E(\gamma) - F'F(\gamma) + E'F(\gamma)\right] \end{aligned}\right\} \quad \text{......(41)}$$

Now ${}^0\Lambda_0^0 = F$, ${}^0\Lambda_0^2$ is given in (40), and ${}^0\Lambda_0^{2n}$ for $n = 2, 3, 4 \ldots$ are then given successively by (37).

Again ${}^0\Lambda_2^0$ is given by (41), and the successive ${}^0\Lambda_2^{2n}$ are given by the general formula (35).

Again (38) and (39) give

$$\begin{aligned} {}^0\Lambda_4^0 &= G\,{}^0\Lambda_2^0 - \tfrac{1}{2}\sec^2\beta\sec^2\gamma\,(\cos^2\gamma\, F + \sin^2\gamma\, E) \\ {}^0\Lambda_6^0 &= \tfrac{3}{2}G\,{}^0\Lambda_4^0 - H'\,{}^0\Lambda_2^0 + \tfrac{1}{4}\sec^2\beta\sec^2\gamma\, F \end{aligned}$$

and by successive applications of the formula (35), we find the successive values of ${}^0\Lambda_4^{2n}$, ${}^0\Lambda_6^{2n}$.

The Ω integrals may apparently be derived by a similar set of formulæ, but since at each step we divide by κ'^2, a small quantity, all accuracy is rapidly dissipated. Although we may safely derive one series of Ω integrals from a preceding one, we cannot so derive a succession of series, and it becomes necessary to find new formulæ.

In order to determine the Ω integrals, consider the group of integrals

$$W_{2m}^{2n} = \int_0^{\frac{1}{2}\pi} \frac{\sin^{2n}\phi}{\Gamma_1^{2m}}\, d\phi$$

If we write $\xi = \dfrac{\cos\gamma\tan\phi}{\cos\beta}$, $\quad a = \dfrac{\cos\gamma}{\cos\beta}$, we find

$$W_{2m}^{0} = \frac{1}{\cos\beta\cos^{2m-1}\gamma}\int_0^\infty \frac{(a^2+\xi^2)^{m-1}}{(1+\xi^2)^m}\, d\xi$$

whence, by some easy integrations,

$$W_2^0 = \frac{\pi}{2\cos\beta\cos\gamma}$$

$$W_4^0 = \frac{\pi}{4\cos\beta\cos\gamma}(\sec^2\beta + \sec^2\gamma)$$

$$W_6^0 = \frac{3\pi}{16\cos\beta\cos\gamma}(\sec^4\beta + \sec^4\gamma + \tfrac{2}{3}\sec^2\beta\sec^2\gamma)$$

On expanding $1/\Gamma$ in powers of κ'^2 we see that

$$\Omega_{2m}^{2n} = W_{2m}^{2n} + \tfrac{1}{2}\kappa'^2 W_{2m}^{2n+2} + \frac{1\,.\,3}{2\,.\,4}\kappa'^4 W_{2m}^{2n+4} + \ldots$$

When $m = 0$ the W integrals are easily determined.

The relationship between successive W integrals is clearly

$$W_{2m}^{2n} = \frac{\cos^2 \beta}{\kappa'^2 \sin^2 \gamma} W_{2m}^{2n-2} - \frac{1}{\kappa'^2 \sin^2 \gamma} W_{2m-2}^{2n-2}$$

I now write for brevity

$$x = \cos \beta, \quad y = \cos \gamma, \quad z = \sin \gamma, \quad \lambda = \frac{z}{1+z}, \quad \sigma = \frac{x}{x+y}$$

It appears that if we write $\gamma_{2i} = \left[\frac{1.3.5 \ldots (2i-1)}{2.4.6 \ldots 2i} \kappa'^i\right]^2$ we may put

$$\Omega_0^{2n} = \tfrac{1}{2}\pi \frac{1.3.5 \ldots (2n-1)}{2.4.6 \ldots 2n} \left\{1 + \frac{1}{n+1}\gamma_2 + \frac{1.2}{(n+1)(n+2)}\gamma_4 + \frac{1.2.3}{(n+1)(n+2)(n+3)}\gamma_6 + \ldots\right\}$$

$$\Omega_2^{2n} = \frac{\pi z}{2xy} F_{2n} + \tfrac{1}{2}\pi \frac{1}{1+z} \frac{1.3 \ldots (2n-1)}{2.4 \ldots 2n} \left\{a_0 + \frac{a_2}{n+1}\gamma_2 + \frac{1.2\, a_4}{(n+1)(n+2)}\gamma_4 + \frac{1.2.3\, a_6}{(n+1)(n+2)(n+3)}\gamma_6 + \ldots\right\}$$

$$\Omega_4^{2n} = \frac{\pi z}{4xy} G_{2n} + \tfrac{1}{2}\pi \frac{1}{1+z} \frac{1.3 \ldots (2n-1)}{2.4 \ldots 2n} \left\{b_0 + \frac{b_2}{n+1}\gamma_2 + \frac{1.2\, b_4}{(n+1)(n+2)}\gamma_4 + \ldots\right\}$$

$$\Omega_6^{2n} = \frac{3\pi z}{16xy} H_{2n} + \tfrac{1}{2}\pi \frac{1}{1+z} \frac{1.3 \ldots (2n-1)}{2.4 \ldots 2n} \left\{c_0 + \frac{c_2}{n+1}\gamma_2 + \frac{1.2\, c_4}{(n+1)(n+2)}\gamma_4 + \ldots\right\}$$

By considering in detail the cases where $n = 0$, I find

$$F_0 = 1, \quad a_0 = 1$$

$$G_0 = 1 + \frac{1}{x^2} + \frac{1}{y^2} = 1 + \frac{1}{\sigma^2 y^2}(1 - 2\sigma + 2\sigma^2), \quad b_0 = 1 - \tfrac{1}{2}\lambda$$

$$H_0 = \frac{1}{x^4} + \frac{1}{y^4} + \tfrac{2}{3}\left(\frac{1}{x^2 y^2} + \frac{1}{x^2} + \frac{1}{y^2}\right) + 1$$

$$= 1 + \frac{2}{3\sigma^2 y^2}(1 - 2\sigma + 2\sigma^2) + \frac{1}{\sigma^4 y^4}(1 - 4\sigma + \tfrac{20}{3}\sigma^2 - \tfrac{16}{3}\sigma^3 + \tfrac{8}{3}\sigma^4)$$

$$c_0 = 1 - \tfrac{7}{8}\lambda + \tfrac{1}{4}\lambda^2$$

By some rather tedious analysis, by considering the manner in which each Ω is derivable from the preceding ones, it may be proved that

$$F_{2n} = \frac{1}{1-2\sigma}\left\{-\sigma^2 F_{2n-2} + \frac{1.3 \ldots (2n-3)}{2.4 \ldots (2n-2)}\sigma(1-\sigma)\right\}$$

$$a_{2i} = \frac{1-\lambda}{1-2\lambda} - \frac{2i}{2i-1}\frac{\lambda^2}{1-2\lambda} a_{2i-2}$$

$$G_{2n} = \frac{1}{1-2\sigma}\left\{-\sigma^2 G_{2n-2} + \frac{2(1-\sigma)^2}{y^2} F_{2n-2} + \frac{1.3 \ldots (2n-3)}{2.4 \ldots (2n-2)}\sigma(1-\sigma)\right\}$$

$$b_{2i} = \frac{1}{1-2\lambda}\left[(1-\lambda)^2 a_{2i} - \frac{2i\lambda^2}{2i-1} b_{2i-2}\right]$$

$$H_{2n} = \frac{1}{1-2\sigma}\left[-\sigma^2 H_{2n-2} + \frac{4(1-\sigma)^2}{3y^2} G_{2n-2} + \frac{1.3\ldots(2n-3)}{2.4\ldots(2n-2)}\sigma(1-\sigma)\right]$$

$$c_{2i} = \frac{1}{1-2\lambda}\left[(1-\lambda)^2 b_{2i} - \frac{2i\lambda^2}{2i-1} c_{2i-2}\right]$$

By successive applications, starting from the values for $n=0$, I find

$$F_0 = 1, \quad F_2 = \sigma, \quad F_4 = \tfrac{1}{2}\sigma(1+\sigma), \quad F_6 = \frac{1.3}{2.4}\sigma(1+\sigma+\tfrac{2}{3}\sigma^2)$$

and generally

$$F_{2n} = \frac{1.3\ldots(2n-3)}{2.4\ldots(2n-2)}\sigma\left\{1 + \frac{2n-4}{2n-3}\sigma^2 + \frac{(2n-6)(2n-8)}{(2n-3)(2n-5)}\sigma^4 + \frac{(2n-8)(2n-10)(2n-12)}{(2n-3)(2n-5)(2n-7)}\sigma^6 + \ldots\right.$$

$$\left. + \sigma + \frac{2n-6}{2n-3}\sigma^3 + \frac{(2n-8)(2n-10)}{(2n-3)(2n-5)}\sigma^5 + \frac{(2n-10)(2n-12)(2n-14)}{(2n-3)(2n-5)(2n-7)}\sigma^7 + \ldots\right\}$$

$$a_0 = 1, \quad a_2 = 1+\lambda, \quad a_4 = 1 + \lambda + \tfrac{2}{3}\lambda^2$$

and generally

$$a_{2i} = 1 + \frac{2i-2}{2i-1}\lambda^2 + \frac{(2i-4)(2i-6)}{(2i-1)(2i-3)}\lambda^4 + \ldots$$

$$+ \lambda + \frac{2i-4}{2i-1}\lambda^3 + \frac{(2i-6)(2i-8)}{(2i-1)(2i-3)}\lambda^5 + \ldots$$

I have not obtained the general law for the G's, H's, b's, and c's. When the suffixes of the G's and H's are greater than 10 or 12 it is easiest to go back to the difference equation with numerical values. For the earlier values we have

$$G_0 = 1 + \frac{1}{x^2} + \frac{1}{y^2}, \quad G_2 = \sigma + \frac{1}{y^2}, \quad G_4 = F_4 + \frac{2\sigma}{y^2}(1-\tfrac{1}{2}\sigma), \quad G_6 = F_6 + \frac{\sigma}{y^2}(1+\sigma-\sigma^2)$$

$$G_8 = F_8 + \frac{3\sigma}{4y^2}(1+\sigma+\tfrac{1}{3}\sigma^2-\sigma^3), \quad G_{10} = F_{10} + \frac{5\sigma}{8y^2}(1+\sigma+\tfrac{3}{5}\sigma^2-\tfrac{1}{5}\sigma^3-\tfrac{4}{5}\sigma^4)$$

$$G_{12} = F_{12} + \tfrac{35}{64}\frac{\sigma}{y^2}(1+\sigma+\tfrac{5}{7}\sigma^2+\tfrac{1}{7}\sigma^3-\tfrac{16}{35}\sigma^4-\tfrac{4}{7}\sigma^5)$$

$$b_0 = 1-\tfrac{1}{2}\lambda, \quad b_2 = 1+\lambda-\lambda^2, \quad b_4 = 1+\lambda+\tfrac{1}{3}\lambda^2-\lambda^3, \quad b_6 = 1+\lambda+\tfrac{3}{5}\lambda^2-\tfrac{1}{5}\lambda^3-\tfrac{4}{5}\lambda^4$$

$$b_8 = 1+\lambda+\tfrac{5}{7}\lambda^2+\tfrac{1}{7}\lambda^3-\tfrac{16}{35}\lambda^4-\tfrac{4}{7}\lambda^5$$

$$H_0 = \frac{1}{x^4} + \frac{1}{y^4} + \tfrac{2}{3}\left(\frac{1}{x^2y^2} + \frac{1}{x^2} + \frac{1}{y^2}\right) + 1, \quad H_2 = \sigma + \frac{2}{3y^2} + \frac{1}{3x^2y^2} + \frac{1}{y^4}$$

$$H_4 = F_4 + \frac{4\sigma}{3y^2}(1-\tfrac{1}{2}\sigma) + \frac{1}{y^4}, \quad H_6 = F_6 + \frac{2\sigma}{3y^2}(1+\sigma-\sigma^2) + \frac{8\sigma}{3y^4}(1-\tfrac{7}{8}\sigma+\tfrac{1}{4}\sigma^2)$$

$$H_8 = F_8 + \frac{\sigma}{2y^2}(1 + \sigma + \tfrac{1}{3}\sigma^2 - \sigma^3) + \frac{4\sigma}{3y^4}(1 + \sigma - 2\sigma^2 + \tfrac{3}{4}\sigma^3)$$

$$H_{10} = F_{10} + \frac{5\sigma}{12y^2}(1 + \sigma + \tfrac{3}{5}\sigma^2 - \tfrac{1}{5}\sigma^3 - \tfrac{4}{5}\sigma^4) + \frac{\sigma}{y^4}(1 + \sigma + 0\,.\,\sigma^2 - 2\sigma^3 + \sigma^4)$$

$$c_0 = 1 - \tfrac{7}{8}\lambda + \tfrac{1}{4}\lambda^2, \quad c_2 = 1 + \lambda - 2\lambda^2 + \tfrac{3}{4}\lambda^3, \quad c_4 = 1 + \lambda + 0\,.\,\lambda^2 - 2\lambda^3 + \lambda^4$$

$$c_6 = 1 + \lambda + \tfrac{2}{5}\lambda^2 - \tfrac{4}{5}\lambda^3 - \tfrac{7}{5}\lambda^4 + \lambda^5, \quad c_8 = 1 + \lambda + \tfrac{4}{7}\lambda^2 - \tfrac{2}{7}\lambda^3 - \tfrac{37}{35}\lambda^4 - \tfrac{5}{7}\lambda^5 + \tfrac{6}{7}\lambda^6$$

By means of these formulæ I then computed a table of the ${}^0\Lambda_{2m}^{2n}$, Ω_{2m}^{2n} integrals corresponding to the critical Jacobian for which $\gamma = 69° \, 49'{\cdot}0$, $\kappa = \sin 73° \, 54'{\cdot}2$.

A little consideration will show that if ${}^0\Lambda_{2m}^0$, ${}^0\Lambda_{2m}^2$, ${}^0\Lambda_{2m}^4$ &c. be a series of the integrals, and that if we form a table of differences, changing the sign of the odd differences we shall find the series of integrals corresponding to ${}^{2p}\Lambda_{2m}^{2q}$. This fact may be stated analytically as follows:—

$$(-)^r \, \Delta^r \, {}^0\Lambda_{2m}^{2p-2r} = {}^{2r}\Lambda_{2m}^{2p-2r}$$

I accordingly tabulate the natural numbers of the ${}^0\Lambda_{2m}$ integrals, and form a table of differences. It seems unnecessary to reproduce these tables of differences, but I will just show how they run in the case of one of the series. We have

${}^0\Lambda_0^0 = 2{\cdot}7024906$			
${}^0\Lambda_0^2 = {\cdot}9505345$	${}^2\Lambda_0^0 = 1{\cdot}7519561$		
${}^0\Lambda_0^4 = {\cdot}6559354$	${}^2\Lambda_0^2 = {\cdot}2945991$	${}^4\Lambda_0^0 = 1{\cdot}4573570$	
${}^0\Lambda_0^6 = {\cdot}5285432$	${}^2\Lambda_0^4 = {\cdot}1273922$	${}^4\Lambda_0^2 = {\cdot}1672069$	${}^6\Lambda_0^0 = 1{\cdot}2901501$
	&c.	&c.	

The tables of differences of the other series of functions Λ_2, Λ_4, Λ_6 are to be formed in exactly the same way.

Before giving the table of results I may remark that it was proved in my paper on Ellipsoidal Harmonics (Paper 10, p. 265, equation 84) that

$$
\begin{aligned}
{}^0\Lambda_0^{2n} = {} & 2^{2n-1}\frac{n!\,\overline{n-1}!}{2n!}\left[1 - \frac{1^2}{2^2(n-1)\,1!}\kappa'^2 + \frac{1^2 . 3^2}{2^4(n-1)(n-2)\,2!}\kappa'^4 \right. \\
& \left. \qquad - \text{\&c. } (n+1) \text{ terms}\right] \\
& + (-)^n \frac{2n!}{2^{2n}(n!)^2}\kappa'^{2n}\left[\left\{L - \tfrac{1}{2}\left(\frac{1}{1} + \frac{1}{2} + \ldots + \frac{1}{n}\right) - \left(\frac{1}{1.1} + \frac{1}{1.2} + \ldots \right.\right.\right. \\
& \qquad \left.\left. + \frac{1}{n(2n-1)}\right)\right\} \\
& + \frac{(2n+1)^2}{2^2(n+1)\,1!}\kappa'^2\left\{L - \tfrac{1}{2}\left(\frac{1}{2} + \frac{1}{3} + \ldots + \frac{1}{n+1}\right) - \left(\frac{1}{1.1} + \frac{1}{1.2} + \ldots \right.\right. \\
& \qquad \left.\left. + \frac{1}{(n+1)(2n+1)}\right)\right\}
\end{aligned}
$$

$$+\frac{(2n+1)^2(2n+3)^2}{2^4(n+1)(n+2)2!}\kappa'^4\left\{L-\tfrac{1}{2}\left(\frac{1}{3}+\frac{1}{4}+\ldots+\frac{1}{n+2}\right)-\left(\frac{1}{1.1}+\frac{1}{1.2}+\ldots\right.\right.$$

$$\left.\left.+\frac{1}{(n+2)(2n+3)}\right\}+\&c.\right]$$

where $L=\log_e\frac{4}{\kappa'}$.

The Λ_0 series of functions was computed also from this formula.

After the earlier values have been computed, it appears that the second portion of the formula becomes negligible. If then we substitute for the factorials the approximate formula

$$n\,!=(2\pi)^{\frac{1}{2}}n^{n+\frac{1}{2}}e^{-n}\left(1+\frac{1}{12n}+\frac{1}{288n^2}-\frac{139}{51840n^3}\ldots\right)^*$$

we find

$${}^0\Lambda_0^{2n}=\tfrac{1}{2}\sqrt{\frac{\pi}{n}}\cdot\left(1+\frac{1}{8n}+\frac{1}{128n^2}-\frac{55}{2^{10}.3^2.n^3}\right)\left(1-\frac{1^2}{2^2(n-1)1!}\kappa'^2+\ldots\right)$$

The calculation of ${}^0\Lambda_0^{22}$ from this formula agrees within unity in the seventh place of decimals with that derived from the rigorous formula.

Even the formula $\tfrac{1}{2}\sqrt{\frac{\pi}{n}}\cdot\left(1+\frac{1}{8n}\right)\left(1-\frac{1}{4(n-1)}\kappa'^2\right)$ only differs by 2 in the fifth place of decimals from the correct result.

I have found other approximate formulæ for the other series of functions, but as no use is made of them they will not be given here.

The following is the table of results, and it may be noted that the Λ_6, Ω_6 series of functions are not required for any higher values than those tabulated.

Table of the Λ functions (natural numbers)†.

n	${}^0\Lambda_0^n$	${}^0\Lambda_2^n$	${}^0\Lambda_4^n$	${}^0\Lambda_6^n$
0	2·7024906	8·034600	30·53878	132·38251
2	·9505345	1·4779482	2·866414	7·149917
4	·6559354	·8294118	1·1590804	1·8826818
6	·5285432	·6160958	·7537022	·9929042
8	·4543269	·5084364	·5844941	·6987687
10	·4044492	·4419004	·4909693	·5582507
12	·3680175	·3958484	·4306317	·4755180
14	·3399181	·3616261	·3878603	·4203234
16	·3173978	·3349328	·3556021	
18	·2988272	·3133708	·3295135	
20	·2831734			
22	·2697454			

* See Boole's *Calc. Fin. Diff.*, Chapter V.

† [In the original paper these functions were found by differencing certain other functions, and the results were not so accurate as I had thought they were. There was rather a bad mistake in

Table of logarithms of the Ω functions.

n	$\log \Omega_0^n$	$\log \Omega_2^n$	$\log \Omega_4^n$	$\log \Omega_6^n$
0	·2047610	1·0302912	1·8667641	2·7138142
2	9·8993673	·7715375	1·6492558	2·5319084
4	9·7729862	·6613000	1·5528790	2·4473549
6	9·6930884	·5897701	1·4886134	2·3893880
8	9·6346685	·5365117	1·4398970	2·3446714
10	9·5886267	·4939849	1·4005014	2·3080689
12	9·5506357	·4585472	1·3673618	2·2769972
14	9·5182995	·4281522	1·3387296	2·2499642
16	9·4901531	·4015325	1·3135070	
18	9·4652355	·3778480	1·2909612	
20	9·4428427	·3565096	1·2705764	
22	9·4226147			

§ 16. *The Integrals σ_2, σ_4, ζ_4**.

In accordance with equation (14) p. 294 of the "Pear-shaped Figure" the third zonal harmonic is defined by

$$S_3 = \mathfrak{P}_3(\mu)\, \mathsf{C}_3(\phi)$$

where
$$\mathfrak{P}_3(\mu) = \sin\theta\,(\kappa^2 \sin^2\theta - q^2)$$
$$\mathsf{C}_3(\phi) = (q'^2 - \kappa'^2 \cos^2\phi)\sqrt{(1 - \kappa'^2\cos^2\phi)}$$

and
$$q^2 = \tfrac{2}{5}\left[1 + \kappa^2 - (1 - \tfrac{7}{4}\kappa^2 + \kappa^4)^{\frac{1}{2}}\right], \quad q'^2 = 1 - q^2$$

The numerical values for the critical Jacobian are

$$\kappa^2 = {\cdot}9231276, \quad q^2 = {\cdot}5746736$$

It may be observed that if $\mathfrak{P}_3(\mu) = f(\kappa^2\cos^2\theta)$, $\mathsf{C}_3(\phi) = \kappa f(-\kappa'^2\sin^2\phi)$; that is to say the form of function is the same, but we have chosen $\mathsf{C}_3(\phi)$ so that it involves the additional factor κ.

We require the squares of both these functions.

Now
$$\mathfrak{P}_3(\mu) = (\kappa^2 - q^2 - \kappa^2\cos^2\theta)\sqrt{(1 - \cos^2\theta)}$$

and if we write

$$\alpha = (\kappa^2 - q^2)^2, \quad \beta = 2\kappa^2(\kappa^2 - q^2) + (\kappa^2 - q^2)^2, \quad \gamma = \kappa^4 + 2\kappa^2(\kappa^2 - q^2), \quad \delta = \kappa^4$$

$$[\mathfrak{P}_3(\mu)]^2 = \alpha - \beta\cos^2\theta + \gamma\cos^4\theta - \delta\cos^6\theta$$

On account of the symmetry of the forms of $\mathfrak{P}_3$ and C_3, if we put

$$\alpha' = \alpha\kappa^2, \quad \beta' = \beta\kappa^2 . \frac{\kappa'^2}{\kappa^2}, \quad \gamma' = \gamma\kappa^2 . \frac{\kappa'^4}{\kappa^4}, \quad \delta' = \delta\kappa^2 . \frac{\kappa'^6}{\kappa^6}$$

we have
$$[\mathsf{C}_3(\phi)]^2 = \alpha' + \beta'\sin^2\phi + \gamma'\sin^4\phi + \delta'\sin^6\phi$$

differencing the functions which gave the Ω_6 functions, and this affected the values from $n = 10$ onwards, but as the Ω functions as tabulated were not actually used in the computations the mistake did not affect the result.]

* [This section has been rewritten.]

The numerical values of these coefficients are as follows:—

$$\log\alpha = 9{\cdot}0843568, \qquad \log\alpha' = 9{\cdot}0496186$$
$$\log\beta = 9{\cdot}8835606, \qquad \log\beta' = 8{\cdot}7693310$$
$$\log\gamma = {\cdot}1748006, \qquad \log\gamma' = 7{\cdot}9810796$$
$$\log\delta = 9{\cdot}9305236, \qquad \log\delta' = 6{\cdot}6573112$$

It may be noted that $\alpha - \beta + \gamma - \delta = 0$.

Before passing on we must give another form of $(\mathfrak{P}_3)^2$; it is

$$[\mathfrak{P}_3(\mu)]^2 = F_0 \sin^6\theta - F_2 \sin^4\theta\cos^2\theta + F_4 \sin^2\theta\cos^4\theta$$

where $$F_0 = (\kappa^2 - q^2)^2, \quad F_2 = 2\kappa^2(\kappa^2 - q^2), \quad F_4 = q^4$$

The logarithms of these are as follows:—

$$\log F_0 = 9{\cdot}0843568$$
$$\log F_2 = 9{\cdot}6026093$$
$$\log F_4 = 9{\cdot}5188031$$

Now by (8) and (6) p. 328 we have

$$\sigma_2 = \iint \Psi (S_3)^2 \, d\theta \, d\phi$$

where

$$\Psi = \frac{6}{\pi}\cos^2\beta\cos^2\gamma\,(\kappa^2\cos^2\theta + \kappa'^2\sin^2\phi)\left[\frac{1}{\Gamma_1^4\Delta_1^2} + \frac{1}{\Gamma_1^2\Delta_1^4} - \frac{G}{\Delta_1^2\Gamma_1^2}\right]\frac{1}{\Delta\Gamma}$$

Let

$$\left.\begin{aligned} f(\Lambda_{2n}^0) &= \alpha\,{}^0\Lambda_{2n}^0 - \beta\,{}^0\Lambda_{2n}^2 + \gamma\,{}^0\Lambda_{2n}^4 - \delta\,{}^0\Lambda_{2n}^6 \\ f(\Lambda_{2n}^2) &= \alpha\,{}^0\Lambda_{2n}^2 - \beta\,{}^0\Lambda_{2n}^4 + \gamma\,{}^0\Lambda_{2n}^6 - \delta\,{}^0\Lambda_{2n}^8 \\ f(\Omega_{2n}^0) &= \alpha'\,\Omega_{2n}^0 + \beta'\,\Omega_{2n}^2 + \gamma'\,\Omega_{2n}^4 + \delta'\,\Omega_{2n}^6 \\ f(\Omega_{2n}^2) &= \alpha'\,\Omega_{2n}^2 + \beta'\,\Omega_{2n}^4 + \gamma'\,\Omega_{2n}^6 + \delta'\,\Omega_{2n}^8 \end{aligned}\right\} \text{ for } n = 1, 2$$

Then clearly

$$\sigma_2 = \frac{6}{\pi}\cos^2\beta\cos^2\gamma\,\{\kappa^2[f(\Lambda_2^2)f(\Omega_4^0) + f(\Lambda_4^2)f(\Omega_2^0) - Gf(\Lambda_2^2)f(\Omega_2^0)]$$
$$+ \kappa'^2[f(\Lambda_2^0)f(\Omega_4^2) + f(\Lambda_4^0)f(\Omega_2^2) - Gf(\Lambda_2^0)f(\Omega_2^2)]\}$$

In order to find σ_4 and ζ_4, S_3 must be raised to the fourth power.

From (8) and (6)

$$\zeta_4 = \iint \Omega\,(S_3)^4 \, d\theta \, d\phi$$

where

$$\Omega = \frac{6}{\pi}\cos^4\beta\cos^4\gamma\,(\kappa^2\cos^2\theta + \kappa'^2\sin^2\phi)\left[-\frac{1}{\Gamma_1^6\Delta_1^6} + G\left(\frac{1}{\Gamma_1^6\Delta_1^4} + \frac{1}{\Gamma_1^4\Delta_1^6}\right)\right.$$
$$\left. - \frac{H}{\Gamma_1^4\Delta_1^4}\right]\frac{1}{\Delta\Gamma}$$

If then we put for $n=2, 3$

$$f(\Lambda_{2n}^0) = \alpha^2\, {}^0\Lambda_{2n}^0 - 2\alpha\beta\, {}^0\Lambda_{2n}^2 + (2\alpha\gamma+\beta^2)\, {}^0\Lambda_{2n}^4 - (2\alpha\delta + 2\beta\gamma)\, {}^0\Lambda_{2n}^6 + (2\beta\delta+\gamma^2)\, {}^0\Lambda_{2n}^8 - 2\gamma\delta\, {}^0\Lambda_{2n}^{10} + \delta^2\, {}^0\Lambda_{2n}^{12}$$

$f(\Lambda_{2n}^2)$ = same coefficients with ${}^0\Lambda_{2n}^2, {}^0\Lambda_{2n}^4, \ldots, {}^0\Lambda_{2n}^{14}$

$f(\Omega_{2n}^0)$ = same in form as $f(\Lambda_{2n}^0)$ but with $\alpha', \beta', \gamma', \delta'$, and Ω's

$f(\Omega_{2n}^2)$ = same in form as $f(\Lambda_{2n}^2)$ but with $\alpha', \beta', \gamma', \delta'$, and Ω's

We have

$$\zeta_4 = \frac{6}{\pi}\cos^4\beta\cos^4\gamma\,\{\kappa^2[-f(\Lambda_6^2)f(\Omega_6^0) + Gf(\Lambda_4^2)f(\Omega_6^0) + Gf(\Lambda_6^2)f(\Omega_4^0) - Hf(\Lambda_4^2)f(\Omega_4^0)] + \kappa'^2[-f(\Lambda_6^0)f(\Omega_6^2) + Gf(\Lambda_4^0)f(\Omega_6^2) + Gf(\Lambda_6^0)f(\Omega_4^2) - Hf(\Lambda_4^0)f(\Omega_4^2)]\}$$

Again from (8) we have

$$\sigma_4 = \frac{6}{\pi}\cos^3\beta\cos^3\gamma\sin\beta\iint(\kappa^2\cos^2\theta + \kappa'^2\sin^2\phi)\left[\tfrac{5}{2}\left(\frac{1}{\Delta_1{}^4\Gamma_1{}^6} + \frac{1}{\Delta_1{}^6\Gamma_1{}^4}\right) - \frac{3G}{\Delta_1{}^4\Gamma_1{}^4}\right]\frac{(S_3)^4}{\Delta\Gamma}\,d\theta\, d\phi$$

Thus with the same functions as for ζ_4 we have

$$\sigma_4 = \frac{6}{\pi}\cos^3\beta\cos^3\gamma\sin\beta\,\{\kappa^2[\tfrac{5}{2}f(\Lambda_4^2)f(\Omega_6^0) + \tfrac{5}{2}f(\Lambda_6^2)f(\Omega_4^0) - 3Gf(\Lambda_4^2)f(\Omega_4^0)] + \kappa'^2[\tfrac{5}{2}f(\Lambda_4^0)f(\Omega_6^2) + \tfrac{5}{2}f(\Lambda_6^0)f(\Omega_4^2) - 3Gf(\Lambda_4^0)f(\Omega_4^2)]\}$$

The computations gave

$$\sigma_2 = \cdot 0136760,\quad \zeta_4 = \cdot 000092343,\quad \sigma_4 = \cdot 000176218$$

These have to be used in a formula which also involves $\mathfrak{A}_3$. Now $\mathfrak{A}_3$ denotes $\mathfrak{P}_3\mathfrak{Q}_3$, or what should be the same thing $\mathsf{P}_1{}^1\mathsf{Q}_1{}^1$. The formulæ in the "Pear-shaped Figure" with $\gamma = 69° 49'\cdot 0$, $\kappa = \sin 73° 54'\cdot 2$ give

$$\mathsf{P}_1{}^1\mathsf{Q}_1{}^1 = \cdot 351697,\quad \mathfrak{P}_3\mathfrak{Q}_3 = \cdot 351663$$

These functions, which should be identical, differ by about one ten-thousandth part of either of them. I think if I had taken $\gamma = 69° 48'\cdot 997$, $\kappa = \sin 73° 54'\cdot 225$ (the actual numerical solution for the critical Jacobian, p. 311) this small discrepancy would have been removed. However, the difference is quite unimportant, and as $\mathfrak{A}_3$ generally means $\mathsf{P}_1{}^1\mathsf{Q}_1{}^1$, I take the former value and put $\log \mathfrak{A}_3 = 9\cdot 5461687$.

With this value I find the required result, namely

$$\mathfrak{A}_3[\tfrac{1}{3}(\sigma_2)^2 + 2\zeta_4] - \tfrac{1}{3}\sigma_4 = -\cdot 00050051 \quad\ldots\ldots\ldots\ldots(43)^*$$

* The discrepancy between this and my computation, in the paper as presented originally, is 4 in the seventh place of decimals.

§ 17. *The Rigorous Expressions for the several Harmonic Functions required.*

In order to obtain the remaining integrals it now becomes necessary to evaluate a number of the harmonic functions.

Rigorous forms have been found in Papers 10 and 11 for all the harmonics of orders up to the third inclusive. For harmonics of the fourth order rigorous algebraic forms may be obtained in all cases except when $s=0, 2, 4$, but these are exactly the cases to be considered in this investigation. We have, then, to show how these functions of higher orders may be evaluated rigorously for an ellipsoid of known ellipticity.

The only case required is that in which both i and s are even, and although all the forms might be evaluated by processes similar to those indicated below, I shall confine myself to this case.

We have seen in "Harmonics" that if β denotes $(1-\kappa^2)/(1+\kappa^2)$ of this paper, and μ denotes $\sin\theta$,

$$\mathfrak{P}_i^s(\mu) = P_i^s(\mu) - \beta q_{s+2} P_i^{s+2}(\mu) + \beta^2 q_{s+4} P_i^{s+4}(\mu) - \ldots + (-)^{\frac{1}{2}(i-s)} \beta^{\frac{1}{2}i} q_i P_i^i(\mu)$$
$$- \beta q_{s-2} P_i^{s-2}(\mu) + \beta^2 q_{s-4} P_i^{s-4}(\mu) - \ldots + (-)^{\frac{1}{2}s} \beta^{\frac{1}{2}s} q_0 P_i(\mu)$$

It is well known that

$$P_i(\mu) = \frac{1}{2^i\, i!} \frac{d^i}{d\mu} (\mu^2-1)^i$$

and

$$P_i^s(\mu) = \frac{(i+s)(i+s-1)\ldots(i-s+1)}{2^s . s!} (1-\mu^2)^{\frac{1}{2}s}$$
$$\times \left[\mu^{i-s} - \frac{(i-s)(i-s-1)}{1!\, 2^2 (s+1)} \mu^{i-s-2}(1-\mu^2)\right.$$
$$\left. + \frac{(i-s)(i-s-1)(i-s-2)(i-s-3)}{2!\, 2^4 (s+1)(s+2)} \mu^{i-s-4} (1-\mu^2)^2 - \ldots\right]$$

Hence we may clearly write $\mathfrak{P}_i^s$ in the form

$$\mathfrak{P}_i^s(\mu) = f_0 \sin^i\theta - f_2 \sin^{i-2}\theta \cos^2\theta + f_4 \sin^{i-4}\theta\cos^4\theta - \ldots$$

Since when s is not zero $P_i^s(1)=0$, and when s is zero $P_i(1)=1$, it follows that $f_0 = (-)^{\frac{1}{2}s} \beta^{\frac{1}{2}s} q_0$. For the zonal harmonics $(s=0)$ this gives $f_0 = 1$. The determination of the other f's depends on that of the q's, which we shall consider later.

Another form of $\mathfrak{P}_i^s(\mu)$ will be useful, viz.:

$$\mathfrak{P}_i^s(\mu) = a - b\cos^2\theta + c\cos^4\theta - d\cos^6\theta + e\cos^8\theta - \ldots$$

It is obvious that

$$a = f_0$$

$$b = f_2 + \frac{i}{2.1!} f_0$$

$$c = f_4 + \frac{i-2}{2.1!} f_2 + \frac{i(i-2)}{2^2.2!} f_0$$

$$d = f_6 + \frac{i-4}{2.1!} f_4 + \frac{(i-2)(i-4)}{2^2.2!} f_2 + \frac{i(i-2)(i-4)}{2^3.3!} f_0$$

&c. &c. &c.

Thus, when the f's are computed it is easy to obtain the a, b, c, d, &c.

We know that $\mathfrak{C}_i^s(\phi)$ (the cosine function of ϕ) is the same function of $-\kappa'^2 \sin^2\phi$ that $\mathfrak{P}_i^s(\mu)$ is of $\kappa^2\cos^2\theta$, except as regards a constant factor.

Hence it follows that

$$\mathfrak{C}_i^s(\phi) = \lambda\left[a + b\frac{\kappa'^2}{\kappa^2}\sin^2\phi + c\frac{\kappa'^4}{\kappa^4}\sin^4\phi + d\frac{\kappa'^6}{\kappa^6}\sin^6\phi + \dots\right]$$

where λ is a constant factor.

Now I desire to define $\mathfrak{P}_i^s(\mu)$ and $\mathfrak{C}_i^s(\phi)$ exactly as in "Harmonics."

This definition has already been adopted as regards $\mathfrak{P}_i^s(\mu)$, but it remains to adjust the constant λ so as to attain the same end as regards $\mathfrak{C}_i^s(\phi)$.

When i and s are even, $\mathfrak{C}_i^s(\phi)$ was defined thus:

$$\mathfrak{C}_i^s(\phi) = \cos s\phi + \beta p_{s+2}\cos(s+2)\phi + \beta^2 p_{s+4}\cos(s+4)\phi + \dots + \beta^{\frac{1}{2}(i-s)} p_i \cos i\phi$$
$$+ \beta p_{s-2}\cos(s-2)\phi + \beta^2 p_{s-4}\cos(s-4)\phi + \dots + \beta^{\frac{1}{2}s} p_0$$

Since

$$\sin^{2r}\phi = \frac{2r!}{2^{2r}(r!)^2} - \frac{2r!}{2^{2r-1}(r-1)!(r+1)!}\cos 2\phi + \frac{2r!}{2^{2r-1}(r-2)!(r+2)!}\cos 4\phi - \dots$$

it follows that the term independent of ϕ in $\mathfrak{C}_i(\phi)$ is

$$\lambda\left[a + \frac{1}{2}b\frac{\kappa'^2}{\kappa^2} + \frac{1.3}{2.4}c\frac{\kappa'^4}{\kappa^4} + \frac{1.3.5}{2.4.6}d\frac{\kappa'^6}{\kappa^6} + \dots\right]$$

The term in $\cos 2\phi$ in $\mathfrak{C}_i^2(\phi)$ is

$$-2\lambda\cos 2\phi\left[\frac{1}{2}\cdot\frac{1}{2}b\frac{\kappa'^2}{\kappa^2} + \frac{1.3}{2.4}\cdot\frac{2}{3}c\frac{\kappa'^4}{\kappa^4} + \frac{1.3.5}{2.4.6}\cdot\frac{3}{4}d\frac{\kappa'^6}{\kappa^6} + \dots\right]$$

The term in $\cos 4\phi$ in $\mathfrak{C}_i^4(\phi)$ is

$$2\lambda\cos 4\phi\left[\frac{1.3}{2.4}\cdot\frac{1.2}{3.4}c\frac{\kappa'^4}{\kappa^4} + \frac{1.3.5}{2.4.6}\cdot\frac{2.3}{4.5}d\frac{\kappa'^6}{\kappa^6} + \frac{1.3.5.7}{2.4.6.8}\cdot\frac{3.4}{5.6}e\frac{\kappa'^8}{\kappa^8} + \dots\right]$$

and so forth.

In accordance with the definition to be adopted, these terms in the three cases $s=0, 2, 4$ respectively are: $1, \cos 2\phi, \cos 4\phi$. Hence λ must be chosen so as to fulfil that condition.

Pursuing only the case of $\mathfrak{C}_i(\phi)$ (where $s=0$) in detail, we have

$$\frac{1}{\lambda} = a + \tfrac{1}{2}b\frac{\kappa'^2}{\kappa^2} + \frac{1.3}{2.4}c\frac{\kappa'^4}{\kappa^4} + \dots$$

If, then, $$\mathfrak{C}_i(\phi) = a' + b'\sin^2\phi + c'\sin^4\phi + \dots$$

we must have $$a' = \lambda a, \quad b' = \lambda b\frac{\kappa'^2}{\kappa^2}, \quad c' = \lambda c\frac{\kappa'^4}{\kappa^4}, \quad \text{\&c.}$$

Thus when f_0, f_2, f_4, &c., are found, it is easy to compute a, b, c, &c., and a', b', c', &c.

Our formulæ tend to involve the differences between large numbers, and this defect becomes more pronounced as the order of harmonics increases. The fault is mitigated by using the forms

$$\mathfrak{P}_i(\mu) = f_0\sin^i\theta - f_2\sin^{i-2}\theta\cos^2\theta + \dots$$

$$\mathfrak{C}_i(\phi) = a' + b'\sin^2\phi + c'\sin^4\phi + \dots$$

In the case of a lower harmonic, however, such as the fourth, we may just as well use the form for $\mathfrak{P}$ involving a, b, c, &c., and powers of $\cos^2\theta$.

We must now show how to complete the evaluation of the f's for the zonal harmonics.

It appears, from p. 212 of "Harmonics," that, when i is even, we have to solve the equation

$$\sigma = \frac{\tfrac{1}{2}\beta^2\{i, 1\}\{i, 2\}}{4.1^2 + \sigma -}\;\frac{\tfrac{1}{4}\beta^2\{i, 3\}\{i, 4\}}{4.2^2 + \sigma - \dots}$$

ending with

$$\frac{-\tfrac{1}{4}\beta^2\{i, i\}\{i, i-1\}}{i^2 + \sigma}, \quad \text{where} \quad \{i, j\} = (i+j)(i-j+1)$$

We are to take that root which vanishes when β vanishes.

Although the equation for σ is of order $\tfrac{1}{2}i + 1$ or $\tfrac{1}{2}(i+1)$, yet at least for such an ellipsoid as we have to deal with, it is very easy to solve it by successive rapid approximations.

It is clear that we may write the equation in the form

$$\sigma^2 + \left[4 - \frac{\tfrac{1}{4}\beta^2\{i, 3\}\{i, 4\}}{4.2^2 + \sigma -}\;\frac{\tfrac{1}{4}\beta^2\{i, 5\}\{i, 6\}}{4.3^2 + \sigma - \dots}\right]\sigma = \tfrac{1}{2}\beta^2\{i, 1\}\{i, 2\}$$

An analytical approximation is found by neglecting the continued fraction in the second term on the left, and we then obtain

$$\sigma = -2 + 2\sqrt{[1 + \tfrac{1}{8}\beta^2(i-1)i(i+1)(i+2)]}$$

If this value of σ is used in computing the first term of the continued fraction, and if the quadratic is solved again, we obtain a closer approximation. We then use the second approximation and include one more term in the continued fraction, and proceed until σ no longer changes.

It is shown on p. 215 of "Harmonics" that

$$\frac{q_2}{q_0} = \frac{1}{4.1^2+\sigma-}\;\frac{\frac{1}{4}\beta^2\{i,3\}\{i,4\}}{4.2^2+\sigma-\ldots}$$

$$\frac{2q_4}{q_2} = \frac{1}{4.2^2+\sigma-}\;\frac{\frac{1}{4}\beta^2\{i,5\}\{i,6\}}{4.3^2+\sigma-\ldots}$$

$$\frac{2q_6}{q_4} = \frac{1}{4.3^2+\sigma-}\;\frac{\frac{1}{4}\beta^2\{i,7\}\{i,8\}}{4.4^2+\sigma-\ldots}$$

....

It may be remarked that the factor 2 occurs in each of these equations on the left, excepting in the first one; also we are to take $q_0 = 1$.

In the course of the successive approximations for the determination of σ, each of these fractions is naturally evaluated. Therefore it is only necessary to extract certain numerical values already found in the course of solving the equation for σ.

As a verification, which shows whether the equation has been correctly solved, we have

$$\frac{q_0}{q_2} = \frac{\frac{1}{2}\beta^2\{i,1\}\{i,2\}}{\sigma}$$

It is now obvious that we are able to find all the q's in terms of q_0, which is unity. We then multiply each q by its appropriate power of β or $\frac{1-\kappa^2}{1+\kappa^2}$, that is to say, we form $\beta^r q_{2r}$ for $r = 1, 2, \ldots, \frac{1}{2}i$, and introduce the results into the formula for $\mathfrak{P}_i(\mu)$.

A closely analogous method enables us to find all the other types of function for an ellipsoid of known ellipticities, but, except for certain harmonics of the fourth order, it is not possible to obtain rigorous analytical solutions. Approximate analytical forms are given in "Harmonics," and the approximation may be carried further if desired.

The following tables give the coefficients in the several functions for the critical Jacobian ellipsoid with which we are dealing:—

i	s	a	b	c	d	e	f
*2	0	0·603374	0·923128				
*2	2	−0·039203	0·923128				
4	0	1·000000	5·442161	4·892138			
4	2	1·769147	−36·154264	−44·93584			
4	4	0·083965	− 7·984389	95·562431			
6	0	1·000000	12·45814	29·55340	18·53561		
†6	2	8·4	−121·8	−439·425	−320·513		
†6	4	3·78	−338·31	3680·303	4482·844		
8	0	1·00000	23·29297	103·90805	155·9554	74·7977	
10	0	1·00000	38·29978	274·94458	721·88640	789·90216	306·12784

i	s	a'	b'	c'	d'	e'	f'
*2	0	0·603374	0·076872				
*2	2	−0·039203	0·076872				
4	0	0·806905	0·365661	0·027371			
4	2	1·065020	−1·812415	−0·187586			
4	4	1·013640	−8·026680	8·000000			
6	0	0·62544	0·64882	0·12816	0·00669		
†6	2	1·1408	−1·5349	−0·4404	−0·0264		
†6	4	1·0305	−7·704	6·944	0·704		
8	0	0·440664	0·854891	0·317488	0·396793	0·001585	
10	0	0·289818	0·924288	0·552512	0·120795	0·011006	0·000355

As it is desirable to use the other form of $\mathfrak{P}$ in the higher zonal harmonics, I give the coefficients f_0, f_2, f_4, &c., in these cases. It will be noticed how much smaller are the numbers involved.

Coefficients of Terms in $\mathfrak{P}_i(\mu)$ when expressed in Sines and Cosines.

i	s	f_0	f_2	f_4	f_6	f_8	f_{10}
6	0	1·00000	9·45814	7·63713	0·44035		
8	0	1·00000	19·29797	40·01416	14·03323	0·45235	
10	0	1·00000	33·29978	131·74546	116·85134	22·76398	0·46728

* The second harmonics are here defined by

$$\mathfrak{P}_2^s(\mu) = \kappa^2 - q_s^2 - \kappa^2 \cos^2\theta, \quad \mathfrak{C}_2^s(\phi) = \kappa^2 - q_s^2 + \kappa'^2 \sin^2\phi \quad (s=0, 2)$$

with $q_s^2 = \frac{1}{3}[1+\kappa^2 \mp \sqrt{(1-\kappa^2\kappa'^2)}]$, with the upper sign for $s=0$ and the lower for $s=2$. In the case of $i=4$, $s=2$, I had in the original paper inadvertently changed the sign of $\mathfrak{P}_4^2$, $\mathfrak{P}_6^2$, and at the same time of $\mathfrak{C}_4^2$, $\mathfrak{C}_6^2$ without, of course, introducing any error.

† These functions are only given in their approximate forms.

§ 18. *The Integrals* ω_i^s, ρ_i^s, ϕ_i^s.

When i and s are both even, any harmonic S_i^s is given by

$$S_i^s = \mathfrak{P}_i^s(\mu)\,\mathfrak{C}_i^s(\phi)$$

where we may write the two factors in the forms

$$\mathfrak{P}_i^s(\mu) = a - b\cos^2\theta + c\cos^4\theta - \ldots$$
$$\mathfrak{C}_i^s(\phi) = a' + b'\sin^2\phi + c'\sin^4\phi + \ldots$$

Each series is, of course, terminable, the number of terms in each being $\frac{1}{2}i+1$.

For the determination of the ω, ρ integrals S_i^s must be multiplied by $(S_3)^2$.

Now we have seen that the squares of the two factors of S_3, which is $\mathfrak{P}_3(\mu)\,\mathrm{C}_3(\phi)$, are given by

$$[\mathfrak{P}_3(\mu)]^2 = \alpha - \beta\cos^2\theta + \gamma\cos^4\theta - \delta\cos^6\theta$$
$$[\mathrm{C}_3(\phi)]^2 = \alpha' + \beta'\sin^2\phi + \gamma'\sin^4\phi + \delta'\sin^6\phi$$

We have also seen that for all the harmonics there are alternative forms for $\mathfrak{P}_i(\mu)$ and $\mathfrak{P}_3(\mu)$, viz.

$$\mathfrak{P}_i^s(\mu) = f_0\sin^i\theta - f_2\sin^{i-2}\theta\cos^2\theta + f_4\sin^{i-4}\theta\cos^4\theta - \ldots$$
$$[\mathfrak{P}_3(\mu)]^2 = F_0\sin^6\theta - F_2\sin^4\theta\cos^2\theta + F_4\sin^2\theta\cos^4\theta$$

From (8) and (6) we have

$$\omega_i^s = \iint \Psi\,(S_3)^2 S_i^s\,d\theta\,d\phi, \quad \rho_i^s = \frac{\cos^2\beta\cos^2\gamma}{\sin\beta}\iint \frac{\Phi}{\Delta_1^2\Gamma_1^2}(S_3)^2 S_i^s\,d\theta\,d\phi$$

where

$$\Psi = \frac{6}{\pi}\cos^2\beta\cos^2\gamma\,(\kappa^2\cos^2\theta + \kappa'^2\sin^2\phi)\left[\frac{1}{\Gamma_1^4\Delta_1^2} + \frac{1}{\Gamma_1^2\Delta_1^4} - \frac{G}{\Gamma_1^2\Delta_1^2}\right]\frac{1}{\Delta\Gamma}$$

$$\Phi = \frac{6}{\pi}\,\frac{\kappa^2\cos^2\theta + \kappa'^2\sin^2\phi}{\Delta\Gamma}$$

It can be seen without actually writing out the intermediate steps of analysis how the several integrals will occur. I write the coefficients as follows:—

$$l_0 = a\alpha, \quad l_2 = a\beta + b\alpha, \quad l_4 = a\gamma + b\beta + c\alpha, \quad l_6 = a\delta + b\gamma + c\beta + d\alpha, \quad \&c.$$
$$m_0 = a'\alpha', \quad m_2 = a'\beta' + b'\alpha', \quad m_4 = a'\gamma' + b'\beta' + c'\alpha', \quad m_6 = a'\delta' + b'\gamma' + c'\beta' + d'\alpha', \quad \&c.$$

In the alternative form for the integrals with respect to θ, I write

$$L_0 = f_0F_0, \quad L_2 = f_0F_2 + f_2F_0, \quad L_4 = f_0F_4 + f_2F_2 + f_4F_0, \quad L_6 = f_2F_4 + f_4F_2 + f_6F_0, \quad \&c.$$

Next let

$$f(\Lambda_{2n}^0) = l_0{}^0\Lambda_{2n}^0 - l_2{}^0\Lambda_{2n}^2 + l_2{}^0\Lambda_{2n}^4 - l_4{}^0\Lambda_{2n}^6 + \ldots \quad (n = 1, 2)$$

or an alternative form of the same function is

$$f(\Lambda_{2n}^0) = L_0^{\,i+6}\Lambda_{2n}^0 - L_2^{\,i+4}\Lambda_{2n}^2 + L_4^{\,i+2}\Lambda_{2n}^4 - \dots \quad (n = 1, 2)$$

Let $$f(\Lambda_{2n}^2) = l_0^{\,0}\Lambda_{2n}^2 - l_2^{\,0}\Lambda_{2n}^4 + l_4^{\,0}\Lambda_{2n}^6 - \dots \quad (n = 1, 2)$$

or an alternative form of the same function is

$$f(\Lambda_{2n}^2) = L_0^{\,i+6}\Lambda_{2n}^2 - L_2^{\,i+4}\Lambda_{2n}^4 + L_4^{\,i+2}\Lambda_{2n}^6 - \dots \quad (n = 1, 2)$$

Let $$f(\Omega_{2n}^0) = m_0\,\Omega_{2n}^0 + m_2\,\Omega_{2n}^2 + m_4\,\Omega_{2n}^4 + \dots \quad (n = 1, 2)$$

$$f(\Omega_{2n}^2) = m_0\,\Omega_{2n}^2 + m_2\,\Omega_{2n}^4 + m_4\,\Omega_{2n}^6 + \dots \quad (n = 1, 2)$$

Then

$$\omega_i^s = \frac{6}{\pi}\cos^2\beta\cos^2\gamma\,\{\kappa^2\,[f(\Lambda_2^2)f(\Omega_4^0) + f(\Lambda_4^2)f(\Omega_2^0) - Gf(\Lambda_2^2)f(\Omega_2^0)]$$
$$+\,\kappa'^2\,[f(\Lambda_2^0)f(\Omega_4^2) + f(\Lambda_4^0)f(\Omega_2^2) - Gf(\Lambda_2^0)f(\Omega_2^2)]\}$$

$$\rho_i^s = \frac{6}{\pi}\,\frac{\cos^2\beta\cos^2\gamma}{\sin\beta}\,\{\kappa^2 f(\Lambda_2^2)f(\Omega_2^0) + \kappa'^2 f(\Lambda_2^0)f(\Omega_2^2)\}$$

I have used the rigorous formulæ for the harmonics in the cases of the zonal harmonics (s equal to zero), because they contribute by far the most important terms, also for the sectorial harmonics $i = 2$, $s = 2$ and $i = 4$, $s = 2$, and $s = 4$. But for the sixth harmonics, where $s = 2$ and 4, I have used the approximate formulæ of "Harmonics" with the parameter β equal to ·039973; the values of a, b, c &c., a', b', c' &c. were given above for all these cases.

It may be well to remark that ρ_0 is needed (but not ω_0), and in this case $a = a' = 1$, and b, c &c., b', c' &c. are all zero.

It seems useless to go in detail through the tedious operations involved in carrying out the processes in the several cases.

Approximate formulæ are given for the ϕ_i^s integrals in § 22 p. 276 of "Harmonics." The $\int p_0 d\sigma$ of that paper is the same as $\frac{4}{3}\pi k^3 \dfrac{\cos\beta\cos\gamma}{\sin^3\beta}\iint \Phi\, d\theta\, d\phi$ of the present one, and the factor there written M is $k^3\cos\beta\cos\gamma/\sin^3\beta$. Hence it follows that

$$\phi_i^s = \frac{3}{4\pi M}\int (\mathfrak{P}_i^s\mathfrak{C}_i^s)^2\, p\, d\sigma \text{ of "Harmonics."}$$

In order to apply this to the harmonics of the second order, it must be borne in mind that a different definition of S_2^s ($s = 0$, 2) is being used here.

If $[\phi_2]$, $[\phi_2^2]$ be the values which would be found from "Harmonics" without making this adaptation, and if ϕ_2, ϕ_2^2 are the required values, we may put

$$\mathfrak{P}_2(\mu) = x\,[\mathfrak{P}_2(\mu)], \quad \mathfrak{C}_2(\phi) = y\,[\mathfrak{C}_2(\phi)]$$

or $$\kappa^2\sin^2\theta - q_0^2 = x\alpha\left(\sin^2\theta + \frac{\gamma}{\alpha}\right), \quad q_0'^2 - \kappa'^2\cos^2\phi = -y\epsilon\left(\cos^2\phi + \frac{\zeta}{\epsilon}\right)$$

where α, ϵ, γ, ζ are the coefficients specified in § 12 p. 237 of "Harmonics."

Then the two definitions will agree if we take $x=\kappa^2/\alpha$, $y=-\kappa'^2/\epsilon$, although γ/α and ζ/ϵ will not be rigorously equal to $-q_0^2$ and $-q_0'^2$ respectively. Hence it follows that

$$\phi_2 = [\phi_2]\frac{\kappa^4\kappa'^4}{\alpha^2\epsilon^2}, \quad \text{and similarly } \phi_2^2 = [\phi_2^2]\frac{\kappa^4\kappa'^4}{\alpha'^2\epsilon'^2}$$

In the table below I have given the approximate values of the ϕ integrals derived in this way in all the several cases, but I have not thought it expedient to trust to the approximation throughout. The approximate values for ϕ_2 and ϕ_2^2 have been corrected, as just explained, so as to agree with the present definitions of the functions.

The ϕ integrals may be determined rigorously in two ways.

First we may find them by our table of integrals as follows:—

From (6) and (8) it appears that

$$\phi_i^s = \frac{6}{\pi}\iint(\kappa^2\cos^2\theta + \kappa'^2\sin^2\phi)\frac{(S_i^s)^2}{\Delta\Gamma}\,d\theta d\phi$$

In this case we have seen that $S_i^s = \mathfrak{P}_i^s(\mu)\,\mathfrak{C}_i^s(\phi)$, where

$$\mathfrak{P}_i^s(\mu) = a - b\cos^2\theta + c\cos^4\theta - d\cos^6\theta + \dots$$

or as an alternative

$$\mathfrak{P}_i^s(\mu) = f_0\sin^i\theta - f_2\sin^{i-2}\theta\cos^2\theta + f_4\sin^{i-4}\theta\cos^4\theta - \dots$$

Also $\qquad \mathfrak{C}_i^s(\phi) = a' + b'\sin^2\phi + c'\sin^4\phi + d'\sin^6\phi + \dots$

Writing

$$\lambda_0 = a^2, \quad \lambda_2 = 2ab, \quad \lambda_4 = 2ac + b^2, \quad \lambda_6 = 2ad + 2bc, \text{ \&c.}$$

or $\qquad L_0 = f_0^2, \quad L_2 = 2f_0f_2, \quad L_4 = 2f_0f_4 + f_2^2, \quad$ &c.

$$\mu_0 = a'^2, \quad \mu_2 = 2a'b', \quad \mu_4 = 2a'c' + b'^2, \quad \text{\&c.}$$

we have

$$[\mathfrak{P}_i^s(\mu)]^2 = \lambda_0 - \lambda_2\cos^2\theta + \lambda_4\cos^4\theta - \dots$$

or $\qquad = L_0\sin^{2i}\theta - L_2\sin^{2i-2}\theta\cos^2\theta + L_4\sin^{2i-4}\theta\cos^4\theta - \dots$

$$[\mathfrak{C}_i^s(\phi)]^2 = \mu_0 + \mu_2\sin^2\phi + \mu_4\sin^4\phi + \dots$$

The results here, as elsewhere, have a tendency to present themselves as the difference between large numbers, and the object of the second alternative form is to obviate this difficulty in a measure.

For the fourth zonal harmonic it is immaterial which form is used, for the sixth zonal the second form is the better one, and for the eighth zonal it is necessary, and for the tenth zonal the use of logarithms of more than seven places would become necessary.

Let us write

$$f(\Lambda_0^0) = \lambda_0{}^0\Lambda_0^0 - \lambda_2{}^0\Lambda_0^2 + \lambda_4{}^0\Lambda_0^4 - \dots$$

or as an alternative form of the same function

$$f(\Lambda_0^0) = L_0{}^{2i}\Lambda_0^0 - L_2{}^{2i-2}\Lambda_0^2 + L_4{}^{2i-4}\Lambda_0^4 - \dots$$

$$f(\Lambda_0^2) = \lambda_0{}^0\Lambda_0^2 - \lambda_2{}^0\Lambda_0^4 + \lambda_4{}^0\Lambda_0^6 - \dots$$

or alternatively

$$f(\Lambda_0^2) = L_0{}^{2i}\Lambda_0^2 - L_2{}^{2i-2}\Lambda_0^4 + L_4{}^{2i-4}\Lambda_0^6 - \dots$$

Also

$$f(\Omega_0^0) = \mu_0\Omega_0^0 + \mu_2\Omega_0^2 + \mu_4\Omega_0^4 + \dots$$

$$f(\Omega_0^2) = \mu_0\Omega_0^2 + \mu_2\Omega_0^4 + \mu_4\Omega_0^6 + \dots$$

We then have

$$\phi_i^s = \frac{6}{\pi}\{\kappa^2 f(\Lambda_0^2) f(\Omega_0^0) + \kappa'^2 f(\Lambda_0^0) f(\Omega_0^2)\}$$

The use of these formulæ as far as the 10th harmonic necessitates the extension of our tables of the series 0 as far as ${}^0\Lambda_0^{22}$, ${}^0\Omega_0^{22}$.

The second alternative process is that given in the next succeeding Paper, No. 13.

It is necessary to evaluate the functions there denoted $[2n, 2m]$ from a sequence equation. The labour of obtaining these functions in an analytical form is practically prohibitive for harmonics higher than the third, but it is easy to evaluate them numerically for the particular ellipsoid with which we are concerned.

The results again present themselves as the differences between large numbers, but by the exercise of great care in computation with seven-figured logarithms it is possible to get a fairly good result as far as the eighth order.

I have computed these integrals by both the methods explained with good agreement of results.

The following table gives the results of the computations which have now been explained.

Table of logarithms of ω_i^s, ρ_i^s, ϕ_i^s.

i	s	$\log \omega_i^s + 10$	$\log \rho_i^s + 10$	$\log \phi_i^s$	Approx. $\log \phi_i^s$ from formula in "Harmonics"
0	0	—	7·6310567	—	—
2	0	7·6714241	7·0286816	9·0051748 − 10	9·00518 − 10
2	2	(−) 5·6818162	(−) 5·0264000	7·0397377 − 10	7·03981 − 10
4	0	8·0332932	7·3558076	9·6886735 − 10	9·68861 − 10
4	2	(−) 8·25158	(−) 7·32157	1·72739	1·72729
4	4	8·30779	7·37092	3·81610	3·81612
6	0	7·96786	7·32449	9·69177 − 10	9·69303 − 10
6	2	(−) 8·72778	(−) 7·94094	—	2·20562
6	4	9·10094	8·13161	—	5·29999
8	0	7·78437	6·96857	9·75611 − 10	9·76872 − 10
10	0	(−) 7·9838	6·6024	9·8473 − 10	9·87800 − 10

Note that ω_{10} is negative while ρ_{10} remains positive.

The calculation of the integrals for $i = 8$ and $i = 10$ was very laborious. and as the results tend to present themselves as the differences between large numbers, it was difficult to obtain accuracy with logarithms of only seven places of decimals. The integrals ϕ were much the most troublesome, and indeed I do not claim close accuracy for ϕ_8. As it appeared to be impossible to compute ϕ_{10} to nearer than ten per cent. by the formula, I computed the several constituent integrals for the tenth harmonic by quadratures and combined them to form ϕ_{10}. The results derived from the approximate formulæ of "Harmonics" are given for the sake of comparison. They clearly give somewhat too large values for the higher harmonics.

I believe ω_{10} and ρ_{10} to be nearly correct. If allowance be made for the difference of definition adopted in this paper from that used in "Harmonics" as regards the second zonal harmonic, it will be found that $\omega_2, \omega_4, \omega_6, \omega_8, \omega_{10}$, when set out graphically, fall into an evenly flowing curve.

The corresponding test for the ρ's is not quite so convincing, but there is nothing which implies a mistake. The values $\rho_0, \rho_2, \rho_4, \rho_6$ fall well into line, and so also do $\rho_4, \rho_6, \rho_8, \rho_{10}$, but there is a gentle elevation in the neighbourhood of ρ_6. In consequence of this test I recomputed the *whole* again independently, after it had been recomputed and verified once, and special attention was devoted to ω_6 and ρ_6.

§ 19. *The Rigorous Expression for the Harmonics of the second kind, and the integrals $\mathfrak{A}_i^s$, $\mathfrak{B}_i^s$.*

The integral $\mathfrak{A}_i^s$ denotes $\mathfrak{P}_i^s(\nu_0)\,\mathfrak{Q}_i^s(\nu_0)$, and $\mathfrak{B}_i^s$ denotes $\mathfrak{Q}_i^s(\nu_0)\dfrac{d}{d\nu_0}\mathfrak{P}_i^s(\nu_0)$. Thus $\mathfrak{A}_i^s$ is, in fact, the harmonic function of the second kind. $\mathfrak{B}_i^s$ is clearly determinable from $\mathfrak{A}_i^s$.

In the original paper $\mathfrak{A}_i^s$ was found by quadrature, which was not, perhaps, a very satisfactory method, and the defect will now be made good by finding these integrals in terms of the F and E elliptic integrals. It appears that my former results were sufficiently near to the truth for practical purposes.

The functions $\mathfrak{P}_i^s(\nu)$ or $\mathsf{P}_i^s(\nu)$ are of eight types, determined by the oddness or evenness of i and s, and the association with a cosine or sine function of ϕ. In "Harmonics" the types are indicated by combinations in groups of three of the four letters E, O, C, S—denoting Even, Odd, Cosine, Sine; for example, OES means i odd, s even, associated with a sine function.

All the roots of the equation $\mathfrak{P}_i^s(\nu)$ or $\mathsf{P}_i^s(\nu) = 0$ are real, and when the form of the function has been determined by the method of § 17 the equation may be solved. Hence these functions are expressible as the products of a number of factors; and it is to be noted that it is not necessary to adopt the same definition as in "Harmonics," because the function may be multiplied by any constant factor, without affecting the result.

For brevity, let

$$\prod_1^n (\kappa^2\nu^2 - q_x^2) = (\kappa^2\nu^2 - q_1^2)(\kappa^2\nu^2 - q_2^2)\dots(\kappa^2\nu^2 - q_n^2)$$

The alternative notation will be needed, in which we write

$$\nu^2 = \frac{1}{\kappa^2 \sin^2\psi}, \quad \Delta_x^2 = 1 - q_x^2 \sin^2\psi, \quad \Delta^2 = 1 - \kappa^2 \sin^2\psi$$

At the surface of the ellipsoid $\nu = \nu_0$, $\psi = \gamma$, and we shall, as before, write $\sin\beta = \kappa \sin\gamma$. At the surface of the ellipsoid we have then

$$\nu = \operatorname{cosec}\beta, \quad \Delta_x^2 = 1 - q_x^2 \sin^2\gamma, \quad \Delta^2 = \cos^2\beta$$

In this notation

$$\prod_1^n \left(\frac{\Delta_x^2}{\sin^2\psi}\right) = \frac{\Delta_1^2\Delta_2^2\dots\Delta_n^2}{\sin^{2n}\psi}$$

A consideration of the eight types of harmonics shows that they may be written as follows:—

Type.

EEC, $\mathfrak{P}_{2n}^{2t}(\nu) = \prod_1^n(\kappa^2\nu^2 - q_x^2) = \operatorname{cosec}^{2n}\psi \prod_1^n(\Delta_x^2)$

EES, $\mathsf{P}_{2n+2}^{2t}(\nu) = (\kappa^2\nu^2 - \kappa^2)^{\frac{1}{2}}(\kappa^2\nu^2 - 1)^{\frac{1}{2}}\prod_1^n(\kappa^2\nu^2 - q_x^2) = \operatorname{cosec}^{2n+2}\psi\Delta\cos\psi\prod_1^n(\Delta_x^2)$

OOC, $\mathsf{P}_{2n+1}^{2t+1}(\nu) = (\kappa^2\nu^2 - 1)^{\frac{1}{2}}\prod_1^n(\kappa^2\nu^2 - q_x^2) = \operatorname{cosec}^{2n+1}\psi\cos\psi\prod_1^n(\Delta_x^2)$

OOS, $\mathfrak{P}_{2n+1}^{2t+1}(\nu) = (\kappa^2\nu^2 - \kappa^2)^{\frac{1}{2}}\prod_1^n(\kappa^2\nu^2 - q_x^2) = \operatorname{cosec}^{2n+1}\psi\Delta\prod_1^n(\Delta_x^2)$

OEC, $\mathfrak{P}_{2n+1}^{2t}(\nu) = \kappa\nu\prod_1^n(\kappa^2\nu^2 - q_x^2) = \operatorname{cosec}^{2n+1}\psi\prod_1^n(\Delta_x^2)$

OES, $\mathsf{P}_{2n+3}^{2t}(\nu) = \kappa\nu(\kappa^2\nu^2 - \kappa^2)^{\frac{1}{2}}(\kappa^2\nu^2 - 1)^{\frac{1}{2}}\prod_1^n(\kappa^2\nu^2 - q_x^2) = \operatorname{cosec}^{2n+3}\psi\Delta\cos\psi\prod_1^n(\Delta_x^2)$

EOC, $\mathsf{P}_{2n+2}^{2t+1}(\nu) = \kappa\nu(\kappa^2\nu^2 - 1)^{\frac{1}{2}}\prod_1^n(\kappa^2\nu^2 - q_x^2) = \operatorname{cosec}^{2n+2}\psi\cos\psi\prod_1^n(\Delta_x^2)$

EOS, $\mathfrak{P}_{2n+2}^{2t+1}(\nu) = \kappa\nu(\kappa^2\nu^2 - \kappa^2)^{\frac{1}{2}}\prod_1^n(\kappa^2\nu^2 - q_x^2) = \operatorname{cosec}^{2n+2}\psi\Delta\prod_1^n(\Delta_x^2)$

Using $\mathfrak{P}$ and $\mathfrak{Q}$ generically for any one of these and for the corresponding function of the other kind, we have

$$\mathfrak{P}(\nu_0)\mathfrak{Q}(\nu_0) = [\mathfrak{P}(\nu_0)]^2 \int_{\nu_0}^{\infty} \frac{d\nu}{[\mathfrak{P}(\nu)]^2(\nu^2 - 1)^{\frac{1}{2}}(\nu^2 - 1/\kappa^2)^{\frac{1}{2}}}$$

Or, changing the variable of integration to ψ,

$$\mathfrak{A} = \kappa[\mathfrak{P}(\nu_0)]^2 \int_0^{\gamma} \frac{d\psi}{[\mathfrak{P}(\nu)]^2\Delta}$$

To effect the integration the reciprocal of the square of $\mathfrak{P}$ must be expressed in partial fractions. Inspection of the eight forms of functions shows that

$$\frac{1}{[\mathfrak{P}(\nu)]^2} = \frac{1}{f}\cdot\frac{1}{\kappa^2\nu^2-\kappa^2} + \frac{1}{g}\cdot\frac{1}{\kappa^2\nu^2-1} + \frac{1}{h}\cdot\frac{1}{\kappa^2\nu^2} + \sum_1^n \frac{1}{A_x}\left[\frac{1}{(\kappa^2\nu^2-q_x^2)^2} - \frac{B_x}{\kappa^2\nu^2-q_x^2}\right]$$

with appropriate values of f, g, h, A_x, B_x to be given hereafter.

In every case but that of OES some or all of f, g, h are infinite.

In terms of ψ

$$\frac{1}{[\mathfrak{P}(\nu)]^2} = \frac{1}{f}\frac{\sin^2\psi}{\Delta^2} + \frac{1}{g}\tan^2\psi + \frac{1}{h}\sin^2\psi + \sum_1^n \frac{1}{A_x}\left[\frac{\sin^4\psi}{\Delta_x^4} - \frac{B_x\sin^2\psi}{\Delta_x^2}\right]$$

This has to be divided by Δ and integrated, and the result will be expressible in terms of the elliptic integrals

$$F(\gamma) = \int_0^\gamma \frac{d\psi}{\Delta}, \quad E(\gamma) = \int_0^\gamma \Delta\, d\psi, \quad \Pi(\gamma, \Delta_x) = \int_0^\gamma \frac{d\psi}{\Delta_x^2\Delta}$$

Accordingly we require certain integrals, which are given on pp. 299, 300, of the "Pear-shaped Figure," but in somewhat different forms. Here and elsewhere κ'^2 denotes $1-\kappa^2$, and $q_x'^2$ denotes $1-q_x^2$.

The integrals needed are as follows:—

$$\left.\begin{aligned}
\int_0^\gamma \frac{\sin^2\psi}{\Delta^3}\,d\psi &= -\frac{1}{\kappa^2}F(\gamma) + \frac{1}{\kappa^2\kappa'^2}E(\gamma) - \frac{\sin\gamma\cos\gamma}{\kappa'^2\cos\beta}\\
\int_0^\gamma \frac{\tan^2\psi}{\Delta}\,d\psi &= \qquad -\frac{1}{\kappa'^2}E(\gamma) + \frac{\tan\gamma\cos\beta}{\kappa'^2}\\
\int_0^\gamma \frac{\sin^2\psi}{\Delta}\,d\psi &= \frac{1}{\kappa^2}F(\gamma) - \frac{1}{\kappa^2}E(\gamma)
\end{aligned}\right\} \quad \text{.....................(44)}$$

$$\int_0^\gamma \frac{\sin^2\psi}{\Delta_x^2\Delta}\,d\psi = \frac{1}{q_x^2}\Pi(\gamma,\Delta_x) - \frac{1}{q_x^2}F(\gamma)$$

$$\int_0^\gamma \frac{\sin^4\psi}{\Delta_x^4\Delta}\,d\psi = \frac{1}{2q_x^2}\left(\frac{1}{q_x'^2} - \frac{1}{q_x^2} + \frac{1}{\kappa^2-q_x^2}\right)\Pi(\gamma,\Delta_x) + \frac{1}{2q_x^2}\left(\frac{1}{q_x^2} - \frac{1}{q_x'^2}\right)F(\gamma)$$
$$- \frac{1}{2q_x^2q_x'^2(\kappa^2-q_x^2)}E(\gamma) + \frac{\sin\gamma\cos\gamma\cos\beta}{2q_x'^2(\kappa^2-q_x^2)\Delta_x^2}$$

The last two of our integrals will occur in the form

$$\int_0^\gamma \frac{\sin^4\psi}{\Delta_x^4\Delta}\,d\psi - B_x\int_0^\gamma \frac{\sin^2\psi}{\Delta_x^2\Delta}\,d\psi$$

and the coefficient of $\Pi(\gamma, \Delta_x)$ in this expression is

$$\frac{1}{2q_x^2}\left(\frac{1}{q_x'^2} - \frac{1}{q_x^2} + \frac{1}{\kappa^2-q_x^2} - 2B_x\right)$$

[I shall now prove that this coefficient vanishes, and at the same time evaluate B_x. For the sake of brevity the suffix x will temporarily be omitted from the symbols q, A, B.

We have seen that the reciprocal of $[\mathfrak{P}]^2$ is expressible in partial fractions; and writing $\rho = \nu^2$, we have

$$\frac{(\kappa^2\rho - q^2)^2}{[\mathfrak{P}(\rho)]^2} = \frac{1}{A} - \frac{B}{A}(\kappa^2\rho - q^2) + L(\kappa^2\rho - q^2)^2$$

where L need not be determined.

If we write $f(\rho)$ for the left-hand side of this equation, it is clear that

$$\frac{1}{A} = f\left(\frac{q^2}{\kappa^2}\right)$$

$$B = -\frac{d}{\kappa^2 d\rho}\log f(\rho), \quad \left(\rho = \frac{q^2}{\kappa^2}\right)$$

Now $$\frac{d}{d\rho}\log f(\rho) = 2\left(\frac{\kappa^2}{\kappa^2\rho - q^2} - \frac{1}{\mathfrak{P}(\rho)}\frac{d\mathfrak{P}(\rho)}{d\rho}\right)$$

$$= 2\frac{\kappa^2\mathfrak{P}(\rho) - (\kappa^2\rho - q^2)\,d\mathfrak{P}(\rho)/d\rho}{(\kappa^2\rho - q^2)\,\mathfrak{P}(\rho)}$$

When $\rho = q^2/\kappa^2$, this expression assumes the form $0^2 \div 0^2$. Accordingly we differentiate the numerator and denominator each twice, and putting $\rho = q^2/\kappa^2$, we find

$$\frac{d}{d\rho}\log f(\rho)\left(\rho = \frac{q^2}{\kappa^2}\right) = -\frac{\mathfrak{P}''}{\mathfrak{P}'}$$

where $\mathfrak{P}'$, $\mathfrak{P}''$ denote the first and second differentials of $\mathfrak{P}$ with $\rho = q^2/\kappa^2$.

Hence $$B = \frac{\mathfrak{P}''}{\kappa^2\mathfrak{P}'}$$

The equation (3) of "Harmonics" (p. 193) is satisfied by $\mathfrak{P}(\nu)$. Now if we replace the independent variable ν by ρ (equal to ν^2) and substitute $(1-\kappa^2)/(1+\kappa^2)$ for β, that equation may be written as follows:—

$$\rho\left(\rho - \frac{1}{\kappa^2}\right)(\rho - 1)\left\{\frac{d^2\mathfrak{P}}{d\rho^2} + \tfrac{1}{2}\left(\frac{1}{\rho} + \frac{1}{\rho - 1} + \frac{1}{\rho - 1/\kappa^2}\right)\frac{d\mathfrak{P}}{d\rho}\right\}$$
$$-\tfrac{1}{4}i(i+1)\left[\rho - \frac{1+\kappa^2}{2\kappa^2}\right]\mathfrak{P} - \tfrac{1}{8}(s^2 - \sigma)\frac{(1+\kappa^2)}{\kappa^2}\mathfrak{P} = 0$$

On substituting for ρ the value q^2/κ^2, which is one of the roots of $\mathfrak{P} = 0$, we find

$$\mathfrak{P}'' + \tfrac{1}{2}\left(\frac{\kappa^2}{q^2} + \frac{\kappa^2}{q^2 - \kappa^2} + \frac{\kappa^2}{q^2 - 1}\right)\mathfrak{P}' = 0$$

Therefore, on reintroducing the suffix x,

$$2B_x = \frac{2\mathfrak{P}''}{\kappa^2\mathfrak{P}'} = \frac{1}{q_x'^2} - \frac{1}{q_x^2} + \frac{1}{\kappa^2 - q_x^2}$$

It follows that the $\Pi(\gamma, \Delta_x)$ integrals disappear, and we may use this value of B_x in the other terms*.]

The coefficient of $F(\gamma)$ in this same combination of integrals is

$$\frac{1}{2q_x^2}\left[\frac{1}{q_x^2} - \frac{1}{q_x'^2} + 2B_x\right]$$

In this we may substitute for B_x its value, and thus find

$$\int_0^\gamma \frac{\sin^4\psi}{\Delta_x^4\Delta}\,d\psi - B_x\int_0^\gamma \frac{\sin^2\gamma}{\Delta_x^2\Delta}\,d\psi = \frac{1}{2q_x^2(\kappa^2 - q_x^2)}F(\gamma) - \frac{1}{2q_x^2q_x'^2(\kappa^2 - q_x^2)}E(\gamma) + \frac{\sin\gamma\cos\gamma\cos\beta}{2q_x'^2(\kappa^2 - q_x^2)\Delta_x^2}$$

This expression together with (44) give all the required integrals, and it only remains to tabulate f, g, h, A_x for the several types of function.

For the sake of brevity I write

$$C_x = (q_x^2 - q_1^2)^2(q_x^2 - q_2^2)^2 \ldots (q_x^2 - q_n^2)^2 \quad (x = 1, 2 \ldots n)$$

the factor which would vanish being in each case omitted.

When there is only one q, C_1 is to be interpreted as unity.

Table of Values of f, g, h, and A_x.

Type	i order of harmonic	s rank of harmonic	f	g	h	A_x
EEC	$2n$	$2t$	∞	∞	∞	C_x
EES	$2n+2$	$2t$	$-\kappa'^2\prod_1^n(\kappa^2-q_x^2)^2$	$\kappa'^2\prod_1^n q_x'^4$	∞	$q_x'^2(\kappa^2-q_x^2)C_x$
OOC	$2n+1$	$2t+1$	∞	$\prod_1^n q_x'^4$	∞	$-q_x'^2C_x$
OOS	$2n+1$	$2t+1$	$\kappa'^2\prod_1^n(\kappa^2-q_x^2)^2$	∞	∞	$-(\kappa^2-q_x^2)C_x$
OEC	$2n+1$	$2t$	∞	∞	$\prod_1^n q_x^4$	$q_x^2C_x$
OES	$2n+3$	$2t$	$-\kappa^2\kappa'^2\prod_1^n(\kappa^2-q_x^2)^2$	$\kappa'^2\prod_1^n q_x'^4$	$\kappa^2\prod_1^n q_x^4$	$q_x^2q_x'^2(\kappa^2-q_x^2)C_x$
EOC	$2n+2$	$2t+1$	∞	$\prod_1^n q_x'^4$	$-\prod_1^n q_x^4$	$-q_x^2q_x'^2C_x$
EOS	$2n+2$	$2t+1$	$\kappa^2\prod_1^n(\kappa^2-q_x^2)^2$	∞	$-\kappa^2\prod_1^n q_x^4$	$-q_x^2(\kappa^2-q_x^2)C_x$

* This proof is adapted from Heine's *Kugelfunctionen*, Vol. I., Part II., Chapter 3, § 100, 62.

We have generally $\mathfrak{B} = \dfrac{\mathfrak{A}}{\mathfrak{P}(\nu_0)} \dfrac{d}{d\nu_0} \mathfrak{P}(\nu_0)$. Hence by logarithmic differentiation of the expressions for the several types of $\mathfrak{P}$ we find results given in the following table:—

Table of the $\mathfrak{B}$ Integrals.

Type	i order of harmonic	s rank of harmonic	$\mathfrak{B}_i^s \div 2 \sin\beta \mathfrak{A}_i^s$
EEC	$2n$	$2t$	$\sum_1^n 1/\Delta_x^2$
EES	$2n+2$	$2t$	$\frac{1}{2}\sec^2\beta + \frac{1}{2}\sec^2\gamma + \sum_1^n 1/\Delta_x^2$
OOC	$2n+1$	$2t+1$	$\frac{1}{2}\sec^2\gamma + \sum_1^n 1/\Delta_x^2$
OOS	$2n+1$	$2t+1$	$\frac{1}{2}\sec^2\beta + \sum_1^n 1/\Delta_x^2$
OEC	$2n+1$	$2t$	$\frac{1}{2} + \sum_1^n 1/\Delta_x^2$
OES	$2n+3$	$2t$	$\frac{1}{2} + \frac{1}{2}\sec^2\beta + \frac{1}{2}\sec^2\gamma + \sum_1^n 1/\Delta_x^2$
EOC	$2n+2$	$2t+1$	$\frac{1}{2} + \frac{1}{2}\sec^2\gamma + \sum_1^n 1/\Delta_x^2$
EOS	$2n+2$	$2t+1$	$\frac{1}{2} + \frac{1}{2}\sec^2\beta + \sum_1^n 1/\Delta_x^2$

In the case of the zonal harmonics ($s = 0$), q_x/κ is always less than unity; for harmonics of rank 2 one of the q_x/κ is greater than unity and the rest are less; for rank 4 two of them are greater than unity and the rest less.

For the zonal harmonics there is some gain in simplicity by putting $\sin^2\theta_x = \dfrac{q_x^2}{\kappa^2}$. We then take the equation

$$\mathfrak{P}_{2n}(\mu) = a - b\cos^2\theta + c\cos^4\theta - \ldots = 0$$

and find all the n roots, say $\theta_1, \theta_2 \ldots \theta_n$.

If we solve the corresponding equation $\mathfrak{P}_{2n}^2(\mu) = 0$ for the tesseral harmonic of rank 2, we find one root for $\cos^2\theta$ to be negative. If this root corresponds to θ_1, we must put $\cos^2\theta_1 = 1 - \dfrac{q_1^2}{\kappa^2}$, so that $q_1^2 = \kappa^2[1 + (-\cos^2\theta_1)]$. Similarly for the harmonics of rank 4, two roots correspond with imaginary angles, and so forth.

Subject to this explanation we may now regard the roots as defined by $\theta_1, \theta_2 \ldots \theta_n$.

Since $\Delta_x^2 = 1 - q_x^2 \sin^2\gamma$, we have $\Delta_x^2 = 1 - \kappa^2 \sin^2\gamma \sin^2\theta_x = 1 - \sin^2\beta \sin^2\theta_x$, and $\mathfrak{P}_{2n}^{2t}(\mu) = \operatorname{cosec}^{2n}\gamma \Delta_1^2 \Delta_2^2 \ldots \Delta_n^2$.

If we write

$$D_x = (\sin^2\theta_x - \sin^2\theta_1)^2 (\sin^2\theta_x - \sin^2\theta_2)^2 \ldots (\sin^2\theta_x - \sin^2\theta_n)^2 \quad (n-1 \text{ factors})$$

our former C_x may be written in the form $\kappa^{4n-4} D_x$, and the several coefficients in the expression for $\mathfrak{A}_i^s$ may be expressed as trigonometrical functions—some of which may, however, be hyperbolic.

We thus have

$$\mathfrak{A}_{2n}^{2t} = \frac{1}{2\kappa^{4n-1}} \prod_1^n (1 - \sin^2\beta \sin^2\theta_x)^2 \left[\sum_1^n \frac{F(\gamma)}{D_x \sin^2\theta_x \cos^2\theta_x} - \sum_1^n \frac{E(\gamma)}{D_x \sin^2\theta_x \cos^2\theta_x (1 - \kappa^2 \sin^2\theta_x)} + \sum_1^n \frac{\kappa^2 \sin\gamma \cos\gamma \cos\beta}{D_x \cos^2\theta_x (1 - \kappa^2 \sin^2\theta_x)(1 - \sin^2\beta \sin^2\theta_x)} \right]$$

This formula agrees with the result given for $\mathfrak{A}_2^s$ ($s = 0, 2$) in § 4, p. 300, of "The Pear-shaped Figure," although the formula is there expressed in terms of q^2, and $1/(\kappa^2 - q^2)$ is replaced by its equivalent $(1 - 2q^2)/q^2 q'^2$.

In the case of the even zonal harmonics of order i, all the θ's are real angles, and it facilitates the solution of the equation for θ to note that, with rough approximation (improving as the order of harmonic increases),

$$\theta_1 = \frac{\pi}{2i}, \quad \theta_2 = \frac{3\pi}{2i}, \quad \theta_3 = \frac{5\pi}{2i}, \ldots, \quad \theta_{\frac{1}{2}i} = \frac{(i-1)\pi}{2i}.$$

The following numerical values apply to the critical Jacobian ellipsoid:—

For the fourth harmonic $\theta_1 = 20° \, 15'$, $\theta_2 = 61° \, 11'$; the rough approximation gives $22° \, 30'$ and $67° \, 30'$.

For the sixth $\theta_1 = 14° \, 1'{\cdot}9$, $\theta_2 = 42° \, 12'{\cdot}2$, $\theta_3 = 71° \, 8'{\cdot}6$; the rough approximation being $15°$, $45°$, $75°$.

For the eighth $\theta_1 = 10° \, 43'{\cdot}1$, $\theta_2 = 32° \, 11'{\cdot}8$, $\theta_3 = 53° \, 51'{\cdot}2$, $\theta_4 = 76° \, 21'{\cdot}8$; the approximation being $11° \, 45'$, $34° \, 15'$, $56° \, 45'$, $79° \, 15'$.

For the tenth $\theta_1 = 8° \, 40'$, $\theta_2 = 26° \, 1'$, $\theta_3 = 43° \, 26'$, $\theta_4 = 61° \, 3'$, $\theta_5 = 79° \, 28'$; the approximation being $9°$, $27°$, $45°$, $63°$, $81°$.

The values of the several 𝔄's found by the quadratures used in this paper in its original form were in every case too small; the correct values are given in the table below. I find that for $\mathfrak{A}_4$ quadratures gave too small a value by a $\frac{1}{300}$th part; for $\mathfrak{A}_6$ by a $\frac{1}{140}$th part; for $\mathfrak{A}_8$ by an $\frac{1}{87}$th part.

The method which I have given above fails for the tenth zonal harmonic, unless we use logarithms of more than seven places; and it is not worth while to undertake so heavy a piece of computation. I conclude by extrapolation that for $\mathfrak{A}_{10}$ quadrature (carried out on exactly the same plan as in all the other cases, but not reproduced here) gives too small a result by a $\frac{1}{70}$th part of itself. I therefore augment in this case the result of the quadratures and find $\mathfrak{A}_{10} = 0{\cdot}11640$; this enables us also to compute $\mathfrak{B}_{10}$.

The following table gives the results of the whole computation:—

Table of Logarithms of $\mathfrak{A}_i^s$ and $\mathfrak{B}_i^s$ Integrals.

i	s	$\log \mathfrak{A}_i^s + 10$	$\log \mathfrak{B}_i^s$
2	0	9·6931231	·0929494
2	2	9·3330037	·4066504
3	0	9·54617	·20462
4	0	9·4332383	·2657402
4	2	9·24250	·39502
4	4	9·04753	·43121
6	0	9·2701270	·3263106
6	2	9·14462	·39512
6	4	9·00632	·42458
8	0	9·15835	·35745
10	0	9·06595	·36897

§ 20. *Synthesis of Numerical results; Stability of the Pear.*

In the following tables and remarks some of the results which occur in the course of the work are collected together.

i	s	(1) $\mathfrak{A}_3 - \mathfrak{A}_i^s$	(2) $\log(\mathfrak{A}_3 - \mathfrak{A}_i^s)\,\phi_i^s = \log C_i^s$	(3) $\mathfrak{A}_i^s\,\omega_i^s + \frac{1}{2}\mathfrak{B}_i^s\rho_i^s$	(4) $\mathfrak{B}_3\rho_i^s - \dfrac{\rho_i^s \sin^2\beta}{2\cos\beta\cos\gamma}$	(3) + (4) B_i^s
2	0	− ·141617	− 8·1562893 − 10	·0029766	− ·0012020	·001776
2	2	·136410	6·1745839 − 10	− ·000024745	− ·000012702	− ·00001204
4	0	·080529	8·5946258 − 10	·0050195	− ·0025529	·0024666
4	2	·176913	·97515	− ·0057115	·0023591	− ·0033524
4	4	·241787	3·19953	·0053573	− ·0026433	·0027140
6	0	·16543	8·91050 − 10	·0039673	− ·0023753	·0015920
6	2	·2122	1·5323	− ·01829	·00982	− ·00847
6	4	·2502	4·6983	·03081	− ·01524	·01557
8	0	·2077	9·07354 − 10	·001936	− ·001047	·000889
10	0	·235	9·21888 − 10	− ·000653	− ·000450	− ·001103

For all harmonics higher than those of the second degree $\mathfrak{A}_3 - \mathfrak{A}_i^s$ is the coefficient of stability. Since in all these cases the expression is positive, the ellipsoid is stable for all such deformations.

If $U + \delta U$ be the energy functions for the pear, whose variations for constant moment of momentum are considered by M. Poincaré, we have in our notation

$$U + \delta U = -\tfrac{1}{2}\int \frac{dm_1 dm_2}{D_{12}} + \tfrac{1}{2}(A_j - A_r)(\omega^2 + \delta\omega^2)$$

It is easy to show from our analysis that for the deformation $f_2 S_2$

$$\delta U = \frac{3M^2}{2k_0}(f_2)^2 \left\{(\mathfrak{A}_3 - \mathfrak{A}_2)\phi_2 + \frac{\omega^2}{2\pi\rho}\frac{\mathfrak{c}^2}{\mathfrak{a}}\right\}$$

and that the corresponding expression with $\mathfrak{b}^2$ in place of $\mathfrak{c}^2$ holds good for the deformation $f_2^2 S_2^2$.

Forestalling the results obtained below, it may be stated that for $f_2 S_2$

$$\delta U = \frac{3M^2}{2k_0}(f_2)^2\{-\cdot 014 + \cdot 040\}$$

and for $f_2^2 S_2^2$

$$\delta U = \frac{3M^2}{2k_0}(f_2^2)^2\{\cdot 00015 + \cdot 00002\}$$

Thus in both cases δU is positive, and this shows that the Jacobian ellipsoid is also stable for the ellipsoidal deformations. The fact that δE (the variation of my function of energy for constant angular velocity) is negative for the deformation S_2, illustrates the truth of M. Poincaré's remark (*Acta Math.*, VII. p. 365): "Si au contraire la rotation de la masse fluide était déterminée par celle d'un axe rigide (comme dans les expériences de Plateau par exemple), tout déplacement produirait une résistance passive et l'ellipsoïde de Jacobi serait toujours instable."

I have written in (32)

$$[i, s] = \frac{1}{\mathfrak{A}_i^s \phi_i^s}\left\{(\mathfrak{A}_i^s \omega_i^s + \tfrac{1}{2}\mathfrak{B}_i^s \rho_i^s)^2 - \frac{\sin^2\beta}{4\cos\beta\cos\gamma}\mathfrak{B}_i^s(\rho_i^s)^2\right\}$$

The following table then gives further stages in the work:—

i	s	(1) $[i, s]$	(2) $(B_i^s)^2/C_i^s$	(1) + (2)	B_i^s/C_i^s
2	0	·000138868	− ·000219736	− ·000080868	− ·12382
2	2	·000000717	·000000970	·000001687	− ·08056
4	0	·000092542	·000154732	·000247274	·06273
4	2	·000001908	·000001190	·000003098	− ·000355
4	4	·000000012	·000000000	·000000012	·0000017
6	0	·000031204	·000031146	·000062350	·019564
6	2	·000003422	·000002107	·000005529	− ·000229
6	4	·000000014	·000000000	·000000014	·0000003
8	0	·000012905	·000006671	·000019576	·007505
10	0	− ·000001030	·000007358	·000006328	− ·00667
			Sum of column (1) + (2)	·000265000	
			$\mathfrak{A}_3[\frac{1}{3}(\sigma_2)^2+2\zeta_4]-\frac{1}{3}\sigma_4=$	− ·000500513	
			Numerator	− ·000235513	

Note that except in the case of the harmonics of the second order $f_i^s = \frac{B_i^s}{C_i^s} e^2$.

The next step is to find

$$L = \frac{\sin\beta}{4q_0^2\cos\beta\cos\gamma}\left(1 - \frac{\cos^2\beta}{D+\kappa^2}\right), \quad M = \frac{\sin\beta}{4q_2^2\cos\beta\cos\gamma}\left(1 + \frac{\cos^2\beta}{D-\kappa^2}\right)$$

where $D^2 = 1 - \kappa^2\kappa'^2$.

The numerical values are

$$\log D = 9{\cdot}9840165, \quad \log L = {\cdot}6454565, \quad \log M = {\cdot}9591963$$

From these we obtain $\mathfrak{c} = L\phi_2$, $\mathfrak{d} = M\phi_2^2$; whence

$$\frac{B_2}{C_2}\mathfrak{c} = -\,{\cdot}0553908$$

$$-\frac{B_2^2}{C_2^2}\mathfrak{d} = \quad {\cdot}0008037$$

$$\mathfrak{b} = \quad {\cdot}0316007$$

$$\text{Denominator} = -\,{\cdot}0229864$$

In accordance with (32) the Numerator divided by the Denominator is $-\delta\omega^2/4\pi\rho e^2$, and I thus find

$$\log\frac{\delta\omega^2}{4\pi\rho e^2} = (-)\,8{\cdot}01054$$

It was found in § 7 of the "Pear-shaped Figure" that the angular velocity of the critical Jacobian was given by $\frac{\omega^2}{2\pi\rho} = {\cdot}1419990$. Accordingly the square of the angular velocity of the pear being $\omega^2 + \delta\omega^2$, we have

$$\omega^2 + \delta\omega^2 = \omega^2\,[1 - {\cdot}1443066e^2]$$

From the formula (31), I then find

$$f_2 = \cdot195979e^2, \quad f_2^2 = \cdot603177e^2$$

The other f_i^s are equal to $\frac{B_i^s}{C_i^s}e^2$, and are given in the preceding table. From (28) and the definitions of $\mathfrak{a}$, $\mathfrak{b}$, $\mathfrak{c}$, $\mathfrak{d}$ it appears that the moment of inertia of the pear is

$$A_j - A_r = \frac{3M^2\mathfrak{a}}{2\pi\rho k_0}\left[1 + \frac{\mathfrak{b}}{\mathfrak{a}}e^2 + \frac{\mathfrak{c}}{\mathfrak{a}}f_2 - \frac{\mathfrak{d}}{\mathfrak{a}}f_2^2\right]$$

With $\log \mathfrak{a} = 9\cdot8559759$, I find

$$A_j - A_r = \frac{3M^2\mathfrak{a}}{2\pi\rho k_0}[1 + \cdot157786e^2]$$

The angular velocity of the pear is

$$\sqrt{(\omega^2 + \delta\omega^2)} = \omega[1 - \cdot072153e^2]$$

Multiplying these last two expressions together, we have the moment of momentum of the pear; it is

$$\frac{3M^2\mathfrak{a}\omega}{2\pi\rho k_0}[1 + \cdot085633e^2]$$

It follows that, while the angular velocity of the pear is less than that of the critical Jacobian, the moment of momentum is greater. This result would afford a rigorous proof of the stability of the pear if the numbers were based on a complete solution of the problem. But as we have not determined an infinite series of harmonic terms, it becomes necessary to consider how the result might differ if the hitherto uncomputed terms were added.

If ϵ denotes the hitherto uncomputed portion of the infinite series $\Sigma\{[i, s] + (B_i^s)^2/C_i^s\}$, and if Δ denotes the addition to be made on that account to any of the results as already computed, we have

$$\Delta\left(\frac{\delta\omega^2}{4\pi\rho}\right) = \frac{\epsilon e^2}{\cdot0229864} \quad \text{and} \quad \Delta\left(\frac{\delta\omega^2}{\omega^2}\right) = \frac{2\epsilon e^2}{\cdot0229864 \times \cdot141999}$$

Whence $$\Delta[\sqrt{(\omega^2 + \delta\omega^2)}] = \tfrac{1}{2}\omega\Delta\frac{\delta\omega^2}{\omega^2} = \omega[1 + 306\cdot367\epsilon e^2]$$

Since $$f_2 = \frac{B_2}{C_2}e^2 + \frac{\delta\omega^2}{4\pi\rho}\frac{\mathfrak{c}}{C_2}, \quad f_2^2 = \frac{B_2^2}{C_2^2}e^2 - \frac{\delta\omega^2}{4\pi\rho}\frac{\mathfrak{d}}{C_2^2}$$

$$\Delta f_2 = \frac{\mathfrak{c}}{C_2}\Delta\frac{\delta\omega^2}{4\pi\rho} = -10^{3\cdot13287}\epsilon e^2$$

$$\Delta f_2^2 = -\frac{\mathfrak{d}}{C_2^2}\Delta\frac{\delta\omega^2}{4\pi\rho} = -10^{3\cdot46288}\epsilon e^2$$

Then

$$\Delta(A_j - A_r) = \frac{3M^2\mathfrak{a}}{2\pi\rho k_0}\left[\frac{\mathfrak{c}}{\mathfrak{a}}\Delta f_2 - \frac{\mathfrak{d}}{\mathfrak{a}}\Delta f_2^2\right] = \frac{3M^2\mathfrak{a}}{2\pi\rho k_0}[-846\cdot302 + 40\cdot349]\epsilon e^2$$

$$= \frac{3M^2\mathfrak{a}}{2\pi\rho k_0}[-805\cdot953]\epsilon e^2$$

Therefore $\sqrt{(\omega^2+\delta\omega^2)}=\omega\,[1-{\cdot}072153e^2+306{\cdot}367\epsilon e^2]$

$$A_j-A_r=\frac{3M^2\mathfrak{a}}{2\pi\rho k_0}[1+{\cdot}157786e^2-805{\cdot}953\epsilon e^2]$$

By multiplication we find that the moment of momentum is

$$\frac{3M^2\mathfrak{a}\omega}{2\pi\rho k_0}[1+{\cdot}085633e^2-499{\cdot}586\epsilon e^2]$$

The coefficient of e^2 will be positive, and the pear stable, provided that

$$499{\cdot}586\epsilon<{\cdot}085633$$

or $$\epsilon<{\cdot}0001714$$

The eighth zonal harmonic gave a contribution of ·00001958 and the tenth of ·00000633. These are respectively 1/8·7 and 1/27 of the critical total ·00017. The pear is then stable unless the residue of the apparently highly convergent series shall amount to more than 27 times the contribution of the last term computed. M. Liapounoff, as explained in the Summary below, claims in effect to prove that this is the case, but to me it seems incredible. I look for the discrepancy between our conclusions in some other direction.

§ 21. *Second Approximation to the Form of the Pear.*

Extracting the numerical values of the f's from our results, we find that the inequality of the critical Jacobian ellipsoid is

$$eS_3+e^2\,[{\cdot}19598\,S_2+{\cdot}60318\,S_2^2+{\cdot}06273\,S_4-{\cdot}000355\,S_4^2+{\cdot}00000017\,S_4^4$$
$$+{\cdot}01956\,S_6-{\cdot}000229\,S_6^2+{\cdot}0000003\,S_6^4-?\,S_6^6+{\cdot}0075\,S_8-{\cdot}0067\,S_{10}]$$

In order to give this expression a clear meaning, it is well to define the several S's.

$$S_3=(\kappa^2\sin^3\theta-q^2\sin\theta)(q'^2-\kappa'^2\cos^2\phi)\sqrt{(1-\kappa'^2\cos^2\phi)}$$

where $$\kappa^2={\cdot}923128,\qquad q^2={\cdot}574647$$
$$\kappa'^2={\cdot}076872,\qquad q'^2={\cdot}425353$$

For the other harmonics we have

$$S_i^s=(a-b\cos^2\theta+c\cos^4\theta-d\cos^6\theta+\ldots)(a'+b'\sin^2\phi+c'\sin^4\phi+d'\sin^6\phi+\ldots)$$

where the values of a, b, &c., a', b', &c. are as given in a table on p. 364.

The surface of the pear is determined by measuring a certain length along the arc of curves orthogonal to the surface of the ellipsoid. By equation (22) it appears that that length measured in the direction of the positive normal is

$$p_0\left[eS_3+\Sigma f_i^s S_i^s+\tfrac{1}{2}e^2\frac{\cos^2\beta\cos^2\gamma}{\Delta_1^2\Gamma_1^2}\left(\frac{1}{\Delta_1^2}+\frac{1}{\Gamma_1^2}-2G\right)(S_3)^2\right]$$

In order to construct a figure it will be convenient to adopt as unit of length c, the greatest axis of the ellipsoid which is deformed. We know that $c=\frac{k}{\sin\beta}$, $b=k\cot\beta$, $a=\frac{k\cos\gamma}{\sin\beta}$, so that $b=c\cos\beta$, $a=c\cos\gamma$, and the mass of the ellipsoid is $\frac{4}{3}\pi\rho c^3\cos\beta\cos\gamma$. But since the mass of the pear is $\frac{4}{3}\pi\rho k_0^3\frac{\cos\beta\cos\gamma}{\sin^3\beta}$ where $k_0^3=k^3(1+e^2\sigma_2)$, it follows that it is

$$\tfrac{4}{3}\pi\rho c^3\cos\beta\cos\gamma\,(1+\cdot0136760e^2)$$

Hence the mass of the pear is a little greater than that of the ellipsoid whose deformations we shall draw, and the protuberances above the surface slightly exceed in volume the depressions below it.

We have

$$p_0=\frac{c\cos\beta\cos\gamma}{\Delta_1\Gamma_1}=\frac{c\cos\beta\cos\gamma}{(1-\sin^2\beta\sin^2\theta)^{\frac{1}{2}}(\cos^2\gamma+\kappa'^2\sin^2\gamma\cos^2\phi)^{\frac{1}{2}}}$$

and the expression for the orthogonal arc, measured from the ellipsoid to the pear, is therefore

$$p_0\left[eS_3+\left(\frac{p_0eS_3}{c}\right)^2\left\{\frac{1}{2(1-\sin^2\beta\sin^2\theta)}+\frac{1}{2(\cos^2\gamma+\kappa'^2\sin^2\theta\cos^2\phi)}\right.\right.$$
$$\left.\left.-\tfrac{1}{2}(1+\sec^2\beta+\sec^2\gamma)\right\}+\Sigma f_i^s S_i^s\right]$$

It appears to me that it will afford a sufficient idea of the corrected form of surface if I draw two principal sections, namely, first, a section through the axis of rotation and the longest axis of the ellipsoid, and, secondly, a section at right angles to the axis of rotation. It is not worth while to consider the third section drawn through the axis of rotation and the mean axis of the ellipsoid, since it will hardly differ sensibly from the uppermost figure shown in the "Pear-shaped Figure," p. 314.

For the sake of brevity I will call the first and second sections the meridian and the equator.

The three ellipsoidal co-ordinates ν, θ, ϕ of any point are connected with x, y, z by the relationships

$$x=c\sin\gamma\,.\,(\kappa^2\nu^2-1)^{\frac{1}{2}}(1-\kappa^2\sin^2\theta)^{\frac{1}{2}}\cos\phi$$
$$y=c\sin\gamma\,.\,\kappa(\nu^2-1)^{\frac{1}{2}}\cos\theta\sin\phi$$
$$z=c\sin\gamma\,.\,\kappa\nu\sin\theta(1-\kappa'^2\cos^2\phi)^{\frac{1}{2}}$$

The equation to the surface of the ellipsoid is $\nu=\frac{1}{\kappa\sin\gamma}=\frac{1}{\sin\beta}$.

The equation to the meridian plane in rectangular co-ordinates is simply $y=0$, that to the equator is $x=0$.

In ellipsoidal co-ordinates the equation to the equator is simply $\phi = \frac{1}{2}\pi$, but the equation to the meridian is peculiar, for it is in part represented by $\theta = \frac{1}{2}\pi$ and in part by $\phi = 0$.

The curve $\theta = \frac{1}{2}\pi$, $\phi = 0$, which defines the limit between the two regions where the equation to the plane has different forms, is clearly the hyperbola

$$z^2 - \frac{x^2}{\kappa'^2} = c^2 \sin^2 \gamma$$

In the region from $z = \infty$ and x small down to this hyperbola the equation is $\theta = \frac{1}{2}\pi$; and between the origin and the hyperbola it is $\phi = 0$.

If we follow the arc of the ellipse from the extremity of the c axis we begin with $\theta = \frac{1}{2}\pi$, $\phi = \frac{1}{2}\pi$, and θ remains constant whilst ϕ falls to zero. Then ϕ maintains a constant zero value whilst θ falls from $\frac{1}{2}\pi$ to zero.

On the side of the origin where z is negative, θ is of course negative and undergoes parallel changes.

The hyperbola $\theta = \frac{1}{2}\pi$, $\phi = 0$ cuts the ellipsoid so near to the extremities of the c axis that an adequate idea of the deformation may be derived from the two extreme values of ϕ, namely, $\frac{1}{2}\pi$ and 0. I have also thought it sufficient to compute the deformations for $\theta = 0$, 30°, 60°, 90°. We thus obtain the following scheme of values of θ, ϕ, together with the corresponding rectangular co-ordinates (with c taken as unity), at which to compute the deformation:—

Meridian ($y=0$).	Equator ($x=0$).
$\theta = 90°$, $\phi = 90°$; $z = 1$, $x = 0$	$\theta = 90°$, $\phi = 90°$; $z = 1$, $y = 0$
$\theta = 90°$, $\phi = 0$; $z = \cdot 961$, $x = \cdot 096$	$\theta = 60°$, $\phi = 90°$; $z = \cdot 866$, $y = \cdot 216$
$\theta = 60°$, $\phi = 0$; $z = \cdot 832$, $x = \cdot 191$	$\theta = 30°$, $\phi = 90°$; $z = \cdot 5$, $y = \cdot 374$
$\theta = 30°$, $\phi = 0$; $z = \cdot 480$, $x = \cdot 303$	$\theta = 0$, $\phi = 90°$; $z = 0$, $y = \cdot 432$
$\theta = 0$, $\phi = 0$; $z = 0$, $x = \cdot 345$	

It did not seem to be worth while to compute the deformations due to the eighth zonal harmonic, since it would be quite impossible to show them on a drawing of any reasonable scale.

In order to exhibit the magnitudes of the contributions of the harmonics of the several orders, I give the normal departures δn at the points $z = \pm 1$, $x = 0$, $y = 0$.

Term of first order	S_3	$\pm\cdot148227e$	
Terms of second order proportional to e^2	$(S_3)^2$		$-\cdot010986$
	S_2	$\cdot080440$	
	$S_2^{\ 2}$		$\cdot000890$
	S_4	$\cdot076544$	
	$S_4^{\ 2}$	$\cdot000596$	
	$S_4^{\ 4}$	$\cdot000000$	
	S_6	$\cdot028139$	
	$S_6^{\ 2}$	$\cdot001657$	
	$S_6^{\ 4}$	$\cdot000001$	
		$\cdot187377$	$-\cdot011876$
		$-\cdot011876$	
		$\cdot175501e^2$	

The following are then the results for the normal departures at the several points whose rectangular co-ordinates are specified *:—

Meridian ($y=0$).

$$z=\pm 1,\quad x=0,\quad \delta n=\pm\cdot1482e+\cdot1723e^2$$
$$z=\pm\cdot961,\quad x=\cdot096,\quad \delta n=\pm\cdot0932e+\cdot0858e^2$$
$$z=\pm\cdot832,\quad x=\cdot191,\quad \delta n=\pm\cdot0189e+\cdot0103e^2$$
$$z=\pm\cdot480,\quad x=\cdot303,\quad \delta n=\mp\cdot0223e-\cdot0033e^2$$
$$z=0,\quad x=\cdot345,\quad \delta n=+\cdot0046e^2$$

Equator ($x=0$).

$$z=\pm 1,\quad y=0,\quad \delta n=\pm\cdot1482e+\cdot1723e^2$$
$$z=\pm\cdot866,\quad y=\cdot216,\quad \delta n=\pm\cdot0300e+\cdot1265e^2$$
$$z=\pm\cdot5,\quad y=\cdot374,\quad \delta n=\mp\cdot0354e-\cdot0220e^2$$
$$z=0,\quad y=\cdot432,\quad \delta n=-\cdot0095e^2$$

In order to draw a figure I take $e=\frac{1}{2}$. Throughout most of the arc of the ellipsoid the approximation is probably good, but at the vertices, which are just the points of most interest, it is pretty clear that we are using a somewhat extreme value for e. The results are:—

* [Down to this point I have corrected the numerical values in consequence of my revision of the computations. But as, even at the two extremities of the axis, the corrections are far too small to be allowed for in a diagram, I have permitted the original computations to stand in the following table and in the figures by which the results are illustrated.]

Meridian ($y=0$).			Equator ($x=0$).		
$z=1$,	$x=0$,	$\delta n=+\cdot117$	$z=1$,	$y=0$,	$\delta n=+\cdot117$
$z=\cdot96$,	$x=\cdot096$,	$\delta n=+\cdot068$	$z=\cdot866$,	$y=\cdot216$,	$\delta n=+\cdot047$
$z=\cdot83$,	$x=\cdot19$,	$\delta n=+\cdot012$	$z=\cdot5$,	$y=\cdot374$,	$\delta n=-\cdot014$
$z=\cdot48$,	$x=\cdot30$,	$\delta n=-\cdot011$	$z=0$,	$y=\cdot432$,	$\delta n=-\cdot002$
$z=0$,	$x=\cdot345$,	$\delta n=+\cdot001$	$z=-\cdot5$,	$y=\cdot374$,	$\delta n=+\cdot003$
$z=-\cdot48$,	$x=\cdot30$,	$\delta n=+\cdot010$	$z=-\cdot866$,	$y=\cdot216$,	$\delta n=+\cdot017$
$z=-\cdot83$,	$x=\cdot19$,	$\delta n=-\cdot007$	$z=-1$,	$y=0$,	$\delta n=-\cdot031$
$z=-\cdot96$,	$x=\cdot096$,	$\delta n=-\cdot025$	*N.B.—For* $z=\pm\cdot866$, δn *is in both cases positive.*		
$z=-1$,	$x=0$,	$\delta n=-\cdot031$			

These numbers are set out graphically in the annexed figure. It will be noticed that whereas the protuberance at the positive end of the z axis is

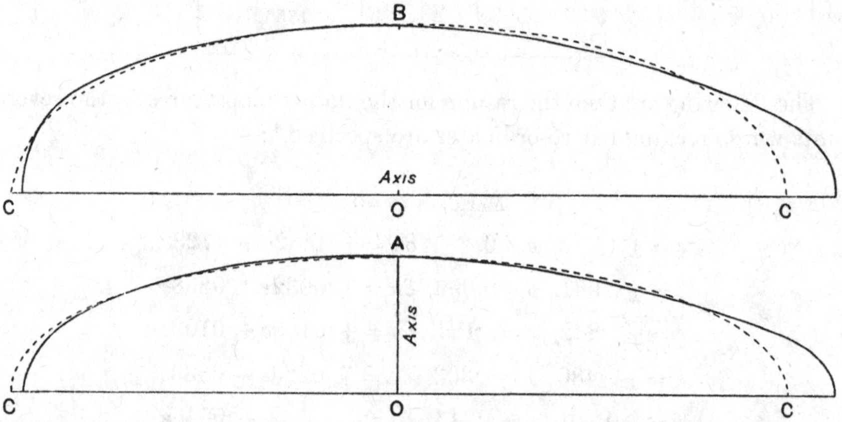

Second approximation to Pear-shaped Figure. Upper section "equatorial," lower "meridional."

great, the deficiency at the negative end is almost filled up. We may describe the general effect by saying that the Jacobian ellipsoid is very little changed, excepting at one end of its longest axis, where it shoots forth a protuberance.

SUMMARY.

If a mass of liquid be rotating like a rigid body with uniform angular velocity, the determination of the figure of equilibrium may be treated as a statical problem, if the mass be subjected to a rotation potential.

The energy, say W, lost in the concentration of a body from a condition of infinite dispersion is equal to the potential of the body in its final configuration at the position of each molecule, multiplied by the mass of the molecule and summed throughout the body. In the system, as rendered statical, it is necessary to add the rotation potential to the gravitation potential before effecting the summation. That portion, say T, of the whole lost energy which arises from the rotation potential is simply the same thing as the kinetic energy of the mass, when the system is regarded as a dynamical one. If we replace $W+T$ by E to denote the whole lost energy of the statical system, the condition that the surface shall be in equilibrium is that the variations of E for constant angular velocity shall be stationary. E must then be a maximum or a minimum, or a maximum for some variations and a minimum for others.

It might appear at first sight that the condition for the secular stability of the figure is that E should be a maximum for all variations, and this is so if certain constraints are introduced; but in the absence of such constraints the figure may be stable although E is a minimax.

It has been shown by M. Poincaré that the stability must be determined from the variations, subject to constancy of angular momentum, of the total energy of the system, both kinetic and potential. The two portions of the total energy, say U, are again W and T; but whereas E involves the lost energy W of the system under the action of the gravitation potential, U involves the potential energy which is equal to $-W$. Thus U is equal to $-W+T$.

The variation of U with constant angular momentum leads to results for the determination of the figure identical with those found from the variation of E with constant angular velocity. But there is this important difference, that to insure secular stability U must be an absolute minimum. It appears, in fact, that, in the case of the pear-shaped figure, while E is actually a maximum for all the deformations but one, it is a minimum for that one, which consists of an ellipsoidal strain of the critical Jacobian ellipsoid from which the pear-shaped figures bifurcate (§ 20).

But M. Poincaré has adduced another consideration which enables us to determine the stability of the pear by means of the function E, without a direct proof that U is a minimum for all variations. For he has shown that if for given angular momentum slightly less than that of the critical Jacobian ellipsoid, the only possible figure is the Jacobian, and if for slightly greater angular momentum there are two figures (namely, the Jacobian and the pear*), then exchange of stability between the two series must occur at the bifurcation. If, on the other hand, the smaller momentum corresponds with the two figures and the larger with only one, one of the two coalescent series must be stable and the other unstable. Now it has been proved that the less elongated Jacobian ellipsoids are stable, so that if the first alternative holds the stability must pass from the Jacobian series to the pear series; and if the second alternative holds the pear series must be unstable throughout. The question of stability is then completely determined by means of the

* For the sake of simplicity we may speak of a single pear, instead of two similar pears in azimuths 180° apart.

angular momentum of the pear; if it is greater than that of the critical Jacobian the pear is stable, and, if less, unstable.

It suffices then to determine the figure by means of the variations of E with constant angular velocity, and afterwards to evaluate the angular momentum.

It was proved by M. Poincaré, and repeated by me in my previous paper, that the first approximation to the pear-shaped figure is given by the third zonal harmonic inequality of the critical Jacobian ellipsoid—zonal with respect to its longest axis. In proceeding to the higher approximation I suppose that the amplitude of the third zonal harmonic is measured by a parameter e, which is to be regarded as a quantity of the first order. We must now also suppose the ellipsoid to be deformed by all and any other harmonics, but with amplitudes of order e^2. In the first approximation the lost energy W is proportional to e^2, but it now becomes necessary to determine W as far as the order e^4. A change in the sign of e means that the figure of equilibrium is rotated in azimuth through 180°. Such a rotation cannot affect the value of the energy, and it thus becomes obvious that the odd powers of e must be absent from the expression for W. We have further to find the moment of inertia of the body as far as the terms of order e^2, and thence to find the kinetic energy T. The function E is equal to $W + T$.

In order to attain the requisite degree of accuracy, it is convenient to regard the pear as being built up in an artificial manner. I construct an ellipsoid similar to and concentric with the critical Jacobian, and therefore itself possessing the same character. The size of this new Jacobian, which I call J, is undefined, and is subject only to the condition that it shall be large enough to enclose the whole pear. The regions between J and the pear being called R, I suppose the pear to consist of positive density throughout J and negative density throughout R (§ 1).

The lost energy of the pear consists of that of J with itself, say $\frac{1}{2}JJ$; of J with R, which is filled with negative density, say $-JR$; and of R with itself, say $\frac{1}{2}RR$. This last contribution to the energy must be broken into several portions. It was the evaluation of $\frac{1}{2}RR$ which baffled me, until M. Poincaré's solution came to my help.

If we imagine the ellipsoid J to be intersected by a family of orthogonal quadrics, and if we suppose for the moment that the region R is filled with positive density, we may further imagine the matter lying inside any orthogonal tube to be transported along the tube, and to be deposited on the surface of J in the form of a condensation of positive surface density $+C$. The mass of $+C$ is equal to that of $+R$, but it is differently arranged. In the actual system R is filled with negative volume density, and we may clearly add to this two equal and opposite surface densities $+C$ and $-C$ on J.

Thus the matter lying in the region R may be regarded as consisting of negative surface density $-C$ on J, together with a double system, namely negative volume density $-R$ in conjunction with equal and opposite surface density $+C$. This double system, say D, is therefore $C-R$. The lost energy $\frac{1}{2}RR$ may be considered as consisting of three parts; first the energy of $-C$ with itself, say $\frac{1}{2}CC$; secondly that of D with itself, say $\frac{1}{2}DD$; thirdly that of $-C$ with D. This third item is obviously equal to $-CC+CR$, and therefore $\frac{1}{2}RR$ is equal to $-\frac{1}{2}CC+CR+\frac{1}{2}DD$.

It follows that the gravitational lost energy of the pear may be written symbolically in the form

$$\tfrac{1}{2}JJ - JR + CR - \tfrac{1}{2}CC + \tfrac{1}{2}DD$$

In this discussion no attention has as yet been paid to the rotation, but fortunately it happens that the introduction of this consideration actually simplifies the problem, for if we suppose $\frac{1}{2}JJ$ and JR to mean the lost energies of J with itself and with R on the supposition that the mass is rotating with the angular velocity of the critical Jacobian, the formulæ become much more tractable than would have been the case otherwise.

The inclusion of part of the angular velocity in this portion of the function E, only leaves outstanding the excess of the kinetic energy of the pear above the kinetic energy, which it would have if it rotated with the angular velocity of the critical Jacobian. If ω denotes the latter angular velocity, and $(\omega^2+\delta\omega^2)^{\frac{1}{2}}$ the actual angular velocity of the pear; if A_j be the moment of inertia of J, and A_r that of R considered as filled with positive density, we have

$$E = \tfrac{1}{2}JJ - JR + CR - \tfrac{1}{2}CC + \tfrac{1}{2}DD + \tfrac{1}{2}(A_j - A_r)\,\delta\omega^2$$

In this statement I have omitted a term which arises from the displacement of the centre of inertia from the centre of the ellipsoid; it is duly considered in the paper, but is shown to vanish to the requisite order of approximation (§§ 2, 14).

The co-ordinates of points are determined by reference to the ellipsoid J, which envelopes the whole pear, and the formula for the internal gravitation of J, inclusive of the rotation ω, is of a simple character. The size of J is indeterminate, and therefore the formulæ must involve an arbitrary constant expressive of the size of J. But the final result E cannot in any way depend on the size of the ellipsoid which is chosen as a basis for measurement, and therefore this arbitrary constant must ultimately disappear. Hence it is justifiable to treat it as zero from the beginning. It appears then that we are justified in using the formula for internal gravity throughout the investigation. If the artifice of the enveloping ellipsoid had not been adopted, it would have been necessary to take note of the fact that the pear is in part protuberant above and in part depressed below the ellipsoid of reference.

M. Poincaré did follow this last plan, and then proceeded to prove the justifiability of using the formula for internal gravity throughout. The argument adduced above seems, however, sufficient to prove the point.

Although the constant expressive of the size of J is put equal to zero—which means that the pear is really partly protuberant above the ellipsoid—I have found that a considerable amount of mental convenience results from always discussing the subject as though the constant were not zero, so that the ellipsoid envelopes the pear, and I shall continue to do so here.

When an ellipsoid is deformed by an harmonic inequality, the volume of the deformed body is only equal to that of the ellipsoid to the first order of small quantities. In the case of the pear, all the inequalities, excepting the third zonal one, are of the second order, and as far as concerns them the volumes of J and of the pear are the same. But it is otherwise as regards the third zonal harmonic term, and the first task is to find the volume of such an inequality as far as e^2. When this is done we can express the volume of J in terms of that of the pear, which is, of course, a constant (§§ 3, 4).

By aid of ellipsoidal harmonic analysis we may now express the first four terms of E in terms of the mass of the pear, and of certain definite integrals which depend on the shape of the critical Jacobian ellipsoid (§§ 5, 6, 7).

The energy $\frac{1}{2}DD$ presents much more difficulty, and it is especially in this that M. Poincaré's insight and skill have been shown. The system D consists of a layer of negative volume density, coated on its outer surface with a layer of surface density of equal and opposite mass.

Two surfaces, infinitely near to one another, coated with equal and opposite surface densities, form together a magnetic layer or a layer of doublets. The change of potential in crossing such a layer is 4π times the magnetic moment at the point of crossing, and is independent of the form of surface. To find the difference between the potential at two points at a finite distance apart, one being on one side and the other on the other side of the layer, we have to add to the preceding difference a term equal to the force on either side of the magnetic layer multiplied by the distance between the two points. This additional term is small compared with that involving the magnetic moment, provided that the distance is small. If the magnetic layer coincided with the surface of an ellipsoid the force in question would be exactly calculable, and if it lies on the surface of a slightly deformed ellipsoid the force remains unchanged by the deformation as a first approximation.

Thus it follows that it is possible to calculate the difference of potential at two points lying on a curve orthogonal to an ellipsoid, when one point is on one side and the other on the other side of a magnetic layer residing on a deformation of the ellipsoid. Further, if the two points lie on the same side of the magnetic layer the term dependent on magnetic moment (which would represent the crossing of the layer) disappears, and only the term dependent on the force remains.

Two equal and opposite layers of matter at a finite distance apart may be built up from an infinite number of magnetic layers interposed between the two surfaces. Hence by the integration of the result for a magnetic layer we may find the change of potential in passing from any one point to any other lying on the same orthogonal curve in the neighbourhood of a finite double layer.

Again, the system D, consisting of $-R$ and $+C$, may be built up by an infinite number of finite double layers. Hence by a second integration we may find the difference between the potential of D at any point inside R and the point lying on J where the orthogonal curve through the first point cuts the surface of J.

Finally, it may be proved that the lost energy $\frac{1}{2}DD$ is equal to half the difference of potentials just determined multiplied by the density and integrated throughout the region R. The required expression for this portion of the energy is found to consist of two parts, of which one depends on magnetic moment and the other on the force (§ 9). The reduction of this part of the energy to calculable forms is not very simple; it is carried out in §§ 11, 12.

The calculation of the moment of inertia of the pear is comparatively easy, since it only involves those harmonic inequalities of J which are expressible by harmonics of the second degree (§ 13). On multiplying the moment of inertia by $\frac{1}{2}\delta\omega^2$, we obtain the last contribution to the expression for E.

The energy function cannot involve e^2, since the vanishing of the coefficient of that term is the condition whence the critical Jacobian was determined. If f denotes the coefficient of any harmonic inequality other than the third zonal one, the part of E independent of $\delta\omega^2$ is found to contain terms in e^2, e^2f and $(f)^2$. The coefficient of $\delta\omega^2$ consists of a constant term, a term in e^2 and terms in f_2 and $f_2{}^2$, where these f's denote the coefficients of the second zonal and sectorial harmonics. This last part does not contain the coefficient of any harmonic of odd degree, and in the first part the coefficient of the term in e^2f for all such harmonics is found to vanish.

The condition for the figure of equilibrium is that the variations of E for variations of e^2 and of each f shall vanish. On differentiating E with respect to the f of any harmonic of odd degree and equating the result to zero, we see that that f must vanish. Hence it follows that the pear cannot involve any odd harmonic excepting the third zonal one. Again, the symmetry of the figure negatives the existence of any even functions involving sine-functions of the quasi-longitude measured from the prime meridian (as I may call it) of symmetry through the axis of rotation. The same consideration negatives the existence of even functions involving cosine functions of odd rank. Accordingly the only functions to be considered are the even ones of even rank, involving the cosine functions of the longitude.

The equation to zero of the variations of E for all the f's, excepting f_2, f_2^2, gives at once all those f's in terms of e^2. The equations to zero of the variations for e^2, f_2, f_2^2 give three equations for the determination of $\delta\omega^2, f_2, f_2^2$ as multiples of e^2. We thus have the means of finding the angular velocity and all the f's in terms of the parameter e, which measures the amount of departure of the pear from the critical Jacobian ellipsoid (§ 14).

It seems unnecessary to give here any explanation of the methods adopted for reducing the analytical results to numbers, and it may suffice to say that the task proved to be a very laborious one.

The harmonic terms included in the computation were those of degree 2 and ranks 0 and 2, of degree 4 and ranks 0, 2, 4, of degree 6 and ranks 0, 2, 4, and of degrees 8 and 10 and rank 0. The sectorial harmonic of degree 6 was omitted because its contribution would certainly prove negligible, and all the tesseral harmonics of degrees 8 and 10 for the same reason.

The expression for $\delta\omega^2$ is found in the form of a fraction, of which the denominator is determinate and the numerator consists of the sum of an infinite series. Eleven terms of this series were computed, namely, a constant term and the contribution of the ten harmonic terms specified above. I found, in fact, that it would only change the numerator by about one-twentieth part of itself, if all the harmonics excepting the zonal ones of degrees 2, 4, 6, 8, 10 had been dropped.

The result shows that the square of the angular velocity of the pear is less than that of the critical Jacobian ellipsoid in about the proportion of $1-\frac{1}{7}e^2$ to 1. On the other hand the angular momentum of the pear is greater than that of the ellipsoid in about the proportion of $1+\frac{1}{12}e^2$ to 1. If this last result were based on a rigorous summation of the infinite series, it would, in accordance with the principle explained above, absolutely prove the stability of the pear. The inclusion of the uncomputed residue of the series would undoubtedly tend in the direction of reducing the coefficient given in round numbers as $\frac{1}{12}$, and if it were to reduce it to a negative quantity, we should conclude that the pear was unstable after all.

The contribution of the eighth zonal harmonic to the series above referred to was about ·00002, and that of the tenth about ·000006, and I find that if the contribution of the uncomputed residue should amount to ·00017, the apparent stability of the pear would just be reversed. The pear is then stable unless the residue of the series shall amount to 27 times the contribution of the last computed term.

Since the convergency of the series is obviously rapid, it seemed incredible to me that the inclusion of the uncomputed residue could materially alter, much less reverse our result. I regarded it then as proved, but by something short of an absolute algebraic argument, that the pear-shaped figure is stable.

[However in Vol. XVII. No. 3 (1905), of the *Memoirs of the Imperial Academy of St Petersburg*, M. Liapounoff has published an abstract of his work on figures of equilibrium of rotating liquid under the title "Sur un Problème de Tchebychef." In this paper he has explained how he has obtained a rigorous solution for the figure and stability of the pear-shaped figure, and he has pronounced it to be unstable.

The stability or instability depends, in fact, on the sign of a certain function which M. Liapounoff calls A, and which is the same as that which I denote $A_0 + \Sigma (B_i^s)^2/C_i^s$, where A_0 is equal to $\mathfrak{A}_3 [\frac{1}{3} (\sigma_2)^2 + 2\zeta_4] - \frac{1}{3}\sigma_4 + \Sigma [i, s]$.

M. Liapounoff tells us that, after having seen my conclusion he repeated all his computations and confirmed his former result. He attributes the disagreement between us to the fact that I have only computed portion of an infinite series, and have only used approximate forms for the elliptic integrals in the several terms. He believes that the inclusion of the neglected residue of the infinite series would lead to an opposite conclusion.

In my computation the function $\mathfrak{A}_3 [\frac{1}{3} (\sigma_2)^2 + 2\zeta_4] - \frac{1}{3}\sigma_4$ is decisively negative, and being numerically greater than $\Sigma \{(B_i^s)^2/C_i^s + [i, s]\}$, which is positive, the sum of the two is negative. As indicated above, the inclusion of the neglected residue undoubtedly tends to make this whole function positive, but after making the revision (incorporated in the present edition of my paper) I still feel unconvinced that the neglected residue can amount to the total needed to invert the sign.

It may be worth mentioning that in revising my work I notice that $\mathfrak{A}_3 [\frac{1}{3} (\sigma_2)^2 + 2\zeta_4] - \frac{1}{3}\sigma_4$ owes its negative sign to the term $-\frac{1}{3}\sigma_4$. This term arises from the energy of the double layer, called $\frac{1}{2}DD$. It comes from the portion of the term $\frac{2}{5}\pi\rho^2 \int \epsilon^3 (1 - \lambda\epsilon)\, d\sigma$, which gives rise to a term in e^4 with a negative sign. This term involves under the integral sign the factor $\frac{5}{2}\left(\frac{1}{\Gamma_1^2} + \frac{1}{\Delta_1^2}\right) - 3G$, all the other factors being positive. If we attribute to θ and to ϕ various values between $\frac{1}{2}\pi$ and zero, we see that in part of the range the factor is positive and in other parts negative. A general inspection does not suffice to determine whether the positive portion outweighs the negative, as in fact it does. Therefore, in order to feel abundantly sure that no gross mistake had been made, I computed by quadratures the eight constituent integrals involved in the final result, and confirmed the correctness of the value found by the rigorous evaluation.

The analysis of the investigation has of course been carefully re-examined throughout, and I have, besides, applied the same method to the investigation of Maclaurin's spheroid, where the solution can be verified by the known exact result*.

As a further check, the formulæ of the present paper have been examined on the hypothesis that the ellipsoid of reference reduces to a sphere. The several terms correctly reproduce the analogous terms in the paper on Maclaurin's spheroid, but in effecting the comparison it is necessary to note that the variable τ of this paper reduces to $\frac{1}{2}(1-r^2/a^2)$, whereas in the paper on Maclaurin's spheroid the corresponding variable τ denotes $\frac{1}{3}(1-r^3/a^3)$, where r is radius vector and a the radius of the sphere.

Dissent from so distinguished a mathematician as M. Liapounoff is not to be undertaken lightly, and I have, as explained, taken especial pains to ensure correctness. Having made my revision, and completed the computations as set forth here, I feel a conviction that the source of our disagreement will be found in some matter of principle, and not in the neglected residue of this series. I can now only express a hope that some one else will take up the question, and I will proceed to discuss the subject further as though my results had not been subjected to this criticism.]

The numbers obtained in the course of the determination of the stability afford the means of giving a second approximation to the form of the pear. The result is shown graphically in the figure of § 21, where the largest value of e is adopted which seemed to secure a fair degree of approximation in the result. I originally called the figure "pear-shaped," because M. Poincaré's conjectural sketch in the *Acta Mathematica* was very like a pear. In the first approximation, shown in my former paper, the resemblance to a pear was not striking, and it needs some imagination to recognise the pear shape in the second approximation shown here; but a distinctive name is so convenient that we may as well continue to call it by that name.

The effects of the new terms now added are almost entirely concentrated at the two ends. All these terms, excepting a very small one arising from the second sectorial harmonic, tend to augment the protuberance at the stalk and to fill up the depression at the blunt end. It is true that there is a small term, arising from the square of the third zonal harmonic, which diminishes the protuberance and increases the depression, but this cannot be regarded as a new term, since it only represents the effect of the fundamental harmonic carried to the second order of small quantities.

The new zonal harmonics furnish by far the most important contributions. The second zonal harmonic denotes that the ellipsoid most nearly resembling the pear is longer and less broad than the Jacobian. The largest contribution of all is that due to the fourth zonal harmonic, and this may be regarded as the octave of the second zonal term. The ratio of the contribution of the

* [Paper 14.]

sixth zonal harmonic to the fourth is about $\frac{1}{4}$; of the eighth to the sixth, and of the tenth to the eighth about $\frac{1}{3}$.

The general effect is that the protuberance at the stalk of the pear is much increased, and the depression at the other end nearly filled up. Over the greater part of the whole surface the depressions and protuberances are less conspicuous than they were. The nodal lines where the surface of the pear cuts that of the ellipsoid are entirely shifted from their former positions. It did not seem worth while to attempt to specify their new positions, because the choice of the ellipsoid to which we refer influences the result so largely. The ellipsoid on which these figures are constructed is that which is called J in this summary. Its volume is a little less than that of the pear, so that the protuberances are a little greater in volume than the depressions.

I think it is hardly too much to say, that in a well-developed "pear" the Jacobian ellipsoid has nearly regained its primitive figure, but that it is subject to a small tidal distortion due to the attraction of a protuberance which it shoots forth at one end. I venture to give here a conjectural sketch of a further stage of the development.

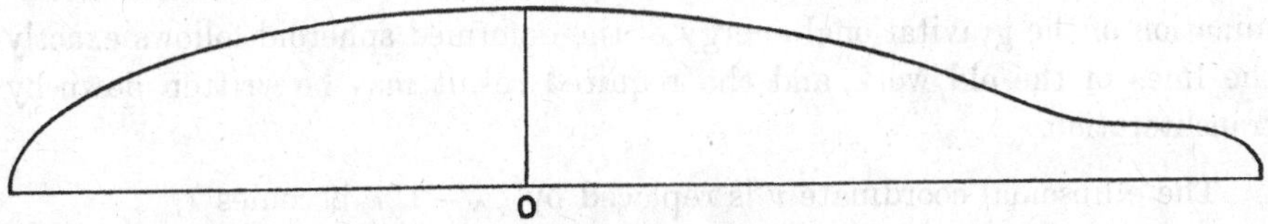

Conjectural Sketch.

If we look at this figure and at those drawn by Mr Jeans in his striking investigation of the parallel changes in the shape of an infinite rotating cylinder (*Phil. Trans. Roy. Soc.*, Vol. 200, A (1902), p. 67), we can hardly fail to be reminded of some such phenomenon as the protrusion of a filament of protoplasm from a mass of living matter.

Notwithstanding the *caveat* which M. Poincaré enters as to the dangers of applying these results to heterogeneous masses and to cosmogony, I cannot restrain myself from joining him in seeing in this almost life-like process a counterpart to at least one form of the birth of double stars, planets, and satellites*.

* [*Sed vide* Paper 15.]

APPENDIX.

NOTE ON THE STABILITY OF JACOBI'S ELLIPSOID.

(An abstract is contained in *Proceedings of Royal Society*, Vol. 82, A (1909), pp. 188—9.)

[It is known that Maclaurin's spheroid of rotating liquid becomes unstable when its eccentricity reaches the value sin 54° 21′ 27″. This is a form of bifurcation, and for increasing momentum the stability passes over to Jacobi's ellipsoids. The discussion of this problem by the method of the preceding paper is valuable, because we are able to see that the method gives correct results, even when only a small number of terms in the infinite series are computed. Moreover we are able to compare in detail the series which gives the result in this case with that found previously. Although it is impossible in this way to prove my conclusion, as against M. Liapounoff's contention, yet I think it tends to indicate that I am right.

In this problem we use spheroidal harmonic analysis applicable to an ellipsoid of revolution of eccentricity $\sin\delta$, where $\delta = 54° 21' 27''$. The determination of the gravitational energy of the deformed spheroid follows exactly the lines of the old work, and the required result may be written down by transliteration.

The ellipsoidal coordinate ν is replaced by $\zeta\sqrt{-1}$, κ becomes 1,

$$\sin\gamma = \sin\beta = \sqrt{(-1)}\tan\delta, \quad \text{and} \quad \cos\gamma = \cos\beta = \sec\delta$$

Also Δ_1^2 or $1 - \sin^2\gamma\sin^2\theta$ becomes $1 + \tan^2\delta\sin^2\theta$, and I will now use Δ^2 to denote $1 + \tan^2\delta\sin^2\theta$; the old Δ becomes $\cos\theta$, and Γ becomes unity.

There is no need for the black-letter notation for $\mathfrak{A}_i^s$, $\mathfrak{B}_i^s$, but it may be retained, $\mathfrak{A}_i^s$ denoting $P_i^s(\zeta_0) Q_i^s(\zeta_0)$, and $\mathfrak{B}_i^s$ being $Q_i^s(\zeta_0)\, dP_i^s(\zeta_0)/d\zeta_0$.

The deformed spheroid is given by

$$\tau = -eS_2^2 - \Sigma f_i^s S_i^s$$

where

$$S_i^s = P_i^s(\mu)\cos s\phi$$

The condition for the limiting stability of Maclaurin's spheroid is

$$\mathfrak{A}_2^2 = \mathfrak{A}_1$$

and the solution of this is $\delta = 54° 21' 27''$.

Thus wherever in the old analysis $\mathfrak{A}_3$ or $\mathfrak{A}_1^1$ occur we must replace them by $\mathfrak{A}_2^2$ or $\mathfrak{A}_1$.

In translating the old work $\sqrt{-1}$ will occur frequently, but these imaginary signs take care of themselves, and we need pay no attention to them.

Thus the old result for the gravitational energy holds good except in two respects; first that the suffix and the affix 2 replace the suffix 3, and the suffix 1 replaces the suffix and affix 1, and secondly that the coefficient $\sin^2\beta/\cos\beta\cos\gamma$ becomes $\sin^2\delta$.

Thus we have for the gravitational energy

$$A_0e^4 + 2\Sigma e^2 f_i^s B_i^s - \Sigma C_i^s (f_i^s)^2$$

where
$$A_0 = \mathfrak{A}_1\left[\tfrac{1}{3}(\sigma_2)^2 + 2\zeta_4\right] - \tfrac{1}{3}\sigma_4 + \sum_2^\infty [i, s]$$

$$[i, s] = \frac{(\mathfrak{A}_i^s\omega_i^s + \frac{1}{2}\mathfrak{B}_i^s\rho_i^s)^2}{\mathfrak{A}_i^s\phi_i^s} - \tfrac{1}{4}\sin^2\delta\,\frac{\mathfrak{B}_i^s(\rho_i^s)^2}{\mathfrak{A}_i^s\phi_i^s}$$

$$B_i^s = \mathfrak{A}_i^s\omega_i^s + \tfrac{1}{2}B_i^s\rho_i^s + (\mathfrak{B}_2^{\,2} - \tfrac{1}{2}\sin^2\delta)\,\rho_i^s$$

$$C_i^s = (\mathfrak{A}_1 - \mathfrak{A}_i^s)\,\phi_i^s$$

The integrals involved in these formulæ may be derived by translation from the old results.

The formulæ for the rotational term in the energy is however quite different, since the rotation takes place round the z axis.

I find that

$$\tfrac{1}{2}C\delta\omega^2 = \tfrac{3}{2}\frac{M^2}{k_0}\frac{\delta\omega^2}{2\pi\rho}\left\{\tfrac{1}{5}\tan\delta + \tfrac{1}{3}e^2(\rho_0\sin^2\delta - \rho_2\sin^2\delta - \omega_2\tan\delta) - \tfrac{1}{3}f_2\phi_2\tan\delta\right\}$$

It appears that all the f's vanish except those for which i is even and s is zero or 4, and all the f's except f_2 may be eliminated.

The equations for finding f_2 and $\delta\omega^2$ become

$$f_2 = \frac{B_2}{C_2} - \frac{\delta\omega^2}{4\pi\rho}\cdot\tfrac{1}{3}\frac{\phi_2\tan\delta}{C_2}$$

$$\left[A_0 + \sum_2^\infty \frac{(B_i^s)^2}{C_i^s}\right]e^2 + \frac{\delta\omega^2}{4\pi\rho}\cdot\tfrac{1}{3}\left[(\rho_0 - \rho_2)\sin^2\delta - \omega_2\tan\delta - \frac{B_2}{C_2}\phi_2\tan\delta\right] = 0$$

Before stating the numerical results I will give the values of the several integrals.

If as already stated we write $\Delta^2 = 1 + \tan^2\delta\sin^2\theta = 1 + \mu^2\tan^2\delta$

$$\omega_i = \tfrac{27}{2}\sec^2\delta\int_0^1\left(\frac{1}{\Delta^4} - \frac{1}{2\Delta^2}\right)(1-\mu^2)^2[\mathrm{P}_i(\mu)]^2\,d\mu$$

$$\rho_i = 27\operatorname{cosec}2\delta\int_0^1\frac{1}{\Delta^2}(1-\mu^2)^2[\mathrm{P}_i(\mu)]^2\,d\mu$$

$$\zeta_4 = \tfrac{729}{16}\sec^4\delta\int_0^1\left(\frac{1}{\Delta^6} - \frac{3}{4\Delta^4}\right)(1-\mu^2)^4\,d\mu$$

$$\sigma_4 = \tfrac{729}{16}\sin\delta\sec^3\delta\int_0^1\left(\frac{5}{\Delta^6} - \frac{3+\cos^2\delta}{\Delta^4}\right)(1-\mu^2)^4\,d\mu$$

The integral σ_2 is the same as ω_0. The integrals ω_i^4, ρ_i^4 only differ from ω_i, ρ_i in having the coefficient outside the integral sign half as large, and $P_i^4(\mu)$ of course replaces $P_i(\mu)$. The integrals ω_i^s, ρ_i^s vanish unless $s = 0$ or 4.

Lastly $$\phi_i = \frac{3}{2i+1}, \quad \phi_i^4 = \frac{3}{2(2i+1)} \frac{i+4!}{i-4!}$$

I only carried the computation as far as the eighth harmonics, and found that

$$f_2 = -8{\cdot}4346e^2, \quad f_4 = 3{\cdot}4700e^2, \quad f_6 = -1{\cdot}1067e^2, \quad f_8 = {\cdot}3358e^2$$

$$f_4^4 = {\cdot}076910e^2, \quad f_6^4 = -{\cdot}0019225e^2, \quad f_8^4 = {\cdot}00013722e^2$$

The coefficients of the sectorial harmonics are very small, but as they are to be multiplied by $P_i^4(\mu)$, which involves a large numerical coefficient, these terms are quite as important as the zonal terms.

I find that $$\frac{\delta\omega^2}{4\pi\rho} = -2{\cdot}3146e^2$$

The angular velocity, say ω_0, of the critical Maclaurin's spheroid is given by $\omega_0^2/4\pi\rho = {\cdot}09356$, and since $\omega^2 = \omega_0^2 + \delta\omega^2$, we have

$$\omega = \omega_0(1 - 12{\cdot}370e^2)$$

The moment of inertia is found to be

$$C = \frac{3M^2}{10\pi\rho k_0} \tan\delta\,(1 + 47{\cdot}124e^2)$$

Hence $$C\omega = \frac{3M^2\omega_0}{10\pi\rho k_0} \tan\delta\,(1 + 34{\cdot}754e^2)$$

Hence the angular velocity of the Jacobian ellipsoid is less, and the angular momentum greater than in the case of the critical Maclaurin spheroid. This shows, as in the case of the pear-shaped figure, that the Jacobian ellipsoid is stable.

In order to test the correctness of the results I have calculated the axes of the ellipsoids given by this analysis, corresponding to $\gamma = 55°, 57°, 60°$ (see Paper 8, p. 130). Even for so long an ellipsoid as the last the result is very nearly exact, for I find

	Computed axes	Correct axes	Error as fraction of correct value
a	1·3747	1·3831	+1/165
b	1·0584	1·0454	−1/80
c	·6873	·6916	+1/161

The chief point of interest is, however, to consider the series which is analogous with that found in investigating the pear-shaped figure.

For Jacobi's ellipsoid there are two series proceeding *pari passu*, viz. the zonal and fourth tesseral series. For these I find the following results:—

i	Zonal harmonics $(B_i)^2/C_i+[i, 0]$	Ratio of each term to the next
2	37·2393	37
4	1·0077	$6\frac{1}{2}$
6	·1554	9
8	·0170	

i	Fourth tesseral harmonics $(B_i^4)^2/C_i^4+[i, 4]$	Ratio of each term to the next
4	3·3307	10
6	·3352	13
8	·0253	

In the case of the pear-shaped figure the corresponding numbers are extremely small fractions, but this is without any signification since it merely depends on the definitions which have been adopted for the harmonic functions. Here we need only consider the zonal terms, since the sectorial terms are clearly insignificant. We had:—

i	$(B_i)^2/C_i+[i, 0]$	Ratio of each term to the next
2	−·000080868	
4	·000247274	3·9
6	·000062350	3·2
8	·000019576	3·1
10	·000006328	

M. Liapounoff contends that if I had taken more terms into account, the conclusion as to the stability of the pear-shaped figure would have been reversed, because the series is so slowly convergent. It would seem, however, as if the uncomputed residue of this series should be about ·000003. It is true that the series is less convergent than those which arise in the case of Jacobi's ellipsoid, but I cannot think that it can converge so slowly as to justify M. Liapounoff. It seems to me then that the present investigation should lead us to look with suspicion on an argument which would show that my former conclusion was incorrect.]

13.

ON THE INTEGRALS OF THE SQUARES OF ELLIPSOIDAL SURFACE HARMONIC FUNCTIONS.

[*Philosophical Transactions of the Royal Society*, Vol. 203, A (1903), pp. 111—137.]

THIS paper forms a sequel to the three preceding papers in the present volume. I shall refer to them as "Harmonics," "The Pear-shaped Figure," and "Stability."

In "Harmonics," the functions being expressed approximately, approximate formulæ are found for the integrals over the surface of the ellipsoid of the squares of all the surface harmonics. These integrals are of course required whenever it is proposed to make practical use of this method of analysis, and the evaluation of them is therefore an absolutely essential step towards any applications.

The analysis used in the determination of some of these integrals was very complicated, and is probably susceptible of improvement. Such improvement might perhaps be obtained by the methods of the present paper, but I do not care to spend a great deal of time on an attempt merely to improve the analysis.

In "Harmonics" the symmetry which really subsists between the three factors of the solid harmonic functions was sacrificed with the object of obtaining convenient approximate forms, and I do not think it would have been possible to obtain such satisfactory results without this sacrifice. But this course had the disadvantage of rendering it difficult to evaluate the integrals of the squares of the surface harmonics.

All the harmonic functions up to the third order inclusive are susceptible of rigorous algebraic expression; and indeed the same is true of some but not of all the functions of the fourth order. Accordingly in these cases rigorous expressions for the integrals should also be obtainable, and the object of the present paper is to complete the preceding investigation in this respect.

It will be well to begin by a restatement of the notation. That used in "Harmonics" was convenient for the approximate and asymmetrical expressions involved, but the notation used in the two later papers seems preferable where the formulæ are rigorous and symmetrical.

In "Harmonics" the squares of the semi-axes of the ellipsoid were

$$a^2 = k^2\left(\nu^2 - \frac{1+\beta}{1-\beta}\right), \quad b^2 = k^2(\nu^2 - 1), \quad c^2 = k^2\nu^2$$

The rectangular coordinates were connected with the ellipsoidal coordinates ν, μ, ϕ by

$$\frac{x^2}{k^2} = -\frac{1-\beta}{1+\beta}\left(\nu^2 - \frac{1+\beta}{1-\beta}\right)\left(\mu^2 - \frac{1+\beta}{1-\beta}\right)\cos^2\phi$$

$$\frac{y^2}{k^2} = -(\nu^2 - 1)(\mu^2 - 1)\sin^2\phi$$

$$\frac{z^2}{k^2} = \nu^2\mu^2\frac{1-\beta\cos 2\phi}{1+\beta}$$

The three roots of the cubic

$$\frac{x^2}{a^2+u} + \frac{y^2}{b^2+u} + \frac{z^2}{c^2+u} = 1$$

were $$u_1 = k^2\nu^2, \quad u_2 = k^2\mu^2, \quad u_3 = k^2\frac{1-\beta\cos 2\phi}{1-\beta}$$

Lastly ν ranges from ∞ to 0, μ between ± 1, ϕ from 0 to 2π.

In the two later papers I put

$$\kappa^2 = \frac{1-\beta}{1+\beta}, \quad \kappa'^2 = 1 - \kappa^2, \quad \nu = \frac{1}{\kappa\sin\gamma}, \quad \mu = \sin\theta$$

and for convenience I introduced an auxiliary constant β (easily distinguishable from the β of the previous notation) defined by $\sin\beta = \kappa\sin\gamma$.

The squares of the semi-axes of the ellipsoid were then

$$a^2 = \frac{k^2\cos^2\gamma}{\sin^2\beta}, \quad b^2 = \frac{k^2\cos^2\beta}{\sin^2\beta}, \quad c^2 = \frac{k^2}{\sin^2\beta}$$

The rectangular coordinates became

$$\frac{x^2}{k^2} = \frac{\cos^2\gamma}{\sin^2\beta}(1 - \kappa^2\sin^2\theta)\cos^2\phi, \quad \frac{y^2}{k^2} = \frac{\cos^2\beta}{\sin^2\beta}\cos^2\theta\sin^2\phi$$

$$\frac{z^2}{k^2} = \frac{1}{\sin^2\beta}\sin^2\theta(1 - \kappa'^2\cos^2\phi)$$

The roots of the cubic were

$$u_1 = \frac{k^2}{\sin^2 \beta}, \quad u_2 = k^2 \sin^2 \theta, \quad u_3 = \frac{k^2}{\kappa^2}(1 - \kappa'^2 \cos^2 \phi)$$

This is the notation which will be used in the present paper.

If $d\sigma$ be an element of surface of the ellipsoid, and p the central perpendicular on to the tangent plane, it appears from the formula on p. 326 of "Stability" that

$$\frac{p\,d\sigma}{d\theta d\phi} = \frac{k^3 \cos\beta \cos\gamma}{\sin^3 \beta} \cdot \frac{\kappa^2 \cos^2\theta + \kappa'^2 \sin^2 \phi}{\Delta\Gamma}$$

where $\Delta^2 = 1 - \kappa^2 \sin^2 \theta$, $\Gamma^2 = 1 - \kappa'^2 \cos^2 \phi$.

In the previous papers I have expressed the two factors of which a surface harmonic consists by $\mathfrak{P}_i^s(\mu)$ or $\mathsf{P}_i^s(\mu)$, and $\mathfrak{C}_i^s(\phi)$, $\mathsf{C}_i^s(\phi)$, $\mathfrak{S}_i^s(\phi)$ or $\mathsf{S}_i^s(\phi)$, one of the two P-functions being multiplied by one of the four cosine or sine functions.

Taking a pair of typical cases, the integrals to be evaluated are

$$\int (\mathfrak{P}_i^s \mathfrak{C}_i^s)^2 p\,d\sigma \quad \text{and} \quad \int (\mathfrak{P}_i^s \mathfrak{S}_i^s)^2 p\,d\sigma$$

As it will be convenient to use an abridged notation, I will write these integrals $I_i^s(\cos)$ and $I_i^s(\sin)$, according to an easily intelligible notation.

These functions involve integrals of even functions, and therefore we may integrate through one octant of space, the limits of θ and ϕ being $\frac{1}{2}\pi$ to 0, and multiply the result by 8.

It is clear then that

$$I_i^s(\cos) = \frac{8k^3 \cos\beta \cos\gamma}{\sin^3\beta} \left[\int_0^{\frac{1}{2}\pi} \frac{\kappa^2 \cos^2\theta\, (\mathfrak{P}_i^s)^2}{\Delta} d\theta \int_0^{\frac{1}{2}\pi} \frac{(\mathfrak{C}_i^s)^2}{\Gamma} d\phi \right.$$
$$\left. + \int_0^{\frac{1}{2}\pi} \frac{(\mathfrak{P}_i^s)^2}{\Delta} d\theta \int_0^{\frac{1}{2}\pi} \frac{\kappa'^2 \sin^2\phi\, (\mathfrak{C}_i^s)^2}{\Gamma} d\phi \right]$$

Similar expressions are applicable to all the other forms of function, but we may proceed with this form as a type of all the others.

This formula shows that the variables are separable, and since we might substitute $\frac{1}{2}\pi - \psi$ for ϕ without changing the result, the ϕ integrals are of the same type as the θ integrals.

It has been stated above that two of the roots of the cubic equation are proportional to $\kappa^2 \sin^2\theta$ and $(1 - \kappa'^2 \cos^2\phi)$. By the nature of the harmonic functions it follows that if $[\mathfrak{P}_i^s(\mu)]^2$ is proportional to a certain function of $\kappa^2 \sin^2\theta$, $[\mathfrak{C}_i^s(\phi)]^2$ is proportional to the same function of $(1 - \kappa'^2 \cos^2\phi)$.

It follows that if $(\mathfrak{P}_i^s)^2 = F(\kappa^2 - \kappa^2 \sin^2\theta) = F(\kappa^2 \cos^2\theta)$,

$$(\mathfrak{C}_i^s)^2 = \alpha F(\kappa^2 - 1 + \kappa'^2 \cos^2\phi) = \alpha F(-\kappa'^2 \sin^2\phi)$$

where α is a constant, which for the present we may regard as being unity.

If then

$$[\mathfrak{P}_i^s(\mu)]^2 = A_0 + A_1\kappa^2\cos^2\theta + A_2\kappa^4\cos^4\theta + A_3\kappa^6\cos^6\theta + \ldots$$

we must have

$$[\mathfrak{C}_i^s(\phi)]^2 = A_0 - A_1\kappa'^2\sin^2\phi + A_2\kappa'^4\sin^4\phi - A_3\kappa'^6\sin^6\phi + \ldots$$

Accordingly if there is a term $A_n\int\kappa^{2n+2}\cos^{2n+2}\theta\frac{d\theta}{\Delta}$ in $\int\kappa^2\cos^2\theta(\mathfrak{P}_i^s)^2\frac{d\theta}{\Delta}$ and a term $(-)^m A_m\int\kappa'^{2m}\sin^{2m}\phi\frac{d\phi}{\Gamma}$ in $\int(\mathfrak{C}_i^s)^2\frac{d\phi}{\Gamma}$, then there must be a term $A_m\int\kappa^{2m}\cos^{2m}\theta\frac{d\theta}{\Delta}$ in $\int(\mathfrak{P}_i^s)^2\frac{d\theta}{\Delta}$, and a term $(-)^n A_n\int\kappa'^{2n+2}\sin^{2n+2}\phi\frac{d\phi}{\Gamma}$. It follows that the coefficient of $(-)^m A_n A_m$ in $I_i^s(\cos)\div\frac{8k^3\cos\beta\cos\gamma}{\sin^3\beta}$ is

$$\int\kappa^{2n+2}\cos^{2n+2}\theta\frac{d\theta}{\Delta}\int\kappa'^{2m}\sin^{2m}\phi\frac{d\phi}{\Gamma}$$
$$-(-)^{n-m+1}\int\kappa^{2m}\cos^{2m}\theta\frac{d\theta}{\Delta}\int\kappa'^{2n+2}\sin^{2n+2}\phi\frac{d\phi}{\Gamma}$$

For the sake of brevity I call this function $[2n+2, 2m]$, and we may state that one term in the required expression is $(-)^m A_n A_m[2n+2, 2m]$, where [ ] indicates the above function of the four integrals. It follows that

$$I_i^s(\cos)\div\frac{8k^3\cos\beta\cos\gamma}{\sin^3\beta}$$
$$\left.\begin{aligned}= \quad & A_0^2[2,0] - A_0A_1[2,2] + A_0A_2[2,4] - \ldots\\ & + A_1A_0[4,0] - A_1^2[4,2] + A_1A_2[4,4] - \ldots\\ & + A_2A_0[6,0] - A_2A_1[6,2] + A_2^2[6,4] - \ldots\\ & \ldots\ldots\ldots\ldots\ldots\ldots\ldots\ldots\ldots\ldots\ldots\end{aligned}\right\}\ldots\ldots\ldots(1)$$

Since

$$[2n, 2m] = \int\kappa^{2n}\cos^{2n}\theta\frac{d\theta}{\Delta}\int\kappa'^{2m}\sin^{2m}\phi\frac{d\phi}{\Gamma}$$
$$-(-)^{n-m}\int\kappa^{2m}\sin^{2m}\theta\frac{d\theta}{\Delta}\int\kappa'^{2n}\sin^{2n}\phi\frac{d\phi}{\Gamma}$$

it is clear that $\quad [2n, 2m] = -(-)^{n-m}[2m, 2n]$

Hence if n and m differ by an odd number $[2n, 2m] = [2m, 2n]$, and if they differ by an even number $[2n, 2m] = -[2m, 2n]$. Also $[2n, 2n] = 0$.

Let us write

$$\{2n\} = \int_0^{\frac{1}{2}\pi}\kappa^{2n}\cos^{2n}\theta\frac{d\theta}{\Delta},\quad \{2n\}' = \int_0^{\frac{1}{2}\pi}\kappa'^{2n}\sin^{2n}\phi\frac{d\phi}{\Gamma}.$$

so that $\quad [2n, 2m] = \{2n\}\{2m\}' - (-)^{n-m}\{2n\}'\{2m\}$

We must now evaluate these functions.

Since $\Delta^2 = 1 - \kappa^2 \sin^2\theta$, we have by differentiation

$$\frac{d}{d\theta}[\Delta \sin\theta \cos^{2n-3}\theta]$$

$$= \frac{1}{\Delta}\{(2n-1)\kappa^2\cos^{2n}\theta - (2n-2)(\kappa^2-\kappa'^2)\cos^{2n-2}\theta - (2n-3)\kappa'^2\cos^{2n-4}\theta\}$$

Integrating between $\frac{1}{2}\pi$ and 0 and multiplying by κ^{2n-2} we have

$$\{2n\} = \frac{2n-2}{2n-1}(\kappa^2-\kappa'^2)\{2n-2\} + \frac{2n-3}{2n-1}\kappa^2\kappa'^2\{2n-4\}$$

and by symmetry

$$\{2n\}' = -\frac{2n-2}{2n-1}(\kappa^2-\kappa'^2)\{2n-2\}' + \frac{2n-3}{2n-1}\kappa^2\kappa'^2\{2n-4\}'$$

Multiplying the first of these by $\{2m\}'$ and the second by $-(-)^{n-m}\{2m\}$ and adding together we have

$$[2n, 2m] = \frac{2n-2}{2n-1}(\kappa^2-\kappa'^2)[2n-2, 2m] + \frac{2n-3}{2n-1}\kappa^2\kappa'^2[2n-4, 2m]$$

By successive applications of this formula we may reduce any function $[2n, 2m]$ until it depends on $[2, 0]$, but the result becomes very complicated after a few successive reductions.

Now $\{0\} = \int_0^{\frac{1}{2}\pi}\frac{d\theta}{\Delta} = \text{F}, \quad \{0\}' = \int_0^{\frac{1}{2}\pi}\frac{d\theta}{\Gamma} = \text{F}'$

$$\{2\} = \int_0^{\frac{1}{2}\pi}\frac{\kappa^2\cos^2\theta}{\Delta}d\theta = \text{E} - \text{F}\kappa'^2, \quad \{2\}' = \int_0^{\frac{1}{2}\pi}\frac{\kappa'^2\sin^2\phi}{\Gamma}d\phi = \text{E}' - \text{F}'\kappa^2$$

Then

$$[2, 0] = \{2\}\{0\}' - (-)^1\{2\}'\{0\}$$
$$= \text{EF}' + \text{F}'\text{F} - \text{FF}'$$

But it is well known that this combination of the complete elliptic integrals with moduli κ and κ' is $\frac{1}{2}\pi$*.

Hence $[2, 0] = \frac{1}{2}\pi$.

It seems unnecessary to reproduce the simple algebra involved in the successive reductions, and I therefore merely give the results, as follows:—

$[0, 2] = [2, 0] = \frac{1}{2}\pi$

$-[0, 4] = [4, 0] = \frac{2}{3}(\kappa^2-\kappa'^2)\cdot\frac{1}{2}\pi$

$[0, 6] = [6, 0] = \frac{1}{3.5}(8 - 23\kappa^2\kappa'^2)\cdot\frac{1}{2}\pi$

$-[0, 8] = [8, 0] = \frac{8}{3.5.7}(\kappa^2-\kappa'^2)(6 - 11\kappa^2\kappa'^2)\cdot\frac{1}{2}\pi$

$[2, 4] = [4, 2] = \frac{1}{3}\kappa^2\kappa'^2\cdot\frac{1}{2}\pi$

$-[2, 6] = [6, 2] = \frac{4}{3.5}(\kappa^2-\kappa'^2)\kappa^2\kappa'^2\cdot\frac{1}{2}\pi$

$[2, 8] = [8, 2] = \frac{1}{3.5.7}(24 - 71\kappa^2\kappa'^2)\kappa^2\kappa'^2\cdot\frac{1}{2}\pi$

$[4, 6] = [6, 4] = \frac{1}{5}\kappa^4\kappa'^4\cdot\frac{1}{2}\pi$

$-[4, 8] = [8, 4] = \frac{6}{5.7}(\kappa^2-\kappa'^2)\kappa^4\kappa'^4\cdot\frac{1}{2}\pi$

$[6, 8] = [8, 6] = \frac{1}{7}\kappa^6\kappa'^6\cdot\frac{1}{2}\pi$

* See for example Durège's *Theorie der Elliptischen Functionen*, p. 293.

These are the only functions of this kind which are needed for the evaluations of integrals in this paper.

When these functions are introduced into (1) and the terms re-arranged, I find:—

$$I_i^s(\cos) \div \frac{4\pi k^3 \cos\beta\cos\gamma}{\sin^3\beta}$$

$$= A_0^2 - \tfrac{1}{3}\kappa^2\kappa'^2 A_1^2 + \tfrac{1}{5}\kappa^4\kappa'^4 A_2^2 - \tfrac{1}{7}\kappa^6\kappa'^6 A_3^2 + \tfrac{1}{9}\kappa^8\kappa'^8 A_4^2 - \dots$$

$$+ 2(\kappa^2 - \kappa'^2)\left[\frac{1}{1.3} A_0 A_1 - \frac{2}{3.5}\kappa^2\kappa'^2 A_1 A_2 + \frac{3}{5.7}\kappa^4\kappa'^4 A_2 A_3 - \frac{4}{7.9}\kappa^6\kappa'^6 A_3 A_4 + \dots\right]$$

$$+ \frac{2}{1.3.5}\left[\tfrac{1}{2}(2.4) - 3^2\kappa^2\kappa'^2\right] A_0 A_2 - \frac{2}{3.5.7}\left[\tfrac{1}{2}(4.6) - 5^2\kappa^2\kappa'^2\right]\kappa^2\kappa'^2 A_1 A_3$$

$$+ \frac{2}{5.7.9}\left[\tfrac{1}{2}(6.8) - 7^2\kappa^2\kappa'^2\right]\kappa^4\kappa'^4 A_2 A_4 - \frac{2}{7.9.11}\left[\tfrac{1}{2}(8.10) - 9^2\kappa^2\kappa'^2\right]\kappa^6\kappa'^6 A_3 A_5 + \dots$$

$$+ 2(\kappa^2 - \kappa'^2)\left\{\frac{2}{1.3.5.7}\left[3.5(1 - \kappa^2\kappa'^2) - 3\right] A_0 A_3 - \frac{3}{3.5.7.9}\left[5.7(1 - \kappa^2\kappa'^2) - 3\right]\kappa^2\kappa'^2 A_1 A_4\right.$$

$$\left. + \frac{4}{5.7.9.11}\left[7.9(1 - \kappa^2\kappa'^2) - 3\right]\kappa^4\kappa'^4 A_2 A_5 - \dots\right\}$$

$$+ \frac{2}{1.3.5.7.9}(192 - 816\kappa^2\kappa'^2 + 525\kappa^4\kappa'^4) A_0 A_4 - \dots \quad \dots\dots\dots\dots (2)$$

In this result a good many terms are added which are not deducible from the table of functions given above, but every term as stated here has actually been computed. The laws governing the succession of terms in the first six lines seem clear, but I do not claim that the proof of the laws is rigorous. I do not perceive how each series is derived from those preceding it, and I have no idea how the series beginning with $A_0 A_4$ would go on. With sufficient patience it would no doubt be possible to determine the general law of the series, but I do not propose to make the attempt at present, since we have more than enough for the immediate object in view.

This result (2) is, of course, equally applicable to the integrals of the type $I_i^s(\sin)$.

In order to effect the required integrations we must define the functions, and I take the definitions (with a few very slight changes) from § 2 of "The Pear-shaped Figure." In order to use the preceding analysis it is necessary that the square of the P-function and the square of the cosine or sine function should be the same functions of $\kappa^2\cos^2\theta$ and of $-\kappa'^2\sin^2\phi$. But as in the definitions to be used this symmetry does not hold good, a difficulty arises, which may, however, be easily overcome. If the P-function be multiplied by any factor f, and the cosine or sine function by any factor g, the integral will be multiplied by f^2g^2. I therefore introduce such factors f and g as will render the residual factors of the squares of the P and cosine or sine-functions symmetrical in the proper manner.

It seems desirable to show how the results found here accord with the approximate integrals as found on pp. 276—7, § 22, of "Harmonics." In this connection I remark that $\dfrac{k^3\cos\beta\cos\gamma}{\sin^3\beta}$, when written in the notation of "Harmonics," is $k^3\nu(\nu^2-1)^{\frac{1}{2}}\left(\nu^2-\dfrac{1+\beta}{1-\beta}\right)^{\frac{1}{2}}$, a factor which I denoted in that paper by M.

It does not seem necessary to give full details of the analysis in the several cases, since it is sometimes tedious, and it merely involves the substitution in the formula of the values of A_0, A_1, A_2, &c.

We will now take the several harmonics successively.

Harmonic of the Order Zero.

This harmonic is simply unity, so that $A_0=1$ and all other A's vanish. The formula is

$$I_0(\cos)=\frac{4\pi k^3\cos\beta\cos\gamma}{\sin^3\beta}\quad\text{.........................(3)}$$

This is obviously right since the integral is $\int p\,d\sigma$, of which this is the known value.

Harmonics of the First Order.

Here we have all the A's zero excepting A_0 and A_1, and when the functions have the proper symmetrical forms, we have from (2),

$$I_i^s\binom{\cos}{\sin}=\frac{4\pi k^3\cos\beta\cos\gamma}{\sin^3\beta}\left[A_0^2-\tfrac{1}{3}\kappa^2\kappa'^2A_1^2+\tfrac{2}{3}(\kappa^2-\kappa'^2)A_0A_1\right],\ (s=0,1)$$

(1) *The Zonal Harmonic.*

I define this thus:—

$$\left.\begin{aligned}\mathfrak{P}_1(\mu)&=\sin\theta=f(\kappa^2-\kappa^2\cos^2\theta)^{\frac{1}{2}}\\ \mathfrak{C}_1(\phi)&=(1-\kappa'^2\cos^2\phi)^{\frac{1}{2}}=g(\kappa^2+\kappa'^2\sin^2\phi)^{\frac{1}{2}}\end{aligned}\right\}\quad\text{..............(4)}$$

where $f=\dfrac{1}{\kappa}$, $g=1$.

On squaring $\mathfrak{P}_1(\mu)$, it is clear that

$$A_0=\kappa^2,\quad A_1=-1$$

Whence I find $\quad I_1(\cos)=\dfrac{4\pi k^3\cos\beta\cos\gamma}{3\sin^3\beta}f^2g^2\kappa^2$

Since with definition (4) $f^2 g^2 \kappa^2 = 1$,

$$I_1(\cos) = \frac{4\pi k^3 \cos\beta \cos\gamma}{3 \sin^3 \beta} \quad \ldots\ldots\ldots\ldots\ldots\ldots\ldots\ldots(5)$$

In "Harmonics" this harmonic is defined by

$$\mathfrak{P}_1(\mu) = P_1(\mu) = \mu; \quad \mathsf{C}_1(\phi) = \sqrt{(1-\beta\cos 2\phi)} \quad \ldots\ldots\ldots\ldots(6)$$

Now we must take for f and g values such as to bring the two definitions into accord. This is the case if

$$f = \frac{1}{\kappa} = \sqrt{\frac{1+\beta}{1-\beta}}, \quad g = \sqrt{1+\beta}$$

and $f^2 g^2 \kappa^2 = 1+\beta$.

Hence $$I_1(\cos) = \tfrac{4}{3}\pi \mathrm{M}(1+\beta) \quad \ldots\ldots\ldots\ldots\ldots\ldots\ldots\ldots(7)$$

agreeing with the result on p. 277 of "Harmonics" for the case $i=1$, $s=0$, type OEC.

(2) *The Sectorial Cosine Harmonic.*

I define this thus:—

$$\left.\begin{aligned} \mathsf{P}_1^1(\mu) &= (1-\kappa^2\sin^2\theta)^{\frac{1}{2}} = f(\kappa'^2 + \kappa^2\cos^2\theta)^{\frac{1}{2}} \\ \mathfrak{C}_1^1(\phi) &= \cos\phi \qquad\qquad\quad = g(\kappa'^2 - \kappa'^2\sin^2\phi)^{\frac{1}{2}} \end{aligned}\right\} \quad \ldots\ldots\ldots\ldots(8)$$

where $f=1$, $g = \dfrac{1}{\kappa'}$.

By symmetry with the last result

$$I_1^1(\cos) = \frac{4\pi k^3 \cos\beta\cos\gamma}{3\sin^3\beta} f^2 g^2 \kappa'^2 = \frac{4\pi k^3\cos\beta\cos\gamma}{3\sin^3\beta} \quad \ldots\ldots\ldots(9)$$

In "Harmonics" I defined the functions thus:—

$$\left.\begin{aligned} \mathsf{P}_1^1(\mu) &= P_1^1(\mu)\left(\frac{\frac{1+\beta}{1-\beta} - \mu^2}{1-\mu^2}\right)^{\frac{1}{2}} = \left(\frac{1+\beta}{1-\beta}\right)^{\frac{1}{2}}\left(1 - \frac{1-\beta}{1+\beta}\sin^2\theta\right)^{\frac{1}{2}} \\ \mathfrak{C}_1^1(\phi) &= \cos\phi \end{aligned}\right\} \quad \ldots\ldots(10)$$

If we take $f = \left(\dfrac{1+\beta}{1-\beta}\right)^{\frac{1}{2}}$, $g\kappa' = 1$, the two definitions agree, and we have

$$I_1^1(\cos) = \tfrac{4}{3}\pi\mathrm{M}\frac{1+\beta}{1-\beta} = \tfrac{4}{3}\pi\mathrm{M}(1+2\beta+2\beta^2) \quad \ldots\ldots\ldots\ldots\ldots(11)$$

This agrees with the result on p. 277 of "Harmonics" with $i=1$, $s=1$, type OOC.

(3) *The Sectorial Sine Harmonic.*

I define this thus:—

$$\left.\begin{aligned}\mathfrak{P}_1^1(\mu) &= \cos\theta = f\kappa\cos\theta\\ \mathfrak{S}_1^1(\phi) &= \sin\phi = g\sqrt{-1}\,.\,\kappa'\sin\phi\end{aligned}\right\}\quad\ldots\ldots\ldots\ldots\ldots\ldots(12)$$

where $f=\dfrac{1}{\kappa}$, $g=\dfrac{1}{\kappa'\sqrt{-1}}$.

On squaring $\mathfrak{P}_1^1$ we find $A_0=0$, $A_1=1$, and

$$I_1^1(\sin)=\frac{4\pi k^3\cos\beta\cos\gamma}{3\sin^3\beta}(-f^2g^2\kappa^2\kappa'^2)=\frac{4\pi k^3\cos\beta\cos\gamma}{3\sin^3\beta}\quad\ldots\ldots(13)$$

In "Harmonics" the definitions were the same, and therefore

$$I_1^1(\sin)=\tfrac{4}{3}\pi\mathrm{M}\quad\ldots\ldots\ldots\ldots\ldots\ldots\ldots\ldots\ldots\ldots(14)$$

This agrees with the result on p. 277 of "Harmonics" with $i=1$, $s=1$, type OOS.

Harmonics of the Second Order.

In these the only coefficients are A_0, A_1, A_2, and (2) becomes

$$I_2^s\binom{\cos}{\sin}=\frac{4\pi k^3\cos\beta\cos\gamma}{\sin^3\beta}\Big[A_0^2-\tfrac{1}{3}\kappa^2\kappa'^2A_1^2+\tfrac{1}{5}\kappa^4\kappa'^4A_2^2$$
$$+\tfrac{2}{3}(\kappa^2-\kappa'^2)A_0A_1-\tfrac{4}{15}(\kappa^2-\kappa'^2)\kappa^2\kappa'^2A_1A_2+\tfrac{2}{15}(4-9\kappa^2\kappa'^2)A_0A_2\Big]$$

with $s=0, 1, 2$.

(1) *and* (4) *The Zonal and Sectorial Cosine Harmonics.*

These are defined thus:—

$$\left.\begin{aligned}\mathfrak{P}_2^s(\mu) &= \kappa^2\sin^2\theta-q^2\\ \mathfrak{C}_2^s(\phi) &= q'^2-\kappa'^2\cos^2\phi,\quad(s=0,\,2)\end{aligned}\right\}\quad\ldots\ldots\ldots\ldots\ldots\ldots(15)$$

where $q^2=\frac{1}{3}[1+\kappa^2\mp(1-\kappa^2\kappa'^2)^{\frac{1}{2}}]$, with upper sign for $s=0$ and lower for $s=2$; and $q'^2=1-q^2$.

Writing $\quad t^2=\kappa^2-q^2=q'^2-\kappa'^2=\frac{1}{3}[\kappa^2-\kappa'^2\pm(1-\kappa^2\kappa'^2)^{\frac{1}{2}}]$

$$\left.\begin{aligned}\mathfrak{P}_2^s(\mu) &= f(t^2-\kappa^2\cos^2\theta)\\ \mathfrak{C}_2^s(\phi) &= g(t^2+\kappa'^2\sin^2\phi),\quad(s=0,\,2)\end{aligned}\right\}\quad\ldots\ldots\ldots\ldots\ldots(15)$$

where $f=1$, $g=1$. It may be noted that t^2 is a symmetrical function in κ^2 and $-\kappa'^2$.

Squaring $\mathfrak{P}_2^s$ we find

$$A_0=t^4,\quad A_1=-2t^2,\quad A_2=1$$

After reduction I find, for $s = 0, 2$,

$$I_i^s(\cos) = \frac{4\pi k^3 \cos\beta \cos\gamma}{3\sin^3\beta}\left[t^8 - \tfrac{4}{3}(\kappa^2 - \kappa'^2)t^6 + \tfrac{2}{15}(4 - 19\kappa^2\kappa'^2)t^4 + \tfrac{8}{15}(\kappa^2 - \kappa'^2)\kappa^2\kappa'^2 t^2 + \tfrac{1}{5}\kappa^4\kappa'^4\right]$$

Now

$$3t^2 = \kappa^2 - \kappa'^2 \pm (1 - \kappa^2\kappa'^2)^{\frac{1}{2}}$$

$$9t^4 = 2 - 5\kappa^2\kappa'^2 \pm 2(\kappa^2 - \kappa'^2)(1 - \kappa^2\kappa'^2)^{\frac{1}{2}}$$

$$27t^6 = (4 - 7\kappa^2\kappa'^2)(\kappa^2 - \kappa'^2) \pm (4 - 13\kappa^2\kappa'^2)(1 - \kappa^2\kappa'^2)^{\frac{1}{2}}$$

$$81t^8 = 8 - 40\kappa^2\kappa'^2 + 41\kappa^4\kappa'^4 \pm 4(2 - 5\kappa^2\kappa'^2)(\kappa^2 - \kappa'^2)(1 - \kappa^2\kappa'^2)^{\frac{1}{2}}$$

Whence on substitution, with $f^2g^2 = 1$,

$$I_2^s(\cos) = \frac{4\pi k^3 \cos\beta \cos\gamma}{5\sin^3\beta}\cdot\frac{2^3}{3^4}\left[(1 - \kappa^2\kappa'^2)^2 \pm (1 + \tfrac{1}{2}\kappa^2\kappa'^2)(\kappa^2 - \kappa'^2)(1 - \kappa^2\kappa'^2)^{\frac{1}{2}}\right] \quad \text{......(16)}$$

The upper sign being taken for the zonal ($s = 0$), the lower for the sectorial harmonic ($s = 2$).

If these expressions be developed in powers of κ' as far as three terms of the series, I find, on reintroducing the factor f^2g^2,

$$I_2(\cos) = \frac{4\pi k^3 \cos\beta \cos\gamma}{5\sin^3\beta}\cdot\frac{2^4}{5^4}(1 - 2\kappa'^2 + \tfrac{21}{16}\kappa'^4)f^2g^2$$

$$= \tfrac{4}{5}\pi M\,.\,\frac{2^4}{5^4}(1 - 4\beta + \tfrac{37}{4}\beta^2)f^2g^2 \quad \text{.....................(17)}$$

$$I_2^2(\cos) = \frac{4\pi k^3 \cos\beta \cos\gamma}{5\sin^3\beta}\cdot\tfrac{1}{3}\kappa'^4(1 - \kappa'^2 + \tfrac{11}{16}\kappa'^4)f^2g^2$$

$$= \tfrac{4}{5}\pi M\,.\,\tfrac{1}{3}\left(\frac{2\beta}{1+\beta}\right)^2(1 - 2\beta + \tfrac{19}{4}\beta^2)f^2g^2 \quad \text{...........(18)}$$

In "Harmonics" I made the following definitions

$$\left.\begin{aligned}\mathfrak{P}_2(\mu) &= P_2(\mu) - \tfrac{1}{4}\beta P_2^2(\mu) = 1 - \tfrac{3}{2}(1 + \tfrac{1}{2}\beta)\cos^2\theta \\ \mathfrak{C}_2(\phi) &= 1 - \tfrac{3}{2}\beta\cos 2\phi \qquad = 1 - \tfrac{3}{2}\beta + \tfrac{3}{2}\beta\sin^2\phi\end{aligned}\right\} \quad \text{.........(19)}$$

In order to make the two forms of definition agree we must take

$$ft^2 = 1, \quad gt^2 = 1 - \tfrac{3}{2}\beta$$

Thus $$f^2g^2 = \frac{1}{t^8}(1 - 3\beta + \tfrac{9}{4}\beta^2)$$

Now on development

$$t^8 = (\tfrac{2}{3})^4(1 - 5\kappa'^2 + \tfrac{81}{8}\kappa'^4) = (\tfrac{2}{3})^4(1 - 10\beta + \tfrac{101}{2}\beta^2)$$

Whence $$f^2g^2 = (\tfrac{3}{2})^4(1 + 7\beta + \tfrac{87}{4}\beta^2)$$

Introducing this I find

$$I_2(\cos) = \tfrac{4}{5}\pi M(1 + 3\beta + 3\beta^2) \quad \text{.....................(20)}$$

agreeing with the result on p. 276 of "Harmonics" with $i=2$, $s=0$, type EEC.

Again in "Harmonics"

$$\left.\begin{aligned}\mathfrak{P}_2^2(\mu) &= 3\beta P_2(\mu) + P_2^2(\mu) = 3\beta + 3(1-\tfrac{3}{2}\beta)\cos^2\theta \\ \mathfrak{C}_2^2(\phi) &= \tfrac{1}{2}\beta + \cos 2\phi \qquad\quad = 1 + \tfrac{1}{2}\beta - 2\sin^2\phi\end{aligned}\right\} \quad \ldots\ldots(21)$$

In order to make the two definitions agree we must take

$$f\kappa^2 = -3(1-\tfrac{3}{2}\beta), \qquad g\kappa'^2 = 2$$

or

$$f = -3(1+\tfrac{1}{2}\beta-\beta^2), \quad g = 2\left(\frac{1+\beta}{2\beta}\right)$$

So that $f^2g^2 = 2^2 . 3^2 \left(\dfrac{1+\beta}{2\beta}\right)^2 (1+\beta-\tfrac{7}{4}\beta^2)$.

Introducing this in (18) we have

$$I_2^2(\cos) = \tfrac{4}{7}\pi M . 12(1-\beta+\beta^2) \ldots\ldots\ldots\ldots\ldots\ldots(22)$$

agreeing with the result on p. 276 of "Harmonics" with $i=2$, $s=2$, type EEC.

(2) *The Cosine Tesseral Harmonic.*

This is defined thus:—

$$\left.\begin{aligned}\mathsf{P}_2^1(\mu) &= \sin\theta\,(1-\kappa^2\sin^2\theta)^{\frac{1}{2}} = f(\kappa^2-\kappa^2\cos^2\theta)^{\frac{1}{2}}(\kappa'^2+\kappa^2\cos^2\theta)^{\frac{1}{2}} \\ \mathsf{C}_2^1(\phi) &= \cos\phi\,(1-\kappa'^2\cos^2\phi)^{\frac{1}{2}} = g(\kappa^2+\kappa'^2\sin^2\phi)^{\frac{1}{2}}(\kappa'^2-\kappa'^2\sin^2\phi)^{\frac{1}{2}}\end{aligned}\right\} \ldots(23)$$

where $f = \dfrac{1}{\kappa}$, $g = \dfrac{1}{\kappa'}$.

Squaring P_2^1 we find

$$A_0 = \kappa^2\kappa'^2, \quad A_1 = (\kappa^2-\kappa'^2), \quad A_2 = -1$$

On substituting in the formula, I find, on putting $f^2g^2\kappa^2\kappa'^2 = 1$ and reducing,

$$I_2^1(\cos) = \frac{4\pi k^3\cos\beta\cos\gamma}{5\sin^3\beta} \cdot \frac{1}{3} \quad \ldots\ldots\ldots\ldots\ldots\ldots(24)$$

In "Harmonics" the definitions were

$$\left.\begin{aligned}\mathsf{P}_2^1(\mu) &= \left(\frac{\frac{1+\beta}{1-\beta}-\mu^2}{1-\mu^2}\right)^{\frac{1}{2}} P_2^1(\mu) = 3\left(\frac{1+\beta}{1-\beta}\right)^{\frac{1}{2}}\sin\theta\left(1-\frac{1-\beta}{1+\beta}\sin^2\theta\right)^{\frac{1}{2}} \\ \mathsf{C}_2^1(\phi) &= (1-\beta\cos 2\phi)^{\frac{1}{2}}\cos\phi = (1+\beta)^{\frac{1}{2}}\cos\phi\left(1-\frac{2\beta}{1+\beta}\cos^2\phi\right)^{\frac{1}{2}}\end{aligned}\right\} \ldots(25)$$

In order to make the two definitions agree we must take

$$f\kappa = 3\left(\frac{1+\beta}{1-\beta}\right)^{\frac{1}{2}}, \quad g\kappa' = (1+\beta)^{\frac{1}{2}}$$

so that $f^2 g^2 \kappa^2 \kappa'^2 = 3^2 \dfrac{(1+\beta)^2}{1-\beta} = 3^2(1+3\beta+4\beta^2)$. On multiplying (24) by this factor, we have

$$I_2^1(\cos) = \tfrac{4}{5}\pi M \,.\, 3(1+3\beta+4\beta^2) \quad \text{.................(26)}$$

agreeing with the result on p. 276 of "Harmonics" with $i=2$, $s=1$, type EOC.

(3) *The Tesseral Sine Harmonic.*

This is defined thus:—

$$\left.\begin{aligned} \mathfrak{P}_2^1(\mu) &= \sin\theta\cos\theta = f\kappa\cos\theta(\kappa^2 - \kappa^2\cos^2\theta)^{\frac{1}{2}} \\ \mathsf{S}_2^1(\phi) &= \sin\phi(1-\kappa'^2\cos^2\phi)^{\frac{1}{2}} = g\sqrt{-1}\,.\,\kappa'\sin\phi(\kappa^2+\kappa'^2\sin^2\phi)^{\frac{1}{2}} \end{aligned}\right\}\text{...(27)}$$

where $f = \dfrac{1}{\kappa^2}$, $g = \dfrac{1}{\kappa'\sqrt{-1}}$.

Squaring $\mathfrak{P}_2^1$ we find

$$A_0 = 0, \quad A_1 = \kappa^2, \quad A_2 = -1$$

whence, on putting $-f^2 g^2 \kappa^4 \kappa'^2 = 1$,

$$I_2^1(\sin) = \frac{4\pi k^3 \cos\beta\cos\gamma}{5\sin^3\beta} \cdot \frac{1}{3} \quad \text{.......................(28)}$$

In "Harmonics" the definitions were

$$\left.\begin{aligned} \mathfrak{P}_2^1(\mu) &= P_2^1(\mu) = 3\sin\theta\cos\theta \\ \mathsf{S}_2^1(\phi) &= \sin\phi(1-\beta\cos 2\phi)^{\frac{1}{2}} = (1+\beta)^{\frac{1}{2}}\sin\phi\left(1 - \frac{2\beta}{1+\beta}\cos^2\phi\right)^{\frac{1}{2}} \end{aligned}\right\}\text{...(29)}$$

Therefore, to make the two definitions agree, we must take

$$f\kappa^2 = 3, \quad g\kappa'\sqrt{-1} = (1+\beta)^{\frac{1}{2}}$$

Therefore $-f^2 g^2 \kappa^4 \kappa'^2 = 3^2(1+\beta)$, and on multiplying (28) by this factor we have

$$I_2^1(\sin) = \tfrac{4}{5}\pi M \,.\, 3(1+\beta) \quad \text{......................(30)}$$

agreeing with the result on p. 276 of "Harmonics" with $i=2$, $s=1$, type EOS.

(5) *The Sectorial Sine Harmonic.*

This is defined thus:—

$$\left.\begin{aligned} \mathsf{P}_2^2(\mu) &= \cos\theta(1-\kappa^2\sin^2\theta)^{\frac{1}{2}} \\ \mathfrak{S}_2^2(\phi) &= \sin\phi\cos\phi \end{aligned}\right\}\text{....................(31)}$$

If in the last integral we had written $\frac{1}{2}\pi - \theta$ for ϕ, and $\frac{1}{2}\pi - \phi$ for θ, and κ' for κ, $\mathfrak{P}_2^1$ would have become $\mathfrak{S}_2^2$, and S_2^1 would have become P_2^2. Therefore the result (28) gives what is needed by merely interchanging κ and κ'.

Therefore $$I_2^2(\sin) = \frac{4\pi k^3 \cos\beta\cos\gamma}{5\sin^3\beta}\cdot\frac{1}{3} \qquad\ldots\ldots(32)$$

For the purpose of comparison I must put

$$\mathsf{P}_2^2(\mu) = f\kappa\cos\theta\,(\kappa'^2+\kappa^2\cos^2\theta)^{\frac{1}{2}},\quad \mathfrak{S}_2^2(\phi) = g\sqrt{-1}\,.\,\kappa'\sin\phi\,(\kappa'^2-\kappa'^2\sin^2\phi)^{\frac{1}{2}} \qquad\ldots\ldots(31)$$

and $$I_2^2(\sin) = \frac{4\pi k^3\cos\beta\cos\gamma}{5\sin^3\beta}\left(-\tfrac{1}{3}f^2g^2\kappa^2\kappa'^4\right) \qquad\ldots\ldots(33)$$

In "Harmonics" the definition was

$$\left.\begin{aligned}\mathsf{P}_2^2(\mu) &= \left(\frac{\frac{1+\beta}{1-\beta}-\mu^2}{1-\mu^2}\right)^{\frac{1}{2}} P_2^2(\mu) = 3\left(\frac{1+\beta}{1-\beta}\right)^{\frac{1}{2}}\cos\theta\left(1-\frac{1-\beta}{1+\beta}\sin^2\theta\right)^{\frac{1}{2}}\\ \mathfrak{S}_2^2(\phi) &= \sin 2\phi \qquad\qquad = 2\sin\phi\cos\phi\end{aligned}\right\}\ldots(34)$$

In order to make the two definitions agree we must take

$$f\kappa = 3\left(\frac{1+\beta}{1-\beta}\right)^{\frac{1}{2}},\qquad g\sqrt{-1}\,.\,\kappa'^2 = 2$$

Thus $-f^2g^2\kappa^2\kappa'^4 = 2^2\,.\,3^2\,\dfrac{1+\beta}{1-\beta}$; introducing this in (33) we have

$$I_2^2(\sin) = \tfrac{4}{5}\pi\mathrm{M}\,12\left(\frac{1+\beta}{1-\beta}\right) = \tfrac{4}{5}\pi\mathrm{M}\,12\,(1+2\beta+2\beta^2) \qquad\ldots\ldots(35)$$

agreeing with the result on p. 276 of "Harmonics" with $i=2$, $s=2$, type EES.

The Harmonics of the Third Order.

In these the only coefficients are A_0, A_1, A_2, A_3, and (2) becomes

$$\begin{aligned}I_3^s\binom{\cos}{\sin} = \frac{4\pi k^3\cos\beta\cos\gamma}{\sin^3\beta}\{&A_0^2 - \tfrac{1}{3}\kappa^2\kappa'^2A_1^2 + \tfrac{1}{5}\kappa^4\kappa'^4A_2^2 - \tfrac{1}{7}\kappa^6\kappa'^6A_3^2\\ &+2(\kappa^2-\kappa'^2)\left[\tfrac{1}{3}A_0A_1 - \tfrac{2}{15}\kappa^2\kappa'^2A_1A_2 + \tfrac{3}{35}\kappa^4\kappa'^4A_2A_3\right]\\ &+\tfrac{2}{15}(4-9\kappa^2\kappa'^2)A_0A_2 - \tfrac{2}{105}(12-25\kappa^2\kappa'^2)\kappa^2\kappa'^2A_1A_3\\ &+\tfrac{4}{35}(\kappa^2-\kappa'^2)(4-5\kappa^2\kappa'^2)A_0A_3\}\quad (s=0,1,2,3)\end{aligned}$$

(1) *and* (4) *The Zonal and Second Tesseral Cosine Harmonics.*

These are defined thus:—

$$\left.\begin{aligned}\mathfrak{P}_3^s(\mu) &= \sin\theta\,(\kappa^2\sin^2\theta - q^2)\\ \mathsf{C}_3^s(\phi) &= (q'^2-\kappa'^2\cos^2\phi)(1-\kappa'^2\cos^2\phi)^{\frac{1}{2}},\quad (s=0,2)\end{aligned}\right\}\ldots\ldots(36)$$

where $q^2 = \frac{2}{5}[1+\kappa^2 \mp (1-\frac{7}{4}\kappa^2+\kappa^4)^{\frac{1}{2}}]$, with the upper sign for $s=0$ and the lower for $s=2$. Writing

$$t^2 = \kappa^2 - q^2 = q'^2 - \kappa'^2$$
$$= \tfrac{1}{5}[3\kappa^2 - 2 \pm (4 - 7\kappa^2 + 4\kappa^4)^{\frac{1}{2}}] = \tfrac{1}{5}[1 - 3\kappa'^2 \pm (1 - \kappa'^2 + 4\kappa'^4)^{\frac{1}{2}}]$$

$$\left.\begin{aligned}\mathfrak{P}_3^s(\mu) &= f(t^2 - \kappa^2\cos^2\theta)(\kappa^2 - \kappa^2\cos^2\theta)^{\frac{1}{2}}\\ \mathrm{C}_3^s(\phi) &= g(t^2 + \kappa'^2\sin^2\phi)(\kappa^2 + \kappa'^2\sin^2\phi)^{\frac{1}{2}}\end{aligned}\right\} \quad \text{............(37)}$$

where $f = \dfrac{1}{\kappa}$, $g=1$.

Squaring $\mathfrak{P}_3^s$ we find

$$A_0 = t^4\kappa^2, \quad A_1 = -(2\kappa^2 + t^2)t^2, \quad A_2 = 2t^2 + \kappa^2, \quad A_3 = -1$$

After some rather tedious reductions I find (for $s = 0, 2$)

$$I_3^s(\cos) = \frac{4\pi k^3 \cos\beta\cos\gamma}{\sin^3\beta}\,\kappa^2\{\tfrac{1}{3}t^8 - \tfrac{4}{15}(1 - 3\kappa'^2)t^6 + \tfrac{2}{105}(4 - 25\kappa'^2 + 33\kappa'^4)t^4$$
$$+ \tfrac{4}{105}(2 - 5\kappa'^2)\kappa^2\kappa'^2 t^2 + \tfrac{1}{35}\kappa^4\kappa'^4\} f^2 g^2$$

Now writing $D = (1 - \kappa'^2 + 4\kappa'^4)^{\frac{1}{2}}$,

$$5t^2 = 1 - 3\kappa'^2 \pm D$$
$$5^2 t^4 = 2 - 7\kappa'^2 + 13\kappa'^4 \pm 2(1 - 3\kappa'^2)D$$
$$5^3 t^6 = 4 - 21\kappa'^2 + 48\kappa'^4 - 63\kappa'^6 \pm (4 - 19\kappa'^2 + 31\kappa'^4)D$$
$$5^4 t^8 = 8 - 56\kappa'^2 + 177\kappa'^4 - 314\kappa'^6 + 313\kappa'^8 \pm (8 - 52\kappa'^2 + 136\kappa'^4 - 156\kappa'^6)D$$

On substituting these in the above expression, and noting that $\kappa^2 f^2 g^2$ will be unity with the definition adopted, I find

$$\left.\begin{aligned}I_3(\cos) &= \frac{4\pi k^3\cos\beta\cos\gamma}{7\sin^3\beta}\cdot\frac{2^3}{5^4}D[(1 - \tfrac{1}{2}\kappa'^2)(1 - \kappa'^2 - \tfrac{8}{3}\kappa'^4) + (1 - \kappa'^2 + \tfrac{2}{3}\kappa'^4)D]\\ I_3^2(\cos) &= \text{the same with the sign of } D \text{ changed}\end{aligned}\right\}$$
$$\text{......(38)}$$

If these expressions be developed in powers of κ', and if the factor $\kappa^2 f^2 g^2$ be reintroduced, I find

$$I_3(\cos) = \frac{4\pi k^3\cos\beta\cos\gamma}{7\sin^3\beta}(\tfrac{2}{5})^4(1 - 2\kappa'^2 + \tfrac{49}{16}\kappa'^4)\,.\,\kappa^2 f^2 g^2$$
$$= \tfrac{4}{7}\pi\mathrm{M}(\tfrac{2}{5})^4(1 - 4\beta + \tfrac{65}{4}\beta^2)\,.\,\kappa^2 f^2 g^2$$
$$I_3^2(\cos) = \frac{4\pi k^3\cos\beta\cos\gamma}{7\sin^3\beta}\cdot\frac{1}{3\,.\,5}\kappa'^4(1 - \kappa'^2 + \tfrac{31}{16}\kappa'^4)\,.\,\kappa^2 f^2 g^2$$
$$= \tfrac{4}{7}\pi\mathrm{M}\,.\,\frac{1}{3\,.\,5}\kappa'^4(1 - 2\beta + \tfrac{39}{4}\beta^2)\,.\,\kappa^2 f^2 g^2$$

In "Harmonics" I defined

$$\mathfrak{P}_3(\mu) = P_3(\mu) - \tfrac{1}{4}\beta P_3^2(\mu) = \sin\theta[\tfrac{5}{2}\sin^2\theta(1 + \tfrac{3}{2}\beta) - \tfrac{3}{2}(1 + \tfrac{5}{2}\beta)] \ldots (39)$$

In order that our previous definition may agree with this we must have

$$f\kappa q^2 = \tfrac{3}{2}(1+\tfrac{5}{2}\beta), \quad f\kappa^3 = \tfrac{5}{2}(1+\tfrac{3}{2}\beta)$$

Now $$q^2 = \tfrac{1}{5}[4 - 2\kappa'^2 - \sqrt{(1-\kappa'^2+4\kappa'^4)}] = \tfrac{3}{5}[1 - \tfrac{1}{2}\kappa'^2 - \tfrac{5}{8}\kappa'^4]$$
$$= \tfrac{3}{5}[1 - \beta - \tfrac{3}{2}\beta^2]$$

whence $f\kappa = \frac{5}{2}(1+\frac{7}{2}\beta+5\beta^2)$, and this value of $f\kappa$ satisfies the second equation.

In "Harmonics" I defined

$$\left.\begin{aligned} \mathrm{C}_3(\phi) &= (1-\beta\cos 2\phi)^{\frac{1}{2}}(1-\tfrac{5}{2}\beta\cos 2\phi) \\ &= (1+\beta)^{\frac{1}{2}}(1+\tfrac{5}{2}\beta - 5\beta\cos^2\phi)\left(1-\frac{2\beta}{1+\beta}\cos^2\phi\right)^{\frac{1}{2}} \end{aligned}\right\} \quad \ldots(40)$$

In order that the previous definition may agree with this we must have

$$gq'^2 = (1+\beta)^{\frac{1}{2}}(1+\tfrac{5}{2}\beta) = 1 + 3\beta + \tfrac{9}{8}\beta^2$$
$$g\kappa'^2 = 5\beta(1+\beta)^{\frac{1}{2}}$$

But $$q'^2 = 1 - q^2 = \tfrac{2}{5}(1 + \tfrac{3}{2}\beta + \tfrac{9}{4}\beta^2)$$

and thence $$g = \tfrac{5}{2}(1+\tfrac{3}{2}\beta - \tfrac{27}{8}\beta^2)$$

This value of g will be found to give the correct value for $g\kappa'^2$.

Then $$fg\kappa = (\tfrac{5}{2})^2(1+5\beta+\tfrac{55}{8}\beta^2)$$

and $$f^2g^2\kappa^2 = (\tfrac{5}{2})^4(1+10\beta+\tfrac{155}{4}\beta^2)$$

Introducing this into the value of $I_3(\cos)$, we find

$$I_3(\cos) = \tfrac{4}{7}\pi\mathrm{M}(1+6\beta+15\beta^2)\ldots\ldots\ldots\ldots\ldots\ldots\ldots(41)$$

agreeing with the result on p. 276 of "Harmonics" for $i=3$, $s=0$, type OEC.

Again in "Harmonics" I defined

$$\mathfrak{P}_3^2(\mu) = 15\beta P_3(\mu) + P_3^2(\mu) = 15\sin\theta[1-\tfrac{3}{2}\beta - (1-\tfrac{5}{2}\beta)\sin^2\theta]\ldots(42)$$

To make the former definition agree with this we must take

$$f\kappa q^2 = -15(1-\tfrac{3}{2}\beta), \quad f\kappa^3 = -15(1-\tfrac{5}{2}\beta)$$

In the present case

$$q^2 = \tfrac{1}{5}[4-2\kappa'^2+\sqrt{(1-\kappa'^2+4\kappa'^4)}] = 1 - \tfrac{1}{2}\kappa'^2 + \tfrac{3}{8}\kappa'^4 + \tfrac{3}{16}\kappa'^6$$
$$= 1 - \beta + \tfrac{5}{2}\beta^2 - \tfrac{5}{2}\beta^3$$

Omitting the term in β^3 we find, with this value of q^2,

$f\kappa = -15(1-\frac{1}{2}\beta - 3\beta^2)$, and that the second equation is satisfied.

Again I defined

$$\begin{aligned} \mathrm{C}_3^2(\phi) &= (\tfrac{3}{2}\beta + \cos 2\phi)(1-\beta\cos 2\phi)^{\frac{1}{2}} \\ &= (1+\beta)^{\frac{1}{2}}(2\cos^2\phi - 1 + \tfrac{3}{2}\beta)\left(1-\frac{2\beta}{1+\beta}\cos^2\phi\right)^{\frac{1}{2}} \ldots(43) \end{aligned}$$

Hence to secure agreement we must take

$$gq'^2 = -(1 - \tfrac{3}{2}\beta)(1+\beta)^{\frac{1}{2}} = -(1 - \beta - \tfrac{7}{8}\beta^2)$$

$$g\kappa'^2 = -2(1+\beta)^{\frac{1}{2}}$$

Now $q'^2 = 1 - q^2 = \beta(1 - \tfrac{5}{2}\beta + \tfrac{5}{2}\beta^2)$, and therefore

$$g = -\frac{1}{\beta}(1 + \tfrac{3}{2}\beta + \tfrac{3}{8}\beta^2)$$

The second equation is satisfied.

We have then

$$f^2 g^2 \kappa^2 \kappa'^4 = 2^2 . 3^2 . 5^2 (1+\beta)(1 - \beta - \tfrac{23}{4}\beta^2) = 2^2 . 3^2 . 5^2 (1 - 0 . \beta - \tfrac{27}{4}\beta^2)$$

Introducing this into the value of $I_3^2(\cos)$, we find

$$I_3^2(\cos) = \tfrac{4}{7}\pi M . 3 . 4 . 5 (1 - 2\beta + 3\beta^2) \quad \text{..............(44)}$$

agreeing with the result on p. 276 of "Harmonics" for $i = 3$, $s = 2$, type OEC.

(2) *and* (6) *First Tesseral Cosine Harmonic and Sectorial Cosine Harmonic.*

These are defined thus:—

$$\left.\begin{aligned} \mathsf{P}_3^s(\mu) &= (\kappa^2 \sin^2\theta - q^2)(1 - \kappa^2 \sin^2\theta)^{\frac{1}{2}} \\ \mathfrak{C}_3^s(\phi) &= \cos\phi\,(q'^2 - \kappa'^2 \cos^2\phi), \quad (s = 1, 3) \end{aligned}\right\} \quad \text{............(45)}$$

where $q^2 = \tfrac{1}{5}[1 + 2\kappa^2 \mp (1 - \kappa^2 + 4\kappa^4)^{\frac{1}{2}}]$, with upper sign for $s = 1$ and lower sign for $s = 3$, and $q'^2 = 1 - q^2$.

Writing $t'^2 = \kappa'^2 - q'^2$,

$$\left.\begin{aligned} \mathsf{P}_3^s(\mu) &= f(t'^2 + \kappa^2\cos^2\theta)(\kappa'^2 + \kappa^2\cos^2\theta)^{\frac{1}{2}} \\ \mathfrak{C}_3^s(\phi) &= g(t'^2 - \kappa'^2\sin^2\phi)(\kappa'^2 - \kappa'^2\sin^2\phi)^{\frac{1}{2}} \end{aligned}\right\} \quad \text{............(46)}$$

where $f = -1$, $g = -\dfrac{1}{\kappa'}$.

It is clear that $[\mathsf{P}_3^s(\mu)\,\mathfrak{C}_3^s(\phi)]^2$ $(s = 1, 3)$ has the same form as $[\mathfrak{P}_3^s(\mu)\,\mathsf{C}_3^s(\phi)]^2$ $(s = 0, 2)$ when in the latter we interchange θ with $\tfrac{1}{2}\pi - \phi$, and κ with κ'. The interchange of the variables of integration clearly makes no difference in the result, and therefore we need only interchange κ and κ', and replace t by t'.

In the present instance

$$t'^2 = \kappa'^2 - q'^2 = q^2 - \kappa^2 = \tfrac{1}{5}[1 - 3\kappa^2 \mp (1 - \kappa^2 + 4\kappa^4)^{\frac{1}{2}}]$$

This shows that t'^2 is the same function of κ^2 that t^2 was of κ'^2, but that $I_3^1(\cos)$ is analogous with $I_3^2(\cos)$, and $I_3^3(\cos)$ with $I_3(\cos)$. Thus we may at once write down the results by interchanging κ and κ' throughout.

Let $$D' = (1 - \kappa^2 + 4\kappa^4)^{\frac{1}{2}}$$

Then putting $\kappa'^2 f^2 g^2 = 1$, we have by symmetry with (38)

$$\left.\begin{aligned} I_3^1(\cos) &= \frac{4\pi k^3 \cos\beta \cos\gamma}{7\sin^3\beta} \cdot \frac{2^3}{5^4} D' [(1 - \kappa^2 + \tfrac{2}{3}\kappa^4) D' - (1 - \tfrac{1}{2}\kappa^2)(1 - \kappa^2 - \tfrac{8}{3}\kappa^4)] \\ I_3^3(\cos) &= \text{the same with the sign of } D' \text{ changed} \end{aligned}\right\} \quad \text{.........(47)}$$

If these expressions be developed in powers of κ' I find, on reintroducing the factor $\kappa'^2 f^2 g^2$,

$$I_3^1(\cos) = \frac{4\pi k^3 \cos\beta \cos\gamma}{7\sin^3\beta} \cdot \frac{2^7}{3 \cdot 5^4}(1 - \tfrac{9}{4}\kappa'^2 + \tfrac{361}{256}\kappa'^4)\kappa'^2 f^2 g^2$$

$$= \tfrac{4}{7}\pi M \cdot \frac{2^7}{3 \cdot 5^4}(1 - \tfrac{9}{2}\beta + \tfrac{649}{64}\beta^2)\kappa'^2 f^2 g^2$$

$$I_3^3(\cos) = \frac{4\pi k^3 \cos\beta \cos\gamma}{7\sin^3\beta} \cdot \frac{1}{2 \cdot 5}\kappa'^4(1 - \tfrac{3}{2}\kappa'^2 + \tfrac{183}{256}\kappa'^4)\kappa'^2 f^2 g^2$$

$$= \tfrac{4}{7}\pi M \cdot \frac{1}{2 \cdot 5}\left(\frac{2\beta}{1+\beta}\right)^2 (1 - 3\beta + \tfrac{375}{64}\beta^2)\kappa'^2 f^2 g^2$$

In "Harmonics" I defined

$$\mathsf{P}_3^1(\mu) = \left(\frac{\frac{1+\beta}{1-\beta} - \mu^2}{1-\mu^2}\right)^{\frac{1}{2}} [P_3^1(\mu) - \tfrac{3}{16}\beta(1 + \tfrac{3}{4}\beta) P_3^3(\mu)]$$

$$= \tfrac{15}{2}\left(1 - \frac{1-\beta}{1+\beta}\sin^2\theta\right)^{\frac{1}{2}} \left(\frac{1+\beta}{1-\beta}\right)^{\frac{1}{2}} [\sin^2\theta(1 + \tfrac{3}{8}\beta + \tfrac{9}{32}\beta^2) - \tfrac{1}{5}(1 + \tfrac{15}{8}\beta + \tfrac{45}{32}\beta^2)] \quad \text{......(48)}$$

But we have defined it above by

$$\mathsf{P}_3^1(\mu) = f(1 - \kappa^2\sin^2\theta)^{\frac{1}{2}}(\kappa^2\sin^2\theta - q^2)$$

Therefore
$$f\kappa^2 = \tfrac{15}{2}\left(\frac{1+\beta}{1-\beta}\right)^{\frac{1}{2}}(1 + \tfrac{3}{8}\beta + \tfrac{9}{32}\beta^2)$$

$$fq^2 = \tfrac{3}{2}\left(\frac{1+\beta}{1-\beta}\right)^{\frac{1}{2}}(1 + \tfrac{15}{8}\beta + \tfrac{45}{32}\beta^2)$$

Now $$q^2 = \tfrac{1}{5}(1 - \tfrac{1}{4}\kappa'^2 - \tfrac{15}{64}\kappa'^4) = \tfrac{1}{5}(1 - \tfrac{1}{2}\beta - \tfrac{7}{16}\beta^2)$$

Whence $$f = \tfrac{15}{2}(1 + \tfrac{27}{8}\beta + \tfrac{189}{32}\beta^2)$$

This value also satisfies the expression for $f\kappa^2$.

Again I defined

$$\mathfrak{C}_3^1(\phi) = \cos\phi - \tfrac{5}{8}\beta(1 + \tfrac{3}{4}\beta)\cos 3\phi$$

$$= \cos\phi[1 + \tfrac{15}{8}\beta + \tfrac{45}{32}\beta^2 - \tfrac{5}{2}\beta(1 + \tfrac{3}{4}\beta)\cos^2\phi] \quad \text{.........(49)}$$

But we have defined it above by

$$\mathfrak{C}_3^1 = g\kappa'\cos\phi(\kappa'^2\cos^2\phi - q'^2)$$

Therefore

$$g\kappa' q'^2 = -(1 + \tfrac{15}{8}\beta + \tfrac{45}{32}\beta^2), \qquad g\kappa'^3 = -\tfrac{5}{2}\beta(1 + \tfrac{3}{4}\beta^2)$$

With the above value for q^2 we have $q'^2 = \frac{4}{5}(1 + \frac{1}{8}\beta + \frac{7}{64}\beta^2)$; whence

$$g\kappa' = -\tfrac{5}{4}(1 + \tfrac{7}{4}\beta + \tfrac{69}{64}\beta^2)$$

Therefore $$f^2 g^2 \kappa'^2 = \frac{3^2 . 5^4}{2^6}[1 + \tfrac{41}{4}\beta + \tfrac{3331}{64}\beta^2]$$

and I find $$I_3^1(\cos) = \tfrac{4}{7}\pi M . 6(1 + \tfrac{23}{4}\beta + \tfrac{257}{16}\beta^2) \quad \text{.................(50)}$$

agreeing with the result on p. 276 of "Harmonics" with $i = 3$, $s = 1$, type OOC.

In "Harmonics" I defined

$$\mathsf{P}_3^3(\mu) = \left(\frac{\frac{1+\beta}{1-\beta} - \mu^2}{1-\mu^2}\right)^{\frac{1}{2}} [\tfrac{5}{4}\beta(1 + \tfrac{3}{4}\beta)P_3^1(\mu) + P_3^3(\mu)]$$

$$= 15\left(1 - \frac{1-\beta}{1+\beta}\sin^2\theta\right)^{\frac{1}{2}}[1 + \tfrac{7}{8}\beta + \tfrac{9}{32}\beta^2 - \sin^2\theta(1 + \tfrac{3}{8}\beta - \tfrac{19}{32}\beta^2)] \quad \text{...(51)}$$

But $$\mathsf{P}_3^3(\mu) = f(1 - \kappa^2\sin^2\theta)^{\frac{1}{2}}(\kappa^2\sin^2\theta - q^2)$$

Therefore

$$f\kappa^2 = -15(1 + \tfrac{3}{8}\beta - \tfrac{19}{32}\beta^2), \qquad fq^2 = -15(1 + \tfrac{7}{8}\beta + \tfrac{9}{32}\beta^2)$$

Now $$q^2 = 1 - \frac{3}{2^2}\kappa'^2 + \frac{3}{2^6}\kappa'^4 + \frac{21}{2^9}\kappa'^6 \dots$$

Therefore $$q^2 = 1 - \tfrac{3}{2}\beta + \tfrac{27}{16}\beta^2$$

and $$f = -15(1 + \tfrac{19}{8}\beta + \tfrac{69}{32}\beta^2)$$

This also gives the correct value to $f\kappa^2$.

Again $$\mathfrak{C}_3^3(\phi) = \tfrac{3}{8}\beta(1 + \tfrac{3}{4}\beta)\cos\phi + \cos 3\phi$$

$$= \cos\phi[4\cos^2\phi - 3(1 - \tfrac{1}{8}\beta - \tfrac{3}{32}\beta^2)] \quad \text{.........(52)}$$

But $$\mathfrak{C}_3^3(\phi) = g\kappa'\cos\phi(\kappa'^2\cos^2\phi - q'^2)$$

Therefore $$g\kappa' = \frac{4}{\kappa'^2}, \quad \text{and} \quad g\kappa' . q'^2 = 3(1 - \tfrac{1}{8}\beta - \tfrac{3}{32}\beta^2)$$

If we eliminate $g\kappa'$, these equations give the correct value for q'^2.

Then $$fg\kappa' = -\frac{60}{\kappa'^2}(1 + \tfrac{19}{8}\beta + \tfrac{69}{32}\beta^2)$$

Therefore $$f^2 g^2 \kappa'^2 = \frac{3^2 . 4^2 . 5^2}{\kappa'^4}(1 + \tfrac{19}{4}\beta + \tfrac{637}{64}\beta^2)$$

Hence we find

$$I_3^3(\cos) = \tfrac{4}{7}\pi M . 360(1 + \tfrac{7}{4}\beta + \tfrac{25}{16}\beta^2) \quad \text{.................(53)}$$

agreeing with the result on p. 276 of "Harmonics" with $i = 3$, $s = 3$, type OOC.

(3) *and* (7) *First Tesseral Sine Harmonic and Sectorial Sine Harmonic.*

These are defined thus:—

$$\left.\begin{aligned}\mathfrak{P}_3^s(\mu) &= \cos\theta\,(\kappa^2\sin^2\theta - q^2)\\ \mathfrak{S}_3^s(\phi) &= \sin\phi\,(q'^2 - \kappa'^2\cos^2\phi),\quad (s=1,3)\end{aligned}\right\}\quad\ldots\ldots\ldots\ldots(54)$$

where $q^2 = \frac{1}{5}[2+\kappa^2 \mp (4-\kappa^2\kappa'^2)^{\frac{1}{2}}]$, with the upper sign for $s=1$, the lower for $s=3$, and $q'^2 = 1-q^2$.

Writing $\qquad t^2 = \kappa^2 - q^2$

$$\left.\begin{aligned}\mathfrak{P}_3^s(\mu) &= f\kappa\cos\theta\,(t^2 - \kappa^2\cos^2\theta)\\ \mathfrak{S}_3^s(\phi) &= g\kappa'\sqrt{-1}\,.\,\sin\phi\,(t^2 + \kappa'^2\sin^2\phi)\end{aligned}\right\}\quad\ldots\ldots\ldots\ldots(55)$$

where $f = \dfrac{1}{\kappa}$, $g = \dfrac{1}{\kappa'\sqrt{-1}}$.

Squaring $\mathfrak{P}_3^s$ we find

$$A_0 = 0,\quad A_1 = t^4,\quad A_2 = -2t^2,\quad A_3 = 1$$

On substitution in the formula for harmonics of the third order I find

$$I_3^s(\sin) = \frac{4\pi k^3\cos\beta\cos\gamma}{\sin^3\beta}\,[\tfrac{1}{3}t^8 - \tfrac{8}{15}(\kappa^2-\kappa'^2)\,t^6 + \tfrac{2}{105}(12 - 67\kappa^2\kappa'^2)\,t^4$$
$$+ \tfrac{12}{35}(\kappa^2-\kappa'^2)\,\kappa^2\kappa'^2 t^2 + \tfrac{1}{7}\kappa^4\kappa'^4]\,(-f^2g^2\kappa^2\kappa'^2)$$

If we write $D = (1 - \frac{1}{4}\kappa^2\kappa'^2)^{\frac{1}{2}}$

$$\frac{5}{2}\,t^2 = 1 - 2\kappa'^2 \pm D$$

$$\frac{5^2}{2^2}\,t^4 = 2 - \tfrac{17}{4}\kappa^2\kappa'^2 \pm 2\,(1-2\kappa'^2)\,D$$

$$\frac{5^3}{2^3}\,t^6 = 4 - \tfrac{51}{4}\kappa^2\kappa'^2 - \tfrac{19}{2}\kappa'^6 \pm (4 - \tfrac{49}{4}\kappa^2\kappa'^2)\,D$$

$$\frac{5^4}{2^4}\,t^8 = 8 - 34\kappa'^2 + \tfrac{897}{16}\kappa'^4 - \tfrac{353}{8}\kappa'^6 + \tfrac{353}{16}\kappa'^8 \pm (8 - 25\kappa'^2 + 51\kappa'^4 - 34\kappa'^6)\,D$$

On substitution I find, on putting $-f^2g^2\kappa^2\kappa'^2 = 1$,

$$\left.\begin{aligned}I_3^1(\sin) &= \frac{4\pi k^3\cos\beta\cos\gamma}{7\sin^3\beta}\cdot\frac{2^3}{3\,.\,5^4}\,[8 - 14\kappa^2\kappa'^2 + 3\kappa^4\kappa'^4\\ &\qquad + (\kappa^2-\kappa'^2)(8+3\kappa^2\kappa'^2)(1-\tfrac{1}{4}\kappa^2\kappa'^2)^{\frac{1}{2}}]\\ I_3^3(\sin) &= \text{the same with the sign of the square root reversed}\end{aligned}\right\}\ldots(56)$$

Developing these expressions in powers of κ'^2, reintroducing the factor $-f^2g^2\kappa^2\kappa'^2$, and reverting to the notation of "Harmonics," I find

$$I_3^1(\sin) = \frac{4\pi k^3\cos\beta\cos\gamma}{7\sin^3\beta}\,\frac{2^7}{3\,.\,5^4}\,(1 - \tfrac{7}{4}\kappa'^2 + \tfrac{169}{256}\kappa'^4)\,(-f^2g^2\kappa^2\kappa'^2)$$

$$= \tfrac{4}{7}\pi\mathrm{M}\,\frac{2^7}{3\,.\,5^4}\,(1 - \tfrac{7}{2}\beta + \tfrac{393}{64}\beta^2)\,(-f^2g^2\kappa^2\kappa'^2)$$

$$I_3^3(\sin) = \frac{4\pi k^3 \cos\beta\cos\gamma}{7\sin^3\beta}\,\frac{1}{2.5}\,\kappa'^4(1-\tfrac{1}{2}\kappa'^2+\tfrac{55}{256}\kappa'^4)(-f^2g^2\kappa^2\kappa'^2)$$

$$= \tfrac{4}{7}\pi M\,\frac{1}{2.5}\,\kappa'^4(1-\beta+\tfrac{119}{64}\beta^2)(-f^2g^2\kappa^2\kappa'^2)$$

In "Harmonics" I defined

$$\mathfrak{P}_3^1(\mu) = P_3^1(\mu) - \tfrac{1}{16}\beta(1-\tfrac{3}{4}\beta)\,P_3^3(\mu)$$
$$= \tfrac{15}{2}\cos\theta\,[\sin^2\theta\,(1+\tfrac{1}{8}\beta-\tfrac{3}{32}\beta^2) - \tfrac{1}{5}(1+\tfrac{5}{8}\beta-\tfrac{15}{32}\beta^2)] \quad\ldots\ldots(57)$$

To make our former definition agree with this we must take

$$f\kappa\,.\,\kappa^2 = \tfrac{15}{2}(1+\tfrac{1}{8}\beta-\tfrac{3}{32}\beta^2); \quad f\kappa\,.\,q^2 = \tfrac{3}{2}(1+\tfrac{5}{8}\beta-\tfrac{15}{32}\beta^2)$$

Hence $f\kappa = \tfrac{15}{2}(1+\tfrac{17}{8}\beta+\tfrac{69}{32}\beta^2)$.

It will be found that $q^2 = \tfrac{1}{5}(1-\tfrac{3}{2}\beta+\tfrac{9}{16}\beta^2)$, and that $f\kappa\,.\,q^2$ has the above form.

Again I defined

$$\mathfrak{S}_3^1(\phi) = \sin\phi - \tfrac{5}{8}\beta(1-\tfrac{3}{4}\beta)\sin 3\phi$$
$$= \sin\phi\,[1-\tfrac{15}{8}\beta+\tfrac{45}{32}\beta^2+\tfrac{5}{2}\beta(1-\tfrac{3}{4}\beta)\sin^2\phi] \ldots\ldots\ldots(58)$$

To make our former definition agree with this we must take

$$g\kappa'\sqrt{-1}\,.\,t^2 = 1-\tfrac{15}{8}\beta+\tfrac{45}{32}\beta^2$$
$$g\kappa'\sqrt{-1}\,.\,\kappa'^2 = \tfrac{5}{2}\beta(1-\tfrac{3}{4}\beta)$$

It will be found that $t^2 = \tfrac{4}{5}(1-\tfrac{17}{8}\beta+\tfrac{151}{64}\beta^2)$.

Whence $g\kappa'\sqrt{-1} = \tfrac{5}{4}(1+\tfrac{1}{4}\beta-\tfrac{27}{64}\beta^2)$, and $g\kappa'\sqrt{-1}\,.\,\kappa'^2$ has the correct form.

Therefore $$f g\kappa\kappa'\sqrt{-1} = \frac{3\,.\,5^2}{2^3}(1+\tfrac{19}{8}\beta+\tfrac{145}{64}\beta^2)$$

$$-f^2g^2\kappa^2\kappa'^2 = \frac{3^2\,.\,5^4}{2^6}(1+\tfrac{19}{4}\beta+\tfrac{651}{64}\beta^2)$$

Whence $$I_3^1(\sin) = \tfrac{4}{7}\pi M\,6\,(1+\tfrac{5}{4}\beta-\tfrac{5}{16}\beta^2) \ldots\ldots\ldots\ldots\ldots\ldots(59)$$

agreeing with the result on p. 276 of "Harmonics" with $i=3$, $s=1$ type OOS.

In "Harmonics" I defined

$$\mathfrak{P}_3^3(\mu) = \tfrac{15}{4}\beta(1-\tfrac{3}{8}\beta)\,P_3^1(\mu) + P_3^3(\mu)$$
$$= 15\cos\theta\,[-(1-\tfrac{15}{8}\beta+\tfrac{45}{32}\beta^2)\sin^2\theta + 1 - \tfrac{3}{8}\beta+\tfrac{9}{32}\beta^2] \ldots(60)$$

In order that this may agree with our former definition we must take

$$f\kappa\,.\,\kappa^2 = -15\,(1 - \tfrac{15}{8}\beta + \tfrac{45}{32}\beta^2), \quad f\kappa\,.\,q^2 = -15\,(1 - \tfrac{3}{8}\beta + \tfrac{9}{32}\beta^2)$$

Whence $f\kappa = -15\,(1 + \frac{1}{8}\beta - \frac{11}{32}\beta^2)$.

It will be found that

$$q^2 = 1 - \tfrac{1}{4}\kappa'^2 + \tfrac{3}{64}\kappa'^4 + \tfrac{3}{512}\kappa'^6 + \dots$$
$$= 1 - \tfrac{1}{2}\beta + \tfrac{11}{16}\beta^2$$

so that $f\kappa\,.\,q^2$ has the correct form.

Again I defined

$$\mathfrak{S}_3{}^3(\phi) = \tfrac{3}{8}\beta\,(1 - \tfrac{3}{4}\beta)\sin\phi + \sin 3\phi$$
$$= 3\sin\phi\,[1 + \tfrac{1}{8}\beta - \tfrac{3}{32}\beta^2 - \tfrac{4}{3}\sin^2\phi] \quad \text{............(61)}$$

In order to make our former definition agree with this we must take

$$g\kappa'\sqrt{-1}\,.\,t^2 = 3\,(1 + \tfrac{1}{8}\beta - \tfrac{3}{32}\beta^2), \quad g\kappa'\sqrt{-1}\,.\,\kappa'^2 = -4$$

Therefore $g\kappa'\sqrt{-1} = -\dfrac{4}{\kappa'^2}$.

It will be found that $t^2 = -\frac{3}{4}\kappa'^2(1 + \frac{1}{8}\beta - \frac{3}{32}\beta^2)$, so that $g\kappa'\sqrt{-1}\,.\,t^2$ has the correct form.

Then
$$fg\kappa\kappa'\sqrt{-1} = -\frac{3\,.\,4\,.\,5}{\kappa'^2}(1 + \tfrac{1}{8}\beta - \tfrac{11}{32}\beta^2)$$

and
$$-f^2g^2\kappa^2\kappa'^2 = \frac{3^2\,.\,4^2\,.\,5^2}{\kappa'^4}(1 + \tfrac{1}{4}\beta - \tfrac{43}{64}\beta^2)$$

Whence
$$I_3{}^3(\sin) = \tfrac{4}{7}\pi M\,.\,360\,[1 - \tfrac{3}{4}\beta + \tfrac{15}{16}\beta^2] \quad \text{..............(62)}$$

agreeing with the result on p. 276 of "Harmonics" with $i = 3$, $s = 3$, type OOS.

(5) *The Second Tesseral Sine Harmonic.*

This is defined thus:—

$$\left.\begin{aligned} \mathsf{P}_3{}^2(\mu) &= \sin\theta\cos\theta\,(1 - \kappa^2\sin^2\theta)^{\frac{1}{2}} = f\kappa\cos\theta\,(\kappa^2 - \kappa^2\cos^2\theta)^{\frac{1}{2}}(\kappa'^2 + \kappa^2\cos^2\theta)^{\frac{1}{2}} \\ \mathsf{S}_3{}^2(\phi) &= \sin\phi\cos\phi\,(1 - \kappa'^2\cos^2\phi)^{\frac{1}{2}} = g\kappa'\sqrt{-1}\,.\,\sin\phi\,(\kappa^2 + \kappa'^2\sin^2\phi)^{\frac{1}{2}}(\kappa'^2 - \kappa'^2\sin^2\phi)^{\frac{1}{2}} \end{aligned}\right\} \quad \text{......(63)}$$

where $f = \dfrac{1}{\kappa^2}$, $g = \dfrac{1}{\kappa'^2\sqrt{-1}}$.

Squaring $\mathsf{P}_3{}^2$ we have

$$A_0 = 0, \quad A_1 = \kappa^2\kappa'^2, \quad A_2 = \kappa^2 - \kappa'^2, \quad A_3 = -1$$

Therefore

$$I_3^2(\cos) = \frac{4\pi k^3 \cos\beta \cos\gamma}{\sin^3\beta} [-\tfrac{1}{3}\kappa^6\kappa'^6 + \tfrac{1}{5}(\kappa^2 - \kappa'^2)^2 \kappa^4\kappa'^4 - \tfrac{1}{7}\kappa^6\kappa'^6$$
$$- \tfrac{4}{15}(\kappa^2 - \kappa'^2)^2 \kappa^4\kappa'^4 - \tfrac{6}{35}(\kappa^2 - \kappa'^2)^2 \kappa^4\kappa'^4$$
$$+ \tfrac{2}{105}(12 - 25\kappa^2\kappa'^2)\kappa^4\kappa'^4].f^2g^2$$

Reducing this expression and putting $-f^2g^2\kappa^4\kappa'^4 = 1$, we have

$$I_3^2(\cos) = \frac{4\pi k^3 \cos\beta\cos\gamma}{7\sin^3\beta}\cdot\frac{1}{3.5} \quad \text{.....................(64)}$$

In "Harmonics" I defined

$$\mathsf{P}_3^2(\mu) = \left(\frac{\frac{1+\beta}{1-\beta} - \mu^2}{1-\mu^2}\right)^{\frac{1}{2}} P_3^2(\mu) = 15\left(\frac{1+\beta}{1-\beta}\right)^{\frac{1}{2}}\left(1 - \frac{1-\beta}{1+\beta}\sin^2\theta\right)^{\frac{1}{2}}\cos\theta\sin\theta \quad (65)$$

To make our former definition agree with this we must take

$$f\kappa^2 = 15\left(\frac{1+\beta}{1-\beta}\right)^{\frac{1}{2}}$$

Again I defined

$$\mathsf{S}_3^2(\phi) = (1 - \beta\cos 2\phi)^{\frac{1}{2}}\sin 2\phi = 2(1+\beta)^{\frac{1}{2}}\left(1 - \frac{2\beta}{1+\beta}\cos^2\phi\right)^{\frac{1}{2}}\sin\phi\cos\phi$$
$$\text{......(66)}$$

To make the former definition agree with this we must take

$$g\kappa'^2\sqrt{-1} = 2(1+\beta)^{\frac{1}{2}}$$

Therefore

$$fg\kappa^2\kappa'^2\sqrt{-1} = 2\,.\,3\,.\,5\,\frac{(1+\beta)}{(1-\beta)^{\frac{1}{2}}}$$

and $$-f^2g^2\kappa^4\kappa'^4 = 2^2\,.\,3^2\,.\,5^2\,\frac{(1+\beta)^2}{1-\beta} = 2^2\,.\,3^2\,.\,5^2(1+3\beta+4\beta^2)$$

Hence in the notation of "Harmonics"

$$I_3^2(\sin) = \tfrac{4}{7}\pi \mathrm{M}\, 3\,.\,4\,.\,5\,(1+3\beta+4\beta^2) \quad \text{..............(67)}$$

agreeing with the result on p. 276 of "Harmonics" with $i = 3$, $s = 2$, type OES.

It may be convenient, as furnishing a kind of index to the foregoing investigation, to state that the $1 + 3 + 5 + 7$ integrals for the harmonics of orders 0, 1, 2, 3 are given in equations 3, 5, 9, 13, 16, 24, 28, 32, 38, 47, 56, 64, corresponding to the definitions contained in 4, 8, 12, 15, 23, 27, 31, 36, 45, 54, 63.

The definitions of the harmonic functions as given in my paper on Harmonic Analysis are repeated in 6, 10, 19, 21, 25, 29, 34, 39, 40, 42, 43, 48, 49, 51, 52, 57, 58, 60, 61, 65, 66. Corresponding to these latter definitions the approximate integrals are given in 7, 11, 14, 20, 22, 26, 30, 35, 41, 44, 50, 53, 59, 62, 67; and the results confirm the correctness of the general approximate formulæ for the integrals given in § 22 of the paper on Harmonic Analysis.

It must be obvious that the method exhibited here may be applied to higher harmonics with whatever degree of accuracy is desired; but it is also clear that the labour of evaluating the integrals increases very much as they rise in order. It is probable that the approximate results of the previous paper will suffice for most practical applications.

POSTSCRIPT.

Mr Hobson has shown me how these integrals may be evaluated by a simpler method of analysis, without the intervention of elliptic integrals. As an example of the method he suggests I take the integral $I_2(\cos)$ evaluated above.

The solid ellipsoidal harmonics are given, except as regards a factor, in § 3 of "The Pear-Shaped Figure."

In (19) of that paper we find

$$S_2 = \mathfrak{P}_2(\nu)\,\mathfrak{P}_2(\mu)\,\mathfrak{C}_2(\phi) = A\left[q^2x^2 + (1-2q^2)\,y^2 - q'^2z^2 + \frac{k^2}{\kappa^2}\,q^2q'^2\right]$$

where A is the factor to be evaluated so as to agree with the definitions

$$\mathfrak{P}_2(\nu) = \kappa^2\nu^2 - q^2, \quad \mathfrak{P}_2(\mu) = \kappa^2\mu^2 - q^2, \quad \mathfrak{C}_2(\phi) = q'^2 - \kappa'^2\cos^2\phi$$

The ellipsoid over which we desire to integrate is defined by $\nu = 1/(\kappa\sin\gamma)$, and the extremity of the c axis is defined by $\mu = \sin\theta = 1$, $\phi = \frac{1}{2}\pi$.

Hence at this point

$$S_2 = (\operatorname{cosec}^2\gamma - q^2)\,(\kappa^2 - q^2)\,q'^2$$

But at the extremity of the c axis

$$x = 0, \quad y = 0, \quad z = c = \frac{k}{\kappa\sin\gamma}$$

Therefore

$$(\operatorname{cosec}^2\gamma - q^2)(\kappa^2 - q^2)\,q'^2 = k^2A\left(-\frac{q'^2}{\kappa^2\sin^2\gamma} + \frac{q^2q'^2}{\kappa^2}\right) = -A\,\frac{k^2q'^2}{\kappa^2}\,(\operatorname{cosec}^2\gamma - q^2)$$

Therefore $A = -\frac{\kappa^2(\kappa^2 - q^2)}{k^2}$, and

$$S_2 = (\kappa^2 - q^2)\left[-q^2\kappa^2\frac{x^2}{k^2} - (1-2q^2)\kappa^2\frac{y^2}{k^2} + q'^2\kappa^2\frac{z^2}{k^2} - q^2q'^2\right]$$

Let us assume

$$x = a\xi = \frac{k\cos\gamma}{\kappa\sin\gamma}\xi, \quad y = b\eta = \frac{k\cos\beta}{\kappa\sin\gamma}\eta, \quad z = c\zeta = \frac{k}{\kappa\sin\gamma}\zeta$$

Then when x, y, z is on the ellipsoid we have

$$\xi^2 + \eta^2 + \zeta^2 = 1$$

Thus we may regard ξ, η, ζ as the coordinates of a point on a sphere of unit radius, or as direction cosines, if it is more convenient to do so. On substituting for x, y, z their values in terms of ξ, η, ζ we find

$$S_2 = (\operatorname{cosec}^2\gamma - q^2)(\kappa^2 - q^2)[-q^2\xi^2 - (1-2q^2)\eta^2 + q'^2\zeta^2]$$

On performing the same operation to the points on the boundary of an element $d\sigma$ of surface of the ellipsoid, we find

$$pd\sigma = abc\,d\omega = \frac{k^3\cos\beta\cos\gamma}{\sin^3\beta}\,d\omega$$

where $d\omega$ is an element of the surface of the sphere of unit radius, or an element of solid angle.

Since on the surface of the ellipsoid, $\mathfrak{P}_2(\nu) = \operatorname{cosec}^2\gamma - q^2$, it follows that

$$\mathfrak{P}_2(\mu)\,\mathfrak{C}_2(\phi) = (\kappa^2 - q^2)[-q^2\xi^2 - (1-2q^2)\eta^2 + q'^2\zeta^2]$$

Hence

$$I_2(\cos) = \frac{k^3\cos\beta\cos\gamma}{\sin^3\beta}(\kappa^2 - q^2)^2\int[-q^2\xi^2 - (1-2q^2)\eta^2 + q'^2\zeta^2]^2\,d\omega$$

It is easy to prove that

$$\int\xi^4 d\omega = \int\eta^4 d\omega = \int\zeta^4 d\omega = \tfrac{4}{5}\pi$$

$$\int\eta^2\zeta^2 d\omega = \int\zeta^2\xi^2 d\omega = \int\xi^2\eta^2 d\omega = \tfrac{4}{15}\pi$$

Therefore

$$I_2(\cos) = \frac{4\pi k^3\cos\beta\cos\gamma}{5\sin^3\beta}\cdot\tfrac{1}{3}(\kappa^2 - q^2)^2[3q^4 + 3(1-2q^2)^2 + 3q'^4 - 2q'^2(1-2q^2) + 2q^2(1-2q^2) - 2q^2q'^2]$$

$$= \frac{4\pi k^3\cos\beta\cos\gamma}{5\sin^3\beta}\cdot\tfrac{4}{3}(\kappa^2 - q^2)^2(1 - 3q^2q'^2)$$

On substituting for q^2 its value, viz., $\frac{1}{3}[1+\kappa^2-(1-\kappa^2\kappa'^2)^{\frac{1}{2}}]$, and effecting reductions we arrive at the result given in (16) above.

It is obvious that this process is considerably simpler and more elegant from the point of view of theory, but to carry these operations through for all the integrals given above would entail a good deal of algebra. I think indeed that the work might not be very much less than what I have already done.

Mr Hobson has further remarked that all the integrations may be avoided by the following theorem:—

If $F_n(\xi, \eta, \zeta)$ be a solid spherical harmonic function of ξ, η, ζ of degree n,

$$\int [F_n(\xi, \eta, \zeta)]^2 d\omega = 4\pi \frac{2^n . n!}{2n+1!} F_n\left(\frac{\partial}{\partial\xi}, \frac{\partial}{\partial\eta}, \frac{\partial}{\partial\zeta}\right) F_n(\xi, \eta, \zeta)$$

Considering, however, how simple are the integrals involved in his first method, it may be doubted whether this would save trouble.

14.

THE APPROXIMATE DETERMINATION OF THE FORM OF MACLAURIN'S SPHEROID.

[*Transactions of the American Mathematical Society*, IV. (1903), pp. 113—133, with which is incorporated a paper entitled "Further Note on Maclaurin's spheroid," *ibid.*, IX. (1908), pp. 34—38.]

PREFACE.

SPHERICAL harmonics render the approximate determination of the figure of a rotating mass of liquid a very simple problem. If ρ be the density, e the ellipticity, and ω the angular velocity of the spheroid, the solution is

$$\frac{\omega^2}{2\pi\rho} = \frac{8}{15}e$$

This result is only correct as far as the first power of the ellipticity, but M. Poincaré has recently shown* how harmonic analysis may be so used as to give results which shall be correct as far as squares of small quantities; and I have myself used his method for the determination of the stability of the pear-shaped figure of equilibrium†.

Both these papers involved the use of ellipsoidal harmonic analysis, and it would be rather tiresome for a reader to extract the method from the complex analysis in which it is embedded. It therefore seems worth while to treat the well-worn subject of Maclaurin's spheroid as an example of the method in question. It will appear below that it would have been possible to obtain a more accurate result than that stated above, even if the rigorous solution of the problem had been beyond the powers of the mathematician.

My own personal reason for undertaking this task was that I desired a sort of collateral verification of the very complicated analysis needed in the case of my previous investigation.

* *Philosophical Transactions of the Royal Society*, Vol. 198, A (1902), pp. 333—373.

† [Paper 12.]

§ 1. *Method of defining the spheroid.*

Let a sphere S be described concentrically with the spheroid, and let it be sufficiently large to enclose the whole of the spheroid. I call R the region between the sphere S and the spheroid; and suppose the density of the liquid in S to be $+\rho$, and that in R to be $-\rho$.

If S_i^s denotes any surface spherical harmonic of colatitude θ and longitude ϕ, it is usual to define the corresponding deformation of a sphere of radius a by the equation $r = a(1 + eS_i^s)$. But in the present investigation it will be found that there is a great saving of labour by defining it by the equation

$$r^3 = a^3(1 + 3eS_i^s)$$

The two forms give identical results as far as the first power of the ellipticity e, but not so when we are to consider the squares of small quantities*.

In general I define S_i^s by one of the two alternative forms $P_i^s(\mu) \begin{matrix}\cos\\ \sin\end{matrix} s\phi$, where $\mu = \cos\theta$. But in the case of the second zonal harmonic ($s = 0$, $i = 2$) it is convenient to write

$$S_2 = \tfrac{1}{3} - \mu^2$$

The fourth zonal harmonic will occur explicitly below, and in accordance with the general definition to be adopted we have

$$S_4 = \tfrac{35}{8}\mu^4 - \tfrac{15}{4}\mu^2 + \tfrac{3}{8}$$

The angular velocity is to be denoted by ω, and the colatitude θ or $\cos^{-1}\mu$ is measured from the axis of rotation.

We must now assume a general form for the equation to the spheroid, and shall subsequently determine the several ellipticities so that the surface may be a figure of equilibrium.

The radius of S being denoted by a, we may write the equation to the surface of the spheroid in the form

$$r^3 = a^3[1 - 3c + 3eS_2 + 3fS_4 + 3\Sigma f_i^s S_i^s]$$

In this expression e is the ellipticity corresponding to the second zonal harmonic, and it represents that term which exists alone in the ordinary approximate solution. Then I suppose that f and f_i^s are quantities of the order e^2, and that there are f_i^s corresponding to all possible harmonics excepting the second and fourth zonal ones. Thus all the f_i^s are of order e^2, excepting f_2 and f_4 which are zero. Lastly c is an arbitrary constant, and is

* [In order that the procedure might be as closely parallel as possible with that adopted in Paper 12, I have also carried out the investigation when the form of the spheroid is defined by $r^2 = a^2(1 + 2eS_i^s)$. If the ellipsoid of Paper 12 be reduced to the sphere, it will be found that the dependent variable τ becomes equal to $(a^2 - r^2)/2a^2$, or $r^2 = a^2(1 - 2\tau)$.]

only subject to the condition that it is greater than the greatest positive value of $eS_2 + fS_4 + \Sigma f_i^s S_i^s$. This condition ensures that S shall envelope the whole spheroid.

It is now convenient to replace the radius vector r by a new variable τ, defined by

$$\tau = \frac{a^3 - r^3}{3a^3} \quad\text{..............................(1)}$$

Thus the equation to the spheroid may be written

$$\tau = c - eS_2 - fS_4 - \Sigma f_i^s S_i^s$$

The problem will be solved by making the energy of the system stationary. It will therefore be necessary to determine the energy lost in the concentration of the spheroid from a condition of infinite dispersion. This will involve the use of the formula for the gravity of S, and since the whole region R is inside S, we only require the formula for internal gravity.

If we were to continue the developments from this point all the formulæ would involve the constant c. But since it is merely needed for defining a sphere of reference of arbitrary size, it cannot finally appear in the formula for the energy. It is useless to encumber the analysis by the introduction of a constant which must disappear in the end, and it is legitimate and much shorter to treat c as zero from the first. It is however easier to maintain a clear conception of the processes if we continue to discuss the problem as though c were not zero, and as if S enveloped the whole spheroid. With this explanation we may write the equation to the spheroid in the form

$$\tau = -eS_2 - fS_4 - \Sigma f_i^s S_i^s \quad\text{...........................(2)}$$

§ 2. *The lost energy of the system.*

If the negative density in R were transported along conical tubes emanating from the centre of S, it might be deposited as surface density on S; I refer to such a condensation as $-C$. I do not, however, suppose the condensation actually effected, but I imagine the surface of S to be coated with equal and opposite condensations $+C$ and $-C$.

The system of masses forming the spheroid may then be considered as being as follows:

Density $+\rho$ throughout S, say $+S$.

Negative condensation on S, say $-C$.

Positive condensation on S and negative volume density $-\rho$ throughout R. This last forms a double system of zero mass, say D, and $D = C - R$.

The lost energy of the system clearly involves the lost energy of each of these three with itself, and the mutual lost energy of the three taken two and two together. Thus the lost energy may be written symbolically

$$\tfrac{1}{2}SS + \tfrac{1}{2}CC + \tfrac{1}{2}DD - SC + SD - CD$$

Since D is $C - R$, the last three terms are equivalent to

$$-SC + (S - C)(C - R) = -SR + CR - CC$$

Thus the gravitational lost energy is

$$\tfrac{1}{2}SS - SR + CR - \tfrac{1}{2}CC + \tfrac{1}{2}DD$$

The lost energy of the system, as rendered statical by the imposition of a rotation potential, is clearly $\tfrac{1}{2}C\omega^2$, where C is the moment of inertia of the spheroid about the axis of rotation.

If C_s denotes the moment of inertia of the sphere S, and C_r the moment of inertia of the region R considered as being filled with positive density $+\rho$, we clearly have

$$C = C_s - C_r$$

Thus if E denotes the lost energy of the system as rendered statical by the imposition of a rotation potential

$$E = \tfrac{1}{2}SS - SR + CR - \tfrac{1}{2}CC + \tfrac{1}{2}DD + \tfrac{1}{2}\omega^2 (C_s - C_r) \quad \text{..........(3)}$$

§ 3. *The energy* $\tfrac{1}{2}SS - SR + CR - \tfrac{1}{2}CC$.

It is in the first place necessary to obtain certain preliminary analytical and numerical results.

If we write $d\sigma$ for $d\mu d\phi$, it is clear that an element dv of volume is given by

$$dv = a^3 d\tau d\sigma = \frac{3M}{4\pi\rho} d\tau d\sigma$$

where M is the mass of the sphere S; for we may obviate the negative sign of dv by taking the limits of τ from τ to zero.

When we integrate throughout the region R the limits of τ are

$$-eS_2 - fS_4 - \Sigma f_i^s S_i^s \text{ to zero}$$

I now define certain integrals, viz.:

$$\phi_i^s = \frac{3}{4\pi}\int (S_i^s)^2 d\sigma, \quad \omega_i^s = \frac{3}{4\pi}\int (S_2)^2 S_i^s d\sigma, \quad \sigma_4 = \frac{3}{4\pi}\int (S_2)^4 d\sigma$$

It is well known that

$$\int_{-1}^{+1} [P_i^s(\mu)]^2 d\mu = \frac{2}{2i+1}\frac{(i+s)!}{(i-s)!}$$

and since in every case, excepting that of S_2, $S_i^s = P_i^s(\mu) \begin{matrix}\cos\\ \sin\end{matrix} s\phi$, we have

$$\phi_i^s = \frac{3}{2i+1}\frac{(i+s)!}{(i-s)!}$$

Since $S_2 = -\frac{2}{3}P_2(\mu)$, the value of ϕ_2 is derivable from the same general formula. Hence we have

$$\phi_0 = 3, \quad \phi_2 = \tfrac{4}{15}, \quad \phi_4 = \tfrac{1}{3} \text{ ....(4)}$$

Since S_i^s involves either $\cos s\phi$ or $\sin s\phi$, ω_i^s vanishes unless $s=0$; hence we need only consider ω_i.

The function $(S_2)^2$ may be expanded in terms of zonal harmonics. Assume then

$$(S_2)^2 = \Sigma \eta_i S_i$$

Multiplying both sides by $3S_i/4\pi$, and integrating throughout angular space, we find

$$\omega_i = \eta_i \phi_i, \quad \text{and} \quad (S_2)^2 = \Sigma \frac{\omega_i}{\phi_i} S_i$$

But, by actual substitution,

$$(S_2)^2 = \tfrac{4}{45}S_0 - \tfrac{4}{21}S_2 + \tfrac{8}{35}S_4$$

Therefore $$\frac{\omega_0}{\phi_0} = \frac{4}{45}, \quad \frac{\omega_2}{\phi_2} = -\frac{4}{21}, \quad \frac{\omega_4}{\phi_4} = \frac{8}{35} \text{ ....(4)}$$

and all the higher ω's vanish.

It will be noticed that $\omega_0 = \phi_2$, hence we have $\omega_0/\phi_0 = \frac{1}{3}\phi_2$. Also

$$\omega_0 = \frac{4}{15}, \quad \omega_2 = -\frac{16}{3^2 . 5 . 7}, \quad \omega_4 = \frac{8}{3 . 5 . 7} \text{ ....(4)}$$

Next we have

$$\sigma_4 = \frac{3}{4\pi}\int (S_2)^4 d\sigma = \frac{3}{4\pi}\int\left[\frac{\omega_0}{\phi_0}S_0 + \frac{\omega_2}{\phi_2}S_2 + \frac{\omega_4}{\phi_4}S_4\right]^2 d\sigma = \frac{(\omega_0)^2}{\phi_0} + \frac{(\omega_2)^2}{\phi_2} + \frac{(\omega_4)^2}{\phi_4}$$

whence $$\sigma_4 = \frac{16}{3^2 . 5 . 7} \text{ ....(4')}$$

It is now necessary to determine the volume of the region R; it is

$$a^3 \iint d\tau\, d\sigma = -a^3 \int [eS_2 + fS_4 + \Sigma f_i^s S_i^s]\, d\sigma = 0$$

From this it follows that the mass of S is equal to that of the spheroid, and therefore M is the mass of the spheroid. It is this result which makes the choice of τ as independent variable so convenient.

We are now in a position to determine the several contributions to the lost energy.

The lost energy of the sphere, denoted by $\frac{1}{2}SS$, is known to be $3M^2/5a$. This is a constant and may be omitted as being of no further interest. The internal potential of S is given by

$$V = \tfrac{2}{3}\pi\rho\,(3a^2 - r^2)$$

But $r^3 = a^3(1 - 3\tau)$, and $r^2 = a^2(1 - 2\tau - \tau^2 - \frac{4}{3}\tau^3 \ldots)$. Therefore as far as the cubes of small quantities,

$$V = \frac{M}{a} + \frac{M}{a}(\tau + \tfrac{1}{2}\tau^2 + \tfrac{2}{3}\tau^3)$$

Since the volume of R is zero the first term of V contributes nothing to the lost energy SR, and the second term of V will give the whole. Therefore to the fourth order

$$SR = \frac{3M^2}{4\pi\rho a}\iint(\tau + \tfrac{1}{2}\tau^2 + \tfrac{2}{3}\tau^3)\,d\tau\,d\sigma$$

$$= \frac{3M^2}{8\pi\rho a}\int[e^2(S_2)^2 + 2efS_2S_4 + 2\Sigma ef_i^s S_2S_i^s + f^2(S_4)^2 + 2\Sigma ff_i^s S_4S_i^s$$
$$+ (\Sigma f_i^s S_i^s)^2 - \tfrac{1}{3}e^3(S_2)^3 - e^2f(S_2)^2S_4 - \Sigma e^2 f_i^s (S_2)^2 S_i^s + \tfrac{1}{3}e^4(S_2)^4]\,d\sigma$$

$$= \frac{M^2}{2a}[e^2\phi_2 + f^2\phi_4 + \Sigma(f_i^s)^2\phi_i^s - \tfrac{1}{3}e^3\omega_2 - e^2f\omega_4 + \tfrac{1}{3}e^4\sigma_4]$$

Thus on rearranging the terms we have

$$\tfrac{1}{2}SS - SR = \frac{M^2}{a}[-\tfrac{1}{2}e^2\phi_2 + \tfrac{1}{6}e^3\omega_2 - \tfrac{1}{6}e^4\sigma_4 + \tfrac{1}{2}e^2f\omega_4 - \tfrac{1}{2}f^2\phi_4 - \tfrac{1}{2}\Sigma(f_i^s)^2\phi_i^s] \quad \ldots(5)$$

We have next to consider the terms depending on the condensation C. Since $dv/d\tau = a^3 d\sigma$ the amount of matter in the region R, if filled with density $+\rho$, which stands on an element of unit area is

$$\rho a\int d\tau = -\rho a\,(eS_2 + fS_4 + \Sigma f_i^s S_i^s)$$

This expression gives the surface density of the condensation $+C$, and it is expressed in surface harmonics.

Now by the usual formula of spherical harmonic analysis the internal potential of surface density $\rho a\epsilon S_i^s$ is

$$\frac{4\pi\rho\epsilon}{2i+1}\frac{r^i}{a^{i-2}}S_i^s = \frac{3M\epsilon}{(2i+1)a}\frac{r^i}{a^i}S_i^s$$

As far as the first power of τ,

$$r^i = a^i(1 - i\tau)$$

but in the case $i = 2$, as far as squares of small quantities,

$$r^2 = a^2(1 - 2\tau - \tau^2)$$

Hence it follows that the internal potential, say V_c, of the condensation $+C$ is given by

$$V_c = -\frac{3M}{a}\left[\tfrac{1}{5}e(1 - 2\tau - \tau^2)S_2 + \tfrac{1}{9}f(1 - 4\tau)S_4 + \Sigma\frac{(1 - i\tau)}{2i+1}f_i^s S_i^s\right] \quad \ldots\ldots(6)$$

On multiplying this by $3M d\tau\, d\sigma/4\pi$ and integrating throughout R we shall obtain the lost energy CR.

Thus

$$CR = \frac{3M}{4\pi}\iint V_c\, d\tau\, d\sigma$$

$$= -\frac{9M^2}{4\pi a}\iint \Big[\tfrac{1}{5}eS_2 + \tfrac{1}{9}fS_4 + \Sigma \frac{1}{2i+1} f_i^s S_i^s$$

$$- \tau\left(\tfrac{2}{5}eS_2 + \tfrac{4}{9}fS_4 + \Sigma\frac{i}{2i+1} f_i^s S_i^s\right) - \tfrac{1}{5}\tau^2 eS_2\Big]\, d\tau\, d\sigma$$

$$= \frac{9M^2}{4\pi a}\int \Big\{\Big[\tfrac{1}{5}eS_2 + \tfrac{1}{9}fS_4 + \Sigma\frac{1}{2i+1} f_i^s S_i^s\Big]\left[eS_2 + fS_4 + \Sigma f_i^s S_i^s\right]$$

$$+ \tfrac{1}{2}\Big[\tfrac{2}{5}eS_2 + \tfrac{4}{9}fS_4 + \Sigma\frac{i}{2i+1} f_i^s S_i^s\Big]\left[e^2 (S_2)^2 + 2efS_2S_4 + 2\Sigma e f_i^s S_2 S_i^s\right]$$

$$- \frac{1}{3\,.\,5}\, e^3 (S_2)^3 \,.\, eS_2\Big\}\, d\sigma$$

Therefore

$$CR = \frac{3M^2}{a}\Big\{\tfrac{1}{5}e^2\phi_2 + \tfrac{1}{9}f^2\phi_4 + \Sigma\frac{1}{2i+1}(f_i^s)^2\phi_i^s$$

$$+ \tfrac{1}{2}\left[\tfrac{2}{5}e^3\omega_2 + \tfrac{4}{9}e^2 f\omega_4 + \tfrac{4}{5}e^2 f\omega_4\right] - \tfrac{1}{15}e^4\sigma_4\Big\}$$

$$= \frac{3M^2}{a}\Big[\tfrac{1}{5}e^2\phi_2 + \tfrac{1}{5}e^3\omega_2 - \tfrac{1}{15}e^4\sigma_4 + \tfrac{28}{45}e^2 f\omega_4 + \tfrac{1}{9}f^2\phi_4 + \Sigma\frac{1}{2i+1}(f_i^s)^2\phi_i^s\Big] \quad \ldots\ldots(7)$$

In order to find $\frac{1}{2}CC$ we have only to deal with surface density. Then the value of V_c at the surface is given by (6) with $\tau = 0$; therefore

$$\tfrac{1}{2}V_c = -\frac{3M}{2a}\Big[\tfrac{1}{5}eS_2 + \tfrac{1}{9}fS_4 + \Sigma\frac{1}{2i+1} f_i^s S_i^s\Big]$$

An element of mass of the surface density $+C$ is

$$-\frac{3M}{4\pi}\left[eS_2 + fS_4 + \Sigma f_i^s S_i^s\right] d\sigma$$

Multiplying these two together and integrating, we find

$$\tfrac{1}{2}CC = \frac{3M^2}{2a}\Big[\tfrac{1}{5}e^2\phi_2 + \tfrac{1}{9}f^2\phi_4 + \Sigma\frac{1}{2i+1}(f_i^s)^2\phi_i^s\Big]$$

And subtracting this from (7)

$$CR - \tfrac{1}{2}CC = \frac{3M^2}{a}\Big[\tfrac{1}{10}e^2\phi_2 + \tfrac{1}{5}e^3\omega_2 - \tfrac{1}{15}e^4\sigma_4 + \tfrac{28}{45}e^2 f\omega_4 + \tfrac{1}{18}f^2\phi_4$$

$$+ \tfrac{1}{2}\Sigma\frac{1}{2i+1}(f_i^s)^2\phi_i^s\Big] \quad \ldots\ldots(8)$$

Again adding this to (5) we have

$$\tfrac{1}{2}SS - SR + CR - \tfrac{1}{2}CC = \frac{M^2}{a}\Big[-\tfrac{1}{5}e^2\phi_2 + \tfrac{23}{30}e^3\omega_2 - \tfrac{11}{30}e^4\sigma_4 + \tfrac{71}{30}e^2 f\omega_4$$
$$- \tfrac{1}{3}f^2\phi_4 - \Sigma\frac{i-1}{2i+1}(f_i^s)^2\phi_i^s\Big] \quad \text{......(9)}$$

It remains to determine the value of the term $\tfrac{1}{2}DD$, and for this end we must apply the theory of double layers, according to the ingenious method devised by M. Poincaré.

§ 4. *Double layers**.

[For the discussion of such a system I refer the reader to § 9, p. 335 of Paper 12.

The result given in equation (21), p. 340 is as follows:—]

$$\tfrac{1}{2}DD = \tfrac{2}{3}\pi\rho^2\int(\epsilon^3 - \lambda\epsilon^4)\,d\sigma + \tfrac{1}{4}\rho\int\epsilon^2\frac{dV}{dn}\,d\sigma$$

§ 5. *The energy $\tfrac{1}{2}DD$.*

The element of surface of the sphere is written $d\sigma$ in § 4, but in order to accord with the notation used elsewhere we must now write it $a^2 d\sigma$.

The first term in $\tfrac{1}{2}DD$ is

$$\tfrac{2}{3}\pi\rho^2\int(\epsilon^3 - \lambda\epsilon^4)\,d\sigma$$

and when the notation for the element of surface is changed we may write it

$$\tfrac{1}{2}\frac{M\rho}{a}\int(\epsilon^3 - \lambda\epsilon^4)\,d\sigma$$

In this expression ϵ is the length measured along a radius from the sphere S to the spheroid. We have denoted the outward normal by n, and therefore to the second order of small quantities,

$$-dn = -dr = a(1+2\tau)\,d\tau$$

The distance measured inward from the sphere to the point defined by τ in the region R is, to the first order of small quantities, $-n = a\tau$. Again

$$\epsilon = -\int dn = a\int(1+2\tau)\,d\tau$$
$$= -a\left[eS_2 + fS_4 + \Sigma f_i^s S_i^s - e^2(S_2)^2\right] \quad \text{..................(10)}$$

* [In the paper as originally printed the investigation of the double layer was reproduced.]

Since $-n$ is what was denoted ϵs in the general investigation referred to in § 4 (see p. 338), we have

$$\begin{aligned} dv &= -a^2(1+\lambda n)\,dn\,d\sigma \\ &= a^3(1-\lambda a\tau)(1+2\tau)\,d\tau\,d\sigma \\ &= a^3[1+(2-\lambda a)\tau]\,d\tau\,d\sigma \end{aligned}$$

But since $dv = a^3 d\tau d\sigma$, we have $\lambda = 2/a$.

Therefore

$$\epsilon^3 = -a^3[e^3(S_2)^3 + 3e^2 f(S_2)^2 S_4 + 3\Sigma e^2 f_i^s (S_2)^2 S_i^s - 3e^4(S_2)^4]$$

$$\lambda\epsilon^4 = 2a^3 e^4 (S_2)^4$$

Whence

$$\epsilon^3 - \lambda\epsilon^4 = -a^3[e^3(S_2)^3 + 3e^2 f(S_2)^2 S_4 + 3\Sigma e^2 f_i^s (S_2)^2 S_i^s - e^4(S_2)^4]$$

This must be multiplied by $\frac{1}{2}M\rho/a$ and integrated throughout angular space. Thus this contribution to the energy becomes

$$\frac{M^2}{a}[-\tfrac{1}{2}e^3\omega_2 + \tfrac{1}{2}e^4\sigma_4 - \tfrac{3}{2}e^2 f\omega_4] \quad \text{.....................(11)}$$

The second term $\frac{1}{4}\rho\int\epsilon^2\frac{dV}{dn}\,d\sigma$

in $\frac{1}{2}DD$ remains for consideration.

In order to evaluate dV/dn it suffices to imagine the volume density $-\rho$ in the region R concentrated on a surface bisecting the space between S and the spheroid. We may then treat the system D as an infinitesimal double layer of thickness $\frac{1}{2}\epsilon$ and with density $+C$ or $-\rho a(eS_2 + fS_4 + \Sigma f_i^s S_i^s)$ on its outer surface. In the present instance it suffices to consider only the leading terms in the density and thickness. Hence by (10) the product denoted $\tau\delta$ in the general investigation becomes

$$(-\tfrac{1}{2}aeS_2)(-\rho aeS_2) = \tfrac{1}{2}\rho a^2 e^2\left[\tfrac{1}{3}\phi_2 S_0 + \frac{\omega_2}{\phi_2}S_2 + \frac{\omega_4}{\phi_4}S_4\right]$$

We thus have $\tau\delta$ expanded in surface harmonics.

Now consider two functions

$$V_e = \Sigma A_i \frac{ia^{i-1}}{r^{i+1}} S_i^s, \quad \text{for space external to } S$$

$$V_i = -\Sigma A_i \frac{(i+1)r^i}{a^{i+2}} S_i^s, \quad \text{for internal space}$$

They are solid harmonics and as such satisfy Laplace's equation throughout space. Hence they are the external and internal potentials of a distribution of matter on S, but since they are not continuous, while their differentials are continuous, that matter constitutes a double layer.

At the surface $r = a$,

$$V_e - V_i = \Sigma\,(2i+1)\frac{A_i}{a^2}S_i^s,$$

But this must be equal to $4\pi\tau\delta$. Hence

$$2\pi\rho a^2 e^2\left[\tfrac{1}{3}\phi_2 S_0 + \frac{\omega_2}{\phi_2}S_2 + \frac{\omega_4}{\phi_4}S_4\right] = \frac{A_0}{a^2}S_0 + \frac{5A_2}{a^2}S_2 + \frac{9A_4}{a^2}S_4$$

Therefore

$$A_0 = \tfrac{3}{2}Mae^2\,.\,\tfrac{1}{3}\phi_2; \quad A_2 = \tfrac{3}{2}Mae^2\,.\,\tfrac{1}{5}\frac{\omega_2}{\phi_2}; \quad A_4 = \tfrac{3}{2}Mae^2\,.\,\tfrac{1}{9}\frac{\omega_4}{\phi_4}$$

Now $$\frac{dV}{dn} = \frac{dV_e}{dr} = \frac{dV_i}{dr} \quad (r=a)$$

$$= -\Sigma i(i+1)\frac{A_i}{a^3}S_i^s$$

$$= -\tfrac{3}{2}\frac{Me^2}{a^2}\left[\tfrac{6}{5}\frac{\omega_2}{\phi_2}S_2 + \tfrac{20}{9}\frac{\omega_4}{\phi_4}S_4\right]$$

Then since

$$\rho\epsilon^2 a^2 = \rho a^4 e^2 (S_2)^2 = \frac{3Ma}{4\pi}e^2\left[\tfrac{1}{3}\phi_2 + \frac{\omega_2}{\phi_2}S_2 + \frac{\omega_4}{\phi_4}S_4\right]$$

we have (on writing $a^2 d\sigma$ for the $d\sigma$ of the general investigation referred to in § 4)

$$\tfrac{1}{4}\rho a^2\int\epsilon^2\frac{dV}{dn}\,d\sigma = -\frac{9M^2}{32\pi a}e^4\int\left[\tfrac{6}{5}\left(\frac{\omega_2}{\phi_2}\right)^2(S_2)^2 + \tfrac{20}{9}\left(\frac{\omega_4}{\phi_4}\right)^2(S_4)^2\right]d\sigma$$

$$= -\tfrac{3}{8}\frac{M^2}{a}e^4\left[\tfrac{6}{5}\frac{(\omega_2)^2}{\phi_2} + \tfrac{20}{9}\frac{(\omega_4)^2}{\phi_4}\right] \quad \text{......................(12)}$$

Adding (9), (11) and (12), we have for the whole gravitational lost energy

$$\frac{M^2}{a}\left[-\tfrac{1}{5}e^2\phi_2 + \tfrac{4}{15}e^3\omega_2 - e^4\left\{\tfrac{9}{20}\frac{(\omega_2)^2}{\phi_2} + \tfrac{5}{6}\frac{(\omega_4)^2}{\phi_4} - \tfrac{2}{15}\sigma_4\right\}\right.$$
$$\left. + \tfrac{13}{15}e^2 f\omega_4 - \tfrac{1}{3}f^2\phi_4 - \Sigma\frac{i-1}{2i+1}(f_i^s)^2\phi_i^s\right] \quad \text{...(13)}$$

§ 6. *Moment of inertia.*

Since ω^2 is of order e, the moment of inertia must be determined to the cubes of small quantities.

We have, to the square of τ,

$$x^2 + y^2 = r^2\sin^2\theta = \frac{3M}{4\pi\rho a}(\tfrac{2}{3} + S_2)(1 - 2\tau - \tau^2)$$

The region R is to be considered as filled with density $+\rho$, and the element of mass is $3M d\tau\, d\sigma/4\pi$.

Hence the moment of inertia of the region R is

$$C_r = \left(\frac{3M}{4\pi}\right)^2 \frac{1}{\rho a} \iint (\tfrac{2}{3} + S_2)(1 - 2\tau - \tau^2)\, d\tau\, d\sigma$$

$$= -\left(\frac{3M}{4\pi}\right)^2 \frac{1}{\rho a} \int [\tfrac{2}{3} + S_2][eS_2 + fS_4 + \Sigma f_i^s S_i^s + e^2 (S_2)^2 + 2efS_2S_4 + 2\Sigma e f_i^s S_2 S_i^s - \tfrac{1}{3} e^3 (S_2)^3]\, d\sigma$$

$$= -\left(\frac{3M}{4\pi}\right)^2 \frac{1}{\rho a} \int [e (S_2)^2 + e^2 (S_2)^3 + 2ef (S_2)^2 S_4 + 2\Sigma e f_i^s (S_2)^2 S_i^s + \tfrac{2}{3} e^2 (S_2)^2 - \tfrac{2}{9} e^3 (S_2)^3 - \tfrac{1}{3} e^3 (S_2)^4]\, d\sigma$$

$$= -\frac{3M^2}{4\pi\rho} [e\phi_2 + e^2 (\omega_2 + \tfrac{2}{3}\phi_2) - e^3 (\tfrac{2}{9}\omega_2 + \tfrac{1}{3}\sigma_4) + 2ef\omega_4]$$

The moment of inertia of S is $\frac{2}{5}Ma^2$. Therefore

$$C_s = \frac{3M^2}{4\pi\rho a} (\tfrac{2}{5})$$

Then

$$C = C_s - C_r = \frac{3M^2}{4\pi\rho a} [\tfrac{2}{5} + e\phi_2 + e^2 (\omega_2 + \tfrac{2}{3}\phi_2) - e^3 (\tfrac{2}{9}\omega_2 + \tfrac{1}{3}\sigma_4) + 2ef\omega_4]$$

Whence

$$\tfrac{1}{2} C\omega^2 = \frac{M^2}{a} \frac{\omega^2}{4\pi\rho} [\tfrac{3}{5} + \tfrac{3}{2} e\phi_2 + e^2 (\tfrac{3}{2}\omega_2 + \phi_2) - e^3 (\tfrac{1}{3}\omega_2 + \tfrac{1}{2}\sigma_4) + 3ef\omega_4] \quad \ldots(14)$$

§ 7. *Solution of the problem.*

Introducing the numerical values of the several integrals we have from (13) and (14)

$$E \div \frac{M^2}{a} = -\frac{4}{3.5^2} e^2 - \frac{64}{3^3.5^2.7} e^3 - \frac{16}{3^3.7^2} e^4 + \frac{104}{3^2.5^2.7} e^2 f - \frac{1}{3^2} f^2 - \Sigma \frac{i-1}{2i+1} (f_i^s)^2 \phi_i^s$$
$$+ \frac{\omega^2}{4\pi\rho} \left[\frac{3}{5} + \frac{2}{5} e + \frac{4}{3.7} e^2 - \frac{8}{3^3.5.7} e^3 + \frac{8}{5.7} ef\right]$$

The conditions for a figure of equilibrium are

$$\frac{dE}{de} = 0, \quad \frac{dE}{df} = 0, \quad \frac{dE}{df_i^s} = 0$$

with ω^2 constant.

The last of these gives at once $f_i^s = 0$. Neglecting all terms in E of higher orders than the second, the equation $dE/de = 0$ gives as a first approximation

$$\frac{\omega^2}{4\pi\rho} = \frac{4}{3.5} e$$

Then the equation $\dfrac{dE}{df} = 0$ gives

$$\frac{104}{3^2.5^2.7} e^2 - \frac{2}{3^2} f + \frac{\omega^2}{4\pi\rho} \cdot \frac{8}{5.7} e = 0$$

Therefore, by means of the first approximation for ω,

$$f = \frac{4}{7} e^2 \quad \text{.................................(15)}$$

Substituting this value of f in the expression for E we have

$$E \div \frac{M^2}{a} = -\frac{4}{3 \cdot 5^2} e^2 - \frac{64}{3^3 \cdot 5^2 \cdot 7} e^3 - \frac{352}{3^3 \cdot 5^2 \cdot 7^2} e^4 + \frac{\omega^2}{4\pi\rho} \left[\frac{3}{5} + \frac{2}{5} e + \frac{4}{3 \cdot 7} e^2 + \frac{808}{3^3 \cdot 5 \cdot 7^2} e^3 \right] \quad \text{......(16)}$$

On equating $\dfrac{dE}{de}$ to zero we find

$$\frac{\omega^2}{4\pi\rho} \left[1 + \frac{20}{3 \cdot 7} e + \frac{404}{3^2 \cdot 7^2} e^2 \right] = \frac{4}{3 \cdot 5} e \left[1 + \frac{8}{3 \cdot 7} e + \frac{176}{3^2 \cdot 7^2} e^2 \right]$$

Whence
$$\frac{\omega^2}{4\pi\rho} = \tfrac{4}{15} e \left(1 - \frac{4}{7} e + \frac{4}{3 \cdot 7^2} e^2 \right) \text{........................(17)}$$

It follows from (15) that the equation to the surface of equilibrium is

$$r^3 = a^3 \left[1 + 3eS_2 + \tfrac{12}{7} e^2 S_4 \right] \quad \text{........................(18)}$$

It remains to verify that the solution given by (17) and (18) is correct.

The equation to an ellipsoid of revolution, whose equatorial and polar radii are a_1 and $a_1(1 - e_1)$, is

$$r^2 = \frac{a_1^2}{\dfrac{\cos^2 \theta}{(1 - e_1)^2} + \sin^2 \theta}$$

If we determine r^3 by developing this expression as far as e_1^3, it appears that the equation to the ellipsoid may be written in the form

$$r^3 = a_1^3 (1 - e_1) \left[1 + 3 \left(e_1 + \tfrac{5}{14} e_1^2 + \tfrac{2}{21} e_1^3 \right) S_2 + \tfrac{12}{7} e^2 \left(1 + \tfrac{9}{11} e \right) S_4 - \tfrac{40}{33} e^3 S_6 \right]$$

Since the volume of this ellipsoid is $\frac{4}{3}\pi a_1^3 (1 - e_1)$ and that of our spheroid of equilibrium was $\frac{4}{3}\pi a^3$ it follows that

$$a^3 = a_1^3 (1 - e_1)$$

If then we write

$$e = e_1 + \tfrac{5}{14} e_1^2 + \tfrac{2}{21} e^3, \qquad e^2 = e_1^2 + \tfrac{5}{7} e_1^3, \qquad e^3 = e_1^3$$

the equation to the ellipsoid of revolution becomes

$$r^3 = a^3 \left[1 + 3eS_2 + \tfrac{12}{7} e^2 \left(1 + \tfrac{8}{77} e \right) S_4 - \tfrac{40}{33} e^3 S_6 \right]$$

and this is the form determined above in (18), but only as far as squares of e.

If the eccentricity of the ellipsoid be denoted $\sin \gamma$, we have

$$\cos \gamma = 1 - e_1$$

$$\sin^2 \gamma = 2e_1 \left(1 - \tfrac{1}{2} e_1 \right)$$

whence

$$\cos\gamma\sin^2\gamma = 2e_1(1-\tfrac{3}{2}e_1+\tfrac{1}{2}e_1^2) = 2e\left(1-\frac{13}{7}e+\frac{509}{2.3.7^2}e^2\right)$$

$$\cos\gamma\sin^4\gamma = 4e_1^2(1-2e_1) = 4e^2(1-\tfrac{19}{7}e)$$

$$\cos\gamma\sin^6\gamma = 8e_1^3 = 8e^3$$

Now it is known that the rigorous solution for the angular velocity of Maclaurin's ellipsoid may be written in the form

$$\frac{\omega^2}{4\pi\rho} = \cos\gamma\sum_1^\infty \frac{(2n-1)!\sin^{2n}\gamma}{(2n+1)(2n+3)[(n-1)!]^2 2^{2n-3}}$$

Taking the first three terms

$$\frac{\omega^2}{4\pi\rho} = \frac{2}{3.5}\sin^2\gamma\cos\gamma + \frac{3}{5.7}\sin^4\gamma\cos\gamma + \frac{5}{2^2.3.7}\sin^6\gamma\cos\gamma$$

$$= \tfrac{4}{15}e\left(1-\frac{4}{7}e+\frac{4}{3.7^2}e^2\right)$$

This agrees with (17), and the solution is found to be correct as far as cubes of small quantities, thus verifying the previous work.

15.

ON THE FIGURE AND STABILITY OF A LIQUID SATELLITE.

[*Philosophical Transactions of the Royal Society,* Vol. 206, A (1906), pp. 161—248.]

TABLE OF CONTENTS.

Preface.

More than half a century ago Édouard Roche wrote his celebrated paper on the form assumed by a liquid satellite when revolving, without relative motion, about a solid planet*. In consequence of the singular modesty of Roche's style, and also because the publication was made at Montpellier, this paper seems to have remained almost unnoticed for many years, but it has ultimately attained its due position as a classical memoir.

The laborious computations necessary for obtaining numerical results were carried out, partly at least, by graphical methods. Verification of the calculations, which as far as I know have never been repeated, forms part of the work of the present paper. The distance from a spherical planet which has been called "Roche's limit" is expressed by the number of planetary radii in the radius vector of the nearest possible infinitesimal liquid satellite, of the same density as the planet, revolving so as always to present the same aspect to the planet. Our moon, if it were homogeneous, would have the form of one of Roche's ellipsoids; but its present radius vector is of course far greater than the limit. Roche assigned to the limit in question the numerical value 2·44; in the present paper I show that the true value is 2·455, and the closeness of the agreement with the previously accepted value affords a remarkable testimony to the accuracy with which he must have drawn his figures.

He made no attempt to obtain numerical solutions except in the case of the infinitely small satellite. In this case the figure is rigorously ellipsoidal, but for a finite satellite this is no longer the case; nor do his equations afford the means of determining exactly the ellipsoid which most nearly represents the truth. These deficiencies are made good below, and we find that even in the extreme case of two equal masses in limiting stability the ellipsoid is a much closer approximation to accuracy than might have been expected.

It is natural that Roche, writing as he did half a century ago, should not have been in a position to discuss the stability of his solutions with completeness, and although he did much in that direction he necessarily left a good deal unsettled.

In 1887 I attempted the discussion of some of the problems to which this paper is devoted, by means of spherical harmonic analysis†. Poincaré's

* "La figure d'une masse fluide soumise à l'attraction d'un point éloigné," *Acad. des Sci. de Montpellier*, Vol. I., 1847–50, p. 243.

† [Paper 9, p. 135.]

celebrated memoir on figures of equilibrium* was published just when my work was finished, and I kept my paper back for a year in order to apply to my solutions the principles of stability enounced by him. The attempt is given in an appendix (p. 180) to my paper, but unfortunately I failed to understand his work completely, and my investigation, as it stood, was erroneous from the fact that one term in the energy was omitted. The erroneous portion of the Appendix is not reprinted in this volume.

The analysis of the present paper is carried out by means of ellipsoidal harmonic analysis. In the course of the work it becomes necessary to refer to previous papers by myself, all republished in this volume, namely, Papers 10, 11, 12, 13. These papers are hereafter referred to by the abridged titles "Harmonics," "The Pear-Shaped Figure," "Stability," and "Integrals."

The analysis involved in the investigation is unfortunately long and complicated, but the subject itself is not an easy one, and the complication was perhaps unavoidable.

The principal inducement to attack this problem was the hope that it might throw further light on the form of the pear-shaped figure in an advanced stage of development when it might be supposed to consist of two bulbs of liquid joined by a very thin neck. The arguments adduced below seem to show that such a figure must be unstable.

M. Liapounoff has recently published a paper in which he states that he is able to prove the instability of the pear-shaped figure even when only infinitesimally furrowed†. In view of my previous work on the stability of this figure, and from other considerations it seems very difficult to accept the correctness of this result.

At the end a summary is given of the conclusions arrived at, and this last subject is discussed amongst others.

* "Sur l'équilibre d'une masse fluide animée d'un mouvement de rotation," *Acta Math.*, 7: 3, 4 (1885), pp. 259—380.

† "Sur un problème de Tchebychef," *Acad. Imp. des Sci. de St. Pétersbourg*, Vol. XVII., No. 3, (1905).

Part I. Analysis.

§ 1. *The Stability of Liquid Satellites.*

This paper deals with two problems concerning liquid satellites which possess so much resemblance that I did not for some time perceive that there is an essential difference between them. One of these is the determination of the figures and of the secular stability of two masses of liquid revolving about one another in a circular orbit without relative motion of their parts. We may refer to this as the problem of "the figures of equilibrium"; the other may be called "Roche's problem," and it differs only from the former in that one of the two masses of liquid is replaced by a particle or by a rigid sphere. However, in the numerical solutions found hereafter, Roche's problem is slightly modified, for the rigid sphere is replaced by a rigid ellipsoid of exactly the same form as that assumed by the other mass of liquid in the problem of the figures of equilibrium. Thus, with this modification, the two problems become identical as regards the shape of the figures; but, as we shall see, they differ widely as to the conditions of secular stability. This difference arises from the fact that in the one case there are two bodies which may be subject to tidal friction, and in the other there is only one.

If in either problem there is no solution when the angular momentum has less than a certain critical value, if for that value there is one solution and for greater values there are two, then the principle of Poincaré shows that the single solution is the starting point of a pair of which one has one fewer degrees of instability than the other. If, then, one of the two solutions is continuous with a solution which is clearly stable, it follows that the determination of minimum angular momentum will give us the limiting stability of that solution; and this is the point of greatest interest in all such problems.

Our two problems differ in the value of the angular momentum of which the minimum has to be found. For, if in Roche's problem the second body is a particle, it has only orbital momentum; if the second body is a sphere, it must be deemed to have no rotation; and, finally, in the modified form of the problem, the rotational momentum of the rigid body must be omitted from the angular momentum, which has to be a minimum for limiting stability.

It will be useful to make a rough preliminary investigation of the regions in which we shall have to look for cases of limiting stability in the two problems. For this purpose I consider the case of two spheres as the analogue of the problem of the figure of equilibrium, and the case of a sphere and a particle as the analogue of Roche's problem.

Let ρ be density, and let the mass of the whole system be $\frac{4}{3}\pi\rho a^3$; let the masses of the two spheres be $\frac{4}{3}\pi\rho a^3\lambda/(1+\lambda)$ and $\frac{4}{3}\pi\rho a^3/(1+\lambda)$, or for Roche's problem let the latter be the mass of the particle.

Let r be the distance from the centre of one sphere to that of the other, or to the particle, as the case may be; and ω the orbital angular velocity.

In both cases we have $\omega^2 r^3 = \frac{4}{3}\pi\rho a^3$

The centre of inertia of the two masses is distant $r/(1+\lambda)$ and $\lambda r/(1+\lambda)$ from their respective centres, and we easily find the orbital momentum to be

$$\tfrac{4}{3}\pi\rho a^3 \omega r^2 \frac{\lambda}{(1+\lambda)^2}$$

In both problems the rotational momentum of the first sphere is

$$\tfrac{2}{5}\left(\tfrac{4}{3}\pi\rho a^3\right) a^2 \left(\frac{\lambda}{1+\lambda}\right)^{5/3} \omega$$

In the first problem the rotational momentum of the second sphere is

$$\tfrac{2}{5}\left(\tfrac{4}{3}\pi\rho a^3\right) a^2 \frac{1}{(1+\lambda)^{5/3}} \omega$$

and in the second problem it is nil.

If, then, we write L_1 for the total angular momentum of the two spheres, and L_2 for that of the sphere and particle, we have

$$\left.\begin{aligned} L_1 &= \tfrac{4}{3}\pi\rho a^5 \omega \left[\tfrac{2}{5}\frac{1+\lambda^{5/3}}{(1+\lambda)^{5/3}} + \frac{\lambda r^2}{(1+\lambda)^2 a^2}\right] \\ L_2 &= \tfrac{4}{3}\pi\rho a^5 \omega \left[\tfrac{2}{5}\frac{\lambda^{5/3}}{(1+\lambda)^{5/3}} + \frac{\lambda r^2}{(1+\lambda)^2 a^2}\right] \end{aligned}\right\} \quad \ldots\ldots\ldots\ldots(1)$$

On substituting for ω its value in terms of r, these expressions become

$$L_1 = \left(\tfrac{4}{3}\pi\rho\right)^{3/2} \frac{a^5}{(1+\lambda)^2}\left[\tfrac{2}{5}(1+\lambda^{5/3})(1+\lambda)^{1/3}\left(\frac{a}{r}\right)^{3/2} + \lambda\left(\frac{r}{a}\right)^{1/2}\right]$$

$$L_2 = \left(\tfrac{4}{3}\pi\rho\right)^{3/2} \frac{a^5}{(1+\lambda)^2}\left[\tfrac{2}{5}\lambda^{5/3}(1+\lambda)^{1/3}\left(\frac{a}{r}\right)^{3/2} + \lambda\left(\frac{r}{a}\right)^{1/2}\right]$$

To determine the minima of these functions, we differentiate with respect to r, and equate to zero.

Then, if r_1, r_2 denote the two solutions, we find

$$\left(\frac{r_1}{a}\right)^2 = \frac{6}{5\lambda}(1+\lambda^{5/3})(1+\lambda)^{1/3}$$

$$\left(\frac{r_2}{a}\right)^2 = \tfrac{6}{5}\lambda^{2/3}(1+\lambda)^{1/3}$$

Whence $\qquad$ Minimum $L_1 = \left(\tfrac{4}{3}\pi\rho\right)^{3/2} a^5 \,.\, 4\left(\tfrac{2}{135}\right)^{1/4} \dfrac{\lambda^{3/4}(1+\lambda^{5/3})^{1/4}}{(1+\lambda)^{23/12}}$

Minimum $L_2 = \left(\tfrac{4}{3}\pi\rho\right)^{3/2} a^5 \,.\, 4\left(\tfrac{2}{135}\right)^{1/4} \dfrac{\lambda^{7/6}}{(1+\lambda)^{23/12}}$

The ratio of the minimum of L_1 to that of L_2 is $\left(\frac{1}{\lambda^{5/3}}+1\right)^{1/4}$. Thus as λ rises from 0 to ∞ this ratio falls from infinity to unity.

All the possible cases of the first problem are comprised between $\lambda=0$ and $\lambda=1$. When $\lambda=0$, $r_1=\infty$; and when $\lambda=1$,

$$\frac{r_1}{a}=\sqrt{(\tfrac{12}{5}\cdot 2^{1/3})}=1{\cdot}738$$

Thus, in the problem of the figures of equilibrium, if one of the two masses is large compared with the other, the two must be far apart to secure secular stability. This is exactly what is to be expected from the theory of tidal friction, for limiting stability is reached when there is coalescence of the two solutions which correspond to the cases where each body always presents the same face to the other*.

The result when the two masses are equal becomes more easily intelligible when it is expressed in terms of the radius of either of them. That radius is $a/2^{1/3}$, so that when $\lambda=1$

$$r=1{\cdot}738a=2{\cdot}191\left(\frac{a}{2^{1/3}}\right)$$

Thus, in the latter case, limiting stability is reached when the two spheres are nearly in contact with one another, for if r were equal to twice the radius of either they would be touching.

When the two bodies are far apart, the solution may be obtained by spherical harmonic analysis, and has comparatively little interest. But when the bodies are equal or nearly equal in mass, limiting stability for the figure of equilibrium would seem, from this preliminary investigation, to occur when they are quite close together. Accordingly, in finding numerical solutions hereafter, I have devoted more attention to this case than to any other.

Turning now to the solution of the analogue of Roche's problem, we see that when $\lambda=0$, $r_2=0$. This would mean that a very small liquid satellite could be brought quite up to its planet without becoming unstable. But we shall see that, when the satellite is no longer constrainedly a sphere, instability first occurs through the variations in the shape of the satellite. This preliminary solution does not, therefore, throw much light on the matter, excepting as indicating that we must consider the cases where the satellite is as near to the planet as possible.

Next, when $\lambda=1$, we have

$$r_2=\sqrt{(\tfrac{6}{5}\cdot 2^{1/3})}\,a=\sqrt{\tfrac{12}{5}}\cdot\left(\frac{a}{2^{1/3}}\right)=1{\cdot}549\left(\frac{a}{2^{1/3}}\right)$$

* See *Roy. Soc. Proc.*, Vol. XXIX., 1879, p. 168 [Paper 5, Vol. II., p. 195], or Appendix G (*b*) to Thomson and Tait's *Natural Philosophy*.

Thus, when the two masses are equal, their distance apart is only about $1\frac{1}{2}$ radii of either, and they will overlap. Here again it would seem as if stability would persist up to contact, but, as before, instability first sets in through variations in the shape of the satellite.

Finally, when λ is large, r_2 also becomes large. This case is the same in principle as that considered in the problem of the figures of equilibrium, for it means that if a large liquid body (formerly called the satellite) be attended by a small rigid body (formerly called the planet), secular stability will be attained when the small rigid body has been repelled by tidal friction to a great distance from the large liquid body. As this case may be adequately treated by spherical harmonic analysis, it need not detain us, and we see that the most interesting cases of Roche's problem are those where λ lies between 0 and 1.

§ 2. *Figures of Equilibrium of a Rotating Mass of Liquid and their Stability.*

A mass of liquid, consisting of either one or more portions, is rotating, without relative motion of its parts, about an axis through its centre of inertia with angular velocity ω. We choose as an arbitrary standard figure one which does not differ very widely from a figure of equilibrium, and we suppose that any departure from the standard figure may be defined by two parameters e and f, which may be called ellipticities. It is unnecessary to introduce more than two ellipticities, because the result for any number becomes obvious from the case of two. We also assume a definite angular velocity for the standard configuration.

Let $V(e, f)$ denote the gravitational energy lost in the concentration of the system from a condition of infinite dispersion into the configuration denoted by e, f.

Let $I(e, f)$, $\omega(e, f)$ denote the moment of inertia and angular velocity about the axis of rotation in the same configuration.

The initial values of these quantities are those for which $e = f = 0$, and are $V(0, 0)$, $I(0, 0)$, $\omega(0, 0)$. These all refer to the arbitrary standard configuration; they are therefore constants, and I shall write them V, I, ω for brevity.

Let ellipticities e, f be imparted to the system, and let the angular velocity be so changed that the angular momentum remains constant.

Then $$I(e, f)\,\omega(e, f) = I(0, 0)\,\omega(0, 0) = I\omega$$

The kinetic energy of the system is half the square of the angular momentum divided by the moment of inertia; and since the angular momentum is constant it is equal to $\frac{1}{2}(I\omega)^2/I(e, f)$.

Thus the whole energy of the system, both kinetic and potential, is equal to

$$-V(e,f)+\frac{\frac{1}{2}(I\omega)^2}{I(e,f)}$$

If $V(e,f)=V+\delta V$, $I(e,f)=I+\delta I$, the expression for the energy as far as squares of small quantities is

$$-(V+\delta V)+\frac{\frac{1}{2}(I\omega)^2}{I+\delta I}=-V+\tfrac{1}{2}I\omega^2-\delta V-\tfrac{1}{2}\omega^2\delta I+\tfrac{1}{2}\omega^2\frac{(\delta I)^2}{I}$$

The first two terms may be omitted as being constant and of no interest, and the energy with the sign changed, so that it is the lost energy of the system, becomes

$$\delta V+\tfrac{1}{2}\omega^2\delta I-\tfrac{1}{2}\omega^2\frac{(\delta I)^2}{I}$$

Since ω is constant, we may write this

$$\delta(V+\tfrac{1}{2}\omega^2 I)-\tfrac{1}{2}\omega^2\frac{(\delta I)^2}{I}$$

On developing this by Taylor's theorem, it becomes

$$\left(e\frac{\partial}{\partial e}+f\frac{\partial}{\partial f}+\tfrac{1}{2}e^2\frac{\partial^2}{\partial e^2}+ef\frac{\partial^2}{\partial e\partial f}+\tfrac{1}{2}f^2\frac{\partial^2}{\partial f^2}\right)(V+\tfrac{1}{2}\omega^2 I)-\tfrac{1}{2}\frac{\omega^2}{I}\left(e\frac{\partial I}{\partial e}+f\frac{\partial I}{\partial f}\right)^2$$

The condition for a figure of equilibrium is that the first differentials of the energy with respect to the ellipticities shall vanish. If, therefore, e_0, f_0 denote the equilibrium ellipticities, the equations for finding them are

$$\left(\frac{\partial}{\partial e}+e_0\frac{\partial^2}{\partial e^2}+f_0\frac{\partial^2}{\partial e\partial f}\right)(V+\tfrac{1}{2}\omega^2 I)-\frac{\omega^2}{I}\left[e_0\left(\frac{\partial I}{\partial e}\right)^2+f_0\frac{\partial I}{\partial e}\frac{\partial I}{\partial f}\right]=0$$

$$\left(\frac{\partial}{\partial f}+e_0\frac{\partial^2}{\partial e\partial f}+f_0\frac{\partial^2}{\partial f^2}\right)(V+\tfrac{1}{2}\omega^2 I)-\frac{\omega^2}{I}\left[e_0\frac{\partial I}{\partial e}\frac{\partial I}{\partial f}+f_0\left(\frac{\partial I}{\partial f}\right)^2\right]=0$$

Multiplying the first of these by e and the second by f, and adding them together, we find

$$\left(e\frac{\partial}{\partial e}+f\frac{\partial}{\partial f}\right)(V+\tfrac{1}{2}\omega^2 I)=-\left[ee_0\frac{\partial^2}{\partial e^2}+(ef_0+e_0f)\frac{\partial^2}{\partial e\partial f}+ff_0\frac{\partial^2}{\partial f^2}\right](V+\tfrac{1}{2}\omega^2 I)$$
$$+\frac{\omega^2}{I}\left[ee_0\left(\frac{\partial I}{\partial e}\right)^2+(ef_0+e_0f)\frac{\partial I}{\partial e}\frac{\partial I}{\partial f}+ff_0\left(\frac{\partial I}{\partial f}\right)^2\right]$$

On substituting this in the expression for the lost energy, it becomes

$$\left[\tfrac{1}{2}e(e-2e_0)\frac{\partial^2}{\partial e^2}+(ef-ef_0-e_0f)\frac{\partial^2}{\partial e\partial f}+\tfrac{1}{2}f(f-2f_0)\frac{\partial^2}{\partial f^2}\right](V+\tfrac{1}{2}\omega^2 I)$$
$$-\tfrac{1}{2}\frac{\omega^2}{I}\left[e(e-2e_0)\left(\frac{\partial I}{\partial e}\right)^2+2(ef-ef_0-e_0f)\frac{\partial I}{\partial e}\frac{\partial I}{\partial f}+f(f-2f_0)\left(\frac{\partial I}{\partial f}\right)^2\right]$$

Now let δe, δf be the excesses of e and f above their equilibrium values e_0, f_0, so that $e = e_0 + \delta e$, $f = f_0 + \delta f$. Then on substitution in the expression for the lost energy it becomes

$$\left[\tfrac{1}{2}\{(\delta e)^2 - e_0^2\}\frac{\partial^2}{\partial e^2} + \{\delta e\,\delta f - e_0 f_0\}\frac{\partial^2}{\partial e\,\partial f} + \tfrac{1}{2}\{(\delta f)^2 - f_0^2\}\frac{\partial^2}{\partial f^2}\right](V + \tfrac{1}{2}\omega^2 I)$$
$$-\tfrac{1}{2}\frac{\omega^2}{I}\left[\{(\delta e)^2 - e_0^2\}\left(\frac{\partial I}{\partial e}\right)^2 + 2\{\delta e\,\delta f - e_0 f_0\}\frac{\partial I}{\partial e}\frac{\partial I}{\partial f} + \{(\delta f)^2 - f_0^2\}\left(\frac{\partial I}{\partial f}\right)^2\right]$$

Since e_0, f_0 are constants, the portion of this involving e_0, f_0 explicitly is constant, and may be dropped.

Thus the variable part of the lost energy may be written

$$\tfrac{1}{2}(\delta e)^2\left[\frac{\partial^2 V}{\partial e^2} + \tfrac{1}{2}\omega^2\left(\frac{\partial^2 I}{\partial e^2} - \frac{2}{I}\left(\frac{\partial I}{\partial e}\right)^2\right)\right] + \delta e\,\delta f\left[\frac{\partial^2 V}{\partial e\,\partial f} + \tfrac{1}{2}\omega^2\left(\frac{\partial^2 I}{\partial e\,\partial f} - \frac{2}{I}\frac{\partial I}{\partial e}\frac{\partial I}{\partial f}\right)\right]$$
$$+\tfrac{1}{2}(\delta f)^2\left[\frac{\partial^2 V}{\partial f^2} + \tfrac{1}{2}\omega^2\left(\frac{\partial^2 I}{\partial f^2} - \frac{2}{I}\left(\frac{\partial I}{\partial f}\right)^2\right)\right]$$

This is a quadratic function of the departures of the ellipticities from their equilibrium values, and the form is obvious which the result would have if there were any number of ellipticities.

Since the condition for secular stability is that the energy shall be a minimum, the lost energy must be a maximum, and therefore this quadratic function of δe, δf, &c., must always be negative in order that the system may possess secular stability.

If F is a quadratic function of n variables, x_1, x_2, x_3, &c., so that

$$\begin{aligned} F = a_{11}x_1^2 &+ 2a_{12}x_1x_2 + 2a_{13}x_1x_3 + \ldots \\ &+ \quad a_{22}x_2^2 \quad + 2a_{23}x_2x_3 + \ldots \\ &\qquad\qquad + \quad a_{33}x_3^2 \quad + \ldots \end{aligned}$$

it is known that the condition that F shall always be negative for all values of the variables is that the series of functions

$$a_{11},\quad \begin{vmatrix} a_{11}, & a_{12} \\ a_{12}, & a_{22} \end{vmatrix},\quad \begin{vmatrix} a_{11}, & a_{12}, & a_{13} \\ a_{12}, & a_{22}, & a_{23} \\ a_{13}, & a_{23}, & a_{33} \end{vmatrix}, \ldots \qquad \ldots\ldots\ldots\ldots(2)$$

shall be alternatively negative and positive.

Since we might equally well begin with any one of the variables, it follows that a_{11}, a_{22}, a_{33} ... must all be negative; also $a_{12}^2 - a_{11}a_{22}$, $a_{13}^2 - a_{11}a_{33}$, $a_{23}^2 - a_{22}a_{33}$ ... must all be negative if F is always to be negative.

Now, suppose that F is the function of lost energy for a system with $n+1$ degrees of freedom, but that a constraint destroys one of the degrees. If the system has secular stability, the n determinants must have their appropriate signs, and when the constraint is removed, the new additional determinant must have its proper sign in order to secure secular stability. It follows that stability can never be restored by the removal of a constraint if the system was unstable when the constraint existed; but stability may be destroyed by the removal of a constraint.

§ 3. *On the Possibility of joining Two Masses of Liquid by a thin Neck.*

This whole investigation was undertaken principally in the hope that it might lead to an approximation to the form of the pear-shaped figure of equilibrium of a rotating mass of liquid at the stage when it should resemble an hour-glass with a thin neck. It seemed probable that such an approximation might be obtained in the following manner:—

Two masses of liquid are revolving in an orbit about one another without relative motion of their parts, so that they form a figure of equilibrium. Imagine them to be joined by a pipe without weight, through which liquid may flow from one part to the other. A flow of liquid will in general take place between the two parts, but there should be some definite partition of masses, corresponding to a given distance apart, at which flow will cease. At this stage we should have an approximation to the hour-glass figure of equilibrium.

In this section a special case of this problem is considered, in which the detached masses, to be joined by a pipe, are constrained to be spheres.

If the notation of § 1 be adopted, it is clear that the system is defined by the two parameters r and λ. In accordance with the notation of § 2 we denote the lost energy of the system by V and the moment of inertia by I. It is easily shown that

$$V = (\tfrac{4}{3}\pi\rho)^2 a^5 \left[\frac{\lambda}{(1+\lambda)^2}\frac{a}{r} + \tfrac{3}{5}\frac{1+\lambda^{5/3}}{(1+\lambda)^{5/3}}\right]$$

$$I = \tfrac{4}{3}\pi\rho a^5 \left[\frac{\lambda}{(1+\lambda)^2}\frac{r^2}{a^2} + \tfrac{2}{5}\frac{1+\lambda^{5/3}}{(1+\lambda)^{5/3}}\right]$$

For brevity write $F = \dfrac{\lambda}{(1+\lambda)^2}$, $G = \dfrac{1+\lambda^{5/3}}{(1+\lambda)^{5/3}}$

and let F', G', F'', G'' denote their first and second differentials with respect to λ.

The equations for determining the configuration of equilibrium are

$$\frac{\partial V}{\partial r} + \tfrac{1}{2}\omega^2\frac{\partial I}{\partial r} = 0, \quad \frac{\partial V}{\partial \lambda} + \tfrac{1}{2}\omega^2\frac{\partial I}{\partial \lambda} = 0$$

The first of these gives at once

$$\omega^2 = \tfrac{4}{3}\pi\rho\left(\frac{a}{r}\right)^3$$

For determining the form of the second we have

$$V = (\tfrac{4}{3}\pi\rho)^2 a^5 \left[F\frac{a}{r} + \tfrac{3}{5}G\right], \quad I = \tfrac{4}{3}\pi\rho a^5 \left[F\frac{r^2}{a^2} + \tfrac{2}{5}G\right]$$

If we differentiate these with respect to λ, substitute in the second equation of equilibrium and give to ω^2 its value in terms of r^2, we find that the result is

$$\frac{a^3}{r^3} + \tfrac{15}{2}\frac{F'}{G'}\frac{a}{r} + 3 = 0$$

Now

$$F' = \frac{1-\lambda}{(1+\lambda)^3}, \quad G' = \tfrac{5}{3}\frac{\lambda^{2/3}-1}{(1+\lambda)^{8/3}}, \quad \text{and} \quad \frac{F'}{G'} = -\tfrac{3}{5}\frac{1+\lambda^{1/3}+\lambda^{2/3}}{(1+\lambda)^{1/3}(1+\lambda^{1/3})}$$

Hence the equation for determining r for a given value of λ is

$$\frac{a^3}{r^3} - \tfrac{9}{2}\frac{1+\lambda^{1/3}+\lambda^{2/3}}{(1+\lambda)^{1/3}(1+\lambda^{1/3})}\frac{a}{r} + 3 = 0 \quad \text{.....................(3)}$$

This cubic has three real roots of which one is negative and has no physical meaning; the second gives so small a value to r that the smaller sphere is either wholly or partially inside the larger one. The third root is the one required.

In order to present the result in an easily intelligible form it may be well to express it also in terms of the radius of the larger of the two spheres, say a_1, where

$$a_1^3 = \frac{a^3}{1+\lambda}$$

The following is a table of solutions for various values of λ:—

$\lambda^{1/3}$	r/a	r/a_1	$r/a_1 - (1+\lambda^{1/3})$
0·0	1·304	1·304	0·304
0·1	1·323	1·323	0·223
0·2	1·368	1·371	0·171
0·3	1·426	1·438	0·138
0·4	1·486	1·517	0·117
0·5	1·543	1·604	0·104
0·6	1·590	1·697	0·097
0·7	1·625	1·793	0·093
0·8	1·649	1·893	0·093
0·9	1·662	1·995	0·095
1·0	1·666	2·099	0·099

The solution is exhibited in fig. 1, the larger sphere being kept of constant size and the successive smaller circles representing the smaller sphere. Many of the circles pass nearly through one point, and it has not been possible to complete them without producing confusion.

The fourth column of the table gives the excess of r above the sum of the two radii of the spheres, and it shows what interval of space is unoccupied by matter. It is remarkable how nearly constant that interval is throughout a large range in the values of λ.

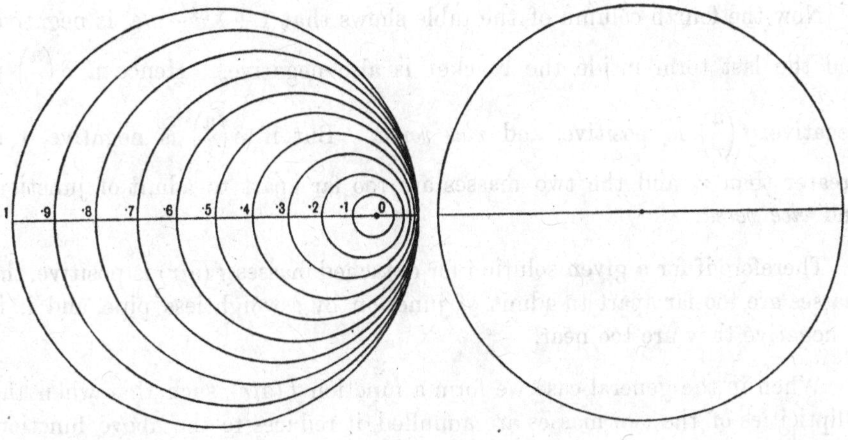

FIG. 1. Solutions for two spheres of liquid joined by a weightless pipe, for successive values of $\lambda^{1/3}$.

In the case where the two bodies are no longer spheres, the equation corresponding to the cubic (3) becomes very complicated. It is therefore desirable to discover whether in any given solution of the figure of equilibrium the two detached masses are too far apart to admit of their being joined by a weightless pipe, or whether they are too near. This may be discovered in the following way:—

Let r_0 be the solution of $f(\mathrm{a}/r_0)=0$, where

$$f\left(\frac{\mathrm{a}}{r}\right)=\frac{\mathrm{a}^3}{r^3}-\frac{9}{2}\frac{1+\lambda^{1/3}+\lambda^{2/3}}{(1+\lambda)^{1/3}(1+\lambda^{1/3})}\frac{\mathrm{a}}{r}+3 \qquad \text{(4)}$$

There is only one solution of $f=0$ between r equal to infinity and the case when the two spheres touch. Hence we can determine on which side of r_0 any given value of r lies by merely considering whether f changes from positive to negative or from negative to positive as r increases through the value r_0.

Now if $\frac{\mathrm{a}}{r_0}+\delta\left(\frac{\mathrm{a}}{r}\right)$ be any neighbouring value of $\frac{\mathrm{a}}{r}$, we have approximately

$$f\left(\frac{\mathrm{a}}{r}\right)=3\left[\frac{\mathrm{a}^2}{r_0{}^2}-\frac{3}{2}\frac{1+\lambda^{1/3}+\lambda^{2/3}}{(1+\lambda)^{1/3}(1+\lambda^{1/3})}\right]\delta\left(\frac{\mathrm{a}}{r}\right)$$

If we express a^2/r_0^2 in terms of r_0/a by means of the equation for r_0, this may be written

$$f\left(\frac{a}{r}\right) = \frac{9}{(1+\lambda)^{1/3}}\left(1 + \lambda^{1/3} - \frac{r}{a_1} - \frac{\lambda^{1/3}}{1+\lambda^{1/3}}\right)\delta\left(\frac{a}{r}\right)$$

where as before
$$a_1^3 = \frac{a^3}{1+\lambda}$$

Now the fourth column of the table shows that $1 + \lambda^{1/3} - r/a_1$ is negative, and the last term inside the bracket is also negative. Hence if $\delta\left(\frac{a}{r}\right)$ is negative, $f\left(\frac{a}{r}\right)$ is positive, and *vice versâ*. But if $\delta\left(\frac{a}{r}\right)$ is negative, r is greater than r_0 and the two masses are too far apart to admit of junction, and *vice versâ*.

Therefore if for a given solution for detached masses $f(a/r)$ is positive, the masses are too far apart to admit of junction by a weightless pipe, and if it is negative they are too near.

When in the general case we form a function $f(a/r)$, such that when the ellipticities of the two masses are annulled, it reduces to the above function, its sign will afford the criterion as to whether the masses are too far or too near to admit of junction by a thin neck of liquid. I return to this subject below in § 13.

The solution of the problem when the two masses are constrainedly spheres is so curious that it seems worth while to consider its stability. This may be done by the method of § 2.

The system depends on two parameters r and λ, and the stability will depend on three functions, which are defined as follows:—

$$\{r, r\} = \frac{\partial^2 V}{\partial r^2} + \tfrac{1}{2}\omega^2 \frac{\partial^2 I}{\partial r^2} - \frac{\omega^2}{I}\left(\frac{\partial I}{\partial r}\right)^2$$

$$\{r, \lambda\} = \frac{\partial^2 V}{\partial r \partial \lambda} + \tfrac{1}{2}\omega^2 \frac{\partial^2 I}{\partial r \partial \lambda} - \frac{\omega^2}{I}\frac{\partial I}{\partial r}\frac{\partial I}{\partial \lambda}$$

$$\{\lambda, \lambda\} = \frac{\partial^2 V}{\partial \lambda^2} + \tfrac{1}{2}\omega^2 \frac{\partial^2 I}{\partial \lambda^2} - \frac{\omega^2}{I}\left(\frac{\partial I}{\partial \lambda}\right)^2$$

These functions correspond to a_{11}, a_{12}, a_{22} of (2) in § 2, and we see that for secular stability $\{r, r\}$ and $\{\lambda, \lambda\}$ must be negative, and

$$\Delta = \{r, r\}\{\lambda, \lambda\} - [\{r, \lambda\}]^2$$

must be positive.

Without giving the details of the several differentiations, I may state that if we write

$$H = (\tfrac{4}{3}\pi\rho)^2 a^3 \frac{F\dfrac{a^3}{r^3}}{F\dfrac{r^2}{a^2} + \tfrac{2}{5}G}$$

so that H is essentially positive, we find

$$\frac{\{r, r\}}{H} = \tfrac{6}{5}G - F\frac{r^2}{a^2}; \quad \frac{\{r, \lambda\}}{H} = -2r\left(F'\frac{r^2}{a^2} + \tfrac{2}{5}G'\right)$$

$$\frac{\{\lambda, \lambda\}}{H} = \frac{a^2}{F}\left\{\left[\tfrac{3}{2}F''\frac{r^2}{a^2} + \tfrac{1}{5}G''\left(1 + 3\frac{r^3}{a^3}\right)\right]\left[F\frac{r^2}{a^2} + \tfrac{2}{5}G\right] - \left[F'\frac{r^2}{a^2} + \tfrac{2}{5}G'\right]^2\right\}$$

From the equation $\quad \dfrac{a^3}{r^3} + \tfrac{15}{2}\dfrac{F'}{G'}\dfrac{a}{r} + 3 = 0$

we find, after some reductions,

$$\tfrac{3}{2}F''\frac{r^2}{a^2} + \tfrac{1}{5}G''\left(1 + 3\frac{r^3}{a^3}\right) = \tfrac{3}{2}\frac{r^2}{a^2}F'\frac{d}{d\lambda}\log\frac{F'}{G'}$$

On substituting for F', G' their values in terms of λ, I find that this expression reduces to

$$\frac{(1 - \lambda^{1/3})^2}{\lambda^{1/3}(1 + \lambda)^4}\frac{r^2}{a^2}$$

an essentially positive quantity.

On substitution in Δ I find

$$\Delta = \frac{H^2 a^2}{F}\left(F\frac{r^2}{a^2} + \tfrac{2}{5}G\right)\left[\frac{(1 - \lambda^{1/3})^2}{\lambda^{1/3}(1 + \lambda)^4}\frac{r^2}{a^2}\left(\tfrac{6}{5}G - F\frac{r^2}{a^2}\right) - 3\left(F'\frac{r^2}{a^2} + \tfrac{2}{5}G'\right)^2\right] \quad \text{.....(5)}$$

The factors outside [ ] are essentially positive and do not affect the sign of Δ, and it is clear that Δ can only be positive if $\tfrac{6}{5}G - Fr^2/a^2$ is positive. But Δ must be positive for secular stability; hence stability can only be secured by $\tfrac{6}{5}G - Fr^2/a^2$ being positive, and it is not necessarily so secured. But if this function is positive, so also is $\{r, r\}$, and if this last is positive the system is unstable. Hence stability is always impossible. As a fact, in all the solutions given above $\{r, r\}$ is positive, and we should have to move the spheres much further apart to make it negative, and therefore on this ground alone the system is always unstable. But Δ is sometimes positive and sometimes negative and vanishes for a certain value of λ. As the vanishing of Δ puzzled me a good deal, I propose to examine the matter further.

Before doing so, however, I will show that the instability of the system may be concluded from other considerations.

It was proved in § 1 that two spheres, unconnected by a pipe, are in limiting stability when their distance apart is given by

$$\frac{r^2}{a^2} = \frac{6}{5\lambda}(1+\lambda^{5/3})(1+\lambda)^{1/3} = \tfrac{6}{5}\frac{G}{F}$$

This is the condition that $\{r, r\}$ should vanish.

When λ is zero the two spheres in limiting stability are infinitely far apart, and when λ is unity they are as near as possible, and $r = 1{\cdot}738a$.

Now the table of solutions in the case where the two are connected by a pipe shows that they are furthest apart when λ is unity, and that then $r = 1{\cdot}666a$.

The removal of the constraint of one degree of freedom may destroy stability, but cannot create it. Hence, when two spheres revolve about one another, the opening of a channel of communication between them may destroy stability, but cannot create it. When two equal spheres revolve about one another at such a distance that they could be connected by a pipe and yet remain in equilibrium, their distance is 1·666; but they are then unstable, because 1·666 is less than 1·738. The opening of a pipe between them, being the removal of a constraint, cannot make the motion stable. *À fortiori* the like is true when the two spheres are unequal in mass.

Hence the system of equilibrium of two spheres joined by a pipe is unstable in all cases.

I will now consider the meaning of the vanishing of Δ.

Having evaluated the angular momentum of the system corresponding to the several solutions tabulated above, I found it had a minimum when $\lambda^{1/3} = 0{\cdot}254$. Such a solution is a critical one and is the starting point of two solutions of which one must have one fewer degrees of instability than the other. The vanishing of Δ must have the same meaning, but it remains to be proved that minimum angular momentum is secured by the vanishing of Δ.

The angular momentum is $I\omega$, and is therefore proportional to μ, where

$$\mu = F\left(\frac{r}{a}\right)^{1/2} + \tfrac{2}{5}G\left(\frac{a}{r}\right)^{3/2}$$

On equating $\dfrac{d\mu}{d\lambda}$ to zero so as to find its minimum, we have

$$2r\left(F'\frac{r^2}{a^2} + \tfrac{2}{5}G'\right) + \left(F\frac{r^2}{a^2} - \tfrac{6}{5}G\right)\frac{dr}{d\lambda} = 0$$

Now since

$$\frac{r^3}{a^3} + \tfrac{1}{3} + \tfrac{5}{2}\frac{F'}{G'}\frac{r}{a} = 0$$

we have

$$\frac{dr}{d\lambda} = \frac{-\tfrac{1}{3}r(1-\lambda^{1/3})^2\lambda^{-1/3}(1+\lambda)^{-4}}{\tfrac{3}{5}G'\dfrac{r}{a} + F'}$$

On substituting this value in the equation $d\mu/d\lambda=0$, I find that the result may be written

$$\frac{(1-\lambda^{1/3})^2}{\lambda^{1/3}(1+\lambda)^4}\frac{r^2}{\mathrm{a}^2}\left(\tfrac{6}{5}G-F\frac{r^2}{\mathrm{a}^2}\right)+6\frac{r^2}{\mathrm{a}^2}\left(F'\frac{r^2}{\mathrm{a}^2}+\tfrac{2}{5}G'\right)\left(\tfrac{3}{5}G'\frac{r}{\mathrm{a}}+F'\right)=0$$

The first term of this is the same as the first term inside the bracket in the expression for Δ in (5). On comparing the two second terms together we see that $\Delta=0$ is the condition for minimum angular momentum, if

$$6\frac{r^2}{\mathrm{a}^2}\left(F'\frac{r^2}{\mathrm{a}^2}+\tfrac{2}{5}G'\right)\left(\tfrac{3}{5}G'\frac{r}{\mathrm{a}}+F'\right)=-3\left(F'\frac{r^2}{\mathrm{a}^2}+\tfrac{2}{5}G'\right)^2$$

that is to say, if $$2\frac{r^2}{\mathrm{a}^2}\left(\tfrac{3}{5}G'\frac{r}{\mathrm{a}}+F'\right)+F'\frac{r^2}{\mathrm{a}^2}+\tfrac{2}{5}G'=0$$

or if $$\frac{\mathrm{a}^3}{r^3}+\tfrac{15}{2}\frac{F'}{G'}\frac{\mathrm{a}}{r}+3=0$$

But this last is true, being the equation (3) determining the figure of equilibrium; hence $\Delta=0$ gives minimum angular momentum.

Since two liquid spheres cannot be joined stably by a pipe, it seems very improbable that two tidal ellipsoids could be so joined as to become stable. Indeed, if the distortion of the surfaces of the two masses into ellipsoidal forms may be regarded as due to the removal of constraints whereby they were previously maintained in a spherical form, stability is impossible.

The question as to whether or not there is an unstable figure with a thin neck will be considered later, for the present we are only concerned with the conclusion that there is no stable figure of this kind.

Mr Jeans has treated an analogous problem in his paper on the equilibrium of rotating liquid cylinders*, and has concluded that the cylinder will divide stably into two portions. The analogy is so close between his problem and the three-dimensional case, that it might have been expected that the analogy would subsist throughout; nevertheless, if we are both correct there must be a divergence between them at some point.

§ 4. *Notation.*

As the solution given below is effected by means of ellipsoidal harmonic analysis, it is well to state the notation employed. It is that used in four previous papers to which references are given in the Preface.

In "Harmonics" the squares of the semi-axes of the ellipsoid were

$$a^2=k^2\left(\nu^2-\frac{1+\beta}{1-\beta}\right),\quad b^2=k^2(\nu^2-1),\quad c^2=k^2\nu^2$$

* *Phil. Trans. Roy. Soc.*, Series A, Vol. 200, pp. 67—104.

The rectangular co-ordinates were connected with ellipsoidal co-ordinates ν, μ, ϕ by

$$\frac{x^2}{k^2} = -\frac{1-\beta}{1+\beta}\left(\nu^2 - \frac{1+\beta}{1-\beta}\right)\left(\mu^2 - \frac{1+\beta}{1-\beta}\right)\cos^2\phi$$

$$\frac{y^2}{k^2} = -(\nu^2-1)(\mu^2-1)\sin^2\phi$$

$$\frac{z^2}{k^2} = \nu^2\mu^2\,\frac{1-\beta\cos 2\phi}{1+\beta}$$

The three roots of the cubic

$$\frac{x^2}{a^2+u} + \frac{y^2}{b^2+u} + \frac{z^2}{c^2+u} = 1$$

were $$u_1 = k^2\nu^2, \qquad u_2 = k^2\mu^2, \qquad u_3 = k^2\,\frac{1-\beta\cos 2\phi}{1-\beta}$$

Lastly ν ranges from ∞ to 0, μ between ± 1, ϕ from 0 to 2π.

In the two later papers, I put

$$\kappa^2 = \frac{1-\beta}{1+\beta}, \qquad \kappa'^2 = 1-\kappa^2, \qquad \nu = \frac{1}{\kappa\sin\gamma}, \qquad \mu = \sin\theta$$

and for convenience I introduced an auxiliary constant β (easily distinguishable from the β of the previous notation) defined by $\sin\beta = \kappa\sin\gamma$.

The squares of the semi-axes of the ellipsoid were then

$$a^2 = \frac{k^2\cos^2\gamma}{\sin^2\beta}, \qquad b^2 = \frac{k^2\cos^2\beta}{\sin^2\beta}, \qquad c^2 = \frac{k^2}{\sin^2\beta}$$

The rectangular co-ordinates became

$$\frac{x^2}{k^2} = \frac{\cos^2\gamma}{\sin^2\beta}(1-\kappa^2\sin^2\theta)\cos^2\phi, \qquad \frac{y^2}{k^2} = \frac{\cos^2\beta}{\sin^2\beta}\cos^2\theta\sin^2\phi$$

$$\frac{z^2}{k^2} = \frac{1}{\sin^2\beta}\sin^2\theta\,(1-\kappa'^2\cos^2\phi)$$

The roots of the cubic were

$$u_1 = \frac{k^2}{\sin^2\beta}, \qquad u_2 = k^2\sin^2\theta, \qquad u_3 = \frac{k^2}{\kappa^2}(1-\kappa'^2\cos^2\phi)$$

The notation employed for the harmonic functions is that defined in "Harmonics."

§ 5. *The Determination of Gravity on Roche's Ellipsoid.*

In Roche's problem a mass of liquid, which assumes approximately the form of an ellipsoid, revolves in a circular orbit about a distant centre of force without any relative motion. In the present section it is proposed to evaluate gravity on the surface of this ellipsoid. I intend to solve the problems of the present paper by means of the principles of energy, and for that purpose it is necessary to determine the law of gravity.

Suppose that the ellipsoid of reference, defined by ν_0, is deformed by a normal displacement defined by the function $pf(\mu, \phi)$, where p is the perpendicular from the centre on to the tangent plane at μ, ϕ. This deformation must be expressible by a series of ellipsoidal harmonic functions, and therefore we may assume

$$f(\mu, \phi) = \Sigma e_i^s \mathfrak{P}_i^s(\mu) \mathfrak{C}_i^s(\phi)$$

The typical term written down must be deemed to include sine-functions as well as cosine-functions, and all those types which I have denoted by P, C, S in "Harmonics."

On multiplying each side of our equation by any harmonic function, and integrating over the surface of the ellipsoid, an element of which surface is denoted by $d\sigma$, we find in the usual way

$$e_i^s = \frac{\int f(\mu, \phi) \mathfrak{P}_i^s(\mu) \mathfrak{C}_i^s(\phi) p d\sigma}{\int [\mathfrak{P}_i^s(\mu) \mathfrak{C}_i^s(\phi)]^2 p d\sigma}$$

Suppose that $f(\mu, \phi)$ is zero everywhere except over a small area $\delta\alpha$ situated at the point μ', ϕ', and that it is there equal to a constant c; also let p' be the value of p at this area $\delta\alpha$.

Then the mass of the inequality is

$$\int \rho f(\mu, \phi) p d\sigma = cp'\rho\delta\alpha$$

where ρ is the density of the solid ellipsoid which is deformed.

Next let us suppose that the mass of the inequality is unity, so that

$$cp'\rho\delta\alpha = 1$$

Then we have

$$\int f(\mu, \phi) \mathfrak{P}_i^s(\mu) \mathfrak{C}_i^s(\phi) p d\sigma = c\mathfrak{P}_i^s(\mu') \mathfrak{C}_i^s(\phi') p'\delta\alpha = \frac{1}{\rho} \mathfrak{P}_i^s(\mu') \mathfrak{C}_i^s(\phi')$$

Hence
$$e_i^s = \frac{\mathfrak{P}_i^s(\mu') \mathfrak{C}_i^s(\phi')}{\rho \int [\mathfrak{P}_i^s(\mu) \mathfrak{C}_i^s(\phi)]^2 p d\sigma}$$

I now write M for the mass of the ellipsoid, and shall subsequently make it equal to $\frac{4}{3}\pi\rho a^3 \frac{\lambda}{1+\lambda}$, while the mass of the distant particle will be $\frac{M}{\lambda}$ or $\frac{4}{3}\pi\rho a^3 \frac{1}{1+\lambda}$.

Since $\int p\rho d\sigma = 3M$, and $\mathfrak{P}_0(\mu)\mathfrak{C}_0(\phi) = 1$, we have $e_0 = \frac{1}{3M}$.

Thus an inequality representing a particle of unit mass at μ', ϕ' on the surface of the ellipsoid is expressed in ellipsoidal harmonics by

$$p\left[\frac{1}{3M} + \Sigma \frac{\mathfrak{P}_i^s(\mu')\mathfrak{C}_i^s(\phi')\mathfrak{P}_i^s(\mu)\mathfrak{C}_i^s(\phi)}{\rho\int[\mathfrak{P}_i^s(\mu)\mathfrak{C}_i^s(\phi)]^2 p d\sigma}\right]$$

By the formula (51) of "Harmonics," the external potential at the point ν, μ, ϕ of the inequality is

$$\frac{1}{k}\mathfrak{P}_0(\nu_0)\mathfrak{Q}_0(\nu_0) + \frac{3M}{k\rho}\Sigma\frac{\mathfrak{P}_i^s(\mu')\mathfrak{C}_i^s(\phi')\mathfrak{P}_i^s(\nu_0)\mathfrak{Q}_i^s(\nu)\mathfrak{P}_i^s(\mu)\mathfrak{C}_i^s(\phi)}{\int[\mathfrak{P}_i^s(\mu)\mathfrak{C}_i^s(\phi)]^2 p d\sigma}$$

But if R is the distance between the point ν_0, μ', ϕ' on the ellipsoid and the external point ν, μ, ϕ, this potential is $1/R$.

If we imagine a particle of mass M/λ situated at ν, μ, ϕ, the above expression multiplied by M/λ is the potential of the particle at the point ν_0, μ', ϕ' on the ellipsoid.

We have no need for the general expression for the potential of a particle situated anywhere in space at the surface of the ellipsoid, because it is only necessary to consider the case where the particle lies on the prolongation of the longest axis of the ellipsoid. In this case

$$\mu = 1, \qquad \phi = \tfrac{1}{2}\pi, \qquad \nu = \frac{r}{k}$$

where r is the distance of the particle from the centre of the ellipsoid.

But it is now no longer necessary to retain the accents to μ', ϕ', since they are only the co-ordinates of a point on the ellipsoid.

Thus the potential of M/λ, lying on the longest axis of the ellipsoid at a distance r from the centre, at the point ν_0, μ, ϕ on the ellipsoid, is

$$\frac{M}{k\lambda}\mathfrak{P}_0(\nu_0)\mathfrak{Q}_0\left(\frac{r}{k}\right) + \frac{3M^2}{k\lambda\rho}\Sigma\frac{\mathfrak{Q}_i^s(r/k)\mathfrak{P}_i^s(1)\mathfrak{C}_i^s(\frac{1}{2}\pi)\mathfrak{P}_i^s(\nu_0)\mathfrak{P}_i^s(\mu)\mathfrak{C}_i^s(\phi)}{\int[\mathfrak{P}_i^s(\mu)\mathfrak{C}_i^s(\phi)]^2 p d\sigma}$$

For the types of functions denoted in "Harmonics" EES, OOS, OES, EOS, $\mathfrak{P}_i^s(1) = 0$, and for EOC, OOC, $\mathfrak{C}_i^s(\frac{1}{2}\pi) = 0$. The only types for which $\mathfrak{P}_i^s(1)\mathfrak{C}_i^s(\frac{1}{2}\pi)$ does not vanish are EEC, OEC; that is to say, cosine-functions of even rank. Accordingly the functions left are $\mathfrak{P}_i^s\mathfrak{C}_i^s$ for i and s even, and

$\mathfrak{P}_i^s \mathsf{C}_i^s$ for i odd and s even; we may however continue to allow $\mathfrak{C}_i^s$ to stand for both types.

For brevity write

$$\mathfrak{T}_i^s = \frac{(2i+1)\rho}{3M} \int [\mathfrak{P}_i^s(\mu)\,\mathfrak{C}_i^s(\phi)]^2\, p\, d\sigma$$

Thence, since $\mathfrak{P}_0(\nu_0) = 1$, the potential may be written

$$\frac{M}{k\lambda}\left\{\mathfrak{Q}_0\left(\frac{r}{k}\right) + \sum_1^\infty \frac{2i+1}{\mathfrak{T}_i^s}\,\mathfrak{Q}_i^s\left(\frac{r}{k}\right)\mathfrak{P}_i^s(1)\,\mathfrak{C}_i^s(\tfrac{1}{2}\pi)\,\mathfrak{P}_i^s(\nu_0)\,\mathfrak{P}_i^s(\mu)\,\mathfrak{C}_i^s(\phi)\right\}$$

for all even values of s.

It must be observed that $\mathfrak{P}_i^s$ and $\mathfrak{C}_i^s$ occur as squares in $\mathfrak{T}_i^s$; they also occur twice in the numerator in the forms $\mathfrak{P}_i^s(1)\,\mathfrak{P}_i^s(\mu)$ and $\mathfrak{C}_i^s(\tfrac{1}{2}\pi)\,\mathfrak{C}_i^s(\phi)$.

Again $\mathfrak{Q}_i^s\left(\frac{r}{k}\right)$ is of dimensions -1 in $\mathfrak{P}_i^s$, and therefore $\mathfrak{Q}_i^s\left(\frac{r}{k}\right)\mathfrak{P}_i^s(\nu_0)$ is of zero dimensions. From these considerations it follows that $\mathfrak{P}_i^s(\mu)$ and $\mathfrak{C}_i^s(\phi)$ may be multiplied by any factors without changing the result, and further that $\mathfrak{P}_i^s(\nu)$ may differ in its mode of definition from $\mathfrak{P}_i^s(\mu)$ without producing any change.

The higher harmonics will be considered later, and for the present it is only necessary to consider the terms defined by $i=1$, $s=0$ and $i=2$, $s=0$ and 2.

The following are the definitions of the several functions, in accordance with "Integrals":—

$$\mathfrak{P}_0(\nu) = 1, \qquad \mathfrak{P}_0(\mu) = 1, \qquad \mathfrak{C}_0(\phi) = 1$$

$$\mathfrak{P}_1(\nu) = \nu, \qquad \mathfrak{P}_1(\mu) = \mu, \qquad \mathsf{C}_1(\phi) = \sqrt{(1-\kappa'^2\cos^2\phi)}$$

$$\mathfrak{P}_2^s(\nu) = \nu^2 - \frac{q_s^2}{\kappa^2}, \qquad \mathfrak{P}_2^s(\mu) = \kappa^2\mu^2 - q_s^2, \qquad \mathfrak{C}_2^s(\phi) = q_s'^2 - \kappa'^2\cos^2\phi \quad (s=0,\,2)$$

where $q_s^2 = \frac{1}{3}[1+\kappa^2 \mp D]$ and $D^2 = 1-\kappa^2\kappa'^2$, with upper sign for $s=0$, and lower for $s=2$.

Hence

$$\mathfrak{P}_1(1)\,\mathfrak{C}_1(\tfrac{1}{2}\pi) = 1, \qquad \mathfrak{P}_2^s(1)\,\mathfrak{C}_2^s(\tfrac{1}{2}\pi) = (\kappa^2 - q_s^2)\,q_s'^2 \quad (s=0,\,2)$$

Then from "Integrals," equations (5) and (6),

$$\mathfrak{T}_1 = 1, \qquad \mathfrak{T}_2^s = \frac{2^3}{3^4}[D^4 \pm (1+\tfrac{1}{2}\kappa^2\kappa'^2)(1-2\kappa'^2)\,D] \quad (s=0,\,2)$$

Thus as far as the second order of harmonics the potential of M/λ at ν_0, μ, ϕ is

$$\begin{aligned}\frac{M}{k\lambda}\Bigg\{\mathfrak{Q}_0\left(\frac{r}{k}\right) &+ 3\mathfrak{Q}_1\left(\frac{r}{k}\right)\mathfrak{P}_1(\nu_0)\,\mathfrak{P}_1(\mu)\,\mathsf{C}_1(\phi)\\ &+ \frac{5}{\mathfrak{T}_2}(\kappa^2-q_0^2)\,q_0'^2\,\mathfrak{Q}_2\left(\frac{r}{k}\right)\mathfrak{P}_2(\nu_0)\,\mathfrak{P}_2(\mu)\,\mathfrak{C}_2(\phi)\\ &+ \frac{5}{\mathfrak{T}_2^2}(\kappa^2-q_2^2)\,q_2'^2\,\mathfrak{Q}_2^2\left(\frac{r}{k}\right)\mathfrak{P}_2^2(\nu_0)\,\mathfrak{P}_2^2(\mu)\,\mathfrak{C}_2^2(\phi)\Bigg\}\end{aligned}$$

We must now express the several solid harmonics involved in this expression in terms of x, y, z co-ordinates of a point on the surface of the ellipsoid.

We have $$\mathfrak{P}_1(\nu_0)\,\mathfrak{P}_1(\mu)\,\mathsf{C}_1(\phi) = \nu\mu\sqrt{(1-\kappa'^2\cos^2\phi)} = \frac{z}{k}$$

By the definition of ellipsoidal co-ordinates the three values of ω^2 which satisfy the equation

$$\frac{x^2}{\omega^2 - 1/\kappa^2} + \frac{y^2}{\omega^2 - 1} + \frac{z^2}{\omega^2} - k^2 = 0 \quad \text{are} \quad \nu^2,\ \mu^2,\ \frac{1}{\kappa^2}(1 - \kappa'^2\cos^2\phi)$$

Hence we have the following identity

$$\frac{x^2}{\omega^2 - 1/\kappa^2} + \frac{y^2}{\omega^2 - 1} + \frac{z^2}{\omega^2} - k^2 = k^2\frac{(\nu_0^2 - \omega^2)(\mu^2 - \omega^2)(1 - \kappa'^2\cos^2\phi - \omega^2\kappa^2)}{(\omega^2 - 1/\kappa^2)(\omega^2 - 1)\,\omega^2\kappa^2}$$

Putting $\omega^2 = \dfrac{q_s^2}{\kappa^2}$ $(s = 0, 2)$ we find

$$\mathfrak{P}_2^s(\nu_0)\,\mathfrak{P}_2^s(\mu)\,\mathfrak{C}_2^s(\phi)$$
$$= q_s^2 q_s'^2(\kappa^2 - q_s^2)\left[-\frac{1}{q_s'^2}\cdot\frac{x^2}{k^2} - \frac{1}{\kappa^2 - q_s^2}\cdot\frac{y^2}{k^2} + \frac{1}{q_s^2}\cdot\frac{z^2}{k^2} - \frac{1}{\kappa^2}\right] \quad (s = 0, 2)$$

Hence the potential of M/λ is

$$\frac{M}{\lambda k}\left\{\mathfrak{Q}_0\left(\frac{r}{k}\right) + 3\mathfrak{Q}_1\left(\frac{r}{k}\right)\cdot\frac{z}{k}\right.$$
$$+\frac{x^2}{k^2}\left[-\frac{5\mathfrak{Q}_2(r/k)}{\mathfrak{T}_2}q_0^2q_0'^2(\kappa^2 - q_0^2)^2 - \frac{5\mathfrak{Q}_2^2(r/k)}{\mathfrak{T}_2^2}q_2^2q_2'^2(\kappa^2 - q_2^2)^2\right]$$
$$+\frac{y^2}{k^2}\left[-\frac{5\mathfrak{Q}_2(r/k)}{\mathfrak{T}_2}q_0^2q_0'^4(\kappa^2 - q_0^2) - \frac{5\mathfrak{Q}_2^2(r/k)}{\mathfrak{T}_2^2}q_2^2q_2'^4(\kappa^2 - q_2^2)\right]$$
$$+\frac{z^2}{k^2}\left[\frac{5\mathfrak{Q}_2(r/k)}{\mathfrak{T}_2}q_0'^4(\kappa^2 - q_0^2)^2 + \frac{5\mathfrak{Q}_2^2(r/k)}{\mathfrak{T}_2^2}q_2'^4(\kappa^2 - q_2^2)^2\right]$$
$$\left.-\frac{1}{\kappa^2}\left[\frac{5\mathfrak{Q}_2(r/k)}{\mathfrak{T}_2}q_0^2q_0'^4(\kappa^2 - q_0^2)^2 + \frac{5\mathfrak{Q}_2^2(r/k)}{\mathfrak{T}_2^2}q_2^2q_2'^4(\kappa^2 - q_2^2)^2\right]\right\}$$

For the object immediately in view we only need the terms involving x, y, z, and may therefore drop the first and last terms.

The expressions for q_s^2 and for $\mathfrak{T}_2^s$ in terms of κ^2 have been given above; by means of these I find that

$$\frac{q_0'^2(\kappa^2 - q_0^2)}{\mathfrak{T}_2} = \frac{9q_2^2}{4D\kappa^2}, \qquad \frac{q_2'^2(\kappa^2 - q_2^2)}{\mathfrak{T}_2^2} = -\frac{9q_0^2}{4D\kappa^2}$$

(note the interchange in the suffixes of the q's).

A common factor $\frac{45}{4}\dfrac{q_0^2q_2^2}{D\kappa^2}$ may be taken from all the coefficients of x^2, y^2, z^2, and since $q_0^2 q_2^2 = \frac{1}{3}\kappa^2$ this common factor is equal to $15/4D$. Hence the terms in x^2, y^2, z^2 inside { } become

$$+\frac{15}{4D}\frac{x^2}{k^2}\left[-\mathfrak{Q}_2\left(\frac{r}{k}\right)q_0'^2(\kappa^2-q_0^2)+\mathfrak{Q}_2^2\left(\frac{r}{k}\right)q_2'^2(\kappa^2-q_2^2)\right]$$

$$+\frac{15}{4D}\frac{y^2}{k^2}\left[-\mathfrak{Q}_2\left(\frac{r}{k}\right)q_0'^2 \qquad +\mathfrak{Q}_2^2\left(\frac{r}{k}\right)q_2'^2\right]$$

$$+\frac{15}{4D}\frac{z^2}{k^2}\left[\mathfrak{Q}_2\left(\frac{r}{k}\right)\frac{q_0'^2}{q_0^2}(\kappa^2-q_0^2)-\mathfrak{Q}_2^2\left(\frac{r}{k}\right)\frac{q_2'^2}{q_2^2}(\kappa^2-q_2^2)\right]$$

On substituting for the several coefficients their values in terms of κ, I find that the potential of M/λ may be written in the following form:—

$$\frac{M}{k\lambda}\left\{3\mathfrak{Q}_1\left(\frac{r}{k}\right)\cdot\frac{z}{k}+\frac{5}{4}\frac{x^2}{k^2}\left[-(2\kappa^2-1)\frac{\mathfrak{Q}_2-\mathfrak{Q}_2^2}{D}-(\mathfrak{Q}_2+\mathfrak{Q}_2^2)\right]\right.$$

$$+\frac{5}{4}\frac{y^2}{k^2}\left[-(2-\kappa^2)\frac{\mathfrak{Q}_2-\mathfrak{Q}_2^2}{D}-(\mathfrak{Q}_2+\mathfrak{Q}_2^2)\right]$$

$$\left.+\frac{5}{4}\frac{z^2}{k^2}\left[(1+\kappa^2)\frac{\mathfrak{Q}_2-\mathfrak{Q}_2^2}{D}+2(\mathfrak{Q}_2+\mathfrak{Q}_2^2)\right]\right\} \quad \text{......(6)}$$

It may be observed that this satisfies Laplace's equation, as it should do.

It remains to obtain approximate expressions for the $\mathfrak{Q}$'s.

The expression for these functions is given by

$$\mathfrak{Q}_i^s(\nu)=\mathfrak{P}_i^s(\nu)\int_\nu^\infty\frac{d\nu}{[\mathfrak{P}_i^s(\nu)]^2(\nu^2-1)^{1/2}(\nu^2-1/\kappa^2)^{1/2}}$$

We require these when ν, which will be put equal to r/k, is large; thus we must develop in powers of $1/\nu$.

Now

$$(\nu^2-1)^{-1/2}(\nu^2-1/\kappa^2)^{-1/2}$$
$$=\frac{1}{\nu^2}\left[1+\frac{1+\kappa^2}{2\kappa^2\nu^2}+\frac{3+2\kappa^2+3\kappa^4}{2^3\kappa^4\nu^4}+\frac{5+3\kappa^2+3\kappa^4+5\kappa^6}{2^4\kappa^6\nu^6}+\dots\right]$$

Since $\mathfrak{P}_0(\nu)=1$, we have by integration

$$\mathfrak{Q}_0\left(\frac{r}{k}\right)=\frac{k}{r}\left[1+\frac{1+\kappa^2}{6\kappa^2}\cdot\frac{k^2}{r^2}+\frac{3+2\kappa^2+3\kappa^4}{40\kappa^4}\cdot\frac{k^4}{r^4}+\frac{5+3\kappa^2+3\kappa^4+5\kappa^6}{112\kappa^6}\cdot\frac{k^6}{r^6}+\dots\right] \quad \text{.........(7)}$$

There is no immediate need for this term, since it has been omitted above, but it will occur again hereafter.

Since $\mathfrak{P}_1(\nu)=\nu$, we have

$$3\mathfrak{Q}_1\left(\frac{r}{k}\right)=\frac{k^2}{r^2}\left[1+\frac{3(1+\kappa^2)}{10\kappa^2}\frac{k^2}{r^2}+\frac{3(3+2\kappa^2+3\kappa^4)}{56\kappa^4}\frac{k^4}{r^4}+\dots\right] \quad \text{......(8)}$$

Lastly, since $\mathfrak{P}_2^s(\nu) = \nu^2 - q_s^2/\kappa^2$, we have

$$[\mathfrak{P}_2^s(\nu)]^{-2} = \frac{1}{\nu^4}\left[1 + \frac{2q_s^2}{\kappa^2\nu^2} + \frac{3q_s^4}{\kappa^4\nu^4} + \ldots\right] \quad (s = 0, 2)$$

so that

$$\frac{1}{[\mathfrak{P}_2^s(\nu)]^2(\nu^2-1)^{1/2}(\nu^2-1/\kappa^2)^{1/2}} = \frac{1}{\nu^6}\left[1 + \left(\frac{1+\kappa^2}{2\kappa^2} + \frac{2q_s^2}{\kappa^2}\right)\frac{1}{\nu^2} + \left(\frac{3+2\kappa^2+3\kappa^4}{8\kappa^4} + \frac{(1+\kappa^2)q_s^2}{\kappa^4} + \frac{3q_s^4}{\kappa^4}\right)\frac{1}{\nu^4} + \ldots\right] \quad (s = 0, 2)$$

If we integrate this, multiply it by $\mathfrak{P}_2^s(\nu)$ and write r/k for ν, we find,

$$5\mathfrak{Q}_2^s\left(\frac{r}{k}\right) = \frac{k^3}{r^3}\left[1 + \left\{\frac{5(1+\kappa^2)}{14\kappa^2} + \frac{3q_s^2}{7\kappa^2}\right\}\frac{k^2}{r^2} + 5\left\{\frac{3+2\kappa^2+3\kappa^4}{72\kappa^4} + \frac{5q_s^2}{126\kappa^4} + \frac{q_s^4}{21\kappa^4}\right\}\frac{k^4}{r^4}\ldots\right] \quad (s = 0, 2) \ldots\ldots(9)$$

On substituting for q_s its value, I find

$$5\mathfrak{Q}_2^s\left(\frac{r}{k}\right) = \frac{k^3}{r^3}\left[1 + \frac{7(1+\kappa^2) \mp 2D}{14\kappa^2}\frac{k^2}{r^2} + \frac{5\{11 + 10\kappa^2 + 11\kappa^4 \mp 4(1+\kappa^2)D\}}{168\kappa^4}\frac{k^4}{r^4}\ldots\right]$$

Whence

$$5[\mathfrak{Q}_2 + \mathfrak{Q}_2^2] = \frac{2k^3}{r^3}\left[1 + \frac{(1+\kappa^2)}{2\kappa^2}\frac{k^2}{r^2} + \frac{5(11+10\kappa^2+11\kappa^4)}{168\kappa^4}\frac{k^4}{r^4}\ldots\right]$$

$$\frac{5}{D}[\mathfrak{Q}_2 - \mathfrak{Q}_2^2] = -\frac{2k^3}{r^3}\left[\frac{1}{7\kappa^2}\frac{k^2}{r^2} + \frac{20(1+\kappa^2)}{168\kappa^4}\frac{k^4}{r^4}\ldots\right]$$

Substituting these values in (6) we have for the potential of M/λ at the surface of the ellipsoid, as far as concerning terms involving x, y, z,

$$\frac{M}{\lambda k}\left\{3\mathfrak{Q}_1\left(\frac{r}{k}\right)\cdot\frac{z}{k} - \frac{x^2}{2k^2}\cdot\frac{k^3}{r^3}\left[1 + \frac{3(3+\kappa^2)}{14\kappa^2}\frac{k^2}{r^2} + \frac{5(5+2\kappa^2+\kappa^4)}{56\kappa^4}\frac{k^4}{r^4}\ldots\right]\right.$$
$$- \frac{y^2}{2k^2}\cdot\frac{k^3}{r^3}\left[1 + \frac{3(1+3\kappa^2)}{14\kappa^2}\frac{k^2}{r^2} + \frac{5(5\kappa^4+2\kappa^2+1)}{56\kappa^4}\frac{k^4}{r^4}\ldots\right]$$
$$\left. + \frac{z^2}{k^2}\cdot\frac{k^3}{r^3}\left[1 + \frac{3(1+\kappa^2)}{7\kappa^2}\frac{k^2}{r^2} + \frac{5(3+2\kappa^2+3\kappa^4)}{56\kappa^4}\frac{k^4}{r^4}\ldots\right]\right\} \ldots(10)$$

If the system be rendered statical by the imposition of a rotation potential, we must add to the above such a potential, and that of the ellipsoid itself.

The expression for the internal potential of an ellipsoid ν_0 is given in (65) of "Harmonics"; it is

$$\frac{3M}{2k}\left\{\frac{\mathfrak{Q}_0(\nu_0)}{\mathfrak{P}_0(\nu_0)} - \frac{x^2}{k^2}\frac{\mathsf{Q}_1^1(\nu_0)}{\mathsf{P}_1^1(\nu_0)} - \frac{y^2}{k^2}\frac{\mathfrak{Q}_1^1(\nu_0)}{\mathfrak{P}_1^1(\nu_0)} - \frac{z^2}{k^2}\frac{\mathfrak{Q}_1(\nu_0)}{\mathfrak{P}_1(\nu_0)}\right\}$$

I will now introduce an abridged notation which was used in some of my previous papers, as follows:—

$$\mathsf{P}_1^1(\nu_0)\,\mathsf{Q}_1^1(\nu_0) = \mathsf{A}_1^1, \quad \mathfrak{P}_1^1(\nu_0)\,\mathfrak{Q}_1^1(\nu_0) = \mathfrak{A}_1^1, \quad \mathfrak{P}_1(\nu_0)\,\mathfrak{Q}_1(\nu_0) = \mathfrak{A}_1$$

Then, since

$$\mathsf{P}_1^1(\nu_0) = \sqrt{(\nu_0^2 - 1/\kappa^2)}, \quad \mathfrak{P}_1^1(\nu_0) = \sqrt{(\nu_0^2 - 1)}, \quad \mathfrak{P}_1(\nu_0) = \nu_0$$

we may, on omitting the term independent of x, y, z, write this potential in the form

$$-\frac{3M}{2k}\left[\frac{x^2}{k^2(\nu_0^2 - 1/\kappa^2)}\;{}_1^1 + \frac{y^2}{k^2(\nu_0^2 - 1)}\mathfrak{A}_1^1 + \frac{z^2}{k^2\nu_0^2}\mathfrak{A}_1\right] \quad\text{.........(11)}$$

The rotation with angular velocity ω takes place about an axis parallel to x through the centre of inertia of the system, which consists of two masses M and M/λ distant r from one another. Hence the rotation potential is

$$\tfrac{1}{2}\omega^2\left[y^2 + \left(z - \frac{r}{1+\lambda}\right)^2\right] = \frac{3M}{2k}\left\{\frac{\omega^2 k^3}{3M}\left(\frac{y^2+z^2}{k^2}\right) - \frac{2}{3}\frac{\omega^2 k^2 r}{(1+\lambda)M}\frac{z}{k}\right\} + \frac{\omega^2 r^2}{2(1+\lambda)^2} \quad\text{.........(12)}$$

The last term, being independent of x, y, z, has no present interest. Then, collecting results from (10), (11), and (12), the whole potential, as far as material, is

$$\frac{3}{2}\frac{M}{k}\left[\frac{2}{3\lambda}\left\{3\mathfrak{Q}_1\left(\frac{r}{k}\right) - \frac{\lambda\omega^2 k^2 r}{(1+\lambda)M}\right\}\frac{z}{k}\right.$$

$$-\frac{x^2}{k^2(\nu_0^2 - 1/\kappa^2)}\left\{\mathsf{A}_1^1 + \frac{(\nu_0^2 - 1/\kappa^2)}{3\lambda}\frac{k^3}{r^3}\left[1 + \frac{3(3+\kappa^2)}{14\kappa^2}\frac{k^2}{r^2} + \frac{5(5+2\kappa^2+\kappa^4)}{56\kappa^4}\frac{k^4}{r^4}\cdots\right]\right\}$$

$$-\frac{y^2}{k^2(\nu_0^2 - 1)}\left\{\mathfrak{A}_1^1 + \frac{(\nu_0^2-1)}{3\lambda}\frac{k^3}{r^3}\left[1 + \frac{3(3\kappa^2+1)}{14\kappa^2}\frac{k^2}{r^2} + \frac{5(5\kappa^4+2\kappa^2+1)}{56\kappa^4}\frac{k^4}{r^4}\cdots\right]\right.$$

$$\left. - \frac{\omega^2 k^3}{3M}(\nu_0^2 - 1)\right\}$$

$$\left.-\frac{z^2}{k^2\nu_0^2}\left\{\mathfrak{A}_1 - \frac{2\nu_0^2}{3\lambda}\frac{k^3}{r^3}\left[1 + \frac{3(1+\kappa^2)}{7\kappa^2}\frac{k^2}{r^2} + \frac{5(3+2\kappa^2+3\kappa^4)}{56\kappa^4}\frac{k^4}{r^4}\cdots\right] - \frac{\omega^2 k^3}{3M}\nu_0^2\right\}\right]$$

The condition that the figure of equilibrium should be the ellipsoid of reference is that this potential when equated to a constant should reproduce the equation to the ellipsoid. The coefficient of z must therefore vanish, and the three coefficients written inside { } must be equal to one another. These conditions give the angular velocity and equations for determining the figure, but as the subject will be reconsidered from a different point of view hereafter, I do not pursue the investigation here.

At present it need only be noted that the coefficient of z vanishes, and that the three coefficients are equal to one another. It is clear then that the potential U of the system, as rendered statical, may be written

$$U = -\frac{3M}{2k}\left\{\mathsf{A}_1^1 + \frac{(\nu_0^2 - 1/\kappa^2)}{3\lambda}\frac{k^3}{r^3}\left[1 + \frac{3(3+\kappa^2)}{14\kappa^2}\frac{k^2}{r^2} + \frac{5(5+2\kappa^2+\kappa^4)}{56\kappa^4}\frac{k^4}{r^4}\cdots\right]\right\}$$

$$\times\left\{\frac{x^2}{k^2(\nu_0^2 - 1/\kappa^2)} + \frac{y^2}{k^2(\nu_0^2-1)} + \frac{z^2}{k^2\nu_0^2}\right\}$$

Now gravity g at the surface of the ellipsoid is $-dU/dn$, where n is the outward normal to the ellipsoid.

Hence $$g = -\left[\frac{px}{k^2(\nu_0^2 - 1/\kappa^2)}\frac{\partial U}{\partial x} + \frac{py}{k^2(\nu_0^2 - 1)}\frac{\partial U}{\partial y} + \frac{pz}{k^2\nu_0^2}\frac{\partial U}{\partial z}\right]$$

But $$\frac{1}{p^2} = \frac{x^2}{k^4(\nu_0^2 - 1/\kappa^2)^2} + \frac{y^2}{k^4(\nu_0^2 - 1)^2} + \frac{z^2}{k^4\nu_0^4}$$

and in our alternative notation

$$\nu_0^2 - \frac{1}{\kappa^2} = \frac{\cos^2\gamma}{\sin^2\beta}$$

Therefore

$$g = \frac{3M}{pk}\left\{\mathsf{A}_1{}^1 + \frac{k^3\cos^2\gamma}{3\lambda r^3\sin^2\beta}\left[1 + \frac{3(3+\kappa^2)}{14\kappa^2}\frac{k^2}{r^2} + \frac{5(5+2\kappa^2+\kappa^4)}{56\kappa^4}\frac{k^4}{r^4} + \dots\right]\right\} \quad \dots(13)$$

As already remarked, we shall put $M = \frac{4}{3}\pi\rho a^3 \dfrac{\lambda}{1+\lambda}$; also, since the three axes of the ellipsoid are $k\cos\gamma\operatorname{cosec}\beta$, $k\cos\beta\operatorname{cosec}\beta$, $k\operatorname{cosec}\beta$, we have

$$\frac{k^3\cos\beta\cos\gamma}{\sin^3\beta} = \frac{\lambda a^3}{1+\lambda}$$

Hence $$\frac{k^3\cos^2\gamma}{3\lambda r^3\sin^2\beta} = \frac{a^3\cos\gamma\tan\beta}{3(1+\lambda)r^3}$$

and the coefficient of the series in the expression for g does not become infinite when λ vanishes; however, it is perhaps more convenient to leave it in the form as written above.

This expression for gravity is the result required, but it is to be noted that it is determined on the hypothesis that the distant body is a particle or sphere instead of being an ellipsoid.

The development ceases with terms of the seventh order, and the harmonic terms of third and higher orders have been neglected. Now the harmonic deformation of Roche's ellipsoid of the third order of harmonics is of order k^4/r^4 in inverse powers of r. This deformation is treated as surface density. If we were to proceed to closer approximation, we should have to take account of the square of the thickness of the layer; such terms would be of order k^8/r^8. Since, then, we are avowedly neglecting terms of this order, it is no use to carry the development higher than terms of the seventh order.

§ 6. *Form of the Expression for the Gravitational Lost Energy of the System.*

The system consists of two ellipsoids, say e and E, with their longest axes co-linear, and each of them is distorted by deformations expressible by ellipsoidal harmonics of orders higher than the second. To the order of approximation to be adopted these deformations may be replaced by layers of surface density, which may be denoted by l and L respectively.

The lost energy of the system may be represented symbolically by

$$V = \tfrac{1}{2}(e+l)^2 + \tfrac{1}{2}(E+L)^2 + (e+l)(E+L)$$

Let s, S denote two spheres of masses equal to e and E and concentric with them respectively.

Then the whole may be written

$$V = \tfrac{1}{2}ee + \tfrac{1}{2}EE + eE + \tfrac{1}{2}ll + \tfrac{1}{2}LL + (e+S)\,l + (E+s)\,L$$
$$+ (e-s)\,L + (E-S)\,l + lL$$

In the term Sl I divide S into two parts, namely S_1, which is to contain all the terms in the potential of S at the surface of e excepting terms expressible by ellipsoidal harmonics of the second order with respect to the ellipsoid e; and S_2, which is to contain the omitted terms of the second order. Similarly, the term sL is to be divided into s_1L and s_2L.

Take the centre of e as origin of co-ordinates x, y, z with the z axis passing through the centre of E, the y axis coincident with the mean axis of e and the x axis coincident with the least axis of e.

Since l is expressible by harmonics higher than the second order, and since y^2+z^2 is expressible by harmonics of orders 0 and 2, it follows that the moment of inertia of the layer l about the axis is zero. If therefore ω is the angular velocity of the system, a contribution to the lost energy of the system which may be written symbolically $[\tfrac{1}{2}\omega^2(y^2+z^2)]\,l$ is zero.

It follows therefore that we may write

$$(e+S)\,l = [e + S_2 + \tfrac{1}{2}\omega^2(y^2+z^2)]\,l + S_1 l$$

Similarly, if the ellipsoid E be referred to a parallel co-ordinate system X, Y, Z through its centre, and such that

$$x = X, \quad y = Y, \quad z = Z + r$$

so that r is the distance between the two origins, we have

$$(E+s)\,L = [E + s_2 + \tfrac{1}{2}\omega^2(Y^2+Z^2)]\,L + s_1 L$$

The problem is already so complicated that it will be convenient to omit certain small terms in the expression for the lost energy, which it would be very troublesome to evaluate.

The term $(e-s)L$ represents the mutual energy of the departure from sphericity of e with the layer of surface density L on E. This term is clearly very small and will be omitted. Similarly $(E-S)l$ will be neglected. It will appear from the results below that these terms are at least of the seventh order in powers of $1/r$. *À fortiori* lL, which is at least of the eighth order, will be omitted.

The whole expression for V will now be divided into several portions.

Let $(eE)_1$ be that portion of eE in which each ellipsoid may be replaced by a particle; it is, in fact, the product of the masses of e and E divided by r.

Let $(eE)_2$ be the rest of eE.

Let (vv) denote that portion of V in which the larger body E may be replaced by a sphere; then

$$(vv) = \tfrac{1}{2}ee + \tfrac{1}{2}ll + [e + S_2 + \tfrac{1}{2}\omega^2(y^2+z^2)]\, l + S_1 l$$

Similarly, let

$$(VV) = \tfrac{1}{2}EE + \tfrac{1}{2}LL + [E + s_2 + \tfrac{1}{2}\omega^2(Y^2+Z^2)]\, L + s_1 L$$

Then $V = (eE)_1 + (eE)_2 + (vv) + (VV) +$ neglected terms.

For Roche's problem, when the second body is a particle, V reduces to $(eE)_1 + (vv)$, but in the modified form of the problem which I am going to solve the whole expression is required.

The evaluation of (eE) is so complicated that I devote a special section to it.

§ 7. *The Mutual Energy of Two Ellipsoids**.

The semi-axes of the ellipsoid of mass e are to be denoted a, b, c, and the corresponding notation for the other ellipsoid is E, A, B, C. The distance between the centres of e and E is r; the axes c of e and C of E are in the same straight line, while a, A and b, B are respectively parallel. For brevity I imagine the densities of the ellipsoids to be unity.

If the external potential of e be U, and if $d\Omega$ be an element of volume or of mass of E, the lost energy to be evaluated is

$$(eE) = \int U d\Omega$$

integrated throughout the ellipsoid E.

* The results of this section were arrived at originally by a longer method. I have to thank one of the referees for showing me the following procedure.

Let us suppose provisionally that the co-ordinates of the centres of e and E are x, y, z and X, Y, Z, and let ξ, η, ζ be the co-ordinates of the element $d\Omega$; the axes being respectively parallel to a, b, c or A, B, C with arbitrary origin.

If $R^2 = (\xi - x)^2 + (\eta - y)^2 + (\zeta - z)^2$, it is well known that the potential U of e at the point ξ, η, ζ is given by

$$U = e\sum_0^\infty \frac{3}{(2n+1)(2n+3)\,2n!}\left(a^2\frac{\partial^2}{\partial\xi^2} + b^2\frac{\partial^2}{\partial\eta^2} + c^2\frac{\partial^2}{\partial\zeta^2}\right)^n \frac{1}{R}$$

Since $\frac{1}{R}$ satisfies Laplace's equation, we may eliminate $\frac{\partial^2}{\partial\zeta^2}$, and observing that $\frac{\partial^2}{\partial x^2}, \frac{\partial^2}{\partial y^2}$ are the same as $\frac{\partial^2}{\partial\xi^2}, \frac{\partial^2}{\partial\eta^2}$ respectively, we have

$$a^2\frac{\partial^2}{\partial\xi^2} + b^2\frac{\partial^2}{\partial\eta^2} + c^2\frac{\partial^2}{\partial\zeta^2} = -(c^2 - a^2)\frac{\partial^2}{\partial\xi^2} - (c^2 - b^2)\frac{\partial^2}{\partial\eta^2}$$

$$= -k^2\left[\frac{1}{\kappa^2}\frac{\partial^2}{\partial x^2} + \frac{\partial^2}{\partial y^2}\right]$$

It follows therefore that

$$U = e\sum_0^\infty \frac{(-)^n\,3}{(2n+1)(2n+3)\,2n!}\,k^{2n}\left(\frac{1}{\kappa^2}\frac{\partial^2}{\partial x^2} + \frac{\partial^2}{\partial y^2}\right)^n \frac{1}{R}$$

Since the operator is independent of ξ, η, ζ, we have

$$(eE) = e\sum_0^\infty \frac{(-)^n\,3}{(2n+1)(2n+3)\,2n!}\,k^{2n}\left(\frac{1}{\kappa^2}\frac{\partial^2}{\partial x^2} + \frac{\partial^2}{\partial y^2}\right)^n \int\frac{d\Omega}{R}$$

But $\int\frac{d\Omega}{R}$ is the potential of the ellipsoid E at the centre of the ellipsoid e, and by an exactly parallel transformation

$$\int\frac{d\Omega}{R} = E\sum_0^\infty \frac{(-)^n\,3}{(2n+1)(2n+3)\,2n!}\,K^{2n}\left(\frac{1}{\mathrm{K}^2}\frac{\partial^2}{\partial x^2} + \frac{\partial^2}{\partial y^2}\right)^n \frac{1}{\rho}$$

where $\rho^2 = (x - X)^2 + (y - Y)^2 + (z - Z)^2$.

Since our co-ordinate axes have a perfectly arbitrary origin, we may at once put $X = 0$, $Y = 0$, $Z = r$, $z = 0$, and after effecting the several differentiations put $x = 0$, $y = 0$.

It follows that, on putting $x = 0$, $y = 0$, after differentiation and writing $\rho^2 = x^2 + y^2 + r^2$,

$$(eE) = eE\sum_0^\infty \frac{(-)^n\,3}{(2n+1)(2n+3)\,2n!}\,k^{2n}\left(\frac{1}{\kappa^2}\frac{\partial^2}{\partial x^2} + \frac{\partial^2}{\partial y^2}\right)^n$$

$$\times \sum_0^\infty \frac{(-)^i\,3}{(2i+1)(2i+3)\,2i!}\,K^{2i}\left(\frac{1}{\mathrm{K}^2}\frac{\partial^2}{\partial x^2} + \frac{\partial^2}{\partial y^2}\right)^i \frac{1}{\rho}$$

If we denote the operator $\frac{1}{\kappa^2}\frac{\partial^2}{\partial x^2}+\frac{\partial^2}{\partial y^2}$ by d^2, and the operator $\frac{1}{\mathrm{K}^2}\frac{\partial^2}{\partial x^2}+\frac{\partial^2}{\partial y^2}$ by D^2, we have

$$(eE)=eE\left[1-\frac{1}{2.5}k^2\mathrm{d}^2+\frac{1}{2^3.5.7}k^4\mathrm{d}^4-\frac{1}{2^4.3^3.5.7}k^6\mathrm{d}^6\ldots\right]$$

$$\times\left[1-\frac{1}{2.5}K^2\mathrm{D}^2+\frac{1}{2^3.5.7}K^4\mathrm{D}^4-\frac{1}{2^4.3^3.5.7}K^6\mathrm{D}^6\ldots\right]\frac{1}{\rho}$$

$$=eE\left[1-\frac{1}{2.5}(k^2\mathrm{d}^2+K^2\mathrm{D}^2)+\frac{1}{2^3.5.7}(k^4\mathrm{d}^4+K^4\mathrm{D}^4)+\frac{1}{2^2.5^2}k^2K^2\mathrm{d}^2\mathrm{D}^2\right.$$

$$\left.-\frac{1}{2^4.3^3.5.7}(k^6\mathrm{d}^6+K^6\mathrm{D}^6)-\frac{1}{2^4.5^2.7}(k^2K^4\mathrm{d}^2\mathrm{D}^4+k^4K^2\mathrm{d}^4\mathrm{D}^2)\ldots\right]\frac{1}{\rho}$$

On effecting the several differentiations, and putting $x=0$, $y=0$, we find

$$\mathrm{d}^2\frac{1}{\rho}=-\frac{1}{r^3}\left(\frac{1}{\kappa^2}+1\right);\qquad \mathrm{d}^4\frac{1}{\rho}=\frac{3}{r^5}\left(\frac{3}{\kappa^4}+\frac{2}{\kappa^2}+3\right)$$

$$\mathrm{D}^2\mathrm{d}^2\frac{1}{\rho}=\frac{3}{r^5}\left(\frac{3}{\kappa^2\mathrm{K}^2}+\frac{1}{\kappa^2}+\frac{1}{\mathrm{K}^2}+3\right)$$

$$\mathrm{d}^6\frac{1}{\rho}=-\frac{3^2.5}{r^7}\left(\frac{5}{\kappa^6}+\frac{3}{\kappa^4}+\frac{3}{\kappa^2}+5\right)$$

$$\mathrm{d}^4\mathrm{D}^2\frac{1}{\rho}=-\frac{3^2.5}{r^7}\left(\frac{5}{\kappa^4\mathrm{K}^2}+\frac{1}{\kappa^4}+\frac{2}{\kappa^2\mathrm{K}^2}+\frac{2}{\kappa^2}+\frac{1}{\mathrm{K}^2}+5\right)$$

and the remaining functions may be found by appropriate changes of small and large letters.

If now we again use ρ to denote the density of the spheroids, and revert to the notation employed elsewhere, namely,

$$e=\tfrac{4}{3}\pi\rho a^3\frac{\lambda}{1+\lambda},\qquad E=\tfrac{4}{3}\pi\rho a^3\frac{1}{1+\lambda}$$

we find

$$(eE)=(\tfrac{4}{3}\pi\rho a^3)^2\frac{\lambda}{(1+\lambda)^2}\left\{1+\frac{1}{2.5r^3}\left[k^2\left(\frac{1}{\kappa^2}+1\right)+K^2\left(\frac{1}{\mathrm{K}^2}+1\right)\right]\right.$$

$$+\frac{3}{2^3.5.7r^5}\left[k^4\left(\frac{3}{\kappa^4}+\frac{2}{\kappa^2}+3\right)+K^4\left(\frac{3}{\mathrm{K}^4}+\frac{2}{\mathrm{K}^2}+3\right)\right]$$

$$+\frac{3}{2^2.5^2r^5}\left[k^2K^2\left(\frac{3}{\kappa^2\mathrm{K}^2}+\frac{1}{\kappa^2}+\frac{1}{\mathrm{K}^2}+3\right)\right]$$

$$+\frac{1}{2^4.3.7r^7}\left[k^6\left(\frac{5}{\kappa^6}+\frac{3}{\kappa^4}+\frac{3}{\kappa^2}+5\right)+K^6\left(\frac{5}{\mathrm{K}^6}+\frac{3}{\mathrm{K}^4}+\frac{3}{\mathrm{K}^2}+5\right)\right]$$

$$+\frac{9}{2^4.5.7r^7}\left[k^2K^4\left(\frac{5}{\kappa^2\mathrm{K}^4}+\frac{1}{\mathrm{K}^4}+\frac{2}{\kappa^2\mathrm{K}^2}+\frac{2}{\mathrm{K}^2}+\frac{1}{\kappa^2}+5\right)\right.$$

$$\left.\left.+k^4K^2\left(\frac{5}{\kappa^4\mathrm{K}^2}+\frac{1}{\kappa^4}+\frac{2}{\kappa^2\mathrm{K}^2}+\frac{2}{\kappa^2}+\frac{1}{\mathrm{K}^2}+5\right)\right]\right\}\ldots(14)$$

The first term in this expression is that which was called above $(eE)_1$, and the rest constitutes $(eE)_2$.

If the body E were a sphere, the only portions of (14) which would remain would be the parts of the expression independent of K.

With the object of effecting certain differentiations hereafter, it is desirable that the formula for (eE) should be expressed in terms of the semi-axes a, b, c and A, B, C.

In accordance with the notation used elsewhere, we have

$$a = \frac{k\cos\gamma}{\sin\beta}, \quad b = \frac{k\cos\beta}{\sin\beta}, \quad c = \frac{k}{\sin\beta}, \quad \text{where } \sin\beta = \kappa\sin\gamma$$

$$A = \frac{K\cos\Gamma}{\sin\mathrm{B}}, \quad B = \frac{K\cos\mathrm{B}}{\sin\mathrm{B}}, \quad C = \frac{K}{\sin\mathrm{B}}, \quad \text{where } \sin\mathrm{B} = \mathrm{K}\sin\Gamma^*$$

The result of the translation into this other notation is as follows:—

$$(eE) = (\tfrac{4}{3}\pi\rho a^3)^2 \frac{\lambda}{(1+\lambda)^2}\left\{1 + \frac{1}{2.5r^3}[2c^2 - a^2 - b^2 + \text{same in } A, B, C]\right.$$

$$+ \frac{3}{2^3.5.7r^5}[3(a^4+b^4) + 8c^4 - 8c^2(a^2+b^2) + 2a^2b^2 + \text{same in } A, B, C]$$

$$+ \frac{3}{2^2.5^2r^5}[2(A^2a^2 + B^2b^2 + C^2c^2) + (A^2+B^2+C^2)(a^2+b^2+c^2)$$

$$- 5C^2(a^2+b^2) - 5c^2(A^2+B^2) + 5C^2c^2]$$

$$+ \frac{1}{2^4.3.7r^7}[16c^6 - 5(a^6+b^6) - 24c^4(a^2+b^2) + 18c^2(a^4+b^4)$$

$$- 3a^2b^2(a^2+b^2) + 12a^2b^2c^2 + \text{same in } A, B, C]$$

$$+ \frac{9}{2^4.5.7r^7}[-A^4(5a^2+b^2-6c^2) - B^4(a^2+5b^2-6c^2)$$

$$- 8C^4(a^2+b^2-2c^2) + 4B^2C^2(a^2+3b^2-4c^2)$$

$$+ 4C^2A^2(3a^2+b^2-4c^2) - 2A^2B^2(a^2+b^2+2c^2)$$

$$\left. + \text{same with small and large letters interchanged}]\right\} \quad \text{......(15)}$$

* The fact that capital β is nearly the same as B must be pardoned; it cannot, I think, cause any confusion.

§ 8. *Remaining Terms in the Expression for the Lost Energy.*

If e be the mass of an ellipsoid of semi-axes a, b, c, the lost energy of its concentration is

$$\tfrac{1}{2}(ee) = \tfrac{3}{10}e^2\psi$$

where
$$\psi = \int_0^\infty \frac{du}{(u+a^2)^{1/2}(u+b^2)^{1/2}(u+c^2)^{1/2}}$$

In the present case
$$e = \tfrac{4}{3}\pi\rho a^3 \frac{\lambda}{1+\lambda}$$

and
$$\psi = \frac{2}{k}\int_{\nu_0}^\infty \frac{d\nu}{(\nu^2-1)^{1/2}(\nu^2-1/\kappa^2)^{1/2}}.$$

where
$$\nu_0 = \frac{1}{\kappa\sin\gamma} = \frac{1}{\sin\beta}$$

Thus
$$\tfrac{1}{2}(ee) = \tfrac{3}{10}(\tfrac{4}{3}\pi\rho a^3)^2 \frac{\lambda^2}{(1+\lambda)^2}\psi \quad \text{.................(16)}$$

By symmetry
$$\tfrac{1}{2}(EE) = \tfrac{3}{10}(\tfrac{4}{3}\pi\rho a^3)^2 \frac{1}{(1+\lambda)^2}\Psi \quad \text{.................(17)}$$

where
$$\Psi = \frac{2}{K}\int_{N_0}^\infty \frac{dN}{(N^2-1)^{1/2}(N^2-1/K^2)^{1/2}}$$

and
$$N_0 = \frac{1}{K\sin\Gamma} = \frac{1}{\sin B}$$

The lost energy (lS_1) is the potential of a particle S, equal in mass to E placed at the centre of E, with the omission of terms of the second order of harmonics, multiplied by the density of the layer l and integrated over the surface of e. This is the same as the potential of the layer l, with the omission of harmonic terms of the second order (and there are none such) at the centre of E multiplied by the mass of E.

A typical term in the surface density representing the layer l is, say,

$$f_i^s \mathfrak{P}_i^s(\mu)\,\mathfrak{C}_i^s(\phi)$$

The external potential corresponding to such a term, at the point ν, μ, ϕ, is by (51) of "Harmonics,"

$$\frac{3}{k}\left(\tfrac{4}{3}\pi\rho a^3 \frac{\lambda}{1+\lambda}\right) f_i^s \mathfrak{Q}_i^s(\nu)\,\mathfrak{P}_i^s(\nu_0)\,\mathfrak{P}_i^s(\mu)\,\mathfrak{C}_i^s(\phi)$$

The co-ordinates of the centre of E are $\nu = \dfrac{r}{k}$, $\mu = 1$, $\phi = \tfrac{1}{2}\pi$; and the mass of E is $\tfrac{4}{3}\pi\rho a^3/(1+\lambda)$.

Hence the contribution to (lS_1) corresponding to this term is

$$\frac{3}{k}(\tfrac{4}{3}\pi\rho a^3)^2 \frac{\lambda}{(1+\lambda)^2} f_i^s \mathfrak{Q}_i^s(r/k)\, \mathfrak{P}_i^s(\nu_0)\, \mathfrak{P}_i^s(1)\, \mathfrak{C}_i^s(\tfrac{1}{2}\pi)$$

Now $\mathfrak{P}_i^s(1)\, \mathfrak{C}_i^s(\tfrac{1}{2}\pi)$ vanishes for all harmonics except cosine-harmonics of even rank. Therefore

$$(lS_1) = \frac{3}{k}(\tfrac{4}{3}\pi\rho a^3)^2 \frac{\lambda}{(1+\lambda)^2} \Sigma f_i^s \mathfrak{Q}_i^s(r/k)\, \mathfrak{P}_i^s(\nu_0)\, \mathfrak{P}_i^s(1) \begin{Bmatrix} \mathfrak{C}_i^s \\ \mathfrak{S}_i^s \end{Bmatrix} (\tfrac{1}{2}\pi) \begin{Bmatrix} i \text{ even} \\ i \text{ odd} \end{Bmatrix} \quad (18)$$

the summation being for all values of i greater than 2, and for all even values of s.

The lost energy (Ls_1) is expressible by the similar function for the other ellipsoid, but I have not adopted a specific notation for the ellipsoidal harmonics for the ellipsoid E, and therefore cannot write down the result.

The lost energy $[e + S_2 + \tfrac{1}{2}\omega^2(y^2+z^2)]\, l$ is the potential of the ellipsoid e, together with the potential of S in as far as it involves harmonics of the second order, and a rotation potential, multiplied by the density of the layer l, and integrated over the surface of e. That is to say it is the potential of gravity on the ellipsoid e integrated throughout the layer l, which it is not permissible to regard as surface density.

If the thickness of the layer be ζ, and if $d\zeta$ be a slice of that part of the layer which is erected normally on an element $d\sigma$ of the surface e, then $\rho d\zeta d\sigma$ is an element of mass of the layer. The potential of gravity is $-g\zeta$. Hence the lost energy is

$$-\rho \iint_0^{\zeta} g\zeta d\zeta d\sigma = -\tfrac{1}{2}\rho \int g\zeta^2 d\sigma$$

Accordingly the lost energy is equal and opposite to the work done in raising the layer, considered as surface density, through half its thickness, against gravity.

We may take as a typical term

$$\zeta = p f_i^s \mathfrak{P}_i^s(\mu)\, \mathfrak{C}_i^s(\phi)$$

and we have shown in (13) that

$$g = \frac{3}{pk}\tfrac{4}{3}\pi\rho a^3 \frac{\lambda}{1+\lambda}\left\{\mathsf{A}_1{}^1 + \frac{k^3}{3\lambda r^3}\frac{\cos^2\gamma}{\cos^2\beta}\left[1 + \frac{3(3+\kappa^2)}{14\kappa^2}\frac{k^2}{r^2} + \frac{5(5+2\kappa^2+\kappa^4)}{56\kappa^4}\frac{k^4}{r^4}\cdots\right]\right\}$$

It should be noted that this expression for gravity takes no account of the change in the ellipticity of e which is due to the fact that E is an ellipsoid and not a sphere. The error introduced thus is however outside the limits of accuracy which have been adopted.

Accordingly this portion of the lost energy is

$$-\frac{3}{2k}\tfrac{4}{3}\pi\rho^2 a^3 \frac{\lambda}{1+\lambda}(f_i^s)^2 \{\mathsf{A}_1{}^1 + \text{series}\} \int [\mathfrak{P}_i^s(\mu)\, \mathfrak{C}_i^s(\phi)]^2 p\, d\sigma$$

Now in § 5 we defined

$$\mathfrak{T}_i^s = \frac{(2i+1)}{4\pi a^3 \lambda/(1+\lambda)} \int [\mathfrak{P}_i^s(\mu)\, \mathfrak{C}_i^s(\phi)]^2 p\, d\sigma$$

Hence

$$[e + S_2 + \tfrac{1}{2}\omega^2(y^2+z^2)]\, l = -\frac{9}{2k}(\tfrac{4}{3}\pi\rho a^3)^2 \frac{\lambda^2}{(1+\lambda)^2} \Sigma \frac{(f_i^s)^2}{2i+1} \{\mathsf{A}_1{}^1 + \text{series}\}\, \mathfrak{T}_i^s \quad \ldots\ldots(19)$$

the summation being for all harmonics.

In determining the lost energy $\frac{1}{2}(ll)$ we may treat the layer as surface density. A typical term in the surface density is $\rho p f_i^s \mathfrak{P}_i^s(\mu)\, \mathfrak{C}_i^s(\phi)$, and the surface value of its potential is

$$\frac{3}{k}(\tfrac{4}{3}\pi\rho a^3) \frac{\lambda}{1+\lambda} f_i^s \mathfrak{P}_i^s(\nu_0)\, \mathfrak{Q}_i^s(\nu_0)\, \mathfrak{P}_i^s(\mu)\, \mathfrak{C}_i^s(\phi)$$

Then, since $\mathfrak{A}_i^s = \mathfrak{P}_i^s(\nu_0)\, \mathfrak{Q}_i^s(\nu_0)$

a typical term of $\frac{1}{2}(ll)$ is

$$\frac{3\rho}{2k}(\tfrac{4}{3}\pi\rho a^3) \frac{\lambda}{1+\lambda} (f_i^s)^2 \mathfrak{A}_i^s \int [\mathfrak{P}_i^s(\mu)\, \mathfrak{C}_i^s(\phi)]^2 p\, d\sigma$$

Thus
$$\tfrac{1}{2}(ll) = \frac{9}{2k}(\tfrac{4}{3}\pi\rho a^3)^2 \frac{\lambda^2}{(1+\lambda)^2} \Sigma \frac{(f_i^s)^2}{2i+1} \mathfrak{A}_i^s \mathfrak{T}_i^s \quad \ldots\ldots\ldots\ldots(20)$$

the summation being made for all harmonics.

The value of $\frac{1}{2}(LL)$ may be written down by symmetry.

§ 9. *Final Expression for the Lost Energy of the System.*

We have
$$V = (eE)_1 + (vv) + (VV) + (eE)_2$$

The several parts are to be collected from (14) or (15), (16), (17), (18), (19), (20), and we have

$$(eE)_1 = (\tfrac{4}{3}\pi\rho a^3)^2 \frac{\lambda}{(1+\lambda)^2 r}$$

$$(vv) = (\tfrac{4}{3}\pi\rho a^3)^2 \frac{\lambda}{(1+\lambda)^2} \Big\{ \tfrac{3}{10}\lambda\psi + \frac{3}{k}\Sigma f_i^s \mathfrak{Q}_i^s(r/k)\, \mathfrak{P}_i^s(\nu_0)\, \mathfrak{P}_i^s(1)\, \mathfrak{C}_i^s(\tfrac{1}{2}\pi)$$
$$(i > 2,\ s \text{ even})$$
$$+ \tfrac{9}{2}\frac{\lambda}{k}\Sigma \frac{(f_i^s)^2 \mathfrak{T}_i^s}{2i+1} \Big[\mathfrak{A}_i^s - \mathsf{A}_1{}^1 - \frac{k^3 \cos^2\gamma}{3\lambda r^3 \sin^2\beta} \Big(1 + \frac{3(3+\kappa^2)}{14\kappa^2}\frac{k^2}{r^2}$$
$$+ \frac{5(5+2\kappa^2+\kappa^4)}{56\kappa^4}\frac{k^4}{r^4} \Big) \Big] \text{(all harmonics)} \Big\}$$

(VV) = symmetrical expression with $1/\lambda$ in place of λ,

$$(eE)_2 = (\tfrac{4}{3}\pi\rho a^3)^2 \frac{\lambda}{(1+\lambda)^2} \left\{ \frac{1}{2 \cdot 5r^3} \left[k^2 \left(\frac{1}{\kappa^2} + 1 \right) + K^2 \left(\frac{1}{\mathrm{K}^2} + 1 \right) \right] \right.$$

$$+ \frac{3}{2^3 \cdot 5 \cdot 7r^5} \left[k^4 \left(\frac{3}{\kappa^4} + \frac{2}{\kappa^2} + 3 \right) + K^4 \left(\frac{3}{\mathrm{K}^4} + \frac{2}{\mathrm{K}^2} + 3 \right) \right]$$

$$+ \frac{3}{2^2 \cdot 5^2 r^5} k^2 K^2 \left(\frac{3}{\kappa^2 \mathrm{K}^2} + \frac{1}{\mathrm{K}^2} + \frac{1}{\kappa^2} + 3 \right)$$

$$+ \frac{1}{2^4 \cdot 3 \cdot 7r^7} \left[k^6 \left(\frac{5}{\kappa^6} + \frac{3}{\kappa^4} + \frac{3}{\kappa^2} + 5 \right) + K^6 \left(\frac{5}{\mathrm{K}^6} + \frac{3}{\mathrm{K}^4} + \frac{3}{\mathrm{K}^2} + 5 \right) \right]$$

$$+ \frac{9}{2^4 \cdot 5 \cdot 7r^7} \left[k^2 K^4 \left(\frac{5}{\kappa^2 \mathrm{K}^4} + \frac{1}{\mathrm{K}^4} + \frac{2}{\kappa^2 \mathrm{K}^2} + \frac{2}{\mathrm{K}^2} + \frac{1}{\kappa^2} + 5 \right) \right.$$

$$\left.\left. + k^4 K^2 \left(\frac{5}{\mathrm{K}^2 \kappa^4} + \frac{1}{\kappa^4} + \frac{2}{\mathrm{K}^2 \kappa^2} + \frac{2}{\kappa^2} + \frac{1}{\mathrm{K}^2} + 5 \right) \right] \right\} \quad \text{.......(21)}$$

§ 10. *Determination of the Forms of the Ellipsoids.*

We have obtained in the last section the expression for V, the lost energy of the system.

The harmonic deformations of the ellipsoids being of orders higher than the second do not enter into the moment of inertia to the order of approximation adopted. Hence the moment of inertia about the axis of rotation, which passes through the centre of inertia of the system, and is parallel to the a and A axes of the ellipsoids, is given by

$$I = \tfrac{4}{3}\pi\rho a^3 \left[\tfrac{1}{5} \frac{\lambda}{1+\lambda} (b^2 + c^2) + \tfrac{1}{5} \frac{1}{1+\lambda} (B^2 + C^2) + \frac{\lambda r^2}{(1+\lambda)^2} \right] \quad \text{....(22)}$$

If f denotes any one of the parameters by which the system is defined, the condition that the figures shall be in equilibrium is

$$\frac{\partial V}{\partial f} + \tfrac{1}{2}\omega^2 \frac{\partial I}{\partial f} = 0$$

The parameters defining the system may be taken as r, the distance between the two centres, $\cos\gamma$, $\cos\beta$ for the smaller ellipsoid and $\cos\Gamma$, $\cos \mathrm{B}$ for the larger one. Besides these we have the coefficients f_i^s, F_i^s of the harmonic inequalities of order i and rank s on the two ellipsoids.

For convenience write

$$\mathfrak{a} = \cos\gamma, \qquad \mathfrak{b} = \cos\beta$$

These letters are chosen on account of the association of $\cos\gamma$, $\cos\beta$ with the semi-axes a, b of the smaller ellipsoid e. It is unnecessary to adopt a corresponding notation for the ellipsoid E, because, when the problem is solved as regards e, it affords the solution for E by symmetry.

Since
$$k^3 \cos\beta \cos\gamma \operatorname{cosec}^3\beta = \lambda \mathfrak{a}^3/(1+\lambda)$$
we have
$$\frac{k}{\mathfrak{a}} = \left(\frac{\lambda}{1+\lambda}\right)^{1/3} \frac{(1-\mathfrak{b}^2)^{1/2}}{(\mathfrak{a}\mathfrak{b})^{1/3}}$$

Hence

$$\frac{a}{\mathfrak{a}} = \frac{k}{\mathfrak{a}}\frac{\cos\gamma}{\sin\beta} = \left(\frac{\lambda}{1+\lambda}\right)^{1/3}\frac{\mathfrak{a}^{2/3}}{\mathfrak{b}^{1/3}}, \quad \frac{b}{\mathfrak{a}} = \frac{k\cos\beta}{\mathfrak{a}\sin\beta} = \left(\frac{\lambda}{1+\lambda}\right)^{1/3}\frac{\mathfrak{b}^{2/3}}{\mathfrak{a}^{1/3}}$$
$$\frac{c}{\mathfrak{a}} = \frac{k}{\mathfrak{a}\sin\beta} = \left(\frac{\lambda}{1+\lambda}\right)^{1/3}\frac{1}{(\mathfrak{a}\mathfrak{b})^{1/3}}$$

Therefore
$$3\frac{da}{a} = 2\frac{d\mathfrak{a}}{\mathfrak{a}} - \frac{d\mathfrak{b}}{\mathfrak{b}}, \quad 3\frac{db}{b} = -\frac{d\mathfrak{a}}{\mathfrak{a}} + 2\frac{d\mathfrak{b}}{\mathfrak{b}}, \quad 3\frac{dc}{c} = -\frac{d\mathfrak{a}}{\mathfrak{a}} - \frac{d\mathfrak{b}}{\mathfrak{b}}$$

Therefore
$$3\mathfrak{a}\frac{\partial}{\partial\mathfrak{a}} = 2a\frac{\partial}{\partial a} - b\frac{\partial}{\partial b} - c\frac{\partial}{\partial c}, \quad 3\mathfrak{b}\frac{\partial}{\partial\mathfrak{b}} = -a\frac{\partial}{\partial a} + 2b\frac{\partial}{\partial b} - c\frac{\partial}{\partial c} \quad \ldots(23)$$

These enable us to differentiate, with respect to $\mathfrak{a}$, $\mathfrak{b}$, functions expressed in terms of a, b, c; the parameters r, f_i^s always occur explicitly.

The equation of condition for the parameter r is
$$\frac{\partial V}{\partial r} + \tfrac{1}{2}\omega^2\frac{\partial I}{\partial r} = 0$$

On differentiating (22) we have
$$\tfrac{1}{2}\omega^2\frac{\partial I}{\partial r} = \tfrac{4}{3}\pi\rho\mathfrak{a}^3\frac{\lambda\omega^2 r}{(1+\lambda)^2}$$

In order to differentiate V we must take separately its several portions as defined in (21).

Now
$$-\frac{\partial}{\partial r}(eE)_1 = (\tfrac{4}{3}\pi\rho\mathfrak{a}^3)^2\frac{\lambda}{(1+\lambda)^2 r^2}$$

$$-\frac{\partial}{\partial r}(vv) = (\tfrac{4}{3}\pi\rho\mathfrak{a}^3)^2\frac{\lambda}{(1+\lambda)^2}\left\{-\frac{3}{k}\Sigma f_i^s \mathfrak{P}_i^s(\nu_0)\,\mathfrak{P}_i^s(1)\,\mathfrak{C}_i^s(\tfrac{1}{2}\pi)\frac{d}{dr}\mathfrak{Q}_i^s\left(\frac{r}{k}\right)\right.$$
$$(i > 2,\ s \text{ even})$$
$$+\tfrac{9}{2}\frac{k^2}{r^4}\Sigma\frac{(f_i^s)^2\mathfrak{T}_i^s}{2i+1}\frac{\cos^2\gamma}{\sin^2\beta}\left[1 + \frac{5(3+\kappa^2)}{14\kappa^2}\frac{k^2}{r^2}\right.$$
$$\left.\left.+\frac{5(5+2\kappa^2+\kappa^4)}{56\kappa^4}\frac{k^4}{r^4}\cdots\right]\text{(all harmonics)}\right\}$$

$$-\frac{\partial}{\partial r}(VV) = \text{symmetrical expression for larger ellipsoid,}$$

$$-\frac{\partial}{\partial r}(eE)_2 = (\tfrac{4}{3}\pi\rho a^3)^2 \frac{\lambda}{(1+\lambda)^2}\left\{\frac{3}{2.5r^4}\left[k^2\left(\frac{1}{\kappa^2}+1\right)+K^2\left(\frac{1}{\mathrm{K}^2}+1\right)\right]\right.$$

$$+\frac{3}{2^3.7r^6}\left[k^4\left(\frac{3}{\kappa^4}+\frac{2}{\kappa^2}+3\right)+K^4\left(\frac{3}{\mathrm{K}^4}+\frac{2}{\mathrm{K}^2}+3\right)\right]$$

$$+\frac{3}{2^2.5r^6}k^2K^2\left(\frac{3}{\kappa^2\mathrm{K}^2}+\frac{1}{\kappa^2}+\frac{1}{\mathrm{K}^2}+3\right)$$

$$+\frac{1}{2^4.3r^8}\left[k^6\left(\frac{5}{\kappa^6}+\frac{3}{\kappa^4}+\frac{3}{\kappa^2}+5\right)+K^6\left(\frac{5}{\mathrm{K}^6}+\frac{3}{\mathrm{K}^4}+\frac{3}{\mathrm{K}^2}+5\right)\right]$$

$$+\frac{9}{2^4.5r^8}\left[k^2K^4\left(\frac{5}{\kappa^2\mathrm{K}^4}+\frac{1}{\mathrm{K}^4}+\frac{2}{\kappa^2\mathrm{K}^2}+\frac{2}{\mathrm{K}^2}+\frac{1}{\kappa^2}+5\right)\right.$$

$$\left.\left.+k^4K^2\left(\frac{5}{\mathrm{K}^2\kappa^4}+\frac{1}{\kappa^4}+\frac{2}{\kappa^2\mathrm{K}^2}+\frac{2}{\kappa^2}+\frac{1}{\mathrm{K}^2}+5\right)\right]\right\}$$

The sum of these last four expressions is equal to $-\partial V/\partial r$, and therefore equal to $\frac{1}{2}\omega^2\partial I/\partial r$.

Now let $$\omega^2 r^3 = \tfrac{4}{3}\pi\rho a^3(1+\zeta)$$
so that ζ represents the correction to Kepler's law of periodic times on account of the ellipticities of the two bodies. Then we have

$$\zeta = \frac{3}{10r^2}\left[k^2\left(\frac{1}{\kappa^2}+1\right)+K^2\left(\frac{1}{\mathrm{K}^2}+1\right)\right]+\frac{3}{56r^4}\left[k^4\left(\frac{3}{\kappa^4}+\frac{2}{\kappa^2}+3\right)\right.$$

$$\left.+K^4\left(\frac{3}{\mathrm{K}^4}+\frac{2}{\mathrm{K}^2}+3\right)\right]$$

$$+\frac{3}{20r^4}k^2K^2\left(\frac{3}{\kappa^2\mathrm{K}^2}+\frac{1}{\mathrm{K}^2}+\frac{1}{\kappa^2}+3\right)+\frac{1}{48r^6}\left[k^6\left(\frac{5}{\kappa^6}+\frac{3}{\kappa^4}+\frac{3}{\kappa^2}+5\right)\right.$$

$$\left.+K^6\left(\frac{5}{\mathrm{K}^6}+\frac{3}{\mathrm{K}^4}+\frac{3}{\mathrm{K}^2}+5\right)\right]$$

$$+\frac{9}{80r^6}\left[k^2K^4\left(\frac{5}{\kappa^2\mathrm{K}^4}+\frac{1}{\mathrm{K}^4}+\frac{2}{\mathrm{K}^2\kappa^2}+\frac{2}{\mathrm{K}^2}+\frac{1}{\kappa^2}+5\right)\right.$$

$$\left.+k^4K^2\left(\frac{5}{\mathrm{K}^2\kappa^4}+\frac{1}{\kappa^4}+\frac{2}{\mathrm{K}^2\kappa^2}+\frac{2}{\kappa^2}+\frac{1}{\mathrm{K}^2}+5\right)\right]$$

$$-\frac{3}{k}\Sigma f_i^s\,\mathfrak{P}_i^s(\nu_0)\,\mathfrak{P}_i^s(1)\,\mathfrak{C}_i^s(\tfrac{1}{2}\pi)\,r^2\frac{d}{dr}\mathfrak{Q}_i^s\left(\frac{r}{k}\right)\quad (i>2,\ s\text{ even})$$

$$+\tfrac{9}{2}\frac{k^2}{r^2}\Sigma\frac{(f_i^s)^2\mathfrak{T}_i^s}{2i+1}\frac{\cos^2\gamma}{\sin^2\beta}\left\{1+\frac{5(3+\kappa^2)k^2}{14\kappa^2}\frac{k^2}{r^2}+\dots\right\}\ \text{(all harmonics)}\dots(24)$$

When $\mathfrak{Q}_i^s(r/k)$ is developed in powers of $1/r$, its first term is one in $r^{-(i+1)}$; hence $r^2\dfrac{\partial}{\partial r}\mathfrak{Q}_i^s$ begins with r^{-i}. Now f_i^s will be determined from terms in the potential of the ellipsoid E of the ith order of harmonics, and will therefore involve $r^{-(i+1)}$. Therefore in the series contained in the last term but one of ζ each term is of order $r^{-(2i+1)}$. Since the lowest value of i

is 3, the term of lowest order in this series is one in r^{-7}, and as I shall not attempt to evaluate ζ beyond r^{-6} the whole of this series is negligible.

Again, since $(f_i^s)^2$ is of order r^{-2i-2}, and since r^{-2} occurs as a factor, each term is of order r^{-2i-4}. Thus the lowest term is of order r^{-10} and is negligible*.

It follows that the only sensible part of ζ arises from the portion of V denoted $(eE)_2$, and the last two terms of (24) may be erased.

We next consider the parameter f_i^s, and, since I does not involve it, the equation reduces to $\partial V/\partial f_i^s = 0$; or, since V only contains f_i^s in the part denoted (vv), it becomes $\partial(vv)/\partial f_i^s = 0$.

This gives, for $i > 2$, s even

$$f_i^s = -\tfrac{1}{3}(2i+1)\frac{\mathfrak{Q}_i^s(r/k)\,\mathfrak{P}_i^s(\nu_0)\,\mathfrak{P}_i^s(1)\,\mathfrak{C}_i^s(\tfrac{1}{2}\pi)}{\lambda\mathfrak{T}_i^s\left\{\mathfrak{A}_i^s - \mathsf{A}_1^1 - \dfrac{k^3}{3\lambda r^3}\dfrac{\cos^2\gamma}{\sin^2\beta}\left[1 + \dfrac{3(3+\kappa^2)}{14\kappa^2}\dfrac{k^2}{r^2} + \dots\right]\right\}} \qquad (25)$$

Since this formula contains λ in the denominator, it would appear at first sight as if f_i^s became infinite when $\lambda = 0$. But this is not so, because when $\mathfrak{Q}_i^s(r/k)$ is developed the first term of the series is one in $(k/r)^{i+1}$; now $k^3 = \dfrac{\lambda}{1+\lambda}\mathrm{a}^3\sin^3\beta\sec\beta\sec\gamma$, and therefore the formula for f_i^s involves the factor $\left(\dfrac{\lambda}{1+\lambda}\right)^{(i+1)/3}\cdot\dfrac{1}{\lambda}$ or $\left(\dfrac{\lambda}{1+\lambda}\right)^{(i-2)/3}\cdot\dfrac{1}{1+\lambda}$. We see then that f_i^s vanishes both when $\lambda = 0$ and $\lambda = \infty$.

This factor is a maximum when $\lambda = \tfrac{1}{3}(i-2)$. Therefore we should expect, *cœteris paribus*, the third harmonics to be most important when $\lambda = \tfrac{1}{3}$, the fourth when $\lambda = \tfrac{2}{3}$, the fifth when $\lambda = 1$, and the higher harmonics when λ is greater than unity. This prevision is partially fulfilled by the numerical results given below, but it was not to be expected that it should be exactly so, because the other conditions are not exactly the same in the solutions for various values of λ.

The formula shows, as stated above, that f_i^s is of order r^{-i-1}. The series in the denominator affects the result but slightly and might be omitted, except, perhaps, in the case of the third zonal harmonic. For all harmonics other than cosine-harmonics of even rank f_i^s is zero.

It is now possible to eliminate f_i^s from (vv) by substituting for it its value. These terms in (vv) become, in this way, equal to

$$(\tfrac{4}{3}\pi\rho\mathrm{a}^3)^2\frac{\lambda}{(1+\lambda)^2}\left\{-\frac{2i+1}{2k\lambda}\frac{[\mathfrak{Q}_i^s(r/k)\,\mathfrak{P}_i^s(\nu_0)\,\mathfrak{P}_i^s(1)\,\mathfrak{C}_i^s(\tfrac{1}{2}\pi)]^2}{\mathfrak{T}_i^s[\mathfrak{A}_i^s - \mathsf{A}_1^1 - \text{series}]}\right\} \dots(26)$$

* It is proper to remark that the terms retained in ζ are really of higher orders than they appear to be. I recur to the neglected portions of ζ hereafter in § 23.

When $i=3$, this term is of order r^{-8}, and is negligible; hence we need no longer pay any attention to the inequalities on the ellipsoid. However, the formula (25) is important as rendering it possible to evaluate the inequalities.

Since for all inequalities, excepting cosine-harmonics of even rank, f_i^s only occurs in the energy function as a square, it is in these cases a principal coordinate, and ($\mathfrak{A}_i^s - \mathsf{A}_1^1$ – series) is a coefficient of stability.

But the like is not true for the cosine-harmonics of even rank, because, when we consider, for example, the harmonics of the third order, we see that $\partial^2 V/\partial f_3^s \partial r$ is of the fifth order and $\partial^2 V/\partial f_3^s \partial \mathfrak{a}$, $\partial^2 V/\partial f_3^s \partial \mathfrak{b}$ $(s=0, 2)$ are of the fourth order.

It is clear that the inequalities on the ellipsoid E are determinable by symmetrical formulæ.

We must now turn to the equations of equilibrium for the parameters $\mathfrak{a}$ and $\mathfrak{b}$. Since differentiation with respect to these parameters is effected most conveniently by means of the formulæ (23), the portion V called $(eE)_2$ should be written in the form (15). After effecting the differentiations it is, however, best to revert to the notation involving k, κ, γ, K, K, Γ; but as an exception to the general rule as to notation, it is most convenient to retain the differentials of $k^2(1+1/\kappa^2)$ and of b^2+c^2 in the forms involving a, b, c. As the algebraic processes involved are rather long, I simply give the results, as follows:—

$$3\mathfrak{a}\frac{d}{d\mathfrak{a}}(b^2+c^2) = -2(b^2+c^2) \quad \text{.............(i)}$$

$$3\mathfrak{b}\frac{d}{d\mathfrak{b}}(b^2+c^2) = 2(2b^2-c^2) \quad \text{.............(i)}'$$

$$3\mathfrak{a}\frac{d}{d\mathfrak{a}}k^2\left(\frac{1}{\kappa^2}+1\right) = 2(b^2-2a^2-2c^2) \quad \text{.............(ii)}$$

$$3\mathfrak{b}\frac{d}{d\mathfrak{b}}k^2\left(\frac{1}{\kappa^2}+1\right) = 2(a^2-2b^2-2c^2) \quad \text{.............(ii)}'$$

The remaining results are expressed in terms of k, κ, γ, &c.

$$3\mathfrak{a}\frac{d}{d\mathfrak{a}}k^4\left(\frac{3}{\kappa^4}+\frac{2}{\kappa^2}+3\right) = \frac{4k^4}{\kappa^4\sin^2\gamma}[-3(3+\kappa^2)+(6+\kappa^2-3\kappa^4)\sin^2\gamma] \text{......(iii)}$$

$$3\mathfrak{b}\frac{d}{d\mathfrak{b}} \quad \text{same} \quad = \frac{4k^4}{\kappa^4\sin^2\gamma}[-3(3\kappa^2+1)+(6\kappa^4+\kappa^2-3)\sin^2\gamma] \text{.....(iii)}'$$

$$3\mathfrak{a}\frac{d}{d\mathfrak{a}}k^6\left(\frac{5}{\kappa^6}+\frac{3}{\kappa^4}+\frac{3}{\kappa^2}+5\right) = \frac{6k^6}{\kappa^6\sin^2\gamma}[-3(5+2\kappa^2+\kappa^4) + (10+3\kappa^2+0\kappa^4-5\kappa^6)\sin^2\gamma] \text{...(iv)}$$

$$3\mathfrak{b}\frac{d}{d\mathfrak{b}} \quad \text{same} \quad = \frac{6k^6}{\kappa^6\sin^2\gamma}[-3(5\kappa^4+2\kappa^2+1) + (10\kappa^6+3\kappa^4+0\kappa^2-5)\sin^2\gamma] \text{...(iv)}$$

$$3\mathfrak{a}\,\frac{d}{d\mathfrak{a}}\,k^2K^2\left(\frac{3}{\kappa^2K^2}+\frac{1}{K^2}+\frac{1}{\kappa^2}+3\right)=\frac{2k^2K^2}{\kappa^2K^2\sin^2\gamma}\left[-(3+K^2)(3-2\sin^2\gamma)\right.$$
$$\left.-\kappa^2(1+3K^2)\sin^2\gamma\right]\ldots(\text{v})$$

$$3\mathfrak{b}\,\frac{d}{d\mathfrak{b}}\qquad\text{same}\qquad=\frac{2k^2K^2}{\kappa^2K^2\sin^2\gamma}\left[-(3K^2+1)(3-2\kappa^2\sin^2\gamma)\right.$$
$$\left.-(K^2+3)\sin^2\gamma\right]\ldots(\text{v})'$$

$$3\mathfrak{a}\,\frac{d}{d\mathfrak{a}}\left[k^2K^4\left(\frac{5}{\kappa^2K^4}+\frac{1}{K^4}+\frac{2}{\kappa^2K^2}+\frac{2}{K^2}+\frac{1}{\kappa^2}+5\right)\right.$$
$$\left.+\text{ same with small and large interchanged}\right]$$
$$=\frac{2k^2K^4}{\kappa^2K^4\sin^2\gamma}\left[-(5+2K^2+K^4)(3-2\sin^2\gamma)-\kappa^2(1+2K^2+5K^4)\sin^2\gamma\right]$$
$$+\frac{4k^4K^2}{\kappa^4K^2\sin^2\gamma}\left[-3(5+\kappa^2)+(10+\kappa^2-\kappa^4)\sin^2\gamma\right.$$
$$\left.-K^2\{3(1+\kappa^2)-(2+\kappa^2-5\kappa^4)\sin^2\gamma\}\right]\ldots(\text{vi})$$

$$3\mathfrak{b}\,\frac{d}{d\mathfrak{b}}\qquad\text{same}$$
$$=\frac{2k^2K^4}{\kappa^2K^4\sin^2\gamma}\left[-(5+2K^2+K^4)\sin^2\gamma-(1+2K^2+5K^4)(3-2\kappa^2\sin^2\gamma)\right]$$
$$+\frac{4k^4K^2}{\kappa^4K^2\sin^2\gamma}\left[-3(1+\kappa^2)+(2\kappa^4+\kappa^2-5)\sin^2\gamma\right.$$
$$\left.-K^2\{3(5\kappa^2+1)-(10\kappa^4+\kappa^2-1)\sin^2\gamma\}\right]\ldots(\text{vi})'$$

On picking out the numerical coefficients of the several terms in $(eE)_2$ as given in (14) or (15), we see that

$$3\mathfrak{a}\,\frac{d}{d\mathfrak{a}}(eE)_2=\frac{(\frac{4}{3}\pi\rho a^3)^2\lambda}{(1+\lambda)^2}\left\{\frac{1\,(\text{ii})}{2.5r^3}+\frac{3\,(\text{iii})}{2^3.5.7r^5}+\frac{3\,(\text{v})}{2^2.5^2r^5}+\frac{1\,(\text{iv})}{2^4.3.7r^7}+\frac{9\,(\text{vi})}{2^4.5.7r^7}\right\}$$

$$3\mathfrak{b}\,\frac{d}{d\mathfrak{b}}(eE)_2=\ldots\ldots\ldots\ldots\ (\text{ii})'\ \ldots\ldots\ (\text{iii})'\ \ldots\ldots\ (\text{v})'\ \ldots\ldots\ (\text{iv})'\ldots\ldots\ldots(\text{vi})'$$

Observe that $\dfrac{k^2}{\kappa^2\sin^2\gamma}=\dfrac{k^2}{\sin^2\beta}=c^2$, and write

$$\left.\begin{aligned}\tau&=\frac{3}{2^3.7r^2}\frac{\kappa^2\sin^2\gamma}{k^2}(\text{iii})+\frac{3}{2^2.5r^2}\frac{\kappa^2\sin^2\gamma}{k^2}(\text{v})+\frac{5}{2^4.3.7r^4}\frac{\kappa^2\sin^2\gamma}{k^2}(\text{iv})+\frac{9}{2^4.7r^4}\frac{\kappa^2\sin^2\gamma}{k^2}(\text{vi})\\ \sigma&=\ldots\ldots\ldots\ldots\ (\text{iii})'\ \ldots\ldots\ldots\ldots\ldots(\text{v})'\ \ldots\ldots\ldots\ldots\ (\text{iv})'\ \ldots\ldots\ldots\ldots\ (\text{vi})'\end{aligned}\right\}\ldots\ldots(27)$$

Then we have

$$\left.\begin{aligned}3\mathfrak{a}\,\frac{d}{d\mathfrak{a}}(eE)_2&=(\tfrac{4}{3}\pi\rho a^3)^2\frac{\lambda}{(1+\lambda)^2}\left[\frac{1}{5r^3}(b^2-2a^2-2c^2)+\frac{1}{5r^3}\cdot\tau c^2\right]\\ 3\mathfrak{b}\,\frac{d}{d\mathfrak{b}}(eE)_2&=(\tfrac{4}{3}\pi\rho a^3)^2\frac{\lambda}{(1+\lambda)^2}\left[\frac{1}{5r^3}(a^2-2b^2-2c^2)+\frac{1}{5r^3}\cdot\sigma c^2\right]\end{aligned}\right\}\ldots(28)$$

The terms in V denoted $(eE)_1$ and (VV) do not contain a, b, c, and their differentials with respect to $\mathfrak{a}$, $\mathfrak{b}$ are zero; also, after omission of the terms in f_i^s, (vv) is reduced to

$$(vv) = (\tfrac{4}{3}\pi\rho a^3)^2 \frac{\lambda}{(1+\lambda)^2} \cdot \tfrac{3}{10}\lambda\psi$$

Hence

$$3\mathfrak{a}\frac{d}{d\mathfrak{a}}(vv) = (\tfrac{4}{3}\pi\rho a^3)^2 \frac{\lambda}{(1+\lambda)^2}\left[\tfrac{3}{10}\lambda\left(2a\frac{\partial\psi}{\partial a} - b\frac{\partial\psi}{\partial b} - c\frac{\partial\psi}{\partial c}\right)\right]$$

$$3\mathfrak{b}\frac{d}{d\mathfrak{b}}(vv) = (\tfrac{4}{3}\pi\rho a^3)^2 \frac{\lambda}{(1+\lambda)^2}\left[\tfrac{3}{10}\lambda\left(2b\frac{\partial\psi}{\partial b} - a\frac{\partial\psi}{\partial a} - c\frac{\partial\psi}{\partial c}\right)\right]$$

Now

$$a\frac{\partial\psi}{\partial a} = -\frac{2}{k}\mathsf{P}_1^1(\nu_0)\,\mathsf{Q}_1^1(\nu_0) = -\frac{2}{k}\mathsf{A}_1^1$$

$$b\frac{\partial\psi}{\partial b} = -\frac{2}{k}\mathfrak{P}_1^1(\nu_0)\,\mathfrak{Q}_1^1(\nu_0) = -\frac{2}{k}\mathfrak{A}_1^1$$

$$c\frac{\partial\psi}{\partial c} = -\frac{2}{k}\mathfrak{P}_1(\nu_0)\,\mathfrak{Q}_1(\nu_0) = -\frac{2}{k}\mathfrak{A}_1$$

Since ψ is homogeneous of degree -1 in a, b, c, the sum of these three is equal to -1, so that

$$\mathsf{A}_1^1 + \mathfrak{A}_1^1 + \mathfrak{A}_1 = \tfrac{1}{2}k\psi$$

Therefore

$$\left.\begin{aligned} 3\mathfrak{a}\frac{d}{d\mathfrak{a}}(vv) &= -(\tfrac{4}{3}\pi\rho a^3)^2 \frac{\lambda}{(1+\lambda)^2} \cdot \frac{3\lambda}{5k}(3\mathsf{A}_1^1 - \tfrac{1}{2}k\psi) \\ 3\mathfrak{b}\frac{d}{d\mathfrak{b}}(vv) &= -(\tfrac{4}{3}\pi\rho a^3)^2 \frac{\lambda}{(1+\lambda)^2} \cdot \frac{3\lambda}{5k}(3\mathfrak{A}_1^1 - \tfrac{1}{2}k\psi) \end{aligned}\right\} \quad \text{.........(29)}$$

On adding together (28) and (29), we find

$$3\mathfrak{a}\frac{dV}{d\mathfrak{a}} = (\tfrac{4}{3}\pi\rho a^3)^2 \frac{\lambda}{(1+\lambda)^2}\left[-\frac{3\lambda}{5k}(3\mathsf{A}_1^1 - \tfrac{1}{2}k\psi) + \frac{1}{5r^3}(b^2 - 2a^2 - 2c^2 + \tau c^2)\right]$$

$$3\mathfrak{b}\frac{dV}{d\mathfrak{b}} = (\tfrac{4}{3}\pi\rho a^3)^2 \frac{\lambda}{(1+\lambda)^2}\left[-\frac{3\lambda}{5k}(3\mathfrak{A}_1^1 - \tfrac{1}{2}k\psi) + \frac{1}{5r^3}(a^2 - 2b^2 - 2c^2 + \sigma c^2)\right]$$

By means of (i) and (i)′ we find the differentials of the moment of inertia I; they are

$$3\mathfrak{a}\frac{dI}{d\mathfrak{a}} = -\tfrac{4}{3}\pi\rho a^3 \frac{\lambda}{1+\lambda} \cdot \tfrac{2}{5}(b^2 + c^2)$$

$$3\mathfrak{b}\frac{dI}{d\mathfrak{b}} = \tfrac{4}{3}\pi\rho a^3 \frac{\lambda}{1+\lambda} \cdot \tfrac{2}{5}(2b^2 - c^2)$$

Then, since $\tfrac{1}{2}\omega^2 = \tfrac{4}{3}\pi\rho a^3 \cdot \dfrac{1}{2r^3}(1+\zeta)$,

$$\tfrac{3}{2}\omega^2\mathfrak{a}\frac{dI}{d\mathfrak{a}} = -(\tfrac{4}{3}\pi\rho a^3)^2 \frac{\lambda}{(1+\lambda)^2} \cdot \frac{1}{5r^3}(1+\lambda)(1+\zeta)(b^2 + c^2)$$

$$\tfrac{3}{2}\omega^2\mathfrak{b}\frac{dI}{d\mathfrak{b}} = (\tfrac{4}{3}\pi\rho a^3)^2 \frac{\lambda}{(1+\lambda)^2} \cdot \frac{1}{5r^3}(1+\lambda)(1+\zeta)(2b^2 - c^2)$$

Now the equations for equilibrium, for the parameters $\mathfrak{a}$ and $\mathfrak{b}$, are

$$\frac{dV}{d\mathfrak{a}} + \tfrac{1}{2}\omega^2 \frac{dI}{d\mathfrak{a}} = 0, \qquad \frac{dV}{d\mathfrak{b}} + \tfrac{1}{2}\omega^2 \frac{dI}{d\mathfrak{b}} = 0$$

Therefore

$$\left.\begin{aligned} &-3\lambda\,(3\mathsf{A}_1{}^1 - \tfrac{1}{2}k\psi) + \frac{k}{r^3}[b^2 - 2a^2 - 2c^2 + \tau c^2 - (1+\lambda)(1+\zeta)(b^2+c^2)] = 0 \\ &-3\lambda\,(3\mathfrak{A}_1{}^1 - \tfrac{1}{2}k\psi) + \frac{k}{r^3}[a^2 - 2b^2 + 2c^2 + \sigma c^2 + (1+\lambda)(1+\zeta)(2b^2 - c^2)] = 0 \end{aligned}\right\} \quad \ldots\ldots(30)$$

Subtracting the second of these from the first and dividing by 9λ, we have

$$\mathfrak{A}_1{}^1 - \mathsf{A}_1{}^1 = \frac{k}{3\lambda r^3}[a^2 - b^2 + b^2(1+\lambda)(1+\zeta) - \tfrac{1}{3}c^2(\tau - \sigma)] \ldots\ldots(31)$$

Since we may write $3\mathsf{A}_1{}^1 - \frac{1}{2}k\psi$ in the form $-(\mathfrak{A}_1{}^1 - \mathsf{A}_1{}^1) - (\mathfrak{A}_1 - \mathsf{A}_1{}^1)$, the first of (30) in combination with (31) gives

$$\mathfrak{A}_1 - \mathsf{A}_1{}^1 = \frac{k}{3\lambda r^3}[a^2 + 2c^2 + c^2(1+\lambda)(1+\zeta) - \tfrac{1}{3}c^2(2\tau + \sigma)] \ldots(32)$$

Referring to the values of τ and σ in (27), I find

$$\begin{aligned} -\tfrac{1}{3}(\tau - \sigma) = {} & \tfrac{3}{14}\frac{k^2\kappa'^2}{r^2\kappa^2}[2 - 3(1+\kappa^2)\sin^2\gamma] \\ & + \tfrac{5}{56}\frac{k^4\kappa'^2}{r^4\kappa^4}[4(1+\kappa^2) - (5 + 6\kappa^2 + 5\kappa^4)\sin^2\gamma] \\ & + \tfrac{3}{10}\frac{K^2}{r^2\mathrm{K}^2}[2\mathrm{K}'^2\cos^2\gamma - \kappa'^2\sin^2\gamma\,(1 + 3\mathrm{K}^2)] \\ & + \tfrac{9}{56}\frac{K^4}{r^4\mathrm{K}^4}[4\mathrm{K}'^2(1+\mathrm{K}^2)\cos^2\gamma - \kappa'^2\sin^2\gamma\,(1 + 2\mathrm{K}^2 + 5\mathrm{K}^4)] \\ & + \tfrac{9}{28}\frac{k^2K^2}{r^4\kappa^2\mathrm{K}^2}[4(1 - \kappa^2\mathrm{K}^2)\cos^2\gamma - \kappa'^2\sin^2\gamma\,(1 + \kappa^2 + \mathrm{K}^2 + 5\kappa^2\mathrm{K}^2)] \\ -\tfrac{1}{3}(2\tau + \sigma) = {} & \tfrac{3}{14}\frac{k^2}{r^2\kappa^2}[7 + 5\kappa^2 - (3 + \kappa^2)\sin^2\gamma] \\ & + \tfrac{5}{56}\frac{k^4}{r^4\kappa^4}[11 + 6\kappa^2 + 7\kappa^4 - (5 + 2\kappa^2 + \kappa^4)\sin^2\gamma] \\ & + \tfrac{3}{10}\frac{K^2}{r^2\mathrm{K}^2}[7 + 5\mathrm{K}^2 - (3 + \mathrm{K}^2)\sin^2\gamma] \\ & + \tfrac{9}{56}\frac{K^4}{r^4\mathrm{K}^4}[11 + 6\mathrm{K}^2 + 7\mathrm{K}^4 - (5 + 2\mathrm{K}^2 + \mathrm{K}^4)\sin^2\gamma] \\ & + \tfrac{9}{28}\frac{k^2K^2}{r^4\kappa^2\mathrm{K}^2}[11 + 3\kappa^2 + 3\mathrm{K}^2 + 7\kappa^2\mathrm{K}^2 - (5 + \kappa^2 + \mathrm{K}^2 + \kappa^2\mathrm{K}^2)\sin^2\gamma] \end{aligned}$$

$$\ldots\ldots(33)$$

In all the cases which we shall have to consider the first of these expressions is small compared with the second, because κ is nearly equal to unity and κ' small, and because $\cos^2\gamma$ is also rather small.

Now let

$$\left.\begin{aligned} \epsilon &= -\tfrac{1}{3}(\tau-\sigma)+\zeta(1+\lambda)\cos^2\beta \\ \eta &= -\tfrac{1}{3}(2\tau+\sigma)+\zeta(1+\lambda) \end{aligned}\right\} \quad \text{..............(33 \textit{bis})}$$

Then, since $a = c\cos\gamma$, $b = c\cos\beta$, the equations (31) and (32) become

$$\left.\begin{aligned} \mathfrak{A}_1{}^1 - \mathsf{A}_1{}^1 &= \frac{kc^2}{3\lambda r^3}(\cos^2\gamma+\lambda\cos^2\beta+\epsilon) \\ \mathfrak{A}_1 - \mathsf{A}_1{}^1 &= \frac{kc^2}{3\lambda r^3}(3+\lambda+\cos^2\gamma+\eta) \end{aligned}\right\} \quad \text{..............(34)}$$

Eliminating $kc^2/3\lambda r^3$, we have

$$(\mathfrak{A}_1-\mathsf{A}_1{}^1)(\cos^2\gamma+\lambda\cos^2\beta+\epsilon) = (\mathfrak{A}_1{}^1-\mathsf{A}_1{}^1)(3+\lambda+\cos^2\gamma+\eta) \quad \text{...(35)}$$

This is the equation to be satisfied by the axes of the ellipsoid. If we treat ϵ and η as zero, it is the same as that found by Roche*.

* The form of this equation is so unlike Roche's, that it may be worth while to prove the identity of the two.

Roche writes his equations in the form

$$\frac{st(t-s)}{(3+\lambda)t-\lambda s}\int_0^\infty \frac{u\,du}{(1+su)(1+tu)R} = \frac{s(1-s)}{s+3+\lambda}\int_0^\infty \frac{u\,du}{(1+su)(1+u)R} = \frac{t(1-t)}{t+\lambda}\int_0^\infty \frac{u\,du}{(1+u)(1+tu)R}$$

where $R^2 = (1+u)(1+su)(1+tu)$, and s is the square of the ratio of the least to the greatest axis, and t the square of the ratio of the least to the mean axis.

In my notation $\quad s = \dfrac{a^2}{c^2} = \cos^2\gamma, \quad t = \dfrac{a^2}{b^2} = \dfrac{\cos^2\gamma}{\cos^2\beta}$

If we write $us+1 = \dfrac{\sin^2\gamma}{\sin^2\psi}$, and change the independent variable from u to ψ, we find

$$\left.\begin{aligned} \text{Roche's first integral} &= \frac{2\cos^3\beta}{\sin^5\gamma\cos^3\gamma}\int_0^\gamma \frac{\sin^2\psi(\sin^2\gamma-\sin^2\psi)}{\Delta^3}d\psi \\ \text{,,}\quad \text{second}\quad \text{,,}\quad &= \frac{2\cos\beta}{\sin^5\gamma\cos\gamma}\int_0^\gamma \frac{\sin^2\psi(\sin^2\gamma-\sin^2\psi)}{\cos^2\psi\,\Delta}d\psi \\ \text{,,}\quad \text{third}\quad \text{,,}\quad &= \frac{2\cos^3\beta}{\sin^5\gamma\cos\gamma}\int_0^\gamma \frac{\sin^2\psi(\sin^2\gamma-\sin^2\psi)}{\cos^2\psi\,\Delta^3}d\psi \end{aligned}\right\} \quad \text{...........(A)}$$

where $\Delta^2 = 1-\kappa^2\sin^2\psi$.

The coefficients are

$$\left.\begin{aligned} \frac{st(t-s)}{(3+\lambda)t-\lambda s} &= \frac{\cos^4\gamma\sin^2\beta}{\cos^2\beta[(3+\lambda)-\lambda\cos^2\beta]} \\ \frac{s(1-s)}{s+3+\lambda} &= \frac{\sin^2\gamma\cos^2\gamma}{3+\lambda+\cos^2\gamma} \\ \frac{t(1-t)}{t+\lambda} &= \frac{\kappa'^2\sin^2\gamma\cos^2\gamma}{\cos^2\beta(\cos^2\gamma+\lambda\cos^2\beta)} \end{aligned}\right\} \quad \text{..........................(B)}$$

Then Roche's equations are equivalent to

1st of (A) × 1st of (B) = 2nd of (A) × 2nd of (B) = 3rd of (A) × 3rd of (B).

It is possible to express this equation in terms of elliptic integrals and to use Legendre's tables for finding the solution, but the method is very tedious, and after finding a few solutions in that way I abandoned it. It may, however, be worth while to mention that

$$\mathfrak{A}_1 - \mathsf{A}_1{}^1 = \frac{\kappa}{\sin^2\gamma}\left[-\frac{1}{\kappa'^2}\sin\gamma\cos\gamma\cos\beta + \frac{F}{\kappa^2} + \frac{\cos^2\beta - 2\kappa'^2}{\kappa^2\kappa'^2}E\right]$$

$$\mathfrak{A}_1{}^1 - \mathsf{A}_1{}^1 = \frac{\kappa}{\sin^2\gamma}\left[-\frac{2}{\kappa'^2}\sin\gamma\cos\gamma\cos\beta - \frac{\cos^2\beta}{\kappa^2}F + \frac{2\cos^2\beta - \kappa'^2}{\kappa^2\kappa'^2}E\right]$$

where $$F = \int_0^\gamma \frac{d\psi}{\sqrt{(1-\kappa^2\sin^2\psi)}}, \qquad E = \int_0^\gamma \sqrt{(1-\kappa^2\sin^2\psi)}\,d\psi$$

When the forms of the ellipsoids have been determined, the radius vector becomes determinable from either of the equations (34).

The conditions that the internal potential of an ellipsoid satisfies Poisson's equation and that ψ is homogeneous in a, b, c of degree -1, give the two following equations:—

$$\frac{\mathsf{A}_1{}^1}{a^2} + \frac{\mathfrak{A}_1{}^1}{b^2} + \frac{\mathfrak{A}_1}{c^2} - \frac{k}{abc} = 0$$

$$\mathsf{A}_1{}^1 + \mathfrak{A}_1{}^1 + \mathfrak{A}_1 - \tfrac{1}{2}k\psi = 0$$

But the two equations are not independent, and I will only pursue the consideration of the form involving the 2nd and 3rd of (A) and (B).

Now $$\frac{\sin^2\psi\,(\sin^2\gamma - \sin^2\psi)}{\cos^2\psi\,\Delta} = \frac{\sin^2\psi}{\Delta} - \cos^2\gamma\,\frac{\tan^2\psi}{\Delta}$$

$$\frac{\sin^2\psi\,(\sin^2\gamma - \sin^2\psi)}{\cos^2\psi\,\Delta^3} = \frac{\cos^2\beta}{\kappa'^2}\frac{\text{siu}^2\psi}{\Delta^3} - \frac{\cos^2\gamma}{\kappa'^2}\frac{\tan^2\psi}{\Delta}$$

and I have proved in (25) of the "Pear-shaped figure, &c." that

$$\mathfrak{A}_1 = \mathfrak{P}_1\mathfrak{Q}_1 = \frac{\kappa}{\sin^2\gamma}\int_0^\gamma \frac{\sin^2\psi}{\Delta}\,d\psi$$

$$\mathsf{A}_1{}^1 = \mathsf{P}_1{}^1\mathsf{Q}_1{}^1 = \frac{\kappa\cos^2\gamma}{\sin^2\gamma}\int_0^\gamma \frac{\tan^2\psi}{\Delta}\,d\psi$$

$$\mathfrak{A}_1{}^1 = \mathfrak{P}_1{}^1\mathfrak{Q}_1{}^1 = \frac{\kappa\cos^2\beta}{\sin^2\gamma}\int_0^\gamma \frac{\sin^2\psi}{\Delta^3}\,d\psi$$

Therefore Roche's second integral is equal to $\dfrac{2\cos\beta}{\kappa\,\text{siu}^3\gamma\cos\gamma}(\mathfrak{A}_1 - \mathsf{A}_1{}^1)$, and his third integral is equal to $\dfrac{2\cos^3\beta}{\kappa\kappa'^2\sin^3\gamma\cos\gamma}(\mathfrak{A}_1{}^1 - \mathsf{A}_1{}^1)$.

Using these transformations of the second and third of (A), and dropping redundant factors, we get

$$(\cos^2\gamma + \lambda\cos^2\beta)(\mathfrak{A}_1 - \mathsf{A}_1{}^1) = (3 + \lambda + \cos^2\gamma)(\mathfrak{A}_1{}^1 - \mathsf{A}_1{}^1)$$

This agrees with the result in the text when ϵ and η are neglected.

Our two equations for $1/r^3$ may be written

$$\mathsf{A}_1{}^1 - \mathfrak{A}_1{}^1 \qquad + \frac{k}{3\lambda r^3}(a^2 + b^2\lambda + c^2\epsilon) = 0$$

$$\mathsf{A}_1{}^1 \qquad - \mathfrak{A}_1 + \frac{k}{3\lambda r^3}[(3+\lambda)c^2 + a^2 + c^2\eta] = 0$$

These four equations afford a determinant by which $\mathsf{A}_1{}^1$, $\mathfrak{A}_1{}^1$, $\mathfrak{A}_1$ may be eliminated. On reduction we find

$$\frac{1}{3\lambda r^3} = \frac{\frac{1}{2}\psi - \dfrac{3/abc}{1/a^2 + 1/b^2 + 1/c^2}}{3c^2 - a^2 + \lambda(b^2 + c^2) + c^2(\eta + \epsilon) - \dfrac{6(1 + \lambda + \frac{1}{2}\eta + \frac{1}{2}c^2\epsilon/b^2)}{1/a^2 + 1/b^2 + 1/c^2}}$$

On putting $a = c\cos\gamma$, $b = c\cos\beta$, and noting that $\psi = \dfrac{2F}{c\sin\gamma}$ and that $\dfrac{1+\lambda}{\lambda}c^3\cos\gamma\cos\beta = \mathrm{a}^3$, I find that

$$\frac{\mathrm{a}^3}{r^3} = \frac{\frac{3}{2}\left[2F\cot\gamma\cos\beta - \dfrac{6}{1+\sec^2\gamma + \sec^2\beta}\right]}{\dfrac{3 - \cos^2\gamma + \lambda(1+\cos^2\beta)}{1+\lambda} - \dfrac{6}{1+\sec^2\gamma+\sec^2\beta} + \delta} \quad \text{......(36)*}$$

* It is by no means obvious how this formula is consistent with results which we know by other means to be true. In the case when $\lambda = \infty$ we have a liquid planet rotating with the same angular velocity as an infinitely small satellite revolving in a circular orbit in its equator.

Let us first consider the value of ζ. In the present case the semi-axes A, B, C pertain to the infinitely small satellite, and are therefore negligible compared with terms in a, b, c. Since the axis denoted by c is that coincident with the satellite's radius vector, and since the equatorial plane of the planet must have a circular section, we have $c = b$.

But since $b = c\cos\beta$, it follows that $\beta = 0$ or $\kappa\sin\gamma = 0$. Now γ does not vanish for $a = c\cos\gamma$, and a is the polar semi-radius of the planet; therefore $\kappa = 0$.

If we consider the formula (24) for ζ, expressing, however, the several terms in the form of (15), we see that for $\lambda = \infty$

$$\zeta = \frac{3}{10r^2}(c^2 - a^2) + \frac{3}{56r^4}\cdot 3(c^2 - a^2)^2 + \frac{1}{48r^6}\cdot 5(c^2 - a^2)^3 \ldots$$

whence

$$\zeta = \frac{3c^2}{10r^2}\sin^2\gamma + \frac{9c^4}{56r^4}\sin^4\gamma + \frac{5c^6}{48r^6}\sin^6\gamma \ldots \quad \text{..............}(a)$$

The factor of correction to Kepler's law of periodic times for a small satellite revolving about an oblate planet, whose equatorial radius is c and whose eccentricity of figure is $\sin\gamma$, is $1+\zeta$, where ζ is expressed by the above series (a).

Now considering the formula (33 *bis*), we see that for $\lambda = \infty$ and $\beta = 0$

$$\frac{\epsilon}{1+\lambda} = \zeta\cos^2\beta = \zeta; \quad \frac{\eta}{1+\lambda} = \zeta$$

The meaning of δ in (36) will be found in the first line of p. 480, and we have

$$\delta = \frac{1}{1+\lambda}\left[\eta + \epsilon - \frac{3(\eta + \epsilon\sec^2\beta)}{1 + \sec^2\beta + \sec^2\gamma}\right] = 2\zeta\left[1 - \frac{3}{3+\tan^2\gamma}\right] = \frac{2\zeta\tan^2\gamma}{3+\tan^2\gamma}$$

where
$$\delta = \frac{1}{1+\lambda}\left[\eta + \epsilon - \frac{3(\eta + \epsilon \sec^2 \beta)}{1 + \sec^2\gamma + \sec^2\beta}\right]$$

However, this is not practically the most convenient form from which to compute the distance between the two ellipsoids.

§ 11. *Solution of the Equations.*

In all the ellipsoids of which we shall have to find the axes, it happens that $\kappa'^2 \tan^2 \gamma$ is fairly small compared with unity. Hence it is possible to expand Δ in powers of that quantity.

When $\kappa = 0$, the elliptic integral F is equal to γ; thus (36) becomes

$$\frac{\mathrm{a}^3}{r^3} = \frac{\frac{3}{2}\left[2\gamma\cot\gamma - \frac{6}{3+\tan^2\gamma}\right]}{2 - \frac{6}{3+\tan^2\gamma} + \frac{2\zeta\tan^2\gamma}{3+\tan^2\gamma}}$$

From this we easily obtain

$$\frac{\omega^2}{2\pi\rho} = \tfrac{2}{3}(1+\zeta)\frac{\mathrm{a}^3}{r^3} = \cot^3\gamma\,[\gamma(3+\tan^2\gamma) - 3\tan\gamma]$$

This is the well-known formula for the angular velocity of Maclaurin's ellipsoid.

It should be remarked that (35) is identically satisfied by $\lambda = \infty$, $\kappa = 0$, for when we use the above values of ϵ and η, the equation becomes divisible by $1+\zeta$.

Since ζ is a symmetrical function of a, b, c and A, B, C, it follows that ζ is the same in form for λ and for $1/\lambda$. Therefore when we consider the case of $\lambda = 0$, the formula (*a*) gives the required result, but c and γ refer to the large body which is throughout most of this paper indicated by capital letters.

Thus for $\lambda = 0$,
$$\zeta = \tfrac{3}{10}\frac{C^2}{r^2}\sin^2\Gamma + \tfrac{9}{56}\left(\frac{C^2}{r^2}\sin^2\Gamma\right)^2 + \tfrac{5}{48}\left(\frac{C^2}{r^2}\sin^2\Gamma\right)^3 \ldots$$
$$= \tfrac{3}{10}\frac{\mathrm{a}^2}{r^2}\frac{\sin^2\Gamma}{\cos^{2/3}\Gamma} + \tfrac{9}{56}\frac{\mathrm{a}^4}{r^4}\left(\frac{\sin^2\Gamma}{\cos^{2/3}\Gamma}\right)^2 + \tfrac{5}{48}\frac{\mathrm{a}^6}{r^6}\left(\frac{\sin^2\Gamma}{\cos^{2/3}\Gamma}\right)^3 \ldots$$

In the case of $\lambda = 0$, k vanishes; K also vanishes and so also does the angle B. Hence we have

$$\epsilon = \zeta\cos^2\beta, \quad \eta = \zeta$$

With these values, equation (35) becomes

$$(\mathfrak{A}_1 - \mathrm{A}_1{}^1)\,[\cos^2\gamma + (\lambda+\zeta)\cos^2\beta] = (\mathfrak{A}_1{}^1 - \mathrm{A}_1{}^1)\,[3 + (\lambda+\zeta) + \cos^2\gamma]$$

Hence ζ plays the part of an augmentation to λ.

With $\lambda = 0$ the equation assumes the form

$$(\mathfrak{A}_1 - \mathrm{A}_1{}^1)(\cos^2\gamma + \zeta\cos^2\beta) = (\mathfrak{A}_1{}^1 - \mathrm{A}_1{}^1)(3 + \zeta + \cos^2\gamma) \quad \ldots\ldots\ldots\ldots\ldots\ldots(b)$$

It follows therefore that an infinitesimal satellite revolving about an oblate planet, whose rotation is the same as the revolution of the satellite, is very nearly identical in form with a small but finite satellite whose mass is a fraction of a *spherical* planet expressed by ζ. This curious conclusion follows from the fact that if we take equation (35) and put ϵ and η zero (which corresponds to a spherical planet and small satellite), we get exactly the equation (*b*) just found, only with λ in place of ζ.

We have $\Delta^2 = 1 - \kappa^2 \sin^2 \gamma = \cos^2 \gamma (1 + \kappa'^2 \tan^2 \gamma)$

and
$$\frac{1}{\Delta} = \frac{1}{\cos \gamma} [1 - \frac{1}{2} \kappa'^2 \tan^2 \gamma + \frac{1.3}{2.4} \kappa'^4 \tan^4 \gamma - \ldots]$$

$$\frac{1}{\Delta^3} = \frac{1}{\cos^3 \gamma} [1 - \frac{3}{2} \kappa'^2 \tan^2 \gamma + \frac{3.5}{2.4} \kappa'^4 \tan^4 \gamma - \ldots]$$

Now from (25) of the "Pear-shaped Figure,"

$$\mathfrak{A}_1 = \frac{\kappa}{\sin^2 \gamma} \int_0^\gamma \frac{\sin^2 \gamma}{\Delta} d\gamma, \quad \mathsf{A}_1{}^1 = \frac{\kappa \cos^2 \gamma}{\sin^2 \gamma} \int_0^\gamma \frac{\tan^2 \gamma}{\Delta} d\gamma, \quad \mathfrak{A}_1{}^1 = \frac{\kappa \cos^2 \beta}{\sin^2 \gamma} \int_0^\gamma \frac{\sin^2 \gamma}{\Delta^3} d\gamma$$

When the Δ's under the integral signs are expanded, all the terms of the series involve integrals of one of two types. If we write

$$\Omega = \frac{1}{\sin \gamma} \log_e \frac{1 + \sin \gamma}{\cos \gamma}$$

the types are

$$\int_0^\gamma \frac{\sin^{2n} \gamma}{\cos^{2n-1} \gamma} d\gamma = \sin \gamma \left[\frac{1}{2n-2} \tan^{2n-2} \gamma - \frac{2n-1}{(2n-2)(2n-4)} \tan^{2n-4} \gamma + \ldots \right.$$
$$\left. + (-)^n \frac{(2n-1)(2n-3)\ldots 5}{(2n-2)(2n-4)\ldots 2} \tan^2 \gamma + (-)^{n+1} \frac{(2n-1)\ldots 3}{(2n-2)\ldots 2} (\Omega - 1) \right]$$

$$\int_0^\gamma \frac{\sin^{2n} \gamma}{\cos^{2n+1} \gamma} d\gamma = \sin \gamma \left[\frac{1}{2n} \tan^{2n} \gamma - \frac{1}{2n(2n-2)} \tan^{2n-2} \gamma \right.$$
$$+ \frac{2n-1}{2n(2n-2)(2n-4)} \tan^{2n-4} \gamma - \ldots - (-)^n \frac{(2n-1)(2n-3)\ldots 5}{2n(2n-2)\ldots 2} \tan^2 \gamma$$
$$\left. - (-)^{n+1} \frac{(2n-1)\ldots 3}{2n(2n-2)\ldots 2} (\Omega - 1) \right]$$

As it is not quite obvious what interpretation is to be put on these formulæ for the smaller values of n, I may mention that when $n = 0, 1, 2$ respectively, the first integral is $\sin \gamma$; $\sin \gamma (\Omega - 1)$; $\sin \gamma [\frac{1}{2} \tan^2 \gamma - \frac{3}{2} (\Omega - 1)]$, and the second is $\sin \gamma \Omega$; $\sin \gamma [\frac{1}{2} \tan^2 \gamma - \frac{1}{2} (\Omega - 1)]$; $\sin \gamma [\frac{1}{4} \tan^4 \gamma - \frac{1}{4.2} \tan^2 \gamma + \frac{3}{4.2} (\Omega - 1)]$. For larger values of n the interpretation is obvious.

If we use these integrals and write

$$\left.\begin{aligned} \frac{\sin \gamma}{\kappa} (\mathfrak{A}_1{}^1 - \mathsf{A}_1{}^1) &= \kappa'^2 [\sigma_0 - \sigma_1 \kappa'^2 + \sigma_2 \kappa'^4 - \sigma_3 \kappa'^6 \ldots] \\ \frac{\sin \gamma}{\kappa} (\mathfrak{A}_1 - \mathsf{A}_1{}^1) &= \tau_0 - \tau_1 \kappa'^2 + \tau_2 \kappa'^4 - \tau_3 \kappa'^6 \ldots \end{aligned}\right\} \quad \ldots\ldots(37)$$

we find

$$\sigma_0 = \tfrac{1}{8}[2\tan^2\gamma + 3 - (3 + \sin^2\gamma)\,\Omega]$$
$$\sigma_1 = \tfrac{3}{32}[\tfrac{4}{3}\tan^4\gamma - \tfrac{8}{3}\tan^2\gamma - 5 + (5 + \sin^2\gamma)\,\Omega]$$
$$\sigma_2 = \tfrac{75}{1024}[\tfrac{16}{15}\tan^6\gamma - \tfrac{8}{5}\tan^4\gamma + \tfrac{10}{3}\tan^2\gamma + 7 - (7 + \sin^2\gamma)\,\Omega]$$
$$\sigma_3 = \tfrac{245}{4096}[\tfrac{32}{35}\tan^8\gamma - \tfrac{128}{105}\tan^6\gamma + \tfrac{28}{15}\tan^4\gamma - 4\tan^2\gamma - 9 + (9 + \sin^2\gamma)\,\Omega]$$
$$\tau_0 = \tfrac{1}{2}[-3 + (3 - \sin^2\gamma)\,\Omega]$$
$$\tau_1 = \tfrac{3}{16}[\tfrac{2}{3}\tan^2\gamma + 5 - (5 - \sin^2\gamma)\,\Omega]$$
$$\tau_2 = \tfrac{15}{128}[\tfrac{4}{15}\tan^4\gamma - \tfrac{4}{3}\tan^2\gamma - 7 + (7 - \sin^2\gamma)\,\Omega]$$
$$\tau_3 = \tfrac{175}{2048}[\tfrac{16}{105}\tan^6\gamma - \tfrac{8}{15}\tan^4\gamma + 2\tan^2\gamma + 9 - (9 - \sin^2\gamma)\,\Omega]$$
$$\tau_4 = \tfrac{2205}{32768}[\tfrac{32}{315}\tan^8\gamma - \tfrac{32}{105}\tan^6\gamma + \tfrac{4}{5}\tan^4\gamma - \tfrac{4}{3}\tan^2\gamma - 11$$
$$+ (11 - \sin^2\gamma)\,\Omega] \quad \text{.........(38)}$$

It would not be difficult to find the general expressions for these functions, but it does not seem worth while to do so.

The equation (35) for determining the form of the ellipsoid involves the factor $\cos^2\gamma + \lambda\cos^2\beta + \epsilon$; if we write

$$M = \frac{\lambda\sin^2\gamma}{(1+\lambda)\cos^2\gamma + \epsilon}$$

this factor may be written in the form $[(1+\lambda)\cos^2\gamma + \epsilon][1 + M\kappa'^2]$. Hence the equation (35) may be written

$$[\tau_0 - \tau_1\kappa'^2 + \tau_2\kappa'^4 - \tau_3\kappa'^6 \ldots][(1+\lambda)\cos^2\gamma + \epsilon][1 + M\kappa'^2]$$
$$= \kappa'^2[\sigma_0 - \sigma_1\kappa'^2 + \sigma_2\kappa'^4 - \sigma_3\kappa'^6 \ldots][3 + \lambda + \cos^2\gamma + \eta]$$

If now we put

$$v_0 = \frac{\tau_0}{\sigma_0}, \qquad v_1 = \frac{\sigma_1}{\sigma_0} - \frac{\tau_1}{\tau_0}, \qquad v_2 = \frac{\sigma_2}{\sigma_0} - \frac{\tau_2}{\tau_0} - \frac{\sigma_1}{\sigma_0}v_1, \qquad v_3 = \frac{\sigma_3}{\sigma_0} - \frac{\tau_3}{\tau_0} - \frac{\sigma_1}{\sigma_0}v_2 - \frac{\sigma_2}{\sigma_0}v_1$$

we have $$\frac{\tau_0 - \tau_1\kappa'^2 + \tau_2\kappa'^4 \ldots}{\sigma_0 - \sigma_1\kappa'^2 + \sigma_2\kappa'^4 \ldots} = v_0[1 + v_1\kappa'^2 - v_2\kappa'^4 + v_3\kappa'^6 - \ldots]$$

Hence our equation may be written

$$\frac{v_0[(1+\lambda)\cos^2\gamma + \epsilon]}{3 + \lambda + \cos^2\gamma + \eta}(1 + v_1\kappa'^2 - v_2\kappa'^4 + v_3\kappa'^6 \ldots)(1 + M\kappa'^2) = \kappa'^2$$

Whence on writing

$$L = \frac{3 + \lambda + \cos^2\gamma + \eta}{(1+\lambda)\cos^2\gamma + \epsilon}$$

we have $$\kappa'^2 = \frac{1 + (Mv_1 - v_2)\kappa'^4 - (Mv_2 - v_3)\kappa'^6 \ldots}{L/v_0 - M - v_1} \quad \text{..............(39)}$$

The determination of L for given value of γ involves that of η and ϵ, and these can only be found from an approximate preliminary solution of the whole problem. But when L is known approximately, the solution of (39) is very simple, for we first neglect the terms in κ'^4 and κ'^6 on the right-hand side, and so determine a first approximation to κ'. As a fact I have not included the term in κ'^6 in my computations, because it would not make so much as 1′ difference in the value of $\cos^{-1}\kappa'$.

For Roche's problem when ϵ and η are neglected the solution is very short, but when these terms are included the computation is laborious.

We now turn to the determination of the radius vector.

We have $$\mathfrak{A}_1 - \mathsf{A}_1{}^1 = \frac{kc^2}{3\lambda r^3}(3 + \lambda + \cos^2\gamma + \eta)$$

Since $c^2 = \dfrac{k^2}{\sin^2\beta}$ and $\dfrac{k^3 \cos\gamma \cos\beta}{\sin^3\beta} = \dfrac{\lambda a^3}{1+\lambda}$, we have

$$\frac{kc^2}{3\lambda r^3} = \frac{1}{3(1+\lambda)} \frac{\kappa \sin\gamma}{\cos\beta\cos\gamma} \cdot \frac{a^3}{r^3}$$

Therefore

$$\frac{\kappa}{\sin\gamma}(\tau_0 - \tau_1\kappa'^2 + \ldots) = \frac{\kappa\sin\gamma}{3(1+\lambda)\cos\beta\cos\gamma}\frac{a^3}{r^3}(3 + \lambda + \cos^2\gamma + \eta)$$

Whence $$\frac{a^3}{r^3} = 3(1+\lambda)\frac{\cos\beta\cos\gamma}{\sin^2\gamma}\,\frac{\tau_0 - \tau_1\kappa'^2 + \tau_2\kappa'^4 - \tau_3\kappa'^6\ldots}{3 + \lambda + \cos^2\gamma + \eta} \quad \text{........(40)}$$

Thus a table of values of τ_0, τ_1, τ_2, τ_3 enables us to compute r for a given value of γ, when κ' has been found.

In the following tables the υ's and τ's were computed for the even degrees of γ and interpolated for the odd degrees. These functions are found as the differences between large numbers, and therefore great care would be required to determine them with a very high degree of accuracy. The differences of the tabulated numbers do not run with perfect smoothness, showing that there are residual errors of one or two units in the last place of decimals. The accuracy is however amply sufficient for the end in view, and it would have been wasteful to spend more time over the computations.

Table of Auxiliary Functions.

γ	Log v_0	v_1	Log v_1	Log v_2	Log v_3	τ_0	Log τ_1	Log τ_2	Log τ_3
°									
30	9·94493	0·14279	9·15470	7·9576	7·15	0·010592	6·85440	5·9805	5·222
31	9·94085	0·15461	9·18925	8·0264	7·25	0·012124	6·94587	6·1063	5·378
32	9·93661	0·16722	9·22329	8·0941	7·36	0·013824	7·03507	6·2288	5·533
33	9·93222	0·18065	9·25684	8·1607	7·47	0·015702	7·12194	6·3478	5·685
34	9·92764	0·19493	9·28988	8·2263	7·58	0·017774	7·20666	6·4639	5·834
35	9·92285	0·21006	9·32234	8·2906	7·67	0·020052	7·28937	6·5780	5·979
36	9·91786	0·22613	9·35436	8·3541	7·76	0·022554	7·37022	6·6899	6·122
37	9·91269	0·24327	9·38608	8·4173	7·87	0·025294	7·44939	6·7994	6·262
38	9·90731	0·26150	9·41747	8·4800	7·97	0·028290	7·52697	6·9069	6·400
39	9·90167	0·28091	9·44857	8·5418	8·06	0·031559	7·60306	7·0127	6·536
40	9·89578	0·30160	9·47943	8·6031	8·14	0·035121	7·67777	7·1167	6·669
41	9·88965	0·32368	9·51011	8·6646	8·23	0·038994	7·75120	7·2192	6·802
42	9·88327	0·34724	9·54063	8·7259	8·32	0·043203	7·82342	7·3202	6·932
43	9·87665	0·37244	9·57105	8·7869	8·42	0·047768	7·89452	7·4200	7·061
44	9·86974	0·39937	9·60138	8·8476	8·51	0·052713	7·96459	7·5186	7·189
45	9·86257	0·42823	9·63168	8·9081	8·60	0·058064	8·03371	7·6162	7·316
46	9·85509	0·45917	9·66197	8·9685	8·69	0·063847	8·10195	7·7128	7·442
47	9·84729	0·49239	9·69231	9·0290	8·79	0·070093	8·16939	7·8087	7·567
48	9·83915	0·52812	9·72274	9·0896	8·87	0·076830	8·23608	7·9036	7·691
49	9·83069	0·56656	9·75325	9·1505	8·96	0·084093	8·30210	7·9981	7·814
50	9·82187	0·60804	9·78393	9·2117	9·05	0·091916	8·36752	8·0920	7·938
51	9·81263	0·65278	9·81477	9·2733	9·15	0·100336	8·43241	8·1855	8·061
52	9·80300	0·70121	9·84585	9·3353	9·24	0·109394	8·49682	8·2786	8·183
53	9·79298	0·75372	9·87721	9·3976	9·33	0·119134	8·56082	8·3716	8·306
54	9·78252	0·81075	9·90889	9·4606	9·43	0·129601	8·62447	8·4643	8·429
55	9·77159	0·87288	9·94095	9·5245	9·52	0·140845	8·68785	8·5570	8·552
56	9·76016	0·94064	9·97342	9·5893	9·62	0·152919	8·75103	8·6500	8·676
57	9·74819	1·01471	·00634	9·6550	9·72	0·165883	8·81407	8·7431	8·801
58	9·73565	1·09590	·03977	9·7216	9·82	0·179801	8·87703	8·8367	8·926
59	9·72252	1·18517	·07378	9·7893	9·92	0·194740	8·93999	8·9306	9·052
60	9·70874	1·28362	·10844	9·8582	·02	0·210779	9·00303	9·0252	9·179
61	9·69426	1·39268	·14385	9·9287	·13	0·227997	9·06626	9·1206	9·308
62	9·67910	1·51349	·17998	·0006	·24	0·246485	9·12966	9·2168	9·439
63	9·66319	1·6480	·21696	·0743	·36	0·266343	9·19331	9·3140	9·572

§ 12. *Determination of the Form of the Second Ellipsoid.*

The parameters Γ and K determine the form of the second ellipsoid in the same way that γ and κ determine the first. It is obvious that a/r is determinable in two ways, and therefore any given value of γ must correspond to a certain definite value of Γ. The fitting together of the two solutions can only be effected with accuracy by interpolation, but it would be so enormously laborious to find by mere conjecture the region in which to begin calculating with assumed values of Γ, that an approximate solution of the problem becomes a practical necessity.

After various trials I find that on neglecting ϵ and η and writing

$$\chi = \frac{\sin^2\gamma}{7(3+\lambda+\cos^2\gamma)}$$

the solution for κ' may be written approximately in the form

$$\kappa'^2 = \frac{(1+\lambda)\cos^2\gamma}{3+\lambda+\cos^2\gamma}[1-(9-7\lambda)\chi-(111+228\lambda-49\lambda^2)\chi^2$$
$$-\tfrac{1}{11}(7875+28185\lambda+42333\lambda^2-3773\lambda^3)\chi^3\ldots]\ldots\ldots(41)$$

If ϵ were added to the numerator of the factor outside the bracket, and η to the denominator, this formula would give nearly as good results as the more accurate method of the last section.

Also I find

$$\frac{a^3}{r^3} = \frac{\frac{2}{5}(1+\lambda)\cos\beta\cos\gamma\sin^2\gamma}{3+\lambda+\cos^2\gamma}[1+\tfrac{9}{2}(5+\lambda)\chi+\tfrac{1}{8}(3055+1718\lambda+175\lambda^2)\chi^2$$
$$+\frac{1}{2^4.3.11}(3533389+3222607\lambda+1021479\lambda^2+60025\lambda^3)\chi^3+\ldots]\ldots(42)$$

In order to obtain the desired approximation, it is necessary to express a^3/r^3 by a series which can be inverted; but this is not possible in the form just given, because $\cos\beta$ depends on κ and therefore involves χ. I find then by means of the above series for κ'^2 that

$$\cos\beta\cos\gamma = 1-\tfrac{7}{2}(7+\lambda)\chi+\tfrac{7}{8}(69-106\lambda-7\lambda^2)\chi^2$$
$$+\tfrac{7}{16}(253+753\lambda-1901\lambda^2-49\lambda^3)\chi^3\ldots$$

On introducing this in the above formula for a^3/r^3, I find

$$\frac{a^3}{r^3} = \tfrac{14}{5}(1+\lambda)\chi[1-(2-\lambda)\chi-(109+67\lambda+0\lambda^2)\chi^2$$
$$-\frac{1}{2^2.3.11}(157712+261395\lambda+97656\lambda^2+1568\lambda^3)\chi^3\ldots]$$

On writing $$\alpha = \tfrac{5}{14}\cdot\frac{1}{1+\lambda}\frac{a^3}{r^3}$$

and inverting the series we find

$$\chi = \alpha+(2-\lambda)\alpha^2+(117+59\lambda+2\lambda^2)\alpha^3$$
$$+\frac{1}{2^2.3.11}(306872+269975\lambda+5739\lambda^2+908\lambda^3)\alpha^4\ldots\ldots(43)$$

This series expresses a function of γ in series proceeding by powers of a^3/r^3, and a similar series must also connect a function of Γ with the radius vector, so as to determine the figure of the second ellipsoid appropriately. This second series may be written down by symmetry.

Since λ must now be replaced by $1/\lambda$, the function corresponding to α is $\frac{5}{14}\frac{\lambda}{1+\lambda}\frac{a^3}{r^3}$ or $\lambda\alpha$, and the function corresponding to χ is $\frac{\lambda\sin^2\Gamma}{7[(3+\cos^2\Gamma)\lambda+1]}$.

If then we write $$X = \frac{\sin^2 \Gamma}{7\left[(3 + \cos^2 \Gamma)\lambda + 1\right]}$$

the symmetrical series for the other ellipsoid is

$$X = \alpha + (2\lambda - 1)\alpha^2 + (117\lambda^2 + 59\lambda + 2)\alpha^3$$
$$+ \frac{1}{2^2 . 3 . 11}(306872\lambda^3 + 269975\lambda^2 + 5739\lambda^3 + 908)\alpha^4 \ldots$$

Now α is easily computed for the first ellipsoid, and then X is computed by the series. Thus we have

$$\sin^2 \Gamma = \frac{(4\lambda + 1) X}{\lambda X + \frac{1}{7}}$$

We obtain in this way a fairly accurate value of Γ corresponding to the value of γ which determines the first ellipsoid. We can then compute K′ by the method of the last section. We may thus obtain a good idea of the values of Γ and K with which it is necessary to work in order to obtain the final solution.

§ 13. *The Equilibrium of Two Ellipsoids joined by a Weightless Pipe.*

In § 3 the problem is considered of the equilibrium of two masses of liquid, each constrainedly spherical, when joined by a pipe without weight. It was shown that the condition determining the ratio of the masses for given radius vector is expressed by a certain equation which was written $f(\mathrm{a}/r) = 0$. Further, it was proved that if $f(\mathrm{a}/r)$ is positive the two spheres of liquid are too far apart to admit of junction, and if it is negative they are too near. Finally we found that all these solutions were unstable.

The solutions for the two spheres showed them to be always very close together, and as all the solutions for two ellipsoids, when they are in limiting stability, made them much further apart than were the two spheres, it seemed somewhat improbable that two ellipsoids could be similarly joined by a pipe, and certain that they would be unstable if such junction were possible. Nevertheless, it seemed conceivable that the additional terms, which must appear in $f(\mathrm{a}/r)$ when the constraint to spherical form is removed, might alter the conditions so that the junction of ellipsoids by a pipe should become possible. It thus became expedient to solve a problem analogous to that of § 3 when the two masses of liquid are ellipsoidal.

The conditions of equilibrium of two ellipsoids unjoined by a pipe are given in § 10, and the additional condition corresponding to junction by a pipe is

$$\frac{dV}{d\lambda} + \tfrac{1}{2}\omega^2 \frac{dI}{d\lambda} = 0$$

In the present investigation I shall neglect the higher ellipticities, denoted f_i^s and F_i^s, and terms of higher order than those in $1/r^5$.

With this degree of approximation we have

$$V = (\tfrac{4}{3}\pi\rho a^3)^2 \left[\frac{\lambda}{(1+\lambda)^2 r} + \tfrac{3}{10}\frac{\lambda^2}{(1+\lambda)^2}\psi + \tfrac{3}{10}\frac{1}{(1+\lambda)^2}\Psi\right] + (eE)_2$$

where $(eE)_2$ is given in (14) (with omission of terms in $1/r^7$).

Also
$$\psi = \frac{2}{k}\int_{\nu_0}^{\infty} \frac{d\nu}{(\nu^2-1)^{1/2}(\nu^2-1/\kappa^2)^{1/2}}$$

and Ψ has a symmetrical form in K and K.

We have besides $\omega^2 = \frac{4}{3}\pi\rho\frac{a^3}{r^3}(1+\zeta)$, where ζ is given in (24); and I is given in (22).

The differentiation with respect to λ and the subsequent re-arrangement of the equation are rather tedious, and I will not give the details of the operations. It may however be well to note that k, K are functions of λ, and that

$$\frac{d\psi}{d\lambda} = -\frac{1}{3\lambda(1+\lambda)}\psi, \qquad \frac{d\Psi}{d\lambda} = \frac{1}{3(1+\lambda)}\Psi$$

I find finally that the equation of condition is $f\left(\frac{a}{r}\right) = 0$, where

$$\begin{aligned}
f\left(\frac{a}{r}\right) = &-\tfrac{9}{2}\frac{1+\lambda^{1/3}+\lambda^{2/3}}{(1+\lambda)^{1/3}(1+\lambda^{1/3})}\frac{a}{r} - \frac{3a(\lambda\psi-\Psi)}{2(1+\lambda)^{1/3}(1-\lambda^{2/3})} \\
&- \frac{1}{(1+\lambda)^{1/3}(1-\lambda^{2/3})}\left[\frac{ac^2}{20r^3}\{48-20\lambda-(19-15\lambda)\cos^2\gamma-(9-25\lambda)\cos^2\beta\}\right. \\
&\quad + \frac{aC^2}{20r^3}\{20-48\lambda-(15-19\lambda)\cos^2\Gamma-(25-9\lambda)\cos^2 \mathrm{B}\} \\
&\quad + \frac{3ac^4}{560r^5}\{288-112\lambda-(260-140\lambda)\cos^2\gamma-(204-196\lambda)\cos^2\beta \\
&\qquad + (87-63\lambda)\cos^4\gamma + (59-91\lambda)\cos^4\beta \\
&\qquad + (30-70\lambda)\cos^2\beta\cos^2\gamma\} \\
&\quad + \frac{3aC^4}{560r^5}\{112-288\lambda-(140-260\lambda)\cos^2\Gamma-(196-204\lambda)\cos^2\mathrm{B} \\
&\qquad + (63-87\lambda)\cos^4\Gamma + (91-59\lambda)\cos^4\mathrm{B} \\
&\qquad + (70-30\lambda)\cos^2\mathrm{B}\cos^2\Gamma\} \\
&\quad + \frac{3ac^2C^2}{40r^5}\{40(1-\lambda)-(18-22\lambda)\cos^2\gamma-(22-18\lambda)\cos^2\Gamma \\
&\qquad -(14-26\lambda)\cos^2\beta-(26-14\lambda)\cos^2\mathrm{B} \\
&\qquad + 15(1-\lambda)(\cos^2\gamma\cos^2\Gamma+\cos^2\beta\cos^2\mathrm{B}) \\
&\qquad \left. + (7-3\lambda)\cos^2\gamma\cos^2\mathrm{B} + (3-7\lambda)\cos^2\Gamma\cos^2\beta\}\right] \qquad (44)
\end{aligned}$$

In this expression c and C are respectively the longest semi-axes of the two ellipsoids, which are pointed at one another.

We may derive ψ and Ψ from Legendre's tables of elliptic integrals for

$$\psi = \frac{2}{c \sin \gamma} F(\kappa, \gamma), \qquad \Psi = \frac{2}{2C \sin \Gamma} F(\mathrm{K}, \Gamma)$$

or we may expand the integrals in powers of $\kappa'^2 \tan^2 \gamma$ and obtain the approximate formula

$$\psi = \frac{2}{c}[1 + (\Omega - 1)(1 + \tfrac{1}{4}\kappa'^2 + \tfrac{9}{64}\kappa'^4 + \tfrac{25}{256}\kappa'^6 \ldots) - \tfrac{1}{4}\kappa'^2 \tan^2 \gamma (1 + \tfrac{3}{16}\kappa'^2 + \tfrac{25}{192}\kappa'^4 \ldots)$$
$$+ \tfrac{3}{32}\kappa'^4 \tan^4 \gamma (1 + \tfrac{5}{36}\kappa'^2 \ldots) - \tfrac{5}{96}\kappa'^6 \tan^6 \gamma (1 + \ldots)] \ldots (45)$$

where $\Omega = \dfrac{1}{\sin \gamma} \log_e \dfrac{1 + \sin \gamma}{\cos \gamma}$. The formula for Ψ is of course symmetrical.

It should be noted that when the two ellipsoids reduce to spheres, we have

$$\gamma = \beta = \Gamma = \mathrm{B} = 0, \quad c = \frac{\lambda^{1/3} \mathrm{a}}{(1 + \lambda)^{1/3}}, \quad C = \frac{1}{(1 + \lambda)^{1/3}} \mathrm{a}$$

$$\psi = \frac{2}{c} = \frac{2}{\mathrm{a}} \frac{(1 + \lambda)^{1/3}}{\lambda^{1/3}}, \quad \Psi = \frac{2}{\mathrm{a}}(1 + \lambda)^{1/3}$$

Thus
$$f\left(\frac{\mathrm{a}}{r}\right) = -\tfrac{9}{2} \frac{1 + \lambda^{1/3} + \lambda^{2/3}}{(1 + \lambda)^{1/3}(1 + \lambda^{1/3})} \frac{\mathrm{a}}{r} + 3 + \frac{\mathrm{a}^3}{r^3}$$

This is the form obtained in the solution of the restricted problem of § 3.

We conclude that if $f(\mathrm{a}/r)$, as expressed in (44), with values derived from any solution of the problem of the equilibrium of two ellipsoids unconnected by a pipe, is positive, the two figures are too far apart to admit of junction, and *vice versâ*. I have in fact always found it positive, although always diminishing as r diminishes, so that junction would seem to be always impossible, at least so long as the approximation retains any validity. This might indicate that there is no figure of equilibrium shaped like an hourglass with a thin neck. However, I return to this subject in discussing numerical solutions, and in the summary of results.

In the case where the two masses are equal, $\lambda = 1$, and the above formula for $f(\mathrm{a}/r)$ fails by becoming indeterminate. As this is a case of especial interest, it must be considered.

Since the two shapes are now exactly alike, we may take κ, γ, β, k to define either of them.

When $\lambda = 1$, the first term of $f(\mathrm{a}/r)$ becomes $-\dfrac{27}{4 \cdot 2^{1/3}} \dfrac{\mathrm{a}}{r}$

The second term becomes

$$-\frac{3\mathrm{a}\psi(\lambda - 1)}{2(1 + \lambda)^{1/3}(1 - \lambda^{2/3})} = \frac{3\psi\mathrm{a}(1 + \lambda^{1/3} + \lambda^{2/3})}{2(1 + \lambda)^{1/3}(1 + \lambda^{1/3})} = \frac{9\psi\mathrm{a}}{4 \cdot 2^{1/3}}$$

All the terms in $1/r^3$ are of one of the two forms

$$F\left[\frac{c^2-\lambda C^2}{1-\lambda^{2/3}}\right] \quad \text{or} \quad G\left[\frac{\lambda c^2-C^2}{1-\lambda^{2/3}}\right]$$

Now,

$$c^2=\frac{\mathrm{a}^2}{(\cos\beta\cos\gamma)^{2/3}}\frac{\lambda^{2/3}}{(1+\lambda)^{2/3}}, \qquad C^2=\frac{\mathrm{a}^2}{(\cos\mathrm{B}\cos\Gamma)^{2/3}}\frac{1}{(1+\lambda)^{2/3}}$$

In the limit $\beta=\mathrm{B}$, $\gamma=\Gamma$, and $\lambda=1$; and we find that the first of these forms becomes $\frac{1}{2}Fc^2$, and the second $-\frac{5}{2}Gc^2$.

Again, of the terms in $1/r^5$, those in c^4 and C^4 are of one of the two forms

$$F\left[\frac{c^4-\lambda C^4}{1-\lambda^{2/3}}\right] \quad \text{or} \quad G\left[\frac{\lambda c^4-C^4}{1-\lambda^{2/3}}\right]$$

In the limit the first of these reduces to $-\frac{1}{2}Fc^4$, and the second to $-\frac{7}{2}Gc^4$.

The last term in $1/r^5$ has a common factor $1-\lambda$, when $\gamma=\Gamma$, $\beta=\mathrm{B}$, and

$$\frac{1-\lambda}{(1+\lambda)^{1/3}(1-\lambda^{2/3})}=\frac{1+\lambda^{1/3}+\lambda^{2/3}}{(1+\lambda)^{1/3}(1+\lambda^{1/3})}=\frac{3}{2\,.\,2^{1/3}}$$

By means of these transformations we find for $\lambda=1$

$$f\left(\frac{\mathrm{a}}{r}\right)=\frac{3}{16^{1/3}}\left[-\frac{9}{2}\frac{\mathrm{a}}{r}+\psi\mathrm{a}+\frac{\mathrm{a}c^2}{30r^3}(-74+47\cos^2\gamma+67\cos^2\beta)\right.$$
$$+\frac{\mathrm{a}c^4}{70r^5}(-272+300\cos^2\gamma+356\cos^2\beta-123\cos^4\gamma-151\cos^4\beta$$
$$\left.-110\cos^2\beta\cos^2\gamma)\right] \quad \ldots(46)$$

If we put $\beta=\gamma=0$, and note that c becomes $\mathrm{a}/2^{1/3}$, this expression reduces as before to the correct form.

§ 14. *Ellipsoidal Harmonic Deformations of the Third Order.*

The two ellipsoids whose forms have been determined are subject to further deformation by harmonic inequalities. The expression for the ellipticity f_i^s corresponding to all the cosine-harmonic functions for i greater than 2 and s even is given in (25), viz. :—

$$f_i^s=-\tfrac{1}{3}(2i+1)\frac{\mathfrak{Q}_i^s(r/k)\,\mathfrak{P}_i^s(\nu_0)\,\mathfrak{P}_i^s(1)\,\mathfrak{C}_i^s(\frac{1}{2}\pi)}{\lambda\mathfrak{T}_i^s\left\{\mathfrak{A}_i^s-\mathsf{A}_1-\dfrac{k^3}{3\lambda r^3}\dfrac{\cos^2\gamma}{\sin^2\beta}\left[1+\dfrac{3(3+\kappa^2)}{14\kappa^2}\dfrac{k^2}{r^2}+\ldots\right]\right\}} \ldots(47)$$

I shall begin by considering the ellipticities f_3 and f_3^2 corresponding to $i=3$ and $s=0$, 2.

I define $\qquad \mathfrak{P}_3^s(\nu)=\nu(\kappa^2\nu-q^2)$, $(s=0, 2)$......................(48)

where $q^2=\frac{2}{5}[1+\kappa^2\mp\sqrt{(1-\frac{7}{4}\kappa^2+\kappa^4)}]$, with upper sign for $s=0$ and lower for $s=2$.

Then $$\mathfrak{Q}_3^s(\nu) = \mathfrak{P}_3^s(\nu)\int_\nu^\infty \frac{d\nu}{[\mathfrak{P}_3^s(\nu)]^2(\nu^2-1)^{1/2}(\nu^2-1/\kappa^2)^{1/2}}$$

Since ν is always greater than unity, the function under the integral sign may be expanded in powers of $1/\nu$, as in (7), (8), (9) of § 5, and the integration may then be effected. In this way I find

$$\mathfrak{Q}_3^s(\nu) = \frac{1}{7\kappa^2\nu^4}\left\{1 + \frac{7(1+\kappa^2)+10q^2}{2\,.\,9\kappa^2\nu^2} + \frac{7[9(3\kappa^4+2\kappa^2+3)+28q^2(\kappa^2+1)+40q^4]}{8\,.\,9\,.\,11\kappa^4\nu^4} + \dots\right\} \quad \dots(49)$$

In all the cases we have to consider κ^2 is nearly unity and κ'^2 is small. Then, since $q^2 = \frac{1}{5}[4 - 2\kappa'^2 \mp \sqrt{(1-\kappa'^2+4\kappa'^4)}]$, and since the function under the square root may be expanded in powers of κ'^2, we may obtain approximate expressions for q^2 in the two cases $s=0$, $s=2$.

When $s=0$, we have

$$q^2 = \tfrac{3}{5}(1 - \tfrac{1}{2}\kappa'^2 - \tfrac{5}{8}\kappa'^4 - \tfrac{5}{16}\kappa'^6 \dots)$$

When $s=2$, $$q^2 = 1 - \tfrac{1}{2}\kappa'^2 + \tfrac{3}{8}\kappa'^4 + \tfrac{3}{16}\kappa'^6 \dots$$

If we substitute these values for q^2 in (49), and express the functions of κ^2 therein in terms of κ'^2, we obtain the following results:—

$$\mathfrak{Q}_3\left(\frac{k}{r}\right) = \frac{k^4}{7\kappa^2 r^4}\left\{1 + \tfrac{10}{9}(1 + \tfrac{1}{2}\kappa'^2 + \tfrac{5}{16}\kappa'^4 + \tfrac{7}{32}\kappa'^6 \dots)\frac{k^2}{r^2} + \tfrac{35}{33}(1 + \kappa'^2 + \kappa'^4 + \kappa'^6 \dots)\frac{k^4}{r^4}\dots\right\}$$

$$\mathfrak{Q}_3^2\left(\frac{k}{r}\right) = \frac{k^4}{7\kappa^2 r^4}\left\{1 + \tfrac{4}{3}(1 + \tfrac{1}{2}\kappa'^2 + \tfrac{21}{32}\kappa'^4 + \tfrac{47}{64}\kappa'^6 \dots)\frac{k^2}{r^2} + \tfrac{49}{33}(1 + \kappa'^2 + \tfrac{45}{28}\kappa'^4 + \tfrac{31}{14}\kappa'^6 \dots)\frac{k^4}{r^4}\dots\right\}$$

Shorter forms may be given to these by making the expansions run in powers of $1/\kappa r^2$; we then have

$$\left.\begin{aligned} \mathfrak{Q}_3\left(\frac{k}{r}\right) &= \frac{k^4}{7\kappa^2 r^4}\left\{1 + \tfrac{10}{9}(1 + 0\kappa'^2 - \tfrac{1}{16}\kappa'^4 - \tfrac{1}{8}\kappa'^6 \dots)\frac{k^2}{\kappa r^2} + \tfrac{35}{33}\frac{k^4}{\kappa^2 r^4}\dots\right\} \\ \mathfrak{Q}_3^2\left(\frac{k}{r}\right) &= \frac{k^4}{7\kappa^2 r^4}\left\{1 + \tfrac{4}{3}(1 + 0\kappa'^2 + \tfrac{9}{32}\kappa'^4 + \tfrac{9}{32}\kappa'^6 \dots)\frac{k^2}{\kappa r^2} + \tfrac{49}{33}(1 + 0\kappa'^2 + \tfrac{17}{28}\kappa'^4 + \tfrac{17}{28}\kappa'^6 \dots)\frac{k^4}{\kappa^2 r^4}\dots\right\} \end{aligned}\right\} \quad (50)$$

It will however suffice for our purposes to take

$$\left.\begin{aligned} \mathfrak{Q}_3\left(\frac{k}{r}\right) &= \frac{k^4}{7\kappa^2 r^4}\left\{1 + \tfrac{10}{9}\frac{k^2}{\kappa r^2}\right\} \\ \mathfrak{Q}_3^2\left(\frac{k}{r}\right) &= \frac{k^4}{7\kappa^2 r^4}\left\{1 + \tfrac{4}{3}\frac{k^2}{\kappa r^2}\right\} \end{aligned}\right\} \quad \dots\dots\dots\dots\dots\dots\dots(51)$$

The next task is to determine the product $\mathfrak{P}_3^s(\nu_0)\,\mathfrak{P}_3^s(1)\,\mathrm{C}_3^s(\tfrac{1}{2}\pi)$ $(s=0,\ 2)$.

The form of the ellipsoid is determined by ν_0, where $\nu_0 = \dfrac{1}{\kappa\sin\gamma} = \dfrac{1}{\sin\beta}$.

If we write $\Delta_1^2 = 1 - q^2\sin^2\gamma$, with the definition of $\mathfrak{P}_3^s$ given above in (48), we have

$$\mathfrak{P}_3^s(\nu_0) = \frac{\Delta_1^2}{\kappa\sin^3\gamma} \qquad \text{.............................(52)}$$

It will be remembered that q has a different value according as $s=0$ or 2.

I now make the following definitions,

$$\mathfrak{P}_3^s(\mu) = \mu(\kappa^2\mu^2 - q^2), \qquad \mathrm{C}_3^s(\phi) = (q'^2 - \kappa'^2\cos^2\phi)\sqrt{(1-\kappa'^2\cos^2\phi)}$$

so that

$$\mathfrak{P}_3^s(1)\,\mathrm{C}_3^s(\tfrac{1}{2}\pi) = q'^2(\kappa^2 - q^2) \qquad \text{....................(53)}$$

It is easy to show that rigorously

$$q'^2(\kappa^2-q^2) = \tfrac{2}{25}\left[1 - \kappa'^2 - \kappa'^4 \pm (1-\tfrac{1}{2}\kappa'^2)\sqrt{(1-\kappa'^2+4\kappa'^4)}\right] \quad (s=0,\ 2)$$

Whence approximately, with the upper sign for $s=0$,

$$q'^2(\kappa^2-q^2) = \tfrac{4}{25}(1 - \kappa'^2 + \tfrac{9}{16}\kappa'^4 + 0\kappa'^6 \ldots) \qquad \text{..............(54)}$$

With the lower sign for $s=2$,

$$q'^2(\kappa^2-q^2) = -\tfrac{1}{4}\kappa'^4(1 + 0\kappa'^2 - \tfrac{9}{16}\kappa'^4 \ldots) \qquad \text{..............(55)}$$

The three last expressions (53), (54), (55), give the two values of $\mathfrak{P}_3^s(1)\,\mathrm{C}_3^s(\tfrac{1}{2}\pi)$.

We now turn to the functions $\mathfrak{T}_3^s$. The general definition of § 5 was

$$\mathfrak{T}_i^s = \frac{(2i+1)\rho}{3M}\int [\mathfrak{P}_i^s(\mu)\,\mathfrak{C}_i^s(\phi)]^2\, p\, d\sigma$$

where M was the mass of the ellipsoid. Hence in the cases under consideration

$$\mathfrak{T}_3^s = \frac{7\sin^3\beta}{4\pi k^3\cos\beta\cos\gamma}\int [\mathfrak{P}_3^s(\mu)\,\mathrm{C}_3^s(\phi)]^2\, p\, d\sigma \quad (s=0,\ 2)$$

These integrals are evaluated in (38) p. 411 of my paper on "Integrals"; whence I find

$$\mathfrak{T}_3^s = \frac{2^3}{5^4} D\left[(1-\tfrac{1}{2}\kappa'^2)(1-\kappa'^2-\tfrac{8}{8}\kappa'^4) + (1-\kappa'^2+\tfrac{9}{8}\kappa'^4)\,D\right]$$

where $D = \pm\sqrt{(1-\kappa'^2+4\kappa'^4)}$, with upper sign for $s=0$ and lower for $s=2$.

It will, however, suffice if we use the development of D in powers of κ'^2. The result is, in fact, given in the equations next below (38) in "Integrals," and they are

$$\left.\begin{aligned} \mathfrak{T}_3 &= (\tfrac{2}{5})^4(1 - 2\kappa'^2 + \tfrac{49}{16}\kappa'^4 \ldots) \\ \mathfrak{T}_3^2 &= \tfrac{1}{8\cdot 5}\kappa'^4(1 - \kappa'^2 + \tfrac{31}{16}\kappa'^4 \ldots) \end{aligned}\right\} \qquad \text{................(56)}$$

From (53), (54), (55), and (56) we now find

$$\left.\begin{aligned}\frac{\mathfrak{P}_3(1)\,\mathsf{C}_3(\tfrac{1}{2}\pi)}{\mathfrak{T}_3}&=\tfrac{25}{4}\,\frac{1-\kappa'^2+\frac{9}{16}\kappa'^4\ldots}{1-2\kappa'^2+\frac{49}{16}\kappa'^4\ldots}=\tfrac{25}{4}(1+\kappa'^2-\tfrac{1}{2}\kappa'^4\ldots)\\ \frac{\mathfrak{P}_3^2(1)\,\mathsf{C}_3^2(\tfrac{1}{2}\pi)}{\mathfrak{T}_3^2}&=-\tfrac{15}{4}\,\frac{1+0\kappa'^2-\frac{9}{16}\kappa'^4\ldots}{1-\kappa'^2+\frac{31}{16}\kappa'^4\ldots}=-\tfrac{15}{4}(1+\kappa'^2-\tfrac{3}{2}\kappa'^4\ldots)\end{aligned}\right\}\quad(57)$$

Thus from (51), (52), (57), we find

$$\left.\begin{aligned}&-\frac{7}{3\lambda\mathfrak{T}_3}\mathfrak{Q}_3\left(\frac{r}{k}\right)\mathfrak{P}_3(\nu_0)\,\mathfrak{P}_3(1)\,\mathsf{C}_3(\tfrac{1}{2}\pi)\\ &\qquad=-\frac{25\Delta_1^2}{12\lambda\sin^3\beta}(1+\kappa'^2-\tfrac{1}{2}\kappa'^4\ldots)\frac{k^4}{r^4}\left(1+\tfrac{10}{9}\frac{k^2}{\kappa r^2}\right)\\ &-\frac{7}{3\lambda\mathfrak{T}_3^2}\mathfrak{Q}_3^2\left(\frac{r}{k}\right)\mathfrak{P}_3^2(\nu_0)\,\mathfrak{P}_3^2(1)\,\mathsf{C}_3^2(\tfrac{1}{2}\pi)\\ &\qquad=\frac{5\Delta_1^2}{4\lambda\sin^3\beta}(1+\kappa'^2-\tfrac{3}{2}\kappa'^4)\frac{k^4}{r^4}\left(1+\tfrac{4}{3}\frac{k^2}{\kappa r^2}\right)\end{aligned}\right\}\quad(58)$$

By (47) f_3^s is equal to the above expressions divided by

$$\mathfrak{A}_3^s-\mathsf{A}_1^1-\frac{k^3}{3\lambda r^3}\frac{\cos^2\gamma}{\sin^2\beta}\left[1+\frac{3(3+\kappa^2)}{14\kappa^2}\frac{k^2}{r^2}\ldots\right]$$

It therefore remains to determine $\mathfrak{A}_3^s$ and A_1^1*.

By § 4 of "The Pear-shaped Figure" (p. 299 and (30) p. 301) we have

$$\mathsf{A}_1^1=\frac{\kappa}{\sin^2\gamma}\left(\frac{1}{\kappa^2}\mathrm{F}-\frac{1}{\kappa^2}\mathrm{E}\right)$$

$$\mathfrak{A}_3^s=\frac{\kappa\Delta_1^4}{\sin^6\gamma}\left[\frac{7q'^2-1}{2\kappa^2q^4q'^2}\mathrm{F}-\frac{2q'^4+5q'^2-1}{2\kappa^2q^4q'^4}\mathrm{E}+\frac{(4q'^2-1)}{2q^4q'^4}\frac{\Delta\sin\gamma\cos\gamma}{\Delta_1^2}\right]$$

where $$q^2=\tfrac{2}{5}[1+\kappa^2\mp(1-\tfrac{7}{4}\kappa^2+\kappa^4)^{1/2}]$$

with the upper sign for $s=0$, and the lower for $s=2$.

§ 15. *The Values of $\mathfrak{Q}_i^s$ for Higher Harmonic Terms.*

For higher harmonic terms it is necessary to adopt the approximate forms of the functions investigated in "Harmonics." The development is there carried out in powers of a parameter β, which I will now write β_0 to avoid confusion; this parameter is equal to $\dfrac{1-\kappa^2}{1+\kappa^2}$, or to $\tfrac{1}{2}\kappa'^2+\tfrac{1}{4}\kappa'^4\ldots$ of the present paper.

* [By a strange oversight I failed to remember that I had evaluated these integrals rigorously, and proceeded to an approximate treatment. The rigorous results are now introduced.]

The functions are here defined by

$$\mathfrak{Q}_i^s(\nu) = \mathfrak{P}_i^s(\nu) \int_\nu^\infty \frac{d\nu}{[\mathfrak{P}_i^s(\nu)]^2 (\nu^2-1)^{1/2} \left(\nu^2 - \frac{1+\beta_0}{1-\beta_0}\right)^{1/2}}$$

but in the notation of § 10 of "Harmonics" this would be called $\mathfrak{Q}_i^s(\nu)/\mathfrak{C}_i^s$. Thus, if $[\mathfrak{Q}_i^s(\nu)]$ denotes that function as defined in "Harmonics," we have

$$\mathfrak{Q}_i^s(\nu) = \frac{[\mathfrak{Q}_i^s(\nu)]}{\mathfrak{C}_i^s}$$

We have for the approximate expression for $\mathfrak{P}_i^s$

$$\mathfrak{P}_i^s(\nu) = P_i^s(\nu) + \beta_0 q_{s-2} P_i^{s-2}(\nu) + \beta_0 q_{s+2} P_i^{s+2}(\nu) + \beta_0^2 q_{s-4} P_i^{s-4}(\nu) + \beta_0^2 q_{s+4} P_i^{s+4}(\nu)$$

The investigation on p. 228 of "Harmonics" shows that the leading term of $[\mathfrak{Q}_i^s(\nu)]$ is

$$(-)^s \frac{2^i\, i\,!}{2i+1\,!} \frac{i+s\,!}{\nu^{i+1}} \left[1 + \beta_0 q_{s-2} \frac{i+s-2\,!}{i+s\,!} + \beta_0 q_{s+2} \frac{i+s+2\,!}{i+s\,!} + \beta_0^2 q_{s-4} \frac{i+s-4\,!}{i+s\,!} + \beta_0^2 q_{s+4} \frac{i+s+4\,!}{i+s\,!}\right]$$

This has to be divided by $\mathfrak{C}_i^s$, the formula for which is given in § 10 of "Harmonics," and we thus obtain the leading term of $\mathfrak{Q}_i^s(\nu)$.

For the second term it will suffice if we take β_0 as zero, so that it is only necessary to consider $Q_i^s(\nu)$, which is equal to $(\nu^2-1)^{\frac{1}{2}s} \dfrac{d^s}{d\nu^s} Q_i(\nu)$.

Since $$Q_i(\nu) = \frac{2^i (i!)^2}{2i+1\,!} \left[\frac{1}{\nu^{i+1}} + \frac{i+2\,!}{2\,.\,1\,!\,i\,!\,(2i+3)\,\nu^{i+3}} + \dots\right]$$

by differentiation, and by expansion of $(\nu^2-1)^{\frac{1}{2}s}$ in powers of $1/\nu^2$ we obtain

$$Q_i^s(\nu) = (-)^s \frac{2^i\,.\,i\,!\,.\,i+s\,!}{2i+1\,!\,\nu^{i+1}} \left[1 + \frac{(i+2)(i+1)+s^2}{2(2i+3)\,\nu^2} + \dots\right]$$

Accordingly, in order to find the second term to the degree of approximation adopted, it is merely necessary to multiply the leading term by

$$1 + \frac{(i+2)(i+1)+s^2}{2(2i+3)\,\nu^2}$$

In order to find the leading term explicitly we have to insert for the q's their values, and after some tedious reductions I find

$$\mathfrak{Q}_i^s\left(\frac{r}{k}\right) = \frac{2^i\,.\,i\,!\,i-s\,!}{2i+1\,!} \frac{k^{i+1}}{r^{i+1}} \{1 + \tfrac{1}{4}\beta_0(\Sigma + 2i) + \tfrac{1}{64}\beta_0^2[-2\Upsilon + \Sigma^2(s^2+3) + 2\Sigma(6i-1+2s^2) + 2i(3i+1) - s^2]\} \times \left\{1 + \frac{(i+2)(i+1)+s^2}{2(2i+3)} \frac{k^2}{r^2}\right\} \quad (61)$$

where $$\Sigma = \frac{i(i+1)}{s^2-1}, \quad \Upsilon = \frac{(i-1)i(i+1)(i+2)}{s^2-4}$$

This formula fails for the cases of $s=0$ and $s=2$, and these cases have to be treated apart. Following a parallel procedure I find

$$\mathfrak{Q}_i^2\left(\frac{r}{k}\right)=\frac{2^i\,.\,i!\,\overline{i-2}!}{\overline{2i+1}!}\frac{k^{i+1}}{r^{i+1}}\{1+\tfrac{1}{4}\beta_0(\Sigma+2i)+\tfrac{1}{512}\beta_0^2[29\Sigma^2+(144i+298)\Sigma$$

$$-48i-40]\}\times\left\{1+\frac{(i+2)(i+1)+4}{2(2i+3)}\frac{k^2}{r^2}\right\}\quad\text{...(62)}$$

$$\mathfrak{Q}_i\left(\frac{r}{k}\right)=\frac{2^i\,(i!)^2}{\overline{2i+1}!}\frac{k^{i+1}}{r^{i+1}}\{1-\tfrac{1}{4}\beta_0 i(i-1)+\tfrac{1}{128}\beta_0^2 i(i-1)(7i^2-3i-6)\}$$

$$\times\left\{1+\frac{(i+2)(i+1)}{2(2i+3)}\frac{k^2}{r^2}\right\}\quad\text{...(63)}$$

The values for i less than 3 are not required, and when $i=3$ these formulæ are found to agree *mutatis mutandis* with those of the last section.

It is pretty clear from general considerations that the higher inequalities corresponding to harmonics other than the zonal ones must be very small. I have, in fact, computed the third tesseral harmonic inequality ($i=3$, $s=2$), and find that it is so very minute compared with the third zonal inequality ($i=3$, $s=0$) as to be negligible. Accordingly it appeared to be a waste of time to develop formulæ for any other than zonal inequalities for values of i greater than 3. Thus of the formulæ just determined the only one of which actual use is made is (63).

§ 16. *The Fourth Zonal Harmonic Inequality.*

In developing the expressions for the higher harmonic inequalities it seems to be most convenient to retain the parameter β_0, which is equal to $\dfrac{1-\kappa^2}{1+\kappa^2}$, instead of developing in powers of κ'^2 as heretofore.

On putting $i=4$ in (63), we find

$$\mathfrak{Q}_4\left(\frac{r}{k}\right)=\frac{8}{5\,.\,7\,.\,9}\frac{k^5}{r^5}\left(1+\tfrac{15}{11}\frac{k^2}{r^2}\right)(1-3\beta_0+\tfrac{141}{16}\beta_0^2)$$

With the notation of "Harmonics" we have

$$\mathfrak{P}_4(\nu)=P_4(\nu)+\tfrac{1}{4}\beta_0 P_4^2(\nu)+\tfrac{1}{128}\beta_0^2 P_4^4(\nu)$$

$$\mathfrak{C}_4(\phi)=1-5\beta_0\cos 2\phi+\tfrac{35}{16}\beta_0^2\cos 4\phi$$

Accordingly

$$\mathfrak{P}_4(1)=P_4(1)=1$$

$$\mathfrak{C}_4(\tfrac{1}{2}\pi)=1+5\beta_0+\tfrac{35}{16}\beta_0^2$$

Again, from § 22 of "Harmonics" for type EEC, $i=4$, $s=0$, we have

$$\int[\mathfrak{P}_4\mathfrak{C}_4]^2 p\,d\sigma=\tfrac{4}{9}\pi k^3\nu_0(\nu_0^2-1)^{1/2}\left(\nu_0^2-\frac{1+\beta_0}{1-\beta_0}\right)^{1/2}[1+10\beta_0+\tfrac{325}{8}\beta_0^2]$$

But $\mathfrak{T}_4$ is this integral multiplied by 9 and divided by 3 times the volume of the ellipsoid.

Therefore $$\mathfrak{T}_4 = 1 + 10\beta_0 + \tfrac{325}{8}\beta_0^2$$

In this formula the coefficients of the powers of β_0 increase with great rapidity, and the approximation may not be very satisfactory; nevertheless it is the best attainable without an enormous increase of labour.

Combining our several results

$$-\frac{9}{3\lambda\mathfrak{T}_4}\mathfrak{Q}_4\left(\frac{r}{k}\right)\mathfrak{P}_4(1)\,\mathfrak{C}_4(\tfrac{1}{2}\pi) = -\frac{8}{105\lambda}\frac{k^5}{r^5}\left(1+\tfrac{15}{11}\frac{k^2}{r^2}\right)(1-8\beta_0+\tfrac{283}{8}\beta_0^2) \quad \text{.........(64)}$$

It remains to find $\mathfrak{P}_4(\nu_0)$, and the denominator in the expression for f_4 which involves $\mathfrak{A}_4 - \mathsf{A}_1^1$.

We have $$\mathfrak{A}_4 = \mathfrak{P}_4(\nu_0)\,\mathfrak{Q}_4(\nu_0)$$

$$= [\mathfrak{P}_4(\nu_0)]^2\int_{\nu_0}^{\infty}\frac{d\nu}{[\mathfrak{P}_4(\nu)]^2(\nu^2-1)^{1/2}(\nu^2-1/\kappa^2)^{1/2}}$$

Inside the integral sign I write $\nu = \dfrac{1}{\kappa\sin\psi}$ and change the independent variable to ψ; I also put $\sin\chi = \kappa\sin\psi$. At the surface of the ellipsoid we have $\psi=\gamma$, $\chi=\beta$, and since $\nu_0 = \dfrac{1}{\kappa\sin\gamma} = \dfrac{1}{\sin\beta}$, these are the values to be used in $\mathfrak{P}_4(\nu_0)$ outside the integral sign.

Then we have

$$\int_{\nu_0}^{\infty}\frac{d\nu}{(\nu^2-1)^{1/2}(\nu^2-1/\kappa^2)^{1/2}} = \kappa\int_0^{\gamma}\sec\chi\,d\psi$$

Since

$$P_4(\nu) = \tfrac{35}{8}(\nu^2-1)^4 + 5(\nu^2-1) + 1 = 1 + 5\cot^2\chi + \tfrac{35}{8}\cot^4\chi$$

$$P_4^2(\nu) = \tfrac{15}{2}(\nu^2-1)[7(\nu^2-1)+6] = \tfrac{15}{2}(6\cot^2\chi + 7\cot^4\chi)$$

$$P_4^4(\nu) = 105(\nu^2-1)^2 = 105\cot^4\chi$$

since further

$$\mathfrak{P}_4(\nu) = P_4(\nu) + \tfrac{1}{4}\beta_0 P_4^2(\nu) + \tfrac{1}{128}\beta_0^2 P_4^4(\nu)$$

we find $$\mathfrak{P}_4(\nu) = 1 + a\cot^2\chi + b\cot^4\chi$$

where $$\left.\begin{aligned} a &= 5(1+\tfrac{9}{4}\beta_0) \\ b &= \tfrac{35}{8}(1+3\beta_0+\tfrac{3}{16}\beta_0^2) \end{aligned}\right\} \quad \text{.................(65)}$$

It must be noted further that when $\nu=\nu_0$ at the surface of the ellipsoid, $\chi=\beta$.

It follows then that $\mathfrak{P}_4(\nu_0) = 1 + a\cot^2\beta + b\cot^4\beta$; and from (64)

$$-\frac{9}{3\lambda\mathfrak{T}_4}\mathfrak{Q}_4\left(\frac{r}{k}\right)\mathfrak{P}_4(\nu_0)\,\mathfrak{P}_4(1)\,\mathfrak{C}_4(\tfrac{1}{2}\pi) =$$

$$-\frac{8}{105\lambda}\frac{k^5}{r^5}\left(1+\tfrac{15}{11}\frac{k^2}{r^2}\right)(1-8\beta_0+\tfrac{283}{8}\beta_0^2)(1+a\cot^2\beta+b\cot^4\beta)$$

It remains to consider the evaluation of $\mathfrak{A}_4$, which now assumes the form

$$\mathfrak{A}_4 = \kappa \int_0^\gamma \left(\frac{1 + a \cot^2 \beta + b \cot^2 \beta}{1 + a \cot^2 \chi + b \cot^2 \chi}\right)^2 \sec \chi \, d\psi$$

It is possible [as shown in the paper on "Stability," § 19, p. 369] to split the subject of integration into partial fractions, and thus obtain an accurate value as was done in the case of $\mathfrak{A}_3$, but it does not seem worth while to undertake so heavy a task, because a sufficiently exact value may be obtained by quadratures.

The method was employed in that paper [as originally presented], and may be explained very shortly.

I divide γ into 10 or 12 equal parts—say 10 for brevity—and let $\delta = \frac{1}{10}\gamma$.

I then compute eleven equidistant values of the subject of integration, say $u_0, u_1, \ldots u_{10}$, corresponding to $\psi = 0, 2\delta, 3\delta, \ldots 10\delta$. As a fact it is unnecessary to compute the first four of these, because they are practically zero.

The equidistant values increase so rapidly that they are very inappropriate for the application of the rules of numerical quadratures. Accordingly I take an empirical and integrable function, say v, such that $v_{12} = u_{12}$ and $v_{11} = u_{11}$, and apply the rules of quadrature only to the differences $u_n - v_n$. The result is a correction to the integral $\int_0^\gamma v \, d\psi$.

The empirical function which satisfies these conditions is

$$v = u_{10} e^{\frac{\psi - \gamma}{\delta} \log_e \frac{u_{10}}{u_9}}$$

When $\psi = \gamma = 10\delta$, $v = u_{10}$; and when $\psi = 9\delta$, $v = u_{10} e^{-\log_e \frac{u_{10}}{u_9}} = u_9$.

Then $\int_0^\gamma v \, d\psi = \dfrac{u_{10}\delta}{\log_e (u_{10}/u_9)} (1 - e^{-10 \log_e (u_{10}/u_9)})$.

In the cases we have to treat $e^{-10 \log_e (u_{10}/u_9)}$ is an extremely small fraction, so that practically $\int_0^\gamma v \, d\psi = \dfrac{u_{10}\gamma}{10 \log_e (u_{10}/u_9)}$; and this is the function to be corrected by the result of quadrature.

For the quadratures we have

$$v_{10} = u_{10}, \quad v_9 = u_9, \quad v_8 = u_{10} \left(\frac{u_9}{u_{10}}\right)^2, \quad v_7 = u_{10} \left(\frac{u_9}{u_{10}}\right)^3, \text{ \&c.}$$

Thus the equidistant values of the function to be integrated (arranged backwards) are

$$0, \quad 0, \quad u_8 - u_{10} \left(\frac{u_9}{u_{10}}\right)^2, \quad u_7 - u_{10} \left(\frac{u_9}{u_{10}}\right)^3, \text{ \&c.}$$

The first two are zero, the next three or four are sensible, and the rest are insensible; thus the quadrature is very short. The correction is found to be very small, and we might perhaps have been content with the empirical integral without material loss of accuracy.

§ 17. *The Fifth Zonal Harmonic Inequality.*

This is treated exactly in the same way as the fourth, and I will only give the results.

We have

$$\mathfrak{Q}_5\left(\frac{r}{k}\right) = \frac{8}{7.9.11}\frac{k^6}{r^6}\left(1 + \tfrac{21}{13}\frac{k^2}{r^2}\right)(1 - 5\beta_0 + \tfrac{385}{16}\beta_0^2)$$

$$\mathfrak{P}_5(\nu) = P_5(\nu) + \tfrac{1}{4}\beta_0 P_5^2(\nu) + \tfrac{1}{128}\beta_0^2 P_5^4(\nu)$$

$$\mathsf{C}_5(\nu) = \sqrt{(1 - \beta_0 \cos 2\phi)}\,.\,[1 - 7\beta_0 \cos 2\phi + \tfrac{63}{16}\beta_0^2 \cos 4\phi]$$

$$\mathfrak{T}_5 = 1 + 15\beta_0 + \tfrac{777}{8}\beta_0^2$$

Whence

$$-\frac{11}{3\lambda\mathfrak{T}_5}\mathfrak{Q}_5\left(\frac{r}{k}\right)\mathfrak{P}_5(1)\,\mathsf{C}_5(\tfrac{1}{2}\pi)$$
$$= -\frac{8}{189\lambda}\frac{k^6}{r^6}\left(1 + \tfrac{21}{13}\frac{k^2}{r^2}\right)(1 - 13\beta_0 + \tfrac{727}{8}\beta_0^2)\sqrt{(1 + \beta_0)}$$

Then $$\mathfrak{P}_5(\nu_0) = \operatorname{cosec}\beta\,(1 + a\cot^2\beta + b\cot^2\beta)$$

where $$a = 7\,(1 + \tfrac{15}{4}\beta_0)$$
$$b = \tfrac{63}{8}(1 + 5\beta_0 + \tfrac{15}{128}\beta_0^2)$$

Finally

$$\mathfrak{A}_5 = \kappa\int_0^\gamma \left[\frac{\operatorname{cosec}\beta\,(1 + a\cot^2\beta + b\cot^4\beta)}{\operatorname{cosec}\chi\,(1 + a\cot^2\chi + b\cot^4\chi)}\right]^2 \sec\chi\, d\psi$$

which is to be evaluated by quadratures as was proposed for the fourth harmonic.

§ 18. *Moment of Momentum and Limiting Stability.*

The moment of momentum of the system is $I\omega$, but when we are determining the configuration of minimum moment of momentum, which is a figure of bifurcation and gives us the configuration of limiting stability, the conditions are different according as whether we are treating the problem of the figures of equilibrium where both masses are liquid, or Roche's problem in which the ellipsoid denoted by capital letters is rigid.

Accordingly I write

$$I = \tfrac{4}{3}\pi\rho a^3\left[\tfrac{1}{5}\frac{\lambda}{1+\lambda}(b^2 + c^2) + \frac{\lambda r^2}{(1+\lambda)^2}\right] + \tfrac{4}{3}\pi\rho a^3\cdot\tfrac{1}{5}\frac{1}{1+\lambda}(B^2 + C^2)$$

and for determining the angular momentum of the figures of equilibrium I take the whole expression for I, but for Roche's problem omit the last term.

Since $\omega^2 = \frac{4}{3}\pi\rho\frac{a^3}{r^3}(1+\zeta)$, I compute for Roche's problem

$$\mu_1 = \frac{a^{3/2}}{r^{3/2}}(1+\zeta)^{1/2}\left[\frac{1}{5}\frac{\lambda}{1+\lambda}\frac{(b^2+c^2)}{a^2}+\frac{\lambda r^2}{(1+\lambda)^2 a^2}\right]$$

and for the figures of equilibrium

$$\mu_2 = \mu_1 + \frac{a^{3/2}}{r^{3/2}}(1+\zeta)^{1/2}\frac{1}{5(1+\lambda)}\cdot\frac{B^2+C^2}{a^2}$$

The moment of momentum is given by

$$I\omega = (\tfrac{4}{3}\pi\rho)^{3/2} a^5 (\mu_1 \text{ or } \mu_2)$$

It will be observed that μ_1 and μ_2 are expressible by numbers for any given solution of the problem.

Suppose now that we have a succession of solutions for equidistant values of γ differing but little from one another. Then if the solutions lie close to the region of limiting stability, we shall find that one of them corresponds to minimum moment of momentum, either of μ_1 or of μ_2, as the case may be. Such a solution is a figure of bifurcation, and of the two coalescent solutions one has one more degree of instability than the other. If one of the two is continuous with a stable solution, and if, moreover, in the passage to the undoubtedly stable solution it passes through no other point of bifurcation, one of our two solutions is secularly stable and the other unstable.

Now, two liquid masses revolving about one another orbitally at an infinite distance are undoubtedly stable, and such a case is also continuous with one of our solutions. Further, Schwarzschild has proved that Roche's ellipsoid has no point of bifurcation from first to last, and as this is true of one such ellipsoid, it is true of two*. Hence we conclude that the minima of μ_1 and μ_2 will afford figures of limiting stability.

§ 19. *Approximate Solution of the Problem.*

It is clear that spherical harmonic analysis is applicable to the case when the two liquid masses are widely distant. When they are so much deformed by their interaction that that method becomes inapplicable, good results may be obtained from the formulæ of the last sections by means of development in powers of $\sin\gamma$, and it is this plan which is especially considered in the present section.

It appears from § 1 that when one of the masses is small compared with the other (λ small), the configuration of limiting stability for the problem of

* Schwarzschild, "Die Poincarésche Theorie des Gleichgewichts," *Neue Annalen der k. Sternwarte München*, Band III., 1896.

figures of equilibrium occurs when the two masses are very far apart. As λ increases, that configuration corresponds with diminishing radius vector. It seemed then probable that at least some of the solutions might be found by means of these series, and if this were so it might, in many cases, prove unnecessary to follow the same laborious procedure as in finding the limiting stability of Roche's ellipsoid. This view was found to be correct, and I therefore think it well to record the methods by which the developments may be obtained, without however giving the full details of the very laborious analysis.

When the masses are far apart, the terms denoted ϵ and η in the equation for κ'^2 and in that for a^3/r^3 are small, and they must be neglected in the developments.

Writing for brevity $g = \sin\gamma$, we may prove that

$$\Omega = \frac{1}{\sin\gamma}\log_e\frac{1+\sin\gamma}{\cos\gamma} = \sum_0^\infty \frac{g^{2n}}{2n+1}$$

$$\tan^2\gamma = -1 + \sum_0^\infty g^{2n}, \qquad \tan^4\gamma = 1 + \sum_0^\infty (n-1)\,g^{2n}$$

$$\tan^6\gamma = -1 + \sum_0^\infty \frac{(n-1)(n-2)}{1\,.\,2}\,g^{2n}$$

$$\tan^8\gamma = 1 + \sum_0^\infty \frac{(n-1)(n-2)(n-3)}{1\,.\,2\,.\,3}\,g^{2n}, \text{ \&c.}$$

Hence the developments may be obtained of the functions $\sigma_0, \sigma_1, \sigma_2 \ldots$ $\tau_0, \tau_1, \tau_2 \ldots$, and thence of $v_0, v_1, v_2 \ldots$ in series proceeding by powers of g^2; and thence we may find κ'^2 in that form.

The result as far as g^4 is

$$\kappa'^2 = \frac{1+\lambda}{4+\lambda}\left[1 - \frac{30}{7(1+\lambda)}g^2 - \frac{12(11+26\lambda)}{7^2(1+\lambda)^2}g^4 \ldots\right]$$

I have also found the term in g^6, but shall make no use of it.

With this value of κ'^2 or of κ^2, which is $1-\kappa'^2$, we develop the expression for a^3/r^3 in the same manner. The result is:—

$$\frac{\mathrm{a}^3}{r^3} = \tfrac{2}{5}\frac{1+\lambda}{4+\lambda}g^2\left[1 + \frac{5+\lambda}{7(4+\lambda)}g^2 - \frac{88+53\lambda}{7^2(4+\lambda)^2}g^4 \ldots\right]$$

By inversion we have

$$g^2 = \sin^2\gamma = \frac{5(4+\lambda)}{2(1+\lambda)}\frac{\mathrm{a}^3}{r^3} - \frac{25(5+\lambda)(4+\lambda)}{2^2\,.\,7(1+\lambda)^2}\frac{\mathrm{a}^6}{r^6} + \frac{125(69+2\lambda)(2+\lambda)(4+\lambda)}{2^3\,.\,7^2(1+\lambda)^3}\frac{\mathrm{a}^9}{r^9} \ldots$$

Since
$$\frac{k^3\cos\beta\cos\gamma}{\sin^3\beta} = \frac{\lambda}{1+\lambda}\mathrm{a}^3$$

it follows that
$$\frac{k^2}{\sin^2\beta} = \left(\frac{\lambda}{1+\lambda}\right)^{2/3}\frac{\mathrm{a}^2}{(\cos\beta\cos\gamma)^{2/3}}$$

The semi-axes of the ellipsoid are given by

$$\frac{c^2}{\text{a}^2} = \frac{k^2}{\text{a}^2 \sin^2 \beta} = \left(\frac{\lambda}{1+\lambda}\right)^{2/3} \frac{1}{(\cos\beta\cos\gamma)^{2/3}}$$

$$\frac{a^2}{\text{a}^2} = \frac{c^2}{\text{a}^2}\cos^2\gamma, \quad \frac{b^2}{\text{a}^2} = \frac{c^2}{\text{a}^2}\cos^2\beta$$

But $\cos^2\gamma = 1 - g^2$, $\cos^2\beta = 1 - \kappa^2 g^2$, and therefore

$$\frac{c^2}{\text{a}^2} = \left(\frac{\lambda}{1+\lambda}\right)^{2/3} (1-g^2)^{-1/3}(1-\kappa^2 g^2)^{-1/3}$$

$$\frac{a^2}{\text{a}^2} = \left(\frac{\lambda}{1+\lambda}\right)^{2/3} (1-g^2)^{2/3}(1-\kappa^2 g^2)^{-1/3}$$

$$\frac{b^2}{\text{a}^2} = \left(\frac{\lambda}{1+\lambda}\right)^{2/3} (1-g^2)^{-1/3}(1-\kappa^2 g^2)^{2/3}$$

Setting apart the factor $[\lambda/(1+\lambda)]^{2/3}$, which is common to all, these three are all expressible in the form, say

$$F = 1 + (a_0 + a_1\kappa^2) g^2 + (b_0 + b_1\kappa^2 + b_2\kappa^4) g^4 + (c_0 + c_1\kappa^2 + c_2\kappa^4 + c_3\kappa^6) g^6 + \ldots$$

where the a_0, a_1, b_0, b_1, &c., have different numerical values according to whichever of the three functions we are treating.

Now the above formula for κ'^2 enables us to write

$$\kappa^2 = A_0 + B_0 g^2 + C_0 g^4 \ldots$$

where the forms of A_0, B_0, C_0 are obvious.

Hence we have

$$F = 1 + (a_0 + a_1 A_0) g^2 + (b_0 + a_1 B_0 + b_1 A_0 + b_2 A_0^2) g^4$$
$$+ (c_0 + c_1 A_0 + c_2 A_0^2 + c_3 A_0^3 + b_1 B_0 + 2b_2 A_0 B_0 + a_1 C_0) g^6 \ldots$$

In this way I find the following expressions for the semi-axes in series proceeding by powers of g^2 or $\sin^2\gamma$:—

$$\left.\begin{aligned}
\frac{c^2}{\text{a}^2} &= \left(\frac{\lambda}{1+\lambda}\right)^{2/3} \left[1 + \frac{7+\lambda}{3(4+\lambda)} g^2 + \frac{524 + 223\lambda + 14\lambda^2}{3^2 . 7 (4+\lambda)^2} g^4 \right. \\
&\qquad \left. + \frac{120926 + 86748\lambda + 19428\lambda^2 + 686\lambda^3}{3^4 . 7^2 (4+\lambda)^3} g^6 \ldots\right] \\
\frac{a^2}{\text{a}^2} &= \left(\frac{\lambda}{1+\lambda}\right)^{2/3} \left[1 - \frac{5+2\lambda}{3(4+\lambda)} g^2 - \frac{64 + 8\lambda + 7\lambda^2}{3^2 . 7 (4+\lambda)^2} g^4 \right. \\
&\qquad \left. - \frac{11122 + 2460\lambda - 1851\lambda^2 + 196\lambda^3}{3^4 . 7^2 (4+\lambda)^3} g^6 \ldots\right] \\
\frac{b^2}{\text{a}^2} &= \left(\frac{\lambda}{1+\lambda}\right)^{2/3} \left[1 - \frac{2-\lambda}{3(4+\lambda)} g^2 - \frac{187 + 110\lambda - 14\lambda^2}{3^2 . 7 (4+\lambda)^2} g^4 \right. \\
&\qquad \left. - \frac{28492 + 36723\lambda + 14160\lambda^2 - 686\lambda^3}{3^4 . 7^2 (4+\lambda)^3} g^6 \ldots\right]
\end{aligned}\right\} \quad \ldots\ldots\ldots(66)$$

It is easy to verify that the product of the three series is unity, as should be the case.

The next step is to substitute for g^2, g^4, g^6... their values in terms of a^3/r^3 and its powers. In this way I find

$$\left.\begin{aligned}
\frac{c^2}{\mathrm{a}^2} &= \left(\frac{\lambda}{1+\lambda}\right)^{2/3}\left[1+\frac{5(7+\lambda)}{6(1+\lambda)}\frac{\mathrm{a}^3}{r^3}+\frac{25(419+187\lambda+11\lambda^2)}{2^2.3^2.7(1+\lambda)^2}\frac{\mathrm{a}^6}{r^6}\right.\\
&\qquad\left.+\frac{125(99848+74769\lambda+16503\lambda^2+488\lambda^3)}{2^3.3^4.7^2(1+\lambda)^3}\frac{\mathrm{a}^9}{r^9}\cdots\right]\\
\frac{a^2}{\mathrm{a}^2} &= \left(\frac{\lambda}{1+\lambda}\right)^{2/3}\left[1-\frac{5(5+2\lambda)}{6(1+\lambda)}\frac{\mathrm{a}^3}{r^3}+\frac{25(11+37\lambda-\lambda^2)}{2^2.3^2.7(1+\lambda)^2}\frac{\mathrm{a}^6}{r^6}\right.\\
&\qquad\left.-\frac{125(23992+17895\lambda+1587\lambda^2+178\lambda^3)}{2^3.3^4.7^2(1+\lambda)^3}\frac{\mathrm{a}^9}{r^9}\cdots\right]\\
\frac{b^2}{\mathrm{a}^2} &= \left(\frac{\lambda}{1+\lambda}\right)^{2/3}\left[1-\frac{5(2-\lambda)}{6(1+\lambda)}\frac{\mathrm{a}^3}{r^3}-\frac{25(157+119\lambda-11\lambda^2)}{2^2.3^2.7(1+\lambda)^2}\frac{\mathrm{a}^6}{r^6}\right.\\
&\qquad\left.-\frac{125(19114+23673\lambda+11577\lambda^2-488\lambda^3)}{2^3.3^4.7^2(1+\lambda)^3}\frac{\mathrm{a}^9}{r^9}\cdots\right]
\end{aligned}\right\} \quad \text{.........(67)}$$

By writing $1/\lambda$ for λ we obtain the formulæ for the axes of the other ellipsoid.

The numerical coefficients increase rather rapidly so that the series are useless unless a/r is small, and accordingly this method fails to give any result for Roche's ellipsoid in limiting stability; it is, however, useful for the problem of figures of equilibrium, as already stated.

If we had relied on spherical harmonic analysis, we should only have obtained the terms in a^3/r^3.

In order to obtain the expression for the angular momentum, which has to be a minimum for limiting stability, we must evaluate ζ. Now, from (24) and (15), we have

$$\begin{aligned}
\zeta &= \frac{3}{10r^2}(2c^2-a^2-b^2+\text{same in } A, B, C)\\
&+\frac{3}{56r^4}[3(a^4+b^4)+8c^4-8c^2(a^2+b^2)+2a^2b^2+\text{same in } A, B, C]\\
&+\frac{3}{20r^4}[2(A^2a^2+B^2b^2+C^2c^2)+(A^2+B^2+C^2)(a^2+b^2+c^2)-5C^2(a^2+b^2)\\
&\qquad -5c^2(A^2+B^2)+5c^2C^2]
\end{aligned}$$

By means of the above series for a^2, b^2, c^2, and of their analogues for A^2, B^2, C^2, I find

$$\zeta = l\frac{\mathrm{a}^5}{r^5}+m\frac{\mathrm{a}^8}{r^8}+n\frac{\mathrm{a}^{10}}{r^{10}}$$

where

$$l = \tfrac{3}{4} \cdot \frac{\lambda^{2/3}(7+\lambda)+7\lambda+1}{(1+\lambda)^{5/3}}$$

$$m = \tfrac{5}{14} \cdot \frac{\lambda^{2/3}(82+38\lambda+\lambda^2)+82\lambda^2+38\lambda+1}{(1+\lambda)^{8/3}}$$

$$n = \tfrac{225}{224} \frac{\lambda^{4/3}(33+10\lambda+\lambda^2)+33\lambda^2+10\lambda+1}{(1+\lambda)^{10/3}} + \tfrac{15}{16} \frac{\lambda^{2/3}(15+102\lambda+15\lambda^2)}{(1+\lambda)^{10/3}}$$

Now we have to evaluate the moment of momentum given above in § 18, viz.,

$$\mu_2 = \left(\frac{\mathrm{a}}{r}\right)^{3/2}(1+\zeta)^{1/2}\left[\tfrac{1}{5}\frac{\lambda}{1+\lambda}(b^2+c^2)+\tfrac{1}{5}\frac{1}{1+\lambda}(B^2+C^2)+\frac{\lambda}{(1+\lambda)^2}r^2\right]$$

When b^2, c^2, B^2, C^2, have their values attributed to them, we find

$$\mu_2 = \frac{\lambda}{(1+\lambda)^2}\left(\frac{\mathrm{a}}{r}\right)^{3/2}(1+\zeta)^{1/2}\left[R+S\frac{\mathrm{a}^3}{r^3}+T\frac{\mathrm{a}^6}{r^6}+U\frac{\mathrm{a}^9}{r^9}\ldots+r^2\right]$$

where

$$R = \frac{2}{5\lambda}(1+\lambda^{5/3})(1+\lambda)^{1/3}$$

$$S = \frac{1}{6(1+\lambda)^{2/3}}\left[\lambda^{2/3}(5+2\lambda)+\frac{1}{\lambda}(5\lambda+2)\right]$$

$$T = \frac{5}{2.3^2.7(1+\lambda)^{5/3}}\left[\lambda^{2/3}(131+34\lambda+11\lambda^2)+\frac{1}{\lambda}(131\lambda^2+34\lambda+11)\right]$$

$$U = \frac{25}{2^2.3^4.7^2(1+\lambda)^{8/3}}\left[\lambda^{2/3}(40367+25548\lambda+2463\lambda^2+488\lambda^3)\right.$$
$$\left.+\frac{1}{\lambda}(40367\lambda^3+25548\lambda^2+2463\lambda+488)\right]$$

In § 1, where the same problem is treated for two spheres, we had l, m, n, S, T, U, all zero.

In order to find the minimum moment of momentum for a given value of λ, I compute l, m, n, R, S, T, U, and assuming several equidistant values of r compute values of $\mu_2(1+\lambda)^2/\lambda$. When the coefficients are computed we very easily find the value of r corresponding to the minimum.

When that value of r is found, we are in a position to compute the axes of the two ellipsoids.

For values of λ less than $\frac{1}{2}$ the results found in this way would be satisfactory, and for $\lambda = \frac{1}{2}$ they are, I think, adequate. Even for the case of $\lambda = 1$ the result is not very remote from the truth, for whereas the correct result for the minimum of angular momentum is $r/\mathrm{a} = 2{\cdot}638$, the result derived from this approximate method is $r/\mathrm{a} = 2{\cdot}51$. But it would have been impossible to foresee that the result would be as good as it is.

Part II. Numerical Solutions.

§ 20. *Roche's Infinitesimal Ellipsoidal Satellite in Limiting Stability.*

We require to find the form of an infinitesimal satellite (so that $\lambda = 0$) revolving in a circular orbit about a *spherical* planet. When this problem is solved we shall be able to see how far the solution will be affected when we allow the spherical planet to become oblate under the influence of a rotation of the same speed as that of the revolution of the infinitesimal satellite. This last is what I have called the modified form of Roche's problem.

The planet being spherical and λ being zero, the small terms ζ, ϵ, η vanish, so that our solution becomes rigorous.

The angular momentum of the planet's axial rotation is to be omitted, and the satellite being infinitesimal the momentum of its axial rotation is zero. Thus the moment of momentum of the system varies as the square root of the satellite's radius vector, and minimum momentum coincides with minimum radius vector.

The solution of the problem has been obtained in two ways: first by Legendre's tables of elliptic integrals, and secondly by means of the auxiliary tables given above. In the first method, I knew with fair approximation by various preliminary computations the values of κ and γ which lay near to the required solution. Now there is a certain function of κ, γ, say $f(\sin^{-1}\kappa, \gamma)$, which vanishes when the ellipsoid is a figure of equilibrium; accordingly I computed by means of Legendre's tables the following eight values of $f(\sin^{-1}\kappa, \gamma)$ for integral degrees of $\sin^{-1}\kappa$ and γ:—

$$f(77°, 57°) = +0{\cdot}0000878, \quad f(77°, 58°) = -0{\cdot}0000624$$
$$f(78°, 59°) = +0{\cdot}0000724, \quad f(78°, 60°) = -0{\cdot}0000785$$
$$f(79°, 61°) = +0{\cdot}0000562, \quad f(79°, 62°) = -0{\cdot}0000939$$
$$f(80°, 63°) = +0{\cdot}0000408, \quad f(80°, 64°) = -0{\cdot}0001046$$

(probably the last significant figure in each of these is inaccurate).

Interpolating from these we find four values satisfying $f(\sin^{-1}\gamma, \kappa) = 0$, namely:—

$$f(77°, 57°{\cdot}5846) = 0, \quad f(78°, 59°{\cdot}4798) = 0$$
$$f(79°, 61°{\cdot}3744) = 0, \quad f(80°, 63°{\cdot}2806) = 0$$

With these solutions I find

$\sin^{-1}\kappa$	r/a
77°	2·467860
78°	2·458191
79°	2·455446
80°	2·460289

By formulæ of interpolation the minimum of r occurs when

$$\sin^{-1}\kappa = 78°{\cdot}8756$$

Then by a second interpolation this value of κ corresponds with $\gamma = 61^\circ{\cdot}1383$, and the minimum value of r is 2·45539. We may take then $\gamma = 61^\circ\ 8'{\cdot}3$, $\kappa = \sin 78^\circ\ 52'{\cdot}5$, whence $\beta = 59^\circ\ 14'{\cdot}5$. Since $\cos\gamma = 0{\cdot}4827$, $\cos\beta = 0{\cdot}5114$, the three axes of the ellipsoid are proportional to 10000, 5114, 4827; Roche gave the ratios 1000, 496, 469, and the radius vector as 2·44, in place of 2·45539.

Turning now to the second solution, I solved the problem by means of the auxiliary tables in two ways, namely, for $\gamma = 60^\circ, 61^\circ, 62^\circ$ and also for $\gamma = 57^\circ, 59^\circ, 61^\circ, 63^\circ$.

They led to virtually identical results, viz., that the minimum of r is 2·45521, corresponding to $\gamma = 61^\circ\ 8'{\cdot}4$, $\sin^{-1}\kappa = 78^\circ\ 52'{\cdot}0$.

Finally the solution for Roche's limit and for the ratio of the axes of the ellipsoid in limiting stability may be taken to be as follows:—

γ	$\sin^{-1}\kappa$	$\cos\gamma$	$\cos\beta$	r/a
$61^\circ\ 8\tfrac{1}{2}'$	$78^\circ\ 52'$	0·4827	0·5114	2·4553

with uncertainty of unity in the last place of decimals in r and of half a minute of arc in $\sin^{-1}\kappa$.

We must next consider the modified form of Roche's problem, in which the large body or planet yields to centrifugal force and becomes an oblate ellipsoid of revolution. The approximate formulæ of § 19 show that when $\lambda = \infty$ or when $\lambda = 0$,

$$l = \tfrac{3}{4}, \quad m = \tfrac{5}{14}, \quad n = \tfrac{225}{224}$$

Hence in this case

$$\zeta = \tfrac{3}{4}\left(\frac{a}{r}\right)^5 + \tfrac{5}{14}\left(\frac{a}{r}\right)^8 + \tfrac{225}{224}\left(\frac{a}{r}\right)^{10} \dots$$

The solution of the modified problem can only differ slightly from that just found when the planet is spherical, and therefore we may compute ζ with sufficient accuracy by means of the values of a/r already found. I accordingly computed ζ for $\gamma = 60^\circ, 61^\circ, 62^\circ$, and found that in each case ζ was very nearly equal to 0·0088.

Now it is proved in the footnote to § 10 that, when $\lambda = 0$ and when the planet yields to centrifugal force, $\epsilon = \eta = \zeta$; as the value of ζ is found with good approximation, it is easy to compute r for these three values of γ. I thus find that in the modified problem, minimum radius vector, and therefore limiting stability, occurs when $r = 2{\cdot}457$, $\gamma = 61^\circ\ 12'$, $\kappa = \sin 78^\circ 50'$, $\beta = 59^\circ 17'$; the axes of the large body are determined by the approximate formulæ of § 10 to be $\frac{C}{a} = \frac{B}{a} = 1{\cdot}0304$, $\frac{A}{a} = 0{\cdot}9418$.

It appears then that the yielding of the planet to centrifugal force makes very little difference, as was to be expected.

These results are included in the table given below of results for solutions of the modified problem of Roche with finite values of λ.

§ 21. *Roche's Ellipsoidal Satellite, of finite mass, in limiting stability, the planet being also ellipsoidal.*

This is the problem which I describe as the "modified" problem of Roche. It seemed unnecessary to carry out the computations for the smaller values of λ, since they are sufficiently represented by the case of the infinitesimal satellite where λ is zero. I therefore begin with the case of $\lambda = 0{\cdot}4$ and pass on to $\lambda = 0{\cdot}5$, $0{\cdot}6$, $0{\cdot}7$, $0{\cdot}8$, $0{\cdot}9$, $1{\cdot}0$.

It seems well to describe the process followed in one case as a type of all. It was, in general, possible either by extrapolation from neighbouring values of λ, or by mere guessing, to begin with some values of ζ, ϵ, η and their corelative functions E, H for the larger body, which were somewhere near the truth. With these we could compute r, κ, Γ, K with fair approximation; thence values of ζ, ϵ, η, E, H could be calculated with close accuracy and the computation could be repeated. It was, of course, a matter of conjecture as to what initial values of γ would be found to embrace the region of minimum angular momentum.

I will now describe the process for $\lambda = 0{\cdot}4$. Passing over the preliminary stages in which fairly good values were found, we begin with the following conjectural values:—

γ	46°	48°	50°	Γ	32°	34°	36°
$\sin^{-1}\kappa$	68° 24′·8	69° 32′·0	70° 41′·3	$\sin^{-1}$K	50° 12′·0	51° 18′·8	52° 34′·9
Γ	33° 13′·9	34° 18′·8	35° 18′·5	γ	43° 50′·9	47° 24′·4	51° 34′·1
$\sin^{-1}$K	50° 52′·0	51° 30′·1	52° 7′·5	$\sin^{-1}\kappa$	67° 15′·0	69° 11′·8	71° 37′·0
$\log r$	0·40245	0·39594	0·39060	$\log r$	0·41071	0·39775	0·38726

whence I compute

ζ	0·056259	0·064863	0·074219	ζ	0·047883	0·062231	0·082010
ϵ	0·045435	0·048102	0·050256	E	0·140179	0·174168	0·218540
η	0·32004	0·36426	0·40838	H	0·40539	0·52564	0·69272

By means of these and the auxiliary tables I find

$\sin^{-1}\kappa$	68° 24′·6	69° 31′·8	70° 40′·5	$\sin^{-1}$K	50° 11′·8	51° 18′·8	52° 33′·8
$\log r$	0·40240	0·39591	0·39046	$\log r$	0·41068	0·39768	0·38693

The computed values are so very close to the conjectural ones, in so far as they have been as yet computed, that we might be content, but in order to illustrate the process when the conjectures are less satisfactory, I proceed to the next stage.

By far the greater part of the discrepancy between assumed and computed values (which in some cases was considerable) arises from error in the assumed values of r. Now assuming κ and K to be correct, it is very easy to correct the results for a changed value of r.

In this case I find

corrected ζ	0·056274	0·064873	0·074271		corrected ζ	0·047890	0·062253	0·082137
" ϵ	0·04545	0·04811	0·05030		" E	0·14020	0·17423	0·21889
" η	0·32012	0·36431	0·40867		" H	0·40546	0·52583	0·69385

Recomputing

	$\sin^{-1}\kappa$ unchanged					$\sin^{-1}$ K unchanged		
corrected $\log r$	0·40240	0·39591	0·39047		corrected $\log r$	0·41069	0·39768	0·38695

By means of these we find two formulæ of interpolation, namely:—

$$\frac{r}{a} = 2{\cdot}4884 - 0{\cdot}0342\left(\frac{\gamma - 48^\circ}{2}\right) + 0{\cdot}0032\left(\frac{\gamma - 48^\circ}{2}\right)^2$$

$$\frac{r}{a} = 2{\cdot}4985 - 0{\cdot}0685\left(\frac{\Gamma - 34^\circ}{2}\right) + 0{\cdot}0075\left(\frac{\Gamma - 34^\circ}{2}\right)^2$$

These two expressions may be equated to one another, and therefore we have the means of finding simultaneous values of γ and Γ, and thence by another formula of interpolation of κ and K. Hence I obtain

γ	46°	48°	50°		Γ	32°	34°	36°
Γ	33° 13′·4	34° 17′·9	35° 17′·4		γ	43° 47′·4	47° 25′·4	51° 34′·4
$\sin^{-1}$ K	50° 52′·2	51° 29′·5	52° 6′·3		$\sin^{-1}\kappa$	67° 12′·1	69° 12′·3	71° 35′·6

Comparison with the initial values shows that the conjectures were very good.

It now remains to compute the moment of momentum, and as we are dealing with Roche's problem the rotational momentum of the larger ellipsoid is not required. It follows that the values of Γ and K are not used, and since they are only required for finding the shape of the larger ellipsoid, there was no necessity for a high degree of accuracy in them. The moment of momentum is represented by the quantity μ_1 of § 18. I find then

γ	46°	48°	50°
μ_1	0·348640	0·348300	0·348519

By formulæ of interpolation the minimum value occurs when $\gamma = 48°\ 12'{\cdot}9$ and $\kappa = \sin 69°\ 39'{\cdot}1$. The corresponding values are $\frac{r}{a} = 2{\cdot}4848$, $\Gamma = 34°\ 25'{\cdot}6$, $K = \sin 51°\ 33'{\cdot}5$. The last step is to compute the axes of the two ellipsoids from the values of κ, γ, K, Γ.

Of course the numbers set out above make no claim to absolute accuracy, but the results tabulated below are, I believe, substantially correct.

The unit of length employed is the radius of a sphere whose mass is equal to the mass of the whole system. If it were preferred to express the results in terms of the mean radius of the larger body, all linear results would have to be multiplied by $(1 + \lambda)^{1/3}$.

We may now collect the results in a tabular form, as follows:—

SOLUTIONS FOR ROCHE'S ELLIPSOID IN LIMITING STABILITY.

The unit of length is the radius of a sphere whose mass is equal to the sum of the masses, *i.e.*,

$$abc + ABC = 1, \text{ and } \frac{abc}{ABC} = \lambda.$$

λ	γ	$\sin^{-1}\kappa$	a	b	c	Γ	$\sin^{-1} K$	A	B	C	r
0	61° 12′	78° 50′	0·482 ÷ ∞	0·511 ÷ ∞	1·0 ÷ ∞	—	—	0·942	1·030	1·030	2·457
0·4	48° 13′	69° 39′	0·562	0·603	0·843	34° 25′	51° 34′	0·815	0·886	0·988	2·485*
0·5	46° 40′	68° 12′	0·597	0·642	0·870	35° 59′	54° 30′	0·792	0·860	0·979	2·484
0·6	45° 5′	66° 43′	0·627	0·674	0·888	37° 14′	56° 41′	0·772	0·836	0·969	2·490
0·7	43° 38′	65° 20′	0·652	0·701	0·901	38° 9′	58° 18′	0·753	0·815	0·958	2·497
0·8	42° 26′	64° 4′	0·673	0·725	0·912	38° 57′	59° 39′	0·737	0·796	0·947	2·502
0·9	41° 25′	62° 58′	0·691	0·744	0·921	39° 40′	60° 47′	0·722	0·778	0·937	2·508
1·0	40° 15′	61° 43′	0·708	0·762	0·927	40° 15′	61° 43′	0·708	0·762	0·927	2·514

* The values $r=2{\cdot}485$ for $\lambda=0{\cdot}4$ and $r=2{\cdot}484$ for $\lambda=0{\cdot}5$ represent 2·4848 and 2·4844 respectively; it is probable that the last significant figure in the former is a little too large and in the latter too small, and that it might have been more correct to invert the 2·485 and 2·484 in the table. I give the result, however, of the computation.

The cases $\lambda = 0{\cdot}4$, $0{\cdot}7$, $1{\cdot}0$ are illustrated by figs. 2, 3, 4. The meaning of the dotted lines near the vertices of the smaller ellipsoid will be explained in the next section.

The distance $r-(c+C)$ is the interval between the vertices of the two ellipsoids; the following are the values, using, however, more places of decimals than are tabulated above:—

λ	$r-(c+C)$	λ	$r-(c+C)$
0	1·030	0·7	0·638
0·4	0·653	0·8	0·643
0·5	0·635	0·9	0·650
0·6	0·633	1·0	0·660

It is remarkable how very nearly constant the intervening space remains throughout a large range in the values of λ.

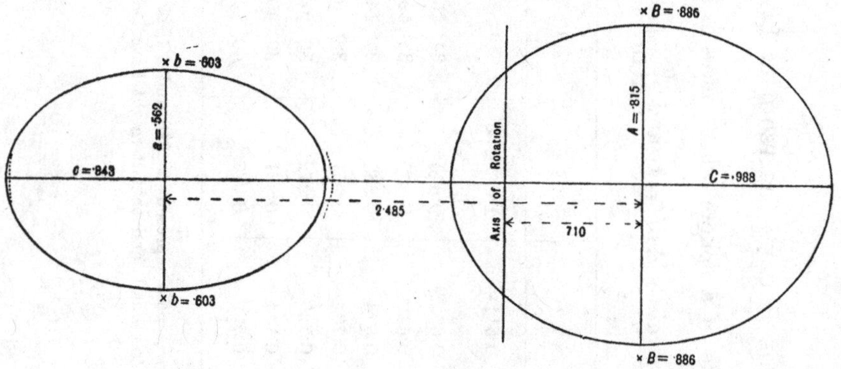

FIG. 2. Roche's ellipsoid in limiting stability, when $\lambda = 0{\cdot}4$.

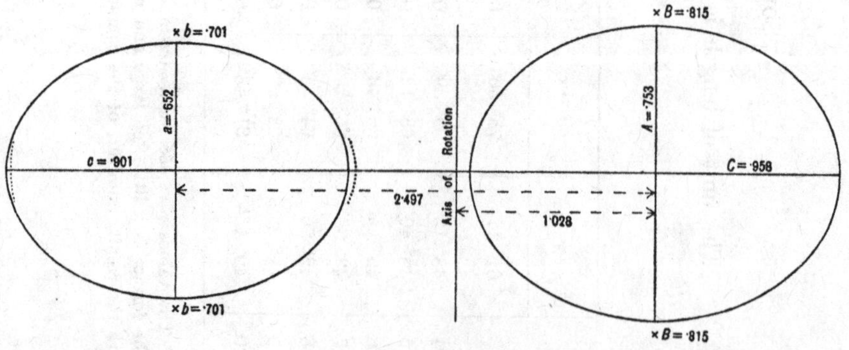

FIG. 3. Roche's ellipsoid in limiting stability, when $\lambda = 0{\cdot}7$.

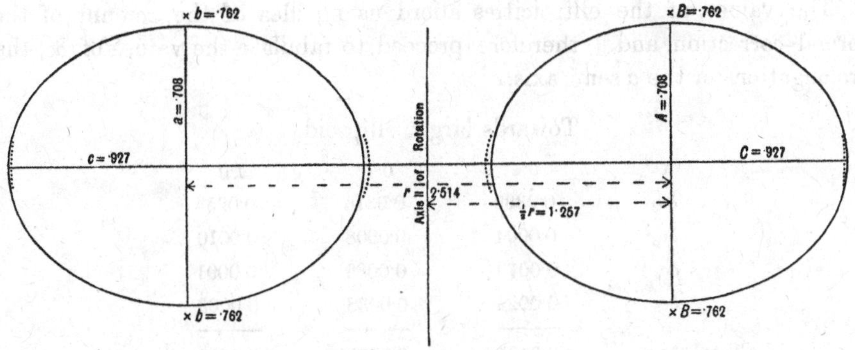

FIG. 4. Roche's ellipsoids in limiting stability for equal masses.

§ 22. *Harmonic Deformations of the Ellipsoids.*

When λ is infinitely small, so that the liquid satellite is infinitely small, the harmonic deformations are evanescent, and the same is true when λ is infinitely great. We saw in § 14 that there was reason to suppose that Roche's ellipsoid in limiting equilibrium might be more markedly deformed for values of λ midway between zero and unity than for the latter value. I therefore determined the harmonic deformations in the three cases $\lambda = 0{\cdot}4$, $0{\cdot}7$, $1{\cdot}0$.

The formulæ for the ellipticities f_3, f_3^2, f_4, f_5 are given in §§ 14, 16, 17 and their values may be found for the ellipsoids in limiting stability tabulated in the last section. It appears that in every case the amount of deformation is small, and therefore it was sufficient to compute the normal deformation at the two extremities of the c semi-axis, that is to say, at the points nearest to and most remote from the other ellipsoid. At these points the normal displacement outwards may be denoted δc, with numbers affixed thereto so as to indicate to which harmonic it is due.

The results may be given in a tabular form, but it may be well to remark that the ellipticities corresponding to the tesseral harmonics of the fourth and fifth orders, viz. $f_4^2, f_4^4, f_5^2, f_5^4$ were not computed, because their effects would be quite negligible. The following are the computed values of the ellipticities*:—

λ	0·4	0·7	1·0
f_3	0·325	0·261	0·214
f_3^2	−0·113	−0·126	−0·089
f_4	0·004	0·007	0·006
f_5	0·001	0·002	0·003

* [A mistake has been corrected in the column $\lambda = 0{\cdot}4$, and certain small corrections have been made in the other columns]

The values of the ellipticities afford us no idea of the amount of the normal correction, and I therefore proceed to tabulate the values of δc, the prolongations of the c semi axis.

Towards larger ellipsoid

λ	0·4	0·7	1·0
δc_3	0·0389	0·0316	0·0254
$\delta c_3{}^2$	0·0004	0·0008	0·0010
δc_4	0·0078	0·0089	0·0091
δc_5	0·0024	0·0038	0·0063
Total δc	0·0495	0·0451	0·0418
c	0·8433	0·9006	0·9270
$c+\delta c$	0·8928	0·9457	0·9688
$\frac{\delta c}{c}$	1/17	1/20	1/23

Away from larger ellipsoid

λ	0·4	0·7	1·0
δc_3	−0·0389	−0·0316	−0·0254
$\delta c_3{}^2$	−0·0004	−0·0008	−0·0010
δc_4	+0·0078	+0·0089	+0·0091
δc_5	−0·0024	−0·0038	−0·0063
Total δc	−0·0339	−0·0273	−0·0236
c	0·8433	0·9006	0·9270
$c+\delta c$	0·8094	0·8766	0·9104
$\frac{\delta c}{c}$	1/25	1/33	1/39

The last line in each division of this table has been given in order to show the relative importance of the total correction. It is clear that the ellipsoid remains a substantially correct solution.

These corrections to the semi-major axes are indicated by dotted lines at the extremities of the longest axis of the smaller mass in figs. 2 and 3, and of both masses in fig. 4.

We have in the last section tabulated $r-(c+C)$, the distance between the two vertices. Now, although I have not calculated the deformations of the larger ellipsoid, it is pretty clear that they must bear to those of the smaller one approximately the ratio of λ to unity. Accepting this conjecture, we have for the δC of the larger ellipsoid towards the smaller one the following values:—

λ	0·4	0·7	1·0
δC	0·020	0·034	0·042

The distance between the two surfaces of liquid is clearly

$$r-(c+\delta c+C+\delta C)$$

Thus we have

λ	0·4	0·7	1·0
$r-(c+C)$	0·653	0·638	0·660
$\delta c+\delta C$	0·070	0·079	0·084
$r-(c+\delta c+C+\delta C)$	0·583	0·559	0·576

§ 23. *Certain Tests and Verifications.*

In order to test how nearly the solution for a^3/r^3 by series in (40) of § 11 would agree with the solution (36) of § 10 in terms of the F elliptic integral, I computed for $\lambda = 0·7$ the value of r/a in the two ways for three values of γ, and found the following results:—

$$\lambda = 0·7$$

γ	42°	44°	46°
r/a by series	2·5355	2·4888	2·4500
r/a by elliptic integrals	2·5359	2·4881	2·4495

The agreement seems to be as close as could be expected when five-figured logarithms are used.

Certain terms in ζ, as expressed in (24) of § 10, were neglected on the ground that they are of higher order than those retained. But it appears from the approximate solution in § 19 that the coefficients of the terms retained are themselves small, so that we are really only retaining terms of the same order as others which are neglected.

The most important of the neglected terms in ζ is

$$-\frac{3}{k} f_3 \mathfrak{P}_3(\nu_0)\, \mathfrak{P}_3(1)\, \mathsf{C}_3(\tfrac{1}{2}\pi)\, r^2 \frac{d}{dr} \mathfrak{Q}_3\left(\frac{r}{k}\right)$$

and this is a term of the seventh order. It seems, therefore, well to compute this in one case, and see how large a proportion it bears to the whole value.

Since $$\mathfrak{Q}_3\left(\frac{k}{r}\right) = \frac{k^4}{7\kappa^2 r^4}\left(1+\tfrac{10}{9}\frac{k^2}{\kappa r^2}\right)$$

$$r^2\frac{d}{dr}\mathfrak{Q}_3\left(\frac{k}{r}\right) = -\frac{4k^4}{7\kappa^2 r^3}\left(1+\tfrac{5}{3}\frac{k^2}{\kappa r^2}\right)$$

Using this value and the other approximate values given in § 14, where f_3 is determined, I find that the neglected term is

$$\frac{-\dfrac{4}{7\lambda}\dfrac{\Delta_1{}^4}{\sin^6\beta}\left(1+0\kappa'^2-\tfrac{15}{16}\kappa'^4\right)\dfrac{k^7}{r^7}\left(1+\tfrac{25}{9}\dfrac{k^2}{\kappa r^2}\right)}{\mathfrak{A}_3-\mathsf{A}_1{}^1-\dfrac{k^3}{3\lambda r^3}\dfrac{\cos^2\gamma}{\sin^2\beta}\left(1+\dfrac{3(3+\kappa^2)}{14\kappa^2}\dfrac{k^2}{r^2}\cdots\right)}$$

The numerical value of this, for the case of Roche's ellipsoid in limiting stability when $\lambda = 0{\cdot}7$, is found to be $+0{\cdot}0016$. Now, the value of ζ, as computed from the terms retained, was found to be $0{\cdot}0677$. Thus the neglected term is about one 42nd of the whole. The neglect then seems fairly justified.

I thought it worth while to discover how far the modification of Roche's problem, whereby the larger body is ellipsoidal, affects the result. I find that whereas it makes but little difference in the solution for any single assumed value of γ, it does make a sensible difference in the incidence of the minimum of angular momentum, and therefore of limiting stability. Thus, when $\lambda = 0{\cdot}5$, I found in one of my preliminary solutions for Roche's modified problem that limiting stability occurs when $r/a = 2{\cdot}49$ (the more correct value is $2{\cdot}484$), but when the larger body is a sphere it occurs when $r/a = 2{\cdot}35$. Thus we see that ellipticity in the larger body induces instability at a greater distance than if it were spherical. This might have been conjectured from general considerations.

§ 24. *Figures of Equilibrium of Two Masses in Limiting Stability.*

In this case both masses are liquid. We saw in § 2 that when one of the masses is infinitely small, stability only exists when the two are infinitely far apart. When λ is less than $0{\cdot}5$ we may obtain fair results from the approximate investigation of § 19, but for greater values of λ it is necessary to employ the laborious method adopted in determining Roche's ellipsoid.

When $\lambda = 0{\cdot}5$ I obtain the following approximate results for the two figures:—

$$r = 2{\cdot}574$$

$$a = 0{\cdot}62, \quad A = 0{\cdot}81$$

$$b = 0{\cdot}66, \quad B = 0{\cdot}87$$

$$c = 0{\cdot}81, \quad C = 0{\cdot}95$$

It is probable that the value of r derived in this way is too small.

For $\lambda = 0{\cdot}4$, I found $r = 2{\cdot}59$, but did not calculate the axes.

The only other case in which the problem has been solved is for equal masses, when $\lambda = 1$. The two ellipsoids are exactly alike, and I find limiting stability occurs for the following values:—

$$\gamma = 36^\circ\, 18', \quad \kappa = \sin 59^\circ\, 33', \quad r = 2{\cdot}638$$

$$a = 0{\cdot}723, \quad b = 0{\cdot}771, \quad c = 0{\cdot}897$$

and

$$r - 2c = 0{\cdot}844$$

This is illustrated in fig. 5.

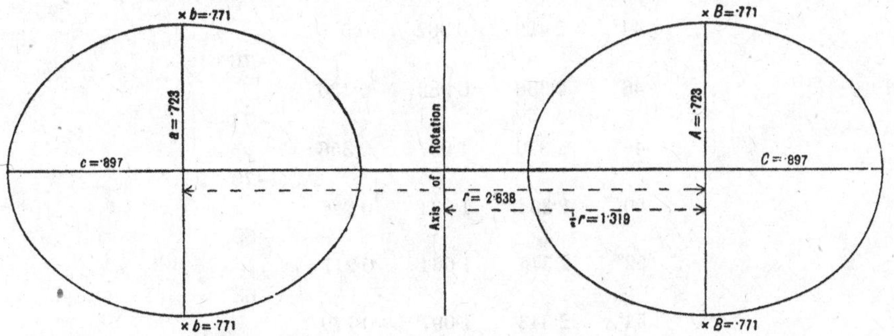

FIG. 5. Two equal masses of liquid in limiting stability.

§ 25. *Unstable Figures of Equilibrium of Two Masses.*

When $\lambda = 0$ the figure of minimum radius vector, when the larger body is rigid, is also that of minimum angular momentum, but for larger values of λ there is an ellipsoid considerably nearer to the larger body than that which possesses limiting stability. I have only determined the ellipsoid of minimum radius vector in two cases, viz., when $\lambda = 0{\cdot}8$ and $1{\cdot}0$.

When $\lambda = 0{\cdot}8$ I find minimum radius vector to be $r = 2{\cdot}36$, whereas limiting stability occurs for $r = 2{\cdot}50$. When $\lambda = 0{\cdot}8$, $r = 2{\cdot}36$, the ellipsoids are determined by the following data:—

$$\gamma = 54^\circ\, 20', \quad \kappa = \sin 71^\circ\, 51', \quad \Gamma = 46^\circ\, 10', \quad K = \sin 64^\circ\, 20'$$

whence

$$a = 0{\cdot}619, \quad A = 0{\cdot}705$$

$$b = 0{\cdot}675, \quad B = 0{\cdot}774$$

$$c = 1{\cdot}063, \quad C = 1{\cdot}018$$

When $\lambda = 1$, the minimum radius vector occurs when γ is about 54° and is then equal to $2{\cdot}343$, whereas limiting stability occurs when $r = 2{\cdot}514$. I have not computed the axes, since it suffices to learn that there is an ellipsoidal solution when the two masses are considerably nearer than is consistent with stability.

As γ increases, the ellipsoids get longer and longer, and it is interesting to inquire whether they increase in length with such rapidity that, notwithstanding the increase of r, the interval between the two vertices continues to decrease, or whether the increase of r annuls the simultaneous increase of c.

The following table of values, computed with fair but not extreme accuracy, affords the answer to this question.

$\lambda = 1$

γ	r	c	$r-2c$	Differences
44°	2·429	0·962	0·506	
				−76
46°	2·396	0·983	0·430	
				−74
48°	2·370	1·007	0·356	
				−70
50°	2·354	1·034	0·286	
				−69
52°	2·345	1·064	0·217	
				−68
54°	2·343	1·097	0·149	
				−67
56°	2·350	1·134	0·082	
				−67
58°	2·367	1·176	0·015	

The differences of $r-2c$ hardly diminish at all, and it is clear that the next entry would be negative, or in other words the two figures would overlap.

These results are obtained on the supposition that our approximation is adequate, but the small terms ζ, ϵ, η, which are really infinite series, show signs of bad convergence as γ increases. I think it probable that when we get to these extreme cases the convergence breaks down. It appears, however, justifiable to argue from these results that the unstable body continually elongates until its end coalesces with the other elongated body. I have no doubt but that the same holds true when the masses are unequal, and that we should always find $r-(c+C)$ diminishing until the two meet. The poorness of the approximation of course would prevent us from making good drawings when coalescence is approaching.

§ 26. *On the Possibility of Joining the Two Masses by a Weightless Pipe.*

This subject is considered in § 13, and it is there shown that if a certain function written $f(\mathrm{a}/r)$, for a given solution of the figures of equilibrium of two detached masses of liquid, is positive, the two masses are too far apart to admit of equilibrium when joined by a pipe without weight—and conversely.

Now I have computed $f(\mathrm{a}/r)$ in a number of cases of Roche's ellipsoids in limiting stable equilibrium, and have found it always to be decisively positive.

The corresponding function for two spheres is given in (3) of § 3, and its first term is $+\mathrm{a}^3/r^3$. When we compute it for two ellipsoids, we find the corresponding term to have become negative, and the additional terms, which are given in (44), § 13, are also negative. Hence $f(a/r)$ is decidedly less for two ellipsoids than it is for two spheres of the same masses with the same radius vector. Thus the deformation of the two bodies tends in the direction of making it possible to join them by a pipe without weight, but it seems certain that in the cases of the Roche's ellipsoids in limiting stability such junction remains impossible.

I also computed $f(a/r)$ for the much elongated ellipsoids which are roughly computed in the last section and finally overlap, and always found $f(a/r)$ to be positive, as far as the approximate formula went. The additional terms tend, however, more and more to cause $f(a/r)$ to vanish, and the approximation becomes very imperfect. Now I believe, although I cannot prove it rigorously, that if we could obtain a more exact evaluation of the forms of these elongated ellipsoids, and if further a more exact value of $f(a/r)$ were calculable, we should find $f(a/r)$ vanishing near the stage when the computations would show the two ellipsoids to overlap. It therefore seems probable that there is a figure of equilibrium consisting of two elongated masses joined by a narrow neck. These ellipsoids are very unstable when detached, and, according to the principles of § 2, it seems inconceivable that junction by a neck of fluid could render them stable.

PART III. SUMMARY.

Since the foregoing investigation may be read by mathematicians, while astronomers and physicists will perhaps wish to learn the nature of the conclusions arrived at, I shall devote this part of the paper to a general discussion of the subject, without reference to the mathematical processes used.

Two problems are solved here simultaneously; for the analysis required for their solutions is almost identical, although the principles involved are very distinct.

We conceive that there are two detached masses of liquid in space which revolve about one another in a circular orbit without relative motion—just as the moon revolves about the earth; the determination of the shapes assumed by each mass, when in equilibrium, is common to both our problems. It is in the conditions which determine secular stability that the problem divides itself into two.

One cause of instability in the system resides in the effect on each body of the reaction on it of the frictionally resisted tides raised by it in the other. If now the larger of the two masses were rigid, while still possessing the same shape which it would have had if formed of liquid, the only effect on the orbital stability of the system would be due to the friction of the tides of the smaller mass generated by the attraction of the larger one. Investigation shows that in this case, as the two masses are brought nearer and nearer together, instability would not supervene from tidal friction until the two masses were almost in contact; but it is clear that the deformation of the figure of the liquid mass presents another possible cause of instability. In fact, instability, as due to the deformation of figure, will set in when the masses are still at a considerable distance apart. It amounts to exactly the same whether we consider the larger mass to be rigid, or whether we treat it as liquid and agree to disregard the instability which arises from the friction of the tides raised in it by the smaller body. Accordingly we may describe the stability just considered as "partial," whilst full secular stability of both bodies will depend on the tidal friction of the larger mass also.

The determination of the figure and partial stability of a liquid satellite (*i.e.*, apart from the effects of the tidal friction of the planet) is the problem of Roche. He, however, virtually regarded the planet as constrainedly a sphere, whilst in general I have treated it as an ellipsoid with the form of equilibrium.

It has already been remarked that, as the radius vector of the satellite diminishes, partial instability first supervenes from the deformation of the smaller body. It therefore hardly seems worth while to consider the partial stability of a system in which the liquid satellite (hitherto described as the smaller body) is greater than the planet. We may merely remark that in this case the problem comes to differ very little from that involved in the determination of the full secular stability of two liquid masses; for if we consider the case of a large liquid mass (the satellite) attended by a small body (the planet), it clearly makes very little difference in the result whether or not the tidal friction of the small body is included amongst the causes of instability.

This being so, I have not thought it worth while to continue the solutions of Roche's problem (modified by allowing the planet to be deformed) to the cases in which the satellite is larger than the planet. The ratio of the masses of satellite to planet is denoted above by λ, and the field examined by means of numerical solutions extends from $\lambda = 0$ to $\lambda = 1$, while the part omitted extends from $\lambda = 1$ to $\lambda = \infty$.

Tidal friction is a slowly acting cause of instability, and from the point of view of cosmical evolution the partial stability of Roche's ellipsoids is of even greater interest than the full secular stability of the system.

The limiting stability of Roche's liquid satellite is determined by the consideration that the angular momentum of the system, exclusive of the rotational momentum of the planet, shall be a minimum. This exclusion of a portion of the momentum of the whole system corresponds with the fact that we are to disregard the tidal friction of the planet as a cause of instability. If all possible cases of the liquid satellite be arranged in order of the corresponding (partial) angular momentum of the system, it is clear that for given momentum there will in general be two forms of satellite; but when the momentum is a minimum the two series coalesce. If then we proceed in order of increasing momentum, the configuration of minimum is the starting point of two series of figures; it is a figure of bifurcation, and one of the two series has one fewer degrees of instability than the other.

One of the two series is continuous with the case of a liquid satellite revolving orbitally at an infinite distance from its planet, and this is a stable configuration. Moreover, M. Schwarzschild has shown* that the whole series of Roche's ellipsoids does not pass through any other form of bifurcation. Hence we conclude that of the two series which start from the configuration of minimum momentum, one is stable and the other unstable.

The unstable series of solutions is continuous with a quasi-ellipsoidal satellite, infinitely elongated along the radius vector of the orbit, and the radius vector itself is infinite. Since two portions of matter cannot occupy the same space, the infinite elongation of the satellite would be physically impossible, unless the order of infinity of the radius vector were greater than that of the longest axis of the satellite. Now it appears from the numerical results of § 25 that this is not the case, and that the satellite becomes more rapidly elongated than the radius vector increases. Hence if the solution of the problem were exact we should reach a stage at which the two masses of liquid would overlap. I shall endeavour hereafter to consider the interpretation which should be put on this result.

A series of solutions for Roche's ellipsoid in limiting stability is tabulated in § 21, and the table gives the radius vector and the three semi-axes of each body. The unit of length adopted is the radius of a sphere whose volume is equal to the sum of the volumes of the two masses. Three of these solutions are illustrated in figs. 2, 3, 4. The section shown is that passing through the axis of rotation and the two centres, but the places are marked which the extremities of the mean axes would reach if the section had been taken at right angles to the axis of rotation.

The table of § 21 shows that the radius vector at which instability sets in only changes from 2·457 to 2·514, whilst λ, the ratio of the mass of the satellite to that of the planet, changes from zero to unity. The distance between the vertices of the two ellipsoids also remains wonderfully nearly constant throughout a wide range of change in the value of λ; for when $\lambda = 0·4$ it is 0·653, and when $\lambda = 1$ it is 0·660, only falling to 0·633 at its minimum.

* See reference in § 18 above.

Thus far I have been speaking of the modified problem of Roche in which the planet assumes the appropriate figure of equilibrium, but I have also obtained the solution of Roche's problem for an infinitely small satellite and a spherical planet. As stated in the Preface, the radius vector of limiting stability, which has been called "Roche's limit," is found to be 2·4553, and the axes of the critical ellipsoid are proportional to the numbers 10000, 5114, 4827. These may be compared with the 2·44 and 1000, 496, 469 determined by Roche himself. When we consider the methods which he employed, we must be struck with the closeness to accuracy to which he attained.

For the infinitely small satellite the modification of Roche's problem hardly introduces any sensible change in the results, but for satellites of finite mass stability will continue to subsist for a slightly smaller radius vector for the spherical than for the ellipsoidal planet. In other words, the ellipticity of the planet induces instability earlier than would be otherwise the case.

Roche did not attempt to investigate how closely his equations were capable of giving the ellipsoid most nearly representative of the truth, nor did he estimate how far the ellipsoid is an accurate solution. These points are considered above, and it was the desirability of making the investigation with a closer degree of accuracy which occasioned many of the difficulties encountered.

For the infinitely small satellite the ellipsoidal solution is exact, and with a spherical planet, but not for an ellipsoidal one, Roche's equations give that ellipsoid exactly. In this case, however, the change introduced by the modification of Roche's problem is quite unimportant.

For finite satellites Roche's equations require sensible modification, and the solution of the modified problem is different from that of the unmodified one, although not to an important extent. But the ellipsoid derived from the corrected equations is deformed by an infinite series of ellipsoidal harmonic deformations, beginning with terms of the third order. Of these, the only ones which have any sensible effect are those which may be described as zonal with respect to the satellite's radius vector.

By far the most important of these is the third zonal harmonic, whereby the satellite assumes a somewhat pear-shaped figure, being sharpened towards the stalk end of the pear pointing towards the planet, and bluntened at the other end. In consequence of this deformation the shape is slightly flattened between the stalk and the middle.

The fifth and successive odd zonal harmonics accentuate the sharpening of the stalk and the bluntening of the remote end. The fourth, sixth, and successive even harmonics also accentuate the protrusion of the stalk, but tend to fill up the deficiency at the remote end.

The general effect must be very like what results from the second approximation to the pear-shaped figure of equilibrium*, for I found that the ellipsoidal form was but slightly changed over the greater part of the periphery, whilst a protrusion occurred at one end—in this present case pointing towards the planet.

In figs. 2, 3, 4, the protrusions at one end and the bluntening at the other, as computed from the third, fourth, and fifth harmonics, are indicated by dotted lines. It appears from these figures that, at least up to the point when instability sets in, the ellipsoid remains surprisingly near to the correct solution.

For an infinitely small satellite minimum radius vector also gives minimum angular momentum, so that the closest possible satellite is also in a state of limiting stability. But this is not the case for finite satellites, and there exists an unstable ellipsoidal satellite with smaller radius vector than is consistent with stability. Thus for a satellite of four-fifths of the mass of the planet the minimum radius vector is 2·36, whilst stability ceases at a distance of 2·50. Again, for equal masses stability ceases at 2·514, whilst the possibility of an ellipsoidal solution extends to 2·343.

If we follow the forms of the more and more elongated satellites, when the radius vector has begun to increase again, we find explicitly in the case of equal masses, and with practical certainty for all ratios of masses, that the distance between the two vertices continues to diminish and finally becomes negative. At this stage the two masses overlap, a conclusion which is, of course, physically impossible. But the calculation is based on the assumed adequacy of the approximations, and it is certain that the harmonic deformations of the ellipsoids increase rapidly, so that each body puts out a protrusion towards the other. The two masses of liquid must therefore really meet before we reach the stage of overlapping ellipsoids. As far as can be seen, the approximation has become very imperfect—perhaps evanescent—before the two ellipsoids cross. It will be best to continue the discussion of the meaning of this result after we have considered the true secular stability of the two masses of liquid.

If a satellite, being a particle, revolves about a rotating planet, whose tides are subject to friction, there are, for given angular momentum, two configurations (if any) in which the planet always presents the same face to the satellite. In one of these, which is unstable, the satellite is close to the planet; in the other, which is stable, it is remote†. If the angular momentum of the system be diminished, the radius vector of the stable configuration diminishes and that of the unstable one increases until the

* See "Stability," referred to in the Preface. [Paper 12, p. 317.]

† See *Roy. Soc. Proc.*, No. 197, 1879, or Appendix G (*b*) to Vol. II. of Thomson and Tait's *Natural Philosophy*. [Paper 5, Vol. II., p. 195.]

two coalesce. For yet smaller angular momentum there is no configuration possible in which the planet shall always present the same face to the satellite. We see then that amongst all possible configurations in which the planet presents the same face to the satellite, that one is in limiting stability, in which the two solutions coalesce with minimum angular momentum.

A rotating liquid planet will continue to repel its satellite so long as it has any rotational momentum to transfer to the orbital momentum of the satellite. Hence an infinitesimal satellite will be repelled to infinity, and the configuration of limiting stability for an infinitesimal satellite attending a planet, which always presents the same face to it, is one with infinite radius vector.

Very nearly the same conditions hold good when both planet and satellite are subject to frictional tides. In § 2 it is proved that when each body is constrainedly spherical, the radius vector of limiting stability is infinite when the ratio of the masses is infinitely small. The radius vector decreases with great rapidity as the ratio of the masses increases, and when the masses are equal, the radius vector of limiting stability is 1·738 times the radius of a sphere whose mass is equal to the sum of the masses, or is 2·19 times the radius of either of the two spheres. Thus, when the ratio of the masses falls from zero to unity (and this embraces all possible cases), limiting stability occurs with a radius vector which falls from infinity until the two spheres are only just clear of one another.

When we pass from the case of the two spheres to that of two masses, each of which is a figure of equilibrium under the attraction of itself and its companion, and subject to centrifugal force, the calculation becomes exceedingly complicated. Since the radius vector of limiting stability in every case must be greater than that of Roche's ellipsoid in limiting stability, and since in the latter case instability sets in through the deformation of the smaller body, it follows that in every case of true limiting secular stability of the system, instability supervenes through tidal friction.

When the ratio of the masses is small, we have seen that limiting stability occurs when the two masses are far apart. In this case the deformations of figure are small, and could easily be computed by spherical harmonic analysis.

For finite values of the ratio of masses, when spherical harmonic analysis would fail, a fair degree of exactness in the result may be obtained from the approximate formula of § 19. There would be no serious error from this formula when the ratio of masses is less than a half, but for greater values of the ratio it seems necessary to have recourse to the laborious processes employed in determining Roche's ellipsoids. I thought, then, that it might suffice to compute the configuration of true secular limiting stability in the case of equal masses. It is illustrated in fig. 5, and we see that the radius vector is 2·638. We found that for a pair of equal spheres, instability only set in when the radius vector, measured in the same unit, was 1·738. Thus the deformations of the two masses forbid them to approach as near to one another as if they were spheres. It should be noted that instability in this case must arise from tidal friction, because Roche's ellipsoid in limiting stability was found to have a radius vector of 2·514.

When Poincaré announced that there is a figure of equilibrium bearing some resemblance to a pear, he also conjectured that the constriction between the stalk and the middle of the pear might become developed until it became a thin neck of liquid joining two bulbs, and that yet further the neck might break and the two masses become detached. References to my own papers on this pear-shaped figure and its stability are given above in the preface, and the present investigation was undertaken in the hope that a revision of Roche's work would throw some light on the figure when the constriction has developed into a thin neck of liquid.

As a preliminary to greater exactness, I have in § 3 considered the motion of two masses of liquid, each constrainedly spherical, and joined to one another by a weightless pipe. Through such a pipe liquid can pass from one sphere to the other, and it will continue to do so until, for given radius vector, the spheres bear some definite ratio to one another; or, to state the matter otherwise, two spherical masses of given ratio, revolving in a circular orbit without relative motion, can be started with some definite radius vector so that liquid will not flow from one to the other.

In this system the ratio of the masses and the radius vector are the only parameters, and I find that the condition of equilibrium is a cubic equation in the radius vector with coefficients which are functions of the ratio of the masses. The cubic has three real roots of which only one has a physical meaning, and the solution is illustrated graphically in fig. 1. The single circle on the right is the larger sphere, and it is maintained of constant size for convenience of illustration. The smaller circles on the left represent the solutions for various ratios of masses, which are the cubes of the numbers written on the successive circles.

The solution of this problem seems to me very curious, but it does not possess much physical interest, since it is proved in § 3 that all the solutions are unstable.

The distance between the two masses is much smaller than is the case with any of Roche's ellipsoids, even with minimum radius vector, and accordingly it did not seem probable that the parallel problem, when the two masses are liquid and deformed, would possess any solution at all; nevertheless, it was worth while to pursue the investigation to the end.

When the masses are ellipsoidal and are joined by a weightless pipe, the solution would become very complicated, but the question may be attacked indirectly. When the masses are spherical there is a certain function of the radius vector and of the ratio of the masses which must vanish when a channel of communication is opened between them. If this function be computed for two given spherical masses with given radius vector, we find that it is negative if the two masses are too close together to admit of junction by a pipe without disturbance of their relative masses, and that it is positive if they are too far apart.

When the figures of equilibrium of two detached masses of liquid are determined, it is possible to form the corresponding function, but part of it consists of an infinite series of which it is only practically possible to give the first few terms. Now I have computed this function in a number of cases of Roche's ellipsoids, and have found that the few terms of the infinite series are small, that the series is apparently rapidly convergent, and that the function is decisively positive. We may conclude then that in none of the cases, for which numerical results have been given, is it even approximately possible to make a junction between the masses; and even if we could do so, the system would be unstable, because removal of a constraint may destroy but cannot impart stability. To find any possible solution we must consider cases where the two masses are much closer together.

I think, however, that there must be a figure of the kind sought, for the following reasons: If the function referred to above be formed for given radius vector and ratio of masses, we find that its value is very much less than if the two masses are spherical. Thus the tendency of liquid to flow from the larger to the smaller mass (when they are too far apart) is much less than if the two masses were spherical. Every increase of ellipticity in the ellipsoids tends to diminish the function, and the series tends to become less convergent; and besides I have made no attempt to evaluate the terms in the function which correspond to the harmonic inequalities of the ellipsoids, and these would tend to diminish the function still further.

It was remarked above that two much elongated ellipsoids seem to coalesce finally, but that the approximations were not satisfactory. I find, however, that even to the end the function, as far as it could be computed, was still positive although much diminished. It appears to me then probable that if we could obtain a more complete expression for the function, we should find that it vanishes before the two ellipsoids overlap. There is then some reason to believe in the existence of a figure of equilibrium consisting of two quasi-ellipsoids joined by a narrow neck; but such a figure must be unstable.

I have, in fig. 6, made a highly conjectural drawing of such a figure where the two bulbs are equal. The data are derived from the computations for the much elongated ellipsoids just before they are found to overlap.

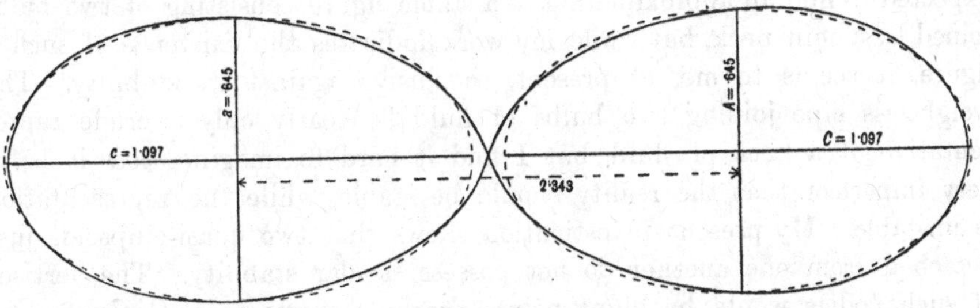

FIG. 6. Conjectural drawing of unstable figure of two equal masses of liquid just in contact.

Mr Jeans has considered the equilibrium and stability of infinite rotating cylinders of liquid. This is the two-dimensional analogue of the three-dimensional problem*. He finds solutions perfectly analogous to Maclaurin's and Jacobi's ellipsoids and to the pear-shaped figure. In consequence of the greater simplicity of the conditions, he is able to follow the development of the cylinder of pear-shaped section until the neck joining the two parts has become quite thin. His analysis, besides, points to the rupture of the neck, although the method fails to afford the actual shapes and dimensions in this last stage of development.

He is able to prove conclusively that the cylinder of pear-shaped section is stable, and it is important in connection with our present investigation to note that he finds no evidence of any break in the stability of that cylinder up to its division into two parts.

The stability of Maclaurin's and of the shorter Jacobian ellipsoids is, of course, well established, and I imagined that the pear-shaped figure with incipient furrowing was also proved to be stable. But M. Liapounoff now states† that he is able to prove the pear-shaped figure to be unstable from the beginning, and he attributes the discrepancy between our conclusions to the fact that my result depended on the supposed rapid convergency of an infinite series, of which only a few terms were computed. The terms computed diminish rapidly, and it seemed to me evident that the rapid diminution must continue, so that I feel unable to accept the hypothesis that the sum of the neglected terms could possibly amount to the very considerable total which would be necessary to reverse my conclusion. I am, therefore, still of opinion that the pear-shaped figure is stable at the

* "On the Equilibrium of Rotating Liquid Cylinders," *Phil. Trans.*, A, Vol. 200, pp. 67—104.

† See reference in Preface.

beginning; and this view receives a powerful confirmation from Mr Jeans's researches. The final decision must await the publication of M. Liapounoff's investigation.

But there is another difficulty raised by the present paper. I had fully expected to find an approximation to a stable figure consisting of two bulbs joined by a thin neck, but while my work indicates the existence of such a figure, it seems to me, at present, conclusive against its stability. The weightless pipe joining two bulbs of fluid is clearly only a crude representative of a neck of fluid, but I find it hard to imagine that it is so very imperfect that the reality should be stable, while the representation is unstable. My present investigation shows that two quasi-ellipsoids just detached from one another do not possess secular stability. The vertices of such bodies would be blunt points nearly in contact; the introduction of a short pipe without weight between these blunt points would differ exceedingly little from two sharp points actually in contact. Is it possible that the difference would produce all the change from great instability even to limiting stability? The opening of a channel between the two masses is the removal of a constraint; the system does not possess true secular stability when the channel is closed, and we should have to believe that the removal of a constraint induces stability; and this is, I think, impossible.

If, then, Mr Jeans is right in believing in the stable transition from the single cylinder to two revolving about one another, and if I am correct now, the two problems must part company at some undetermined stage. M. Liapounoff will no doubt contend that it is at the beginning of the pear-shaped series, but for the present I should disagree with such an opinion.

I have no suggestion to make as to the stage at which the pear-shaped figure may become unstable, or as to the figure which must be coalescent with it when instability supervenes. These points must await the elucidation which they will no doubt receive from future investigations.

One question remains: If my present conclusions are correct, do they entirely destroy the applicability of this group of ideas to the explanation of the birth of satellites or of double stars? I think not, for we see how a tendency to fission arises, and it is not impossible that a period of turbulence may naturally supervene in the process of separation. Finally, as Mr Jeans points out, heterogeneity of density introduces new and important differences in the conditions.

INDEX TO VOLUME III.

CAMBRIDGE: PRINTED BY JOHN CLAY, M.A. AT THE UNIVERSITY PRESS.

Printed in the United States
By Bookmasters